Complete Engine Performance and Diagnostics

by Robert Scharff
& the Editors of
MOTOR SERVICE

WE ENCOURAGE
PROFESSIONALISM

THROUGH TECHNICIAN
CERTIFICATION

Delmar Publishers Inc.®

NOTICE TO THE READER

Publisher does not warrant or guarantee any of the products described herein or perform any independent analysis in connection with any of the product information contained herein. Publisher does not assume, and expressly disclaims, any obligation to obtain and include information other than that provided to it by the manufacturer.

The reader is expressly warned to consider and adopt all safety precautions that might be indicated by the activities described herein and to avoid all potential hazards. By following the instructions contained herein, the reader willingly assumes all risks in connection with such instructions.

The publisher makes no representations or warranties of any kind, including but not limited to, the warranties of fitness for particular purpose or merchantability, nor are any such representations implied with respect to the material set forth herein, and the publisher takes no responsibility with respect to such material. The publisher should not be liable for any special, consequential or exemplary damages resulting, in whole or in part, from the readers' use of, or reliance upon, this material.

Text, Design, and Production by
Scharff Associates, Ltd.
RD 1 Box 276
New Ringgold, PA 17960

Scharff Staff
Production Manager: Marilyn Strouse-Hauptly
Staff Writers: Dave Caruso, John Corinchock, and Keith Mullen
Cover Design: Eric Schreader
Logo Design: Ed Foulk

Delmar Staff
Editor-in-Chief: Mark W. Huth
Associate Editor: Joan Gill

For more information address Delmar Publishers Inc.
3 Columbia Circle, Box 15-015
Albany, New York 12212-5015

Printed in the United States of America
Published simultaneously in Canada
by Nelson Canada,
a division of International Thomson Limited

10 9 8 7 6 5

Library of Congress Cataloging-in-Publication Data

Scharff, Robert.
Complete engine performance and diagnostics.

Includes index.
1. Automobiles—Motors—Maintenance and repair.
I. Motor service (Chicago, Ill. : 1951) II. Title.
TL210.S35 1989 629.25'04 89-7832
ISBN 0-8273-3570-2
ISBN 0-8273-3581-4 (shop manual)
ISBN 0-8273-3580-6 (instructor's guide)

CONTENTS

Complete Engine Performance and Diagnostics

PREFACE

The past two decades have been an era of rapid change in automotive design and technology. The average car today is smaller, lighter, and more aerodynamic in design than vehicles of the past. Rear-wheel automobiles have given way to front-wheel drive, with subsequent changes in suspensions and drivetrains. Engines are smaller; the age of the large V-8 is gone. Turbocharged, multivalve, four- and six-cylinder engines are common. The carburetor has been replaced with fuel injection, ignition timing is electronically controlled, and computers monitor and control an increasing number of functions from the volume and mass of the incoming air to the operation of the windshield wipers.

These technological changes and refinements place a tremendous burden of training on the automotive repair industry and its technicians. The new technician as well as the seasoned mechanic require extensive, on-going education as new technology and repair procedures are introduced, refined, and updated. Meeting that need is the purpose of this book.

As explained in Chapter 1, the nature and scope of engine diagnosis and tune-up have radically changed in the past twenty years. Adjustments and replacements that were once common tune-up procedures are no longer possible or even necessary. More than ever, a thorough tune-up must ensure that all engine systems are functioning to specifications. A misadjustment in one system will adversely affect the operation of others. The tune-up technician must understand the interrelationships between various engine systems and how a malfunction in one can affect the others.

Although many test procedures and tools that were used in diagnosis and tune-up twenty years ago are still applicable to late-model vehicles, new tools and procedures have been developed to troubleshoot electronic ignition systems, computer control systems, and new fuel injection systems. Some of the more important tools are described in Chapter 2.

It is impossible to properly tune an engine that has mechanical problems. Chapter 3 describes some basic engine performance tests that should be performed prior to making adjustments to related engine systems. Performance problems can also be caused by a malfunction in the battery, charging, and starting systems. Diagnosis procedures for these systems are given in Chapters 4, 5, and 6.

In the early 1970s, a basic tune-up consisted of replacing point and plugs, adjusting the ignition timing and dwell, and setting the idle speed. Although the only task still applicable to many modern vehicles is replacing the spark plugs, Chapter 7 covers the basic replacement adjustment procedures used to tune conventional ignition systems.

Conventional ignition systems were replaced with electronic systems in the late 1970s. Although the basic operating principles are the same for most systems, there is a great variety in system components and testing procedures. Changes in early electronic ignition systems were made almost yearly, and most manufacturers have several electronic ignition systems. These systems are covered in Chapter 8.

In today's vehicle, ignition, fuel delivery, air induction, and emissions systems are integrated under a computerized control system. Chapter 9 explains these electronic engine control systems and their diagnostic procedures. As automotive electronics becomes increasingly sophisticated and is applied to more components and functions, diagnosis becomes more and more the function of the computer. Current systems have self-diagnostic circuits and functions that detect component malfunction and system misadjustments.

In addition to spark, internal engine combustion requires the right combination of air and fuel. Air and fuel delivery systems are covered in Chapters 10 through 13. Methods of metering air and fuel have changed as radically as ignition systems. Most vehicles today are equipped with a fuel injection system and are often turbocharged, if not supercharged. The tune-up technician must know how to service these systems.

The last chapter, Chapter 14, deals with emission systems. Most of the recent innovations in engine design, ignition systems, and fuel systems are partly the result of the manufacturers' efforts to produce cleaner running vehicles. For many years, most emission controls simply attempted to remove emissions from the exhaust. Today's engines are reaching a stage of refinement in which emission control systems are being simplified and, in some cases, even eliminated. However, emission system components must be operating as intended or the vheicle will suffer driveability problems. Chapter 14 explains how to diagnose emission system problems.

ACKNOWLEDGMENTS

To organize a book requires the help of a great number of sources of information and the aid of many people and organizations. The cooperation of the staff of *Motor Service* and its editorial director, Jim Halloran, played a great part in the preparation of the text and furnishing of illustrations. In addition we would like to thank the big three automotive manufacturers—Chrysler, Ford, and General Motors—for permission to use information and illustrations from their training programs and service manuals.

Certain photographs were used with permission from Robert Scharff's *Fuel Systems and Emission Controls*, copyright © 1989 by Delmar Publishers Inc.

Grateful acknowledgment is also made to the following companies, which provided research material, photographs, and/or illustrations for use in preparing this book.

Actron Manufacturing Company
Allen Testproducts Division
All-Test, Inc.
Babcox Publications
Bear Automotive Service Equipment Company
Carter Automotive Division, ACF Industries, Inc.
Chrysler Corporation
Electro Specialties
Ford Motor Company
Ford Parts and Service Division (Training and Publications Department)
Garrett Aftermarket Division
General Motors Corporation
Hamilton Test Products
Holley Carburetor Division, Colt Industries Operating Corporation
Johnson Controls, Inc. (Battery Division)
Kleer-Flo Company
MAC Tools, Inc.
Michael Ede Management, Inc.
Micro-Processors, Inc.
OTC Tool & Equipment Division of SPX Corporation
Penray Company
Snap-On Tools Corporation
Sun Electric Corporation
The John Fluke Manufacturing Company
TIF Instruments, Inc.

We would like to thank the following for reviewing the manuscript and for their helpful comments:

Mr. William Demtsey
ASE
1920 Association Drive
Reston, VA 22091-1502

Mr. Reginald Ludwig
American Educational Complex
Central Texas College
US Highway 190 West
PO Box 1800
Killen, TX 76540-9399

Mr. Richard Neet
Spokane Community College
North 1810 Greene Street
Spokane, WA 99212

CHAPTER ONE

INTRODUCTION TO AUTOMOTIVE TUNE-UPS

Objectives

After reading this chapter, you should be able to:

- State the objectives of a tune-up.
- Explain the changes that have affected tune-up procedures.
- Outline the basic steps in a quality tune-up.
- List the sources of information needed when performing a tune-up.
- Explain safe tune-up practices.

WHAT IS A TUNE-UP?

Not too many years ago, the average car owner brought his or her car in for a tune-up twice a year—usually each spring and fall. The tune-up was designed to keep the car operating according to factory specifications. Usually, the spark plugs, ignition contact breaker points, and condenser were replaced. The dwell, idle mixture and speed, and ignition timing were adjusted, if necessary, to keep the engine operating as it was designed to.

A thorough tune-up has a two-fold purpose. First, a tune-up is a preventive maintenance effort. It identifies wearing parts and replaces them before the components fail. It also maintains operating parameters—such as idle speed and ignition timing—so that performance, fuel efficiency, and emission levels do not suffer. Secondly, a tune-up diagnoses existing performance problems, identifies the fault, and corrects it.

Keeping a vehicle operating according to factory specifications is still the goal of today's tune-up (Figure 1-1). People are bringing their vehicles in for tune-ups at a rate of over 60 million a year, according to a recent survey. However, much has changed since the days when changing "points and plugs" twice a year kept a car running efficiently.

FIGURE 1-1 A complete tune-up requires the use of specialized equipment to analyze engine and ignition system performance. *(Courtesy of OTC Tool and Equipment, Division of SPX Corp.)*

AN AUTOMOTIVE REVOLUTION

The changes began in the early 1960s when the United States Congress passed legislation establishing emission standards for American-made vehicles. The exhaust produced by the powerful, gas-guzzling engines of that day were polluting the nation's air with carbon monoxide, oxides of nitrogen, unburned hydrocarbons and other gases. The situation was especially critical in major cities, such as Los Angeles, California. In fact, California established its own emissions standards in 1961 for cars sold in that state and still today has stiffer standards than the federally mandated limits.

EMISSION CONTROL DEVICES

The new laws required car manufacturers to equip their vehicles with emission-controlling devices. The following is a synopsis of the major developments in emission controls that resulted from early regulations:

- *Positive Crankcase Ventilation (PCV) Systems.* In 1963, all domestic car manufacturers voluntarily equipped their vehicles with a PCV system that draws crankcase fumes into the intake manifold to be burned. PCV systems operate by manifold vacuum to lower hydrocarbon (HC) emissions.
- *Air Injector Reactor (AIR).* First introduced in 1968, the air injector reactor (or thermactor) pumps outside air into the exhaust manifold to oxidize any unburned HC and carbon monoxide (CO) in the exhaust. As shown in Figure 1-2, the AIR pump consists of a belt-driven pump and vacuum-controlled check valve. Other air injection systems use the scavenging vacuum of the exhaust stream to draw air into the exhaust system.

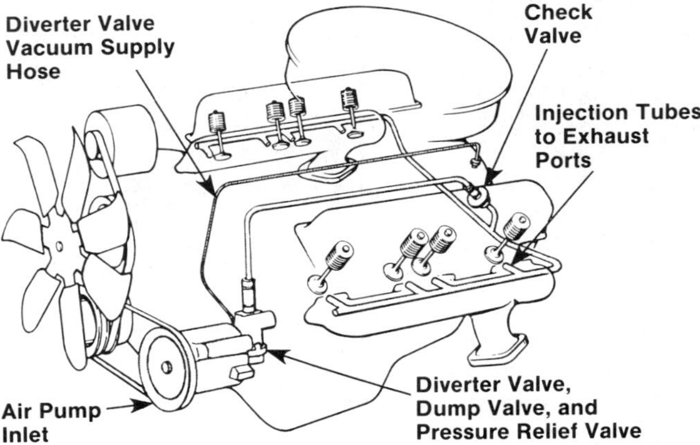

FIGURE 1-2 An air injection system oxidizes unburned hydrocarbons in the exhaust system.

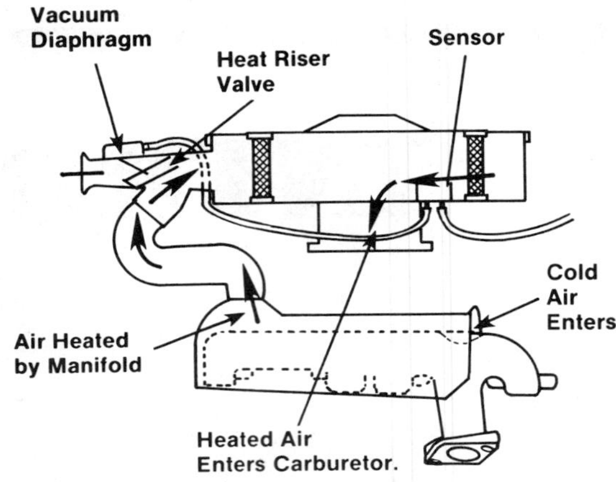

FIGURE 1-3 The heat riser valve opens to draw warm air into the carburetor until the engine warms up.

- *Lean Carburetors.* In 1968, carburetors were recalibrated to deliver a leaner air/fuel mixture in all ranges. This lowered CO emissions, but when made too lean it actually raised HC levels.
- *Retarded Timing.* In 1968, the dual-diaphragm distributor was introduced. The dual diaphragm allows the timing to be retarded during idle and part-throttle operation to give more complete combustion of leaner air/fuel mixtures. It lowers both HC and oxide of nitrogen (No_x)emissions.
- *Thermostatic Air Cleaner.* To help vaporize gasoline more efficiently on cold start-ups, the thermostatic air cleaner was introduced in 1966. The air cleaner has a vacuum-controlled heat riser valve that draws intake air across the exhaust manifold and into the air cleaner (Figure 1-3). It reduces both HC and CO emissions.
- *Evaporative Emission Control (EEC) System.* Introduced in 1971, the EEC system prevents raw gas vapors from escaping from the gas tank and carburetor and lowers HC emissions.
- *Exhaust Gas Recirculation (EGR) System.* The EGR system (Figure 1-4) introduces exhaust gases into the intake manifold to reduce oxides of nitrogen during acceleration and full throttle operation. This system was developed in 1973 and operates virtually all the time, except during idle and full throttle.
- *Catalytic Converter.* The catalytic converters introduced in 1975 chemically converted hydrocarbons and carbon monoxide into harmless carbon dioxide and water. They also reduce NO_x emissions.

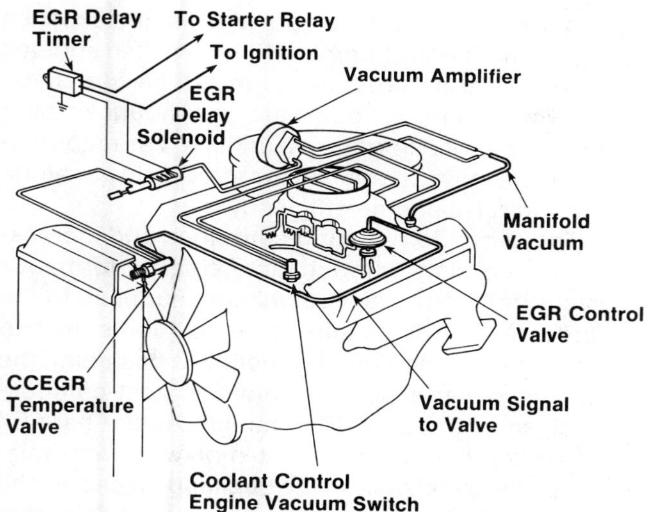

FIGURE 1-4 In addition to the EGR valve, a typical exhaust gas recirculation system contains many other components.

In addition to these developments, car manufacturers also reworked their engines to make them more efficient. Combustion chambers and intake manifolds were redesigned, compression ratios were lowered, and camshaft valve overlap was increased in an effort to lower emission levels.

These developments did lower harmful emissions. However, the combination of a leaner air/fuel mixture, retarded timing, exhaust gas recirculation, and lower compression ratios drastically hurt engine performance and fuel economy. Because the early systems were mechanically or vacuum operated, many people simply disconnected them, restoring power and economy but frustrating the original intent of the federal laws.

Federal legislation continued to tighten emission control requirements. In 1970, Congress passed the Clean Air Act, which required a 90 percent reduction in exhaust emissions by 1976. Table 1-1 shows the progressively stricter limits imposed from 1966 to present.

TABLE 1-1: AIR POLLUTION LEGISLATION

Year	Regulations	HC	CO	NO_x
Before Controls		850 ppm (16.8 g/mi)*	3.4% (125.0 g/mi)*	1000 ppm (4.0 g/mi)*
1966-67	Calif.	275 ppm	1.5%	none
	U.S. Federal	none	none	none
1968-69	Calif.	275 ppm	1.5%	none
	U.S. Federal	275 ppm	1.5%	none
1970	Calif. and U.S. Federal	4.6 g/mi	46.0 g/mi	none
1971	Calif.	4.6 g/mi	46.0 g/mi	4.0 g/mi
	U.S. Federal	4.6 g/mi	46.0 g/mi	none
	Canadian	2.2 g/mi**	23.9 g/mi**	none
1972	Calif.	3.2 g/mi	39.0 g/mi	3.2 g/mi
	U.S. Federal	3.4 g/mi	39.0 g/mi	none
	Canadian	3.4 g/mi	39.0 g/mi	none
1973	Calif.	3.2 g/mi	39.0 g/mi	3.0 g/mi
	U.S. Federal	3.4 g/mi	39.0 g/mi	3.0 g/mi
	Canadian	3.4 g/mi	39.0 g/mi	3.0 g/mi
1974	Calif.	3.2 g/mi	39.0 g/mi	2.0 g/mi
	U.S. Federal	3.4 g/mi	39.0 g/mi	3.0 g/mi
	Canadian	3.4 g/mi	39.0 g/mi	3.0 g/mi
1975-76	Calif.	0.9 g/mi	9.0 g/mi	2.0 g/mi
	U.S. Federal	1.5 g/mi	15.0 g/mi	3.1 g/mi
	Canadian	2.0 g/mi	25.0 g/mi	3.0 g/mi
1977-79	Calif.	0.41 g/mi	9.0 g/mi	1.5 g/mi
	U.S. Federal	1.5 g/mi	15.0 g/mi	2.0 g/mi
	Canadian	2.0 g/mi	25.0 g/mi	3.0 g/mi
1980	Calif.	0.39 g/mi	9.0 g/mi	1.0 g/mi
	U.S. Federal	0.41 g/mi	7.0 g/mi	2.0 g/mi
1981	Calif.	0.41 g/mi	3.4 g/mi	1.0 g/mi
1981-85	U.S. Federal	0.41 g/mi	3.4 g/mi	1.0 g/mi
1982-Present	Calif.	0.41 g/mi	7.0 g/mi	0.4 g/mi

*approximate estimations of uncontrolled levels
**theoretical flow rate

Note: ppm = parts per million
g/mi = grams per mile

FUEL EFFICIENT VEHICLES

A second catalyst for change occurred in 1973. The Arab oil embargo of that year created a worldwide oil shortage, escalating the costs of gasoline. In response to this crisis, Congress passed laws requiring domestic manufacturers to increase their fleet's average miles per gallon. By 1985, federal law required an average of 27.5 mpg per fleet for each American auto manufacturer.

The engineers in Detroit responded to the demand by

- Designing Smaller, Lighter Cars. The average full-size American passenger car lost 1500 pounds in 10 years.
- Building Smaller Engines. Four- and six-cylinder engines with a small displacement consumed less fuel per combustion cycle.
- Reducing Final Drive Ratios. A numerically lower (taller) rear axle gear ratio allows the engine to turn more slowly at a given speed, thereby using less gas per mile.

Gas mileage improved but again at the expense of performance and comfort. The smaller engines and taller gearing produced sluggish performance. The lighter vehicles rode hard and bounced over bumps. And inspite of the sacrifices, car manufacturers were falling behind in the effort to lower emissions and raise engine efficiency.

The secret to reducing emissions and increasing fuel economy without sacrificing performance was controlling the air/fuel mixture supplied to the cylinders. The ideal ratio of air to fuel is 14.7 to 1—called the stoichiometric point. At this point, the catalytic converter best performs its post-combustion cleanup function. This minimizes the harmful emissions produced by combustion while maximizing the energy released. The challenge faced by the automobile industry was finding a way to control the air/fuel charge so as to constantly maintain a stochiometric ratio—regardless of the engine operating demands.

Electronics had found its way under the hood of the automobile as early as 1964. In that year, AMC began using electronic voltage regulators. By 1970,

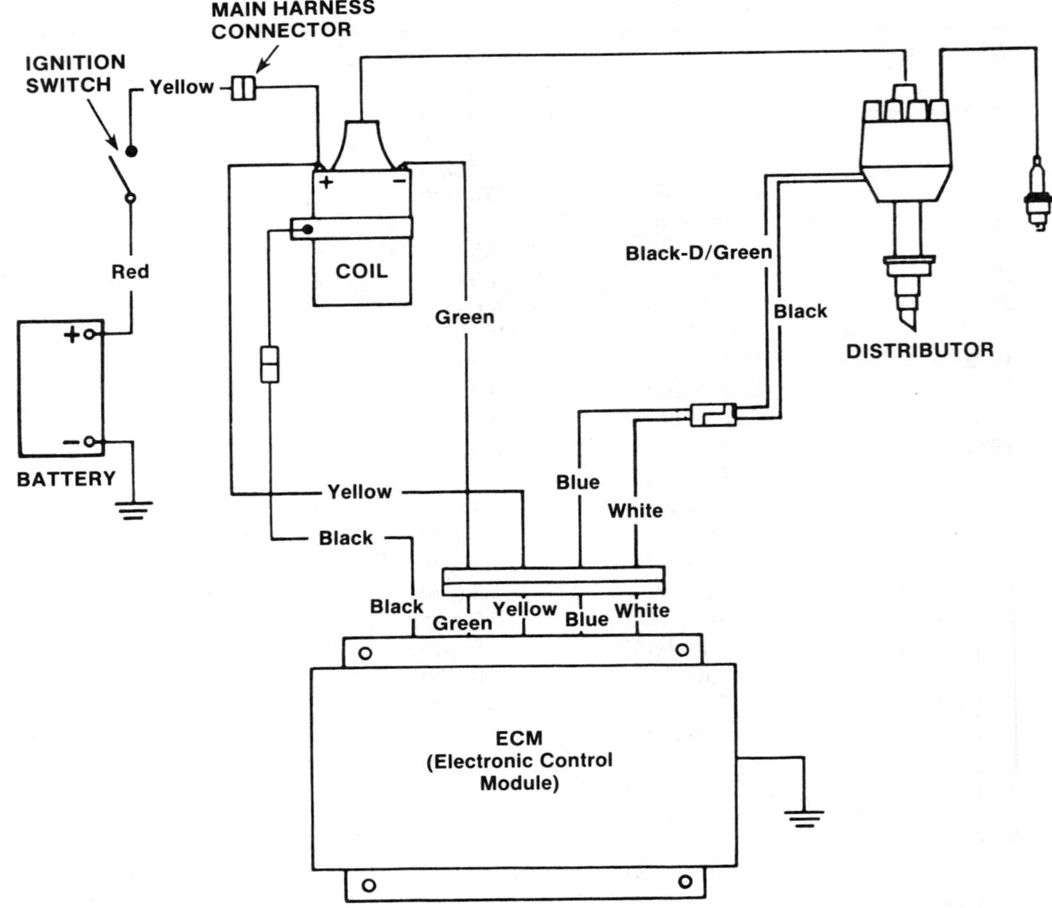

FIGURE 1-5 An electrical diagram of an electronic ignition system

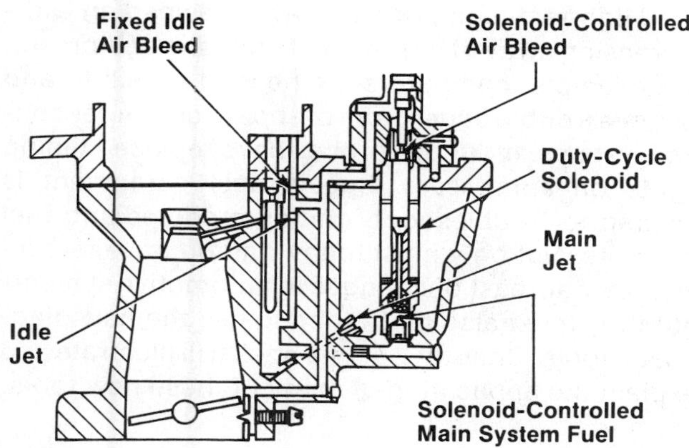

FIGURE 1-6 A duty-cycle solenoid in this feedback carburetor constantly regulates the air/fuel mixture according to feedback from the O_2 sensor.

both Chrysler and GM were using electronic regulators.

The first big advance in using electronics to control engine operation came in 1971. Chrysler introduced an electronic ignition, replacing the age-old ignition breaker points with a transistorized switching device. Breaker points were limited in the amount of voltage they could carry without excessive wear. An electronic system, however, could easily handle much higher secondary voltages. This allowed the spark plugs to generate a much hotter spark, which in turn permitted the use of a leaner fuel mixture than was previously possible.

Other domestic car manufacturers quickly followed Chrysler's lead, developing their own breakerless ignition systems (Figure 1-5). The electronic signals generated by the various breakerless distributors were processed by an electronic control module (ECM).

The control module turned the primary circuit on and off to generate the secondary voltage. Early GM control modules also varied the dwell according to engine speed. Soon, advances in timing control systems eliminated mechanical and vacuum advance mechanisms from the distributor. The development of microprocessor technology allowed for total control of the ignition timing in response to operating parameters such as coolant temperature, air temperature, engine rpm, engine load, and emissions levels.

Electronic control of the air/fuel mixture was made possible by the introduction of feedback carburetors in the late 1970s. The feedback carburetor worked in conjunction with an electronic control module (ECM), an oxygen (O_2) sensor, and a three-way catalytic converter. The O_2 sensor located in the exhaust system, produced a voltage signal based on the amount of oxygen in the exhaust. This voltage signal was used by the electronic control module to determine the air/fuel ratio. Based on this input, the ECM varied the duty cycle of solenoid valves (or stepper motors) that controlled fuel-metering pins and/or air bleeds in the carburetor (Figure 1-6). Under certain engine operating conditions, this effectively kept the air/fuel mixture close to the stoichiometric ratio.

Yet even the sophisticated feedback carburetor could not satisfy the demands of the ever shrinking emissions levels mandated by Congress and the Environmental Protection Agency. Fuel and air supply was still controlled by mechanical means—fuel metering rods and/or air bleeds. The process was not always precise, and the reaction time was not fast enough for sudden performance demands. Besides these drawbacks, the feedback carburetor was complex and expensive, both to manufacture and to repair.

A simpler, less expensive alternative was found in fuel injection. Fuel injection systems were introduced on most domestic passenger cars in the early 1980s. Since then they have almost entirely replaced the clumsier carburetor. Fuel injectors are electrical devices that respond instantly to electronic commands. Fuel can be delivered in precise amounts that can be constantly varied according to changing performance demands.

By the mid to late 1980s, fuel and ignition systems were integrated into a comprehensive electronic engine control system. Not only did computer modules regulate spark timing and air/fuel delivery, but the computer constantly monitored the performance of all engine systems, alerting the driver of malfunctions and providing the technician with precise analysis of the engine systems performances. The age of the electronic car had dawned (Figure 1-7).

CHANGES IN TUNE-UP TECHNOLOGY

These changes in automotive systems have had a profound effect on the automotive repair industry. Mechanics—or technicians as they are now referred to—who fail to keep up with the fast-changing technology are less and less able to correctly troubleshoot and repair new cars. Certain diagnostic procedures used on pre-electronic vehicles are obsolete. Some of the older diagnostic tools are useless and might even be damaging to sensitive electronic components.

FIGURE 1-7 Performing a tune-up on an electronically controlled, fuel-injected engine.

High-tech components have spawned an ever-increasing array of high-tech diagnostic equipment. For example, computers are needed to talk to and retrieve trouble codes from on-board computer control systems. Inductive probes have replaced timing lights on some cars. Special test equipment is needed to troubleshoot ignition modules and fuel injectors. Not only must today's automotive technician stay abreast of changes in automotive systems but they must also be trained to use the new diagnostic tools. Chapter 2 of this book will illustrate and explain the application of many of these new tools.

MAINTENANCE SCHEDULES

Theoretically, today's electronic vehicles need less attention than their older counterparts. Whereas a tune-up was needed every six months on vehicles with a conventional ignition system, today's

TABLE 1-2: TYPICAL SERVICE INTERVALS

	1968		1984	
Components	**Months**	**Miles**	**Months**	**Miles**
Replace				
Crankcase oil	4	6,000	12	7,500
Crankcase oil filter	8	12,000		15,000
PCV valve	12	12,000	As Needed	
Ignition points		12,000	None	
Spark plugs		12,000		30,000
Fuel filter	12	12,000	As Needed	
Coolant	24		24	30,000
Air cleaner filter				
Rotate		12,000	Not Required	
Replace		24,000	36	30,000
PCV filter	Not Required		36	30,000
Adjust				
Valves		12,000	Not Required	
Idle speed		12,000	Not Required	
Rotate distributor cam lubricator		12,000	Not Required	
Ignition timing		12,000		30,000
Check				
Ignition condenser		12,000	None	
PCV system	4	6,000		30,000
Ignition cables		12,000		30,000
Carburetor choke	Not Required			30,000
Vacuum advance		12,000		30,000
Evaporator fuel control system	None			30,000
EGR system	None			30,000
THERMAC	None			30,000

vehicle can travel 30,000 miles on a set of spark plugs. Table 1–2 compares service intervals for 1968 and 1984 automobiles. Almost every tune-up service interval has been substantially increased, many due to the removal of lead additives from the fuel.

Just as important are the service items that are no longer necessary. There is no longer any ignition points or condenser to replace. The idle speed and valves need no adjustment. On vehicles without a distributor, it is even impossible to adjust the timing.

However, service checks have been added to the emission control systems. This reflects the federally mandated five-year, 50,000 mile warranty on emission control devices. A warranty, of course, is only of value if it is used. Therefore, responsibility for maintaining emission systems falls on the car owner. And as Chapter 14 will demonstrate, faults in emission control systems have an adverse effect not just on emission levels but on the overall performance of the vehicle. Verification of the proper operation of each emission control system is an integral part of a thorough tune-up.

STEPS IN A GOOD TUNE-UP

A great many improvements have occurred in automotive technology. Unfortunately, many car owners have not kept up with the changes. Too many—maybe more than 50 percent—still think their cars have breaker point ignition systems, and believe a change of plugs and points will solve their performance problems. Part of the job of the tune-up technician is to educate the consumer in just what is involved in tuning a late-model, electronically controlled vehicle. Table 1–3 is one tune-up checklist that might be shown to a prospective customer both to inform and to sell the garage's services.

In spite of the changes that have occurred in automotive systems in the past two decades, there are still some basic procedures that apply to every tune-up. The following is a basic outline of diagnostic procedures that should be followed to thoroughly tune-up any car.

GATHERING INFORMATION

The first step in performing any tune-up is: asking questions. Ask why the customer is bringing in his or her vehicle for a tune-up. If the vehicle has a specific performance problem such as hard-starting, hesitation, "pinging," etc.—carefully question the customer as to the conditions under which the trouble occurs. If possible, test drive the vehicle with the customer present. Also make sure the customer understands that the tune-up diagnosis might uncover problems not correctable by simply tuning the engine. Also, carefully explain what tasks the tune-up does and does not include.

TEST THE BATTERY, STARTER, AND CHARGING SYSTEMS

For the ignition and electronic control systems to work properly, there must be sufficient voltage available. Always check the battery charge and clean the dirty terminals. Make sure that the starter is drawing sufficient current and that the alternator current output is up to specifications.

TEST OVERALL ENGINE PERFORMANCE

Before installing new plugs and adjusting idle speed and timing, make sure that the engine is sound enough to be tuned. Check the compression in each cylinder; check manifold vacuum. Perform cylinder power balance test. These tests can identify problems that should be corrected before a tune-up can be completed.

TABLE 1-3: A TUNE-UP CHECKLIST
Perform the following:
☐ Ask customer about any driveability problems.
☐ Road test car before tune-up.
☐ Hook up car to analyzer/scope for complete analysis.
☐ Check plugs, caps, and rotor; replace, if necessary.
☐ Replace air filter.
☐ Replace fuel filter.
☐ Check ignition wires; replace, if necessary.
☐ Check EGR valve; replace, if necessary.
☐ Check PCV valve; replace, if necessary.
☐ Test fuel injectors; clean, if necessary.
☐ Clean or adjust carburetor.
☐ Check, then adjust timing, if necessary.
Also, check the following:
☐ Belts
☐ Hoses
☐ Battery
☐ Battery cables
☐ Tires
☐ Fuel (test fuel sample for alcohol, RVP, water, etc.)
☐ Oil, coolant, brake fluid, ATF, windshield fluid levels
☐ Emission components
☐ Road test after the tune-up

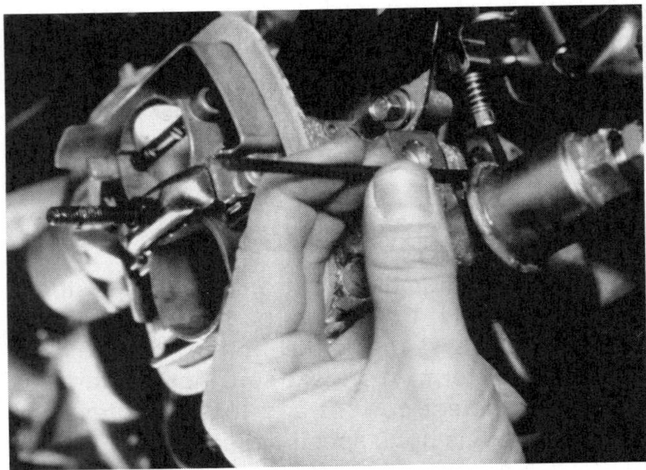

FIGURE 1-8 Adjusting the idle speed on a non-feedback carburetor

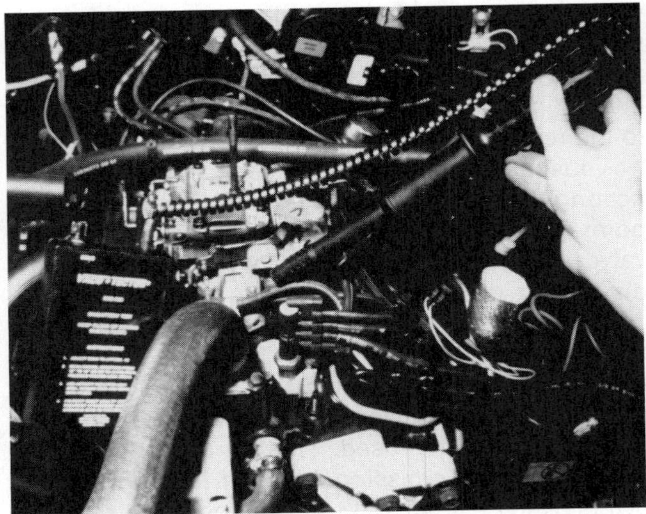

FIGURE 1-9 Checking for vacuum leakage, using an ultrasonic leak detector

CHECK AIR AND FUEL DELIVERY SYSTEMS

It will be impossible to tune the engine if the air and fuel systems are not functioning properly. The problem may be as simple as a dirty air filter or a clogged fuel filter. Low fuel pressure will adversely affect fuel injection systems. The air and fuel delivery systems must operate as intended to avoid performance problems (Figure 1-8).

CHECK THE IGNITION SYSTEM

Adequate current must be present in both the primary and secondary circuits to generate a spark of sufficient power and duration to provide optimum combustion. And this spark must be correctly timed to ensure maximum power transfer. Both dwell and timing must be calibrated to factory specifications during the whole range of operating conditions if the engine is to function properly.

CHECK THE COMPUTER DIAGNOSTICS

Most computers have a self-diagnostic feature. The computer constantly monitors the feedback of various input components and stores this information in its memory. Some computers flash related trouble codes when it is programmed to do so. Other computers relay performance information to diagnostic equipment specially designed to receive and interpret that information. No car can be successfully tuned when trouble is present in the electronic control system.

CHECK THE EMISSION SYSTEMS

Poor start-up, hesitation, and a host of other performance problems can be traced to a vacuum leak or a defective valve in an emission system. A thorough tune-up will always include verifying the operation of the PCV valve, the EGR valve, the thermostatic air cleaner, the EEC system, spark control, and air injection system, as well as checking all hoses and fittings for leaks (Figure 1-9).

DOUBLE-CHECK YOUR WORK

When the tune-up is complete, check the basic performance once more on an engine analyzer. Recheck overall ignition performance, cylinder power balance, and HC and CO levels. These final performance checks allow comparison to earlier tests, and when printed out provide a written form of assurance for the customer that his or her car is operating according to specifications.

ROAD TEST THE VEHICLE

Drive the vehicle to make sure it performs well and that the customer's specific performance complaints have been corrected. Put the vehicle through a full range of operating conditions: acceleration, wide-open throttle, deceleration, stop-and-go traffic, cruising speed, etc. Make sure that the vehicle not only starts easily, but also operates properly when cold and when running at normal operating temperatures. A good test drive can take 10 to 15 minutes.

PERFORMANCE TROUBLESHOOTING

Because of the lengthy service intervals required on late-model cars, many times a car owner will not bring his or her vehicle in for a tune-up until the vehicle develops a specific performance problem. That means, in addition to performing the regular tune-up procedure, the tune-up technician must be familiar with troubleshooting techniques and more specific diagnostic procedures.

The complexity of the modern automobile demands a thorough knowledge of the engine and its related systems and how these systems are controlled and relate to overall engine performance. This understanding is essential to helping to identify and locate a specific problem. Troubleshooting also requires an orderly process of investigation. All good troubleshooting is a process of elimination, working from the most likely to the least likely cause. Use the following ten steps to locate even the most elusive fault.

1. Interview the owner or driver. When someone brings in a vehicle, complaining of a specific performance problem, ask some pertinent questions to try to better define the problem. Here are some questions to ask:
   - When does the problem occur (cold idle, hot idle, cruising mode, etc.)?
   - Was the problem accompanied by any unusual noises, smells, or other signs?
   - Has the problem occurred before and if so what repairs were made to correct it?
   - What repairs or work has been done to the car recently?

   This initial interview is often conducted by the service writer for larger repair shops. The service writer then prepares a repair order (RO) with a description of the performance complaints and the specific work to be done.
2. Acquaint yourself with the car. Either look under the hood or check out the appropriate service manual to discover what equipment is on the car. What kind of ignition system, fuel system, emission systems, etc., does it have? Determine if any of these systems could be contributing to the problem. Also, be sure to check the emissions label.
3. Observe the problem. Equipped with a description of the problem and an understanding of the engine and its various systems, start the car and observe the problem firsthand. Look for related symptoms and pinpoint the conditions under which the problem occurs. Related symptoms and operating conditions will help eliminate potential causes and pinpoint the trouble area.
4. List the possible causes. From the observations made in steps 1 through 3, the problem should be evident. For example, if the problem is rough idle, the cause might be one or more of the following:
   - Percolation and vapor lock
   - Incorrect ignition timing
   - Faulty ignition system part, such as spark plugs
   - Loose or leaky vacuum hoses
   - Incorrect carburetor idle speed or fast-idle adjustment
   - Defect in the automatic choke system
   - Partially clogged fuel filter
   - Weak fuel pump
   - Dirty carburetor
   - Plugged PCV valve
   - Stuck thermostatic air-cleaner assembly damper valve
   - Leaking EGR valve
   - Stuck early fuel evaporation (EFE) valve or manifold heat-control valve

   Make a mental or written list of the potential problems.
5. Isolate the problem by following the manufacturer's recommended diagnostic procedure. Check the suspected component or system by performing the appropriate tests: compression, vacuum, leakage, electrical, etc. Sometimes by simply disconnecting or isolating the suspected component, the problem area can be pinpointed. An infrared exhaust gas analyzer can also isolate problems quickly. After eliminating systems that are operating properly, confine troubleshooting efforts to the most likely trouble area.
6. Dig deeper. After determining the general source of trouble, examine the system in detail. Diagrams and shop manuals are crucial for identifying components and tracing electrical connections and vacuum routing.
7. Test systematically. Identify the direction that power, vacuum, fuel, air, etc., flows; then, test components from the source through the various components in the di-

rection of flow until the problem is found. Do not jump back and forth within the system in search of a fault. Test components in the order of flow. This orderly process of elimination will avoid oversights that can be frustrating and time consuming.

8. Verify the test results. If testing exposes the possibility of a faulty solenoid or thermoswitch, for example, make sure that its electrical connections are clean and tight before replacing the component. If possible, test the component's operation before replacing it.
9. After narrowing the problem to a specific component or cause, make the necessary repairs or replace the defective component.
10. Test the repair. After making the repair, start the engine and verify that the repair has actually corrected the problem. There might be more than one contributing cause. Always recheck work carefully before returning the repaired vehicle to the customer.

TUNE-UP INFORMATION

This book will explain diagnostic procedures that generally apply to all vehicles. In some cases, specific models are used as examples of procedures common to all manufacturers. However, some vehicles have unique variations to the general procedures. For example, some vehicles require that certain electrical connections be made before testing the ignition timing. Others require the air conditioner, lights, or heater fan to be activated before testing the idle speed.

SERVICE MANUALS

These model-specific instructions must be obtained from the manufacturer's service manuals or by service manuals published independently by other parts manufacturers or publishers.

In addition to these specific diagnostic procedures, the service manuals are also an essential source of factory specifications applicable to the vehicle. The specifications tell the technician how the vehicle is supposed to operate when properly tuned. Without these specifications, it will be impossible to tune the engine.

Some typical specifications needed to tune up a vehicle would include some of the following:

- Spark plug type and gap
- Distributor gap
- Distributor dwell
- Slow idle speed
- Fast idle speed
- Ignition timing
- Cylinder numbering sequence and firing order
- Voltage, ohms, and amperage values for electrical components
- Compression pressure
- Fuel pump discharge pressure
- HC and CO levels in the exhaust (at idle)

However, factory service manuals are published every year and the specifications contained in them apply only to that particular model year. Another disadvantage of using service manuals to look up specs is that the specifications are distributed throughout the manuals. One section will give engine specs, another section will list specs for the fuel system, and so on.

TUNE-UP SPECIFICATION BOOKS

A faster source of specifications is found in specification books published by aftermarket manufacturers. Such books contain tune-up specifications for all American and import cars for several model years.

VECI DECALS

The most excessible source of tune-up information is the Vehicle Emission Control Information label or decal (Figure 1–10). All late-model cars are required by law to have a VECI label somewhere under the hood. The label lists some of the basic specifications and other pertinent information concerning the engine and emission systems. The VECI label might include some or all of the following:

- Engine identification
- Slow-idle speed
- Fast-idle speed
- Enriched-idle speed
- Emission system
- Exhaust emission limits
- Initial timing setting
- Breaker-point dwell
- Spark plug type and gap
- Carburetor adjustment instructions
- Vacuum routing diagrams

The information contained on the VECI decal is specific to that car. Any changes or modifications authorized by the manufacturer will also be marked on the label by the technician making the modification. The technician might also record the changes of a specially authorized modification decal and post the decal near the VECI label.

ANALYZER EQUIPMENT

A fourth source of specification and tune-up data is manufacturers of engine analyzer equipment. Computerized analyzers have the specifications programmed into their software packages that accompany the testers. Other analyzers require that the specifications be keyed into the machine by the operator. The analyzer will then run through a programmed sequence of tests comparing the actual vehicle readings to the specifications it has stored in its memory.

TROUBLESHOOTING AIDS

In addition to procedures and specifications, service manuals provide electrical wiring diagrams, vacuum diagrams, and troubleshooting trees. Without these diagnostic aids, engine diagnosis and tune-up would often be very confusing and frustrating.

ELECTRICAL DIAGRAMS

Electrical diagrams are "road maps" of the electrical systems in the vehicle. They identify the wiring (by color) between components and alert the technician to the presence of fuses, fusible links, circuit breakers, etc., in the circuits. They will also identify loads and grounds in a circuit. Figure 1–11 shows some typical symbols used in electrical diagrams. Tracing the electrical path taken by current within a circuit is an essential element in troubleshooting electronic systems prevalent on today's cars. A typical wiring diagram is shown in Figure 1–12.

VACUUM DIAGRAMS

Vacuum diagrams are just as important in locating trouble in emissions systems. Vacuum diagrams identify components in the systems (such as valves, solenoids, and amplifiers); the hoses that connect them (usually these are identified by color); and the source of the vacuum (intake manifold, venturi vacuum, ported carburetor vacuum, etc.). Shown in Figure 1–13, a vacuum system can be very complicated, with many of the parts interconnected. Without the aid of a vacuum system diagram, finding a vacuum leak can be difficult.

It is important to note that emission systems can vary on a single model, depending on the intended destination of the vehicle. Cars destined for the California market may have a somewhat different vacuum system than ones intended for the general American market. A vehicle destined for the Canadian market or for a low altitude environment might also have a different vacuum system. These small differ-

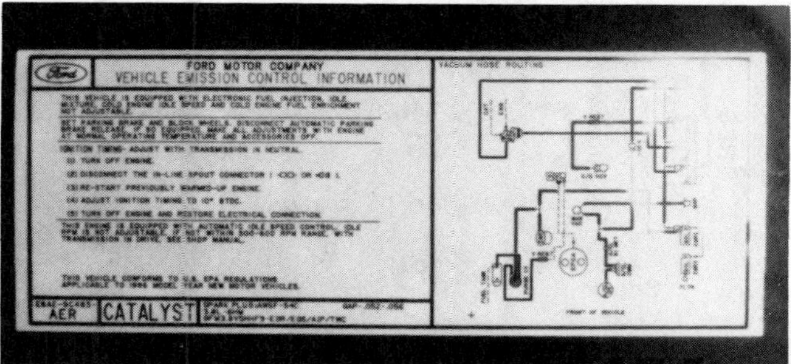

FIGURE 1–10 Vehicle emission control information decals contain vacuum routing diagrams and procedures for checking ignition timing, fast-idle speed, and so on.

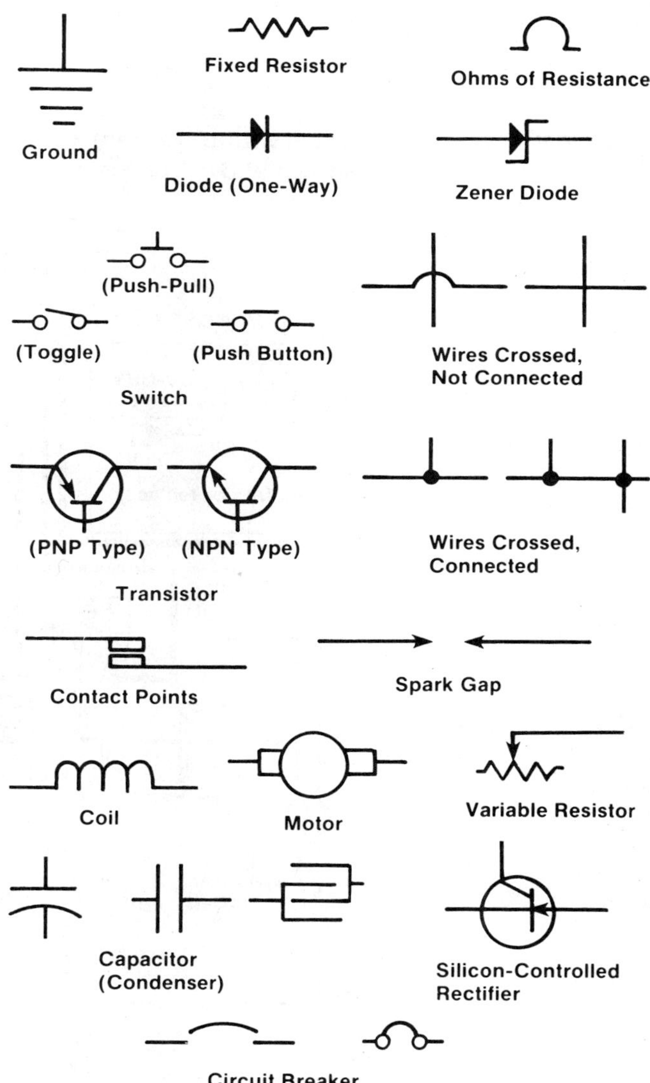

FIGURE 1–11 Common electrical diagram symbols

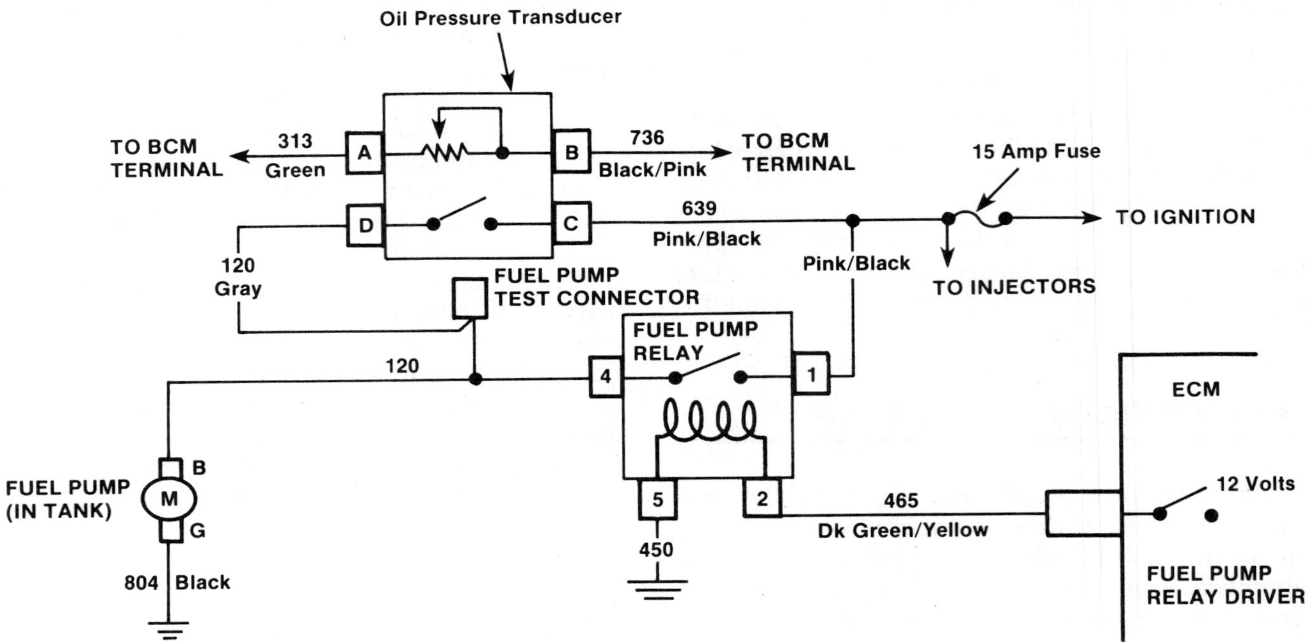

FIGURE 1-12 An electrical wiring diagram will show all the components, connectors, fuses, switches, etc., contained in a circuit. Shown above is a fuel pump circuit for an Oldsmobile fuel injection system.

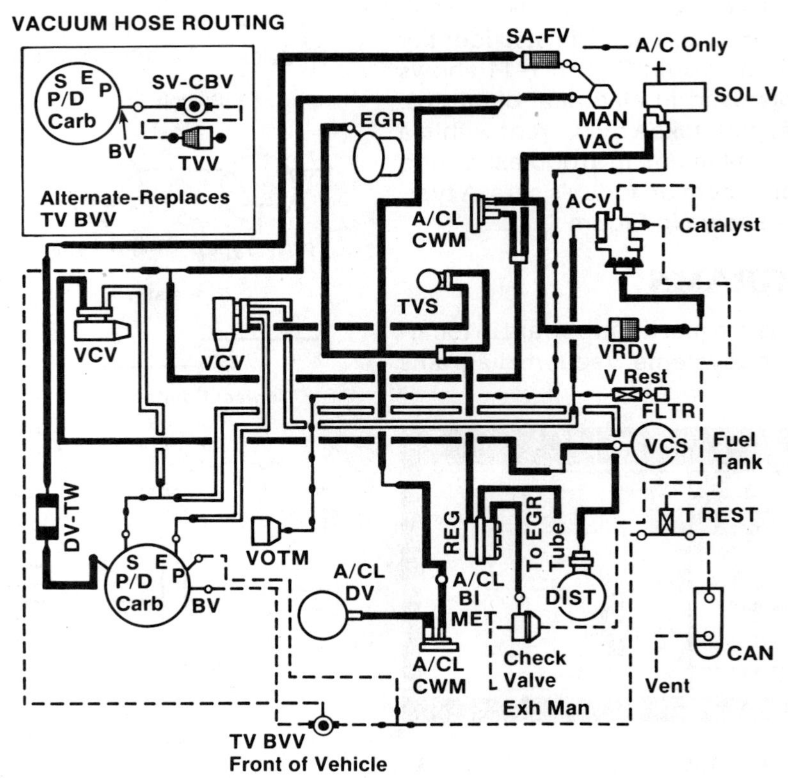

FIGURE 1-13 A vacuum system can be very complicated, as shown here on this vacuum system diagram for 1985 Ford 1.6 liter 4-cylinder engines (California only).

ences can be frustrating to even the experienced technician when he or she must work without the aid of a vacuum system diagram.

TROUBLESHOOTING TABLES

Troubleshooting tables are another valuable aid found in factory service manuals. The troubleshooting tables, or trees as they are sometimes called, take the technician step by step through a diagnostic procedure. For example, a troubleshooting table designed to discover the reason for a no-start condition will guide a technician through a battery of tests that logically eliminate potential causes until the fault is found. Other tables relate to troubleshooting codes that are output by the self-diagnostic feature of the on-board computers. A troubleshooting tree outlining the test procedure for testing the idle air control valve is shown in Figure 1-14.

GENERAL SHOP SAFETY

The most important considerations in any automotive repair shop should be accident prevention and safety. Carelessness and the lack of safety habits cause accidents. Accidents have a far-reaching effect, not only on the victim, but also on the victim's family and society in general. More important, accidents can cause serious injury, temporary or permanent, or even death. Therefore, it is the obligation

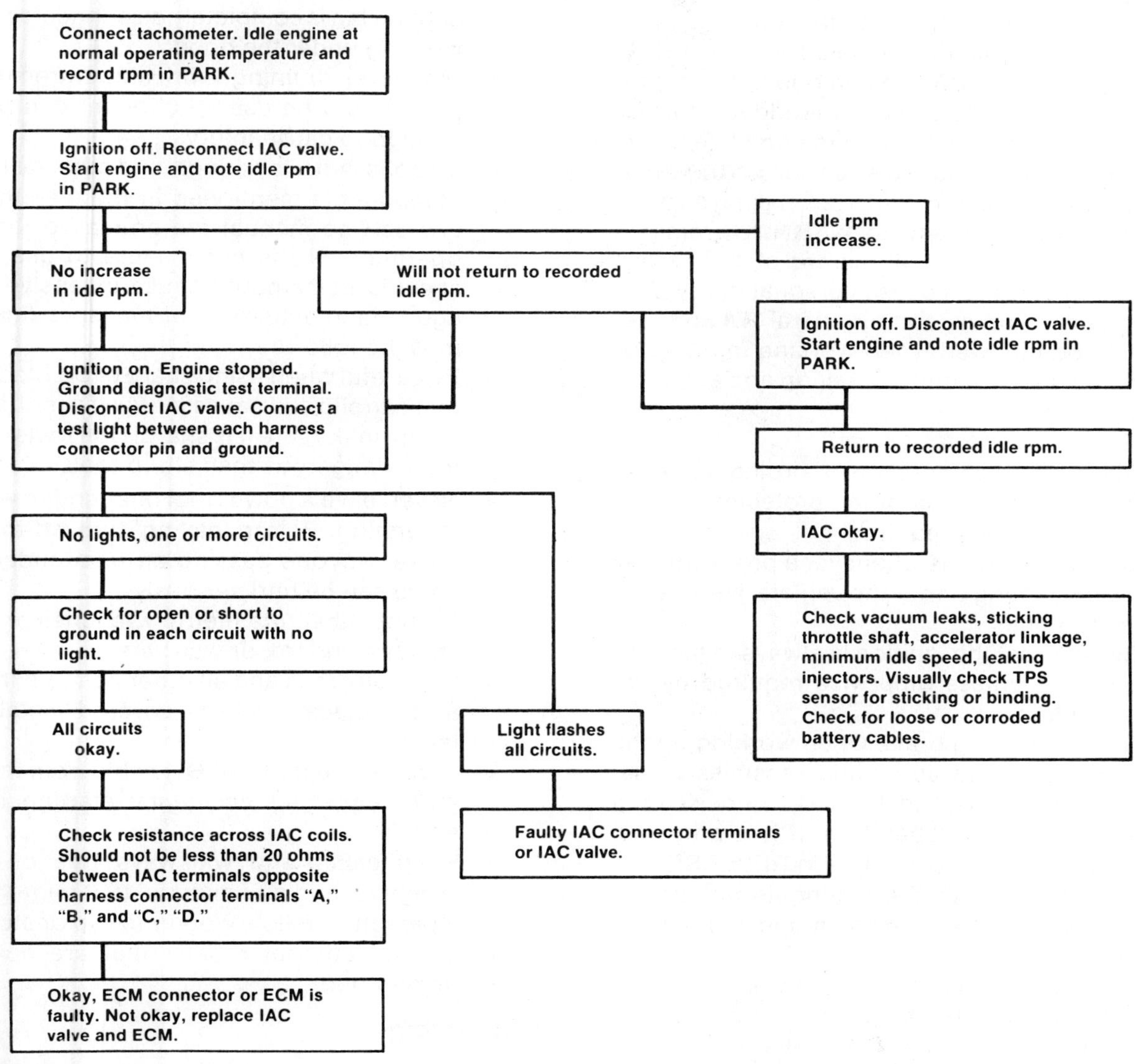

FIGURE 1-14 A typical troubleshooting tree

of all shop employees and the employer to develop a safety program to protect the health and welfare of those involved.

For example, extreme caution should be used while working with any of the components of the fuel system. Gasoline is a very volatile substance. Do not expose it to an open flame, spark, or high heat. Disconnect the negative terminal of the battery before doing any task that will release gasoline from any part of the system. Use containers to catch the gasoline and cloths to wipe up minor spills. Use a flashlight rather than a trouble light or droplight. Gasoline spilled on a hot bulb could cause the bulb to explode and ignite the gasoline. Always keep a Class B fire extinguisher or one capable of fighting type A, B, and C fires nearby to deal with problems that may occur. The Class B type is specially intended for use on gasoline fires.

In the following chapters of this book, the text contains special notations labeled **SHOP TALK, CAUTION,** and **WARNING.** Each one is there for a specific purpose. **SHOP TALK** gives added information that will help the technician to complete a particular procedure or make a task easier. **CAUTION** is given to prevent the technician from making an error that could damage the vehicle. **WARNING** reminds the technician to be especially careful of those areas where carelessness can cause personal injury. The following text contains some general **WARNINGs** that should be followed when working in an automobile repair shop on fuel inspection and emission control systems:

- Always wear safety glasses, protective clothing, respirator, or other protective equipment whenever required.
- Use safety stands whenever a procedure requires getting under the vehicle. Never trust jacks alone.
- Be sure that the ignition is always in the OFF position, unless otherwise required by the procedure.
- Set the parking brake when working on the vehicle. If it is an automatic transmission, set it in PARK unless instructed otherwise for a specific service operation. If it is a manual transmission, it should be in REVERSE (engine off) or NEUTRAL (engine on) unless instructed otherwise for a specific service operation.
- Operate the engine only in a well-ventilated area to avoid the danger of carbon monoxide. Most shop exhaust systems carry the exhaust away from a vehicle directly to the outside of the shop.
- Keep clothing away from moving parts when the engine is running, especially the fan and belts.
- To prevent serious burns, avoid contact with hot metal parts such as the radiator, exhaust manifold, tail pipe, catalytic converter, and muffler.
- Do not smoke while working on the vehicle.
- To avoid injury, always remove rings, watches, loose hanging jewelry, and loose clothing before beginning to work on a vehicle. Tie long hair securely behind the head.
- Keep hands and objects clear of the radiator fan blades. Electric cooling fans can start to operate at any time by an increase in underhood temperatures, even though the ignition is in the OFF position. Therefore, care should be taken to ensure that the electric cooling fan is completely disconnected when working under the hood.
- Whenever draining lubricant, extreme caution should be used. Lubricant can be hot enough to cause injury.
- All bolts, nuts, lock rings, and other fastening components mentioned in the manufacturer's service manual are crucial to the safe operation of the car. Failure to use those specific items could cause extensive damage. Manufacturer's torque specifications must be followed.
- Be careful when using sharp or pointed tools that can slip and cause injury. If a tool is to be sharp, make sure it is sharp. Dull tools can be more dangerous than sharp tools.
- Never leave a power tool unattended when it is running. Before leaving, turn off the machine. Anyone passing an unattended machine can be hurt seriously.
- Do not substitute metric wrenches or sockets for standard, or vice versa.
- Store oily rags and all other combustibles in a safe place, such as covered metal containers.
- Remove the ground (B–) cable from the battery to prevent accidental starting of the engine.
- Keep aisles and walkways free of tools, creepers, and any material that might cause a person or fellow worker to trip or stumble.
- Do not attempt repairs that are not fully understood.

Lift Safety

Some undercar work is necessary to raise the car to repair fuel, emission, and exhaust systems.

Raising a vehicle on a lift or a hoist requires specific care. For example, it is important to make sure that vehicles equipped with a catalytic converter have enough clearance between the hoist and exhaust system components before driving the vehicle onto the ramps.

Adapters and hoist plates must be positioned correctly on twin post and rail type lifts to prevent damage to the underbody of the vehicle. The tie-rod, rod bracket, and shock absorbers are some of the undercar components that could be damaged if the adapters and hoist plates are incorrectly placed.

There are specific contact points to use where the weight of the vehicle is evenly supported by the adapters or hoist plates. The correct lifting points can be found in the vehicle's service manual. Figure 1-15 shows typical locations for both frame and unibody cars. These diagrams are for illustration only; always use the manufacturer's instructions as to the safe lifting procedure.

- Before operating any lift or hoist, carefully read the owner's manual and understand all the operating and maintenance instructions.

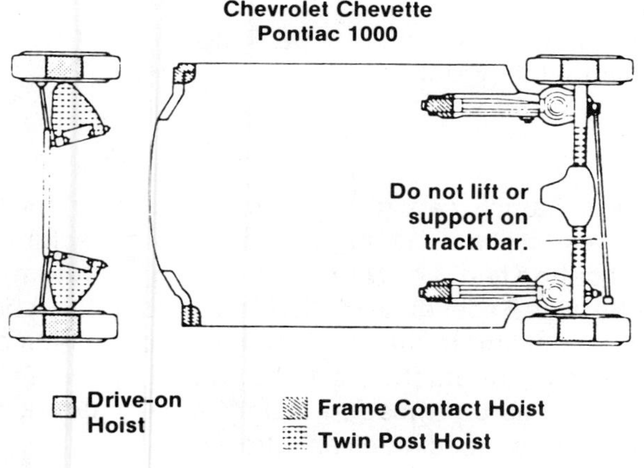

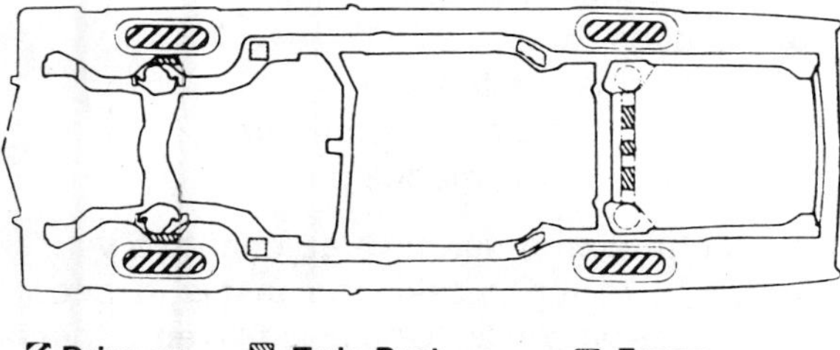

FIGURE 1-15 Lift points for unibody and frame cars

WARNING: Always use the rated tonnage of a lift jack for tons specified. If a jack is rated for 2 tons, do not attempt to use it for a job requiring 20 tons. It is dangerous for both the technician and the vehicle.

- Before driving a vehicle over a lift, position the arms and supports to provide unobstructed clearance. Do not hit or run over lift arms, adapters, or axle supports. This could damage the lift or vehicle.
- Load the vehicle on the lift carefully. Check to make sure adapters or axle supports are in secure contact with the vehicle, according to manual instructions, before raising it to the desired working height. Remember that unsecured loads can be dangerous.
- Make sure the vehicle's doors, hood, and trunk are closed prior to raising the vehicle. Never raise a car with passengers inside.
- Position the lift supports to contact at the vehicle manufacturer's recommended lifting points. Raise the lift until the supports contact the vehicle. Check supports for secure contact with the vehicle and raise the lift to the desired working height.

WARNING: When working under a car, the lift should be raised high enough for the locking device to be engaged.

- After lifting a vehicle to the desired height, be sure to always lower the unit onto mechanical safeties.
- Note that with some vehicles, the removal (or installation) of components can cause a critical shift in the center of gravity and result in raised vehicle instability. Refer to the vehicle manufacturer's service manual for recommended procedures when vehicle components are removed.
- Make sure tool trays, stands, and so forth are removed from under the vehicle. Release locking devices as per instructions before attempting to lower the lift.
- Before removing the vehicle from the lift area, position the arms, adapters, or axle supports to assure that the vehicle or lift will not be damaged.
- Inspect the lift daily. Never operate it if it malfunctions or if it has broken or damaged

parts. It should be removed from service and repaired immediately. A lift requires immediate attention if it
—Jerks or jumps when raised
—Slowly settles down after being raised
—Slowly rises when not in use
—Slowly rises when in use
—Comes down very slowly
—Blows oil out of the exhaust line
—Leaks oil at the packing gland
- Repairs should be made with original equipment parts only.

EMISSION SYSTEM SAFETY

1. Never run a vehicle for emission testing in an enclosed area. Exhaust system leaks should be repaired immediately.
2. Catalytic converter housings get very hot. Do not park a converter-equipped vehicle over dry grass, leaves, or other flammable materials. To avoid burns, do not touch a catalytic converter until the engine has been off for at least 45 minutes.
3. Any situation that allows unburned fuel vapors to enter a catalytic converter can be dangerous. Correct any over-rich air/fuel condition or ignition system problems immediately. Do not try to push-start a converter-equipped car. If the engine must be tested with one or more cylinders not firing, run the engine with all cylinders firing for 30 seconds between tests so that the converter can cool. After testing a cylinder, give the engine a minute or two to recover and let the test speed stabilize before moving on to the next cylinder.

ASE CERTIFICATION

Just as doctors, nurses, accountants, electricians, and other professionals are licensed or certified in order to practice their profession, so the mechanic can be certified. Certification protects the general public and the practitioner or professional. It assures the general public and the prospective employer that certain minimum standards of performance have been met. Many employers now expect their mechanics or technicians to be certified. The certified technician is recognized as a professional by the public, employer, and peers. For this reason, the certified technician usually receives higher pay than the noncertified employee.

FIGURE 1-16 Certified ASE Automobile Technician shoulder patch

Mechanics can get certified in one or more technical areas by taking and passing a mechanic certification test. The National Institute for **A**utomotive **S**ervice **E**xcellence (ASE) offers a voluntary certification program that is recommended by the major vehicle manufacturers in the United States. This program has certification tests that fall into eight categories or specialties as follows:

1. Engine repairs
2. Automatic transmissions/transaxle
3. Manual drivetrain and rear axle
4. Suspension and steering
5. Brakes
6. Electrical systems
7. Heating and air-conditioning
8. Engine performance

The technician must have at least two years of experience in order to take the test and become certified. Those who wish to be certified as a Specialist Technician need to pass only the examination that pertains to the area of specialty. To be certified as a General Technician requires passing examina-

FIGURE 1-17 Certified ASE Master Automobile Technician shoulder patch

tions in all eight areas. A nominal fee is charged for each test taken.

A technician who passes one examination receives an Automobile Technician shoulder patch (Figure 1-16). A Master Automobile Technician meeting minimum standards in all eight categories receives the shoulder patch shown in Figure 1-17.

To help prepare for the ASE examinations the test questions at the end of each chapter are similar in design and content to that used by the ASE. For further information on the ASE certification program, write: National Institute for Auto Service Excellence, 1920 Association Drive, Reston, Virginia 22091. Remember that an ASE certified technician (or specialist) is usually better paid than the general line mechanic.

REVIEW QUESTIONS

1. Which of the following is an objective of a tune-up?
   a. to keep the vehicle running to factory specifications
   b. to replace wearing components before they fail
   c. to repair or replace defective components
   d. all of the above

2. Which of the following was one of the factors leading to today's electronic car?
   a. federally mandated emissions levels
   b. fuel shortages in the 1970s
   c. the development of microprocessors
   d. all of the above

3. Which of the following states was the first to develop emissions standards?
   a. New York
   b. New Jersey
   c. Texas
   d. California

4. According to service schedules for most late-model cars, how often should spark plugs be changed?
   a. after 7,500 miles
   b. after 10,000 miles
   c. after 15,000 miles
   d. after 30,000 miles

5. Which one of the following adjustments might not be necessary on computer-controlled cars?
   a. idle mixture
   b. ignition breaker point gap
   c. ignition timing
   d. all of the above

6. Technician A looks under the hood to discover what equipment is on a car. Technician B checks out the appropriate service manual. Who is right?
   a. Technician A
   b. Technician B
   c. Both A and B
   d. Neither A nor B

7. Which of the following emission gasses are not controlled by federal legislation?
   a. CO_2
   b. HC
   c. NO_x
   d. CO

8. Which of the following is not true of electronic ignitions?
   a. experiences less wear than older breaker point ignitions
   b. generates a much hotter spark
   c. requires readjustment often
   d. all of the above

9. Keeping the air/fuel mixture close to the stoichiometric ratio ____________.
   a. increases the emissions levels
   b. decreases fuel efficiency
   c. increases power delivery and performance
   d. all of the above

10. When working on a manual transmission vehicle with the engine off, Technician A sets it in PARK. Technician B sets it in REVERSE. Who is right?
    a. Technician A
    b. Technician B
    c. Both A and B
    d. Neither A nor B

11. The first step in any tune-up is to ____________.
    a. test drive the vehicle
    b. check the battery
    c. ask questions
    d. change the spark plugs

12. To discover trouble with the engine and its systems ____________.
    a. check engine performance with an oscilloscope
    b. check engine exhaust using an analyzer
    c. check the on-board computer diagnostics
    d. all of the above

13. Which of the following is not a step in troubleshooting a performance problem?
    a. discussing the problem with the owner or driver
    b. observing the problem
    c. listing the probable causes
    d. replacing suspected components with new ones

14. Which of the following is not a source of tune-up specifications?
    a. factory service manuals
    b. tune-up specification books
    c. engine analyzer equipment software
    d. this textbook

15. Which of the following is included on vehicle emission control information (VECI) decals?
    a. electrical wiring diagrams
    b. troubleshooting trees
    c. vacuum routing diagrams
    d. all of the above

CHAPTER TWO

ENGINE TUNE-UP TOOLS AND EQUIPMENT

Objectives

Upon completion of this chapter, you should be able to:
- Name the diagnostic tools and equipment used in engine tune-ups.
- Describe the basic application and use of diagnostic tools.
- Explain the importance of trained technicians skilled in the use of state-of-the-art electronic and computerized test equipment.

As the trend toward electronic integration of ignition, fuel, and emission systems progresses, diagnostic test equipment must also reflect these changes. New tools and techniques have developed to support diagnosis of electronic ignition systems, fuel injection systems, and computer engine controls. Other tools are no longer useful on late-model vehicles.

Today's automotive technician must not only keep abreast of changes in automotive technology, but he or she must also stay informed on new testing procedures and the use of specialized diagnostic equipment. The successful garage must invest in the testing equipment and technical training necessary to troubleshoot today's electronic engine systems.

However, most testing devices used on pre-electronic controlled vehicles are still important diagnostic tools. Not all performance problems are related to electronic control systems, and time-proven test procedures designed to identify electrical and mechanical faults are still a very vital part of a thorough tune-up. The following are some of the basic tools used in engine tuneup and diagnostics.

COMPRESSION AND VACUUM TESTERS

Compression and vacuum are two important indicators of engine condition. Manifold vacuum is also used to control emission devices. The following test equipment is used for measuring compression and vacuum, and for locating pressure and vacuum leaks.

COMPRESSION GAUGE

Internal combustion engines depend on compression of the air/fuel mixture to maximize the energy potential of the air/fuel charge. The upward movement of the piston on the combustion stroke of a four stroke engine compresses the air within the combustion chamber. The air/fuel mixture gets hotter as it is compressed. The hot mixture is easier to ignite than the same mixture at room temperature, and when ignited, it will generate much more power.

If the combustion chamber has leaks in it, some of the air/fuel mixture will escape when compressed, resulting in a loss of efficiency and power. The leaks could be the fault of burned valves, a blown head gasket, worn rings, a slipped timing chain, worn valve seats, a cracked head, and more.

An engine with poor compression cannot be successfully tuned to factory specifications. That is why a compression check of each cylinder should be performed when the spark plugs are removed.

A compression gauge is used to check cylinder compression. The dial face on a typical compression gauge indicates pressure in both pounds per square inch (psi) and metric kilopascals (kPa). The range is usually 0 to 300 psi and 0 to 2100 kPa.

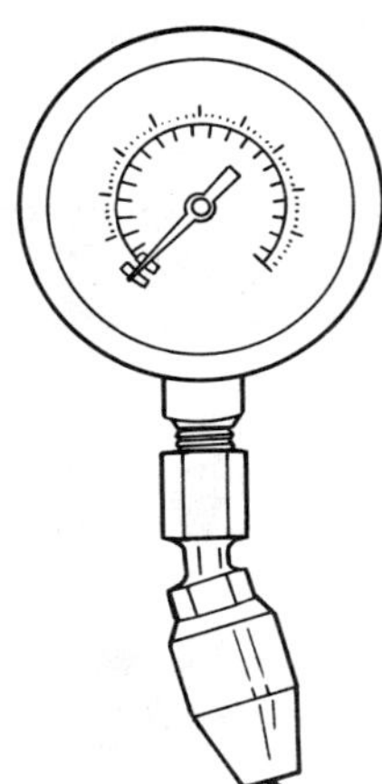

FIGURE 2-1 Push-in type compression gauge

There are two basic types of compression gauges: the push-in gauge (Figure 2-1) and a screw-in gauge (Figure 2-2).

The push-in type has a short stem that is either straight or bent at a 45-degree angle. The stem ends in a tapered rubber tip that fits any size spark plug hole. The rubber tip is placed in the hole and held there by hand while the engine is cranked through several compression cycles. Although simple to use, the push-in gauge will give an inaccurate reading if it is not held tightly in the hole.

The second type of compression gauge—the screw-in type—has a long flexible hose that ends in a threaded adapter. The flexible hose can reach into areas that are inaccessible with a push-in type. The adapters are changeable and come in several thread sizes to fit 10 mm, 12 mm, 14 mm, and 18 mm diameter holes. The adapters screw into the spark plug holes in place of the spark plugs. When installed properly, the gauge is leakproof and the technician's hands are free.

The better compression gauges have a vent valve that holds the highest pressure reading on the dial. Opening the valve releases the pressure when the test is complete.

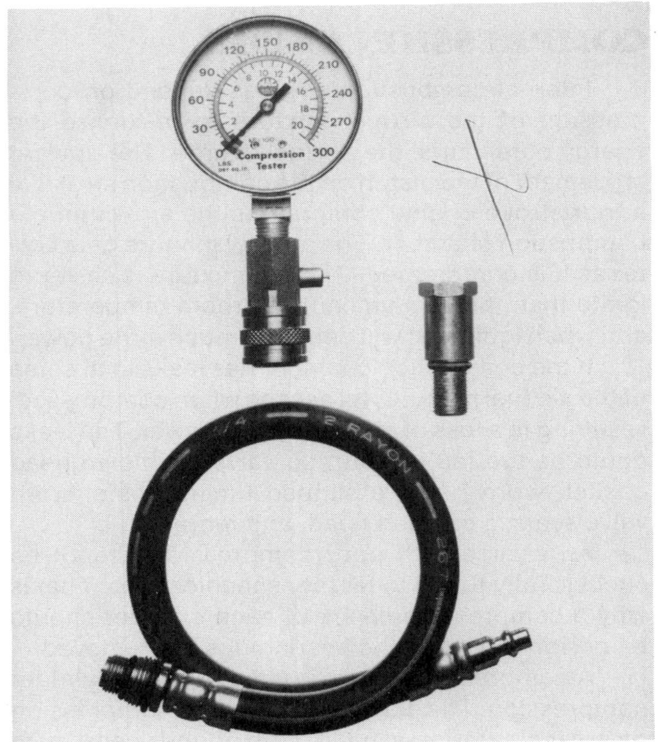

FIGURE 2-2 Screw-in type compression gauge with adapters *(courtesy of MAC Tools, Inc.)*

CYLINDER LEAKAGE TESTER

If a compression test shows any of the cylinders to be leaking, a cylinder leakage test can be performed to measure the percentage of compression lost and help locate the source of leakage.

A cylinder leakage tester (Figure 2-3) applies compressed air to the cylinder through the spark plug hole. A threaded adapter on the end of the air pressure hose screws into the spark plug hole. Compressed air is provided from the shop compressed air system. A pressure regulator in the tester controls the pressure applied to the cylinder. An analog gauge registers the percentage of compression lost from the cylinder when the compressed air is applied. The scale on the dial face reads 0 to 100 percent.

A zero reading means that there is no leakage from the cylinder. Readings of 100 percent would indicate that the cylinder will not hold any pressure.

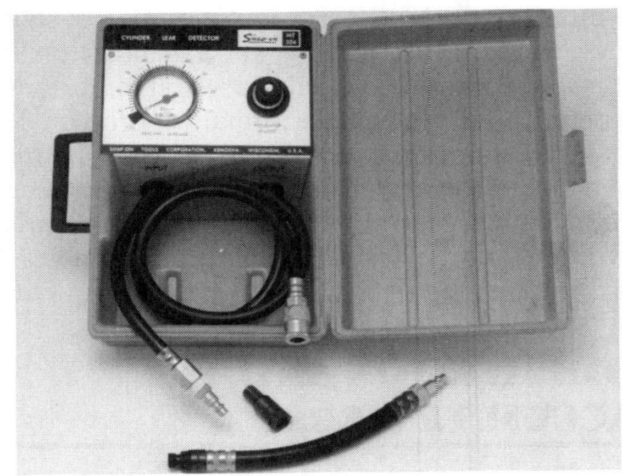

FIGURE 2-3 Cylinder leakage tester *(courtesy of Snap On Tools)*

Most vehicles, even new cars, experience some leakage around the rings. Up to 20 percent is considered acceptable during the leakage test. When the engine is actually running, the rings will seal much better. However, there should be no leakage around the valves or the head gasket.

VACUUM GAUGE

Measuring intake manifold vacuum is another way to diagnose the condition of an engine. Manifold vacuum is tested with a vacuum gauge (Figure 2-4).

The vacuum gauge measures the difference in pressure between the intake manifold and the outside atmosphere. If the manifold pressure is lower than the atmospheric pressure, a vacuum exists. Vacuum is measured in inches of mercury (in./Hg) and in kiloPascals (kPa) or millimeters of mercury (mm/Hg). The atmospheric pressure at sea level is 30 in./Hg. For every 1000 feet increase in altitude, the atmospheric pressure drops 1 inch.

A flexible hose on the gauge must be connected to a source of manifold vacuum, either on the manifold or the carburetor. Sometimes this requires removing a plug from the manifold and installing a special fitting.

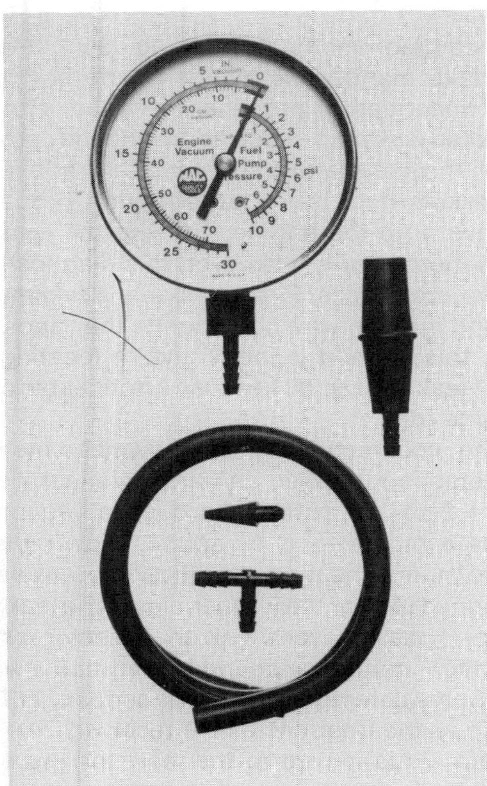

FIGURE 2-4 Vacuum gauge kit *(courtesy of MAC Tools, Inc.)*

The test is made with the engine cranking or running. A good vacuum reading is 15 to 20 inches of mercury (50 to 65 kPa). However, be sure to consult the factory service manual for specifications.

Low or fluctuating readings can indicate many different problems. For example, a low, steady reading might be caused by retarded ignition or valve timing. A sharp vacuum drop at regular intervals might be caused by a burned intake valve.

Other conditions that can be revealed by vacuum readings are as follows:

- Stuck or burned valves
- Improper valve or ignition timing
- Weak valve springs
- Leaking intake manifold
- Uneven compression
- Worn rings or cylinder walls
- Leaking head gaskets
- Incorrect carburetor adjustments
- Restricted exhaust system
- Ignition defects

The vacuum gauge is also used to test vacuum-operated components, such as a distributor vacuum advance mechanism or a thermal vacuum valve. A "tee" connector is used to connect the gauge in line with the system's vacuum hoses.

VACUUM PUMPS

Another tool used to test vacuum actuated components is the vacuum pump (Figure 2-5). The vacuum pump consists of a hand pump, a vacuum gauge, and a length of rubber hose used to attach the pump to the component being tested. Tests with the vacuum pump can usually be performed without removing the component from the car.

When the handles of the pump are squeezed together, a piston in the pump body draws air out of the component being tested. The partial vacuum created by the pump is registered on the pump's vacuum gauge. The vacuum level needed to actuate a given component should be compared to the specifications given in the factory service manual.

The vacuum pump is also commonly used to locate vacuum leaks. This is done by connecting the vacuum pump to a suspect vacuum hose or component and applying vacuum. If the needle on the vacuum gauge beings to drop after the vacuum is applied, a leak exists somewhere in the system.

The following list shows some of the systems and components that can be tested with a vacuum pump:

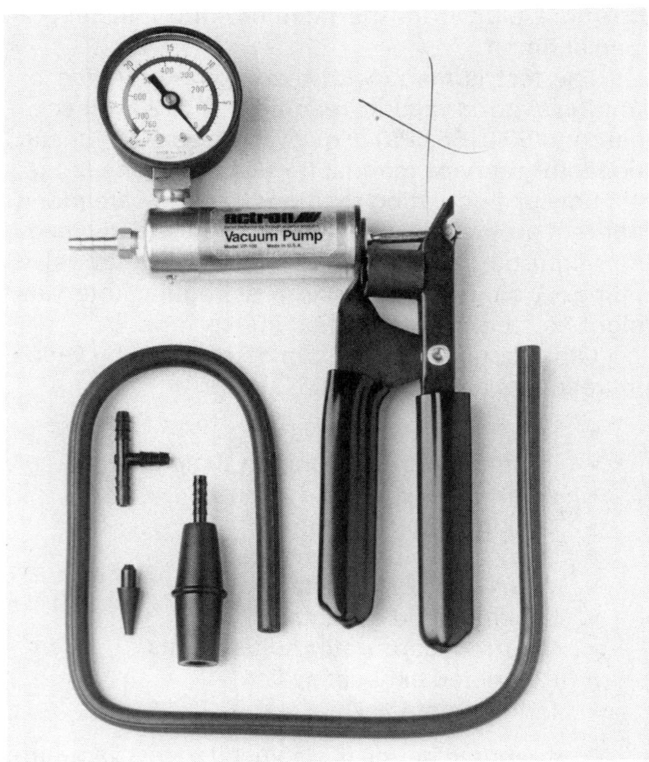

FIGURE 2-5 Vacuum pump with accessories *(courtesy of Actron Manufacturing Co.)*

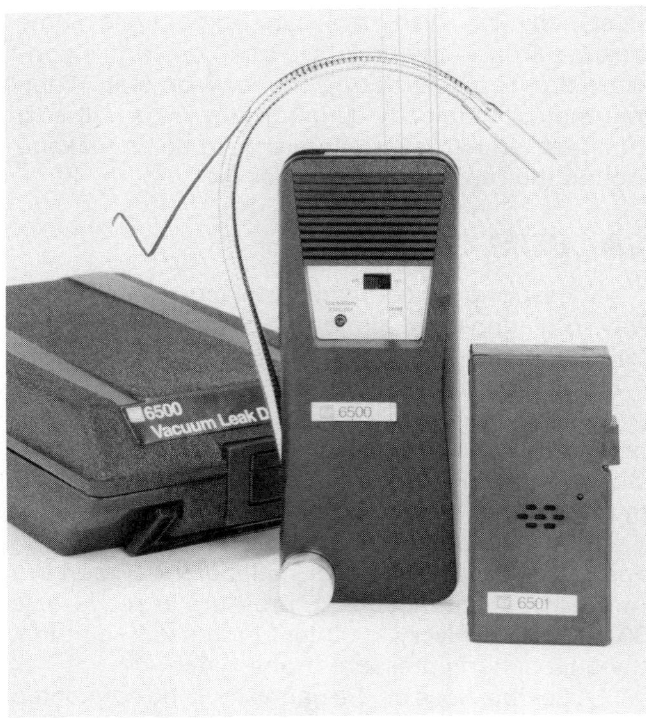

FIGURE 2-6 Ultrasonic vacuum leak detector *(courtesy of TIF Instruments, Inc.)*

1. Carburetor Service
   - Choke pull-off diaphragm
   - Vacuum break diaphragm
2. Ignition System Service
   - Vacuum advance unit
   - Vacuum retard unit
3. Computerized Engine Control Systems
   - Vacuum transducer MAP sensor
4. Emission Control Systems
   - Thermal vacuum switches
   - Vacuum delay valves
   - Heated air intake systems
   - EGR (Exhaust Gas Recirculation) valves
5. Automatic Transmission Modulator Valves

VACUUM LEAK DETECTOR

The existence of a vacuum leak or compression leak might be revealed by a compression check, a leak down test, or a manifold vacuum test. However, finding the location of the leak can be very difficult.

A simple, but time-consuming way to find leaks in a vacuum system is to check each component and vacuum hose with a vacuum pump. Simply apply vacuum to the suspected area and watch the gauge for any loss of vacuum.

A common method of finding vacuum leaks in the intake manifold is with propane enrichment. Small quantities of propane are sprayed over the suspected leak point, such as around the carburetor gasket, throttle shafts, EGR stems, and intake manifold gaskets. If the leak is large enough, propane will be drawn into the engine, causing the rpm to increase momentarily. However, this diagnostic trick has two drawbacks. First, propane is flammable. A sparking ignition wire could ignite the vapors. Secondly, this method is ineffective in locating small leaks—leaks too small to cause a noticeable change in engine rpm.

The most technologically advanced method of leak detection is called an ultrasonic leak detector (Figure 2-6). Air rushing through a vacuum leak creates a high-frequency sound, higher than the range of human hearing. An ultrasonic leak detector is designed to hear the frequencies of the leak. When the tool is passed over a leak, the detector responds to the high-frequency sound by emitting a warning beep. Some detectors also have a series of LEDs that light up as the frequencies are received. The closer the detector is moved to the leak, the more LEDs light up or the faster the beeping occurs. This allows the technician to zero in on the leak. An ultrasonic leak detector can sense leaks as small as 1/500 inch and accurately locate the leak to within 1/16 inch.

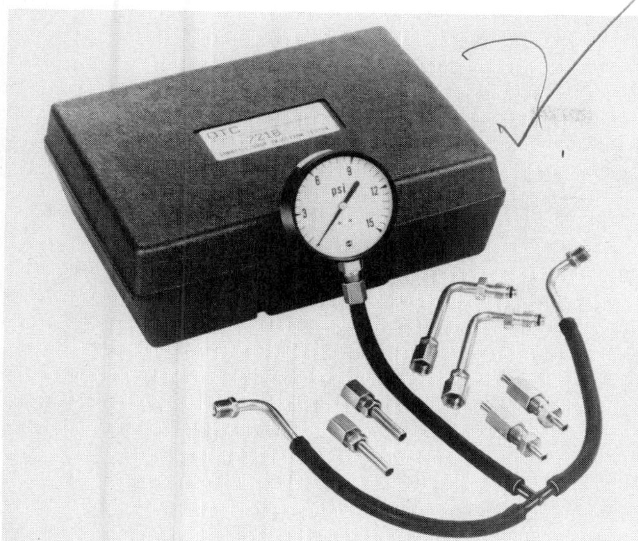

FIGURE 2-7 Throttle body injection tester *(courtesy of OTC Tool and Equipment Division of SPX Corp.)*

An ultrasonic leak detector can also be used to find compression leaks, bearing wear, and electrical arcing and can be used to diagnose fuel injector operation.

FUEL SYSTEM TEST EQUIPMENT

Misadjusted or misfunctioning fuel system components will adversely affect engine performance. The following tools are used to test fuel system components.

PRESSURE GAUGE

A pressure gauge (Figure 2-7) is needed to measure the pressure in the fuel system. This is particularly true in fuel-injected systems in which the discharge pressure of the fuel pump may be 35 to 70 psi. A drop in fuel pressure will reduce the fuel delivered to the injectors and result in a lean air/fuel mixture.

The pressure gauge can be used to check discharge pressure of fuel pumps, the regulated pressure of fuel injection systems, and injector pressure drop.

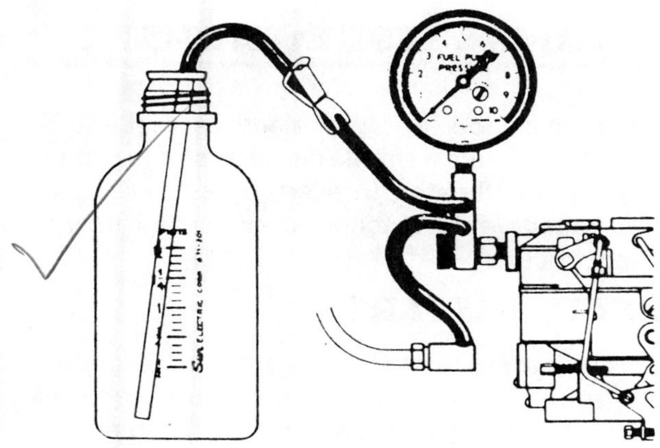

FIGURE 2-8 Fuel pressure tester

SHOP TALK

The fuel system on fuel-injected vehicles is highly pressurized. The pressure must be relieved before servicing any component in the fuel system. Follow the procedures given in Chapter 13 for relieving fuel pressure.

Some fuel pressure gauges also have a valve and outlet hose for testing fuel pump discharge volume (Figure 2-8). The manufacturer's specification for discharge volume will be so many pints or liters of fuel in a certain number of seconds. Discharge volume testing is particularly applicable to carbureted engines. Fuel-injected systems that are computer controlled have specific procedures for relieving pressure. Be sure to consult the applicable service manuals.

FUEL INJECTOR CLEANER

Clogged fuel injectors are becoming a much more common problem as the use of fuel injection systems increases. Clogged injectors are the result of inconsistencies in fuel gasoline detergent levels and high sulfur content. When these sensitive injection systems become partially clogged, fuel flow is restricted, and spray patterns are altered, causing poor performance and reduced fuel economy.

The solution to a sulfated and plugged fuel injector is to clean it, not replace it. There are two kinds of fuel injector cleaners. One is a pressure tank (Figure 2-9). A mixture of solvent and unleaded gasoline is placed in the tank, following the manufacturer's instructions for mixing, quantity, and safe handling. The vehicle's fuel pump must be disabled and, on some vehicles, the fuel line must be blocked between the pressure regulator and the return line. Then, the hose on the pressure tank is connected to the service port in the fuel system. The in-line valve is then partially opened and the engine is started. It should run at approximately 2000 rpm for about 10 minutes to clean the injectors thoroughly.

An alternative to the pressure tank is a pressurized canister (Figure 2-10) in which the solvent solu-

FIGURE 2-9 Pressurized tank used for cleaning fuel injectors *(courtesy of OTC Tool and Equipment Div. of SPX Corp.)*

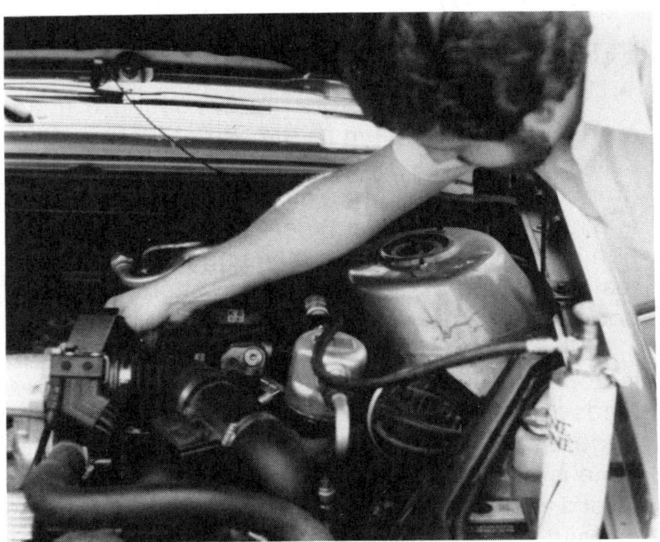

FIGURE 2-10 Throwaway canister of fuel injector cleaner *(courtesy of Penray Co.)*

tion is premixed. Use of the canister-type cleaner is similar to the procedure described above, but does not require mixing or pumping.

The canister is connected to the injection system's service fitting and the valve on the canister is

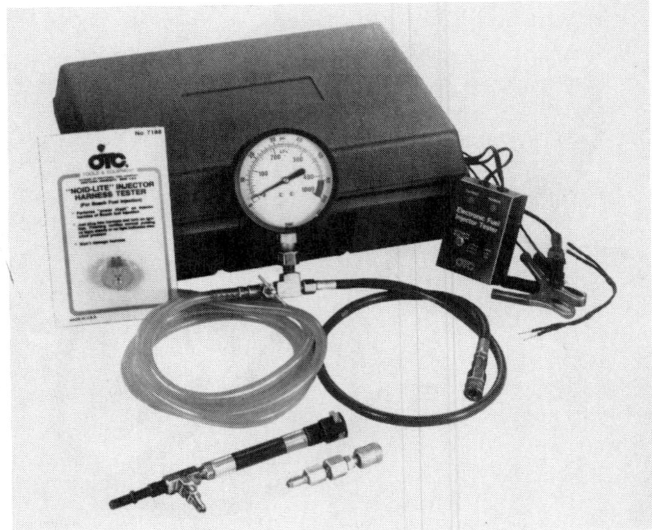

FIGURE 2-11 Fuel pressure gauge and electronic fuel injector tester *(courtesy of OTC Tool and Equipment Div. of SPX Corp.)*

opened. The engine is cranked and allowed to run until it dies. Then, the canister can be discarded.

FUEL INJECTOR PULSE TESTER

The only accurate execution of a cylinder balance test when injectors are suspect in engine performance problems is with a fuel injector pulse tester (Figure 2-11). This electronic tool fires individual fuel injectors in 1/2-second increments in three different ranges: 1 pulse of 500 milliseconds, 50 pulses of 10 milliseconds, and 100 pulses of 5 milliseconds. While the injector is being pulsed, the technician will monitor pressure drop. Ideally each injector should create the same pressure drop when activated for the same amount of time. Little or no pressure drop indicates a plugged or defective injector. Excessive pressure drop indicates an overly rich condition.

ELECTRICAL TEST EQUIPMENT

There is a wide variety of electrical test equipment—ranging from a simple circuit tester to a complex, digital multimeter. In every case, these tools test for voltage, resistance, and/or current flow in electrical circuits.

CIRCUIT TESTERS

Circuit testers (Figure 2-12) are used to identify shorts, grounds, and open circuits in any electrical circuit. Low-voltage testers are used to troubleshoot 6- to 12-volt circuits. High-voltage circuit testers diagnose primary and secondary ignition circuits.

A circuit tester looks like a stubby ice pick. Its handle is transparent and contains a light bulb. A probe extends from one end of the handle and a ground clip and wire from the other end. When the ground clip is attached to a good ground and the probe touched to a live connector, the bulb in the handle will light up. If the bulb does not light, voltage is not available at the connector.

A self-powered circuit tester is called a continuity tester (Figure 2-13). It is used with the power off in the circuit being tested. It looks like a regular circuit tester, except that it has a small internal battery. When the ground clip is attached to the ground terminal of a component and the probe touched to the feed wire, the light will be illuminated if there is continuity in the circuit. If an open circuit exists, the light will not be illuminated.

ELECTRICAL TEST METERS

There are several meters that are essential for testing and troubleshooting electrical systems. These are the voltmeter, ohmmeter, and ammeter.

Voltmeter

The voltmeter (Figure 2-14) measures voltage available at any point in an electrical system. For example, it can be used to measure the voltage available at the battery. It can also be used to test the voltage available at the terminals of any component connector. The voltage meter can also be used to test voltage drop across a relay, switch, connector etc.

A voltmeter usually has two leads: a red positive lead and a black negative lead. The red lead should be connected to the positive (voltage side) circuit and the black should be connected to ground or to the negative side of the component. Voltage meters should always be connected parallel with the circuit being tested. When connected in series, the high resistance of the meter will prevent proper operation of the component.

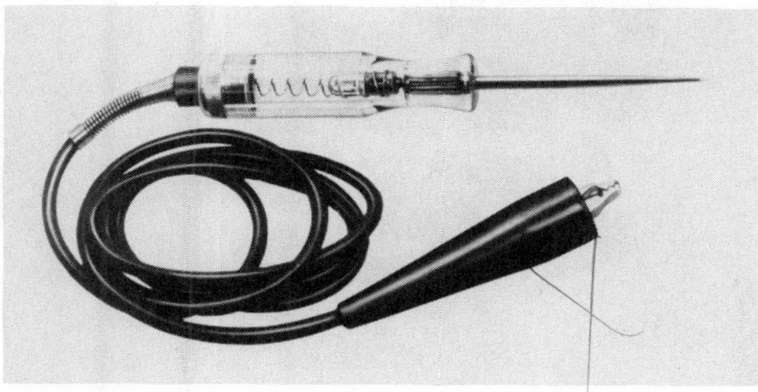

FIGURE 2-12 Circuit tester *(courtesy of MAC Tools)*

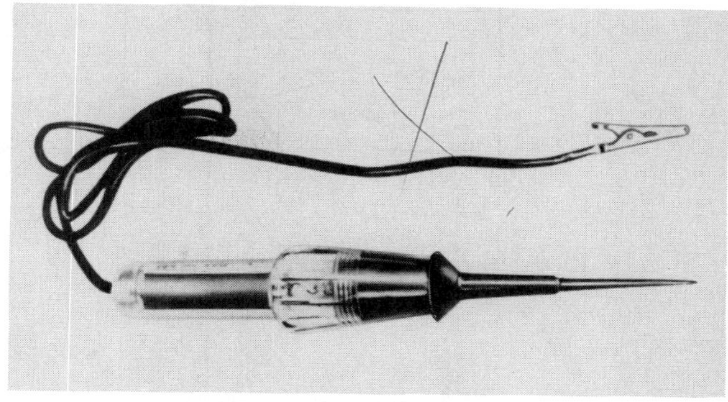

FIGURE 2-13 Continuity tester *(courtesy of MAC Tools)*

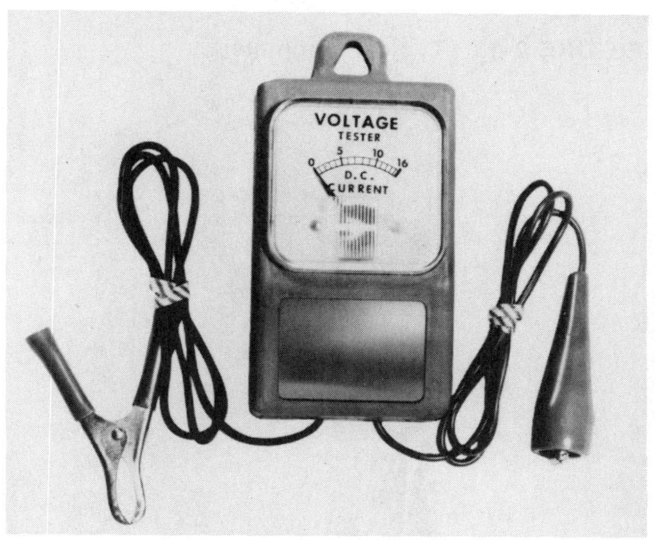

FIGURE 2-14 Voltage meter *(courtesy of MAC Tools)*

Ohmmeter

An ohmmeter (Figure 2-15) measures resistance to current flow in a circuit. In contrast to the voltmeter, which uses the voltage available in the circuit being tested, the ohmmeter is battery powered. The circuit being tested must be open; if power is on in the circuit, the ohmmeter will be damaged.

The two leads of the ohmmeter are placed parallel with the circuit being tested. The red lead is placed on the positive side of the circuit and the black lead is placed on the ground or negative side of the circuit. The meter supplies voltage to the circuit and compares the voltage available on the positive side of the circuit with the voltage available at

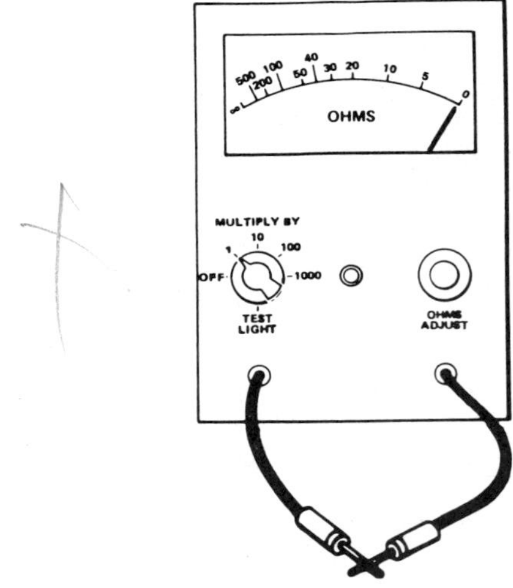

FIGURE 2–15 Typical ohmmeter

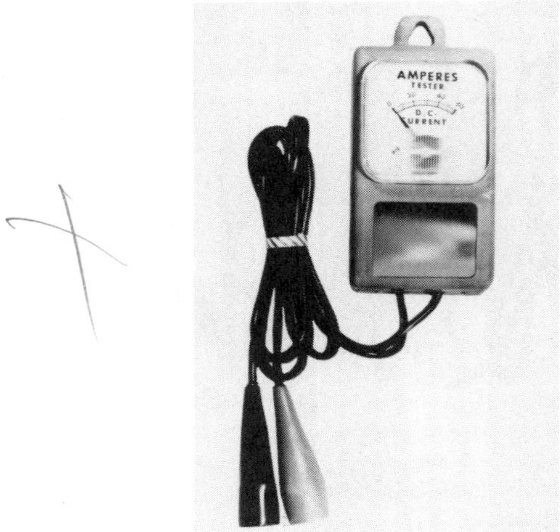

FIGURE 2–16 Ammeter *(courtesy of MAC Tools)*

the negative side of the circuit. The voltage drop is used to determine the resistance in the circuit or component. The ohmmeter scale will read from 0 to infinity (∞). A 0 reading means there is no resistance in the circuit and may indicate a short in a component that should show a specific resistance. An infinity reading indicates a number higher than the meter can measure. This usually is an indication of an open circuit.

Ammeter

An ammeter (Figure 2–16) measures current flow in a circuit. Current measures in amperes, or amps (Figure 2–17). Unlike the voltmeter and ohmmeter, the ammeter must be placed in series with the circuit being tested. Normally, this requires disconnecting the positive wire from the component being tested and connecting the ammeter between them. With some meters, a shunt is used to connect the meter parallel with the circuit.

It is much easier to test current using an ammeter with an inductive pickup. The pickup clamps around the wire or cable being tested (Figure 2–18). The ammeter determines amperage based on the magnetic field created by the current flowing through the wire. No time-consuming wiring arrangements must be made.

Volt/Amp Tester

A volt/amp tester (VAT), shown in Figure 2–19, is used to test batteries, starting systems, and charging systems. The tester contains a voltmeter, ammeter, and carbon pile. The carbon pile is a variable resistor. A knob on the tester allows the technician to vary the resistance of the pile. When the tester is attached to the battery, the carbon pile will draw current out of the battery. The ammeter will read the amount of current flowing. The resistance of the carbon pile must be adjusted to match the size of the battery. The smaller the battery, the lower the load. Chapter 3 will explain how to determine the proper load for the battery being tested and the test being performed.

Figure 2–20 shows a small battery tester. This battery tester does not have a carbon pile. However, it can be used to test open circuit voltage and to perform battery load tests on 6- and 12-volt batter-

FIGURE 2–17 Inductive probe used to measure amperage *(courtesy of All-Test, Inc.)*

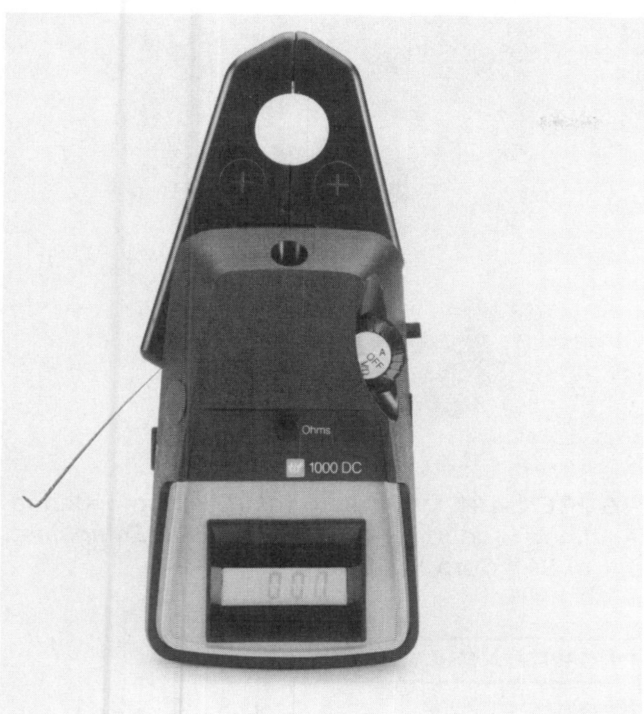

FIGURE 2-18 Inductive clamp used to test ohms, amperage, and voltage *(courtesy of TIF Instruments, Inc.)*

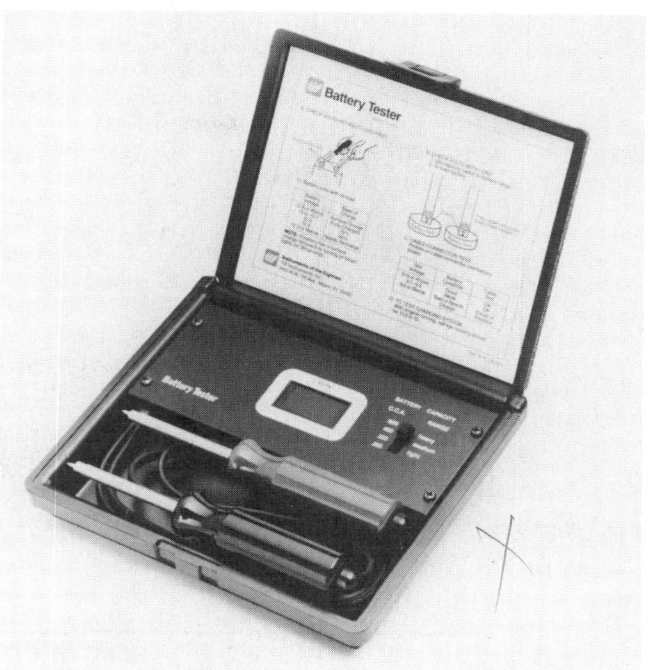

FIGURE 2-20 Battery tester *(courtesy of TIF Instruments, Inc.)*

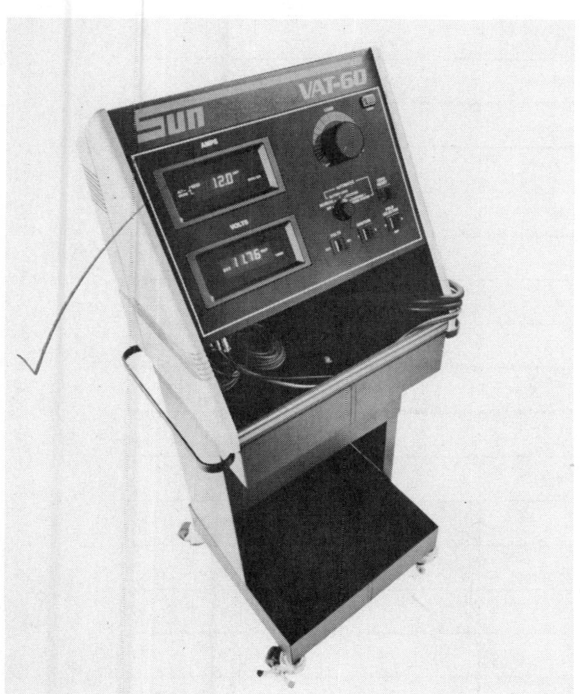

FIGURE 2-19 Volt-amp tester *(courtesy of Sun Electric Corporation)*

ies. It can also be used to check the condition of battery cables and connectors, and to check running voltage.

Multimeter

It is not necessary for a technician to own separate voltmeters, ohmmeters, and ammeters. These tools are available combined in a single tool called a multimeter. A multimeter is one of the most versatile tools used in diagnosing engine performance.

The most basic multimeter is the DVOM—the digital volt/ohmmeter. As its name implies, the DVOM tests both voltage and resistance. Many popular multimeters combine the functions of a DVOM with those of a tach/dwellmeter (Figure 2-21).

Top-of-the-line multimeters are truly multifunctional. Most test—DC and AC—volts, ohms, and amps. Usually there are several test ranges provided for each of these functions. In addition to these basic electrical tests, multimeters also test engine rpm, ignition dwell, diode condition, distributor conditions, frequency, and even temperature (Figure 2-22). Table 2-1 shows some of the many test that can be performed with a multimeter.

Multimeters are available with either analog or metric displays. Analog meters (Figure 2-23) enjoyed wide popularity prior to the advent of electronic control systems. They are still favored by techni-

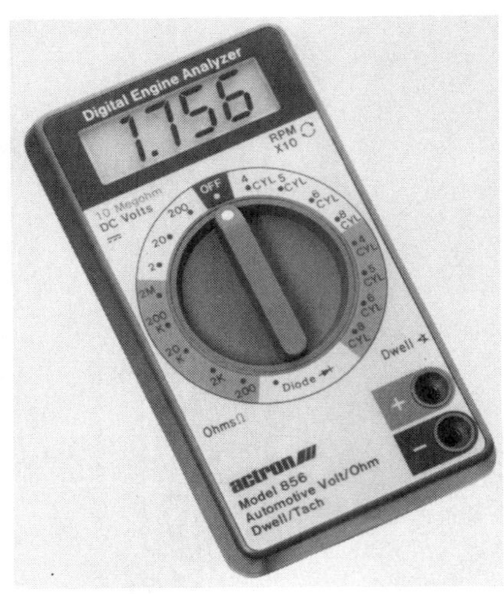

FIGURE 2-21 Digital multimeter volt/ohm *(courtesy of Actron Manufacturing Co.)*

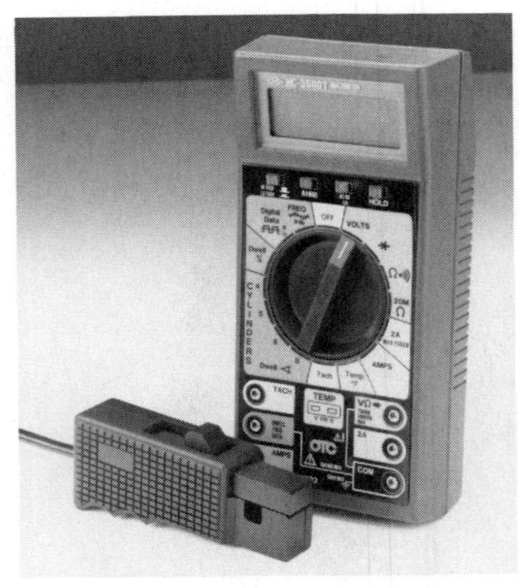

FIGURE 2-22 Multifunctional, low impedance multimeter *(courtesy of OTC Tool and Equipment Div. of SPX Corp.)*

TABLE 2-1: ELECTRICAL TESTING WITH A MULTIMETER

System/Component	Measurement Type			
	Voltage Presence & Level	Voltage Drop	Current	Resistance
Charging System				
Alternators	•		•	
Regulators	•			
Diodes		•		•
Connectors	•	•		•
Starting System				
Batteries	•	•	•	
Starters		•	•	
Solenoids	•	•		•
Connectors		•		•
Interlocks	•			
Ignition System				
Coils	•			•
Connectors	•	•		•
Condensers				•
Contact set (points)	•			•
Distributor caps				•
Plug wires				•
Rotors				•
Magnetic pick-up	•			•

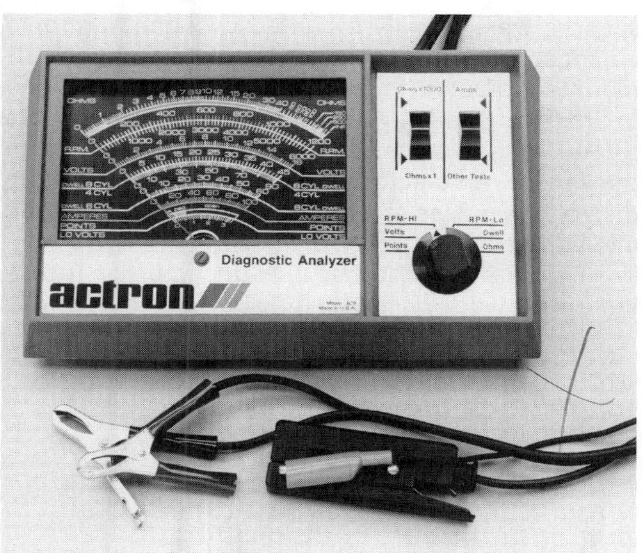

FIGURE 2-23 Analog-type multimeter *(courtesy of Actron Manufacturing Co.)*

cians who have used them for years and prefer the visual reference provided by a moving needle.

However, there are several drawbacks to using most analog-type meters for testing electronic control systems. Many electronic components require very precise test results. Digital meters can measure volts, ohms, or amps in tenths and hundredths. Some have multiple test ranges that must be manually selected. Others are auto ranging. The multimeter shown in Figure 2-24 has 18 test ranges. For example, it can measure DC voltage readings as small as 1/10 millivolt or as high as 1000 volts.

Another problem with analog meters is their low internal resistance (input impedance). The low input impedance allows too much voltage to flow through circuits being tested to permit the meter's use on delicate electronic devices.

Digital meters, on the other hand, have a high input impedance, usually 10 megohms (10 million ohms). Metered voltage for resistance tests is well below 5 volts, reducing the risk of damage to sensitive components and delicate computer circuits.

FIGURE 2-24 This multimeter has many test ranges. *(courtesy of TIF Instruments Inc.)*

IGNITION SYSTEM TEST EQUIPMENT

Some of the tools used to troubleshoot ignition systems have been in use for years. Others are relatively new, having been developed in response to electronic ignition systems.

TACH-DWELLMETER

A tool very commonly used for tune-up and engine diagnosis is the tach-dwellmeter (Figure 2-25).

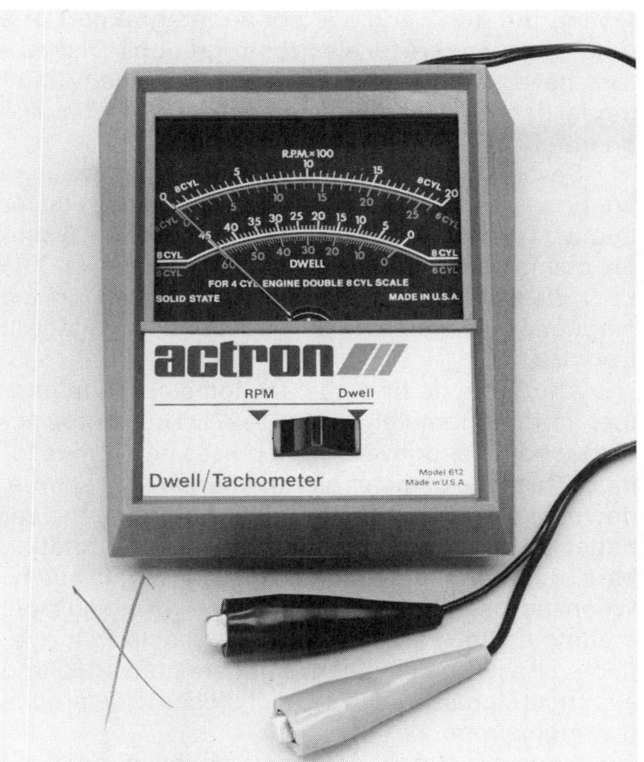

FIGURE 2-25 Meter for testing engine rpm and dwell *(courtesy of Actron Manufacturing Co.)*

This meter is a combination tachometer and dwellmeter. A tachometer measures engine revolutions per minute (rpm). A dwellmeter measures the degree of crankshaft rotation that occurs while the ignition points are closed. On electronic ignition systems, the dwell time is the degree of crankshaft rotation during which the primary circuit is on. On older GM vehicles with the C-4 or Computer Command Control carburetors, the dwellmeter can be used to measure the type of signal (rich or lean) going to the carburetor mixture control solenoid.

A typical tach-dwellmeter has three leads. Two heavy-duty leads connect the meter to the battery terminals. The third lead usually connects the meter to the distributor or tach terminal on the coil. As the breaker points or electronic control module turns the primary ignition system on and off, the tach-dwellmeter uses this signal to determine engine speed and dwell angle.

DIGITAL TACHOMETERS

The advances made in automotive electronics in recent years have diminished the usefulness of the tach-dwellmeter. On computer-controlled ignition systems, dwell is not adjustable. Although a dwell reading that is not to specifications will indicate mechanical problems, such as a worn distributor shaft, testing the dwell angle is not an integral part of a tune-up procedure for electronic ignitions. Distributors have even been eliminated from many late-model cars and are difficult to make connections to on others.

Several types of inductive pick-up tachometers are now available that simplify rpm testing. The inductive type tachometer shown in Figure 2–26 simply clamps over the number 1 spark plug wire. The LED display gives the engine rpm based on the magnetic pulses created by the secondary voltage in the wire.

Another type of digital tachometer is the photoelectric tachometer (Figure 2–27). The photoelectric tachometer converts light pulses into rpm readings. This type of meter has an internal light source, powered either by an internal battery or by the car battery. Reflective tape is applied to any rotating part of the engine, such as the crankshaft pulley. When the meter light is pointed at the revolving tape, a photoelectric "eye" in the meter reacts to the reflected light. The flashes of light are converted into electrical signals, which the meter uses to determine the engine rpm.

When electronic ignitions replaced breaker-point ignitions, magnetic pulse sensors were used to determine crankshaft position and rotation. The sensors were usually mounted over the harmonic balancer, the flywheel, or beside a crankshaft-mounted reluctor (these components are explained in Chapters 8 and 9). As the mechanical components rotate, voltage pulses are magnetically induced in the sensor. An electronic control module monitors the voltage pulses to determine crankshaft position and engine speed.

Figure 2-28 illustrates a magnetic probe receptacle is provided on the sensing device. This permits

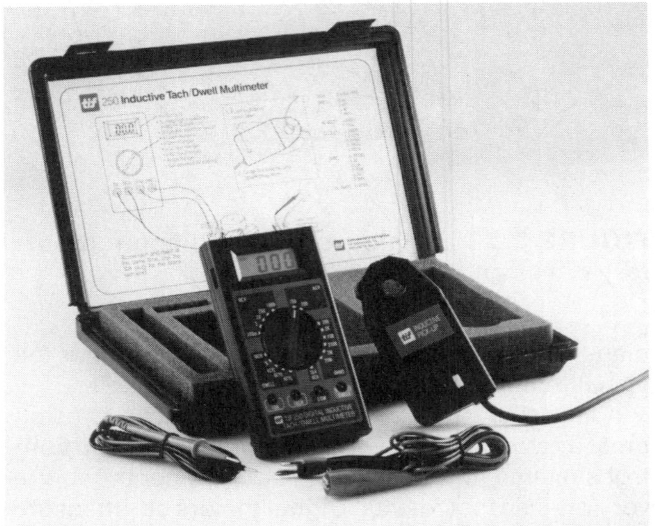

FIGURE 2–26 This inductive pickup clamps over any spark plug wire for testing engine rpm. *(courtesy of TIF Instruments, Inc.)*

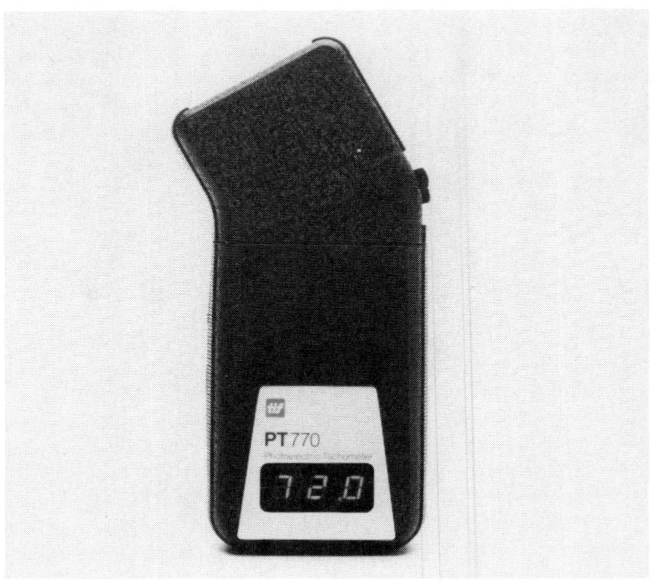

FIGURE 2–27 Photoelectric tachometer *(courtesy of TIF Instruments, Inc.)*

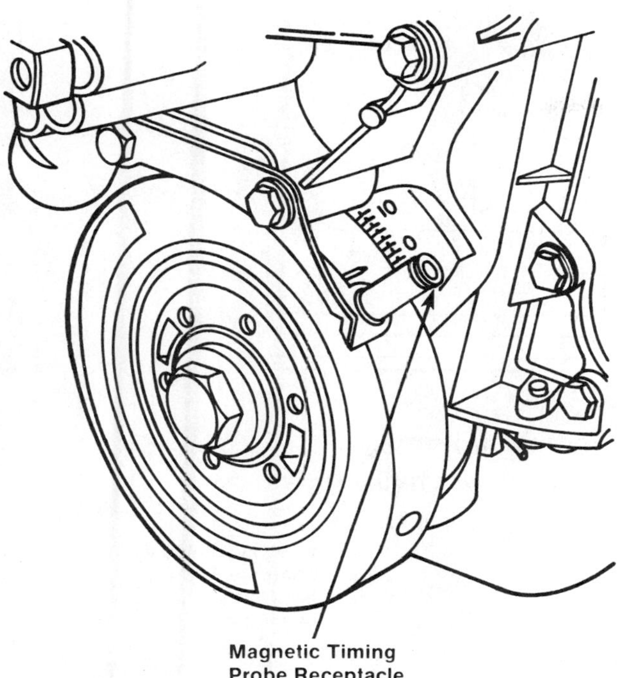

FIGURE 2-28 Magnetic probe receptacle for checking ignition timing

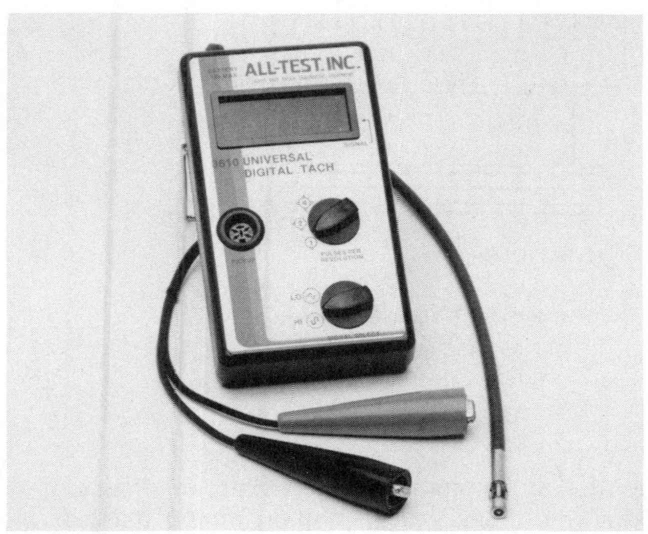

FIGURE 2-29 Digital tachometer with inductive probe *(courtesy of All-Test Inc.)*

a probe to be placed in the receptacle and the electrical pulses to be remotely monitored. Figure 2-29 shows a digital tachometer with a magnetic coil pickup. The tachometer counts the number of pulses and, based on the number of pulses generated per minute by the vehicle's crankshaft position sensor, determines the revolutions per minute.

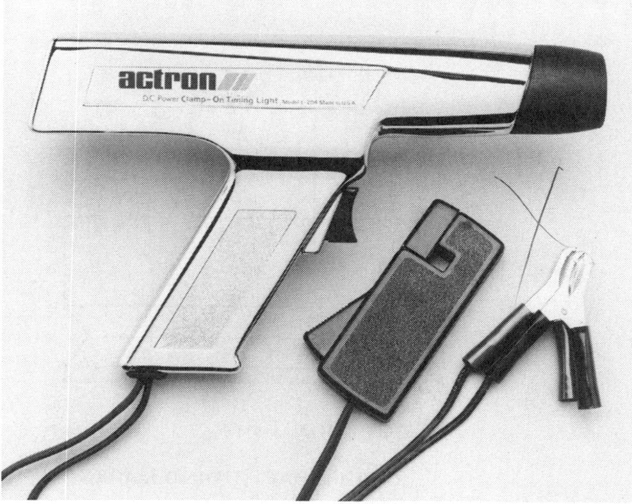

FIGURE 2-30 Timing light with inductive probe *(courtesy of Actron Manufacturing Co.)*

TIMING LIGHT

A timing light (Figure 2-30) is an essential tool in tuning up most vehicles. A timing light is used to reveal when the spark plugs are firing in relation to the piston position. A timing light normally consists of a strobe light and three leads. Two of the leads connect to the car battery to power the strobe. The third lead (pick-up lead) connects to the number 1 spark plug wire. The pick-up lead on most newer timing lights have an inductive pickup that clamps over the spark plug wire. Older-type timing lights have an adapter that must be placed between the spark plug wire and the spark plug. When the engine is turned on, the timing light will flash every time the number 1 plug fires.

CAUTION: Never pierce the spark plug wire to connect the pick-up lead when using a noninductive timing light.

To check the ignition timing, the technician will point the flashing timing light at the ignition timing marks. The timing marks (Figure 2-31) are located either in front of the engine on the harmonic balancer or crank shaft pulley or at the rear of the engine on the flywheel.

The timing marks consist of a pointer, line, or notch and a scale. The scale indicates degrees of crankshaft rotation. The scale usually has a zero reference mark that refers to the crankshaft position when the number 1 piston is at top dead center

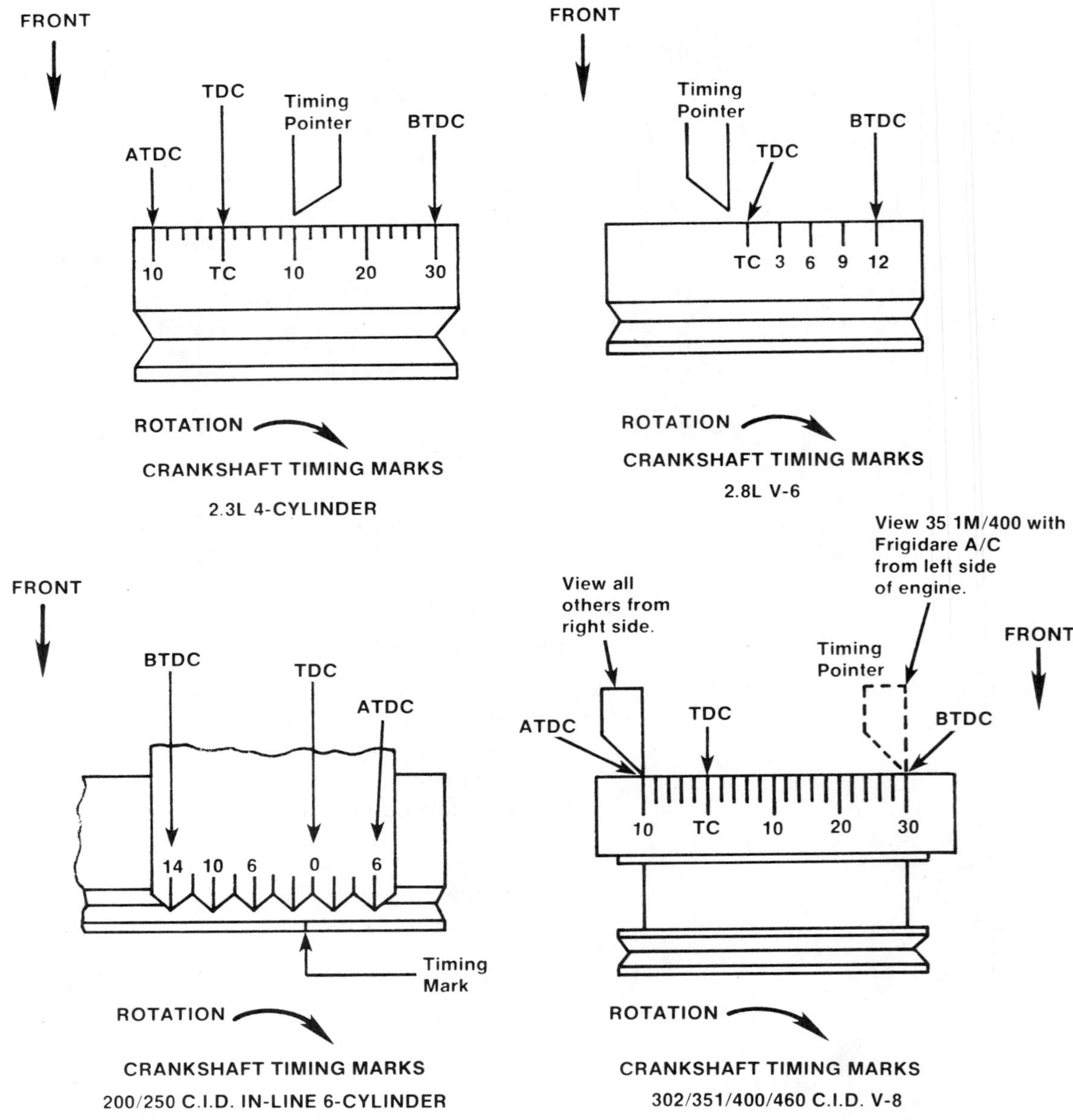

FIGURE 2–31 Typical timing marks

(TDC) on the compression stroke. On either side of the zero reference mark are marks spaced every 2 or 3 degrees. These marks indicate degrees of crankshaft rotation before top dead center (BTDC) and after top dead center (ATDC).

When the number 1 piston is at top dead center on the compression stroke, the line or notch will line up with the zero reference mark on the timing plate. Usually an engine is timed so that the number 1 spark plug fires several degrees BTDC.

When the timing light is connected to the battery and the number 1 spark plug wire, it will flash every time the number 1 spark plug fires. When pointed at the timing marks, the strobe effect of the light will "freeze" the spinning timing mark as it passes the timing scale. In this way the technician can determine the ignition timing by observing the degree of crankshaft rotation before or after TDC when the spark plug fires.

Base timing is checked at base, or curb, idle. Advanced timing is checked at a higher rpm specified by the vehicle manufacturer. As engine speed increases, the timing is advanced (either mechanically, by vacuum, or electronically). The actual amount of timing advance can be measured with an advance timing light (Figure 2–32).

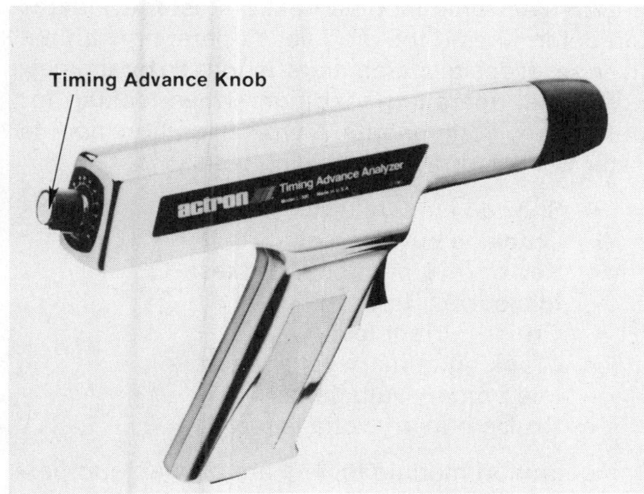

FIGURE 2-32 Timing light with advance knob *(courtesy of Actron Manufacturing Co.)*

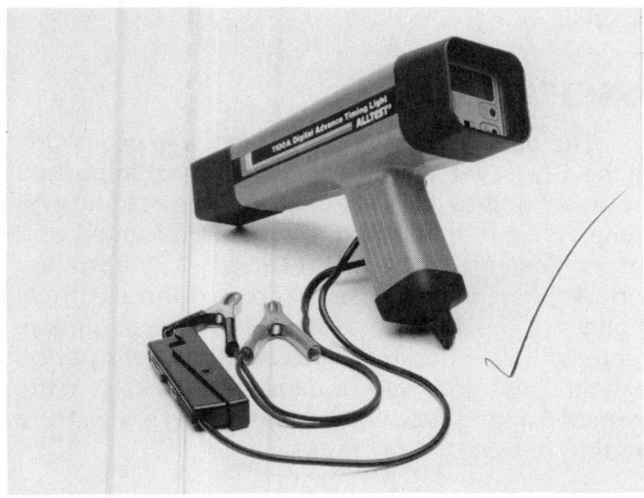

FIGURE 2-33 Digital timing light with LED display *(courtesy of All-Test, Inc.)*

The advance timing light has either a scale or meter that indicates degrees of timing advance. The advance knob on the light can be turned to delay the flash of the light. By pointing the flashing light at the timing marks and turning the dial until the timing marks appear to line up again at the base timing position, the dial will indicate the amount of timing advance.

A more versatile advanced timing light is the digital timing light. The timing light shown in Figure 2-33 electronically measures timing advance as the engine rpm is increased and displays timing advance on the LED display. This light flashes only when a trigger is squeezed, unlike most timing lights

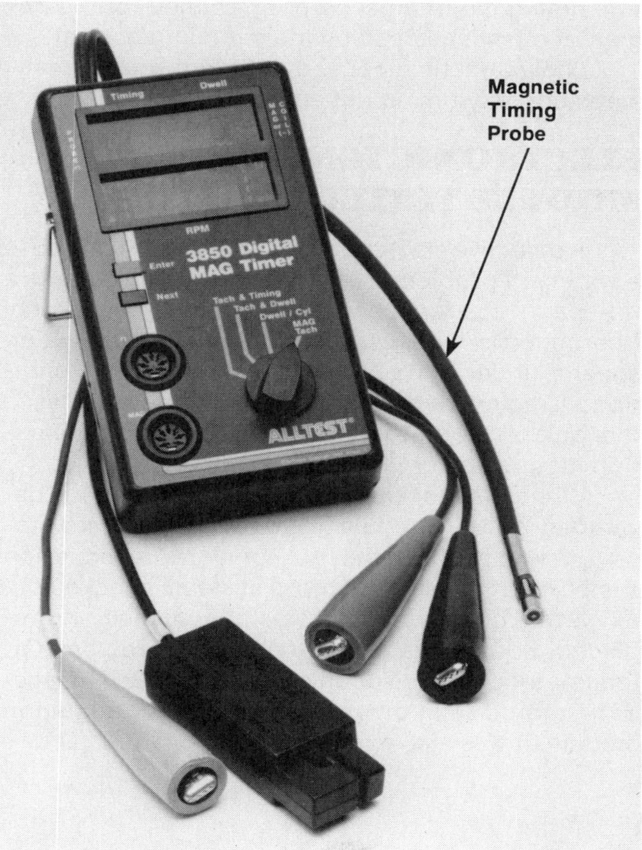

FIGURE 2-34 Ignition timing tester with magnetic timing probe attachment *(courtesy of All-Test, Inc.)*

which flash all the time the engine is running. When the trigger is released, the LED displays engine rpm. This combined feature eliminates the need for a separate tachometer when setting timing.

MAGNETIC TIMING PROBE

The timing light is no longer necessary when tuning many late-model cars. The magnetic probe receptacle on the crankshaft position sensor allows the ignition timing to be electronically monitored with a magnetic timing probe (Figure 2-34). Changes in the magnetic field create electrical pulses in the tip of the timing probe. These pulses are monitored by the timing meter to determine crankshaft position and ignition timing.

The only trick to using a magnetic timing probe is correcting the offset of the probe receptacle. The receptacle is usually situated on either side of the vehicle's coil pickup. The degree of offset must be factored into the timing probe's reading or the timing readout will be inaccurate by that many degrees.

The timing meter must be programmed for the degree of offset specified by the car manufacturer.

The magnetic timer shown in Figure 2–34 measures not only timing but engine rpm as well.

ELECTRONIC IGNITION MODULE TESTER

A defective electronic ignition module can cause a variety of problems. Some may be as obvious as a no-spark, no-start condition. Others may be only intermittent problems, such as a misfire at cruising speeds under certain load or temperature conditions. Tracing the source of intermittent problems to the ignition module can be difficult without an ignition module tester (Figure 2–35).

When the first cars with electronic ignitions developed ignition module troubles, the standard procedure was to check the fuel, ignition, and emission systems for problems. If these systems checked out okay, the ignition module was condemned and replaced. Sometimes this corrected the problem. Oftentimes it did not. Not only was this "R & R" procedure imprecise in diagnosing faults in the ignition module, it was also expensive.

FIGURE 2–35 Digital electronic ignition tester *(courtesy of All-Test, Inc.)*

An electronic module ignition tester evaluates and determines if the module is operating within a given set of design parameters. It does so by simulating normal operating conditions while looking for faults in key components. A typical ignition module tester will perform the following tests:

- Shorted module test
- Cranking current test
- Key on/engine off current test
- Idle current test
- Cruise current test
- Cranking primary voltage test
- Idle primary voltage test
- Cruise primary voltage test

Some ignition module testers are also able to perform an ignition coil-spark test (actually firing the coil) and a distributor pick-up test.

Test selection is made by pushing the appropriate button. The module tester usually responds to these tests with a "pass" or "fail" indication.

OSCILLOSCOPE

The oscilloscope (Figure 2–36) was developed in the 1950s; yet, it is still one of the most important troubleshooting tools for diagnosing electrical systems. Some of the tests that can be performed with an oscilloscope are shown in Table 2–2. The primary job of the oscilloscope is to convert the electrical signals of the ignition system into a visual image representing voltage changes over a specific period of time. This information is displayed on a CRT in the form of a continuous voltage line called a waveform pattern or trace.

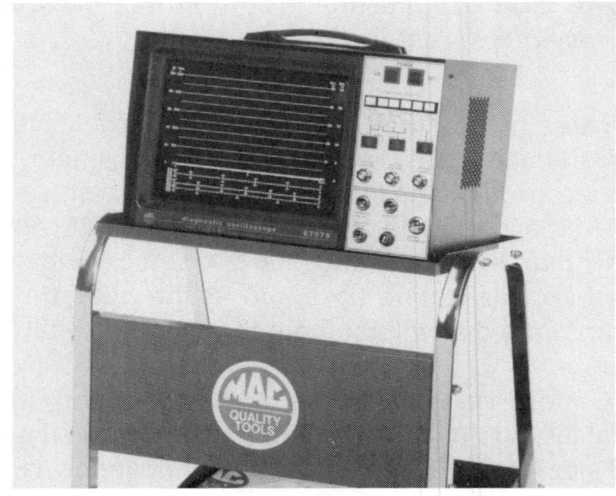

FIGURE 2–36 Oscilloscope *(courtesy of Electro Specialties)*

TABLE 2-2: IGNITION SYSTEM SCOPE TESTS

Conventional Systems	Electronic Systems
Cranking coil output	Cranking spark duration
Coil polarity	Coil polarity
Spark plug firing voltage	Spark plug firing voltage
Rotor air gap voltage drop	Rotor air gap voltage drop
Rotor register	Rotor register
Secondary circuit resistance	Secondary circuit resistance
Maximum coil output	Running spark duration
Secondary insulation	Secondary insulation
Spark plugs under load	Spark plugs under load
Coil and condenser	Coil condition
Cylinder timing accuracy	Cylinder timing accuracy
Breaker point dwell	Not applicable
Breaker point condition	Not applicable
Dwell variation	Dwell and dwell variation
Battery voltage	Battery voltage
Charging voltage	Charging voltage
Alternator condition	Alternator condition

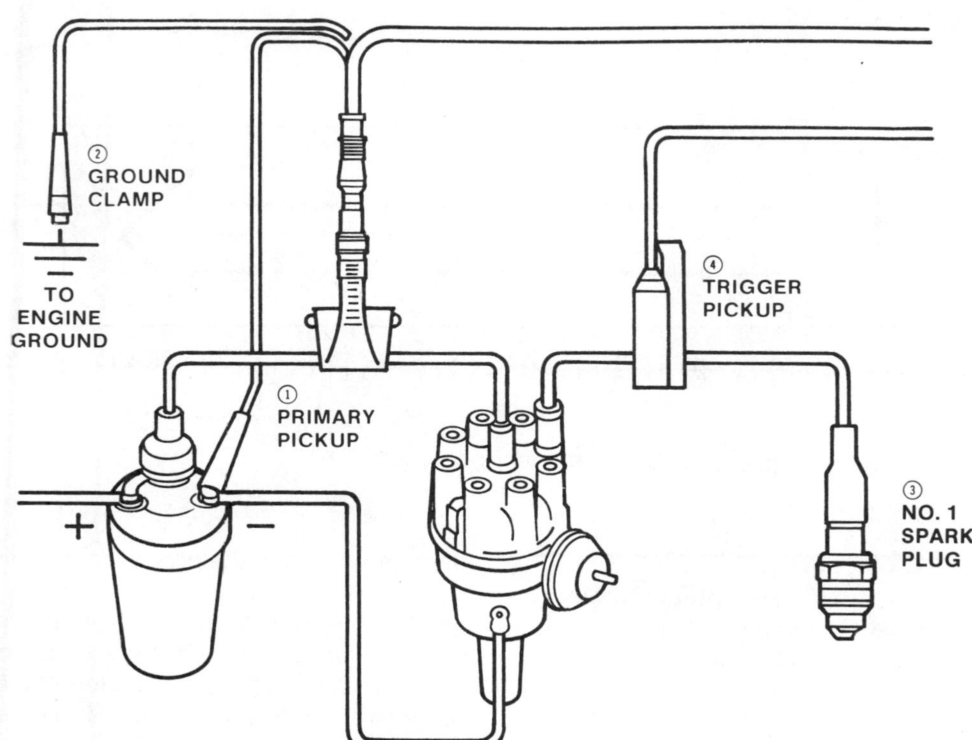

FIGURE 2-37 Typical oscilloscope connections to an ignition system

The oscilloscope normally has four leads (Figure 2-37): the primary pickup that connects to the primary, or negative, terminal of the ignition coil; the ground lead which connects to a good engine ground; the secondary pickup, which clamps around the coil's high tension wire; and, the trigger pickup, which clamps around the spark plug wire of the number 1 cylinder. The information received from these leads is translated into the scope patterns on the CRT screen.

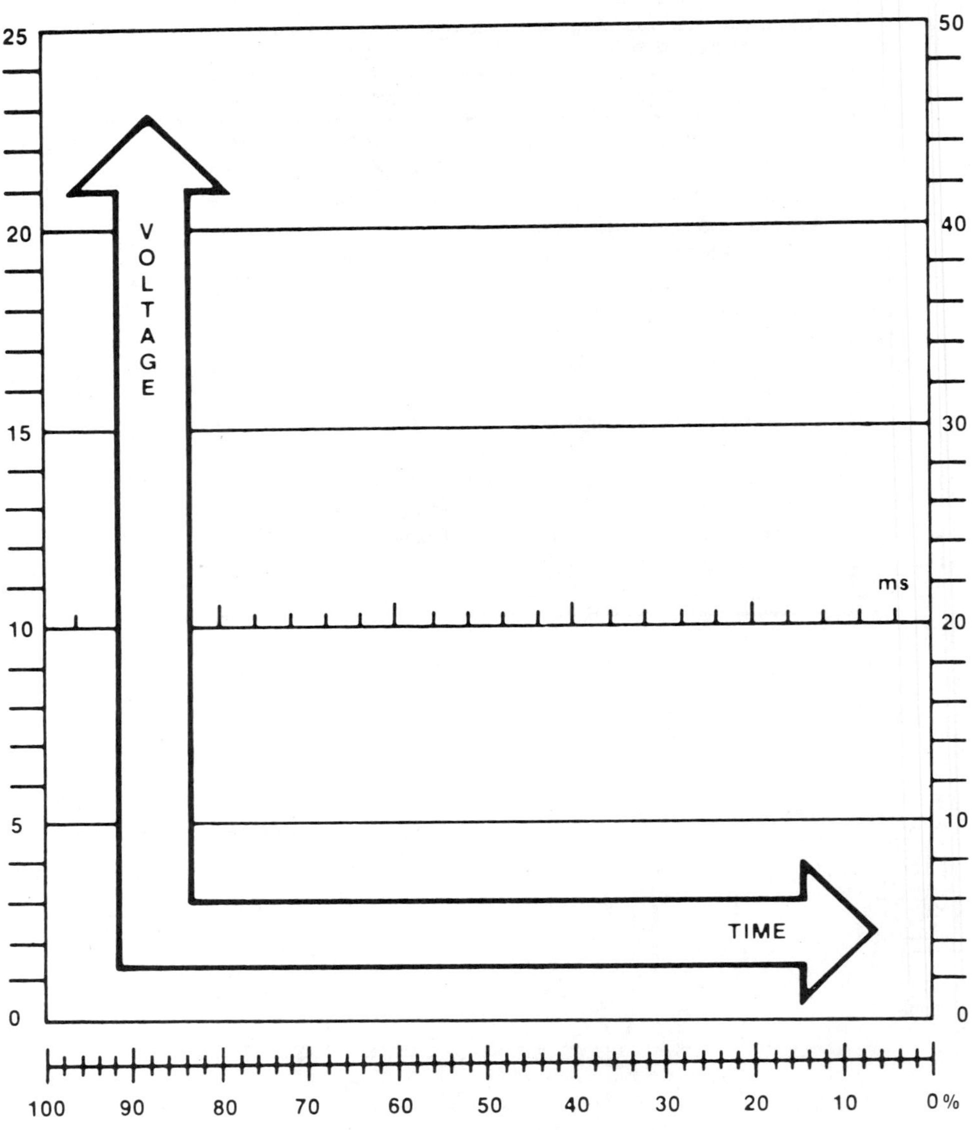

FIGURE 2-38 Voltage vs. time graph on the oscilloscope screen

---

CAUTION: Always follow the manufacturer's instructions for connecting test leads.

Scales

On the face of the CRT screen is a voltage versus time graph (Figure 2-38). The waveform patterns are displayed on the graph. The graph charts the changes in voltage and the time span in which the changes occur. The voltage versus time graph has four scales: two vertical scales (one on the left and one on the right), a horizontal *percent of dwell* scale, and a horizontal *millisecond* scale. The technician must select the proper scale for the test being conducted.

The left and right vertical scales measure voltage. The vertical scale on the left side of the graph is divided into increments of 1 kilovolt (1000 volts) and ranges from 0 to 25 kilovolts (kV). This scale is useful for testing secondary voltage. It can also be used to measure primary voltage by interpreting the scale in volts rather than kilovolts.

The vertical scale on the right side of the graph is divided into increments of 2 kilovolts and has a range of 0 to 50 kV. This scale is also used for testing secondary voltage. This scale can also be used to measure primary voltage in the 0 to 500 V range.

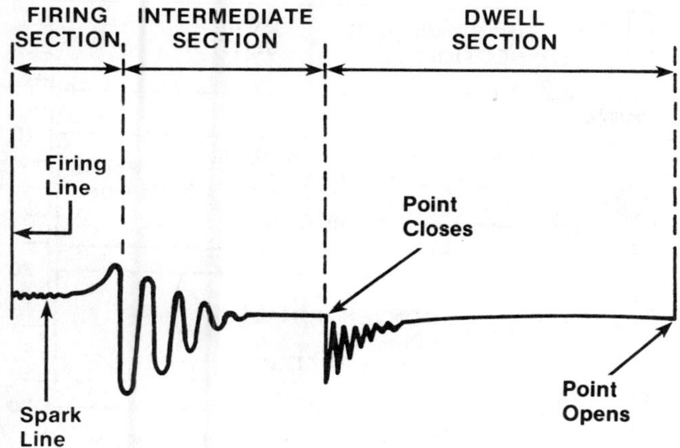

FIGURE 2-39 Typical pattern for a conventional secondary system

The horizontal percent of dwell scale is located at the bottom of the scope screen. This scale is used for checking the dwell angle in both the primary and secondary circuits. The dwell scale is divided into increments of 2 percentage points and ranges from 0 to 100 percent.

The fourth scale—the millisecond scale—is a horizontal line that runs along the center of the voltage versus time graph. Depending on the test mode selected by the technician, the millisecond scale shows a range of 0 to 5 milliseconds (ms) or 0 to 25 milliseconds. The 5 ms scale is often used to measure the duration of the spark.

Functions

When performing any test on the oscilloscope, a technician must select the function to be tested: either the primary circuit or the secondary circuit. A typical conventional secondary pattern is shown in Figure 2-39. From this pattern, a technician can observe the

- Firing voltage
- Duration of the spark
- Coil and condenser oscillations
- Contact breaker point condition and action
- Dwell and cylinder timing accuracy
- Secondary circuit resistance

A typical primary circuit pattern is shown in Figure 2-40. The primary scope pattern is used when secondary circuit connections are not possible. It is also useful for observing contact breaker point condition and action on conventional systems and for providing a means to observe dwell and cylinder timing problems.

The scope patterns for electronic ignition systems are similar to those of conventional systems. A scope pattern of an electronic ignition system will not show condenser oscillations, and the switching action of the transistor, rather than the opening and closing on breaker points, is displayed. Figure 2-41 and 2-42 compare scope patterns of a conventional system and a Chrysler electronic ignition system (EIS).

Patterns

The oscilloscope has several ways to display the voltage patterns of the primary and secondary circuits. When the *display* pattern is selected, the oscilloscope will display the patterns of all the cylinders in a row from left to right as shown in Figure 2-43. Each cylinder's ignition cycle is displayed in the engine's firing order. In the example shown in Figure 2-43 the firing order is 1, 8, 4, 3, 6, 5, 7, 2. The pattern will begin with the spark line of the No. 1 cylinder and will end with the firing line for the No. 1 cylinder. This display pattern allows the technician to compare the individual patterns for each cylinder.

A second choice of patterns is the *raster* pattern (Figure 2-44). A raster pattern stacks the voltage patterns of the cylinders one above the other. The number 1 pattern is displayed at the bottom of the screen and the rest of the cylinder's firing patterns are arranged above it in the engine's firing order.

In a raster pattern, the patterns for each cylinder are displayed across the width of the graph, also beginning with the spark line and ending with the firing line. This allows for a much closer inspection of the voltage patterns than is possible with the display pattern.

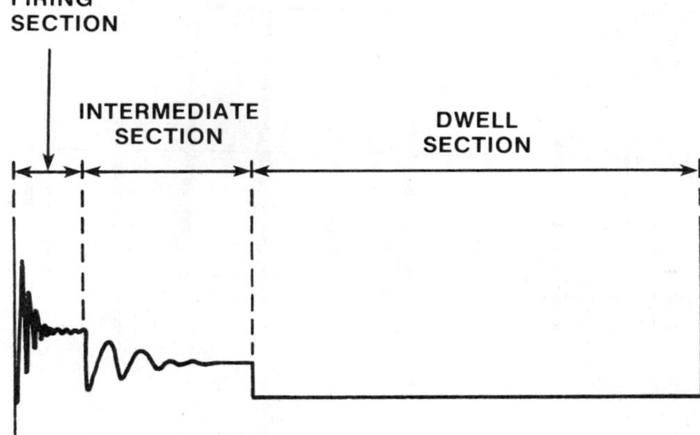

FIGURE 2-40 Typical pattern for a conventional primary system

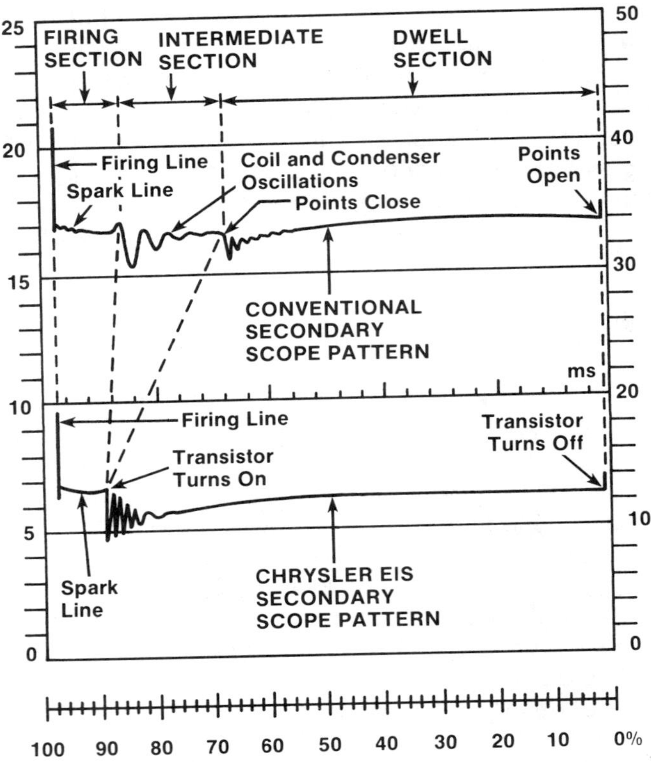

FIGURE 2-41 Typical secondary patterns for a conventional system and an electronic system

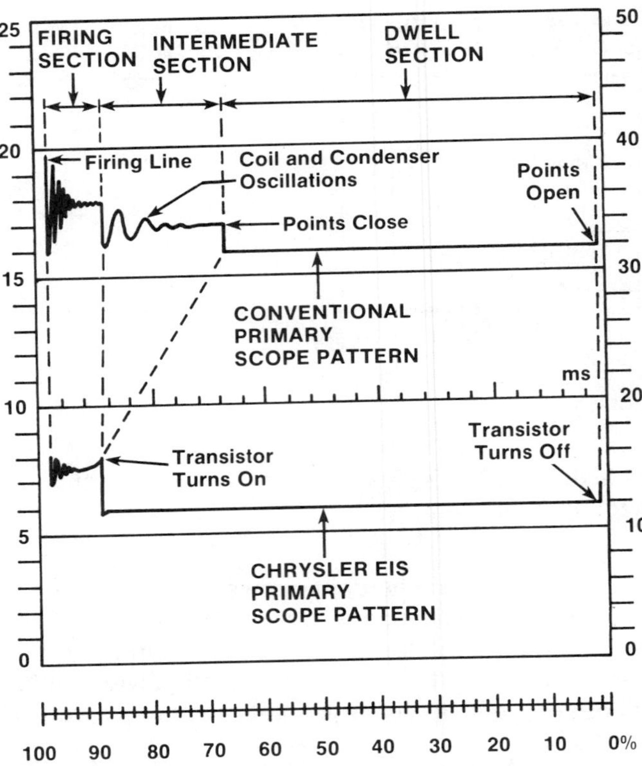

FIGURE 2-42 Typical primary patterns for a conventional system and an electronic system

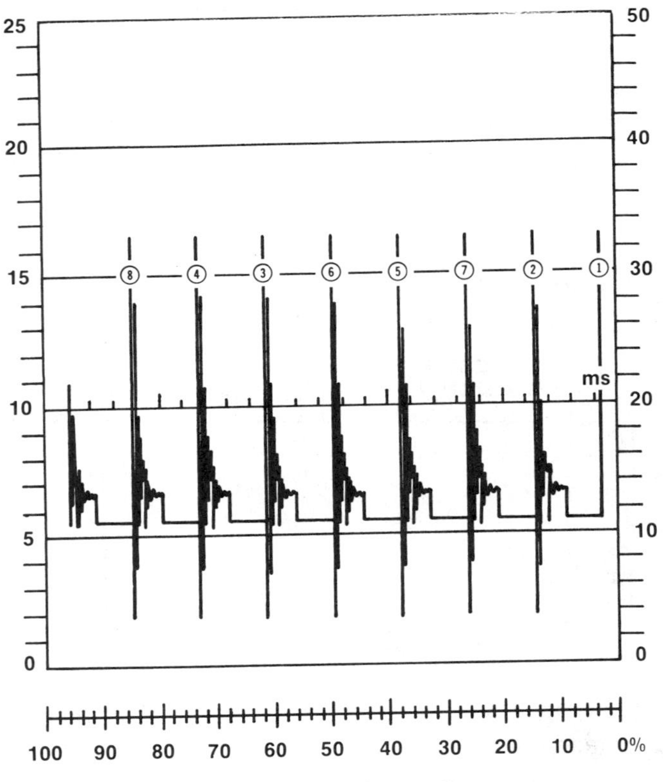

FIGURE 2-43 Primary display pattern

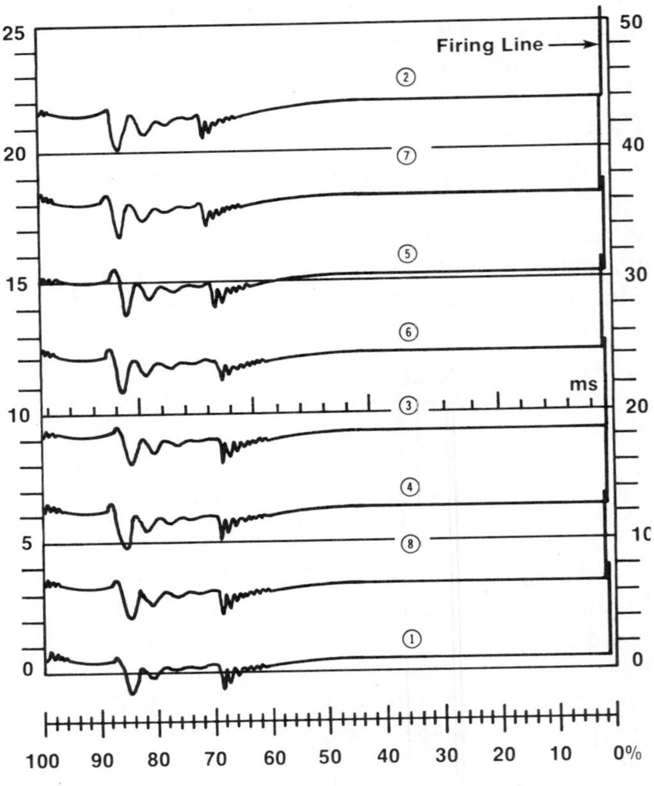

FIGURE 2-44 Secondary raster pattern

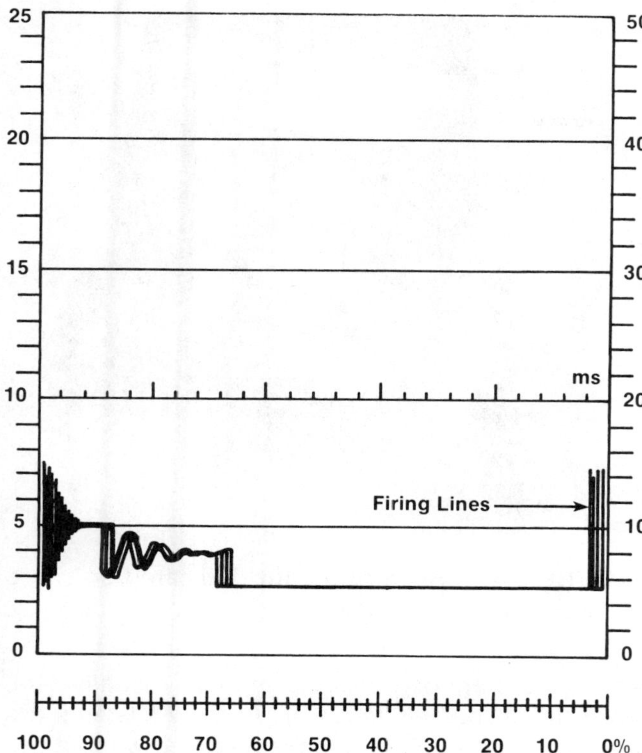

FIGURE 2-45 Primary superimposed pattern

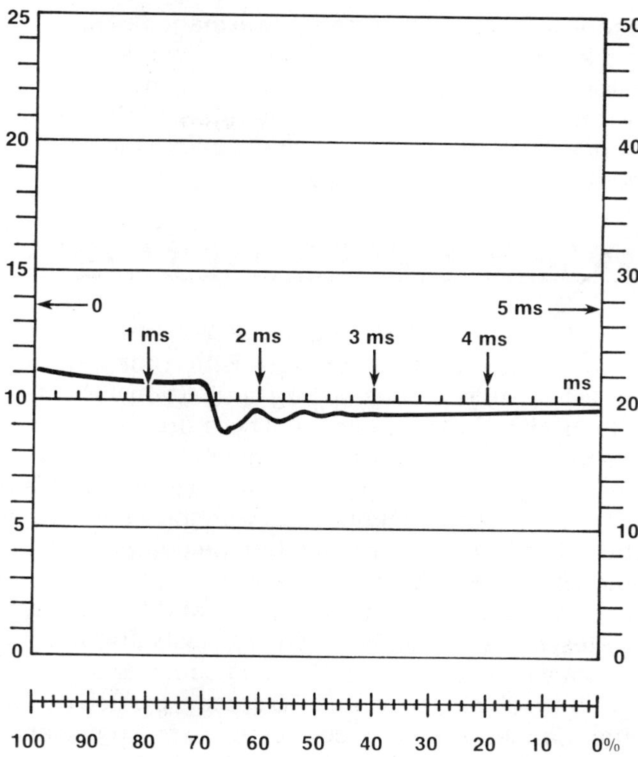

FIGURE 2-46 5-millisecond pattern

All the circuits for voltage patterns for the cylinders can also be displayed in a *superimposed* pattern. A superimposed pattern displays all the patterns one on top of the other. Like the raster pattern, the superimposed voltage patterns are displayed the full width of the screen, beginning with the spark line and ending with the firing line.

A superimposed pattern allows a technician to detect variations of one cylinder's pattern from the others. A superimposed pattern for the primary circuits is shown in Figure 2-45.

The *millisecond* pattern is available on oscilloscopes with a millisecond scale. One of two modes can be selected—the 5 millisecond mode and the 25 millisecond mode. In the 5 millisecond mode (Figure 2-46), the first 5-millisecond segment of the voltage pattern will be displayed the full width of the screen. This mode is used to measure the spark time. If the 25-millisecond mode is selected, 25 milliseconds of the voltage pattern will be displayed (Figure 2-47). In this mode, the complete voltage pattern will be displayed.

In addition to the selection of function, scale, and pattern, the technician must follow specific instruction procedures for obtaining the desired test patterns. These procedures are given in the oscillo-

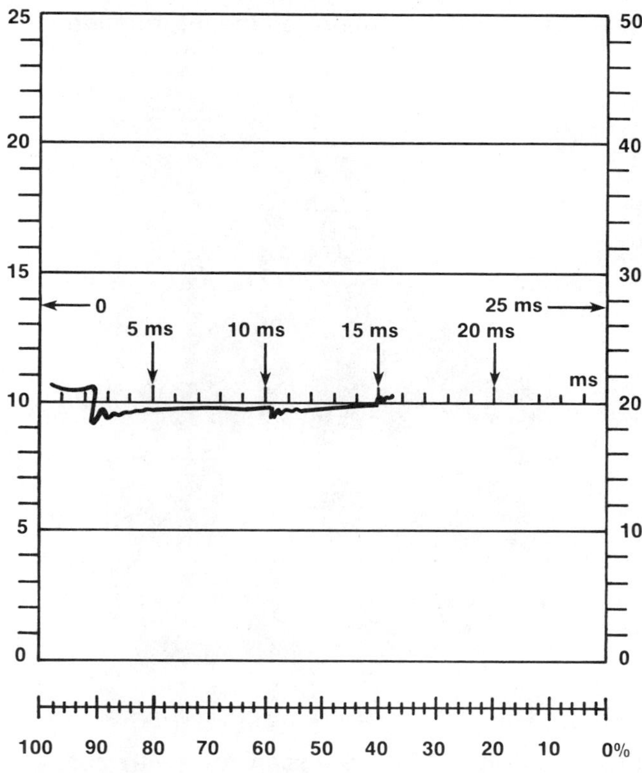

FIGURE 2-47 25-millisecond scale

scope manufacturer's instruction manual. The technician must be familiar with the oscilloscope's operation and the procedures and precautions that apply to the vehicle being tested. Additional information concerning the use of oscilloscopes will be presented in Chapters 7 and 8.

MICROCOMPUTER SCAN TOOLS

The introduction of computer-controlled ignition and fuel systems brought with it the need for tools capable of diagnosing and troubleshooting electronic control systems. There are a variety of microcomputer scan tools available today that do just that. A scan tool (Figure 2–48) is a microprocessor designed to communicate with the vehicle's onboard computer. Connected to the computer module via model specific connectors, a scan tool will access trouble codes, run key-on and key-off test-to-test system operations, and run tests that activate various engine controls. Trouble codes and test results are displayed on an LED screen, printed out on the scanner printer, or connected to an engine analyzer or PC monitor for data display.

A computer scan tool may derive its intelligence from one of several sources, depending on its design. Some scan tools, such as the one in Figure 2–48, have a programmable read-only memory (PROM) chip that contains all the information necessary to diagnose specific models (usually GM, Ford, and Chrysler models). As new systems are introduced, the PROM chip is replaced with an updated version.

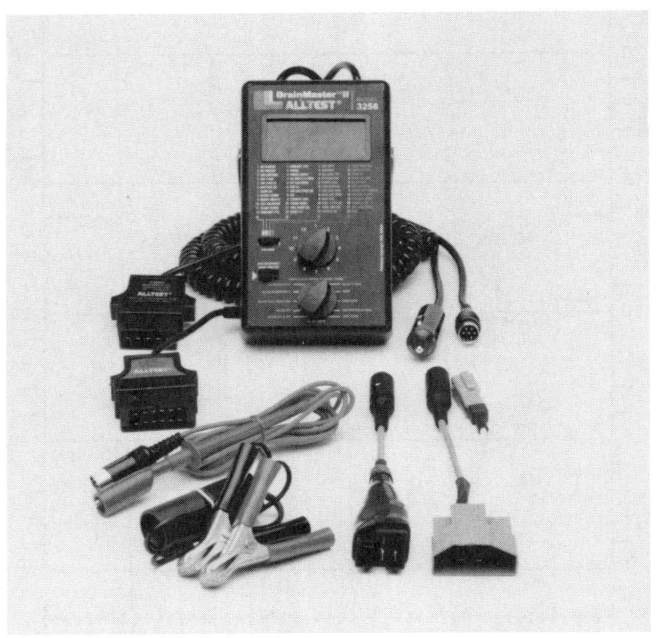

FIGURE 2–48 Computer scan tool with connectors *(courtesy of All-Test, Inc.)*

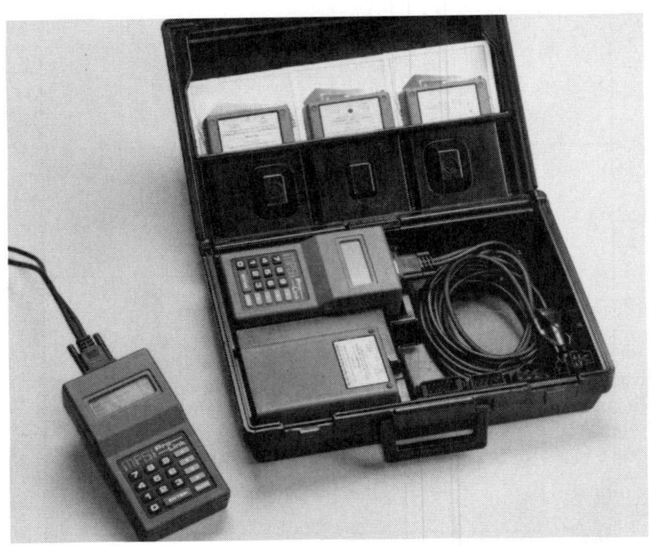

FIGURE 2–49 Scan tool with dedicated software cartridges *(courtesy of Micro-Processors, Inc.)*

Other scan tools use software cartridges that contain model specific information (Figure 2–49). A software package is needed for each make and model and as new systems are introduced, new software is developed by the scan tool manufacturer.

The scan tool in Figure 2–50 derives its diagnostic knowledge from plastic ID cards. The manufac-

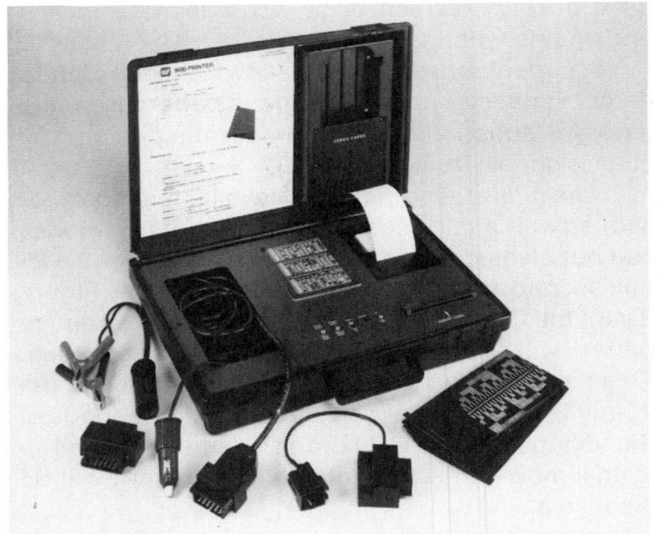

FIGURE 2–50 This scan tool has a coded card for each car model and a built-in printer. *(courtesy of TIF Instruments, Inc.)*

FIGURE 2-51 This scan tool and accessory can be used to capture intermittent faults. *(courtesy of OTC Tool and Equipment Div. of SPX Corp.)*

turer furnishes a card for each model car or truck. The card is inserted into the scanner and is optically read. As new systems are introduced on new models, new cards are developed and made available to the scan tool owner.

A computer scan tool does the same job that a larger engine analyzer does but costs many times less. It is also portable; during a road test (Figure 2-51), some scan tools can be activated to capture the entire data stream generated by the on-board computer at the time an intermittent fault is detected. When activated in this capacity, most scan tools store 30 seconds of data stream prior to the fault occurrence and 30 seconds after. This information can be retrieved later from the scan tool's memory for analysis.

The small size of the scan tool is a handicap as well as an advantage. The LED display is generally only large enough to display four short lines of information. This feature limits the technician's ability to compare test data. However, most scan tools overcome this inadequacy by storing the test data in a random access memory (RAM). The scan tool then can be interfaced with a printer, personal computer, or larger engine analyzer that can retrieve the information stored in the memory and produce a hard copy of the test results. The scan tool in Figure 2-50 has a built-in printer so that the test results are printed as soon as they are entered into the scan tool's memory.

EMISSIONS ANALYZER

Chapter 1 discussed the federal laws that require new cars and light trucks to meet specific emissions levels. State governments also passed laws requiring that car owner's maintain their vehicles so that the emissions remained below an acceptable level. Most states require an annual emissions inspection to meet that goal. Therefore, most garages have an emission analyzer (Figure 2-52).

Early emission analyzers measured the amount of hydrocarbons (HC) and carbon monoxides (CO) in the exhaust. Hydrocarbons in the exhaust is raw, unburned fuel. Emissions analyzers measure HC in parts per million (ppm) or grams per mile (g/mi). Carbon monoxide is an odorless, toxic gas that is the product of incomplete combustion. CO is measured as a percent of the total exhaust.

FIGURE 2-52 Four-gas emission analyzer *(courtesy of Bear Automotive Service Equipment Company)*

An emissions analyzer measures the emissions in the exhaust by infrared refraction. Two air samples—one from the exhaust pipe and one from the shop—are drawn into glass tubes in the analyzer. (Some analyzers have a sealed reference cell and do not require a fresh air sample.) An infrared beam shines through both tubes. As the beam passes through the exhaust sample, it bends, or refracts. The amount of refraction is determined by the level of impurities in the sample. The analyzer measures the amount of refraction and converts it into electrical signals to activate the HC and CO meters.

The level of HC and CO in the exhaust is an indication of engine performance. A high level of hydrocarbons could indicate a fouled spark plug, a defective spark plug wire, or a burned valve allowing unburned fuel vapor to enter the exhaust system. A high CO level indicates an excessively rich air/fuel mixture. A low level of carbon monoxide is the result of an excessively lean air/fuel mixture.

However, the advances in electronic spark control, fuel and air metering, and electronic integration of related systems, along with the development of very efficient catalytic converters, has reached a point where there is little HC and CO remaining in the exhaust. On most late-models cars, HC and CO levels are no longer a useful indicator of engine operation.

The manufacturers of emissions analyzers have responded by producing the four-gas emissions analyzer. In addition to measuring HC and CO levels, a four-gas emissions analyzer also monitors carbon dioxide (CO_2) and oxygen (O_2) levels in the exhaust. This diagnostic capability will permit a technician to use the four-gas analyzer to diagnose the following conditions:

- Rich or lean mixtures (Table 2-3)
- Air pump malfunctions
- Catalytic converter malfunction
- Blown head gaskets
- Faulty carburetors or injectors
- Intake manifold leaks
- Leaking EGR valves
- Excessive misfire
- Excessive spark advance
- Leaks or restrictions in the exhaust system

TABLE 2-3: ANALYSIS OF 4-GAS EMISSIONS ANALYZER RESULTS

Gas Condition		With Catalytic Converter	Without Catalytic Converter
HC*	Over 110 ppm	Rich or lean	—
	Over 250 ppm	—	Rich or lean
CO	Over 3%	—	Rich
	Over 1%	Rich	—
O_2	Over 2%	Lean Mixture	
	Under 1%	Rich Mixture	
CO_2	Less than 10%	Rich Mixture or Misfire	

*HC readings can vary significantly depending on the age and condition of the engine, and the temperature of the engine.

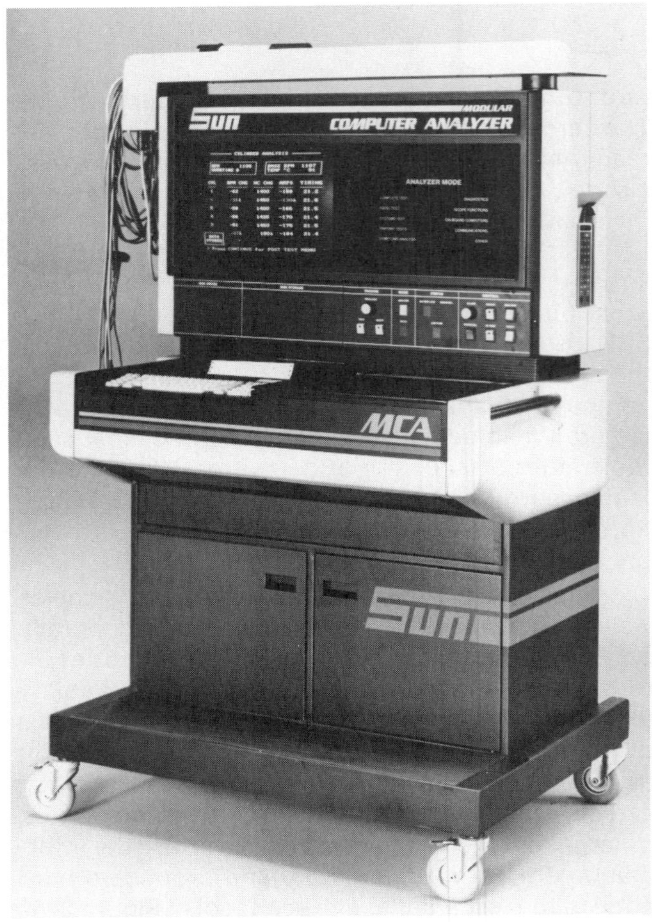

FIGURE 2-53 Computerized engine analyzer *(courtesy of Sun Electric Corporation)*

ENGINE ANALYZER

When performing a complete engine performance analysis, an engine analyzer keeps all of the necessary test equipment within easy reach. Although the term "engine analyzer" is often loosely

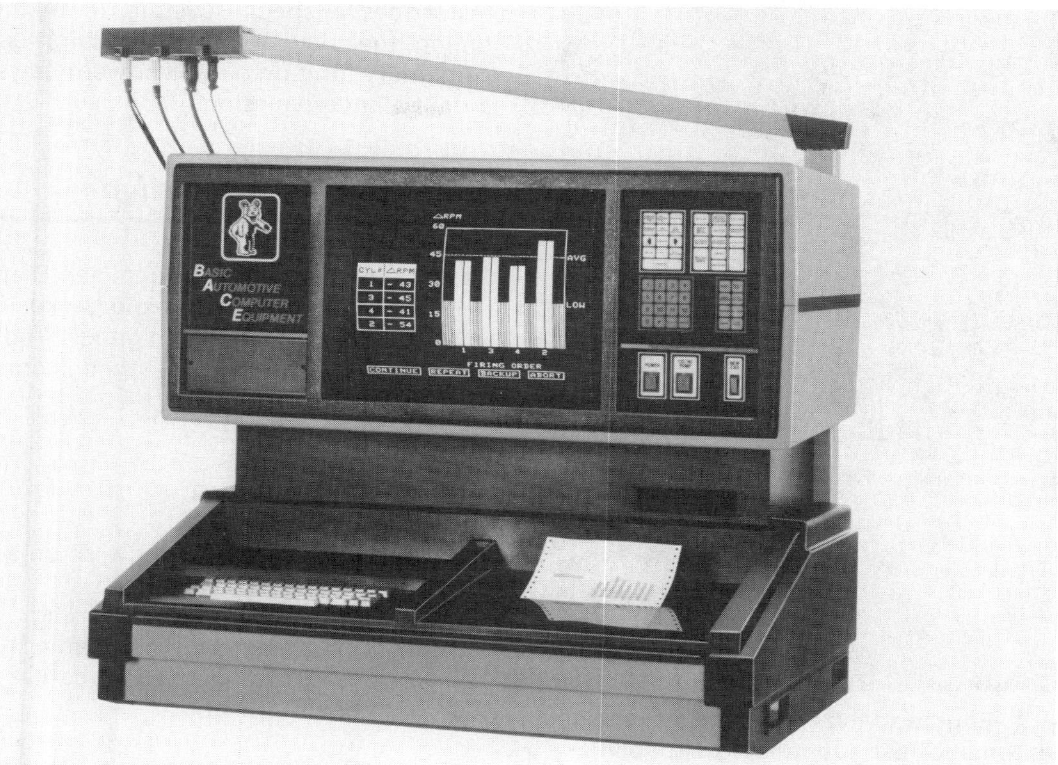

FIGURE 2-54 This engine analyzer has graphic display capabilities, a computer keyboard, and a built-in printer. *(courtesy of Bear Automotive Service Equipment Company)*

applied to any multipurpose test meter, a complete engine analyzer (Figure 2-53) will incorporate most, if not all, of the tools mentioned in this chapter. A state-of-the-art analyzer will do the work of the following tools:

- Compression gauge
- Pressure gauge
- Vacuum gauge
- Vacuum pump
- Tachometer
- Timing light/probe
- Voltmeter
- Ohmmeter
- Ammeter
- Oscilloscope
- Computer scan tool
- Emissions analyzer

With an engine analyzer, one can perform tests on the battery, starting system, charging system, primary and secondary ignition circuits, electronic control systems, the fuel system, the emissions systems, and the engine block. The analyzer is connected to these systems by a variety of leads, inductive clamps, probes, connectors, etc. The data received from these connections is processed by several microprocessors (computers) within the analyzer.

The microprocessors in some computerized engine analyzers are programmed with specifications pertaining to specific model vehicles. Diagnostic trouble codes have been loaded into the analyzer's memory circuits. Based on the input from the leads and connectors, the microprocessors will identify worn, misadjusted, or malfunctioning components in all major engine systems. The analyzer will logically deduce the probable cause of specific performance complaints and will prompt, or guide, the technician step by step through a troubleshooting procedure designed to verify and correct the problem.

Commands and specifications can be entered into the analyzer on the computer-like keyboard. Specifications, commands, test results, etc., are displayed on the CRT screen. Some analyzers have adapted PC technology and display test results graphically on the CRT screen (Figure 2-54). The analyzer's printer will print out hard copies of the information that appears on the screen.

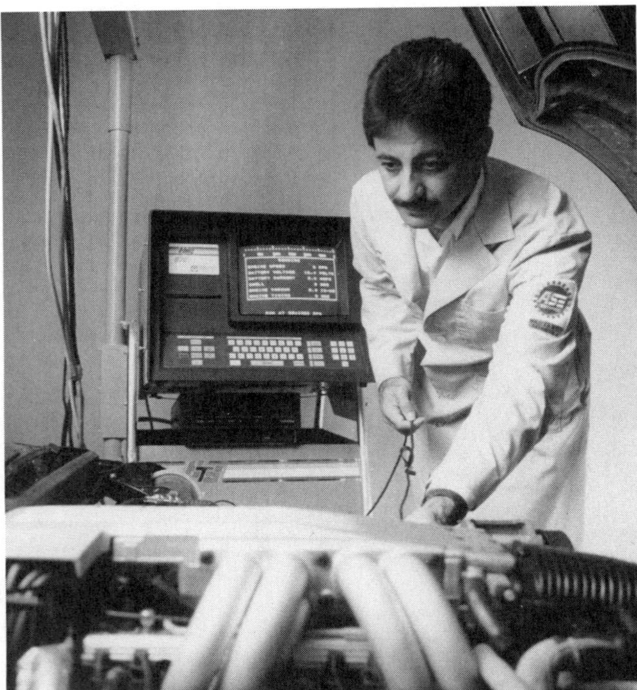

FIGURE 2–55 Engine analyzers like the one shown here have automatic test sequencing capabilities. *(courtesy of Hamilton Test Products)*

Most engine analyzers have both manual and automatic test modes. In the manual modes, any single test, such as cylinder compression or alternator current, can be performed. The manual test mode is useful when troubleshooting for a specific performance problem. The automatic test mode is useful when performing a general tune-up. When the automatic test mode is selected (usually by striking the appropriate command key on the analyzer's keyboard), specific tests are performed in a specific sequence automatically by the analyzer. The engine analyzer in Figure 2–55 performs the following tests in the sequence shown when the analyzer is placed in the automatic test mode:

- Cranking voltage
- Relative compression
- Charging voltage
- Running voltage
- Primary circuit voltage
- Secondary circuit kilovoltage
- Dwell per cylinder
- Cylinder power balance

These tests give a basic performance analysis of the engine, starting system, charging system, and ignition system. Other engine analyzers have automatic test modes for specific systems (cranking, charging, timing, fuel system, primary ignition circuit/ dwell, secondary ignition circuit/power analysis, and cylinder balance/emissions).

REVIEW QUESTIONS

1. Technician A and Technician B are testing for a leaking intake manifold gasket. Technician A uses a compression gauge. Technician B uses a vacuum gauge. Who is correct?
   a. Technician A
   b. Technician B
   c. Both A and B
   d. Neither A nor B

2. A burned valve will show up on a vacuum gauge as a ____________.
   a. low, steady vacuum reading
   b. high, steady vacuum reading
   c. sharp vacuum drop at regular intervals
   d. none of the above

3. Which of the following tools are used to locate a vacuum leak?
   a. compression tester
   b. vacuum pump
   c. oscilloscope
   d. digital multimeter

4. A fuel injector pulse test reveals that an injector is not firing. Technician A says that the problem could be that the fuel injector is plugged. Technician B says the problem could be a short in the injector connector. Who was correct?
   a. Technician A
   b. Technician B
   c. Both A and B
   d. Neither A nor B

5. Which of the following tools can be used to test for resistance in a circuit?
   a. voltmeter
   b. ohmmeter
   c. ammeter
   d. tachometer

6. Which of the following tests cannot be made with a digital multimeter?
   a. diode condition
   b. engine rpm

c. circuit resistance
d. cylinder compression

7. Which of the following tools is least likely to be used when tuning up late-model vehicles?
a. dwellmeter
b. tachometer
c. timing light (or probe)
d. all of the above

8. On which of the following tools can you find an inductive pickup?
a. ammeter
b. tachometer
c. timing light
d. all of the above

9. An electronic ignition module tester can be used to test ____________ .
a. coil spark
b. distributor pickup
c. ignition module
d. all of the above

10. Which of the following oscilloscope patterns would one choose when measuring spark duration?
a. raster
b. 5 millisecond
c. superimposed
d. none of the above

11. Technician A and Technician B are performing a tune-up on a car with breaker point ignition. Technician A says the dwell could be checked with a dwellmeter. Technician B says the dwell could be checked on the oscilloscope. Who is correct?
a. Technician A
b. Technician B
c. Both A and B
d. Neither A nor B

12. Which of the following tests cannot be performed using the oscilloscope?
a. battery voltage
b. cranking coil output
c. spark plug firing voltage
d. cylinder compression

13. To retrieve trouble codes from a vehicle's onboard computer, one should use a ____________ .
a. scan tool
b. emissions analyzer
c. fuel injector tester
d. ammeter

14. Which of the following gasses is not measured by a four-gas emissions analyzer?
a. O_2
b. HC
c. NO_x
d. CO

15. A high CO level on an emissions analyzer test convinces Technician A that the air/fuel mixture is rich. Technician B says that a burned valve is causing the rich condition. Who is probably correct?
a. Technician A
b. Technician B
c. Both A and B
d. Neither A nor B

16. Leaks in the combustion chamber can be caused by ____________ .
a. worn rings
b. a slipped timing chain
c. worn valve seats
d. all of the above

17. The dial face on a typical compression gauge indicates pressure in ____________ .
a. kPa
b. psi
c. both a and b
d. neither a nor b

18. On most vehicles, how much leakage is acceptable during a cylinder leakage test?
a. up to 10 percent
b. up to 20 percent
c. between 20 and 30 percent
d. between 5 and 10 percent

19. Which of the following conditions can be revealed by vacuum readings?
a. air pump malfunctions
b. retarded ignition timing
c. misfiring spark plugs
d. all of the above

20. The solution to a sulfated and plugged fuel injector is to ____________ .
a. run the engine at high rpm

b. clean it
c. replace it
d. none of the above

21. Which of the following statements is incorrect?
    a. An ammeter should be connected parallel with the circuit being tested.
    b. A voltmeter measures voltage available at any point in an electrical system.
    c. On a volt/amp tester, the carbon pile is a variable resistor.
    d. An infinite ohmmeter reading is usually an indication of an open circuit.

22. Timing marks are located ____________.
    a. at the rear of the engine on the flywheel
    b. in front of the engine on the harmonic balancer
    c. both a and b
    d. neither a nor b

23. An oscilloscope normally has ____________ leads.
    a. five
    b. two
    c. three
    d. four

CHAPTER THREE

ENGINE PERFORMANCE TESTS

Objectives

Upon completion of this chapter, you should be able to:

- Explain the principle of compression ratio and how it relates to engine torque and horsepower.
- Prepare an engine for a dry compression test, perform the test, and interpret the results.
- Explain when a wet compression test is needed, perform the test, and interpret the results.
- Prepare an engine for a cylinder leakage test, perform the test, and interpret the results.
- Perform an engine vacuum test and interpret the results.
- Use a vacuum gauge to test a PCV system and an exhaust system.
- Explain the significance of engine noise as a diagnostic aid.
- Perform a cylinder power balance test.
- Describe the three ways that the valve clearance is adjusted.
- Perform a valve adjustment on a running engine and a stationary engine.

Contrary to what many car owners expect of a tune-up, it is not a guaranteed cure-all for an ailing engine. Some vehicles simply are not good candidates for tuning because their engines suffer from mechanical defects. Attempting to tune an engine that is not mechanically sound can be extremely difficult and time-consuming—if not downright impossible.

In order to determine if an engine can be successfully tuned, a series of performance tests are necessary: compression, leakage, vacuum, and noise diagnosis. It is also possible to conduct a power check of the cylinders using diagnostic equipment such as engine analyzers. However, while this test does reveal whether any of the cylinders are weak, it does not pinpoint the exact cause of the problem. A weak cylinder can be caused by any number of mechanical defects; therefore, it is important to do all of the performance tests in their entirety. Keep in mind that these tests apply to the engine block only (Figure 3–1); problems dealing with ignition, fuel, and other systems are covered in later chapters.

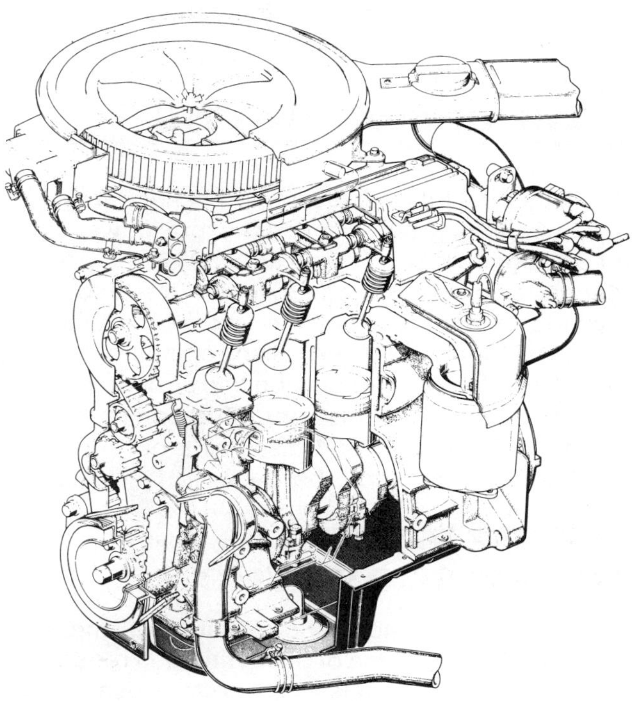

FIGURE 3–1 Engine performance tests help determine whether a tune-up should be attempted.

An important thing to keep in mind before doing any performance tests is this: Do not underestimate

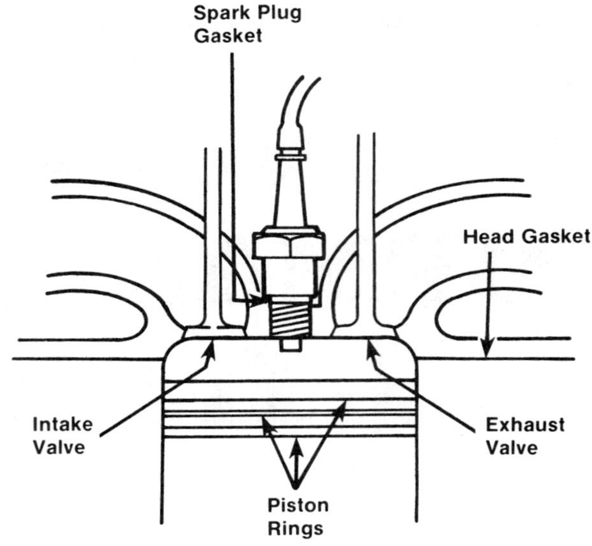

FIGURE 3-2 The amount of compression depends on how well the cylinder is sealed at various points.

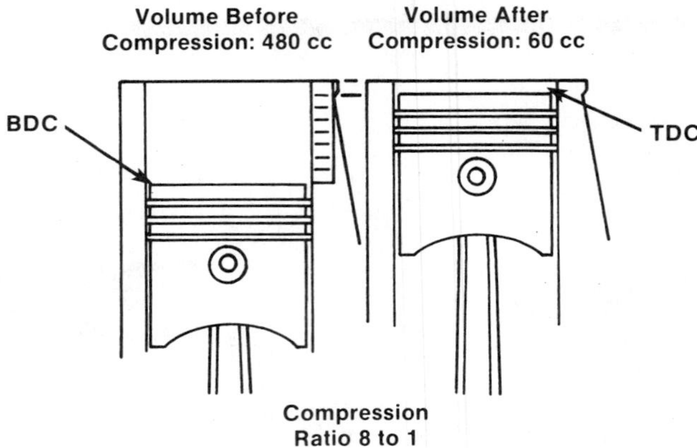

FIGURE 3-3 Compression ratio measures how much the cylinder volume is reduced as the piston moves from bottom dead center (BDC) to top dead center (TDC).

the value of oil in reducing engine wear. Good lubrication is a vital part of engine performance. When checking the lube system, this is also a good time to do the necessary preventive maintenance. First, examine the cylinder head bolts and retorque to specifications, if needed. Second inspect the timing belt; it should be replaced every 50,000 to 60,000 miles. Consult a service manual for details concerning these procedures.

COMPRESSION TESTING

The amount of compression generated in each cylinder has a great effect on the total power output of an engine. Likewise, the amount of compression depends on how well the cylinder is sealed by the cylinder head gasket, piston rings, valves, and the gasket (Figure 3-2). A leak at any of these points will result in a loss of compression and poor driveability. A compression test measures the compression in each individual cylinder. It also points out any significant pressure variations between all of the cylinders.

UNDERSTANDING COMPRESSION RATIO

The compression ratio (Figure 3-3) measures how much the volume of the cylinder is reduced by the piston as it travels from bottom dead center (BDC) to top dead center (TDC). When the piston is at TDC, this is known as the clearance volume. In other words, the compression ratio is simply: the total volume of the cylinder divided by the clearance volume. For example, if the total cylinder volume is 50 cubic inches, and the clearance volume is 5 cubic inches, the compression ratio is 50/5 or 10:1.

The higher the compression ratio, the greater the engine torque and horsepower. This is because the piston forces the air/fuel mixture into a smaller area, which increases the combustion pressure in the engine. An engine with an 8:1 compression ratio has a compression pressure of approximately 150 psi and a combustion pressure of 600 psi; in comparison, an engine with a 7:1 compression ratio has a compression pressure of about 125 psi and a combustion pressure of 500 psi.

TEST SPECIFICATIONS

Every car manufacturer provides compression-pressure specifications for its engines. Always compare the compression test results with these figures. The specifications are usually given one of two ways: percentage or minimum.

- A typical percentage specification might state that the compression pressure be 150 psi plus or minus 10 percent. This plus or minus tolerance is crucial because it limits the pressure variation between cylinders. Another percentage specification might specify that the lowest compression reading of any cylinder can be no less than 75 percent of the highest reading. For example, if the highest reading is 200 psi, the lowest must be at least 150 psi.

- A minimum figure specification generally also includes an allowable pressure variation between cylinders. For example, the minimum compression pressure for an engine might be 150 psi, with a variation of no more than 30 psi. In this case, each cylinder must have a compression pressure of at least 150 psi, but no higher than 180 psi.

PREPARING FOR THE TEST

The two special tools needed to do a compression test are the compression gauge and remote starter switch. Always be sure that the tools used are high quality and in good condition; otherwise, the test results will be invalid.

Compression Gauge

For gasoline engines, the typical compression gauge can measure pressures up to 300 psi. Diesel engines require specialized gauges that measure higher pressures. A popular compression gauge design features a threaded adapter that fits into the spark plug hole; a hose connects the adapter to the gauge. The vent valve traps the pressure in the gauge until the reading is taken.

Another type of compression gauge utilizes a rubber tip that fits on the end of a metal stem; the stem threads into the gauge (Figure 3-4). The tips fit any size spark plug hole, making this gauge very adaptable. The stems come in various lengths and shapes, a very useful feature for working around obstructions. A vent valve is also used on this gauge.

FIGURE 3-4 Holding the rubber tip in the spark plug opening is necessary with some compression gauges.

To take a reading, the rubber tip must be manually held in the spark plug opening, which can be rather difficult when working with a high-compression engine.

Remote Starter Switch

A remote starter switch can be an individual tool or part of a compression gauge. Since it permits the engine to be cranked from under the hood, it is an indispensable tool when working alone. In addition to checking the compression, tasks such as adjusting valves and setting ignition points can be done without the aid of a second person.

Readying the Engine

Use the following procedure to prepare an engine for a compression test:

1. Run the engine until it reaches its normal operating temperature, then turn it off.
2. Disconnect the spark plug cables from the spark plugs (Figure 3-5).
3. Use compressed air to clean all dirt and other foreign matter out of the spark plug wells (Figure 3-6).
4. Remove all of the spark plugs.

SHOP TALK

When evaluating the condition of spark plugs, keep in mind that a plug badly fouled with oil, combined with a low compression reading, probably means that the piston rings are worn.

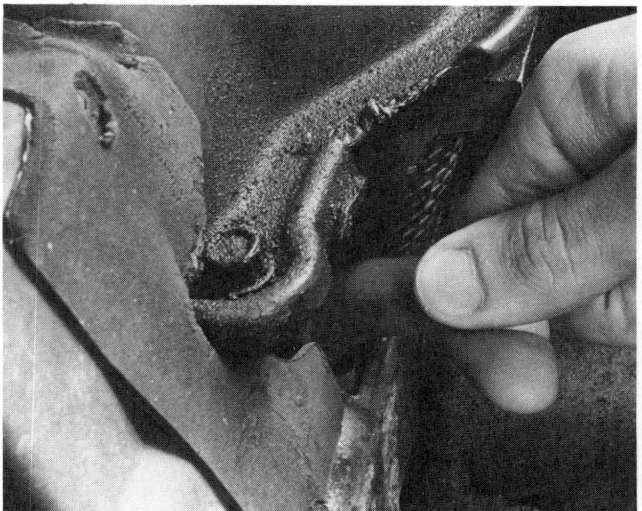

FIGURE 3-5 Disconnecting spark plug cables

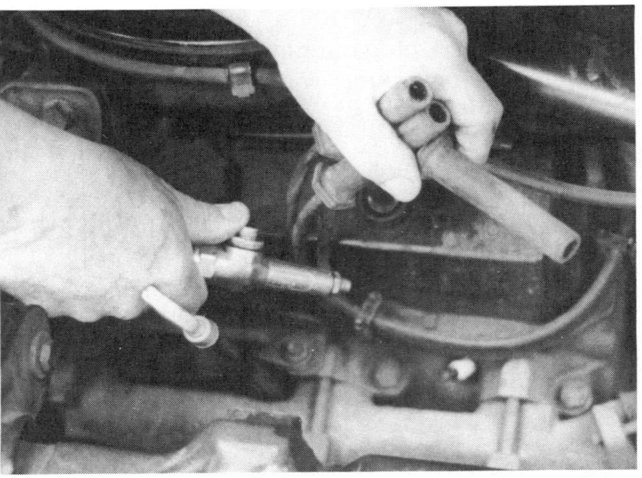

FIGURE 3-6 Using compressed air to clean out the spark plug walls

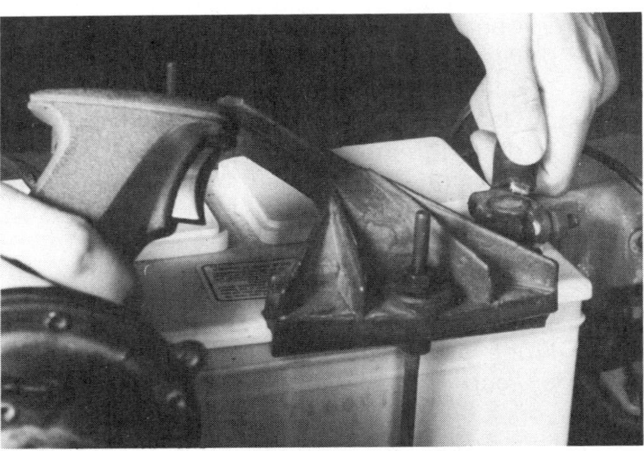

FIGURE 3-7 Connecting a remote starter switch

5. Be sure to remove all spark plug gaskets or tubes from the cylinder head.
6. Remove the air cleaner from the carburetor and block the choke and throttle plate in the wide-open position.
7. Remove the ignition coil's high-tension wire from the distributor cap and ground it. By disabling the ignition system in this manner, the chance of electrical shock or fire is greatly reduced. (The technique can damage certain electronic and HEI systems; consult the service manual for information.)
8. Connect a remote starter switch to the starter relay or solenoid (Figure 3-7). Attach one of its leads to the positive battery terminal and the other to the S-terminal on the relay or solenoid.

DRY COMPRESSION TEST

Use the following procedure to perform a dry compression test:

1. Depending on the type of gauge being used, either thread the adapter into the spark plug hole, fingertighten, and connect the gauge to the adapter, or insert the rubber tip into the hole and hold it firmly in place by hand.
2. Turn the engine over four full compression strokes, keeping an eye on the gauge needle at all times.
3. Record the first and fourth gauge readings. The needle should rise steadily with each subsequent stroke. The fourth and final reading should be well within specifications.
4. Open the vent valve to release the compression pressure.
5. Disconnect and remove the gauge.
6. Repeat the procedure on the remaining cylinders.

To interpret the results of the dry compression test, use the following guidelines:

- If any reading is low on the first stroke, but gradually gets higher without reaching the specified pressure, the piston rings are probably worn badly.
- If any reading is low on the first stroke, and gets only a little higher during subsequent strokes, the cause may be a sticking or burned valve.
- If the reading is equally low on two adjacent cylinders, the probable cause is a leaking head gasket.
- If the total gauge reading is higher than specified, excessive carbon deposits in the combustion chamber are the likely cause.

Any cylinders that record a low pressure reading should be given a wet compression test.

WET COMPRESSION TEST

A wet compression test should be performed before disassembling the engine for inspection and repair. Perform a wet compression test as follows:

1. Squirt one or two ounces of medium-viscosity oil into the cylinder through the spark plug hole (Figure 3-8).
2. Turn the engine over several times; this will allow the oil to work its way down around the rings.

3. Connect the compression gauge to the cylinder and conduct the compression test as described earlier for the dry test procedure.

Guidelines for the wet compression test are as follows:

- If the readings are now normal or very close to normal, either the rings, pistons, or cylinders need servicing.
- If the readings do not improve at all, the leakage is probably at the valves or head gasket.
- If the readings improve only slightly, the rings and valves should both the looked at carefully.

SHOP TALK

The wet compression test should not be performed on horizontally opposed cylinder engines such as those found on Subarus or older Volkswagens. For this engine, a leakage test is the most accurate method of discovering the source of low compression.

CYLINDER LEAKAGE TESTING

While the compression test gives a good indication of the amount of pressure in the cylinders, the leakage test provides a more accurate method of testing engine condition. Even minute leaks in the valves, rings, or head gasket can be detected and measured. The leakage test also points out leaks around the exhaust and intake valves, leaks between cylinders, leaks into the water jacket, and any other cause of compression loss.

FIGURE 3-8 Oil must be added to the cylinder prior to the wet compression test.

LEAKAGE TESTER

The leakage tester contains a precision gauge for extremely accurate readings. Its scale ranges from zero to 100 percent; zero means that the cylinder is perfectly sealed, while 100 indicates that it is holding no air whatsoever. To receive as accurate a reading as possible, the gauge should be calibrated before every leakage test. To do this, first turn the control regulator knob counterclockwise until it rotates freely. Connect a 70 to 200 psi air supply to the tester's air input fitting. Now turn the knob clockwise until the gauge reads zero. Connect and disconnect a test adapter to the tester's cylinder connection fitting. The gauge should rise to 100 percent, then return to zero; if it does not, the control regulator knob must be readjusted.

PREPARING FOR THE TEST

To prepare an engine for a cylinder leakage test, do the following:

1. Check the coolant level; fill if needed.
2. Run the engine until it reaches its normal operating temperature, then shut it off.
3. Disconnect the spark plug cables from the spark plugs.
4. Use compressed air to clean all dirt and other foreign matter out of the spark plug wells.
5. Remove all the spark plugs, as well as any gaskets or tubes that may have been used.
6. Remove the air cleaner from the carburetor. Use a long screwdriver or similar tool to block the choke and throttle plate in the wide-open position.
7. Disconnect the PCV hose from the crankcase.

PERFORMING THE TEST

In addition to the tester, a TDC indicator and indicator light are also needed to conduct a cylinder leakage test. Always begin at the number 1 cylinder. Use the following procedure:

1. Install the proper test adapter hose in the number 1 cylinder spark plug hole. Connect the tester whistle to the adapter hose (Figure 3-9).
2. Use a wrench on the crankshaft pulley nut or bolt to slowly rotate the engine in the normal direction until the whistle blows,

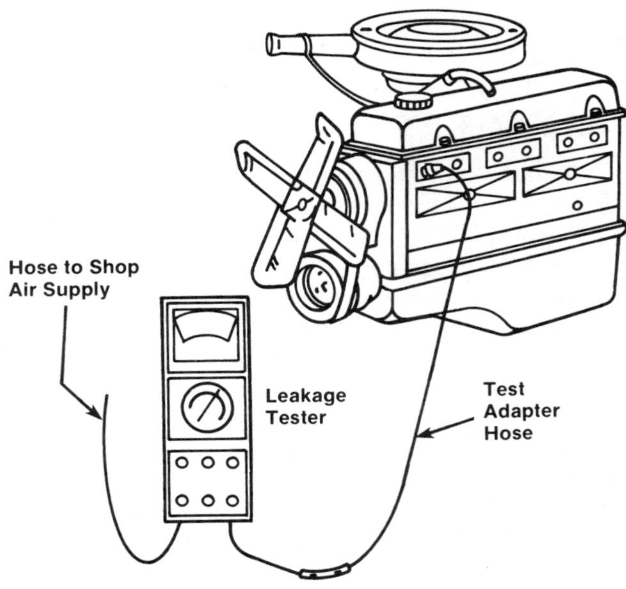

FIGURE 3-9 Leakage tester and whistle connected to the number 1 spark plug hole via test adapter hose

indicating the beginning of a compression stroke.

3. Continue the rotation until the timing mark on the crankshaft pulley lines up with the engine-timing pointer on the timing chain cover. Remove the whistle from the adapter.
4. Use a jumper lead to connect the coil-to-distributor secondary cable to a good ground.
5. Remove the distributor cap and rotor, then mount a TDC indicator on the distributor shaft. Mark a chalk reference point on the engine that lines up with the appropriate cylinder marking on the TDC indicator.
6. Connect an indicator light to the ignition system (Figure 3-10). To do this, attach one of its leads to the distributor's primary terminal on the coil; attach the other lead to a good ground. Turn the ignition switch on.
7. Connect the tester to the adapter hose. If the gauge shows more than 20 percent leakage, air might be escaping through the carburetor, tailpipe, or crankcase. Check also for air bubbles in the radiator.
8. Disconnect the tester from the adapter hose. Resume rotating the engine until the next appropriate cylinder mark on the TDC indicator lines up with the chalk mark on the engine. The indicator light should glow, meaning that the piston is in firing position.
9. Remove the adapter from the previously tested cylinder. Install it in the spark plug hole of the next cylinder in the engine's firing order. (The piston in this cylinder is at top dead center.)
10. Repeat steps 7, 8, and 9 on all of the cylinders.

CAUTION: Remember that the engine could rotate suddenly while a cylinder is under pressure.

All readings should be relatively even and less than 20 percent. Any reading of 30 percent or higher indicates a definite problem. Use a hydrosonic leak detector and the following guidelines to determine where the leak is and its cause:

- Air escaping from the exhaust pipe means that an exhaust valve is leaking.
- Air escaping through the carburetor means that an intake valve is leaking.
- Air bubbles in the radiator indicate a crack in the engine block or head, or a leaking head gasket.
- Leakage from two adjacent cylinders might also indicate a cracked block or head or a leaking head gasket.

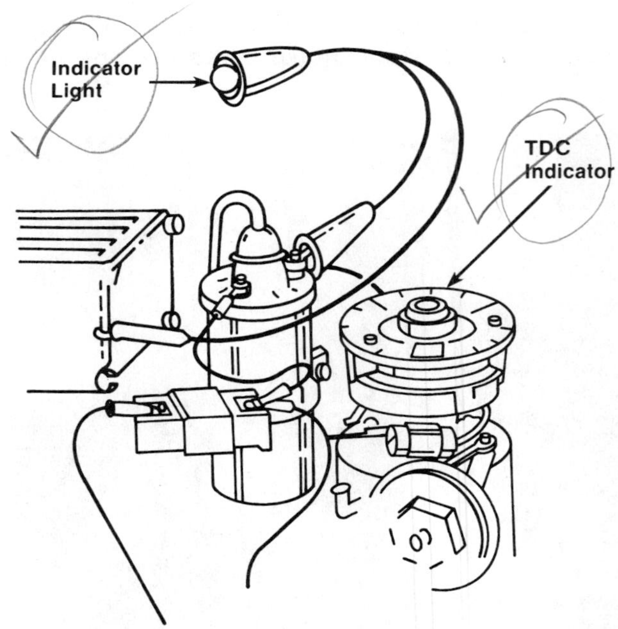

FIGURE 3-10 TDC indicator and indicator light set up for cylinder leakage test

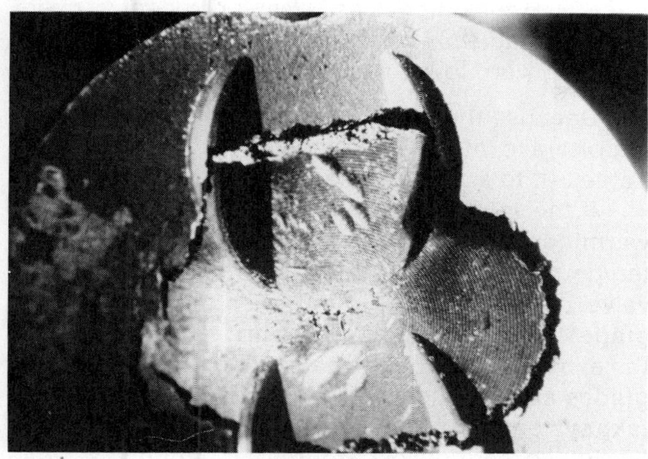

FIGURE 3–11 A cracked piston can cause crankcase leakage.

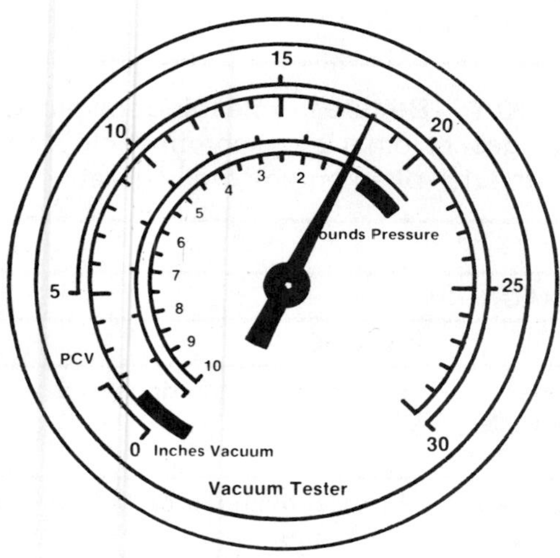

FIGURE 3–12 A normal vacuum gauge reading

- Crankcase leakage could mean worn piston rings or cylinder walls, or a cracked piston (Figure 3–11). However, keep in mind the age of the engine; if it is relatively new, the problem could just be that the rings are stuck or not yet seated.

In many cases, it is useful to compare results of compression and cylinder leakage tests in order to best analyze a difficult or unusual mechanical engine problem. For example, an older engine whose compression reading is within specifications but has air escaping from the crankcase opening is probably just showing signs of old age. High-mileage engines typically suffer from excessive blowby and poor gas mileage, as well as heavy carbon deposits in the combustion chambers. These deposits can raise the compression readings, thus creating the impression that the engine is more mechanically sound than it is.

Low compression combined with minimal cylinder leakage usually means a valve-related problem. An improperly installed timing chain or gears, a loose timing chain, or the wrong type of camshaft can also be the cause of this situation. In all of these cases, the valves are not opening and closing at the correct point in the engine cycle.

VACUUM TESTING

An engine vacuum test is one of the quickest and easiest ways to test an engine. Since any piston-driven engine is basically a combination vacuum pump and heat exchanger, it is the higher pressure on the outside of the cylinders that causes air to be pushed into the cylinders. When an engine loses the ability to create this pressure differential, its performance suffers.

CAUTION: Like compression and leakage tests, vacuum testing alone should not be used to locate the exact source of a problem. Always perform it as part of a series of diagnostic tests.

The vacuum gauge itself may be a separate tool or it can be part of an engine analyzer. Increments are in inches of mercury (Hg), and range from zero to 30. The "normal" range is between 18 and 20 inches Hg, with most of today's emission-controlled cars falling in the low end of this scale (Figure 3-12). It is important to keep in mind that all vacuum gauge readings are dependent on altitude: for every 1,000 feet above sea level, the reading will be low by 1 inch. Therefore, add 1 inch Hg to the reading for every 1,000 feet above sea level.

PERFORMING THE TEST

Use the following procedure to perform a vacuum test:

1. Use a length of hose to connect the vacuum gauge to a nonrestricted port on the intake manifold. The hose should be at least 3 feet long in order to dampen vibrations from the needle.
2. In some cases, it may be necessary to further dampen the needle by clamping the hose; this will slightly restrict its passageway.

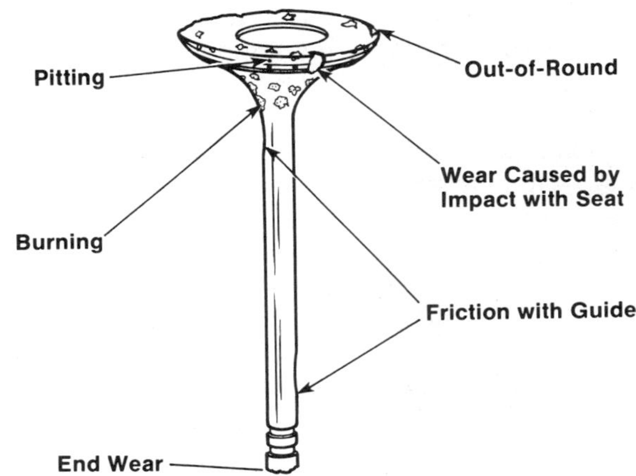

FIGURE 3–13 Examples of valve guide wear

3. Run the engine until it reaches its normal operating temperature. With a few exceptions, vacuum tests must always be performed with the engine at idle rpm; consult a service manual for details.
4. If the needle remains constant between 15 and 20 inches Hg, the engine vacuum is good. In addition to engines with emission control devices, new or recently overhauled engines also have a tendency toward lower vacuum readings.

Because the results can be so varied, it is very important to interpret the vacuum test correctly. Use Table 3–1 to arrive at a diagnosis.

If the valve guides are suspect, check them by warming up the engine to its normal operating temperature. Turn off the engine, then remove the valve covers and squirt oil over the tops of the guides. Turn the engine back on; if blue smoke exits the exhaust pipe and the gauge needle steadies, the guides are definitely worn (Figure 3–13). As for intake system leaks, they are usually caused by defective intake manifold or carburetor mounting gaskets, or defective vacuum hoses. To test for vacuum leaks at the gaskets, run the engine at idle rpm and squirt cleaning solvent along the gasket joints. If the vacuum increases and the idle smoothes out, the leak has been found.

CAUTION: Be sure the cleaning solvent used on the gasket joints is noncombustible. Otherwise, the risk of an engine fire is great.

TABLE 3–1: VACUUM TEST DIAGNOSIS

Reading	Possible Cause	Remedy
1. Low but steady, between 12 and 15 inches Hg	1. Leakage around piston rings, late ignition timing, or late valve timing	1. Replace piston rings or reset timing.
2. Needle oscillates slowly, then rapidly, between 12 and 18 inches Hg	2. Ignition timing too far advanced or carburetor idle mixture too lean	2. Reset timing or carburetor idle mixture.
3. Regular needle drop between 1 and 2 inches Hg	3. Burned or leaking valve or spark plug in one of the cylinders is not firing	3. Replace valve or spark plug.
4. Irregular needle drop between 1 and 2 inches Hg	4. Sticking valve, carburetor out of adjustment, or intermittent spark plug misfire	4. Replace valve, adjust carburetor, or replace spark plug.
5. Normal at idle speed, but excessive vibrations at higher rpm	5. Weak valve springs	5. Replace valve springs.
6. Excessive vibrations at idle speed, but steadies at higher rpm	6. Worn valve guides	6. Replace valve guides.
7. Excessive vibration at all rpm	7. Leaky head gasket	7. Replace head gasket.
8. Needle oscillates between 3 and 9 inches Hg lower than normal	8. Intake system leak	8. Replace faulty component.
9. Normal at idle speed, but drops to near zero and rises to lower than normal	9. Restriction in exhaust system	9. Repair or replace exhaust system.

FIGURE 3-14 On in-line engines like this, the normal method of finding intake leaks often cannot be used.

If the lower edges of the intake manifold gasket are not accessible, this method of finding intake leaks might not work; this is often the case with V-type and in-line engines (Figure 3-14). Instead, check the exhaust for excessive smoke, a sign that there is a problem.

CRANKING VACUUM TEST

To perform a cranking vacuum test, follow this procedure:

1. Run the engine until it reaches normal operating temperature, then turn it off.
2. Connect the vacuum gauge hose to a source of engine manifold vacuum (Figure 3-15).
3. Disable the ignition system. On a vehicle with a separate ignition coil, do this by removing the coil wire from the distributor cap and connecting a jumper lead from the coil wire to an engine ground. On a vehicle without a separate ignition coil, disconnect the electrical connector that supplies battery voltage to the system.

CAUTION: Make sure the vacuum gauge hose is well away from the belts, pulley, and fan before cranking the engine.

---

4. Use a remote starter switch to crank the engine, but do not depress the accelerator pedal.
5. Note the vacuum reading. An engine in good condition should produce a reading of at least 5 inches.
6. If the reading is less than 5 inches, check for external leakage. Inspect the intake manifold gasket and carburetor gasket, and look for broken or disconnected vacuum lines.

TESTING THE EXHAUST SYSTEM

A vacuum gauge can also be used to test the exhaust system for restriction, as follows:

1. Attach the gauge to the intake manifold as described earlier, then connect a tachometer to the engine.
2. Run the engine until it reaches its normal operating temperature.
3. Accelerate the engine slowly. After it reaches 2,000 rpm, note the reading on the gauge; it should drop a little, then rise sharply.
4. Close the throttle quickly. The needle should return to the normal idle reading as quickly as it rose.
5. If the needle gives a normal reading at idle speed and at 2,000 rpm, drops to near zero, and rises to a below normal reading, there is a restriction in the exhaust system. Check the muffler and tailpipe for damage, and inspect the manifold heat control valve to see if it is stuck or frozen.

If working with a large displacement engine, it will probably be necessary to do this test while driving the vehicle. Proceed as follows:

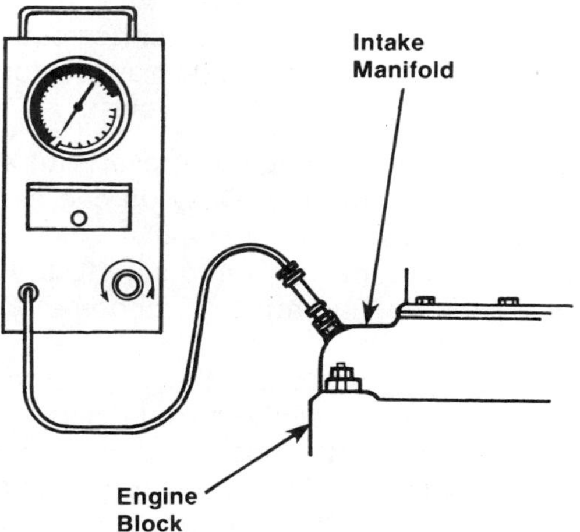

FIGURE 3-15 Making the vacuum gauge connection

FIGURE 3-16 Connecting a tachometer to the engine

1. Connect the gauge to the intake manifold. Use a hose long enough so that the gauge is inside the vehicle.
2. Connect a tachometer to the engine; route its wires so that it is also entirely inside the vehicle (Figure 3-16).
3. Drive the vehicle until it reaches the speed where the engine loses power. The gauge reading should be approximately the same each time a power loss occurs, with the needle dropping toward zero. The greater the restriction, the closer to zero the needle will drop.

VACUUM TESTING FOR LOSS OF COMPRESSION

A vacuum gauge can even be used to test for compression loss due to leakage around the pistons. However, this test should not be performed unless normal readings were produced on all previous vacuum tests. Conduct the test as follows:

1. Make sure the engine oil level is full and the oil is not too old. Dirty oil can cause an incorrect vacuum reading.
2. Connect the vacuum gauge to the intake manifold, and attach a tachometer to the engine.
3. Quickly accelerate the engine to 2,000 rpm, then close the throttle fast.
4. As the throttle closes, the needle should rise 5 inches Hg or more above the normal reading. An increase of less than 5 inches Hg means there is a compression loss around the pistons, rings, or cylinder walls.

NOISE DIAGNOSIS

More often than not, a malfunction in the engine will reveal itself first as an unusual noise. This can happen before the problem affects the driveability of the vehicle. Problems such as loose pistons, badly worn rings or ring lands, loose piston pins, worn main bearings and connecting rod bearings, loose vibration damper or flywheel (Figure 3-17), and worn or loose valve-train components all produce telltale sounds. Of course, unless the technician has experience in listening to and interpreting engine noises, it can be very hard to distinguish one from the other.

When correctly interpreted, engine noise can be a very valuable diagnostic aid. For one thing, a costly and time-consuming engine tear-down might be avoided. *Always* make a noise analysis before doing any repair work; this way, there is a much greater likelihood that only the necessary repair procedures will be done. Careful noise diagnosis also reduces the chances of ruining the engine by continuing to use the vehicle despite the problem.

USING A STETHOSCOPE

While some engine sounds can be easily heard without using a listening device, others are impossible to hear unless magnified. A stethoscope is very helpful in locating engine noise by magnifying the sound waves; it can also distinguish between normal and abnormal noise. The procedure for using a stethoscope is simple: Use the metal prod to trace the sound until it reaches its maximum intensity. Once the precise location has been discovered, the sound can be better evaluated. A sounding stick,

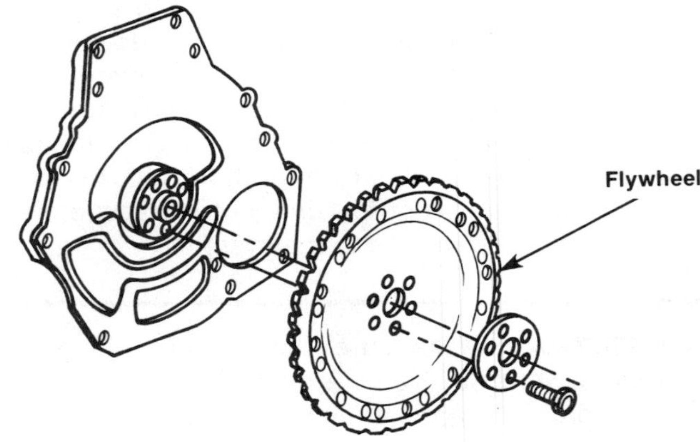

FIGURE 3-17 The flywheel is one of many components that produces telltale sounds when a problem exists.

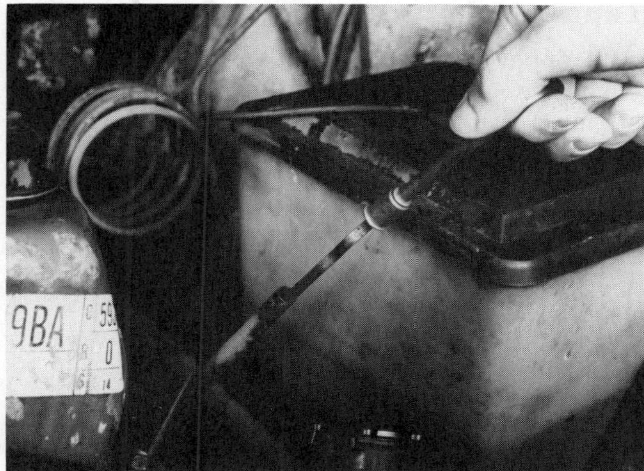

FIGURE 3-18 Always check the oil level first when trying to locate the source of engine noise.

which is nothing more than a long, hollow tube, works on the same principle, though a stethoscope gives much clearer results.

COMMON NOISES

Following are examples of abnormal engine noises, including a description of the sound, its likely cause and ways of eliminating it. An important point to keep in mind is that everyone has his or her own way of describing a particular noise; one person's "rattle" can be another person's "thump." While the owner's descriptions can be helpful, the final diagnosis is in the hands of the technician. It should also be noted that insufficient lubrication is the most common cause of engine noise. For this reason, always check the oil level first before moving on to other areas of the vehicle (Figure 3-18).

Ring Noise

This sound can be heard during acceleration as a high-pitched rattling or clicking in the upper part of a cylinder. It can be caused by worn rings or cylinders, broken piston ring lands, or insufficient ring tension against the cylinder walls. Ring noise is corrected by replacing the rings, pistons, or sleeves, or reboring the cylinders. Shorting out the spark plug of the affected cylinder usually will not help to eliminate ring noise.

Piston Slap

This is a common sound when the engine is cold, and often intensifies when the vehicle accelerates. When a piston slaps against the cylinder wall, the result is a hollow, bell-like sound. Piston slap is caused by worn pistons or cylinders, collapsed piston skirts, misaligned connecting rods, excessive piston-to-cylinder wall clearance, or lack of lubrication, resulting in worn bearings. Correction requires either replacing the pistons, reboring the cylinder, replacing or realigning the rods, or adding oil to the engine and replacing the bearings. Shorting out the spark plug of the affected cylinder might quiet the noise, but this is only a stop-gap measure; it does not correct the problem.

Piston-Pin Knock

Piston-pin knock is a sharp, metallic rap that can sound more like a rattle if all the pins are loose. It originates in the upper portion of the engine, and is most noticeable when the engine is idling. Piston-pin knock is caused by a worn piston pin, piston-pin boss, or piston-pin bushing, or lack of lubrication, resulting in worn bearings. To correct it, either install oversize pins, replace the boss or bushings, or replace the bearings.

Ridge Noise

This noise is less common, but very distinct. As a piston ring strikes the ridge at the top of the cylinder, the result is a high-pitched rapping or clicking that becomes louder during deceleration (Figure 3-19).

There can be more than one reason for the ridge interfering with the ring's travel. For one thing, if new rings are installed without removing the old ridge,

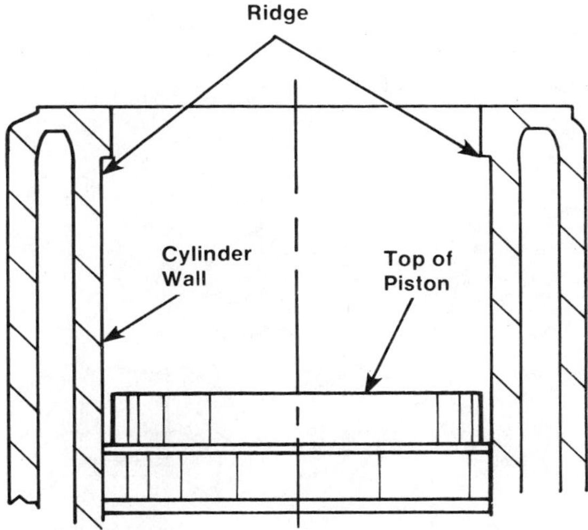

FIGURE 3-19 As the piston strikes the ridge at the top of the cylinder, a high-pitched rapping or clicking sound is made.

the new rings will contact the ridge and make a noise. Also, if the piston pin is very loose or the connecting rod has a loose or burned-out bearing, the piston will go high enough in the cylinder for the top ring to contact the ridge. Thus, in order to eliminate ridge noise, remove the old ring ridge and replace the piston pin or connecting rod bearings.

Rod Bearing Noise

The result of worn or loose connecting-rod bearings, this noise is heard at speeds over 35 mph. Depending on how badly the bearings are worn, the noise can range from a light tap to a heavy knock or pound. Shorting out the spark plug of the affected cylinder can lessen the noise, unless the bearing is totally burned out; in this case, shorting out the plug will have no effect. Rod-bearing noise is caused by a worn bearing or crankpin, a misaligned rod, or lack of lubrication, resulting in worn bearings. To correct it, service or replace the crankshaft, realign or replace the connecting rods, or replace the bearings.

Main or Thrust Bearing Noise

A loose crankshaft main bearing produces a dull, steady knock, while a loose crankshaft thrust bearing produces a heavy thump at irregular intervals. The thrust bearing noise may only be audible on very hard acceleration. Both of these bearing noises are usually caused by worn bearings or crankshaft journals (Figure 3-20). To correct the problem, replace the bearings and/or crankshaft.

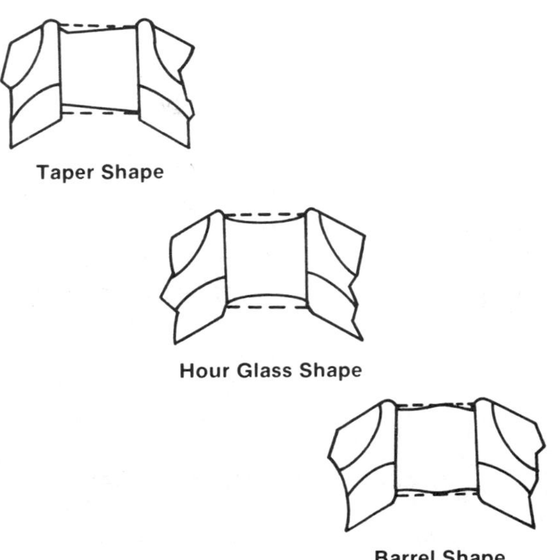

FIGURE 3-20 Examples of worn crankshaft journals

Tappet Noise

Tappet noise is characterized by a light, regular clicking sound that is more noticeable when the engine is idling. It is the result of excessive clearance in the valve train. The clearance problem area is located by inserting a feeler gauge between each lifter and valve, or between each rocker arm and valve tip, until the noise subsides. Tappet noise can be caused by improper valve adjustment, worn or damaged parts, dirty hydraulic lifters, or lack of lubrication. To correct the noise, adjust the valves, replace any worn or damaged parts, or clean or replace the lifters.

Detonation Knock

Also known as spark knock or just plain "ping," this noise is most noticeable during acceleration with the engine under load and running at normal temperature. Excessive detonation knock can be very harmful to the engine. It is often caused by advanced ignition timing, or by substantial carbon build-up in the combustion chambers that increase the combustion pressure. Carbon deposits that get so hot they glow will also preignite the air/fuel mixture, causing detonation. Another possible cause is too low an octane fuel. Detonation knock is cured by removing the carbon deposits from the combustion chambers with a rotary wire brush (Figure 3-21), as well as recommending use of a higher octane gasoline.

Damper or Flywheel Noise

A loose vibration damper causes a heavy rumble or thump in the front of the engine that is more

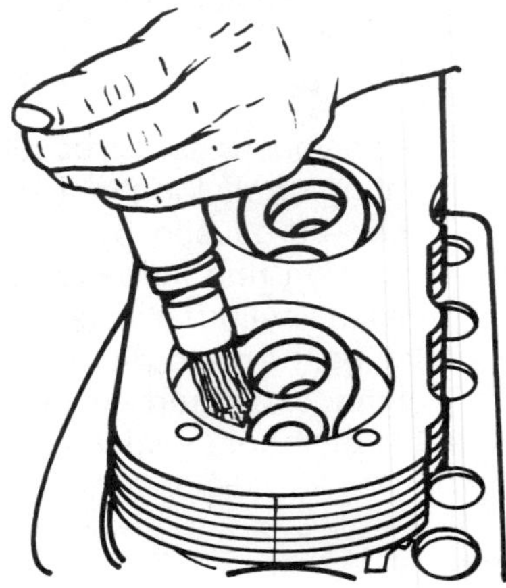

FIGURE 3-21 Removing carbon deposits from the combustion chamber

apparent when the vehicle is accelerating from idle under load or is idling unevenly. A loose flywheel causes a heavy thump or light knock at the back of the engine, depending on the amount of play and the type of engine. Both of these problems are corrected either by tightening or replacing the damper or flywheel.

CYLINDER POWER BALANCE TESTING

The cylinder power balance test is useful in determining if a cylinder or bank of cylinders is producing its share of engine power. Ideally, all the cylinders should be doing the same amount of work, and changes in engine rpm should be about equal as each cylinder is shorted out. Unequal cylinder power can mean a problem in the cylinders themselves, as well as the rings, valves, intake manifold, head gasket, fuel system, or ignition system.

The power balance test is performed quickest and easiest using an engine analyzer, because the spark plugs can be controlled with push buttons. Changes are measured in "rpm drop." Keep in mind that the push-button numbers refer to the cylinder firing order, not the cylinder number designation. For example, when testing an engine with a firing order of 1-3-4-2, pushing the first button shorts out the number 1 cylinder, pushing the second button shorts out the number 3 cylinder, and so on.

TEST PRECAUTIONS

On some computer-controlled or fuel-injected engines, certain components must be disconnected before attempting the power balance test. Because of the wide variation from manufacturer to manufacturer, consult the vehicle's service manual for specific instructions.

If the engine being tested has an exhaust gas recirculation (EGR) system, added precautions must be taken. If the system is valve controlled, disconnect the vacuum or electrical connection to the EGR valve. This will prevent the valve from cycling due to vacuum changes when the cylinders are shorted out. For engines with a floor jet EGR system, the power balance test cannot be performed accurately because of the possibility of the unburned fuel mixture being sent back into the cylinders. The compression test is the recommended alternative in such cases.

Care must be taken when performing the power balance test on vehicles with catalytic converters.

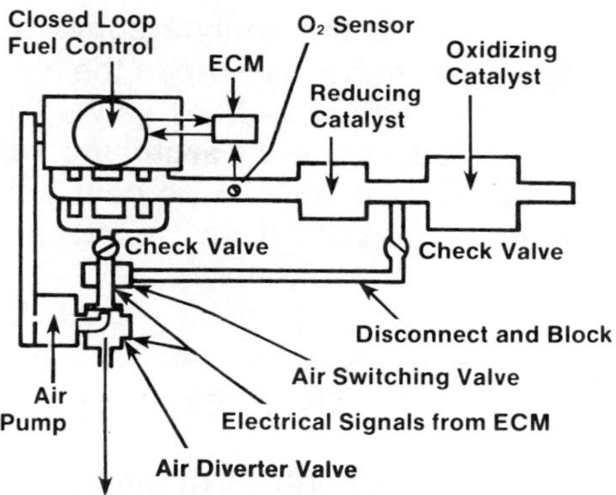

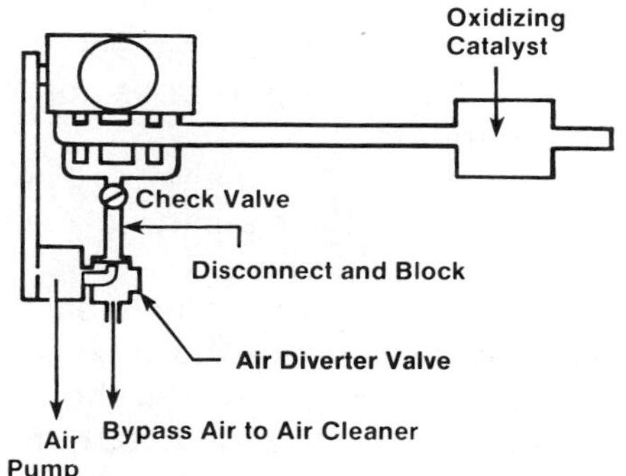

FIGURE 3-22 A cylinder power balance test differs, depending on whether the engine (A) has an air/fuel mixture feedback control or (B) does not.

To prevent unburned fuel from building up in the converter, short each spark plug for less than 15 seconds, then wait another 30 seconds before shorting another one.

SHOP TALK

The old style power balance testers, on which a single knob controls individual spark plugs, might not be safe for use on vehicles with electronic ignitions. The ignition system or solid state components could be damaged, not to mention the possibility of inaccurate test results. Make certain the tester being used is compatible with electronic ignition systems.

PERFORMING THE TEST

The standard power balance test is fairly simple. Use the following procedure:

1. If the engine has an air/fuel mixture feedback control or O_2 sensor (Figure 3-22), disconnect and plug either the air pump

hose going to the catalytic converter or the downstream hose between the air switching valve and the check valve. On some Ford models, the air switching valve can routinely have up to 10 percent leakage, in which case both hoses must be disconnected and plugged.

2. If the engine does not have an air/fuel mixture feedback control or O_2 sensor, disconnect and block the air pump on the valve side.
3. Override the controls of the electric cooling fan by jumper wiring the controls so that the fan runs constantly. If the fan cannot be bypassed, disconnect it, but be careful that the engine does not overheat during the test.
4. Connect the engine analyzer's leads, referring to the vehicle's service manual for specific instructions.
5. Turn on the engine and let it reach its normal operating temperature before beginning the test. Engine speed should be stabilized at approximately 1,000 rpm. When a cylinder is shorted, note any drop in rpm or manifold vacuum.

As each cylinder is shorted out, a noticeable drop in engine speed should occur. Little or no decrease in rpm indicates a weak cylinder. If all the readings are fairly close to each other, the engine is in sound mechanical condition; if a reading in one or more cylinders differs greatly from the rest, there is a problem. Further testing should be done to determine if the problem is purely mechanical, or if it is in the ignition or fuel system.

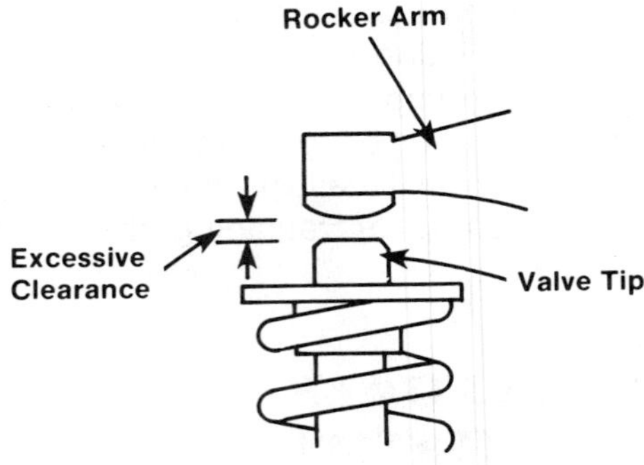

FIGURE 3-23 Wear on the valve-train parts creates excessive clearance.

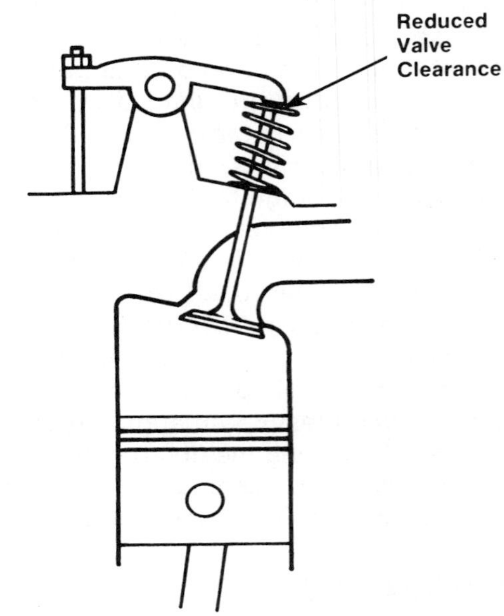

FIGURE 3-24 Reduced valve clearance leads to rough idle and valve burning.

VALVE ADJUSTMENT

Valve adjustment is a must on small engines with mechanical valve lifters. The hydraulic lifters (also called zero lash adjusters) that are usually found on larger engines do not require any adjustment. Generally, this procedure is recommended at the same mileage interval as spark plug replacement.

There are two major reasons why periodic valve adjustment is so important. First, normal wear on the valve-train parts creates excessive clearances in the valve train. This clearance refers to the gap between the cam lobe and the tappet or follower (on overhead cam engines) or between the valve tip and the rocker arm (on I-head engines). Such clearances, in turn, lead to noisy vehicle operation as the rocker arm or lifter repeatedly hammers the valve tip (Figure 3-23). This hammering only serves to increase the wear on the affected components. Excessive clearance also lessens overall engine efficiency by reducing the actual distance the valves lift off their seats.

The other reason valve adjustment is so important is to check for reduced clearance, or clearance that is less than specifications. This problem, the result of a stretched valve stem or other component expansion, can cause rough idle and valve burning (Figure 3-24). It can also cause the valves to open too soon, thus reducing engine efficiency and increasing exhaust emissions.

METHODS OF VALVE ADJUSTMENT

Valve clearance is adjusted one of three ways: with an adjustable rocker arm, an adjustable tappet,

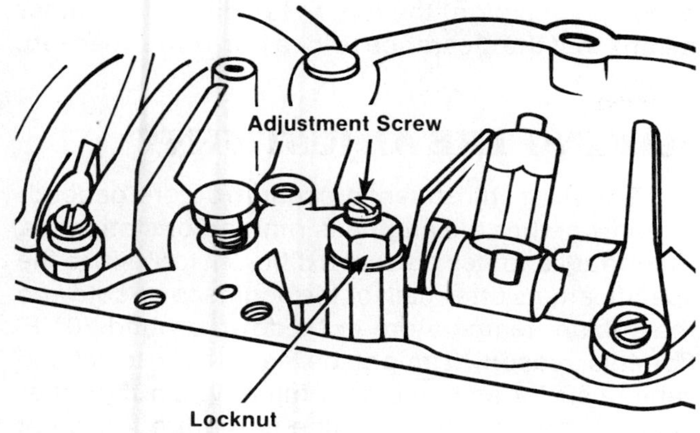

FIGURE 3-25 Adjustment screw and locknut on the pushrod end of a rocker arm

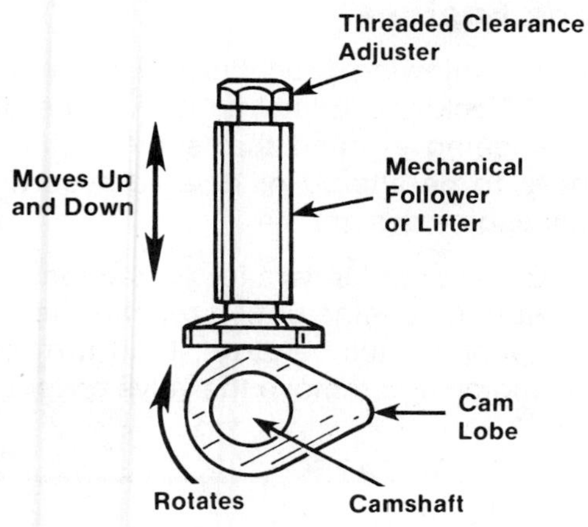

FIGURE 3-26 An early adjustable tappet design

or adjustment discs or shims. Adjustable rocker arms are found on overhead valve engines with the camshaft in the engine block. On some designs the entire rocker arm adjusts up and down on its support stud; other designs utilize an adjustment screw on the pushrod end of the rocker arm (Figure 3-25). The screw can be a self-locking capscrew or an adjusting screw with a locknut.

Adjustable tappets, or cam followers, have been on the automotive scene for years. An early design featured the tappet sitting on top of the cam lobe, with a threaded adjusting device directly under the valve tip (Figure 3-26). Most of today's overhead-cam engines utilize an adjustment screw; one side of the screw is flat and rests against the valve tip, and the other side is tapered and fits underneath the follower (Figure 3-27). By threading the screw in or out of the follower, the gap between the cam lobe surface and the top of the follower is made larger or smaller. The screw must be adjusted one complete turn at a time to keep the flat side against the valve tip; if necessary, install a different size adjustment screw in order to get proper clearance for turning.

Some overhead-cam engines have an adjustment disc or shim between the cam lobe surface and the follower (Figure 3-28). To adjust clearance, a special tool and magnet must be used. To correct excessive clearance, a thicker disc or shim is added; if reduced clearance is the problem, a thinner disc or shim must be installed.

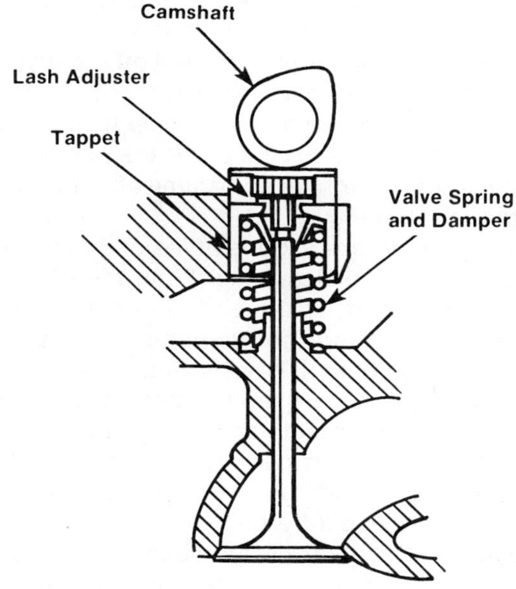

FIGURE 3-27 Some overhead-cam engines feature a cam follower with an adjustment screw.

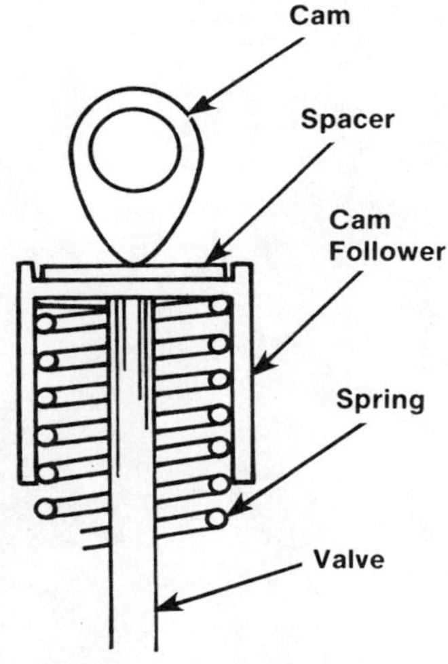

FIGURE 3-28 Adjustment discs or shims can also be used to control valve clearance.

VALVE CLEARANCE SPECIFICATIONS

Whenever valve clearance is specified, the measurement should always be taken with the valve completely closed. All clearance checks and adjustments are made with the engine either cold or hot, depending on the particular vehicle; consult the service manual. When setting the valves cold, the coolant temperature should be as close to the ambient air temperature as possible.

When the vehicle manufacturer calls for the engine to be *hot,* this means at its normal operating temperature. If a clearance check and adjustment are in progress with the engine hot but not running, the valves and rocker arms can cool off rapidly, making it necessary to reinstall the valve covers and run the engine for a few minutes to keep it warmed up. This is done despite the fact that the coolant temperature remains high for some time after the vehicle has stopped running.

There are different kinds of feeler gauges available to check valve clearance (Figure 3-29). Those with the strips bent at the ends are easiest to use when working around hot manifolds. The stepped type is also convenient; it has strips in two different sizes. The thin strip is used for measuring the gap, while the thick strip acts as a no-go gauge. If the thick strip also passes between the gap the clearance is too much and adjustment is necessary.

There is often a difference in the clearances for intake and exhaust valves, with exhaust clearances slightly higher. This is because exhaust valves expand more during engine operation, thus requiring greater clearance. When checking any valve clearance specification, the letter H behind the number means that the clearance is for a hot, or warmed-up, engine.

FIGURE 3-29 Checking valve clearance

MAKING THE ADJUSTMENT

The valve adjustment procedure differs, depending on whether the engine is running or stationary. Stationary engines present a different look since the specifications often call for the engine to be cold and the coolant temperature to be down around 70° F. For this reason, it might be necessary to let the vehicle sit for as much as a full day. On the other hand, if the stationary engine is required to be at normal operating temperature, it will have to be operated periodically throughout the adjustment to keep the temperature of the valves up.

Running Engines

Use the following procedure to make a valve clearance check and adjustment with the engine running, keeping in mind that some of the steps might have to be altered or deleted according to individual engine design.

1. Cover both fenders for protection.
2. Run the engine until it reaches its normal operating temperature, then turn off the engine and remove the valve covers.

SHOP TALK

It may be necessary to remove the air cleaner, PCV hoses, and other accessories before the valve covers can be reached for removal. Be sure to carefully identify each hose so they can be replaced correctly.

3. Inspect the valve springs and valve stems for signs of damage.
4. Make sure that the oil-drain passages in the cylinder heads are open. Clean off any sludge built up on the valve-train components; engine flush is ideal for this purpose. Be sure to completely drain the oil and add a fresh supply after using engine flush.
5. Turn on the engine. Check the valves for clearance, using a front-to-rear or rear-to-front pattern to avoid skipping any valves. Be sure to use the correct type and size feeler gauge strips to make the measurements.
6. With the engine at its slowest idle, insert the strip between the end of the rocker arm

and the valve tip (Figure 3-30). It should pass through the gap with a slight drag. If the strip must be forced, or if the engine starts to miss, the clearance is too small.

7. If the strip passes through the gap easily, or if a choppy, jerking sensation is felt as it passes through, the clearance is too large.
8. If adjustment is needed, turn the adjustment screw in or out until a slight drag is felt as the strip is passed through the gap. If the screw has a separate locknut, recheck the clearance after tightening the locknut.
9. Replace the valve covers (using new gaskets) and any other accessories that might have been removed.
10. Turn on the engine and check the valve covers for oil leakage. Make sure that the engine oil is up to the recommended level.

Stationary Engines

Since the engine is not running during this procedure, each piston must be clearly at its top dead center (TDC) position on the compression stroke in order to determine that the valves are completely closed. The best way to resolve this is with a test adapter, whistle, TDC indicator, and indicator light, using the same method used during the cylinder leakage test. The valve check and adjustment procedure for stationary engines is as follows:

1. Cover both fenders for protection.
2. If called for, turn on the engine and let it warm up to its normal operating temperature.

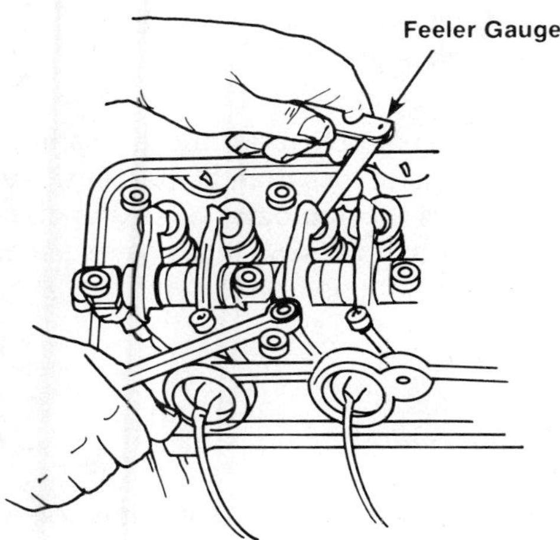

FIGURE 3-30 Checking the clearance between the end of the rocker arm and the valve tip

3. Turn off the engine and remove the valve covers, as well as any components that are blocking access to the valve covers.
4. Remove the number 1 spark plug. Insert the test adapter hose and whistle in the spark plug hole.
5. Place a wrench on the crankshaft nut or bolt, and slowly rotate the engine in the normal direction until the whistle sounds.
6. Continue rotating the engine until the timing mark on the crankshaft pulley lines up with the engine-timing pointer on the timing-chain cover.
7. Remove the distributor cap and ground the coil-to-distributor secondary lead wire using a jumper lead.
8. Mount the TDC indicator on the distributor shaft. Mark a chalk reference point on the engine that lines up with the appropriate cylinder marking on the TDC indicator.
9. Connect the indicator light to the ignition system. To do this, attach one of its leads to the distributor's primary terminal on the coil, and attach the other lead to a good ground. Turn the ignition switch on.
10. Insert the appropriate feeler gauge strip between both the intake valve and exhaust valve gaps of the number 1 cylinder. Adjust as needed to get a slight drag on the strip as it is moved in and out.
11. Resume rotating the engine until the next appropriate cylinder mark on the TDC indicator lines up with the chalk mark on the engine. The indicator light should glow, meaning that the piston of the next cylinder in firing order is at its top dead center position.
12. Measure the gaps on the intake and exhaust valves of this cylinder, and adjust as needed.
13. Crank the engine over again until the next cylinder in the firing order is at its top dead center position. Check and adjust its valves. Continue in this manner until all the valves have been checked and their gaps adjusted (if necessary).
14. Remove the test adapter hose and whistle from the number 1 spark plug hole. Reinstall the plug.
15. Remove the TDC indicator and light from the distributor. Reinstall the cap and secondary lead wire.
16. Replace the valve covers (using new gaskets) and any other components that were removed.

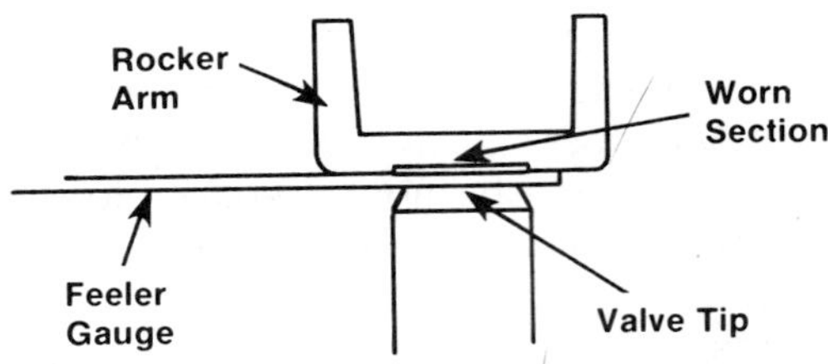

FIGURE 3-31 If the rocker arm is worn above the valve tip, use a feeler gauge strip that is slightly narrower than the tip.

17. Turn on the engine and check the valve covers for oil leakage. Make sure that the engine oil is up to the recommended level.

If the valves are still noisy, or if consistent measurements cannot be achieved with the engine running, the rocker arms might be excessively worn above the valve tip. In this case, use a feeler gauge strip that is slightly narrower than the diameter of the tip, being careful to place the strip exactly in the worn area and in line with the rocker arm (Figure 3-31). If this does not eliminate the noise, the rocker arms must be either resurfaced or replaced.

SHOP TALK

Do not tighten the clearance on a valve in an attempt to quiet it. This will create rough idle and lead to valve burning.

REVIEW QUESTIONS

1. The amount of compression in a cylinder depends on how well it is sealed by the ____________.
   a. piston rings and valves
   b. cylinder head gasket
   c. area surrounding the spark plug
   d. all of the above

2. When the piston is at top dead center, this is known as ____________.
   a. compression ratio
   b. clearance ratio
   c. clearance volume
   d. minus tolerance

3. A dry compression test yields a higher than specified gauge reading. Technician A says the likely cause is a sticking or burned valve. Technician B says the likely cause is excessive carbon deposits in the combustion chamber. Who is right?
   a. Technician A
   b. Technician B
   c. Both A and B
   d. Neither A nor B

4. If a compression test reveals equally low readings on two adjacent cylinders, the probable cause is ____________.
   a. a leaking head gasket
   b. excessive carbon deposits in the combustion chamber
   c. worn piston rings
   d. a sticking or burned valve

5. A cylinder leakage test can detect leaks ____________.
   a. between cylinders
   b. in the valves, rings, or head gasket
   c. around the exhaust and intake valves
   d. all of the above

6. An older engine whose compression reading is within specifications but has air escaping from the crankcase opening ____________.
   a. might just be showing signs of old age
   b. probably has a valve-related problem
   c. probably has a cracked block or head
   d. none of the above

7. The normal engine vacuum range is between ____________.
   a. 18 and 20 inches Hg
   b. 28 and 30 inches Hg
   c. 15 and 30 inches Hg
   d. 10 and 20 inches Hg

8. A cylinder leakage test produces relatiely even readings of less than 20 percent. Technician A says this indicates a leaking exhaust valve. Technician B says it indicates a leaking head gasket. Who is right?
   a. Technician A
   b. Technician B
   c. Both A and B
   d. Neither A nor B

9. The most common cause of engine noise is ____________.
   a. worn pistons

b. bad gas
c. shorted out spark plugs
d. lack of lubrication

10. Detonation knock or "ping" ____________.
a. can be very harmful to the engine
b. is sometimes caused by using too low an octane fuel
c. is most noticeable during acceleration with the engine under load
d. all of the above

11. The old style cylinder power balance testers are usually not safe for use on vehicles with ____________.
a. electronic ignitions
b. fuel injection
c. a floor jet EGR system
d. all of the above

12. A vehicle is experiencing piston slap. Technician A says a possible cause is misaligned connecting rods. Technician B says a possible cause is insufficient ring tension against the cylinder walls. Who is right?
a. Technician A
b. Technician B
c. Both A and B
d. Neither A nor B

13. Which of the following statements is incorrect?
a. The valve adjustment procedure differs, depending on whether the engine is running or stationary.
b. There should be no difference in the clearances for intake and exhaust valves.
c. Normal wear creates excessive clearances in the valve train.
d. Valve adjustment is required on all small engines with mechanical valve lifters.

14. When a vacuum test produces a normal reading at idle speed, but excessive vibrations at higher rpm, a possible cause is ____________.
a. worn valve guides
b. a leaky head gasket
c. late ignition timing
d. weak valve springs

15. Adjustable tappets are also known as ____________.
a. hydraulic lash tappets
b. lifters
c. cam followers
d. none of the above

16. The higher the compression ratio, ____________.
a. the greater the engine torque and horsepower
b. the lesser the engine torque and horsepower
c. the lesser the combustion pressure in the engine
d. none of the above

17. For gasoline engines, a typical compression gauge can measure pressures up to ____________.
a. 500 psi
b. 400 psi
c. 300 psi
d. 200 psi

18. What is the first step in preparing an engine for a cylinder leakage test?
a. Run the engine until it reaches its normal operating temperature.
b. Disconnect the spark plug cables from the spark plugs.
c. Check the coolant level.
d. Remove the air cleaner from the carburetor.

19. Which of the following tasks can be done by one person using a remote starter swtich?
a. adjusting valves
b. checking compression
c. setting ignition points
d. all of the above

20. Air escaping through the carburetor usually means that the ____________ is leaking.
a. intake valve
b. exhaust valve
c. head gasket
d. exhaust pipe

21. Which of the following is not done during a cranking vacuum test?
a. Connect the vacuum gauge hose to a source of engine manifold vacuum.
b. Depress the accelerator pedal.

c. Run the engine until it reaches its normal operating temperature.
d. All of the above.

22. Tappet noise is characterized by ______________.
    a. a pinging sound during acceleration with the engine under load.
    b. a light clicking sound that is more noticeable when the engine is idling.
    c. a heavy knock at speeds in excess of 35 mph
    d. a heavy thump at irregular intervals

23. Valve adjustment is generally recommended ______________.
    a. every five years or 50,000 miles
    b. at the same mileage interval as oil filter replacement
    c. every three years or 36,000 miles
    d. at the same mileage interval as spark plug replacement

24. A spark plug is observed badly fouled with oil after a low reading was produced on a compression test. Technician A says this probably means that the piston rings are worn. Technician B says this probably means that there is leakage at the valves or head gasket. Who is right?
    a. Technician A
    b. Technician B
    c. Both A and B
    d. Neither A nor B

CHAPTER FOUR

BATTERY TESTING AND SERVICING

Objectives

Upon completing this chapter, you should be able to:

- Perform safety and working precautions that must be taken when inspecting, testing, and charging storage batteries.
- Perform the proper steps for routine battery inspections, cleaning, testing, and replacement.
- Describe the function and operation of battery fast and slow chargers and how they differ.
- Properly jump-start vehicle with a discharged battery using a booster battery.

The storage battery (Figure 4–1) is the heart of a vehicle's electrical system. It plays an important role in the operation of the starting, charging, ignition, and accessory circuits.

Contrary to what many people believe, a battery has no electricity in it. Its technical name is a "storage battery" because it converts electrical current generated by the alternator or generator into chemical energy, then stores that energy until it is needed. When switched into an external electrical circuit, the battery's chemical energy is converted back to electrical energy.

A vehicle's battery has three main functions:

1. Produce voltage and a source of current for starting, lighting, and ignition.
2. Act as a voltage stabilizer for the entire electrical system of the vehicle.
3. Provide current whenever the vehicle's electrical demands exceed the output of the charging system.

In order for the battery to maintain these functions, the alternator or generator must replace these current withdrawals. If the system's output exceeds the input, the battery's chemical energy is drained, and it is no longer able to function as a source of voltage and current.

Testing and servicing the battery is always a preliminary step in any thorough tune-up. A battery that does not generate sufficient voltage can affect the charging, ignition, and fuel systems. Correct battery output is especially important for today's electronically controlled engines. Low battery voltage can cause false signals to be generated by various sensors and switches in the electronic control system, adversely affecting engine performance.

FIGURE 4–1 A typical 12-volt storage battery *(courtesy of Johnson Controls, Inc.)*

BASIC BATTERY OPERATION

A typical 12-volt storage battery (Figure 4–2) used in the modern automobile has six cells. Each cell is made up of positive and negative plates surrounded by an electrolyte. A positive plate consists of lead peroxide. A negative plate is made of sponge lead. The electrolyte is composed of water and sulfuric acid. The sulfur reacts with the lead compounds so that negatively charged ions (with an excess of electrons) accumulate on the negative plate and positively charged ions (lacking in electrons) gather on the positive plate. When an electrical circuit connects the two plates, electrons pass from the negative plate to the positive plate.

While the battery is discharging electrons, a chemical reaction takes place that converts some of sulfuric acid to water. This weakens electrolyte, reducing the battery's ability to produce an electrical charge. However, by reversing the flow of electrons through the battery, this chemical reaction is reversed and the battery is "recharged." This is one of the functions of an alternator.

The open circuit voltage of a fully charged battery is roughly 12.6 to 13.2 volts and the specific gravity of the electrolyte is approximately 1.280. However, as the battery discharges, its specific gravity decreases. That is why measuring the specific gravity is a good indication of how much charge a battery has lost. Table 4-1 shows the percent of charge that normally corresponds to the specific gravity of a battery.

SHOP TALK

Batteries intended for use in tropical climates where water never freezes have a specific gravity of 1.210 to 1.230. Batteries used in arctic climates have a higher specific gravity than 1.280—usually 1.290 to 1.300. The higher the specific gravity, the less likely the electrolyte is to freeze. Table 4-2 gives the approximate freezing points of electrolyte at various levels of specific gravity.

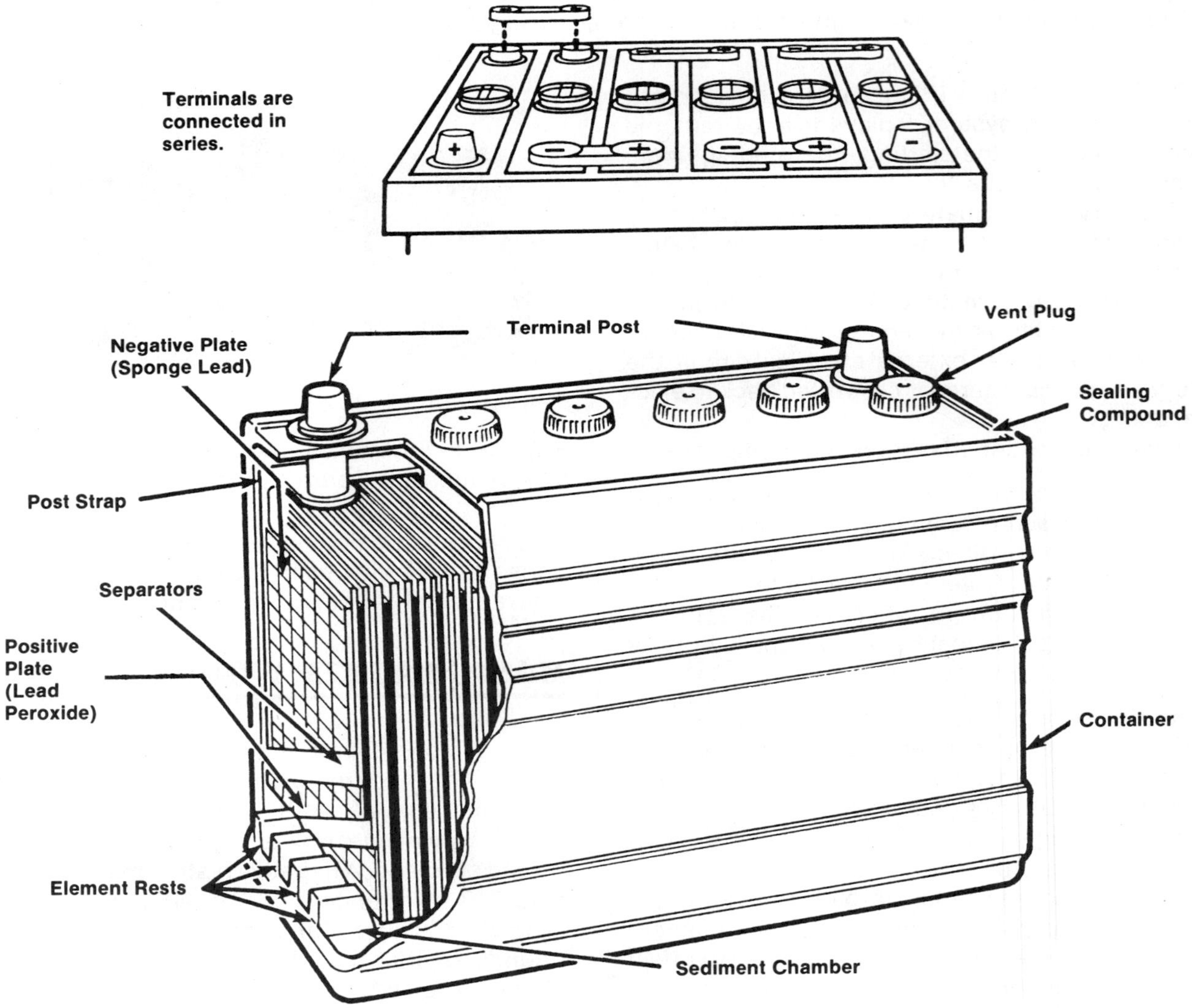

FIGURE 4-2 Component parts and features of a typical battery

TABLE 4-1: ELECTROLYTE SPECIFIC GRAVITY AS RELATED TO CHARGE

Specific Gravity	Percent of Charge
1.265	100%
1.225	75%
1.190	50%
1.155	25%
1.120 or lower	discharged

TABLE 4-2: FREEZING TEMPERATURE OF ELECTROLYTE AS CONTROLLED BY SPECIFIC GRAVITY

Specific Gravity	Freezing Temperature (°F)
1.100	18
1.120	13
1.140	8
1.160	1
1.180	-6
1.200	-17
1.220	-31
1.240	-50
1.260	-75
1.280	-92

BATTERY RATING METHODS

A battery must provide good cranking capability and sufficient reserve power for emergency situations. To measure a battery's ability to meet these standards the Battery Council International (BCI) and the Society of Automotive Engineers (SAE) have revised the methods used for rating batteries.

When replacing a battery, always refer to an application chart to select a battery with the correct BCI group number. Vehicle options, such as air conditioning and a number of major electrical accessories, may indicate the need for an optional heavy-duty battery with a higher rating. Remember, to handle cranking power and the vehicle's other electrical needs, the replacement unit should never have a lower rating than the original battery.

Reserve Capacity

A reserve capacity rating represents the approximate time in minutes it is possible to travel at night with battery ignition and minimum electrical load, but without a charging system in operation. The time in minutes is based on a current draw of 25 amperes while maintaining a minimum battery terminal voltage of 10.2 volts (12-volt batteries) or 5.1 volts (6-volt batteries) at 80° F. This rating replaces the previous 20-hour capacity (ampere-hour) rating and more accurately represents the electrical load that must be supplied by the battery in the event of a charging system failure.

Cold Cranking

A cold-cranking amps (CCA) rating specifies the minimum amps available at 0° F and at -20° F. This rating replaces the old method of relating voltage and time as measures of cranking and starting ability. It is much more accurate because it allows cranking capability to be related to such significant variables as engine displacement, compression ratio, temperature, cranking time, condition of engine and electrical system, and lowest practical voltage for cranking and ignition (the old tests do not take these factors into account). The new test relates a discharge rating in amperes that a fully charged battery will maintain for 30 seconds without the terminal voltage falling below 7.2 volts for a 12-volt battery or 3.6 volts for a 6-volt battery.

To provide enough starting power under adverse conditions, a 12-volt system generally requires 1 ampere for each cubic inch of engine displacement. In other words, a 350 cubic inch engine requires a battery with a cold cranking rating of at least 350 amps.

A rule of thumb that can be used when the amp-hour rating cannot be determined is: Divide the cold cranking rating at 0° F by 2. The figure obtained in this manner equals the load that should be applied when making a battery capacity test, and is almost equal to the value of three times the amp-hour rating of batteries that use the earlier rating system.

FACTORS AFFECTING BATTERY LIFE

All storage batteries have a limited service life, but many conditions can decrease service life.

IMPROPER ELECTROLYTE LEVELS

With nonsealed batteries, water should be the only portion of the electrolyte lost due to evaporation during hot weather and gassing during charging. Maintaining an adequate electrolyte level is the basic step in extending battery life for these designs.

When adding water to the cells, use distilled water when available or clean, soft water. Hard water

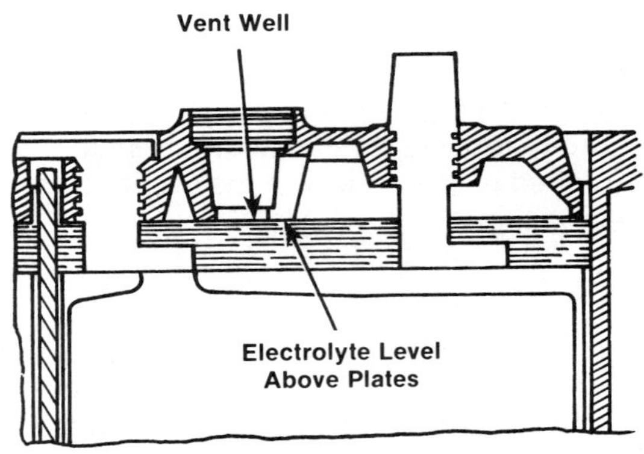

FIGURE 4-3 Add water to the cells until the electrolyte completely covers the plates.

that is high in mineral content will cause reactions in the cells, which will shorten battery life.

Fill each cell to just above the top of the plates (Figure 4-3). Underfilling causes a greater concentration of acid, which deteriorates the plates' grids more rapidly. This low level also exposes the tops of the plates, which causes the active materials to harden and become chemically inactive.

Overfilling the battery is also harmful. The additional water weakens the concentration of sulfuric acid (reduces the electrolyte's specific gravity), which reduces the efficiency of the battery. Any overflow of electrolyte will cause corrosion around the battery terminals, increasing resistance to current flow.

CORROSION

Battery corrosion is commonly caused by spilled electrolyte or electrolyte condensation from gassing. In either case, the sulfuric acid from the electrolyte corrodes, attacks, and can destroy not only connectors and terminals but holddown straps and the carrier box as well.

Corroded connections increase resistance at the battery terminals, which reduces the applied voltage to the vehicle's electrical system. The corrosion in the battery cover can also create a current leakage path that can allow the battery slowly to discharge. Finally, corrosion can lead to mechanical failure of the holddown straps and carrier box, which can result in physical damage to the battery.

OVERCHARGING

Batteries can receive an overcharge from either the vehicles charging system or from a battery charging unit. In either case, the result is a violent chemical reaction within the battery that causes a loss of water in the cells by separating the electrolyte into hydrogen and oxygen gas bubbles. This can push active materials off the plates permanently reducing the capacity of the battery. Overcharging can also cause excessive heat, which can oxidize the positive plate grid material and even buckle the plates, resulting in a loss of cell capacity and early battery failure.

UNDERCHARGE/SULFATION

The vehicle's charging system might not fully recharge the battery due to excessive battery output, stop-and-go driving, or a fault in the charging system. In all cases, the battery operates in a partially discharged condition. A battery in this condition will become sulfated when the sulfate normally formed in the plates becomes dense, hard, and chemically irreversible. This occurs because the sulfate has been allowed to remain in the plates for a long period.

Sulfation of the plates causes two problems. First, it lowers the specific gravity levels and increases the danger of freezing at low temperatures. Secondly, in cold weather, a sulfated battery often fails to crank the engine because of its lack of reserve power.

POOR MOUNTING

Loose holddown straps or covers will allow the battery to vibrate or bounce during vehicle operation. This can have several adverse effects. It can shake the active materials off of the grid plates, severely shortening battery life, loosen the plate connections to the plate strap, loosen cable connections, and even crack the battery case.

CYCLING

Cycling is simply the discharge and recharging of the battery. Heavy and repeated cycling can cause the positive plate material to break away from its grids and fall into the sediment chambers at the base of the case. This problem reduces battery capacity and can lead to premature short circuiting between the plates. Fortunately, the new envelope design of many batteries reduces this problem.

SAFETY PROCEDURES

The potential dangers caused by the sulfuric acid in the electrolyte and the explosive gasses generated during battery charging requires that battery service and troubleshooting be conducted under absolute safe working conditions. According to the National Society to Prevent Blindness, 14,238 Americans suffered serious eye damage from acid in wet-cell batteries in a recent year.

WARNING: Always wear safety glasses or goggles when working with batteries no matter how small the job.

Sulfuric acid can also cause severe skin burns. If electrolyte contacts the skin or eyes, flush the area with water for several minutes. When eye contact occurs, force the eyelid open. Always have a bottle of neutralizing eyewash on hand and flush the affected areas with it. Do not rub the eyes or skin.

WARNING: Receive prompt medical attention if electrolyte contacts the skin or eyes.

When a battery is charging or discharging, it gives off quantities of highly explosive hydrogen gas. Some hydrogen gas is present in the battery *at all times.* Any flame or spark can ignite this gas, causing the battery to explode violently, propelling the vent caps at a high velocity and spraying acid in a wide area. To prevent this dangerous situation, practice the following preventions:

- Keep sparks, open flames, and smoking materials far away from batteries at all times.
- Remove wristwatches and rings before servicing any part of the electrical system. This will help prevent the possibility of electrical arcing and burns.
- Always disconnect the battery's ground cable when working on the electrical system or engine. This prevents sparks from short circuits and prevents accidental starting of the engine.
- Never "short" across cable connections or battery terminals.
- Always operate charging equipment in well-ventilated areas.
- Never connect or disconnect charger leads when the charger is turned on. This will always generate a dangerous spark.
- Never lay metal tools or other objects on the battery because a short circuit across the terminals can result.

Other battery and electrical system safety precautions are as follows:

- Always disconnect the battery ground cable before fast-charging the battery in the vehicle, especially if the system is equipped with an alternator. Improper connection of charger cables to the battery can reverse the current flow and damage the alternator.
- When removing a battery from a vehicle, always disconnect the battery ground cable first. When installing a battery, connect the ground cable last.
- Never attempt to polarize an alternator after reconnecting a battery. Polarization is not needed, and any attempt to do so might damage the alternator, regulator, or circuits.
- Never reverse the polarity of the battery connections. Generally, all vehicles use a negative ground. Reversing this polarity will damage the alternator and circuit wiring.
- Never attempt to use a fast charger as a boost to start the engine.
- Before charging or boost-starting a battery in cold temperatures, always check the electrolyte in all cells for signs of freezing. Passing current through a frozen battery can cause it to rupture or explode. If ice or slush is visible or the electrolyte level cannot be seen, allow the battery to thaw at room temperature before servicing. Do not take chances with sealed batteries. If there is any doubt, allow them to warm to room temperature before servicing.
- Always use a battery carrier or lifting strap to make moving and handling batteries easier and safer (Figure 4–4). Batteries are heavy and often awkward to handle. Dropping a battery can crack it and spill electrolyte. It can also cause damage to the vehicle and/or personal injury.
- Acid from the battery will damage a vehicle's paint and metal surfaces and harm shop equipment. To neutralize any electrolyte spills during servicing or activating of dry charge units, use 1 pound of baking soda in 1 gallon of water or 1 pint of household ammonia in 1 gallon of water.

FIGURE 4–4 Typical battery carrier or strap in use *(courtesy of Johnson Controls, Inc.)*

ROUTINE INSPECTIONS

As part of any tune-up procedure or electrical system work, always check the battery as follows:

1. Visually inspect the battery cover and case for dirt and grease that could create an electrical pathway to ground, discharging the battery. Clean as needed.
2. If possible, check the electrolyte level, adding water as needed.
3. Inspect the unit for cracks, loose terminal posts, and other signs of physical damage. These problems require replacement of the battery.
4. Check for missing cell plug covers and caps. Replace any damaged or missing caps.
5. Inspect all cables for broken or corroded wires, frayed insulation, or loose or damaged connectors. Replace any failed parts.
6. Visually check battery terminals, cable connectors, metal parts, holddowns, and trays for corrosion damage or buildup. Clean as needed (see below). Tighten or replace loose or damaged parts.
7. Check the heat shield for proper installation on vehicles so equipped.

ROUTINE CLEANING

Before removing battery connectors or the battery itself for cleaning or other service, always neutralize accumulated corrosion on terminals, connectors, and other metal parts by applying a solution of baking soda and water or ammonia and water (Figure 4–5).

FIGURE 4-5 Cleaning and neutralizing acid corrosion using a solution of baking soda and water

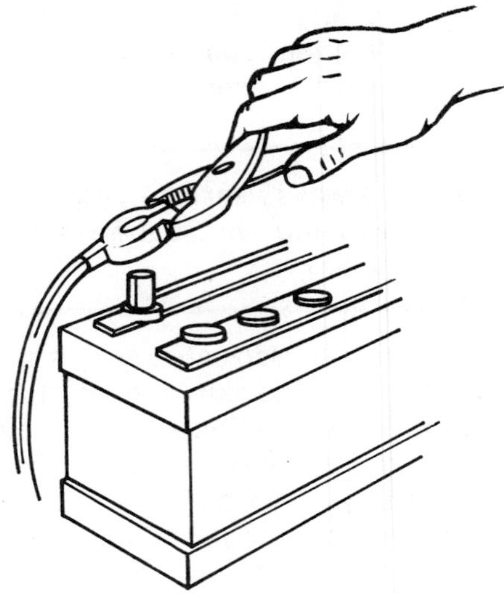

FIGURE 4-6 Removing spring-type battery cables using wide jaw pliers

Do not splash the corrosion onto vehicle paint, metal or rubber parts, or onto the hands and face. Be sure the solution cannot enter the battery cells. A stiff bristle brush is ideal for removing heavy buildup. Dirt and accumulated grease can be removed with a detergent solution or solvent.

After cleaning, rinse the battery and cable connections with clean water and dry the components with a clean rag or low-pressure compressed air.

To clean the inside surfaces of the connectors and the battery terminals, remove the cables, always beginning with the ground cable. Spring-type cable connectors are removed by squeezing the ends of their prongs together with wide-jaw, vise-gripping, channel lock, or battery pliers. This pressure expands the connector so it can be lifted off the terminal post (Figure 4–6).

For connectors tightened with nuts and bolts, loosen the nut about 3/8 inch using a box wrench or cable-clamp pliers. Using ordinary pliers or an open-end wrench can cause problems. These tools might slip off under pressure with enough force to break the cell cover or damage the casing.

Always grip the cable while loosening the nut. This will eliminate unnecessary pressure on the terminal post that could break it or loosen its mounting in the battery. If the connector does not lift easily off the terminal when loosened, use a clamp puller to free it (Figure 4–7). Prying with a screwdriver or bar strains the terminal post and the plates attached to it.

This can break the cell cover or pop the plates loose from the terminal post.

Once the connectors have been removed, open the connector using a connector-spreading tool, and neutralize any remaining corrosion by dipping it in a baking soda or ammonia solution. Next, clean the inside of the connectors and the posts using a wire brush with external and internal bristles (Figure 4-8).

Felt washers treated with corrosion-resistant compound can be installed over the terminals before reinstalling the cable connectors.

Begin reinstallation by placing the positive connector on its post. The connector must be seated on the post as shown in Figure 4-9. Do not overtighten any nuts or bolts since this could damage the post or connector. Finally, coat the connectors with petroleum jelly or battery anticorrosion paste or paint.

BATTERY TESTING

Testing batteries is an important part of electrical system service. Poor and inaccurate tests can lead to serious problems and expensive and unneeded repairs. Depending on the design of the battery, state of charge and capacity can be determined by several different test methods: specific gravity tests, built-in hydrometers, capacity test, alternative capacity test, cadmium-probe test, and three-minute charge test.

FIGURE 4-7 Battery pullers will remove the cable without damage to the terminal post.

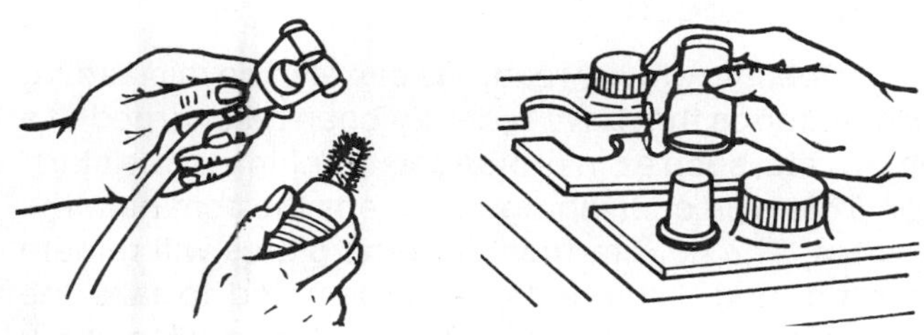

FIGURE 4-8 A combination external/internal wire brush is excellent for cleaning both battery terminals and the inside of cable connectors.

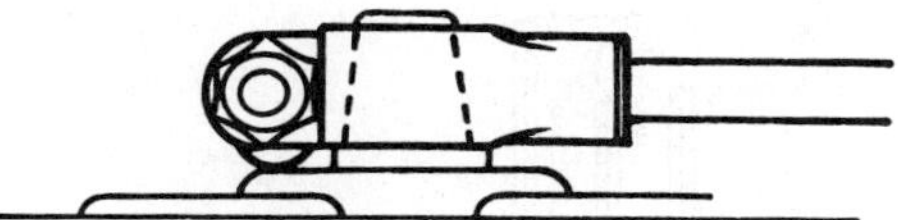

FIGURE 4-9 When installing cable connectors, be certain that they seat fully down on the terminal post.

SPECIFIC GRAVITY TESTS

On unsealed battery designs, the specific gravity of the electrolyte can be measured to give a fairly good indication of the battery's state of charge. A hydrometer is used to perform this test (Figure 4-10). A basic battery hydrometer consists of a glass tube or barrel, rubber bulb, rubber tube, and a glass float or hydrometer with a scale built into its upper stem. The glass tube encases the float and forms a reservoir for the test electrolyte. Squeezing the bulb pulls electrolyte into the reservoir.

When filled with test electrolyte, the sealed hydrometer float bobs in the electrolyte. The depth to which the glass float sinks in the test electrolyte indicates its relative weight compared to water. The reading is taken off of the scale by sighting along the level of the electrolyte. Disregard the curvature of

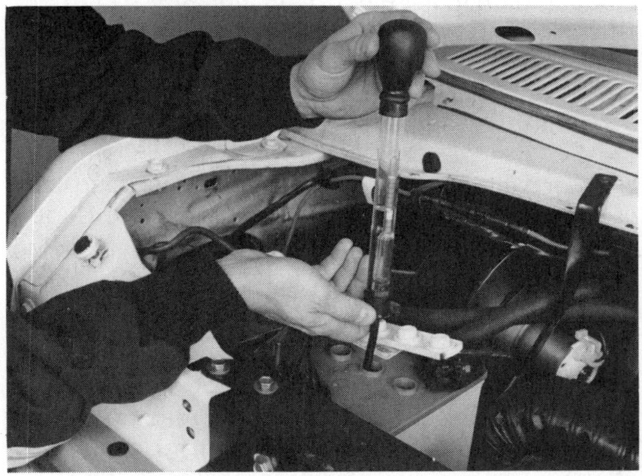

FIGURE 4-10 Using a typical hydrometer to check the specific gravity of battery electrolyte

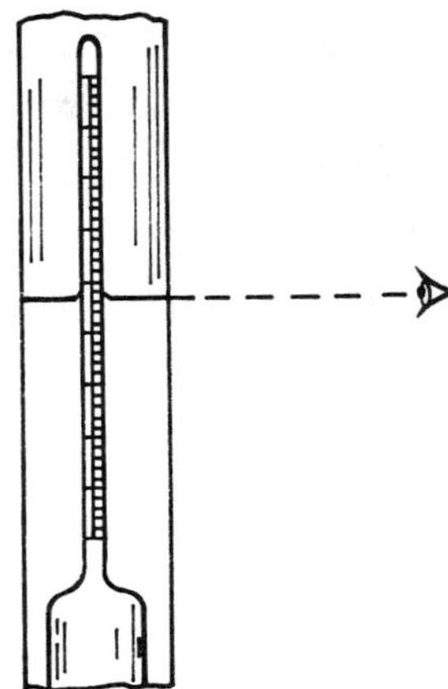

FIGURE 4-11 To accurately read a hydrometer, sight along the level of the electrolyte until it contacts the scale. Disregard the curvature of the liquid.

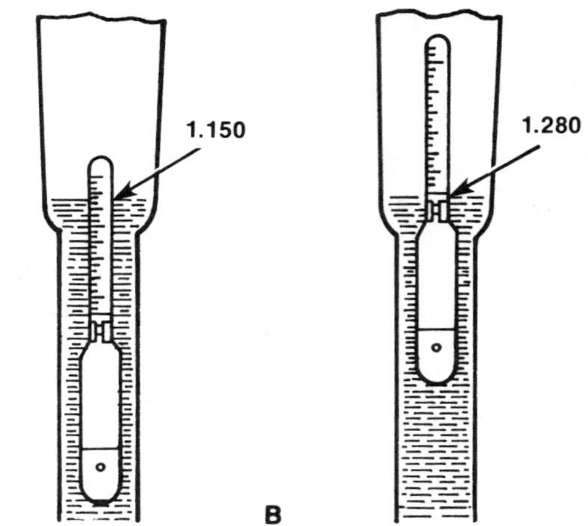

FIGURE 4-12 (A) When the scale sinks deep in the electrolyte, the specific gravity reading is low. (B) Floating high in the electrolyte indicates a high specific gravity.

the electrolyte against the scale and glass tube (Figure 4-11).

For example, if the hydrometer floats deep in the electrolyte, the specific gravity is low (Figure 4-12A). If the hydrometer floats shallow in the electrolyte, the specific gravity is high (Figure 4-12B).

Temperature Correction

At extremely high and low electrolyte temperatures, it is necessary to correct the reading by adding or subtracting 4 points (0.004) for each 10° F above or below the standard of 80° F. Most hydrometers have a built-in thermometer to measure the temperature of the electrolyte (Figure 4-13). Hydrometer reading can be misleading if not adjusted. For example, a reading of 1.260 taken at 20° F would be 1.260 - (6 × 0.004 or 0.024) = 1.236. This lower reading means the cell has less charge than indicated.

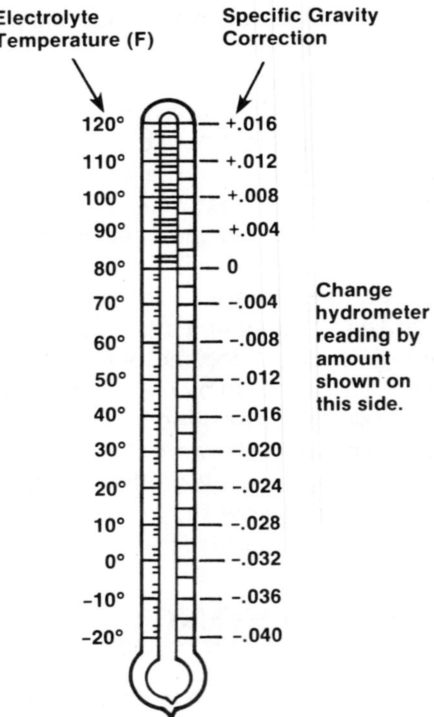

FIGURE 4-13 Many hydrometers are equipped with a thermometer and correction table so the reading can be easily adjusted to account for electrolyte temperature.

The same is true at high electrolyte temperatures. A reading of 1.245 at 100° F is actually 1.245 + (2 × 0.004 or 0.008) = 1.253. This indicates more charge than the reading. It is extremely important to make these adjustments at high and low temperatures to determine the battery's true state of charge.

Mixing Electrolyte

Specific gravity readings can also be misleading on batteries that have recently been discharged at a high rate, such as in prolonged cranking of the starter. This type of discharge weakens the acid near the plates. The acid farther from the plates will remain strong, and if this is the acid sampled to take the specific gravity test, readings will be misleadingly high. To mix the strong and weak acid, allow the battery to stand at idle for several hours. Charging the battery at a high rate for about 10 minutes will also mix the electrolyte. The gasses created during charging mix the electrolyte more rapidly. If water must be added to the cells before a reading can be taken, the electrolyte must be mixed using either of these methods before an accurate reading can be taken.

Interpreting Results

The specific gravity of the cells of a fully charged battery should be between 1.260 and 1.280 when corrected for electrolyte temperature. Table 4-1 (page 69) shows the relationship between specific gravity and the state of charge of the cell.

Recharge any battery if the specific gravity drops below an average of 1.230. A specific gravity difference of more than 50 points between cells is a good indication of a defective battery in need of replacement.

BUILT-IN HYDROMETERS

On many sealed maintenance-free batteries a special temperature-compensated hydrometer is built into the battery cover. A quick visual check will indicate the batteries state of charge. The hydrometer has a green ball within a cage that is attached to a clear plastic rod. The green ball will float at a predetermined specific gravity of the electrolyte that represents about a 65 percent state of charge. When the green ball floats, it rises within the cage and positions itself under the rod. Visually, a green dot then shows in the center of the hydrometer (Figure 4-14A). The built-in hydrometer provides a guide for battery testing and charging.

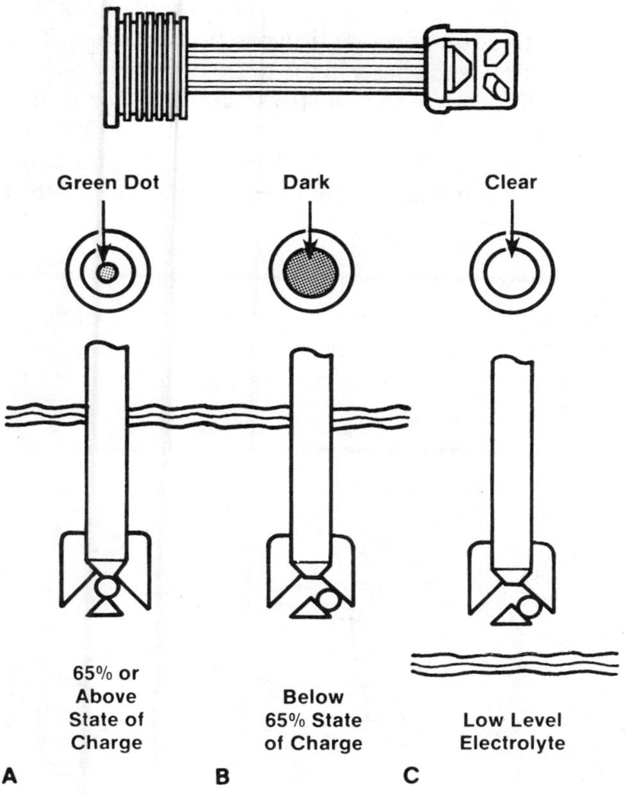

FIGURE 4-14 Typical design and operation of built-in hydrometers on maintenance-free sealed batteries

In testing, the green dot means the battery is charged enough for testing. If the green dot is not visible, and has a dark appearance (Figure 4-14B), it means the battery must be charged before the test procedure is performed.

In charging, the appearance of the green dot means that the battery is sufficiently charged. Charging can be stopped to prevent overcharging.

The hydrometer on some batteries may be clear or light yellow (Figure 4-14C). This means the fluid level might be below the bottom of the rod and attached cage. This might have been caused by excessive or prolonged charging, a broken case, excessive tipping, or normal battery wearout. Whenever this clear or light yellow appearance is present while looking straight down on the hydrometer, always tap the hydrometer lightly with a small screwdriver to dislodge any gas bubbles that might be giving a false indication of low electrolyte level. If the clear or light yellow appearance remains, and if a cranking complaint exists that is caused by the battery, replace it.

It is important when observing the hydrometer that the battery have a clean top to see the correct indication. A flashlight may be required in some poorly-lit areas. Always look straight down when viewing the hydrometer.

On some special applications, some hydrometers feature a red dot indication in addition to the green dot, dark, and clear appearances. The red dot means the battery is nearing complete discharge and must be charged before being used in service.

Complete hydrometer information on most batteries is printed on the label located on the top of the battery. By referring to this label, an accurate interpretation of the hydrometer appearance can be made.

OPEN CIRCUIT VOLTAGE TEST

An open circuit voltage check can be used as a substitute for the hydrometer specific gravity test on maintenance-free sealed batteries with no built-in hydrometer. As the battery is charged or discharged, slight changes occur in the battery's voltage. So

TABLE 4-3: BATTERY OPEN CIRCUIT VOLTAGE AS AN INDICATOR OF STATE OF CHARGE

Open Circuit Voltage	State of Charge
12.6 or greater	100%
12.4 to 12.6	75–100%
12.2 to 12.4	50–75%
12.0 to 12.2	25–50%
11.7 to 12.0	0–25%
11.7 or less	0%

battery voltage with no load applied can give some indication of the state of charge.

The battery's temperature should be between 60° to 100° F and the voltage must be allowed to stabilize for at least 10 minutes with no load applied. On vehicles with high parasitic drains (computer controls, clocks, and accessories that always draw a small amount of current), it may be necessary to disconnect the battery ground cable. On batteries that have just been recharged, apply a 300 amp load for 15 seconds to remove the surface charge and then allow the battery to stabilize. Once voltage has stabilized use a digital voltmeter to measure the battery voltage to the nearest one-tenth of a volt (Figure 4–15). Use Table 4-3 to interpret the results. As you can see, minor changes in battery open circuit voltage can indicate major changes in state of charge.

Checking open circuit voltage across the terminals is simple and fast, but obtaining a reasonable reading does not automatically indicate the battery has the ability to supply a useful amount of amperage or current. For example, if the meter indicates less than the 12.6 or more volts that indicate a full charge, two conditions can be present: (1) all cells might have a low charge, or (2) five cells might have sufficient charge (2.1) and one may have substantially less (1.8).

If the open circuit voltage test indicates a charge of below 75 percent of full charge, recharge the battery and perform the capacity test outlined below to determine battery condition.

FIGURE 4–15 Measuring open circuit voltage across battery terminals using a voltmeter

CAPACITY TEST

The load or capacity test determines how well any type of battery, sealed or unsealed, functions under a load. In other words, it determines the battery's ability to furnish starting current and still maintain sufficient voltage to operate the ignition system.

The load or capacity test can be performed with the battery either in or out of the vehicle, but the battery must be at or very near a full state of charge. Use the specific gravity test or open circuit voltage test to determine charge, and recharge the battery if needed before proceeding. For best results, the electrolyte should be as close to 80° F as possible. Cold batteries will show considerably lower capacity. Never load test a sealed battery if its temperature is below 60° F.

On some batteries designed with side terminals, obtaining a sound connection can be a problem. The best solution is to screw in the appropriate manufacturer's adapter (Figure 4–16). If an adapter is not available, use a 3/8-inch coarse bolt with a nut on it. Bottom out the bolt, back it off a turn, and then tighten the nut against the contact. Now attach the lead to the nut.

CAUTION: Simply turning in a bolt will not do the job. The thread contact area is too small to carry enough current for the load test (or for battery charging).

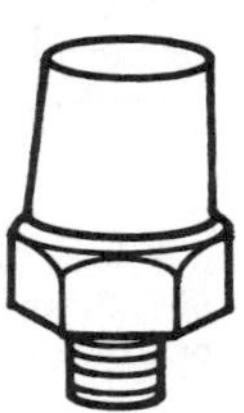

FIGURE 4–16 Several types of adapters that may be needed to test and/or charge side-terminal mount batteries

TABLE 4-4: MINIMUM LOAD TEST VOLTAGES AS AFFECTED BY TEMPERATURE

BATTERY TEMPERATURE (F)	MINIMUM TEST VOLTAGE
70°	9.6 volts
60°	9.5 volts
50°	9.4 volts
40°	9.3 volts
30°	9.1 volts
20°	8.9 volts
10°	8.7 volts
0°	8.5 volts

Follow these steps when performing a load test:

1. Make sure the battery can accept a charge by charging at 20 to 25 amps for 3 minutes. Allow the battery to rest for 10 minutes.
2. On unsealed units, check the specific gravity in each cell (Figure 4–17). If it is below 1.225, or there is a 0.50 difference between cells, charge the battery at 20 to 25 amps for 15 minutes and retest. If the 0.50 difference remains between cells, discontinue the load test and replace the battery.

   On sealed units, check the no-load voltage across the terminals and charge if the reading is below 12.4 volts.
3. If charging was needed, remove the surface charge of the freshly charged battery by applying a 300-amp load for 15 seconds, or by disabling the ignition and cranking the engine for that length of time.

SHOP TALK

Without applying a load it takes 2 hours for a surface charge to dissipate.

4. Adjust all mechanical settings on the tester to zero. Rotate the load increase control fully counterclockwise to the OFF position.
5. Connect the capacity tester to the battery. An inductive pickup must surround all the wires from the negative terminal. Observe correct polarity and be sure the test leads contact the battery posts.
6. If the tester is equipped with an adjustment for battery temperature, turn it to the proper setting. For best results, use a thermometer to check electrolyte temperature in one of the cells. On sealed units, estimate the temperature as closely as possible.
7. Refer to the battery specifications to determine its cold cranking amps (CCA) rating or its amp-hour rating. Either can be used to determine the proper test load.
8. Turn the load control knob on the tester to draw battery current at a rate equal to one-half (1/2) the battery's cold cranking amps (CCA) rating or three times (3X) the battery's amp-hour rating. For example, if a battery's CCA rating is 440, the load should be set at 220 amperes.

SHOP TALK

On fixed load testers, set the battery size selector to the appropriate position.

9. Maintain proper load for 15 seconds while observing the voltmeter of the tester. Turn the control knob off immediately after 15 seconds of current draw.
10. At 70° F or above or on testers that are temperature corrected, the voltage at the end of the 15 seconds should not fall below 9.6 volts. If the tester is not temperature corrected, use Table 4-4 to determine the adjusted minimum voltage reading for the battery temperature.

Interpreting Results

If the voltage reading exceeds the specification by a volt or more, the battery is supplying sufficient

FIGURE 4–17 Performing a specific gravity test

current with a good margin to safety. If the reading is right on the spec, the battery might not have the reserve necessary to handle cranking under tough conditions, such as low temperatures. Keep in mind that this varies with the state of charge determined before the load test was made. If the battery was at 75 percent charge and fell right on the load spec, it is probably in good shape.

If the voltage reads below the temperature corrected minimum, continue to observe the voltmeter of the tester after removing the load. If it rises above 12.4, the battery is bad—it can hold a charge, but has insufficient cold cranking amps. The battery can be recharged and retested, but the results are likely to be the same.

If the voltage tests below the minimum and the voltmeter does not rise above 12.4 when the load is removed, the problem may only be a low state of charge. Recharge the battery and load test again.

ALTERNATIVE CAPACITY TEST

If a volt-amp tester is not available, the starter motor can be used as a loading device to conduct a capacity test as follows:

1. Make certain the battery is at or near full charge and the starting circuit and starter are in good condition.
2. Disable the ignition system so the engine will not start.

CAUTION: Electronic ignition systems might sustain damage during prolonged engine cranking unless properly disconnected. Refer to the vehicle's specific instructions if needed.

---

3. Connect the voltmeter to the battery terminals with its positive lead to the positive (+) terminal and the negative lead to the negative (-) terminal.
4. Crank the engine over continuously for 15 seconds. Observe the voltmeter reading at the end of this cranking period.

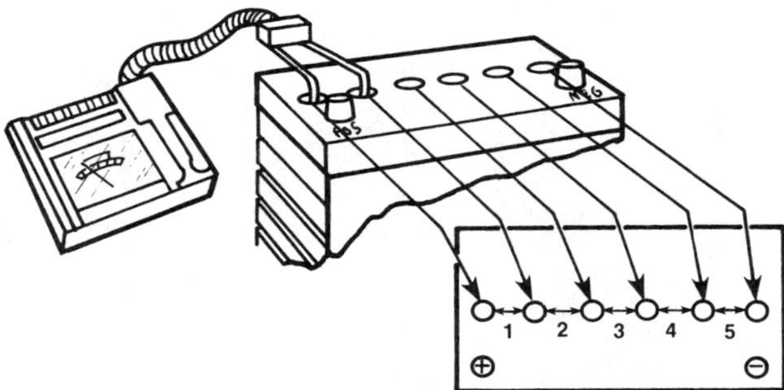

FIGURE 4-18 Performing a cadmium-probe test to check open circuit voltage of individual battery cells

Interpreting Results

If the voltmeter reading is above the minimum listed in Table 4-4, the battery and cranking circuit are in good condition. If the voltage reading drops below the levels listed, the battery might be in poor condition or the starting circuit might be drawing too much current. If this is the case, perform both the 3-minute charge test (next page) and the cranking current test on page 97.

CADMIUM-PROBE TEST

The cadmium-probe test measures the open circuit voltage of the battery's individual cells. It can be performed in or out of the vehicle on nonsealed batteries.

The test uses a special tester designed with two probes connected to a voltmeter. The probe tops are cadmium tubes measuring about 1 inch long. The tubes contain an absorbent material that keeps the cadmium moist. When inserted into adjacent plug openings in the battery, the probes contact the electrolyte in adjacent cells and react to it in much the same way as the active materials on the battery plates. This chemical reaction causes a voltage in the tester (Figure 4-18).

The meter face of the tester is graduated to indicate the efficiency of the cell's electrolyte in creating voltage in the tester. If the reading of any two tested cells varies by five divisions or more, the battery is near the point of failure.

To perform the test:

1. Remove the vent caps and check the electrolyte level in each cell. If adding water is necessary, charge the battery at a slow rate for 5 to 10 minutes to mix the electrolyte.
2. Remove the surface charge from the battery by turning the headlights on for 1 minute before testing the battery. If the vehicle has not been operated 8 hours prior to testing, the battery will not have a surface charge and this step is not needed.
3. Be sure the headlights and all accessories are turned off before testing.
4. Place the red probe in the positive cell next to the (+) terminal and the black probe in the second cell. Record the reading on the meter.

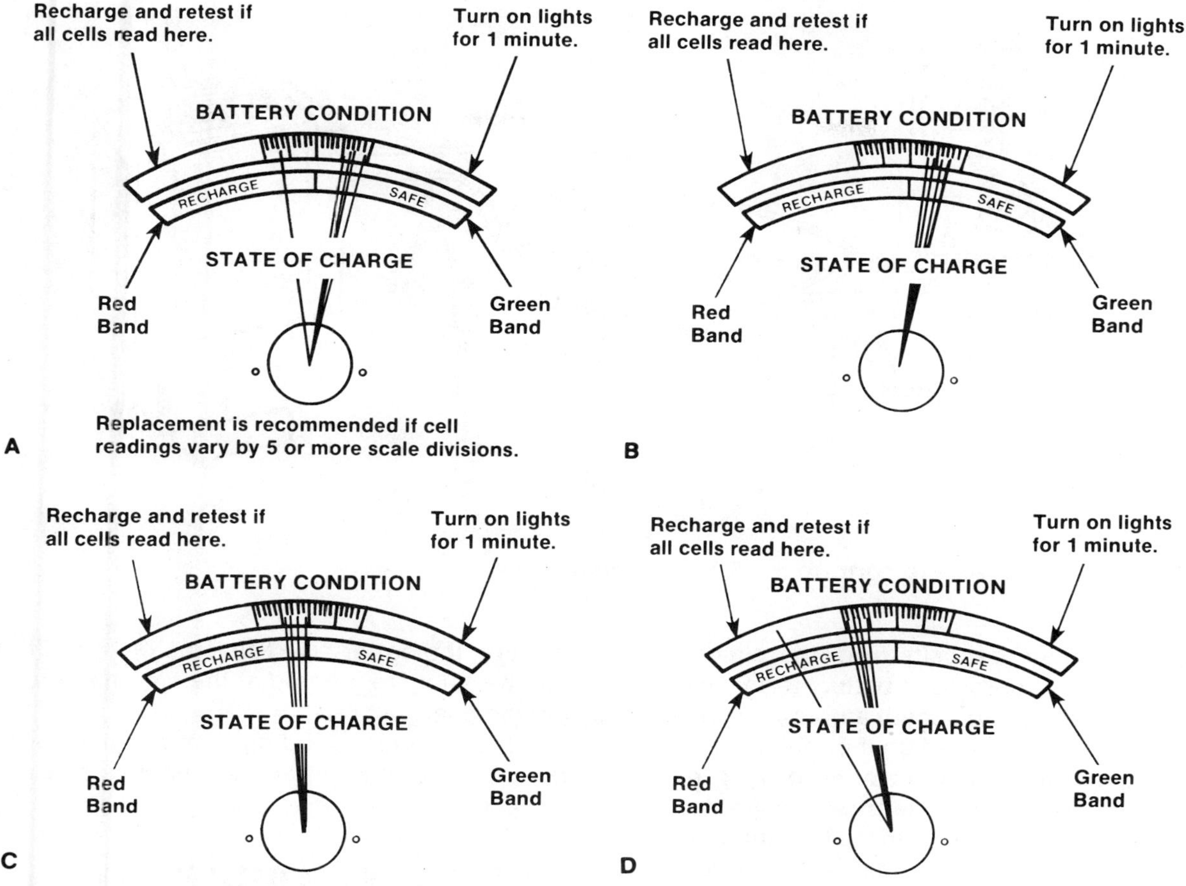

FIGURE 4-19 Typical results of a cadmium-probe test

SHOP TALK

If the probes are reversed, there will be no meter reading at this time.

5. Move the red probe to the second cell and the black probe to the third cell, and record the reading.
6. Move the red probe to the third cell and the black probe to the fourth cell, and record the reading. Continue this process until all cells are tested in this way. Record all readings for comparison.

Interpreting Results

If the reading of any two cells varies five divisions or more on the top scale as shown in Figure 4-19A, regardless of the colored section into which they may fall on the lower scale, the battery is near failure and replacement is needed.

If all cells vary less than five divisions on the top scale and are in the green section on the lower scale (Figure 4-19B), the battery is in good condition and sufficiently charged.

If all cells vary less than five divisions on the top scale but if any of the cells test in the red section on the lower scale (Figure 4-19C), the battery is in good condition but has a low charge.

If any cell readings are in the recharge-retest section of the top scale and the balance of the readings are within the first four scale divisions as shown in Figure 4-19D, the battery has too low a charge for testing. It must be recharged and the surface charge removed before it can be properly tested.

THREE-MINUTE CHARGE TEST

This test determines if conventional unsealed batteries are too badly sulfated to accept a charge. The test is not accurate on maintenance-free batteries.

A battery fast-charger is needed to pass a high charging current through the battery for a three-minute period (Figure 4-20). If the battery is not very sulfated, this high current dislodges the sulfate deposits from the plates. If the high charging does not

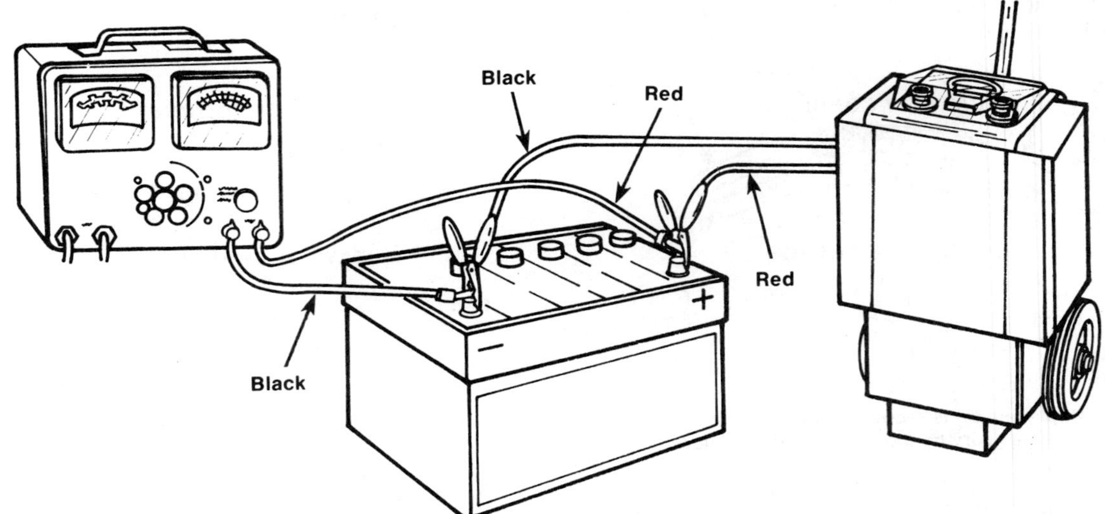

FIGURE 4–20 Setup used to perform the 3-minute charge test

dislodge the sulfate deposits from the plates and high voltage occurs across the battery terminals, the battery is too sulfated to accept a normal charge.

If the battery is charged on the vehicle, disconnect the negative battery connector in order to avoid damaging the alternator or electrical system. Also, if high voltage is recorded early in the test, immediately stop the procedure. This indicates high internal resistance due to sulfation or poor internal connections that will cause excessive heat and electrolyte boiling. If this is the case, the battery must be replaced.

The steps in performing the 3-minute charge test are as follows:

1. Connect the battery charger's positive and negative leads to the corresponding positive and negative battery terminals.
2. Connect a voltmeter across the battery with its positive lead to the positive (+) terminal and negative lead to the negative (−) terminal.
3. Turn the charger's power switch on. Set the timer switch to the 3-minute mark.
4. Adjust the charger's output switch to the highest possible rate not exceeding 40 amperes for 12-volt batteries and 75 amperes for 6-volt batteries.
5. When the timer switch cuts off after 3 minutes, turn the timer switch to the fast-charge position and read the voltmeter.

Interpreting Results

If the voltmeter reading does not exceed 15.5 volts for a 12-volt battery or 7.75 volts for a 6-volt battery, the battery is in satisfactory condition and can be safely recharged at the manufacturer's suggested charging rate.

If the voltage reading is more than the values given above, the battery is defective and must be replaced.

BATTERY CHARGING

To charge a battery, a given charging current is passed through the battery for a period of time. For example, a 10-ampere charging rate for 4 hours will result in a 40-ampere-hour charging input to the battery.

Both fast and slow charging units (Figure 4–21) are used to supply this input and each has its advantages. Fast-chargers are the most popular. They charge batteries at a higher rate or charge—usually 40 amperes for 12-volt batteries and 70 amperes for 6-volt batteries. At this rate fast-chargers can recharge most batteries in about 1 hour. However, batteries must be in good condition to accept a fast charge. Sulfation on the plates of the battery can lead to excessive gassing, boiling, and heat buildup during fast charging. Never fast-charge a battery that has failed the 3-minute charge test (page 79) or shows evidence of sulfation buildup or separator damage.

Slow types of trickle chargers provide low charging currents of about 5 to 15 amperes. Slow charging may require 12 to 24 hours but is the only safe way of charging sulfated batteries. In general, almost any battery can be charged at any current rate as long as excessive electrolyte gassing does not occur and the electrolyte temperature does not exceed 125° F. However, when time is available, slow charging is the safest and easiest method to use. In fact, many fast-chargers can be adjusted to provide slow charging.

Regardless of fast or slow charging rates, always begin by checking the electrolyte level when possible. Add water as needed. Charging with electrolyte levels below the separators can damage the battery.

Be certain the power outlet is delivering full power to the charging unit. Connecting the charger to a heavily loaded circuit in the shop will reduce the charger's output. For example, a 20 percent power drop in the circuit will reduce charger output by 35 to 40 percent. If extension cords are needed to set up the charger, use 14 gauge or heavier cords not exceeding 25 feet in length. Be sure the cord has a three-pronged plug to provide ground. As with battery testing, special adapters may be needed to connect the charging leads to batteries with side terminals.

Perform charging in a well-ventilated area away from sparks and open flames. Always be sure the charger is off before connecting or disconnecting the leads to the battery. Remember to wear eye protection, and never attempt to charge a frozen battery.

All battery chargers have manufacturer specific characteristics and operating instructions that must be followed. The following material is a general overview and should be used as specific instruction. When charging a battery in the vehicle, always disconnect the battery cables to avoid damaging the alternator or other electrical components. Also follow these steps for any recharging:

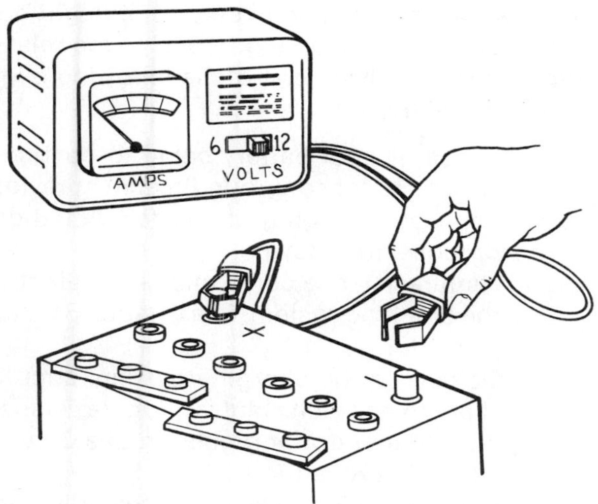

FIGURE 4-21 Typical battery charger in use. Follow all manufacturer's guidelines.

1. When connecting the charger leads to the battery, check for correct polarity.
2. After making terminal connections, turn on the charger and set it at the desired rate.
3. Check specific gravity and electrolyte temperature (if possible) periodically during charging.

CAUTION: Stop charging immediately if electrolyte temperature exceeds 125° F.

---

4. Check the voltage across the terminals periodically. If it rises above 15.5 volts, lower the charging rate until voltage drops below this value.
5. The battery has reached full charge when all cells are gassing freely and specific gravity has not increased for 3 hours. When fully charged, specific gravity should be 1.260 to 1.280 or the manufacturer's specification.
6. When charging a maintenance-free sealed battery, follow the battery manufacturer's specifications for the charging rate and time.

CAUTION: Do not exceed the manufacturer's battery charging limits.

If the battery has a built-in hydrometer, charge until the indicator shows a full charge (green ball indicator).

SHOP TALK

Tipping or shaking the battery might be needed to make the green ball appear.

CAUTION: Never charge the battery if the built-in hydrometer registers clear or light yellow. Replace the battery.

---

7. After charging, flush the battery top with baking soda or ammonia solution to remove any acid deposited because of gassing. Dry the battery top and check and adjust the electrolyte level if needed.

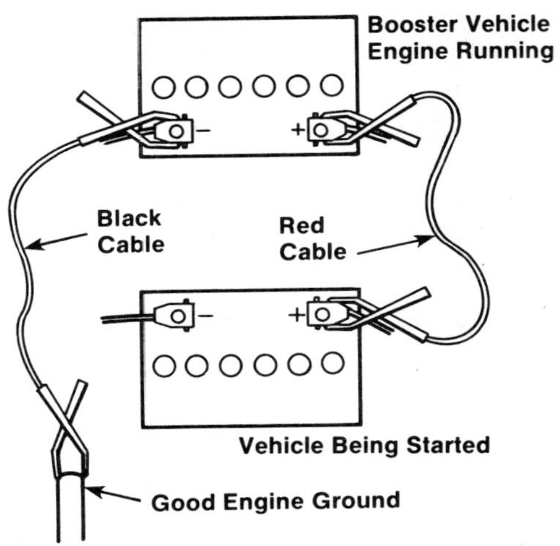

FIGURE 4-22 Proper setup and connections for boost starting a vehicle with a drained battery

JUMP STARTING

When it is necessary to jump-start a car with a discharged battery using a booster battery and jumper cables, follow this procedure to avoid damaging the charging system or creating a hazardous situation (Figure 4-22). Always wear eye protection when making or breaking jumper cable connections. To jump-start the vehicle:

CAUTION: Consult the manufacturer's service manual for procedures and precautions when jump starting late-model vehicles with electronic control systems. Excessive battery voltages can damage sensitive electronic components.

1. Set the hand brake, turn off all accessories and the ignition switch, and place the transmission in NEUTRAL or PARK.
2. Connect one end of the first cable to the positive (+) terminal of the discharged battery.
3. Connect the other end of the first cable to the positive (+) terminal of the booster battery.
4. Connect one end of the second cable to the ground (negative) terminal of the booster battery. Connect the other end of this second cable to a good ground on the engine block of the disabled car as far from the battery as possible.

SHOP TALK

Do not connect this cable directly to the negative (-) terminal of the battery.

5. Turn on the ignition and starter of the disabled vehicle. If it does not start immediately and the booster battery is in another vehicle, turn that vehicle on to avoid excessive drain on its battery.

SHOP TALK

Make certain the vehicles are not touching.

6. After the disabled vehicle starts, remove the cable connection to the engine block first. Then remove the other end of that cable from the booster battery.
7. Finally, remove the other cable by first disconnecting it at the booster battery.

CAUTION: Never use a 24-volt booster charger to jump-start late-model vehicles. Gasoline computer engine controls and diesel engine glow plug systems will suffer immediate, severe, and costly damage.

TROUBLESHOOTING

If a battery tests good and then does not maintain a charge when placed in service, it might be a sign that something is wrong in one of the related electrical systems. If so, one of the following might be the cause of trouble:

1. Loose or poor battery cable-to-post connections, loose ground cable, previous improper charging of a rundown battery, or loose hold-downs.
2. High resistance connections or defects in the cranking system. See Chapter 5 for test procedures.
3. Defects in the charging system, such as slipping fan belt, high wiring resistance, faulty generator or regulator. See Chapter 6 for test procedures.
4. A vehicle electrical load exceeding the generator capacity, with the addition of electrical devices, such as radio equip-

ment, air-conditioning, window defoggers or light systems.
5. Defects in the electrical system, such as shorted wires.

REVIEW QUESTIONS

1. Which of the following statements concerning car batteries is untrue?
   a. A function of car batteries is to act as a voltage stabilizer for the entire electrical system of the vehicle.
   b. Car batteries generate electricity.
   c. A car battery is technically a storage battery.
   d. All of the above.

2. When installing a battery, Technician A connects the ground cable first. Technician B connects it last. Who is correct?
   a. Technician A
   b. Technician B
   c. Both A and B
   d. Neither A nor B

3. What is the ability to deliver a given amount of current flow over a period of time?
   a. specific gravity
   b. battery capacity
   c. rate of discharge
   d. cell rate

4. When replacing a battery, Technician A always refers to an application chart to select a battery with the correct BCI group number. Technician B uses the zero degree cranking ability rating. Who uses the better method for selecting a battery?
   a. Technician A
   b. Technician B
   c. Both A and B
   d. Neither A nor B

5. What is the discharge and recharging of a battery called?
   a. cycling
   b. sulfation
   c. overcharging
   d. corrosion

6. Which of the following influences the voltage of a cell during discharge?
   a. cell size
   b. rate of discharge
   c. electrolyte temperature
   d. all of the above

7. To neutralize any electrolyte spills, use ____________.
   a. a mixture of baking soda and water
   b. a mixture of ammonia and water
   c. either A and B
   d. clear water

8. What is the first step when removing an old battery?
   a. Disconnect the ground cable.
   b. Remove the battery holddown straps and cover.
   c. Inspect and clean the area.
   d. Remove the heat shield.

9. The specific gravity of the cells of a fully charged battery should be ____________.
   a. between 1.230 and 1.240
   b. between 2.260 and 2.280
   c. between 1.260 and 1.280
   d. any of the above, depending on the type of battery

10. Technician A tested a battery with a hydrometer. Technician B used a capacity tester. Who got the most reliable indication of the battery condition?
    a. Technician A
    b. Technician B
    c. Both A and B
    d. Neither A nor B

11. Which of the following statements are true?
    a. Water should be added periodically to sealed batteries.
    b. Wrist watches and rings should be removed before servicing the battery.
    c. Batteries are vented to prevent them from exploding.
    d. All of the above.

12. Before fast charging a battery on the vehicle, Technician A checks the electrolyte level in each cell. Technician B disconnects the positive battery cable. Who is correct?
    a. Technician A
    b. Technician B
    c. Both A and B
    d. Neither A nor B

13. After charging, a battery failed to keep a charge when placed in service. Technician A thinks the alternator could be faulty. Technician B suspects a loose ground connection. Which technician might be correct?
    a. Technician A
    b. Technician B
    c. Both A and B
    d. Neither A nor B

14. Before performing a capacity test on a battery with side terminals, Technician A measures the temperature of the electrolyte. Technician B installs a test adapter. Who is correct?
    a. Technician A
    b. Technician B
    c. Both A and B
    d. Neither A nor B

15. When performing a load test on a battery with 400 CCAs, Technician A sets the test load at 400 amps. Technician B removes the surface charge by applying a 300-amp load for 5 minutes. Who is correct?
    a. Technician A
    b. Technician B
    c. Both A and B
    d. Neither A nor B

16. Technician A performs an open voltage test before performing a load test. Technician B performs an open voltage test before recharging the battery. Who is correct?
    a. Technician A
    b. Technician B
    c. Both A and B
    d. Neither A nor B

CHAPTER FIVE

STARTING SYSTEM TESTING AND DIAGNOSIS

Objectives

Upon completing this chapter, you should be able to:

- List components of the starter system, starter circuit, and control circuit.
- Explain the different types of magnetic switches and starter drive mechanisms.
- Explain how a starter motor operates.
- Describe the operation and function of an overriding clutch.
- Explain how to perform and interpret the following tests: battery load test, cranking voltage test, cranking current test, insulated circuit resistance test, starter relay bypass test, ground circuit resistance test, and control circuit voltage and resistance tests.
- Explain how to replace a starter.

There are many reasons why a vehicle might be hard to start or even refuse to start. Sometimes the problem might even be in the starter system. When a customer complains of a no-start or hard start condition, the smart technician will check the starter and its related circuits before beginning a diagnosis of the ignition and fuel systems.

SYSTEM DESIGN AND COMPONENTS

The vehicle's starting system is designed to turn or "crank" the engine over until it can operate under its own power. To do this, the starter motor receives electrical power from the storage battery. The starter motor then converts this energy into mechanical energy, which it transmits through the drive mechanism to the engine's flywheel.

The only function of the starting system is to crank the engine fast enough to run. The vehicle's ignition and fuel systems provide the spark and fuel for engine operation, but they are not considered components of the basic starting system.

A typical starting system has five basic components and two distinct electrical circuits. The components are

1. Battery
2. Ignition switch
3. Battery cables
4. Magnetic switch (either electrical relay or solenoid)
5. Starter motor

The starter motor draws a great deal of electrical current from the battery. A large starter motor might require 300 to 400 amperes of current. This current flows through the heavy-gauge cables that connect the battery to the starter.

The driver controls the flow of this current using the ignition switch mounted (usually) on the steering column. However, if the cables were routed from the battery to the ignition switch and then on to the starter motor, the voltage drop caused by resistance in the cables would be too great. To avoid this problem, the system is designed with two connected circuits: the starter circuit and the control circuit. The starter circuit, indicated by the solid lines in Figure 5-1, carries the heavy current flow from the battery to the starter motor by way of the magnetic switch or solenoid. The control circuit, shown by the dashed lines in Figure 5-1, ties the ignition switch to the battery and magnetic switch so this heavy current flow can be conveniently controlled.

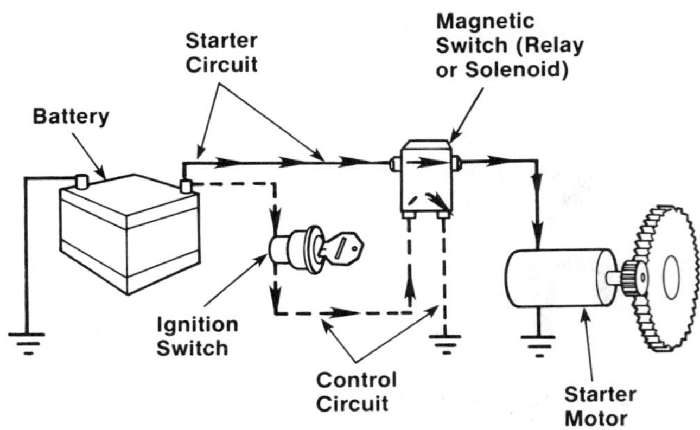

FIGURE 5-1 The major components and circuits of a typical starting system. The starter circuit is shown in a solid line. The control circuit is indicated by a dashed line.

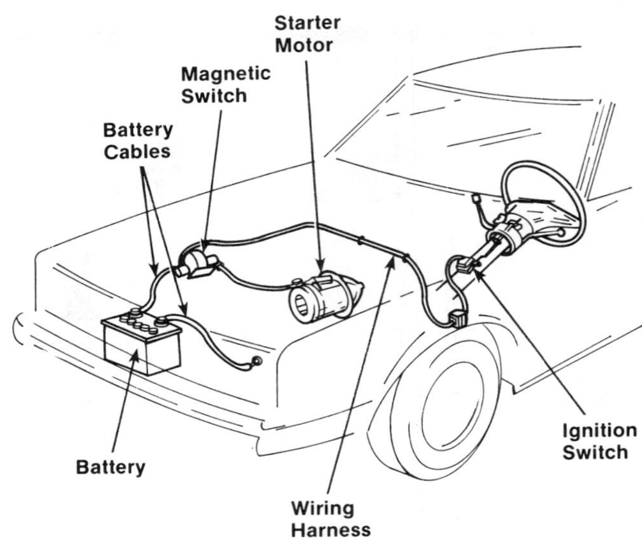

FIGURE 5-2 Basic battery cable and wiring connections in a typical starting system

THE STARTER CIRCUIT

The starter circuit carries the high current flow within the system and supplies power for the actual engine cranking. Components of the starting circuit are the battery, battery cables, magnetic switch or solenoid, and the starter motor.

Battery and Cables

Chapter 4 details the important role the battery plays in a vehicle's electrical system. Many problems associated with the starting system can be solved by troubleshooting the battery and its related components.

Any diagnosis of starter system problems must begin with an inspection of the battery. Be sure the battery terminal posts are clean and free of corrosion and that the terminal connections are tight. Check the battery itself. It must be fully charged and in good condition. Perform any needed battery testing, service, and recharging before beginning any starter system diagnosis. Refer to Chapter 4 for full details.

The cold cranking amp (CCA) capacity of the battery is also extremely important. It must match the ampere requirements of the starter motor or it will not be able to start the engine in cold weather. The general rule of one CCA for each cubic inch of displacement does not always hold true. For example, some GM vehicles use an "MX high torque" starter that draws roughly 100 amperes more than a standard starter. If the battery does not have this capacity, the engine might be difficult to crank.

The starting circuit requires two or more heavy-gauge cables, usually of No. 1 or No. 0 gauge (Figure 5-2). Two of these cables attach directly to the battery. One of these cables connects between the battery's negative terminal and an excellent ground (for negatively grounded battery systems). The other cable connects the battery's positive terminal with the magnetic switch or solenoid. On vehicles where the solenoid does not mount directly on the starter motor, two cables are needed. One runs from the positive battery terminal to the solenoid and the second from the solenoid to the starter motor terminal. In any case, these cables carry the necessary heavy current from the battery to the starter and from the starter back to the battery.

All cables must be in good condition. Cables can be corroded by battery acid and contact with engine parts and other metal surfaces can fray the cable insulation. Frayed insulation can cause a dead short that can seriously damage some of the electrical units of the vehicle. A short to ground is the cause of many dead batteries, and can result in fire.

Cables must also be heavy enough to comfortably carry the required current load. Cranking problems can be created when undersized cables are installed. Some replacement cables use smaller-gauge wire encased in thick insulation, so the small gauge cables can look as heavy as the original equipment. In warm weather, with good connections and no extra current drawn from lights or accessories, an engine will start with smaller cables. But many starts must be done under conditions that are less than ideal. With undersized cables, the starter motor will not develop its greatest turning effort and even a fully charged battery might be unable to start the engine. Always check cable size during the general inspection of the system.

One trick that is sometimes used to aid starting is to install 6-volt battery cables. Although difficult to locate, 6-volt cables can carry twice the current load of standard 12-volt cables.

When checking cables and wiring, always check any fusible links in the wiring. Almost all vehicles are equipped with fusible links to protect the wiring from overloading. These links are different in construction than a fuse, but operate in much the same way (Figure 5-3). The most common type is made of a wire with a special nonflammable insulation. Wire used to make fusible links is ordinarily four sizes smaller than the wire in the circuit they are designed to protect. Often, when a fusible link is subjected to a current overload, the insulation will become charred and give the appearance of a failed link. This is not always a true indication. The best test is to connect an ohmmeter across the link to check for continuity.

The largest fusible link is usually located at the starter solenoid battery terminal. From this terminal, current is distributed to all parts of the vehicle. Another large fusible link joins this battery terminal to the main body harness and protects the complete wiring of the vehicle. This link may take several forms, from a wire to a small piece of metal with terminal connections on each end. It may be located in its own special holder or be attached directly to the starter relay terminal. When a link has failed, always troubleshoot the system and locate the cause before replacing the link.

Magnetic Switches

Every starting system contains some type of magnetic switch that enables the control circuit to open and close the starter circuit. This magnetic switch can be one of several designs.

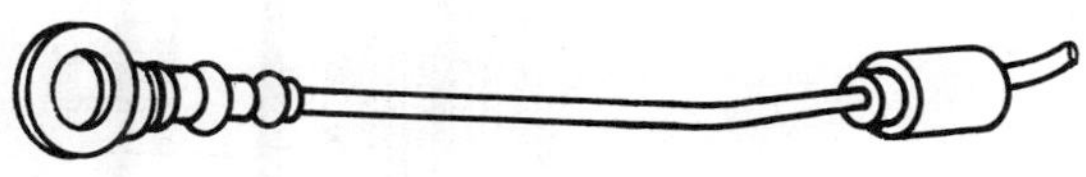

Fusible Link Before Short Circuit

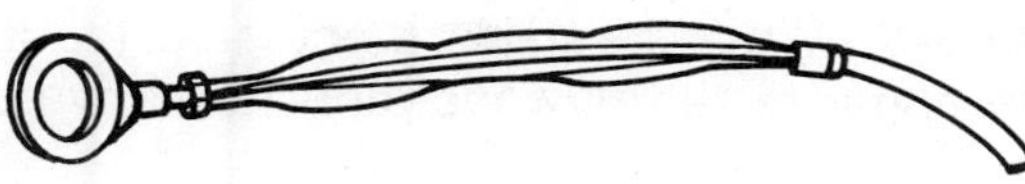

Fusible Link After Short Circuit

FIGURE 5-3 Example of a "good" and "short-circuited" fusible link

Relays (Figure 5-4) are designed with a hinged armature. When the relay receives power through the control circuit, its coil generates an electromagnetic field that attracts the armature and closes the contact points. This allows full battery current to flow to the starter motor.

Solenoids are the second major type of magnetic switches used. Solenoids use the electromagnetic field generated by their coil to pull a plunger into the coil, closing the contact points and allowing full current flow from battery to starter (Figure 5-5).

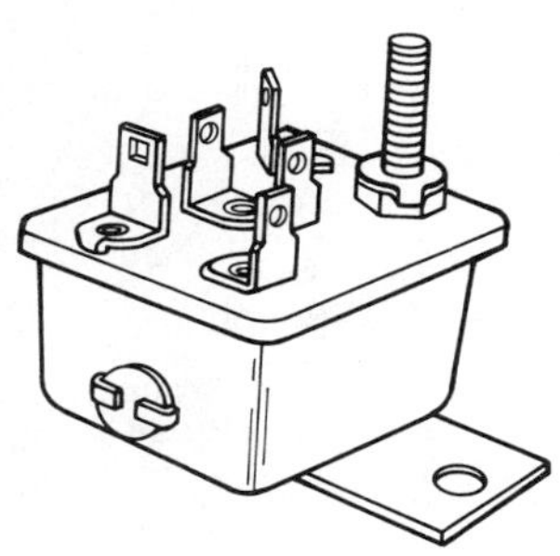

FIGURE 5-4 Example of a relay used to close the starting circuit. This type of relay performs no mechanical action.

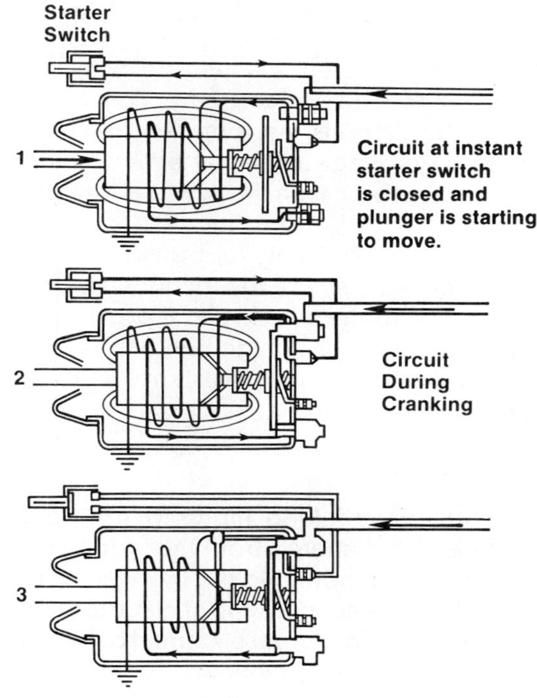

FIGURE 5-5 The operation of a typical solenoid, the second type of magnetic switch used to control current flow in the starting circuit. It also performs a mechanical action.

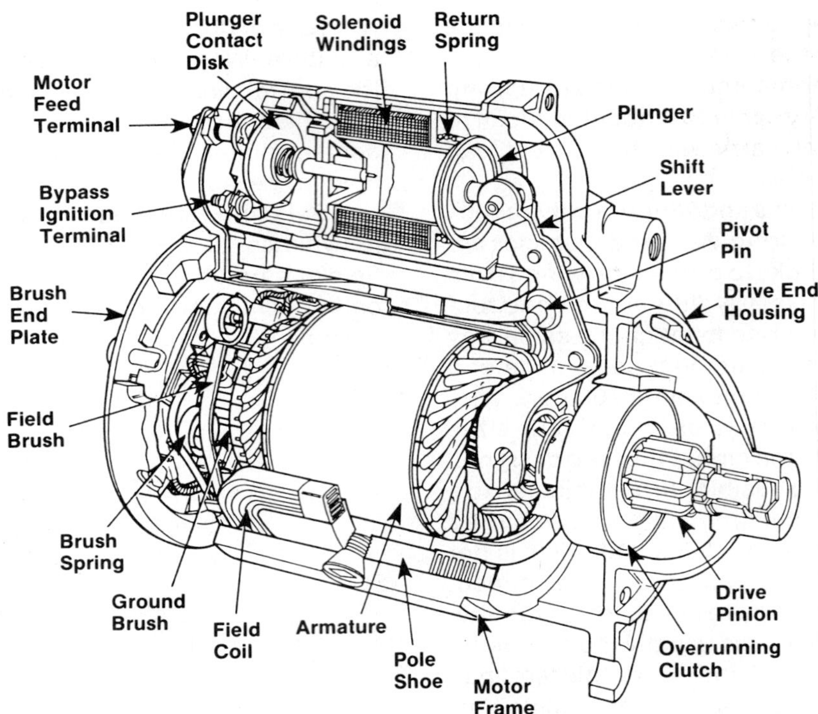

FIGURE 5-6 Example of a solenoid-actuated starter where the solenoid mounts directly to the starter motor

Some solenoids are mounted directly to the starter; others are mounted on the fender apron near the battery.

Starter-mounted solenoids (Figure 5-6), like those used on most GM, Delco-Remy, Bosch, and Japanese starting systems, perform two functions: routing full battery voltage to the starter and moving the pinion gear into mesh with the flywheel ring gear (Figure 5-7).

A remotely mounted solenoid, also called a starter relay (Figure 5-8) has two functions: it allows full battery voltage to the starter and it provides an alternate electrical path to the ignition coil in order to bypass the ballast resistor or resistance wire during startup.

Starting Motors

The starting or cranking motor (Figure 5-6) converts the electrical energy from the battery into mechanical energy for cranking the flywheel or the engine. The starter is a special type of electric motor designed to operate under great electrical overloads and to produce very high horsepower.

Because of these design features, the starter can only operate for short periods of time without rest. The high current needed to operate the starter creates considerable heat, and continuous operation for any length of time will cause serious heat damage to the unit. The starter must never operate for more than 30 seconds at a time and should rest for 2 minutes between these extended crankings. This permits the heat to dissipate without damage to the unit.

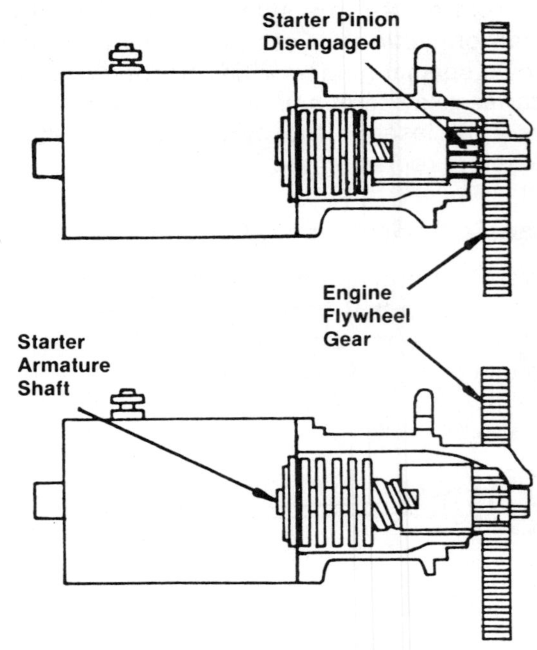

FIGURE 5-7 When the starter is actuated, the starter drive engages the flywheel and turns over the engine

All starting motors are generally the same in design and operation. Basically the starter motor consists of a housing, field coils, an armature, a commutator and brushes, and end frames. The main difference between designs is in the drive mechanism used to engage the flywheel. The armature is the only rotating part of the starter (Figure 5-9). Its movement produces the mechanical energy that is used to crank the engine.

Operating Principle

The starter motor converts electric current into torque or twisting force through the interaction of magnetic fields. It has a stationary magnetic field (created by passing current through the field coils) and a current-carrying conductor (the armature windings). When the armature windings are placed in this stationary magnetic field and current is passed through the windings, a second magnetic field is generated with its lines of force wrapping around the wire (Figure 5-10). Since the lines of force in the stationary magnetic field flow in one direction across the winding, they combine on one side of the wire increasing the field strength, but are opposed on the other side, weakening the field strength. This creates an unbalanced magnetic force, pushing the wire in the direction of the weaker field.

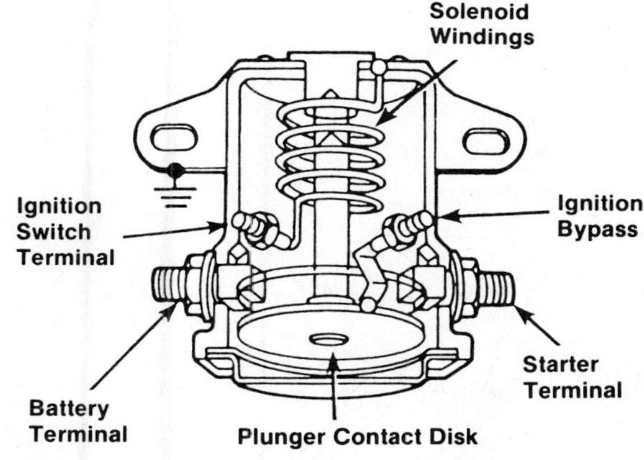

FIGURE 5-8 Cross-section view of a starter-relay, which is actually a solenoid used to close an electrical circuit.

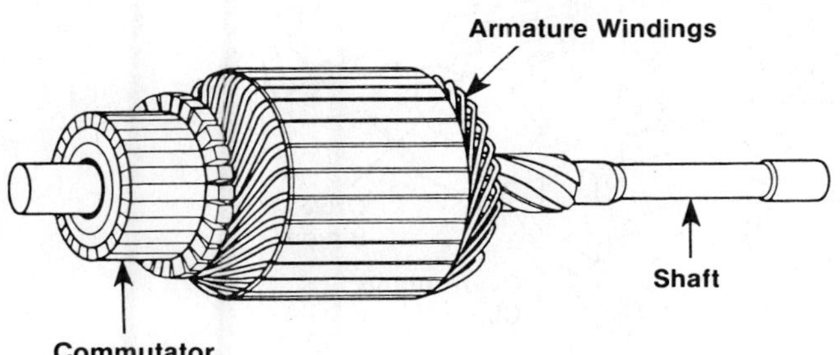

FIGURE 5-9 The armature windings are wound lengthwise through a laminated iron core to strengthen and concentrate their magnetic fields.

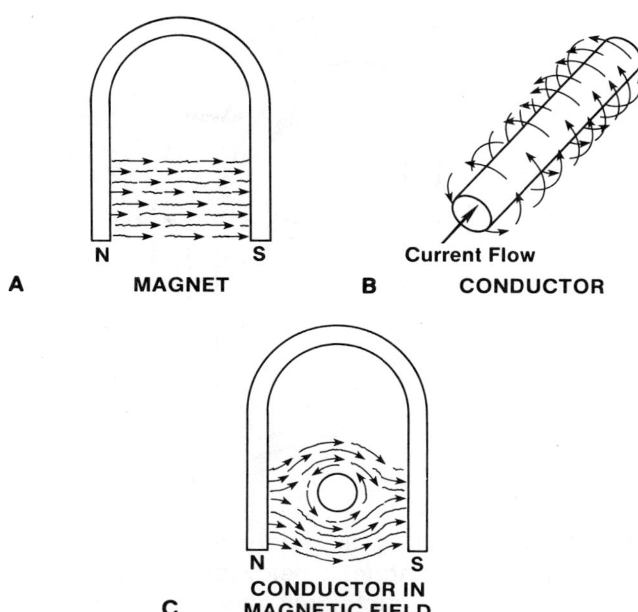

FIGURE 5-10 (A) Magnetic lines of force flow from the north pole to the south pole of a magnet. (B) A magnetic field is generated whenever electrical current is passed through a conductor. (C) If an energized conductor is placed in a second magnetic field, the two fields will interact.

Since the armature windings are formed in loops or coils, current flows outward in one direction and returns in the opposite direction. Because of this, the magnetic lines of force are oriented in opposite directions in each of the two segments of the loop. When placed in the stationary magnetic field of the field coils, one part of the armature coil is pushed in one direction while the other part is pushed in the opposite direction. This causes the coil, and the shaft to which it is mounted, to rotate (Figure 5-11).

DRIVE MECHANISMS

Four different types of drive designs are used on late-model vehicles. The solenoid-actuated direct drive system has already been explained. The following are the other types of designs commonly used.

Movable Pole Shoe Drive

Movable pole shoe drive starters (Figure 5-12) are used only by Ford. In this design, the drive mechanism is an integral part of the motor, and the

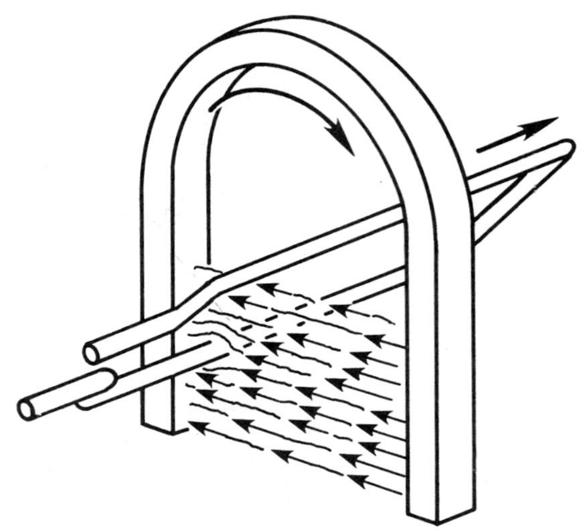

FIGURE 5-11 When an energized conductor is placed within a magnetic field, the magnetic field created by the conductor will cause it to rotate.

drive pinion is engaged with the flywheel before the motor is energized.

When the ignition switch is moved to the start position, current runs through the winding of the movable pole shoe and through a set of contacts to ground. This generates a magnetic force that pulls down the movable pole shoe. It also forces the drive pinion to engage the flywheel ring gear using a lever action and opens the contacts. A small holding coil keeps the movable shoe and lever assembly engaged during cranking. When the engine starts, an overrunning clutch disengages the drive pinion.

Solenoid-Actuated Reduced Drive

Chrysler solenoid-actuated reduction-drive starters use a solenoid to engage the pinion with the flywheel and to close the motor circuit (Figure 5-13). The starter armature does not drive the pinion directly; instead it drives a small gear that is permanently

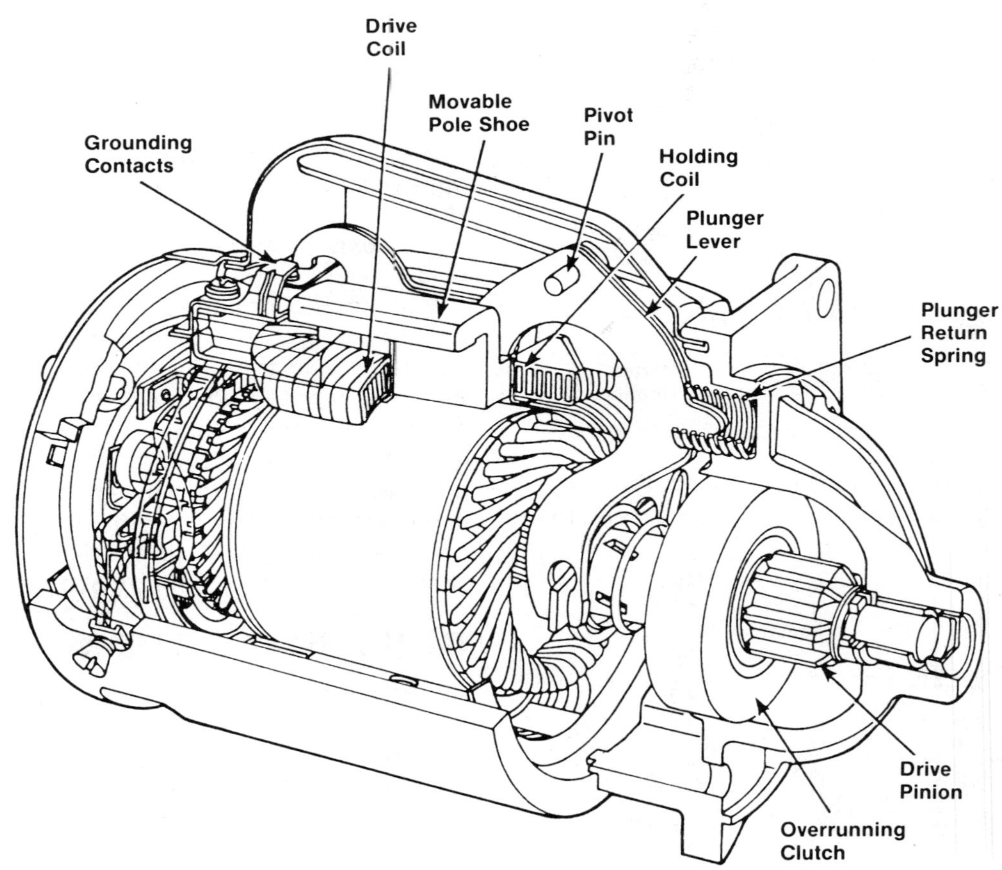

FIGURE 5-12 Typical movable pole shoe starter

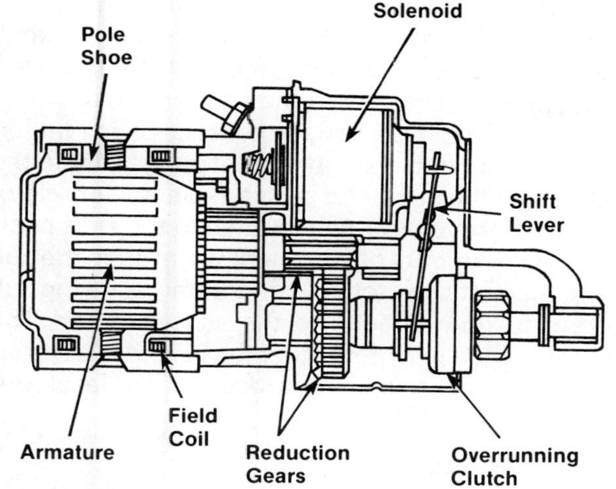

FIGURE 5-13 Typical reduction gear starter design

meshed with a larger gear having a reduction ratio of 2:1 to 3.5:1, depending on the engine size. This design allows a small, high-speed motor to develop increased turning torque at a satisfactory cranking rpm. The solenoid and starter drive operation is basically the same as in solenoid-actuated direct drive systems.

Permanent Magnet Planetary Drive

Available on some late-model Chrysler, AMC, and GM vehicles, the permanent magnet planetary drive starter has a number of unique features (Figure 5-14). First, the field coils and pole shoes are replaced with a permanent magnet made of a new iron and rare earth alloy. This eliminates the field coil circuit, simplifying design and service.

Secondly, the starter is a high-speed, low-torque motor that operates the drive mechanism

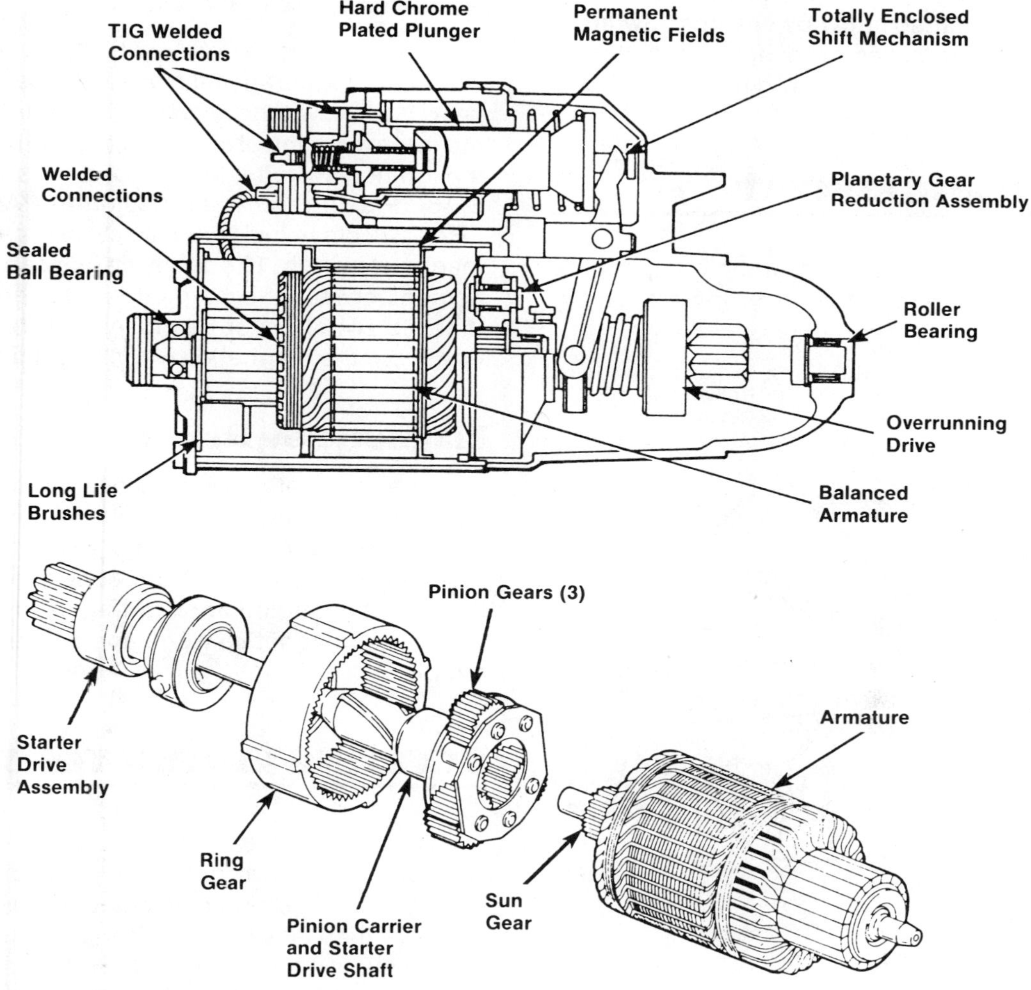

FIGURE 5-14 Construction details of the newer permanent magnet, planetary drive starters

through gear reduction provided by a simple planetary gearset instead of the two spur gears as in the older Chrysler reduction designs.

Although lightweight and easier to service than traditional designs, these units require special handling. The permanent magnet material is quite brittle and can be destroyed with a sharp blow or if the starter is dropped.

OVERRIDE CLUTCHES

The override clutch performs a very important job in protecting the starter motor. When the engine starts and runs, its speed increases. If the starter motor remained connected to the engine through the flywheel, the starter motor would spin at very high speeds, destroying the armature winding.

To prevent this, the starter must be disengaged from the engine as soon as the engine turns more rapidly than the starter has cranked it. But with a movable pole shoe or solenoid-actuated drive, the pinion remains engaged until power stops flowing to the starter. In these cases an overriding clutch is used to disengage the starter. A typical overriding clutch is shown in Figure 5–15.

THE CONTROL CIRCUIT

With a clear understanding of the components of the starting circuit and how they operate, take a closer look at the element of the control circuit. The control circuit allows the driver to use a small amount of battery current to control the flow of a large amount of current in the starting circuit.

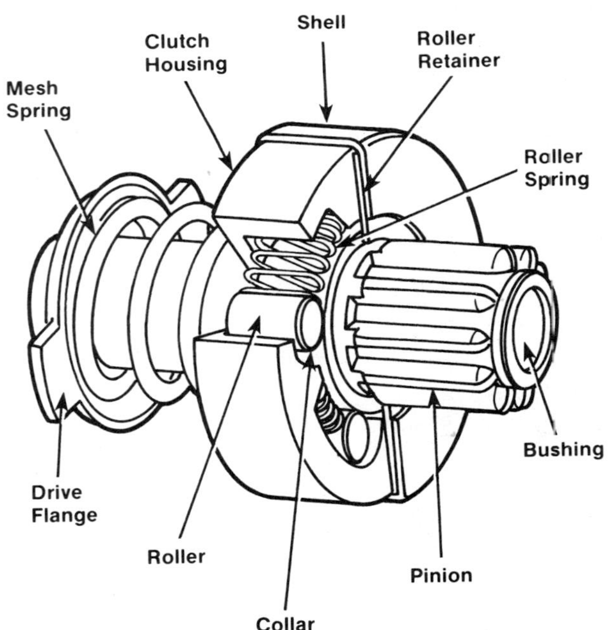

FIGURE 5–15 A sectional view of a typical overriding clutch

The entire circuit usually consists of an ignition switch connected through normal gauge wire to the battery and the magnetic switch (solenoid or relay). When the ignition switch is turned to the start position, a small amount of current flows through the coil of the magnetic switch, closing it and allowing full current to flow directly to the starter motor. The ignition switch performs other jobs besides controlling the starting circuit. It normally has at least four separate positions:

1. Accessory
2. Off
3. On (run)
4. Start

Starting Safety Switch

The starting safety switch prevents vehicles with automatic transmissions from being started in gear. Safety switches can be located in either of two places in the control circuit. One position is between the ignition switch and the magnetic switch (Figure 5–16). Placing the transmission in PARK or NEUTRAL will close the switch so current can flow to the magnetic switch. The safety switch can also be connected between the magnetic switch and ground so that the switch must be closed before current can flow from the magnetic switch to ground.

Magnetic Switches

The magnetic switch covered earlier in the chapter is actually the point in the starting system where the control circuit and starter circuit come together. Low current in the control circuit passes through the ignition and starting safety switches to trigger the magnetic switch and activate the starter circuit.

STARTING SYSTEM TESTING

As mentioned earlier, the starter motor is a special type of electrical motor designed for intermittent use only. During testing, it should never be operated for more than 30 seconds without resting for 2 minutes to allow it to cool. Because the cranking motor is designed to operate under great overload for short periods of time, it provides high horsepower output for its small size.

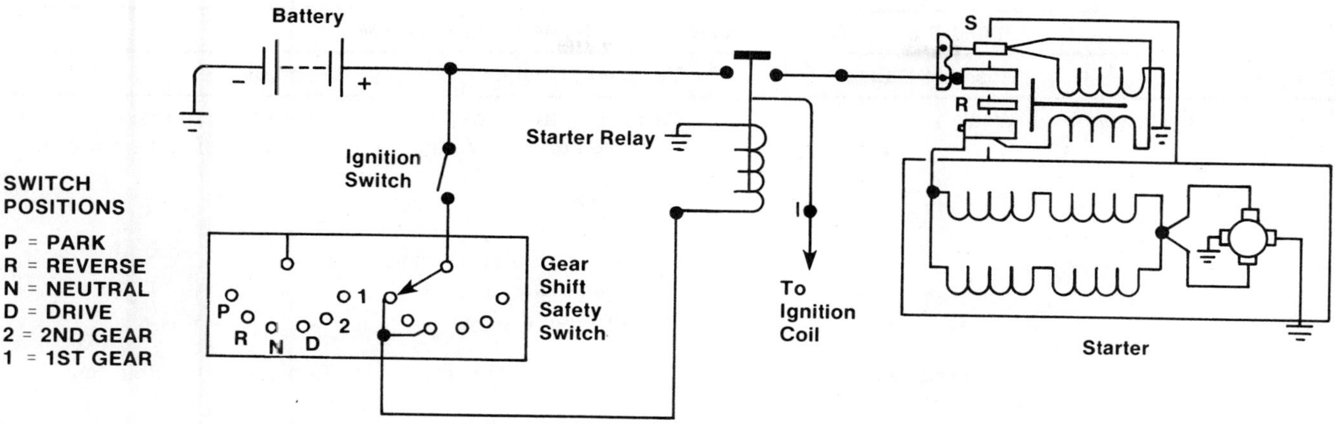

FIGURE 5-16 Schematic drawing of a typical safety switch installed on automatic transmission vehicles. Switch allows starting in NEUTRAL and PARK only.

PRELIMINARY CHECKS

The cranking output obtained from the motor is also affected by the condition and charge of the battery, the wiring circuit, and the engine's cranking requirements.

The battery should be checked and charged as needed before testing. Be sure the battery is rated to meet or exceed the vehicle's manufacturer's recommendations. The voltage rating of the battery must also match the voltage rating of the starter motor.

Check the wiring for clean, tight connections. The starter motor may draw several hundred amperes or more during cranking. Loose or dirty connection will cause excessive voltage drop. Cables should also be checked for correct size.

Also make certain the engine crankcase is filled with the proper weight oil as recommended by the vehicle manufacturer. Heavier than specified oil when coupled with low operating temperatures will drastically lower cranking speed to the point where the engine will not start.

Check the ignition switch for loose mounting, damaged wiring, sticking contacts, and loose connections. Check the wiring and mounting of the safety switch, if so equipped, and make certain the switch is properly adjusted. Check the mounting, wiring, and connections of the magnetic switch and starter motor. Also be sure that the starter pinion is properly adjusted.

SAFETY PRECAUTIONS

Almost all starting system tests must be performed while the starter motor is cranking the engine. However, the engine must not start and run during the test or the readings will be inaccurate.

To prevent the engine from starting, the ignition switch can be bypassed with a remote starter switch that will allow current to flow to the starting system but not to the ignition system. On vehicles with the ignition starting bypass in the ignition switch or the starter relay, the ignition must be disabled.

On standard ignition systems, disable the ignition by removing the secondary lead from the center of the distributor cap and grounding the lead (Figure 5-17). With electronic ignition systems or engine control systems, disconnect the wiring harness connector from the distributor.

During testing, be sure the transmission is out of gear during cranking. A remote starter switch can bypass the safety switch. When servicing the bat-

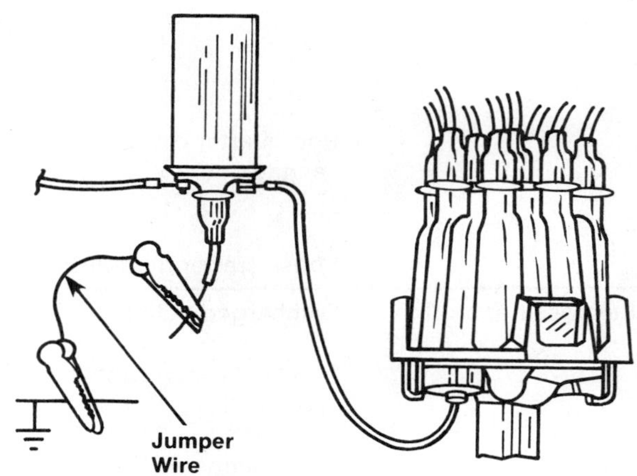

FIGURE 5-17 To disable the ignition, remove the secondary lead from the distributor center tower and ground the lead using a jumper wire.

TABLE 5-1: TROUBLESHOOTING STARTING SYSTEM

Problem	Possible Cause	Tests and Checks	Remedy
Engine cranks slowly or unevenly.	Weak battery	Perform battery open circuit voltage and load voltage tests. Perform battery load tests (capacity). Check capacity and voltage ratings against engine requirements.	Service, recharge, or replace defective battery.
	Undersized or damaged cables	Perform visual inspection.	Replace as needed.
	Poor starter circuit connections	Perform visual inspection for corrosion and damage.	Clean and tighten. Replace worn parts.
	Defective starter motor caused by high internal resistance.	Perform cranking voltage test.	If cranking voltage is within specs, replace motor. If voltage is under specs, proceed with follow-up testing.
	Engine oil too heavy for application	Check oil grade.	Change oil to within specs.
	Ignition timing incorrectly set	Check timing.	Adjust as needed.
	Seized pistons or bearing	Check compression and cranking torque.	Repair as needed.
	Overheated solenoid or starter motor	Check for missing or damaged heat shields. Troubleshoot cooling system.	Replace shield. Service as needed.
	High resistance in starter circuit	Use cranking current test. Insulated circuit test, and ground circuit test to pinpoint area of high resistance.	Replace defective components.
	Poor starter drive/flywheel engagement	Perform visual inspection of drive and flywheel components.	Replace damaged items.
	Loose starter mounting	Perform visual inspection.	Tighten as needed.
Engine does not crank.	Discharged battery	As listed above	As listed above
	Poor or broken cable connections	As listed above	As listed above
	Seized engine components	As listed above	As listed above
	Loose starter mounting	As listed above	As listed above

TABLE 5-1: TROUBLESHOOTING STARTING SYSTEM (CONTINUED)

Problem	Possible Cause	Tests and Checks	Remedy
Engine does not crank.	Open in control circuit	Perform control circuit test to determine "open" condition or areas of high resistance.	Repair or replace components as needed.
	Defective relay, starter relay, or solenoid	Perform starter relay bypass test.	Replace relay, starter relay, or solenoid if engine cranks when these components are bypassed.
	Defective starter motor caused by internal motor malfunction	Perform starter relay bypass test.	Replace starter motor if engine will not crank when relay, starter relay, or solenoid is bypassed.
Starter motor spins but does not crank engine	Defective starter drive	Perform starter drive test.	Replace starter drive.
	Worn or damaged pinion gear	Perform visual inspection of components.	Replace starter drive.
	Worn or damaged flywheel gears	Perform visual inspection of flywheel	Replace as needed.
Starter does not operate. or Movable pole shoe starter chatters or disengages before engine has started.	Battery discharged	Perform battery load test.	Recharge or replace.
	High resistance in starting circuit	Perform cranking current test, insulated circuit test, and ground circuit tests to pinpoint area of high resistance.	Replace defective components.
	Open in solenoid or movable pole shoe hold-in winding	—	Replace solenoid or movable pole shoe starter.
	Worn solenoid unable to overcome return spring pressure	—	Replace solenoid. Install lighter return spring.
	Defective starter motor	—	Replace motor.
Noisy starter cranking	Poor mounting	Perform visual inspection.	Tighten mounts. Correct alignment.

tery, always follow the safety precautions outlined in Chapter 4. Always disconnect the battery ground cable before making or breaking connections at the system's relay, solenoid, or starter motor.

TROUBLESHOOTING PROCEDURES

A systematic troubleshooting procedure is essential when servicing the starting system. Consider the fact that nearly 80 percent of "defective" starters returned on warranty claims work perfectly when tested. This is often the result of poor or incomplete diagnosis of the starting and related charging systems. Table 5-1 summarizes a systematic approach to starting system diagnosis. Testing for the starting system can be divided into area tests which check voltage and current in the entire system and more detailed pinpoint tests which target one particular component or segment of the wiring circuit.

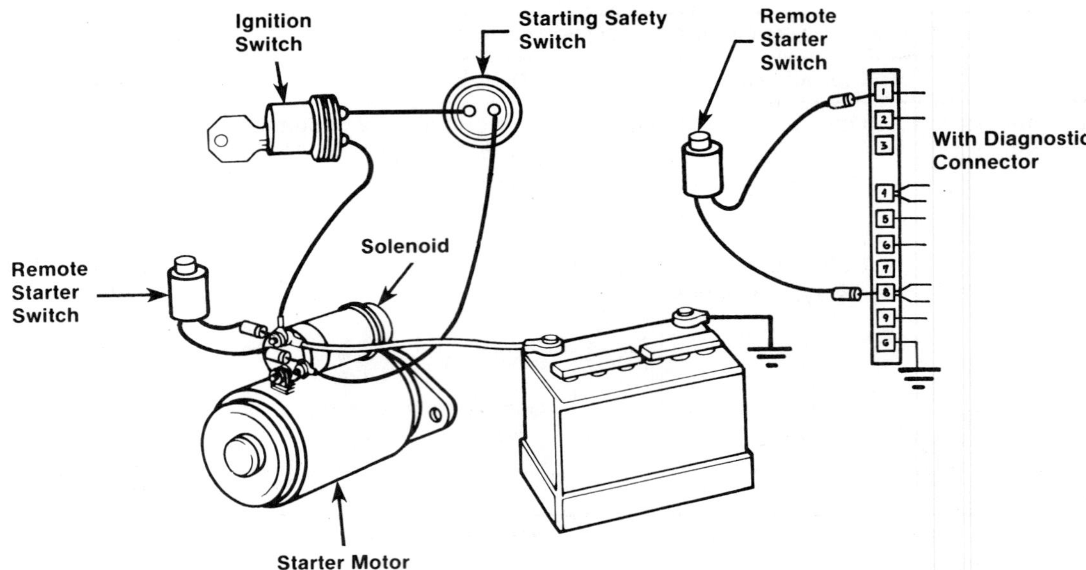

FIGURE 5–18 Using a remote starter switch to bypass the ignition system on typical GM vehicles

BATTERY LOAD TEST (AREA TEST)

An engine that turns sluggishly when cranked is often not receiving sufficient current (amperes) from the battery. The battery must be able to crank the engine under all load conditions while maintaining enough voltage to supply ignition current for starting.

As described in detail in Chapter 4, the open circuit voltage of a charged battery should test at 12.4 volts. If it is less than this, the battery should be charged prior to battery load testing. Details of the battery load test are given in Chapter 4. A battery that shows 10 volts or more during a load test is considered good. A result of 9.6 to 9.9 volts is considered fair and can be used to perform starting system tests. However, these batteries should be completely tested to determine whether the battery is defective or merely needs charging. In this way marginal batteries will be detected and replaced, preventing a road failure.

A load voltage of less than 9.6 volts (4.8 or 6-volt batteries) means that the battery is not serviceable in its present condition. These batteries must be tested and replaced or recharged before performing starter system tests.

CRANKING VOLTAGE TEST (AREA TEST)

The cranking voltage test measures the battery voltage available to the starter during cranking. To perform the test:

1. Disable the ignition or bypass the ignition switch with a remote starter switch. On GM vehicles, the remote starter switch is connected between the ignition switch terminal and the starting safety switch terminal on the starter solenoid. On GM models equipped with diagnostic connectors, the remote starter switch is connected between terminals 1 and 8 (Figure 5–18). On Ford, Chrysler, and AMC cars, connect the switch leads between the battery terminal at the starter relay and the terminal shown in Figure 5–19.
2. Connect the negative (–) lead of the voltmeter to ground.
3. Connect the positive (+) lead of the voltmeter to the proper test point as illustrated: GM vehicles, Figure 5–20; Ford and AMC vehicles, Figure 5–21; and Chrysler vehicles, Figure 5–22.
4. Crank the engine and observe the reading on the voltmeter.
5. Compare the volt reading to vehicle specifications. For cranking voltage, most vehicles are rated at 9.6 volts although some are rated at 9.0 volts. On 6-volt systems, 4.8 volts is the minimum.

Test Conclusions

If the reading is above specifications but the motor still cranks poorly, the problem is in the starter motor itself. If the reading is below specifications even with a fully charged battery, the cranking cur-

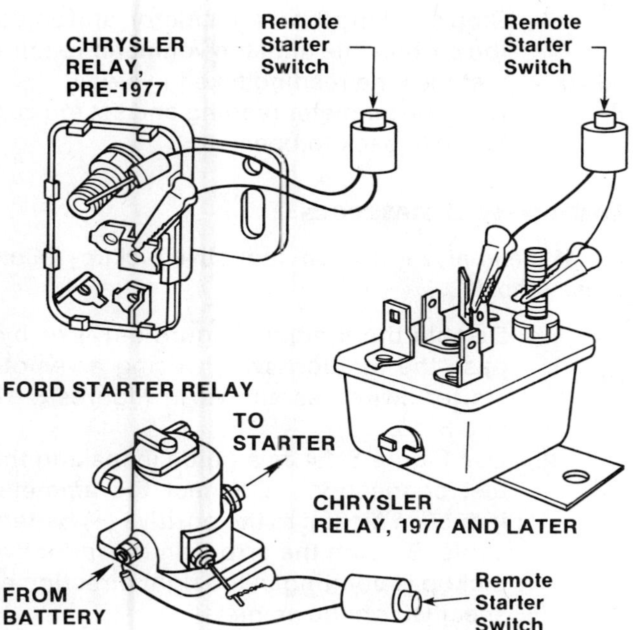

FIGURE 5-19 Using a remote starter switch to bypass the ignition system on typical Ford and AMC, and Chrysler vehicles

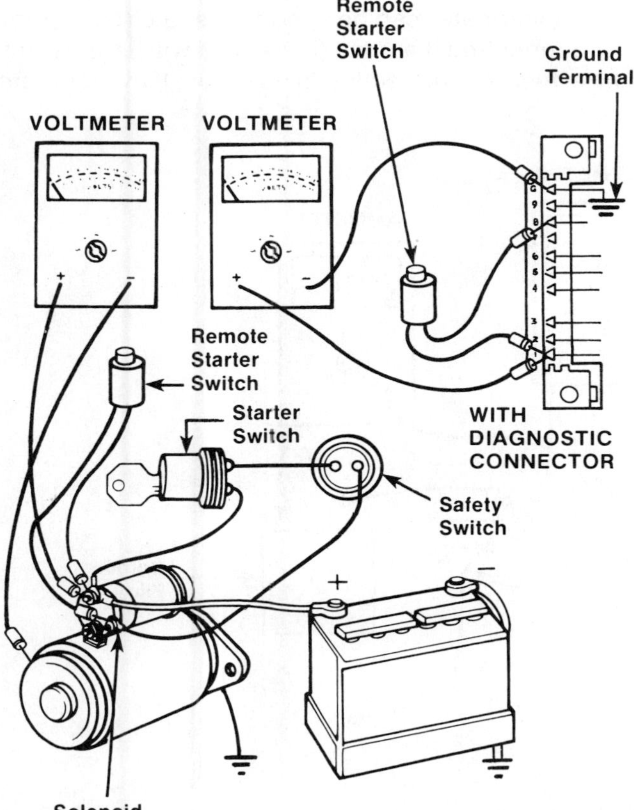

FIGURE 5-20 Measuring cranking voltage on GM vehicles

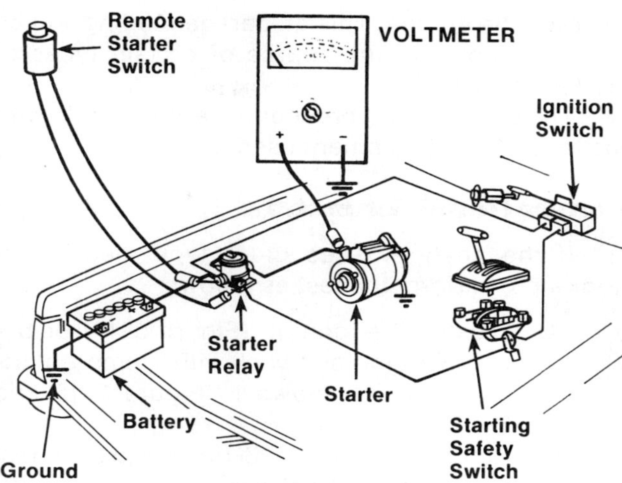

FIGURE 5-21 Setup for measuring cranking voltage on Ford and AMC vehicles

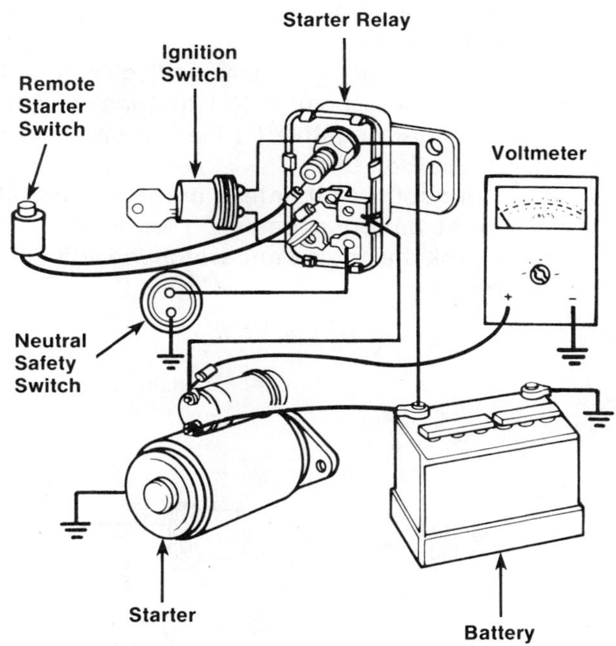

FIGURE 5-22 Setup for measuring cranking voltage on a typical Chrysler vehicle

rent test given below should be performed to determine if the problem is high resistance in the starter circuit, or heavy overloading of the starter by the engine due to seizing, dragging, or preignition.

CRANKING CURRENT TEST (AREA TEST)

The cranking current test measures the amount of current, in amperes, that the starter circuit draws to crank the engine. This amperage reading will be useful in isolating the source of certain types of starter problems.

The exact testing procedure varies slightly with the type of test equipment used.

Conventional Ammeters

If the analyzer uses older type mechanical hook-ups, perform the test as follows:

1. Disable the ignition (Figure 5-17) or bypass the ignition switch with a remote starter switch as shown in Figure 5-18 and 5-19.
2. Refer to Figure 5-23 for the proper test connections. As shown, connect the voltmeter positive (+) lead to the battery positive (+) terminal. Connect the voltmeter negative (-) lead to the battery negative (-) terminal.
3. Set the carbon pile to its maximum resistance (open).
4. Connect the ammeter positive (+) lead to the battery positive (+) terminal and the ammeter negative (-) lead to one lead of the carbon pile.
5. Connect the other lead of the carbon pile to the battery negative (-) terminal.
6. Crank the engine and watch the voltmeter reading.
7. Stop cranking the starter motor, and adjust the carbon pile until the voltmeter reading matches the reading taken in step 6.
8. Note the ammeter reading and set the carbon pile back to open.

Inductive Ammeters

If the analyzer uses an inductive pickup, follow these steps:

1. Disable the ignition (Figure 5-17) or bypass the ignition switch using a remote starter switch as shown in Figures 5-18 and 5-19.
2. Use Figure 5-24 as a guide in making the test connections. Connect the ammeter inductive pickup to the positive (+) battery cable. Be sure the arrow on the inductive pickup is pointing in the right direction as specified on the ammeter.
3. Crank the engine for 15 seconds and observe the ammeter reading.
4. Compare the ammeter reading to specifications.

Voltage Drop Testing

A voltmeter can be used to test a circuit simply by connecting it across (in parallel with) the switch, wire, fuse, or connection in question. If the voltmeter

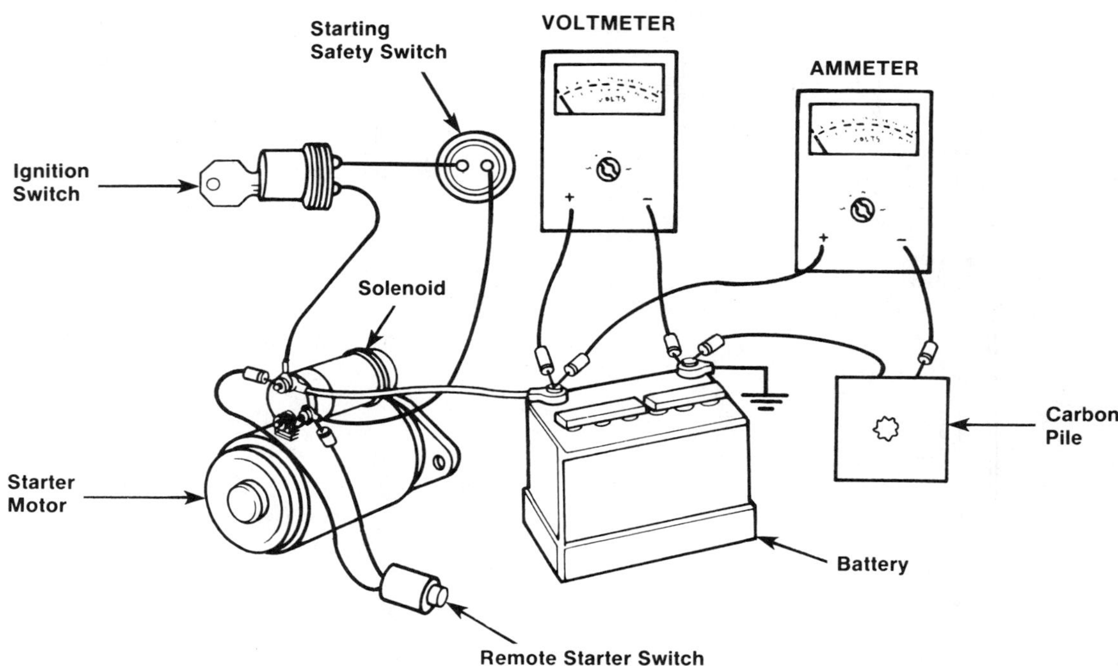

FIGURE 5-23 The test meter setup to perform a starter current draw test.

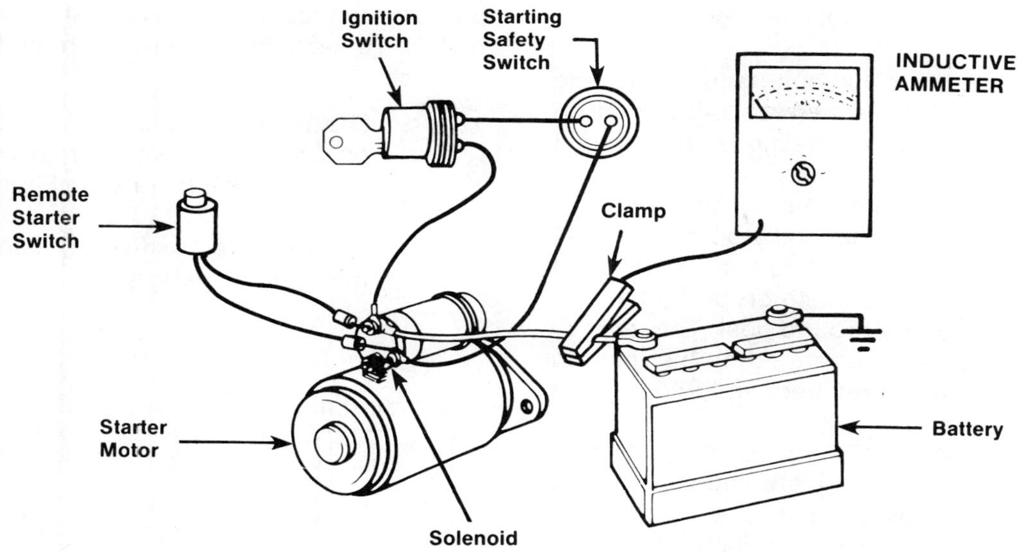

FIGURE 5-24 Using an inductive ammeter to test the starter current draw

reading reveals low voltage, the component being tested is normal. If the reading is equal to the battery voltage, the component being tested is open. If the reading reveals more than the normal voltage drop, there is excessive resistance in the circuit.

Test Conclusions

Compare the reading obtained during testing to the manufacturer's specifications. As a rule of thumb, reading for 8-cylinder engines should be approximately 280 amps. Reading for 6-cylinder engines should be approximately 240 amps and 2 to 5 cylinder engines, 210 amps. Table 5-2 summarizes the most probable causes of current draw that is too high or too low. If the problem appears to be caused by too much resistance in the system, test the starting system resistance as covered in the circuit resistance test in the next sections.

Excessive current draw in the starter can be caused by high resistance within the starter itself. This can be the result of worn brushes, or ground or opens in the armature or coil windings. It can also be the result of increased internal friction due to shaft bushings that bind or an armature that is rubbing against the housing. Noise in the starter is a good indication that it may be dragging.

However, noise can also be caused by starter misalignment. A high-pitched whine after starting may signal a starter drive that is hanging up, often due to variations in the machining of the bell housing. It may be necessary to shim the starter by placing washers between it and the bell housing.

TABLE 5-2: RESULT OF CRANKING CURRENT TESTING

Problem	Possible Cause
Low current draw	Undercharged or defective battery
	Excessive resistance in circuit due to faulty components or connections.
High current draw	Short in starter motor
	Mechanical resistance due to binding engine or starter system component failure or misalignment.

If the starter teeth are too close, they too can bind when the starter becomes hot. The recommended clearance is 0.020 inch. When properly aligned, the drive teeth should engage the flywheel teeth about three quarters of the way down from the top of each tooth. To check this clearance, push a starter drive out by inserting a screwdriver through a hole in the bottom of the starter housing and use a hooked wire gauge to check the clearance between the teeth.

A loosely mounted starter may also cause slow or no cranking conditions. Loose mounting bolts can cause a weak ground connection. The starter may also shift, slip, chatter, or fail to engage. To prevent this, always use lock washers or locking compound on the starter bolts and torque them to specifications.

In normal testing situations, failures indicate that more detailed tests should be performed on individual components, connections, and circuit links in the area.

As you learned earlier, the starting system's main electrical circuit is generally a series circuit from the battery-insulated post, to a starter solenoid or relay, to the starter motor, to ground (chassis), and return to the battery ground post. The relay or solenoid is controlled or operated by the ignition switch. This control circuit frequently contains a safety switch.

The starting system will function properly only when these circuits and electrical components are in good condition. All switches involved must also make good electrical connections when closed. The following three pinpoint tests are designed to locate the point of high resistance in the circuit that is causing excessive voltage loss. The resistance usually occurs at one of the circuit connections, but can also be caused by internal faulty wiring or cables. A general rule for voltage loss in a starting system is 0.1 volt per connection. Losses greater than this will usually cause problems.

INSULATED CIRCUIT RESISTANCE TEST

The complete starter circuit is made up of the insulated circuit and the bare ground circuit. The insulated circuit includes all of the high current cables and connections from the battery to the starter motor. To test the insulated circuit for high resistance:

1. Disable the ignition (Figure 5-17); or bypass the ignition switch with a remote starter switch as illustrated in Figures 5-18 and 5-19
2. Connect the positive (+) lead of the voltmeter to the battery's positive (+) terminal

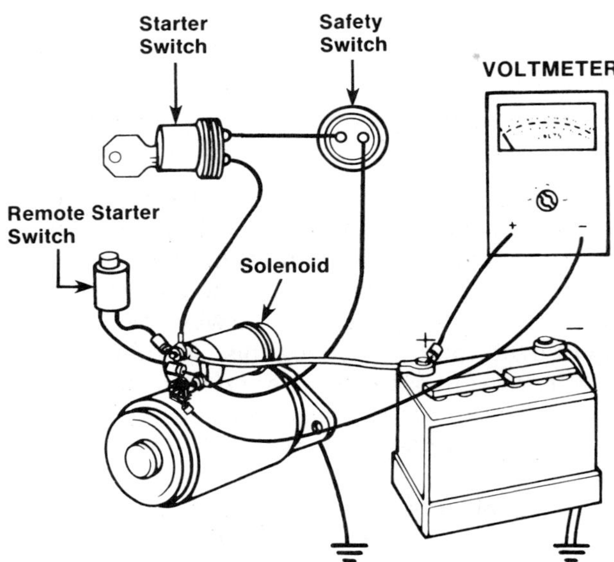

FIGURE 5-25 Setup for the testing of resistance in the insulated circuit of the starting system

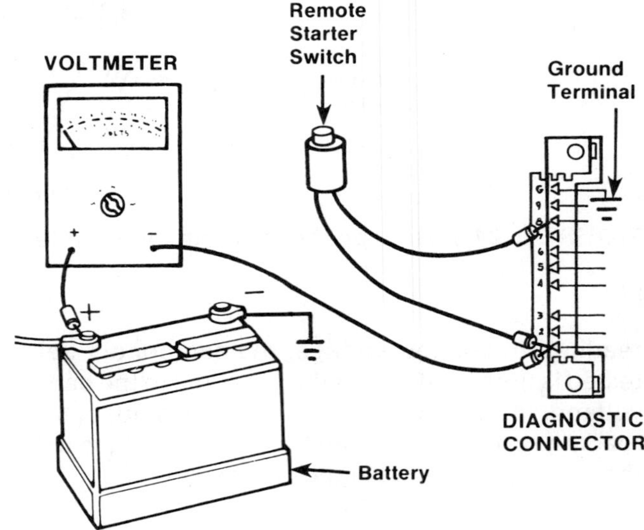

FIGURE 5-26 Setup for using the GM diagnostic connector to test insulated circuit resistance

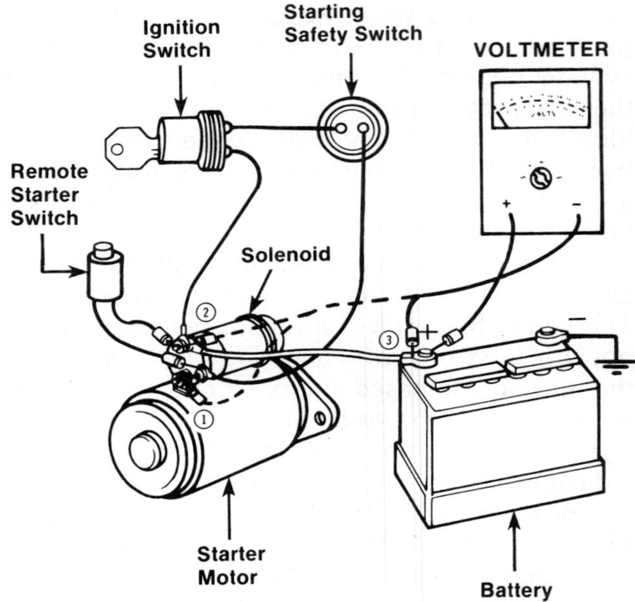

FIGURE 5-27 Specific test points to use when checking insulated circuit resistance on GM vehicles

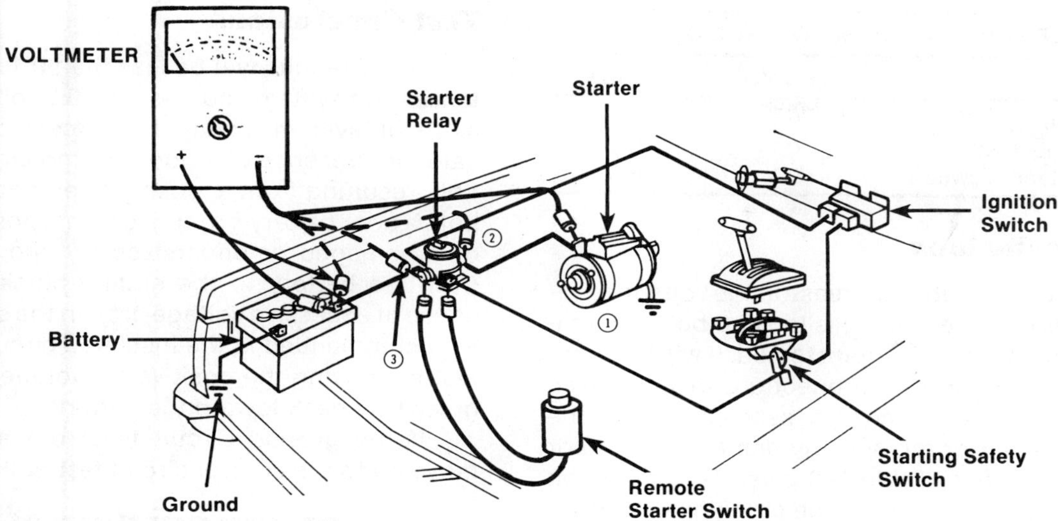

FIGURE 5-28 Specific test points to use when checking insulated circuit resistance on Ford and AMC vehicles

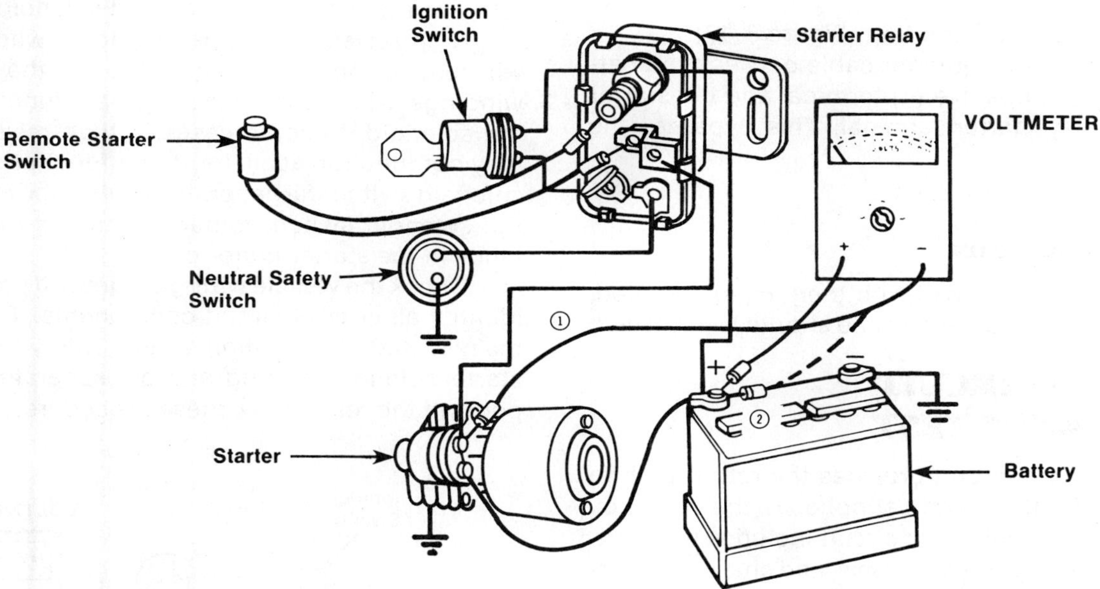

FIGURE 5-29 Specific test points to use when checking insulated circuit resistance of Chrysler vehicles

post or nut. By connecting the lead to the cable, it might be possible to bypass a point of high resistance in the cable-to-post connection.

3. Connect the negative (-) lead of the voltmeter to the terminal at the starter (Figure 5-25). For GM diagnostic connectors, use the hookup shown in Figure 5-26.
4. Crank the engine and record the voltmeter reading. If the reading is within specifications (usually 0.2 to 0.6 voltage drop), the insulated circuit does not have excessive resistance. Proceed to the ground circuit resistance test outlined in the next section.
5. If the reading indicates a voltage loss above specifications, move the volt lead on the starter progressively toward the battery, cranking the engine at each test point indicated in the following illustrations: GM models (Figure 5-27), Ford and AMC models (Figure 5-28), and Chrysler models (Figure 5-29).

TABLE 5-3: MAXIMUM VOLTAGE DROPS

	6-volt	12-volt
Each cable	0.1	0.2
Each connection	0.1	0.1
Starter solenoid switch	0.3	0.3

Test Conclusions

When a noticeable decrease in the voltage reading is observed, the trouble is located between that point and the preceding point tested. It will be either a damaged cable or poor connection, an undersized wire, or possibly a bad contact assembly within the solenoid. Repair or replace any damaged wiring or faulty connections. Table 5-3 gives 6- and 12-volt maximum voltage drops for the cranking circuit.

STARTER RELAY BYPASS TEST

The starter relay bypass test is a simple method of determining if the relay is operational.

1. Disable the ignition (Figure 5-17).
2. Connect a jumper cable between the battery's positive (+) terminal and the starter relay starter terminal. This bypasses the relay.
3. Crank the engine.

Test Conclusions

If the engine cranks with the jumper installed, the starter relay is defective and should be replaced.

GROUND CIRCUIT RESISTANCE TEST

The ground circuit provides the return path to the battery for the current supplied to the starter by the insulated circuit. This circuit includes the starter-to-engine, engine-to-chassis, and chassis-to-battery ground terminal connections. To test the ground circuit for high resistance, follow these procedures.

1. Disable the ignition (Figure 5-17), or bypass the ignition switch with a remote starter switch (Figures 5-18 and 5-19).
2. Connect the positive (+) lead of the voltmeter to the starter housing.
3. Connect the negative (-) lead of the voltmeter to the battery's negative (-) terminal post or nut to avoid bypassing any resistance in the cable-to-post connection (Figure 5-30).
4. Crank the engine and record the voltmeter reading.

Test Conclusions

Good results will be less than 0.1 voltage drop for a 6-volt system and less than 0.2 voltage drop for a 12-volt system. Voltages in excess of these indicate the presence of a poor ground circuit connection, resulting from a loose starter motor mounting bolt, a poor battery ground terminal post connector, or a damaged or undersized ground system wire from the battery to the engine block. Isolate the cause of excessive voltage drop in the same manner as recommended in the insulated circuit resistance test by moving the positive (+) voltmeter lead progressively back toward the battery.

If the ground circuit tests out satisfactorily, move on to the control circuit test outlined below.

CONTROL CIRCUIT TEST

The control circuit test examines all the wiring and components used to control the magnetic switch, whether it is a relay, solenoid acting as a relay, or a starter motor-mounted solenoid.

High resistance in the solenoid switch circuit will reduce the current flow through the solenoid windings, which can cause improper functioning of the solenoid. In some cases of high resistance, it may not function at all. Improper functioning of the solenoid switch will generally result in the burning of the solenoid switch contacts, causing high resistance in the starter motor circuit.

Check the vehicle wiring diagram, if possible, to identify all control circuit components. These normally include the ignition switch, safety switch, the starter solenoid winding, and/or a separate relay. To perform the test, follow these procedures.

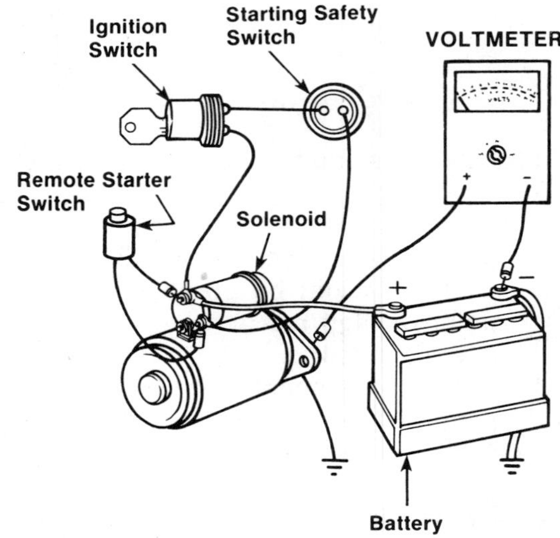

FIGURE 5-30 Setup for testing the resistance of the ground circuit.

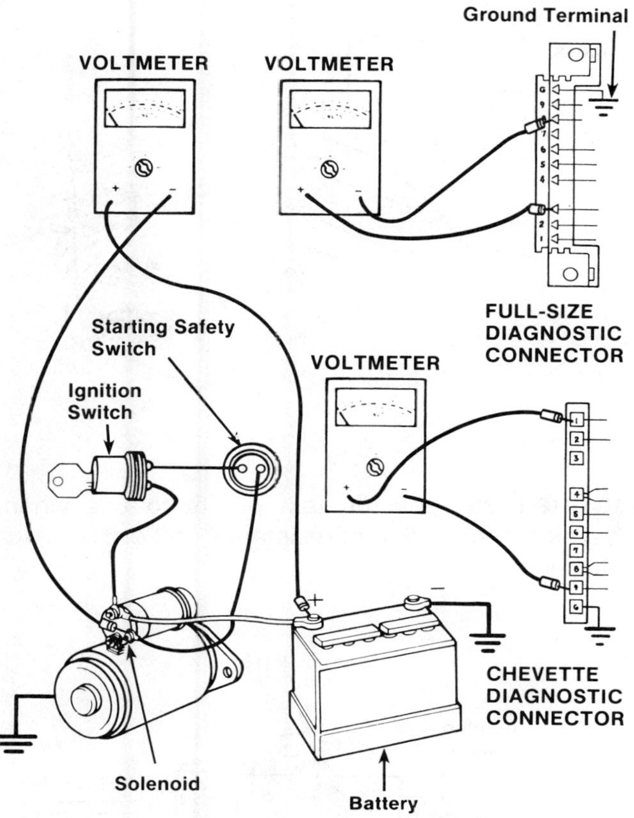

FIGURE 5-31 Setup for testing the control circuit resistance on GM vehicles

1. Disable the ignition system as shown in Figure 5-17.
2. Connect the positive (+) lead of the voltmeter to the battery's positive (+) terminal.
3. Connect the negative (-) lead of the voltmeter to the switch terminal at the relay or solenoid. Exact setups for GM, Ford and AMC, and Chrysler models are illustrated in Figures 5-31, 5-32, and 5-33 respectively.
4. Crank the engine and record the voltmeter reading.

Test Conclusions

Generally, good results will be less than 0.5 volt, indicating that the circuit condition is good. If starter current draw was previously tested to be high or the cranking speed is slow, a faulty starter motor is likely at fault. In this case the motor should be removed for repair or replacement.

If voltage reads more than 0.5 volts, it is usually an indication of excessive resistance. However, on certain vehicles, a slightly higher voltage loss may be normal.

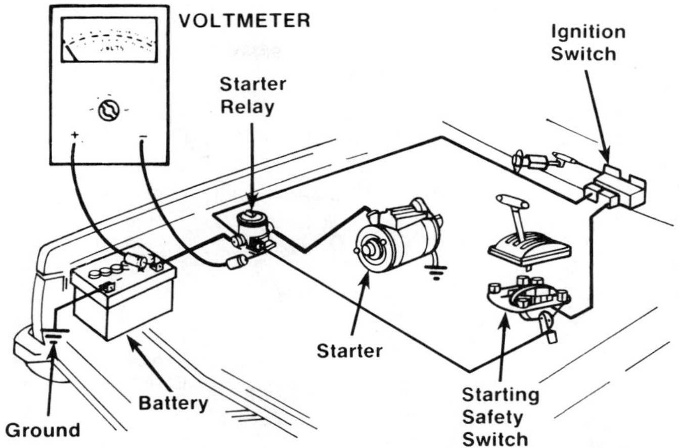

FIGURE 5-32 Setup for testing the control circuit resistance on Ford or AMC vehicles

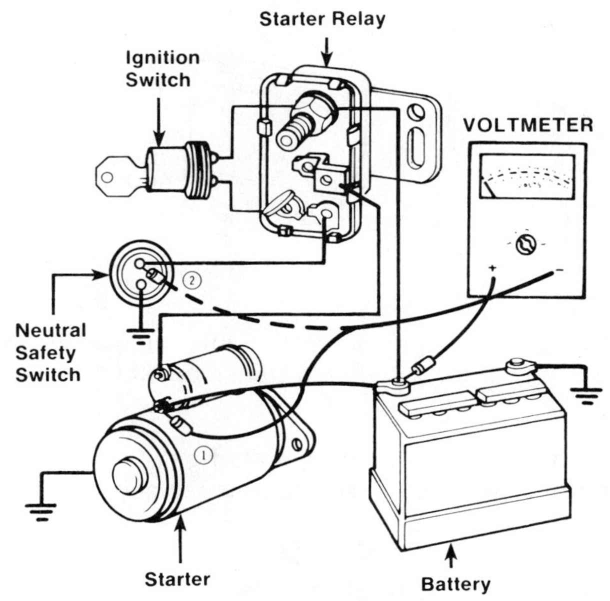

FIGURE 5-33 Setup for testing the control circuit resistance on Chrysler vehicles

Isolate the point of high resistance by placing the voltmeter's positive lead to the points indicated in Figures 5-34, 5-35, and 5-36, for GM, Ford and AMC, and Chrysler models respectively.

A reading of more than 0.1 volt across any one wire or switch is usually an indication of trouble. If a high reading is obtained across the safety switch used on automatic transmissions, check the adjustment of the switch according to the manufacturer's service manual.

Mechanical blocking devices, such as those used on GM models, should not require service dur-

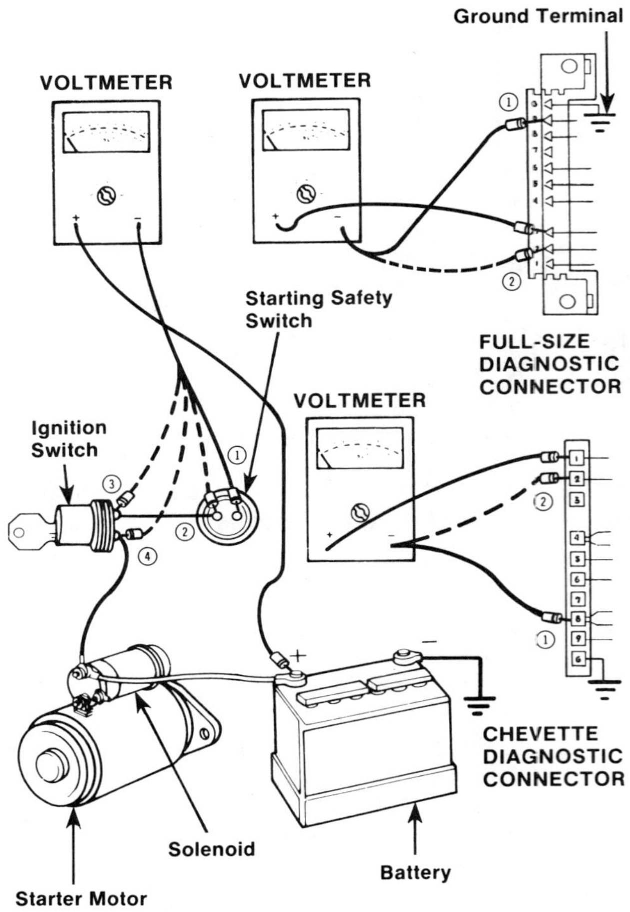

FIGURE 5-34 Specific test points to use when checking control circuit resistance on GM vehicles

ing normal use. Clutch-operated safety switches cannot be adjusted. They must be replaced.

TESTING STARTER DRIVE COMPONENTS

This test will detect a slipping starter drive without removing the starter from the vehicle.

1. Disable the ignition (Figure 5-17) or bypass the ignition switch with a remote starter switch (Figures 5-18 and 5-19).
2. Turn the ignition switch to start and hold it in this position for several seconds.
3. Repeat the procedure at least three times to detect an intermittent condition.

Test Conclusions

If the starter cranks the engine smoothly, this is an indication of acceptable starter drive. If the engine stops cranking and the starter spins noisely at high speed, the drive is slipping and should be replaced.

If the drive is not slipping, but the engine is not being cranked, inspect the flywheel for missing and/or damaged teeth. Remove the starter from the vehicle and check its drive components. Inspect the pinion gear teeth for wear and damage (Figure 5-37). Test the overriding clutch mechanism. If good, the

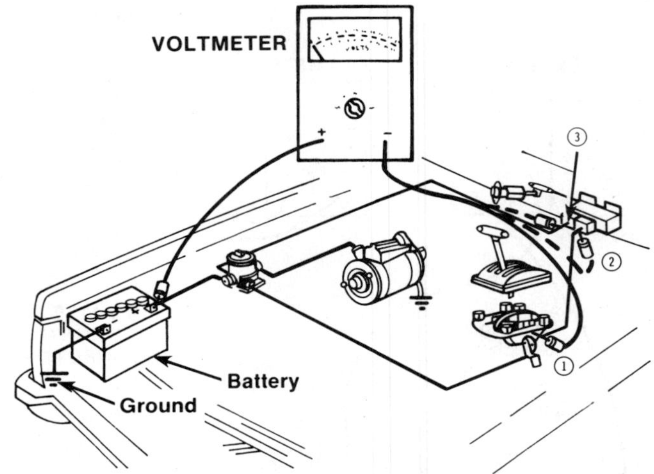

FIGURE 5-35 Specific test points to use when checking control circuit resistance on Ford or AMC vehicles

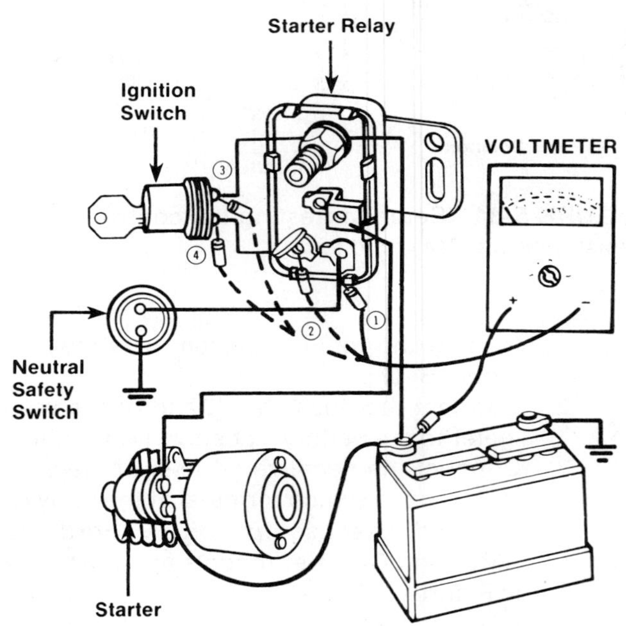

FIGURE 5-36 Specific test points to use when checking control circuit resistance on Chrysler vehicles

overriding clutch should turn freely in one direction, but not in the other. A bad clutch will turn freely in both directions or not at all. If a drive locks up, it can over-rev and destroy the starter.

The weak point in sliding pole starter is the pole shoe that pulls in toward the armature to engage the starter. This starter requires a minimum of 10.5 volts and about 300 to 400 amperes to operate. Otherwise, it will simply click and not engage.

As a sliding pole starter wears, the pivot bushing sometimes hangs up and prevents the engagement pole shoe from being pulled down. When this happens, the starter will spin but fail to engage the flywheel.

A similar problem can occur on solenoid-actuated starters. If the solenoid is too weak to overcome the force of the return spring, the starter will not operate. If replacing the solenoid does not work, installing a lighter return spring may be needed.

STARTER MOTOR REMOVAL AND REPLACEMENT

When removing a starter motor from the vehicle, it may be necessary to remove or loosen heat shields, support brackets, or exhaust pipes. It is always helpful to tag or otherwise note nuts, bolts, washers, and all disconnected wires (Figure 5-38).

Follow all electrical and shop safety precautions. A basic outline of the typical steps involved is given here:

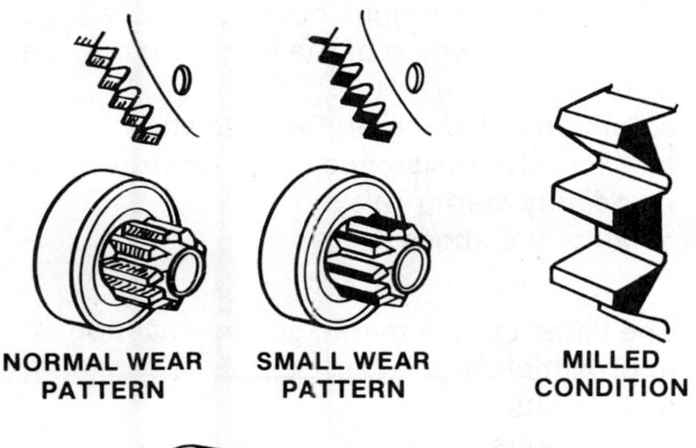

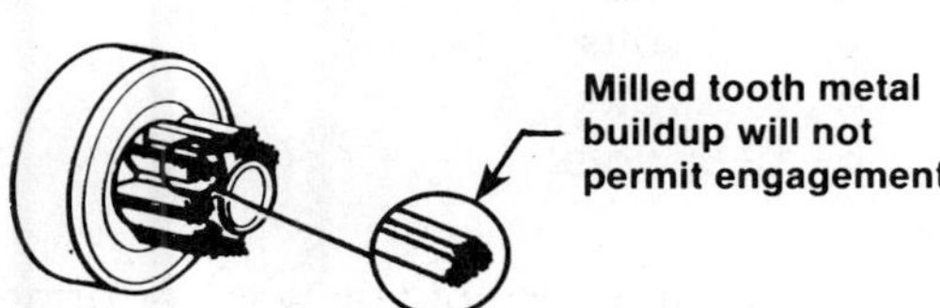

FIGURE 5-37 Typical wear patterns that occur on overriding clutch teeth and flywheel teeth

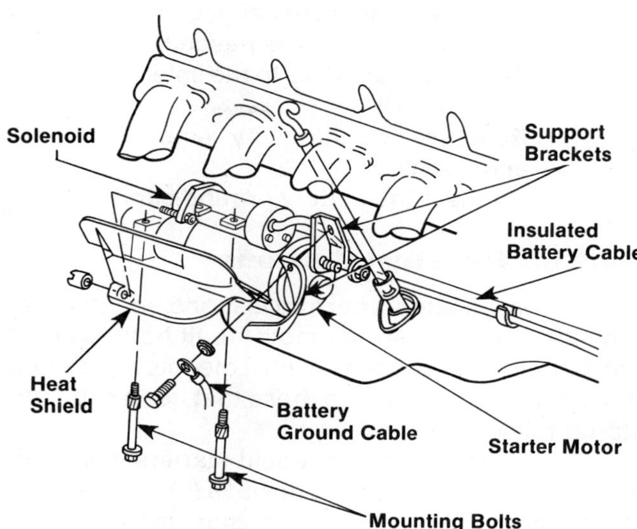

FIGURE 5-38 Many components such as heat shields, brackets, and wires must be carefully removed and reinstalled when replacing a starter.

1. Disconnect the battery ground cable.
2. Raise and support the vehicle for easy access to the starter area. Use safety stands or a hoist.
3. On some vehicles it may be necessary to turn the front wheels and disconnect the tie-rods for easy access to the starter. The wheel may also have to be removed.
4. Loosen or remove exhaust pipes or other components that block access to the starter.
5. Disconnect and tag all wires from the starter motor or solenoid.
6. Remove the heat shields and support brackets.
7. Remove all mounting bolts and shims used to secure the starter to the engine. Lift out the starter.

To reinstall the unit:

1. Attach the support brackets (if used) to the new or serviced motor. It may also be possible to attach the heat shield before mounting.
2. Install the motor on the engine with all mounting bolts.
3. Check the flywheel engagement. GM vehicles may require shims to produce the correct engagement of the starter pinion gear with the flywheel. Shims are placed between the starter drive housing and the engine block. Remove the flywheel cover

and check for proper engagement, adding or removing shims as needed.

4. Reconnect all wires to the solenoid or motor terminals.
5. Remount or tighten any removed or loosened parts.
6. Connect the battery ground cable.

Note on Ford Installations

To consolidate starter applications, some starter rebuilders use one starter motor to fit both types of Ford applications—those with solenoids mounted directly to the starter, and those that use a separate starter relay.

On the back of these rebuilt starters is a small metal buss bar between the positive (+) terminal and "S" terminal. On applications that use a starter-mounted solenoid, the bar must be removed or the starter will not work. Those that use a fender-mounted starter relay need the bar in place.

REVIEW QUESTIONS

1. Which of the following is not part of the starter circuit?
   a. battery
   b. starting safety switch
   c. starter motor
   d. relay solenoid

2. Which of the following could result in a hard starting condition?
   a. corroded battery cables
   b. excessive CCA capacity
   c. heavy gauge battery cables
   d. all of the above

3. What is the only rotating part of the starter?
   a. commutator
   b. armature
   c. brushes
   d. both a and b

4. Which of the following drive mechanisms is found on only Ford vehicles?
   a. permanent magnet planetary drive
   b. solenoid-actuated reduced drive
   c. movable pole shoe drive
   d. none of the above

5. Which of the following is not a member of the control circuit?
   a. ignition switch
   b. starter relay
   c. ballast resistor
   d. starting safety switch

6. Technician A and Technician B were preparing to perform a cranking voltage test. Technician A says they ought to disable the ignition system first. Technician B says that is unnecessary because the car has an electronic ignition system. Which technician is correct?
   a. Technician A
   b. Technician B
   c. Both A and B
   d. Neither A nor B

7. Excessive starter current draw might be caused by ______________.
   a. worn shaft bushings
   b. broken armature windings
   c. improper clearance of starter teeth
   d. all of the above

8. Which of the following tests would be performed to check for high resistance in the battery cables?
   a. cranking voltage test
   b. insulated circuit resistance test
   c. starter relay bypass test
   d. ground circuit resistance test

9. When the starter spins but does not engage the flywheel, which of the following may be true?
   a. binding bushing on the pole shoe
   b. excessive resistance in the control circuit
   c. a faulty starter relay
   d. all of the above

10. The usual cranking voltage specification is approximately ______________.
    a. 9.6 volts
    b. 10.5 volts
    c. 11.0 volts
    d. 12.65 volts

11. A cranking current test was performed, and the amperage was found to be less than spec-

ification. Technician A says that the starter is bad and that it should be replaced. Technician B insists on testing the resistance of the cables, grounds, and connections. Who is correct?
   a. Technician A
   b. Technician B
   c. Both A and B
   d. Neither A nor B

12. A control circuit test may uncover which of the following conditions?
    a. high resistance in the solenoid switch circuit
    b. defective starter relay
    c. loose battery cable connection
    d. short in the starter armature windings

13. An engine cranks slowly. Technician A says that a possible cause of the problem is poor starter circuit connections. Technician B says that a possible cause is incorrectly set ignition timing. Who is right?
    a. Technician A
    b. Technician B
    c. Both A and B
    d. Neither A nor B

14. If a ground circuit test reveals a voltage drop of more than 0.2 volts, the problem may be ______________.
    a. a loose starter motor mounting bolt
    b. a poor battery ground terminal post connector
    c. a damaged battery ground cable
    d. all of the above

15. Which of the following is not a safety hazard that ought to be followed when performing a cranking current test?
    a. Disconnect the negative battery cable.
    b. Disable the ignition system.
    c. Make sure the transmission is out of gear.
    d. All of the above.

16. Which of the following tools should be used to perform a voltage drop test?
    a. ammeter
    b. ohmmeter
    c. voltmeter
    d. all of the above

17. Which of the following is not a function of starting systems?
    a. to turn the engine until it starts
    b. to bypass the ignition ballast resistor
    c. to recharge the battery
    d. all of the above

18. Which of the following is true about starting systems?
    a. Starting systems draw heavy electrical current from the battery.
    b. All starter solenoids are mounted on the starter.
    c. The ignition switch is part of the starter circuit.
    d. All of the above.

19. The battery CCA rating must match engine demands. The rule of thumb is ______________ CCA for each cubic inch of displacement.
    a. 1
    b. 2
    c. 10
    d. 15

20. To check the condition of a fusible link, ______________.
    a. use an ohmmeter to check for continuity
    b. use a voltmeter to perform a voltage drop test.
    c. perform a batter load test
    d. all of the above

21. Which of the following is used to engage the starter pinion gear and the flywheel?
    a. starter relay
    b. starter-mounted solenoid
    c. ballast resistor
    d. fender-mounted solenoid

22. Cranking the starter for longer than 30 seconds at a time will result in ______________.
    a. dangerously high compression in engine
    b. an overheated coil
    c. damage to the starter
    d. all of the above

23. Which of the following components protects the clutch from excessive turning speeds?
    a. sun gear
    b. drive pinion

c. override clutch
d. none of the above

24. If a starter will not run, where is a likely spot to look for an open in the electrical system?
a. starter relay
b. ignition switch
c. battery ground
d. all of the above

25. Which of the following tools would not be used to perform a cranking current test?
a. voltmeter
b. remote starter switch
c. inductive pickup
d. ammeter

26. The maximum voltage drop across a connection electrical in the starter system is ______________ volts.
a. 0.1
b. 0.2
c. 1
d. 1.2

CHAPTER SIX

OPERATION AND TESTING OF CHARGING SYSTEMS

Objectives

Upon completion of this chapter, you will be able to:

- Explain how AC systems work, and describe their advantages over DC systems.
- Explain half- and full-wave rectification, and how they relate to alternator operation.
- Identify the different types of AC voltage regulators.
- Perform a charging system inspection.
- Perform a general charging system test using a charging-starting-battery analyzer.
- Describe the five basic charging system tests.

The charging system converts the mechanical energy of the engine into electrical energy. During cranking, the battery supplies all of the vehicle's electrical energy. However, once the engine is running, the charging system is responsible for producing enough energy to meet the demands of all the loads in the electrical system, while also recharging the battery. With more and more cars featuring electronic steering, brakes, suspension, transmissions, and air-conditioning systems—not to mention traditional items like wipers, lights, and horn—the demands on the charging system are great.

For many years, the automotive charging system was DC, or direct current. This system consisted of a belt-driven generator (Figure 6–1), a regulator, and the necessary wiring and switches; energy was produced by a series of rotating conductors within a fixed magnetic field. By the early 1960s, cars were being built with more electrical accessories than ever before. At the same time, greatly increased traffic congestion brought about the need for engines that could handle slower speed, stop-and-go driving. These two factors led to the development of AC, or alternating current, charging systems. By replacing the generator with an alternator, this system can produce a higher charging rate at slow speeds and while idling.

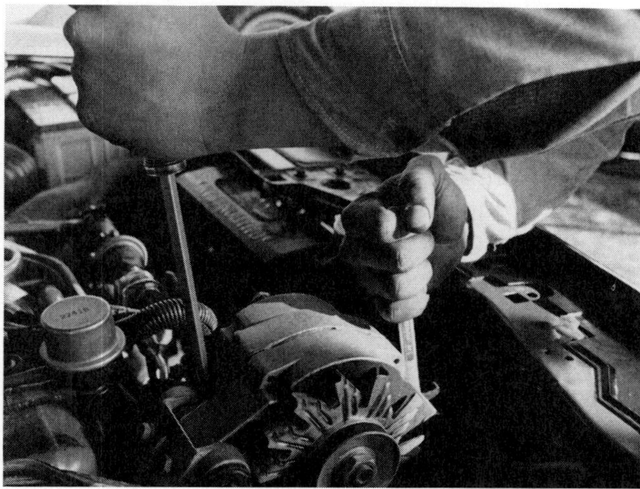

FIGURE 6–1 The belt-driven generator was once standard equipment in all automobiles.

AC SYSTEMS

Alternating current charging systems are used today because they have several advantages over DC systems. In addition to producing more current at low rpm, AC systems also feature

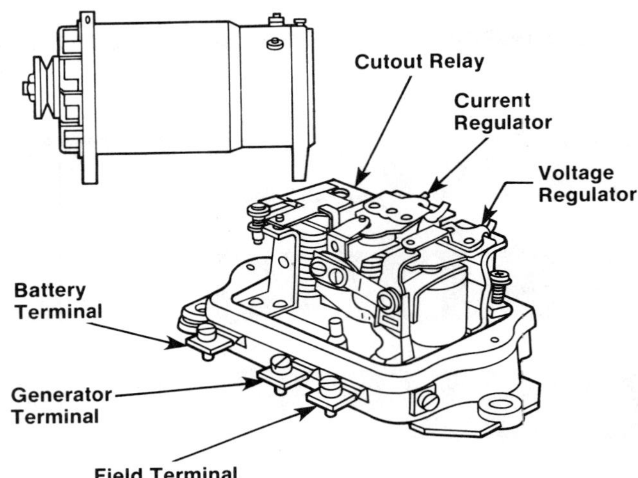

FIGURE 6-2 A typical direct current regulator

- Less weight per ampere of output.
- Efficient operation at much higher speeds.
- Less current travel through the brushes, thus reducing brush wear.
- No need for external current regulation.
- Rotational speeds up to 15,000 rpm without danger of physical damage.

Alternator Construction

For an alternator to produce the necessary voltage output, two things must be present: a rotating magnetic field of sufficient strength, and multiple conductors in which the voltage can be induced. The basic construction of the alternator limits the maximum current output of the AC system. To change the output as needed for particular applications, manufacturers vary the design of the system components. Following are descriptions of the major alternator parts: rotor, slip rings and brushes, stator, and end-frame assembly.

Rotor

As the only moving component within the alternator, the rotor is responsible for producing the rotating magnetic field. The rotor consists of a coil, two pole pieces, and a shaft, as shown in Figure 6-3. The magnetic field is the result of the current flow through the coil; this coil is simply a series of windings wrapped around an iron core. Altering the current flow through the coil varies the strength of the magnetic field, which in turn affects alternator output.

The coil is located between the interlocking pole pieces. As current flows through the coil, the

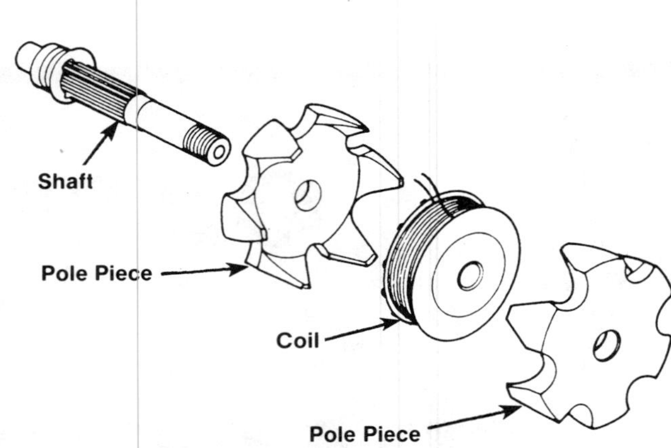

FIGURE 6-3 The rotor is made up of a coil, pole pieces, and a shaft.

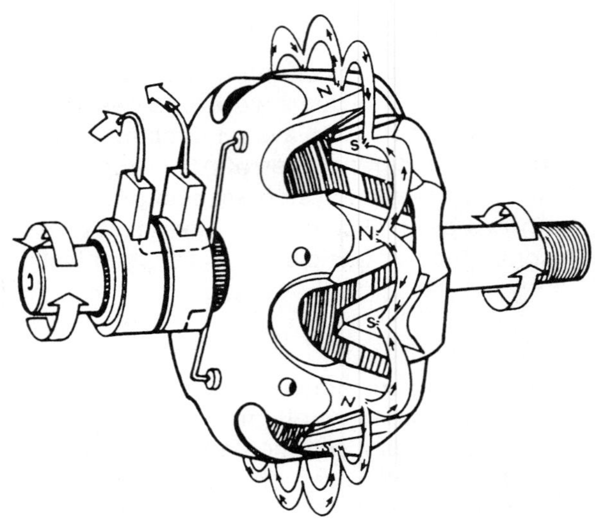

FIGURE 6-4 The magnetic field moves from the N poles, or fingers, to the S poles.

core essentially becomes a magnet, with the pole pieces assuming the magnetic polarity of the end of the core that they touch. Thus, one pole piece has a north polarity and the other has a south polarity. The extensions on the pole pieces, known as the fingers, form the actual magnetic poles. A typical rotor has 14 poles, seven north and seven south, with the magnetic field beween the pole pieces moving from the N poles to the adjacent S poles (Figure 6-4). The more poles in a rotor, the higher the alternator output.

Slip Rings and Brushes

The slip rings and brushes conduct current to the rotor. Most alternators have two slip rings mounted directly on the rotor shaft; they are insulated from the shaft and from each other. A carbon brush is located on each slip ring to carry the current to and from the rotor windings (Figure 6–5). Because the brushes carry only 1.5 to 3.0 amperes of current, they do not require frequent maintenance. This is in direct contrast to generator brushes, which conduct all of the generator's current output and therefore must be serviced often.

Stator

The stator is made up of a number of conductors, or wires, into which the magnetic field induces the voltage. The conductors are wound into slots in a cylindrical frame, with each conductor forming several coils spaced evenly around the frame (Figure 6–6). There are as many coils in each conductor as there are pairs of north/south rotor poles. One end of each conductor attaches to a separate pair of diodes, one positive and the other negative.

End Frame Assembly

The end frame assembly, or housing, is made of two pieces of cast aluminum and contains the bearings for the end of the rotor shaft where the drive pulley is mounted. Each end frame also has built-in air ducts so the air from the rotor shaft fan can pass through the alternator. A heat sink containing three positive diodes is attached to the rear end frame; heat can pass easily from these diodes to the moving air (Figure 6–7). Three negative diodes are contained in the end frame itself. Because the end frames are bolted together and then bolted directly to the engine, the end frame assembly is part of the electrical ground path. This means that anything connected to the housing that is not insulated from the housing is grounded.

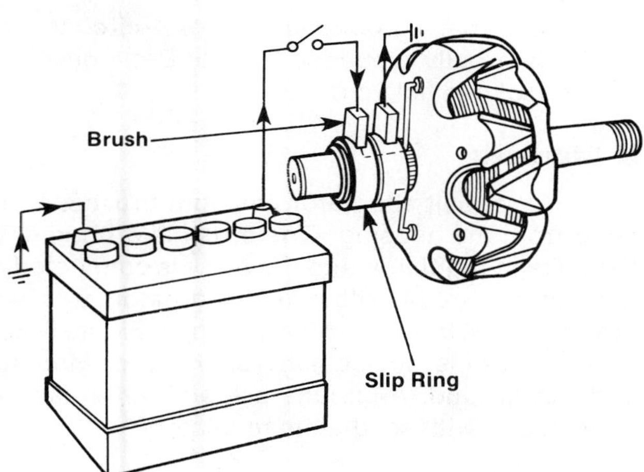

FIGURE 6–5 Current is carried by the brushes to and from the rotor windings via the slip rings.

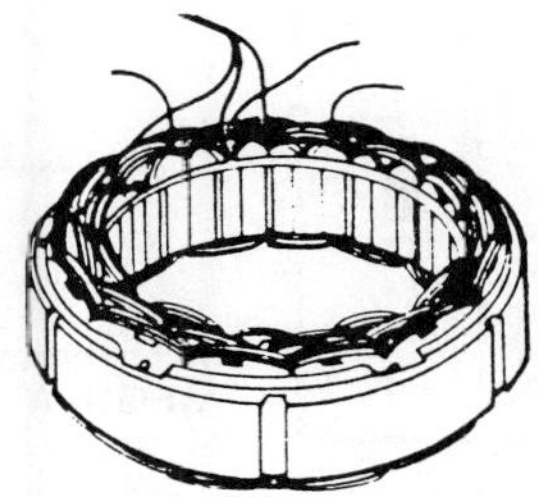

FIGURE 6–6 A typical stator

ALTERNATOR OPERATION

Both generators and alternators produce alternating current. In order to be used in an automotive electrical system, the AC must be converted, or rectified, to DC. While generators accomplish this by means of a mechanical commutator and brushes, alternators do it electronically with diodes. Simply put, a diode is an electrical valve that permits current flow in one direction only.

When a diode is placed in a simple AC circuit, one-half of the voltage is blocked. In other words, the current can flow from X to Y as shown in Figure 6–8, but it cannot flow from Y to X. When the voltage reverses at the start of the next rotor revolution, the current again can pass from X to Y, but not back. Obviously, this type of output is hardly efficient since current is available only half of the time. Because only 50 percent of the AC voltage produced by the alternator is being converted to DC this is known as half-wave rectification.

By adding more diodes to the circuit, more of the AC is rectified. In Figure 6–9A, the current flows from X to Y. It flows from X, through diode 2, through

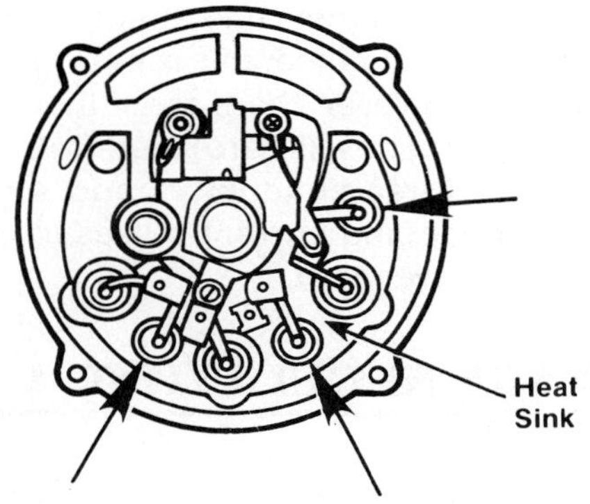

FIGURE 6–7 Three positive diodes mount directly onto the heat sink.

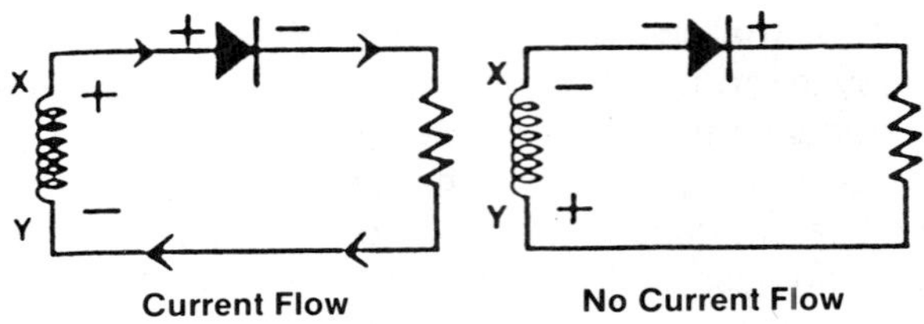

FIGURE 6-8 One diode in an AC circuit results in half-wave rectification.

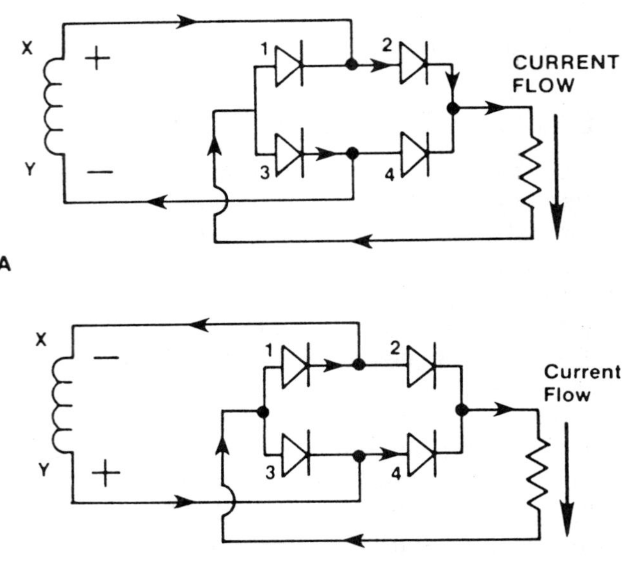

FIGURE 6-9 By adding more diodes, the current flows (A) from X to Y and (B) from Y to X, resulting in full-wave rectification.

the load, through diode 3, and then to Y. In Figure 6-9B, current flows from Y to X. It flows from Y, through diode 4, through the load, through diode 1, and back to X. In both cases, the current flowed through the load in the same direction. Because all of the AC is now rectified, this is known as full-wave rectification. However, there are still brief moments when the current flow is zero; therefore, most alternators utilize three conductors and six diodes to produce overlapping current pulses to ensure that the output is never zero.

VOLTAGE REGULATION

As pointed out earlier, the maximum current output of an AC system is limited by the design of the alternator. As the induced voltage produces current flow in the stator conductors, a simultaneous countervoltage is also produced that prevents any increased current flow. The more current the alternator puts out, the greater the countervoltage. When the countervoltage reaches the point where it stops any further current increase, the alternator has reached its maximum current output.

Alternator output is directly related to the strength of the magnetic field and the speed of the rotor; an increase or decrease in either will increase or decrease the voltage output. And while rotor speed is controlled by the engine and cannot be changed to accommodate the alternator, field strength can be changed, namely by controlling the field current flow through the rotor windings. By lowering the field current, alternator output can be kept at a constant maximum—even when rotor speed increases.

TYPES OF CIRCUITS

As is the case with generators, alternators are designed with different types of field circuits. The type of circuit used depends on where the voltage regulator is connected and where the field current is being drawn from. The most common circuit types are

- Externally grounded field, or A-circuits
- Internally grounded field, or B-circuits
- Isolated field circuits

A-Circuits

An A-circuit alternator has both brushes insulated from the housing. One brush is grounded to the voltage regulator, and the other is connected to the output circuit within the alternator, where it draws current from the rotor windings (Figure 6-10). The regulator is located between the rotor windings and the ground. A-circuits are ordinarily used in conjunction with solid-state regulators.

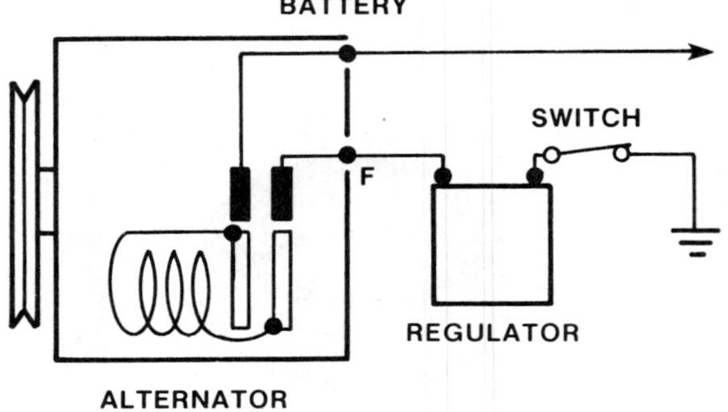

FIGURE 6-10 A typical A-circuit

B-Circuits

A B-circuit alternator has one brush insulated from the housing; it is connected through the insulated voltage regulator terminal to the alternator output circuit (Figure 6–11). The other brush is grounded within the housing. The rotor windings are located between the regulator and the ground. B-circuits are often used with electromagnetic voltage regulators.

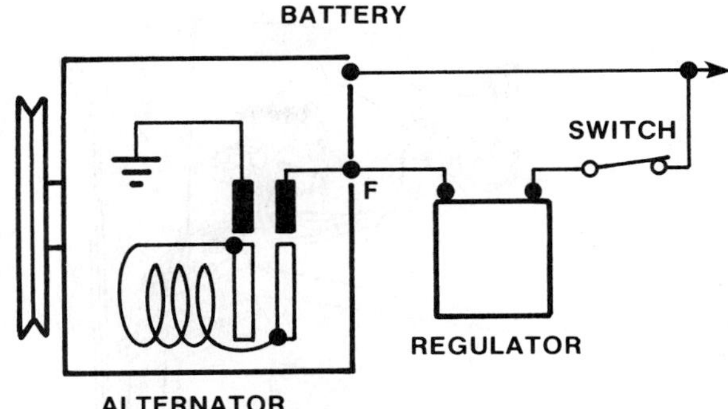

FIGURE 6–11 A typical B-circuit

Isolated Field Circuits

This is actually a variation of the A-circuit. The rotor windings are grounded to the voltage regulator, with current drawn from the alternator output circuit outside the housing flowing through a third terminal on the housing (Figure 6–12). Current flow through this third terminal is regulated by an external switch or field relay. Isolated field circuits are found mainly in Chrysler vehicles.

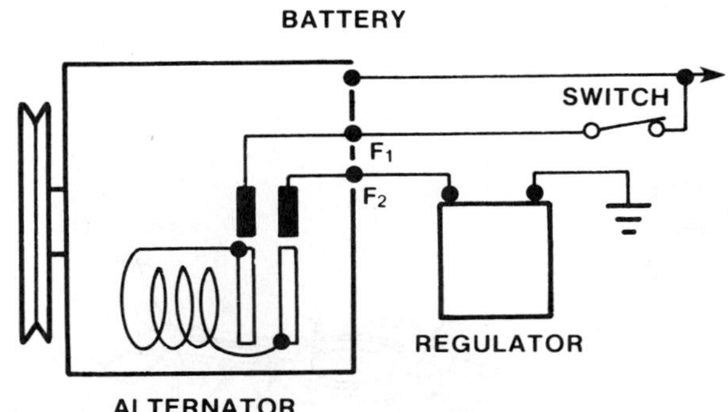

FIGURE 6–12 A typical isolated field circuit

VOLTAGE REGULATORS

A voltage regulator can be either electromagnetic, transistorized, or an integrated circuit. Electromagnetic regulators are always remote mounted, while transistorized regulators can be remote mounted or connected to the alternator housing. Integrated circuit regulators are integral with the alternator; on some late-model cars, they can be part of the electronic engine control system. Most AC systems also include a field relay along with the regulator. It can be part of the regulator or a separate unit. The field relay is responsible for energizing the field circuit when the ignition switch is turned on.

Electromagnetic Regulators

The electromagnetic coil of this regulator is connected from the ignition switch to a ground. The system voltage is received from the alternator output circuit or directly from the battery. The magnetic field of the coil opens and closes contact points that control current flow to the field.

Most electromagnetic voltage regulators are double-contact units. This is because at high rotor speeds too much current can be forced through a single-contact regulator, thus surpassing the desired output. This condition is known as voltage creep, or voltage drift. Double-contact regulators offer more consistent regulation over a wide range of alternator speeds.

Transistorized Regulators

This type of regulator replaces the mechanical contacts of the electromagnetic variety with solid-state switches. Though some early designs retained the electromagnetic field relay, today all transistorized regulators are fully solid-state. Alternator output is controlled by simply switching the field current on and off. One drawback with this type of regulator is that no adjustments are possible; if it becomes defective in any way, it must be replaced.

Integrated Circuit Regulators

This is the most compact regulator, with all of the control circuitry and components located on a single silicon chip. The chip is sealed in a plastic module and mounted either inside or on the back of the alternator. Integrated circuit regulators are also nonserviceable and must be replaced if defective.

COMPUTER REGULATION

On many late-model, four-cylinder Chrysler vehicles, a separate regulator is no longer used; instead, voltage is controlled by means of an engine control computer. This computer-controlled charging system is used with both Chrysler and Bosch alternators. Its circuit is very similar to the isolated field type, with current controlled by integrated cir-

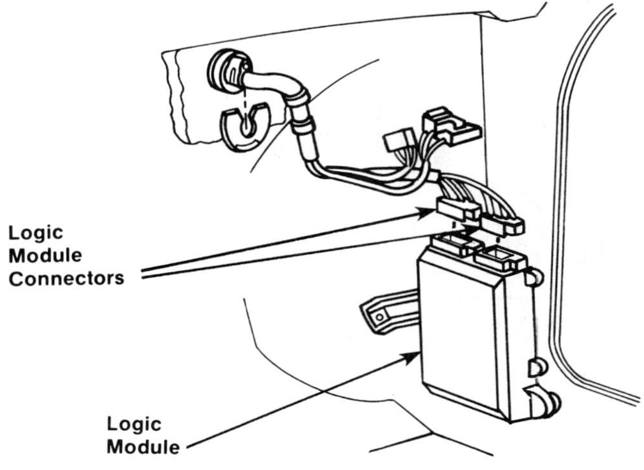

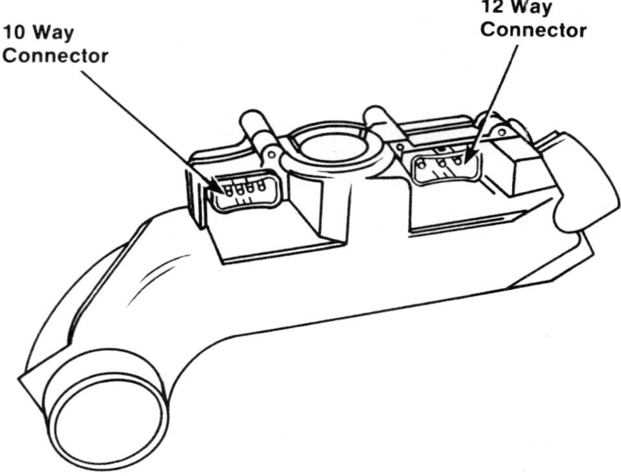

FIGURE 6-13 A computer-regulated system controls voltage by means of (A) the logic module and (B) the power module.

cuitry in the logic and power control modules (Figure 6-13). A significant feature of this system is its ability to vary the amount of voltage according to vehicle requirements and ambient temperatures, with the logic module switching the field current on and off in a duty cycle.

INDICATORS

It is very important to monitor charging system performance during the course of vehicle operation. The three pieces of equipment used are the ammeter, voltmeter, and indicator light.

Ammeter

The ammeter is placed in series with the circuit to monitor current flow into or out of the battery. When the alternator is delivering current to the battery, the ammeter will show a positive (+) indication. When not enough current (or none at all) is being supplied, the result is a negative (-) indication; this is because the current flow is in the opposite direction, from the battery to the electrical system. Once the battery has been recharged after starting the vehicle, no more current will flow into it through the ammeter. Thus, the ammeter is mainly a battery condition monitor.

Voltmeter

Voltmeters indicate the amount of voltage across an electrical circuit. A fully charged battery will indicate just over 12 volts. As energy is taken from it, the battery's voltmeter reading will drop; at the same time, the voltage regulator senses this and raises the charging voltage output to 14 volts or more. When the battery has been recharged, the alternator cuts back. The voltmeter will now show the fully charged battery voltage of just over 12 volts

FIGURE 6-14 A typical charging-starting battery analyzer *(courtesy of Bear Automotive Service Equipment Company)*

again. This on again, off again action of the alternator is constantly monitored by the voltmeter.

Indicator Light

This is the simplest and most common method of monitoring alternator performance. When the battery fails to supply sufficient current, the red light turns on. However, when the ignition switch is first activated, the light also comes on. This is due to the fact that the alternator is not providing power to the battery and other electrical circuits. Thus, the only current path is through the ignition switch, indicator light, voltage regulator, part of the alternator, and ground, then back through the battery. Once enough power is provided, the contact in the regulator changes and only the battery, regulator, and alternator are in the circuit. With no current flowing through the indicator light, it goes out.

With the engine running, the indicator light will again come on if the electrical load is more than the alternator can supply. In this case, the alternator is not turning fast enough to produce voltage as high as that of the battery; this occasionally happens when the engine is idling. If there are no problems, the light should go out as the engine speed is increased. If it does not, either the alternator or regulator is not working properly.

SHOP TALK

On late-model cars, the indicator light is often combined with the oil pressure warning light, and is usually labeled ENGINE. If this light turns on while the engine is running, either the alternator is not charging or the oil pressure is low—or both.

PRELIMINARY CHECKS

The key in solving charging system problems is getting to the root of the trouble the first time. Once a customer drives away with the assurance that his or her problem is solved, another case of a dead battery will be very costly—both in terms of a free service call and a damaged reputation. Add to this the many possible hours of labor trying to figure out why the initial repair failed, and the importance of a correct initial diagnosis becomes all too clear.

When it comes to choosing equipment, the best tool to use for diagnosing the full spectrum of charging system problems is a charging-starting battery (CSB) analyzer (Figure 6–14). While a voltmeter will help in making intelligent decisions, the analyzer will pinpoint the exact source of the trouble. Choose an analyzer that combines the functions of the carbon pile and separate meters to test the battery, alternator, regulator, and wiring.

SAFETY PRECAUTIONS

Before beginning any work on the charging system, keep the following safety precautions in mind:

- Disconnect the battery ground cable before removing *any* leads from the system. Do not reconnect the battery ground cable until *all* wiring connections have been made.
- Avoid contact with the alternator output terminal. This terminal is hot at all times when the battery cables are connected.
- Always make sure the ignition switch is off, except during actual test procedures.
- The alternator is not made to withstand a lot of force; only the front housing is relatively strong. When adjusting belt tension, apply pressure to the front housing only to avoid damaging the stator and rectifier.
- Never operate the alternator without first connecting an external load to it. Failure to do so can cause extremely high voltage and possible alternator burnout.
- When installing a battery, be careful to observe the correct polarity (Figure 6–15). Reversing the cables will destroy the diodes.

FIGURE 6–15 Proper polarity must be observed when installing a battery.

TABLE 6-1: CHARGING SYSTEM TROUBLESHOOTING

Symptom	Possible Cause	Remedy
• Meter needle flutters • Indicator light flickers	1. Loose connections in system wiring 2. Loose or worn brushes 3. Oxidized regular points	1. Repair system wiring. 2. Repair or replace alternator. 3. Replace regulator.
• Ammeter needle reads discharge • Voltmeter needle shows low system voltage • Indicator light stays on • Battery discharged	1. Faulty alternator drive belt 2. Corroded battery cables 3. Loose system wiring 4. Defective field relay 5. Defective battery 6. Wrong battery 7. Alternator output low	1. Check and adjust belt. 2. Replace battery cables. 3. Repair system wiring. 4. Replace field relay. 5. Replace battery. 6. Replace battery. 7. Repair or replace alternator.
• Ammeter needle reads charge • Voltmeter needle shows high system voltage • Battery is overcharged	1. Loose system wiring 2. Poor regulator ground 3. Burned regulator points 4. Incorrect regulator setting 5. Defective regulator	1. Repair system wiring. 2. Tighten regulator ground. 3. Replace regulator. 4. Adjust regulator. 5. Replace regulator.
• Indicator light stays on when ignition switch is off	1. Shorted positive diode	1. Repair or replace alternator.
• Alternator makes squealing noise	1. Loose or damaged drive belt 2. Worn or defective rotor shaft bearing 3. Defective stator 4. Loose or misaligned pulley	1. Adjust or replace drive belt. 2. Repair or replace alternator. 3. Repair or replace alternator. 4. Adjust pulley.
• Alternator makes whining noise	1. Shorted diode	1. Repair or replace alternator.

Proper polarity must also be observed when connecting a booster battery: positive to positive and negative to ground.

- Never attempt to polarize an alternator in the same manner used on a DC generator.
- Always disconnect the battery ground cable before charging the battery. This will greatly lessen the chance of flying sparks that could cause the battery to explode.
- Keep the carbon pile off at all times, except during actual test procedures.
- Make sure all hair, clothing, and jewelry are kept away from moving parts.

INSPECTION

In addition to observing the ammeter, voltmeter, or indicator light, there are some common warning signs of charging system trouble. For example, a low state of battery charge often signals a charging problem, as does a noisy alternator. Other symptoms, causes, and cures are listed in Table 6–1.

Many charging system complaints stem from easily repairable problems that reveal themselves during a visual inspection of the system. Remember to always look for the simple solution before performing more involved diagnostic procedures. Use the following inspection procedure when a problem is suspected.

1. Before doing any belt adjustments, check for proper pulley alignment. This is especially critical with serpentine belts.
2. Inspect the alternator drive belt; loose drive belts are a major source of charging problems. If a belt does not have the proper tension, it might produce a loud squealing sound. Use a belt tension gauge to check the tension as shown in Figure 6–16, but keep in mind that different belts (V, cogged V, and multi-ribbed V) require different tensions. Always consult the manufacturer's specifications. In addition, check the belt for signs of fraying, cracking, or glazing, and replace if needed (Figure 6–17).

SHOP TALK

Some late-model Chrysler engines require a torque reading to be taken when tension is applied to the alternator drive belt. This is especially important on the longer, multi-ribbed V belts.

FIGURE 6–16 Using a belt tension gauge

FIGURE 6–17 Replacing the alternator drive belt on a Chevrolet X-Car.

3. Inspect the battery (see Chapter 3). Check the electrolyte level and specific gravity. It might be necessary to charge the battery in order to restore it to a fully charged state. If the battery cannot be charged, it must be replaced. Also, make sure the posts and cable clamps are clean and tight, since a bad connection can cause reduced current flow.
4. Inspect all system wiring and connections; an inductive probe is a useful tool in this regard (Figure 6–18). Many automotive electrical systems contain fusible links to protect against overloads; these are wires that melt at a lower temperature than the current-carrying wires they protect. Fusible links can blow like a fuse without being noticed. Also, look for a short circuit, an open ground, or high resistance in any of the circuits that could cause a problem that would appear to be in the charging system.
5. Inspect the alternator and regulator mountings for loose and/or missing bolts. Replace or tighten as needed.

If all preliminary visual checks come out positive, take a few moments to listen for noisy belts, bad bearings, or the whirring sound of a bad diode. If these checks also fail to produce anything, it is time to test the charging system.

USING CSB ANALYZERS

A charging-starting battery, or CSB, analyzer greatly simplifies charging system diagnosis. Due to the wide variety of these analyzers on the market, it is impossible to give a single set of instructions that will apply to all of them. The following general procedures illustrate how a CSB analyzer is used to test a charging system:

1. Connect the analyzer to the vehicle according to the directions in the analyzer operator's manual. Typical hookups, as shown in Figure 6–19, are as follows:

   Green lead—rpm measurement and disabling the ignition
   Secondary clip—rpm measurement (instead of green lead)
   Yellow lead—voltage measurement
   Black lead—chassis ground
   Current probe—charging amps reading and diode pattern display
2. Set the test selector switch to the charging test position.

FIGURE 6–18 Using an inductive probe

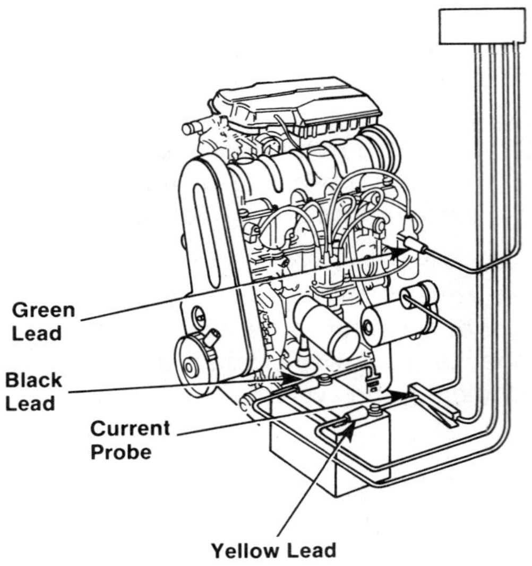

FIGURE 6–19 Typical hookups for testing a charging system with an analyzer

3. Turn on the ignition switch and note the ammeter reading. This is the current required to power the ignition system and accessories—it is *not* included in the maximum output reading.
4. Start the engine. Adjust it to approximately 2000 rpm, or whatever test speed is called for by the manufacturer.
5. Adjust the carbon pile to achieve the maximum ammeter reading, while keeping the system voltage above 12 volts.
6. If the ammeter reading is within 10 percent of the manufacturer's specifications, proceed to the next step. If it is not, skip to step 11.
7. Set the test selector switch to the regulator test position.
8. Again adjust the engine speed to 2000 rpm, or the manufacturer's specified test speed.
9. When the voltmeter stops rising, note the reading. The time it takes to stabilize will vary, depending on the battery's state of charge.
10. If the voltmeter reading is within specifications, the charging system is in good condition. If the reading is not within specifications, replace the regulator and repeat steps 7 to 10.
11. Turn off the engine. Disconnect the field wire from the alternator.
12. Connect the field lead to the alternator field terminal.
13. Start the engine. Adjust the speed to 2000 rpm, or the manufacturer's specified test speed.
14. Adjust the carbon pile to obtain a voltmeter reading approximately 2 volts less than the system voltage.
15. Switch the field selector switch on the analyzer to the A or B position, as specified by the manufacturer.
16. Adjust the carbon pile to obtain a voltmeter reading within the specified range.
17. Note the ammeter reading, return the carbon pile to the open position, then turn off the engine.
18. Add the ammeter reading made in step 3 to the reading from the previous step. If it is within 10 percent of specifications, proceed to the next step. If it is not, remove the alternator for more testing.
19. Check the wiring between the alternator and the regulator. If repairs are necessary, make them; then repeat steps 13 to 18. If the wires are in good condition, replace the regulator.

GENERAL TEST PROCEDURE

On-car AC system tests are all basically the same, whether the vehicle being tested is a Colt or a Cadillac. The major differences from model to model are the meter test points and the specifications. Keep in mind that not all vehicle manufacturers require all of these tests to be performed, while others suggest even more. During *all* of these tests, it is very important to refer to the vehicle manufacturer's specifications; even the most accurate test results are no good if they are not matched against the correct specs. Before beginning, obtain a copy of the vehicle service manual and keep it handy for reference. Also, while these tests are done easiest and most accurately using a CSB analyzer, separate voltmeters, ammeters, and so on can be used if access to an analyzer is not possible.

CAUTION: All of the meter readings that follow apply to 12-volt AC systems. While it is possible to test 6-, 24-, and 32-volt systems, do not attempt to do so without first checking the specifications provided by the manufacturer of the vehicle or the charging system.

Circuit Resistance Test

This test measures the voltage drop in the charging system. They are usually done with the engine running at 1500 to 1800 rpm and the alternator producing a specified amperage. The insulated circuit test measurement is taken from the positive battery terminal to the battery, or output, terminal on the alternator. If the drop exceeds specifications, the connections within the circuit must be tested to locate the area of resistance.

A voltage drop test of the ground circuit is made from the alternator frame to the battery ground terminal; a voltage drop test of the field circuit is made at the regulator. In order to get the most accurate test results, any loose or corroded connections or damaged wiring must be repaired or replaced before beginning.

Current Output Test

This test is conducted in two stages, though some car manufacturers recommend doing only one or the other. The first stage consists of connecting a carbon pile across the battery, connecting the voltmeter between the positive battery terminal and a ground, and connecting the ammeter between the positive battery terminal and the battery terminal at the alternator. While running the engine at a specified speed, the carbon pile is adjusted either to maintain a steady 15 volts or to achieve the greatest possible voltage reading. Compare the ammeter reading to the specifications.

The second stage involves bypassing the regulator; it is performed if the ammeter reading did not meet specifications. If the regulator is remote mounted, a jumper wire must be used. For integral regulators, different methods of bypassing can be used. In either case, bypassing the regulator connects the field winding to full battery voltage. Run the engine at test speed, adjust the carbon pile, and compare the ammeter reading to specifications. If it is within range, the regulator might be faulty. If it still does not meet the specs, the alternator must be removed from the vehicle for further testing.

Voltage Output Test

This is similar to, and often a recommended substitute for, the current output test. (Some car manufacturers feel that it is far less likely to damage the electronic components.) It is performed with the engine running at fast idle, with either the headlights or a carbon pile used to load the battery. If voltage measured at the positive battery terminal is less than 13 volts, bypass the regulator and repeat the test. If the voltage increases to around 16 volts, the regulator might be faulty. If, on the other hand, the voltage stays low, the alternator should be removed for further testing. The system voltage must not be allowed to exceed 16 volts at any time during this test.

Field Current Draw Test

Most car manufacturers recommend bypassing the ignition switch or warning lamp circuit, or both, when performing this test. Regardless, it must always be done with the engine off. Connect the ammeter either between the positive battery terminal and the field, or between the field and a ground. By turning on the ignition switch without starting the engine, or bypassing the ignition switch and warning lamp with a jumper wire, the ammeter will measure the field current draw. If the reading is not within specifications, the alternator must be removed for more testing. If the reading is within specifications, try turning the alternator pulley by hand; if the ammeter reading fluctuates, the slip rings and brushes might need to be serviced.

Voltage Regulator Test

This test can be made by connecting the voltmeter either between the alternator output terminal and a ground, or between the positive battery terminal and a ground. Run the engine at fast idle, with a load on the battery, for 10 to 15 minutes; this will get the regulator up to operating temperature. At this point, the voltmeter reading is the setting of the normally closed, or series, contacts of an electromechanical regulator.

By reducing the load and increasing rpm, the voltmeter reading becomes the setting of the normally open, or grounding, contacts of an electromechanical regulator; this is also the setting of a solid-state regulator. If the readings do not meet the specifications, the regulator must be either adjusted or replaced.

REVIEW QUESTIONS

1. Which of the following does not characterize AC systems?
   a. efficient operation at high speeds
   b. increased current travel through the brushes
   c. rotational speeds up to 15,000 rpm
   d. no need for external current regulation

2. What part of the alternator produces the rotating magnetic field?
   a. stator
   b. rotor
   c. brushes
   d. poles

3. A typical rotor contains ______________.
   a. 14 north poles and 14 south poles
   b. 14 north poles or 14 south poles
   c. 7 north poles and 7 south poles
   d. a pair of poles, one north and one south

4. Alternating current is produced by ______________.
   a. alternators
   b. generators
   c. both a and b
   d. neither a nor b

5. Which type of circuit is ordinarily used with solid-state regulators?
   a. isolated field
   b. A-circuit
   c. B-circuit
   d. none of the above

6. What is responsible for energizing the field circuit when the ignition switch is turned on?
   a. field relay
   b. coil
   c. logic control module
   d. windings

7. An alternator is making a whining noise. Technician A says a shorted diode could be the cause. Technician B says a defective stator could be the cause. Who is right?
   a. Technician A
   b. Technician B
   c. Both A and B
   d. Neither A nor B

8. What is used to measure the amount of voltage across an electrical circuit?
   a. ammeter
   b. heat sink
   c. diode
   d. voltmeter

9. Which of the following safety precautions is incorrect?
   a. Disconnect the battery ground cable after removing any leads from the charging system.
   b. Make sure the ignition switch is off, except during test procedures.
   c. Avoid contact with the alternator output terminal.
   d. Disconnect the battery ground cable before charging the battery.

10. The indicator light remains on when the ignition switch is off. Technician A says loose or worn brushes could be the cause. Technician B says a shorted positive diode could be the cause. Who is right?
    a. Technician A
    b. Technician B
    c. Both A and B
    d. Neither A nor B

11. A circuit resistance test is usually performed with the engine ______________.
    a. running at idle speed
    b. running at 1500 to 1800 rpm
    c. running at a minimum 2000 rpm
    d. shut off

12. What is often used as a substitute for the current output test?
    a. circuit resistance test
    b. field current draw test
    c. voltage regulator test
    d. voltage output test

13. Which of the following is not a possible cause of a squealing alternator?
    a. loose or misaligned pulley
    b. loose or damaged drive belt
    c. defective stator
    d. poor regulator ground

14. On a typical oscilloscope, the yellow lead is used for ______________.
    a. voltage measurement
    b. chassis ground
    c. yam measurement
    d. none of the above

15. Most electromagnetic voltage regulators are ______________.
    a. double-contact units
    b. single-contact units
    c. transistorized units
    d. both b and c

CHAPTER SEVEN

TROUBLESHOOTING BREAKER POINT IGNITION SYSTEMS

Objectives

Upon completion of this chapter, you should be able to:

- Name the components of the primary and secondary ignition system circuits, and describe the function of each.
- Explain dwell and its significance to breaker point ignitions.
- Explain the importance of reach, heat range, and air gap to spark plugs.
- Describe the difference between centrifugal and vacuum spark advance.
- Perform the full range of primary and secondary circuit tests using an oscilloscope, voltmeter, and ohmmeter.
- Perform breaker point maintenance, including aligning and setting points.
- Remove, inspect, regap, and install spark plugs.
- Explain the procedural differences involved in setting static and dynamic timing.

The purpose of any ignition system is to supply properly timed, high-voltage surges to the spark plugs. This, in turn, ignites the air/fuel mixture within each of the combustion chambers. There are two main types of ignition systems: breaker point and electronic. The breaker point, or conventional, ignition system has been used on automobiles for more than 60 years. The electronic, or solid-state, type is relatively new, not having appeared on domestic passenger cars until the early 70s. This system has now become the norm, due mainly to stricter emission control standards; it is discussed in detail in the next chapter.

An ignition system is comprised of two interconnected circuits: the primary (low-voltage) circuit, and the secondary (high-voltage) circuit (Figure 7-1). When the ignition switch is turned on, battery current flows through the switch and the primary resistor, to the coil primary winding, through the breaker points or other switching device, to the grounded terminal of the battery. The low-voltage current flow in the coil primary winding creates a magnetic field. When this flow is interrupted by the switching device, the magnetic field collapses and a high-voltage surge is induced in the coil secondary winding. As seen in Figure 7-2, the current then flows from the coil to the distributor via an ignition cable, through the distributor cap, rotor, rotor air gap, and another ignition cable, to the proper spark plug. From there, it arcs to a ground.

This chapter deals specifically with the operation and troubleshooting of breaker point ignition systems. However, it is important to note that any ignition system theory contained in this chapter applies equally to electronic ignition systems. Also, while the number of vehicles with breaker point ignitions has dwindled greatly in recent years, much of the oscilloscope diagnostic procedures discussed in this chapter is applicable with the ignition systems of today.

PRIMARY CIRCUIT

The primary circuit consists of the following:

- Battery
- Ignition switch
- Primary resistor
- Starting bypass

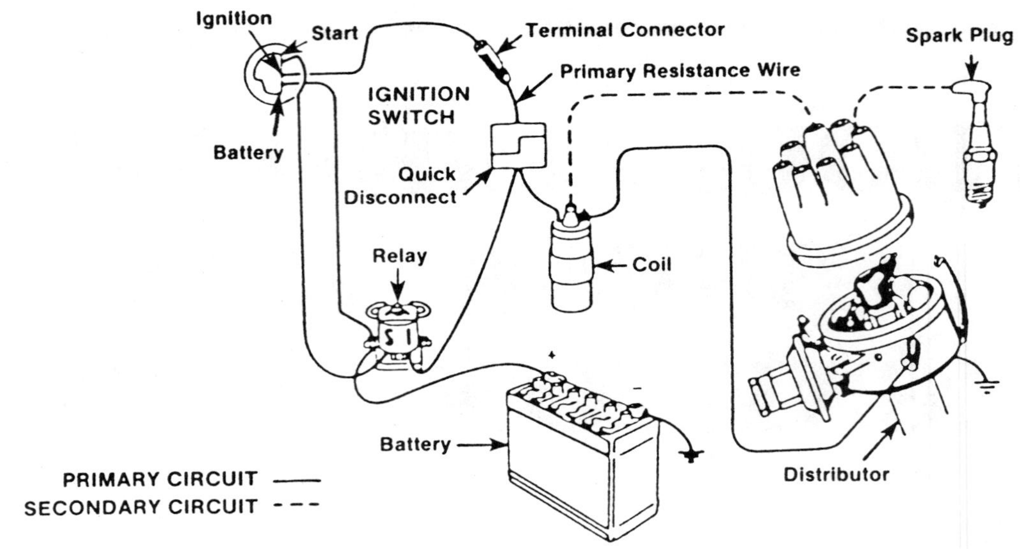

FIGURE 7-1 A typical breaker point ignition system

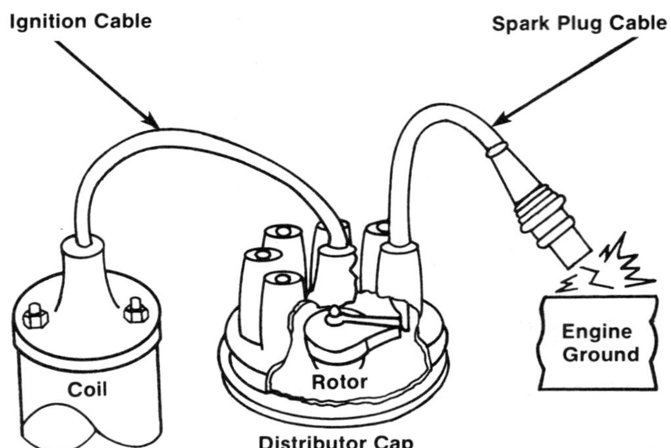

FIGURE 7-2 Tracing high-voltage current flow

- Coil primary winding
- Switching device (in the distributor)

As the supplier of low-voltage current to the primary circuit, the battery is essentially the heart of the entire ignition system (see Chapter 4). Descriptions of the other components follow.

IGNITION SWITCH

This is a simple on/off switch that controls the flow of low-voltage current from the battery to the coil primary winding. Placing the ignition switch in either the START or RUN position will start the current flow; any other positions will route the current to the accessory circuits and lock the steering wheel. Full battery voltage is always present at the ignition switch.

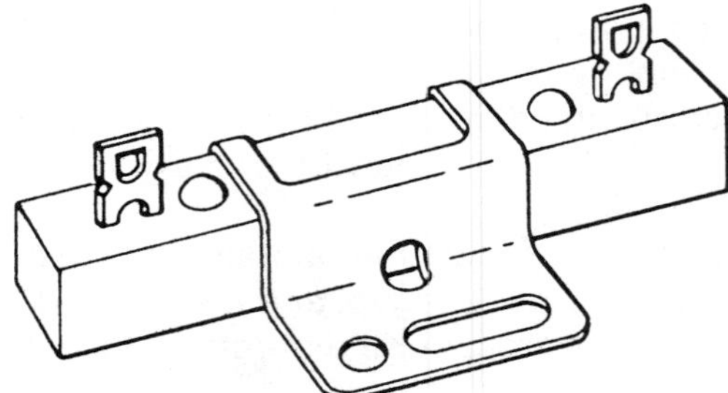

FIGURE 7-3 Ballast primary resistors are contained in a ceramic holder.

PRIMARY RESISTOR

The primary resistor can be either a resistance wire extending from the ignition switch to the coil primary winding, part of the coil itself, or a ballast resistor. The latter is simply a small coil of high-resistance wire in a ceramic holder and mounted on the firewall or an inner fenderwell (Figure 7-3). In any case, the primary resistor is located in series between the battery and the primary coil winding, and is responsible for maintaining the primary circuit voltage at the desired level (about 9 or 10 volts), thus protecting the contact points.

STARTING BYPASS

The primary resistor must be bypassed during cranking because the starter motor already drops the voltage to the desired level; any additional reduction caused by the resistor would affect the ignition system's ability to produce an arc at the spark plugs.

The bypassing is accomplished by means of a parallel circuit; it is controlled by the ignition switch (Figure 7-4) or a starter relay or solenoid (Figure 7-5).

CAUTION: Do not bypass the resistor at any time other than when the vehicle is being started. Doing so would burn the contact points and could also damage the coil primary winding.

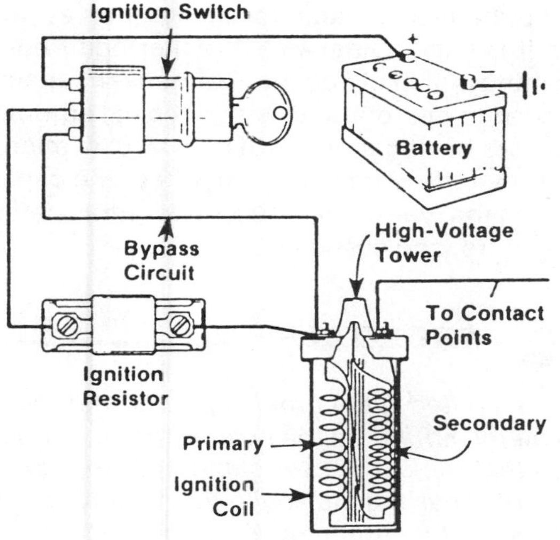

FIGURE 7-4 A bypass circuit controlled by the ignition switch

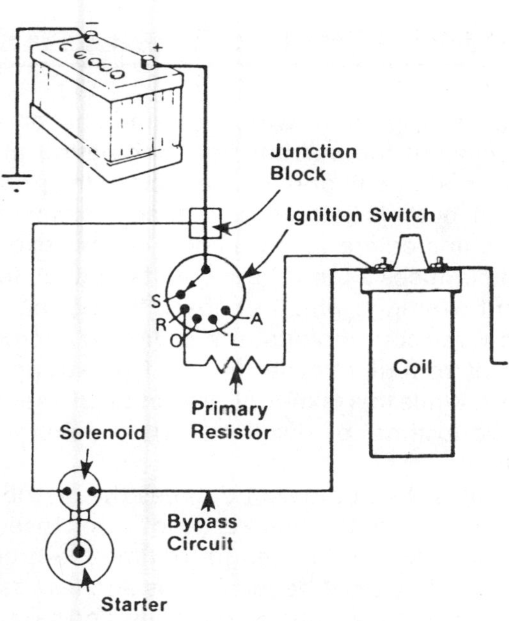

FIGURE 7-5 A bypass circuit controlled by the starter solenoid

COIL PRIMARY WINDING

The coil primary winding consists of 100 to 150 turns of heavy copper wire, insulated from each other by a thin coat of enamel. Each end of the winding is connected to a primary terminal on the top of the coil (Figure 7-6). In a negative-ground electrical system, the positive coil terminal is connected through the primary circuit wiring to the positive battery terminal; the negative coil terminal is attached to the ignition breaker points, and through the points to a ground.

SWITCHING DEVICE

In order to induce the necessary voltage in the secondary winding, the magnetic field of the coil primary winding must collapse totally. This can only happen if the current flow through the primary winding stops abruptly; repeated stopping and starting of the current is needed to induce subsequent high-voltage discharges. Thus, the switching device is responsible for continually breaking and completing the circuit.

Breaker Points

For many years, mechanical breaker, or contact points, have been used as the primary circuit switching device. The breaker point assembly, which is mounted on the breaker plate inside the distributor, consists of a fixed contact, movable contact, mov-

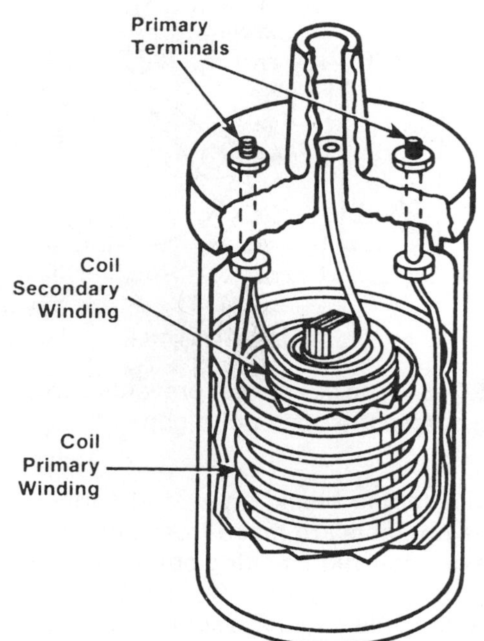

FIGURE 7-6 A cutaway view of a typical ignition coil

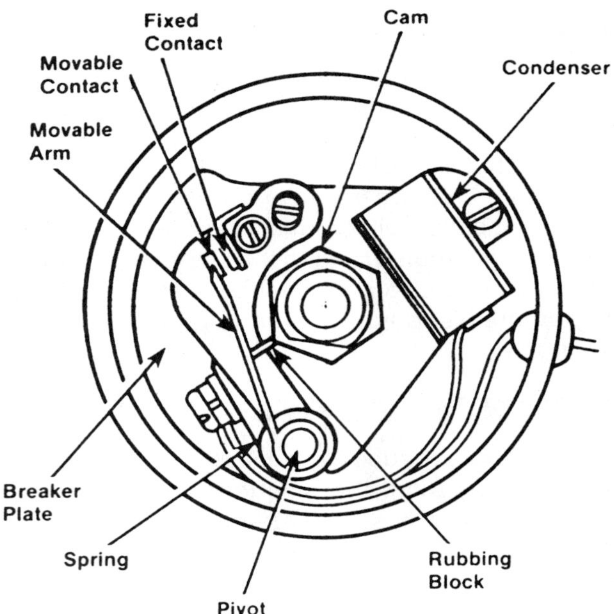

FIGURE 7-7 Mechanical breaker points are the switching device for the primary circuit.

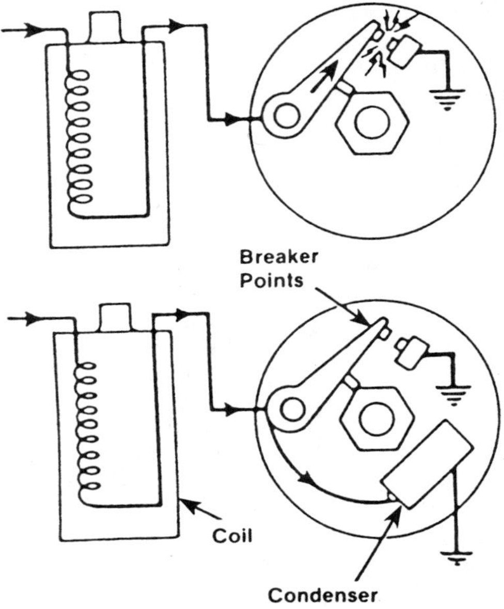

FIGURE 7-8 The condenser prevents current from jumping across the open point gap.

able arm, rubbing block, pivot, and spring (Figure 7-7). The contacts are made from tungsten, which has an extremely high melting point. The fixed contact is grounded through the distributor housing, while the movable contact is connected to the negative terminal of the coil primary winding. The movable contact is known as the positive point, and the fixed contact is the negative point.

Condenser

When the breaker points are open, voltage is still present at the movable arm. This can cause current to jump across the open point gap and damage the points. To prevent this from occuring, a condenser (Figure 7-8) is attached to the movable arm. Made up of two conducting surfaces separated by an insulator, the condenser stores voltage without using any of it. In this way, the voltage at the movable arm is retained instead of being transmitted across the gap.

The condenser capacity is very important; it must be the proper value for the ignition system in which it is used. Otherwise, breaker point burning and pitting will likely occur. Condenser capacity is measured in microfarads, with a typical automotive condenser having a rating of 0.18 to 0.32 microfarads. Many different factors can affect the capacity, including the type of coil used, distributor shaft rpm, temperature, and even altitude.

SHOP TALK

If a pitted set of breaker points has a built-up metal mound on the positive point, this is a sign that the condenser capacity is too low. If the metal has transferred to the negative point, the capacity is too high. Keep in mind, however, that pitted points can also be caused by excessive resistance in either the primary or secondary circuit.

PRIMARY CIRCUIT OPERATION

With the ignition switch on and the breaker points closed, the current flow originating at the battery creates a magnetic field around the primary windings, but this field does not reach its maximum intensity immediately. This is because the expanding field induces a temporary countervoltage in the adjacent windings which opposes the normal current flow; the countervoltage must first be overcome by circuit voltage in order for the field strength to increase. While this takes only a fraction of a second, the buildup time, or dwell period, is nonetheless important.

Dwell is the number of degrees the distributor cam rotates while the breaker points are closed, as seen in Figure 7-9. The length of time the breaker points remain closed becomes less and less as engine rpm increases. Thus, current flow and its resulting magnetic field buildup have a more difficult time reaching full potential at high engine speeds.

When the breaker cam reopens the points, the second phase of the primary circuit operation begins. The current flowing through the windings begins to decrease rapidly, which causes the magnetic field to shrink back toward the center of the coil. The collapsing field induces high voltage in the primary winding to try to maintain the current flow across the separating breaker points.

It is at this point that the condenser receives the charge of induced current from the primary windings. By the time the condenser has been charged, the breaker points have opened far enough so that very little current can arc across the gap. As shown in Figure 7-10, the condenser discharges its stored energy in the form of current, which immediately "bounces" and flows back through the primary coil until it reaches the positive terminal, which contains the same charging potential as the opposite side of the condenser. The current then comes back again, and this oscillating action continues until all of the electrical energy has been dissipated as heat. The complete cycle is over in a short time, but not before a dozen or more current reversals take place.

Because the condenser promotes fast and complete breaking of the primary circuit, the magnetic field collapses approximately 20 times faster than if there were no condenser. However, despite the condenser action, there is inevitably still some arcing at the breaker points. For this reason, vehicles driven consistently at low speed experience more rapid point failure than those driven faster.

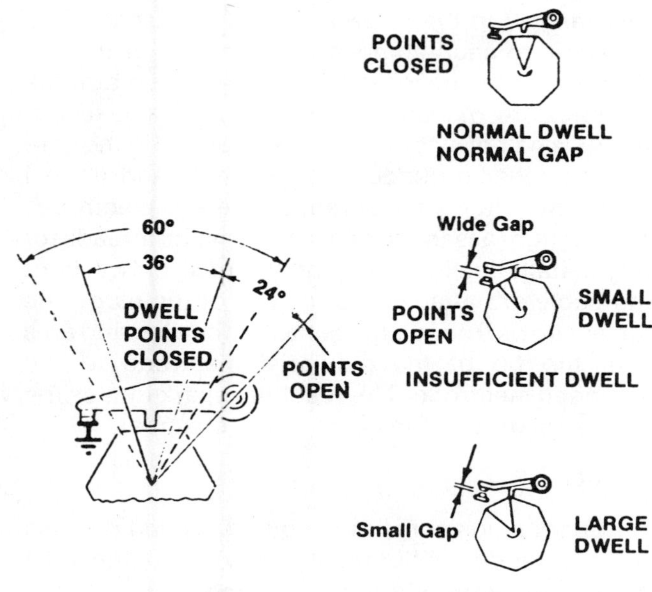

FIGURE 7-9 Normal vs abnormal dwell comparison

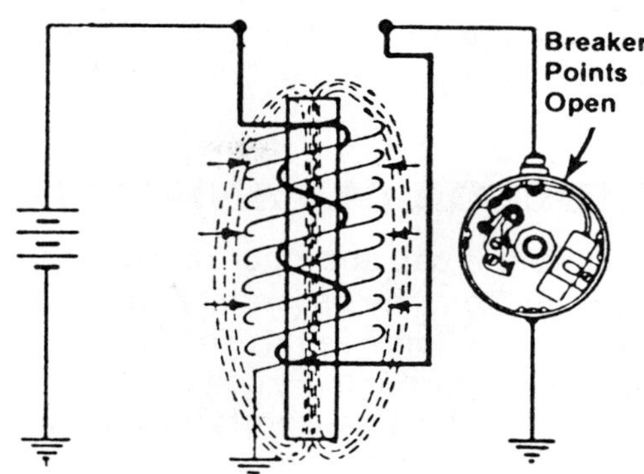

FIGURE 7-10 As the breaker points open, the condenser discharges its stored energy in the form of current.

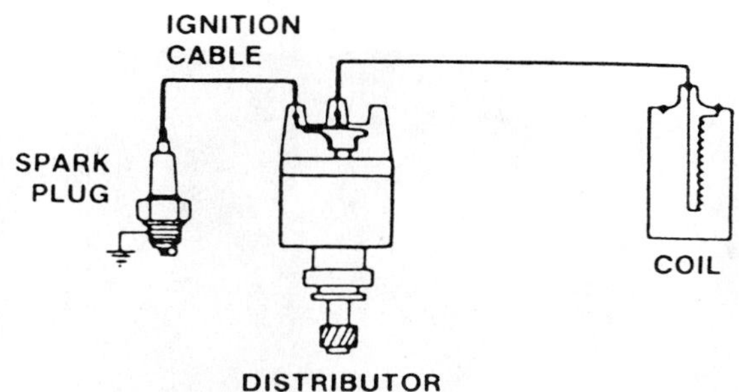

FIGURE 7-11 A typical secondary circuit

SECONDARY CIRCUIT

The secondary circuit (Figure 7-11) consists of the following:

- Coil secondary winding
- Distributor cap and rotor
- Ignition cables
- Spark plugs

The job of the secondary circuit is twofold: it increases the battery voltage enough to produce an arc at the spark plugs, and it delivers this high-voltage surge to the plugs. Descriptions of the secondary circuit components follow.

COIL SECONDARY WINDING

The secondary winding is formed from 15,000 to 30,000 turns of very fine insulated copper wire wrapped around a soft-iron core; a laminated shell surrounds the windings and core. One end of the secondary winding is connected to the primary winding, and the other is connected to the high-voltage terminal on the top of the coil. The windings, core, and shell are encased in a metal container, which is filled with oil or insulating material and sealed with a cap.

DISTRIBUTOR CAP AND ROTOR

The distributor cap and rotor receive the high-voltage current from the secondary winding via a high-tension wire (Figure 7-12). The current enters the distributor cap through the coil tower, or center terminal. The rotor then transports the current from the coil tower to the spark plug electrodes in the rim of the cap. The rotor mounts on the upper portion of the distributor shaft and rotates with it.

The distributor cap (Figure 7-13) is made from silicone plastic or similar material that offers protection from chemical attack. It is attached to the distributor housing with screws or spring-loaded clips. The coil tower contains a carbon insert that carries the current from the high-tension coil lead to the raised portion of the electrode on the rotor. Spaced evenly around the coil tower are the spark plug towers, one for each plug.

FIGURE 7-12 The distributor cap and rotor receive the high-tension current from the secondary winding.

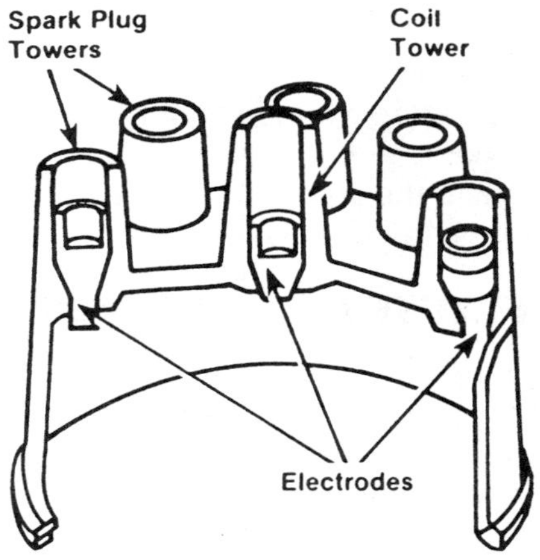

FIGURE 7-13 A cutaway view of a distributor cap

An air gap of a few thousandths of an inch exists between the tip of the rotor electrode and the spark plug electrode in the cap. This gap is necessary in order to prevent the two electrodes from making contact; if they did, both would wear out rapidly. It cannot be measured when the distributor is assembled, so the gap is usually described in terms of the voltage needed to create an arc between the electrodes.

IGNITION CABLES

Also known as spark plug cables, the ignition cables conduct the high-voltage current. Unlike the solid metal cores used in the past, today's ignition cables contain fiber cores that act as a resistor in the secondary circuit; they cut down on radio and television interference and reduce spark plug wear. Insulated boots on the ends of the cables strengthen the connections to the coil, distributor cap, and plugs, as well as prevent dust and water infiltration and voltage loss.

SPARK PLUGS

The spark plugs provide the crucial air gap through which the high-voltage current from the coil flows across in the form of an arc. One spark plug threads into each combustion chamber of the engine. The three main parts of a spark plug are: the ceramic core, or insulator, which acts as a heat conductor; a pair of electrodes, one insulated in the core and the other grounded on the shell; and a steel shell, which holds the ceramic core and electrodes in a gas-tight assembly and also contains threads for plug installation in the engine (Figure 7-14). A boot and cable are attached to the top of the plug. The current flows through the center of the plug and arcs from the tip of the insulated electrode to the grounded electrode. The resulting spark ignites the air/fuel mixture in the combustion chamber.

Reach

Spark plugs come in various sizes and designs to accommodate different engines. One of the most important design differences among spark plugs is the reach, illustrated in Figure 7-15; this refers to the length of the shell from the contact surface at the seat to the bottom of the shell, including both

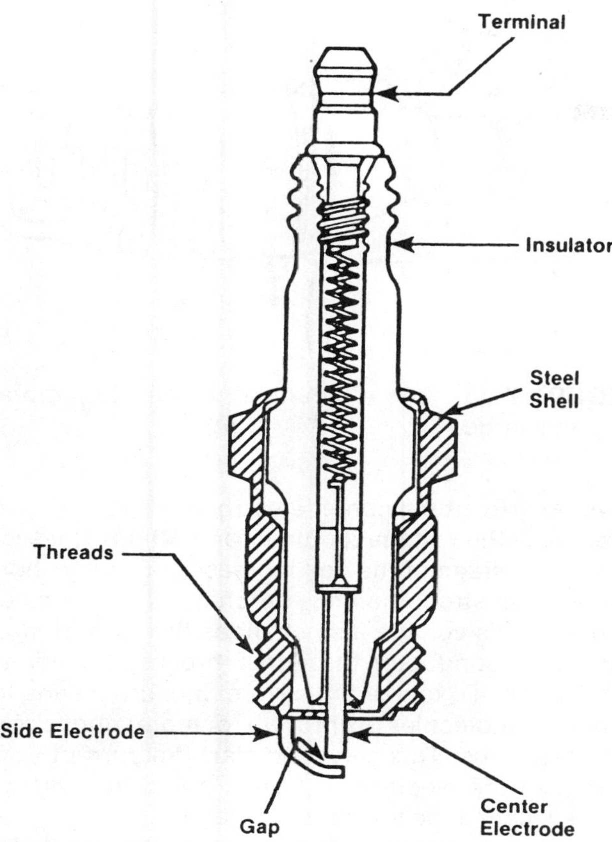

FIGURE 7–14 The parts of a typical spark plug

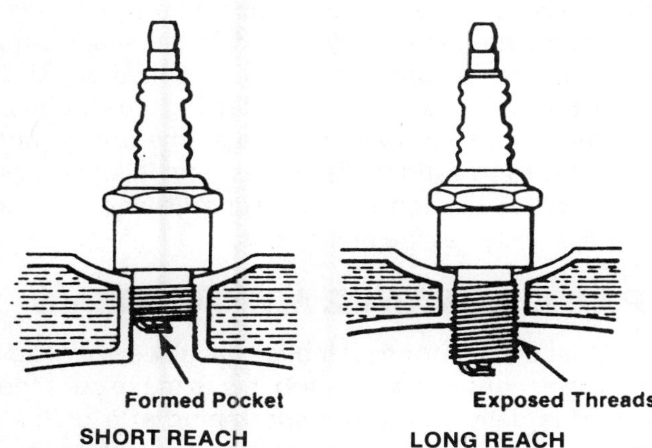

FIGURE 7–15 Spark plug reach: long vs short

threaded and nonthreaded sections. Reach is crucial so that the plug's air gap is in the correct position to produce a spark. Installing plugs with too short a reach means that the electrodes will be in a pocket and the arc will not be able to ignite the air/fuel mixture; in addition, the exposed threads in the cylinder head will accumulate carbon deposits. If the reach is too long, the exposed plug threads could

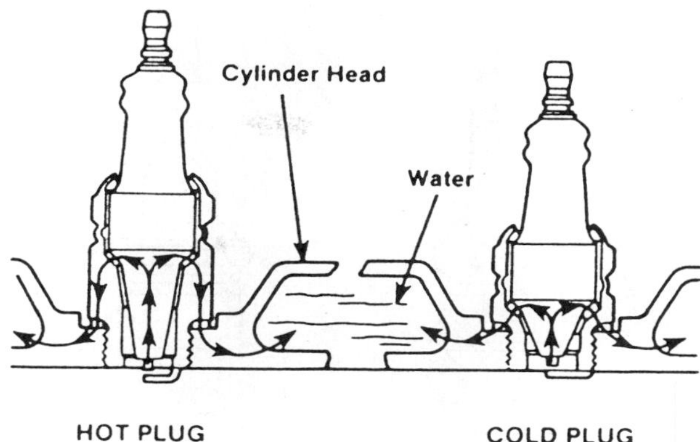

FIGURE 7–16 Spark plug heat range: hot vs cold

get so hot that they ignite the air/fuel mixture at the wrong time, causing preignition.

Heat Range

The heat range determines a spark plug's ability to transfer heat from the firing end to the engine's cooling system. This is very important because the firing end must run hot enough to burn away fouling deposits while the engine is idling, yet cool enough at higher speeds to prevent preignition. Manufacturers specify plugs as being cold or hot, as illustrated in Figure 7–16.

- *Cold Spark Plug.* This plug has a short core tip, which gives the heat a shorter path to travel. The heat can dissipate rapidly to maintain a lower firing tip temperature.
- *Hot Spark Plug.* This plug has a long core tip that permits a slower heat transfer, and thus a higher firing tip temperature.

SHOP TALK

When spark plugs must be replaced, note the plug type specifications as recommended by the engine and plug manufacturers. These are recommendations only; for example, continuous heavy driving demands for a much cooler plug than if the vehicle does a lot of stop-and-go driving.

Threads and Seats

Spark plugs are available in either 14- or 18-millimeter diameters. All 18-millimeter plugs feature tapered seats that match similar seats in the cylinder head and therefore need no gaskets. The 14-millimeter variety can have either a flat seat that requires a gasket, or a tapered seat that does not (Figure

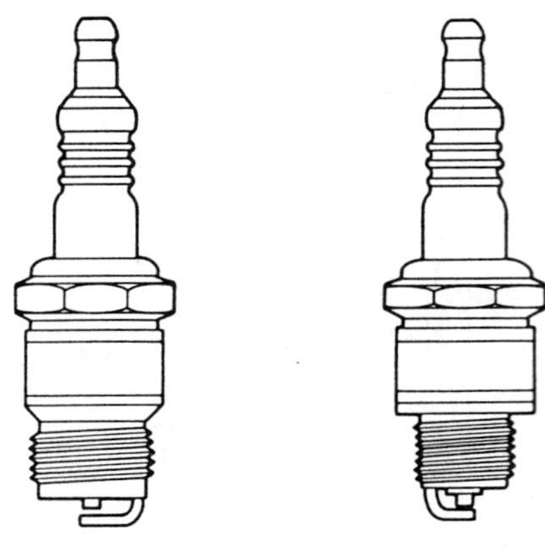

FIGURE 7-17 Spark plug seats: tapered vs flat

7-17); the latter is found mainly in late-model applications. All spark plugs have a hex-shaped shell that accommodates a socket wrench for easy installation and removal. The 14-millimeter, tapered seat plugs have shells with a 5/8-inch hex; 14-millimeter gasketed and 18-millimeter tapered seat plugs have shells with a 13/16-inch hex.

Spark Plug Air Gap

The correct spark plug air gap is essential to achieve optimum engine performance and long plug life. A gap that is too wide (Figure 7-18) requires higher voltage to jump the gap; if the required voltage is greater than what is available, the result will be misfiring. On the other hand, a gap that is too narrow increases the current flow and leads to rough idle and prematurely burned electrodes. Always set the gap in accordance with the vehicle manufacturer's specifications. Electronic ignition systems utilize wider air gaps than breaker point systems due to the leaner air/ fuel mixtures of late-model engines.

SPARK PLUG OPERATION

The discharge across the spark plug air gap can be one of two types: capacitive or inductive. When the coil sends the initial surge to the plug's center electrode, the air/fuel mixture within the gap is not yet ready to conduct the arc; the mixture is literally acting as an insulator. At the same time, the spark plug acts as a capacitor or condenser, with the center electrode storing the negative charge and the grounded electrode storing the less negative charge. The air gap works in a similar fashion to the insulating material found between the two conducting strips of a condenser.

Though it is opposite from what is expected in a typical negative-ground electrical system, a nega-

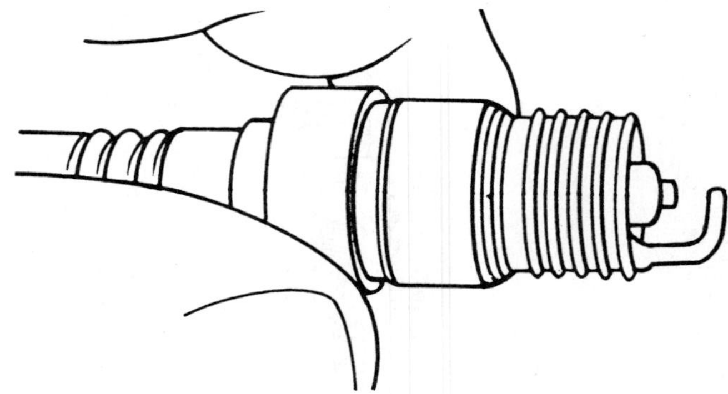

FIGURE 7-18 It is difficult for voltage to jump a gap this large.

tive polarity at the center electrode is necessary to decrease the voltage required for ignition. The secondary voltage of the coil increases in a very short time, thus strengthening the charges at the electrodes. This voltage also changes the air/fuel mixture to a conductor through a process known as ionization. To ionize the air/fuel mixture means to break its molecules up into two (or more) oppositely charged ions. This serves to start the current flow between the electrodes. At this point, the voltage level is called the ionization voltage.

The current now flowing across the gap is the capacitive discharge part of the spark as it moves from the more negatively charged center electrode to the less negatively charged grounded electrode. This flow uses the energy stored in the spark plug while it acts as a condenser prior to ionization. This part of the arc is responsible for starting the combustion process. The remaining voltage (the voltage not needed for ionization) dissipates; this is known as the inductive portion of the discharge and causes the visible flash at the electrodes.

SPARK ADVANCE MECHANISMS

Most distributor assemblies have two spark advance mechanisms whose job it is to make sure the current is delivered to the spark plug at a precise moment in the engine's cycle: the centrifugal advance and the vacuum advance.

Centrifugal Advance

The centrifugal advance mechanism (Figure 7-19) times the high-voltage current surges so that it can fire the air/fuel mixture at the correct instant, as determined by engine rpm. For example, at idle the timing of the arc usually occurs just before the piston reaches top dead center, or TDC. At higher engine speeds, however, less time is available for this process. Thus, in order to achieve maximum power from the air/fuel mixture at higher rpm, the spark must be delivered to the cylinder much earlier in the cycle.

In order to change the timing of the arc in relation to rpm, the centrifugal advance mechanism consists of two weights attached to the distributor shaft with springs; a cam assembly mounts on a second shaft that fits over the distributor shaft like a sleeve. The movement of the weights transfers the motion of the distributor shaft to the outer shaft. As engine speed increases, the distributor shaft speeds up, and the weights move outward. This movement shifts the outer shaft and cam ahead of the distributor shaft rotation. In this way, the cam opens the points earlier and the spark is produced sooner.

Vacuum Advance

During part-throttle engine operation (see Chapter 13), high vacuum is present within the intake manifold, with the result being that less air and fuel enter the cylinder. In this situation, additional spark

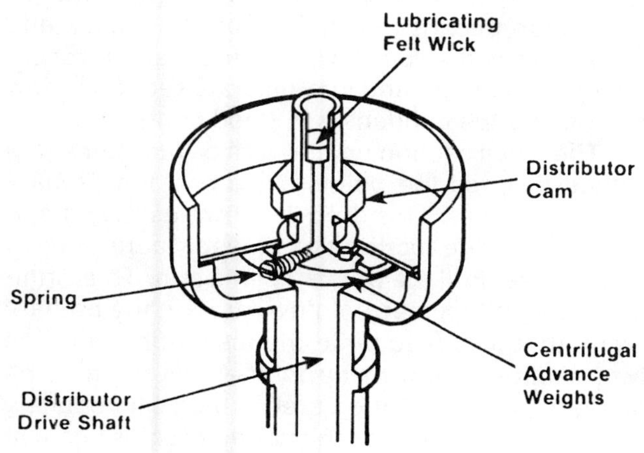

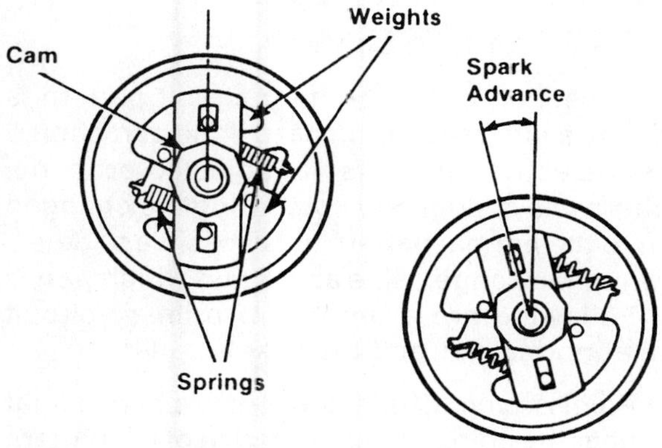

FIGURE 7-19 A typical centrifugal advance mechanism

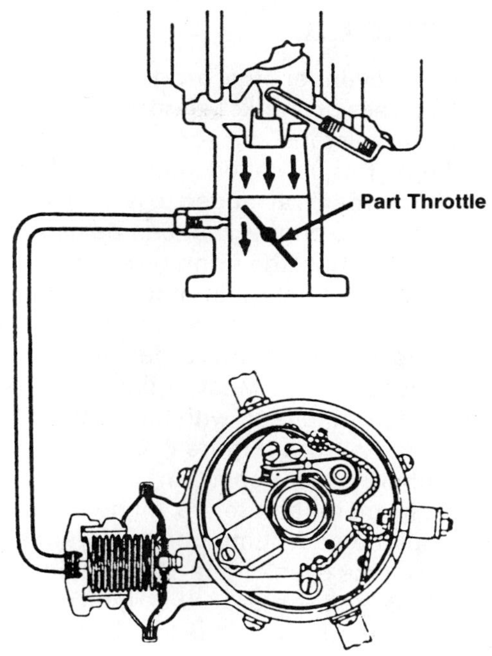

FIGURE 7-20 A typical vacuum advance mechanism

advance serves to increase fuel economy. In order to attain the highest possible power and economy, the arc must occur at the plugs even earlier in the cycle than is needed with a centrifugal advance mechanism.

The heart of the vacuum advance mechanism (Figure 7-20) is the spring-loaded diaphragm, which fits inside a metal housing and connects to the distributor breaker plate. When the throttle valve is in the idle position, there is no vacuum acting on the diaphragm; the housing chamber on the other side of the diaphragm is open to the atmosphere. Any throttle-valve movement past the idle position will cause the valve to swing past the vacuum opening. The intake manifold vacuum draws air from the airtight portion of the vacuum advance chamber; this causes the atmospheric pressure on the opposite side of the diaphragm to push it over against the spring tension. This motion is then transmitted to the distributor breaker plate assembly, which begins rotating. The vacuum advance mechanism governs the amount of rotation by monitoring the amount of vacuum in the intake manifold.

Next, the breaker plate assembly and points move around the breaker cam to an advanced position. This causes the cam to contact the rubbing block of the movable point and open it earlier in the cycle. The spark advance varies, depending on the compression pressures in the cylinder, permitting greater economy of engine operation.

FIRING ORDER

Each engine cylinder must produce power once in every 720 degrees of crankshaft rotation. To make this possible, the pistons and rods are arranged in precise fashion; this is called the engine's firing order. The firing order is arranged to reduce rocking and imbalance in the engine that can be caused by the piston strokes; thus, the firing order varies from engine to engine. Vehicle manufacturers simplify cylinder identification by numbering each cylinder. This numbering system is the order in which the cylinders produce power. Most in-line engines are numbered from front to rear, with the number 1 cylinder located at the front (Figure 7-21).

The numbering sequence on V-type engines can differ from manufacturer to manufacturer. In some cases, all the odd-numbered cylinders are on one bank and all the even on the other. Other times, numbers 1, 2, 3, and 4 are on the right bank, and numbers 5, 6, 7, and 8 are on the left. Regardless of the particular firing order used, the number 1 cylinder always starts the firing procedure, with the rest of the cylinders following in sequence.

SHOP TALK

The firing of the fuel charge within the number 1 cylinder is the basis for setting the engine's basic ignition timing, as well as checking the advance curves of the distributor.

OSCILLOSCOPE TESTING

The oscilloscope, or scope, is an analyzer that is ideal for the task of observing and testing the ignition system. It displays easy-to-interpret graphic images, known as the scope pattern, on its screen. The scope pattern supplies information on any voltage changes in the ignition cycle as they occur; each component can be analyzed while the engine is running. Always follow the manufacturer's instructions when using an oscilloscope.

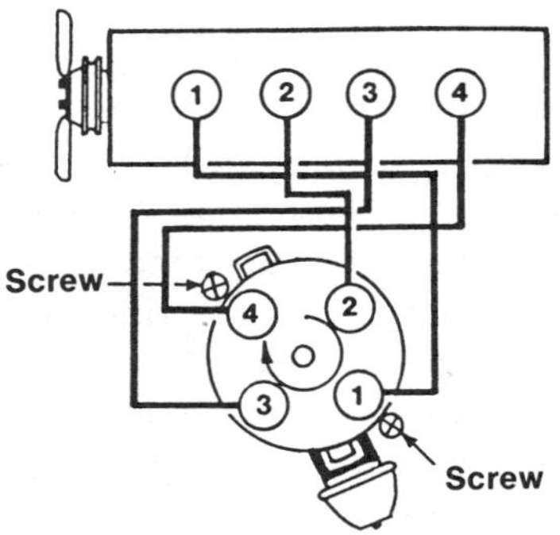

FIGURE 7-21 Most in-line engines are numbered from front to rear, beginning with the number 1 cylinder.

NORMAL PRIMARY CIRCUIT PATTERNS

The typical primary circuit pattern can be divided into three distinct trace sections: firing, intermediate, and dwell. Both positive and negative voltage waves, or oscillations, occur in the firing section. This takes place as the breaker points open the primary circuit and high-voltage current flows through the secondary circuit to arc across the spark plug air gap. Since the primary circuit is open, the current neither flows normally nor stops entirely; instead, it oscillates through the condenser and coil.

The intermediate section begins when the spark plug arc is extinguished, and ends when the breaker points close. It has shorter oscillations than the firing section because of the current flow back and forth between the coil and the condenser. The oscillations of the firing and intermediate sections show the coil and the condenser in good condition.

The dwell section begins with a short but sharp voltage drop as the points close and the primary current resumes. This drop is known as the *points-close* signal. The section continues as more or less a straight line until the points open again. This signifies a low-voltage current flow through the primary circuit that serves to build the magnetic field. The dwell section ends with the *points-open* signal at the far right portion of the screen, then immediately goes into a sharp rise as the pattern begins again at the left side.

ABNORMAL PRIMARY CIRCUIT PATTERNS

Each section of the primary circuit pattern is affected by a different circuit part. If the part should become defective, or if it is not adjusted properly, the pattern will change. By comparing the changed pattern to the normal pattern, it is possible to determine why the change came about and which part is faulty. Following are examples of primary circuit pattern abnormalities and their likely causes.

- *Point Arcing.* Characterized by a blip of light near the point-opening section of the pattern (Figure 7-22). Can be caused by a defective condenser; misaligned, burned, pitted, or dirty points; excessively high charging voltage; or a faulty primary resistor.

- *Point Bounce.* Characterized by hash marks in the points-closing section (Figure 7-23). If the condition is particularly bad, it will show up as oscillations in the same section. Can be caused by a weak breaker point spring; a binding breaker point pivot; or loose or misaligned points.

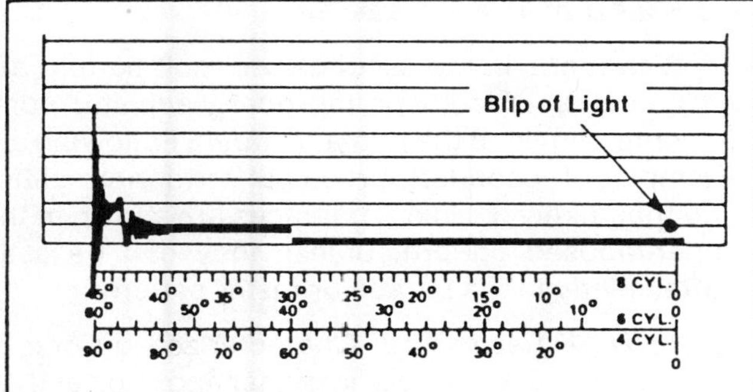

FIGURE 7-22 Point arcing

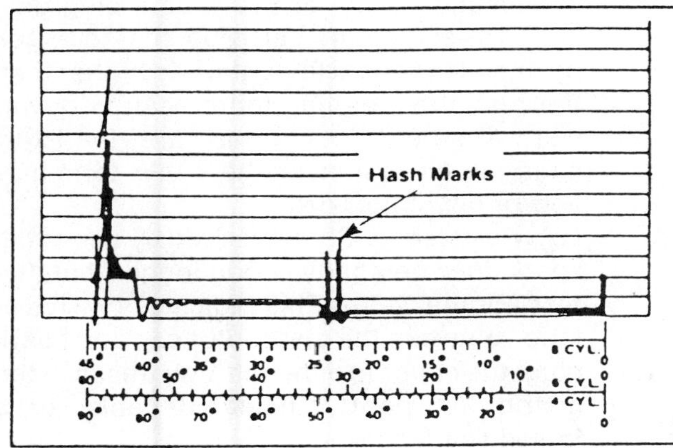

FIGURE 7-23 Point bounce

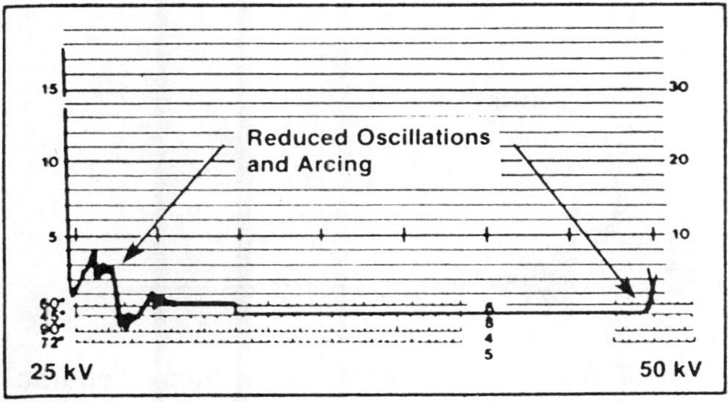

FIGURE 7-24 Condenser series resistance

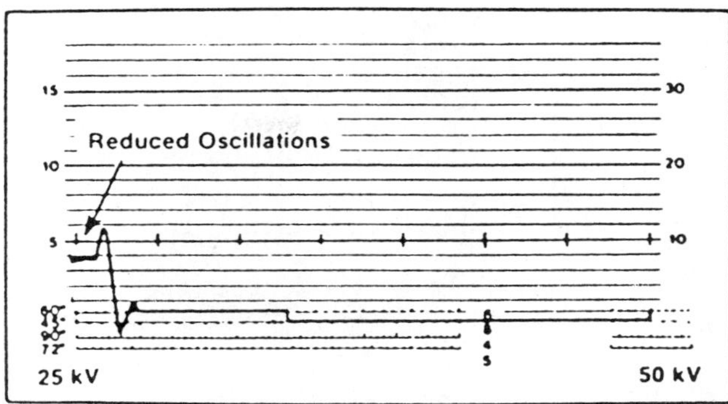

FIGURE 7-25 An intermittently open condenser

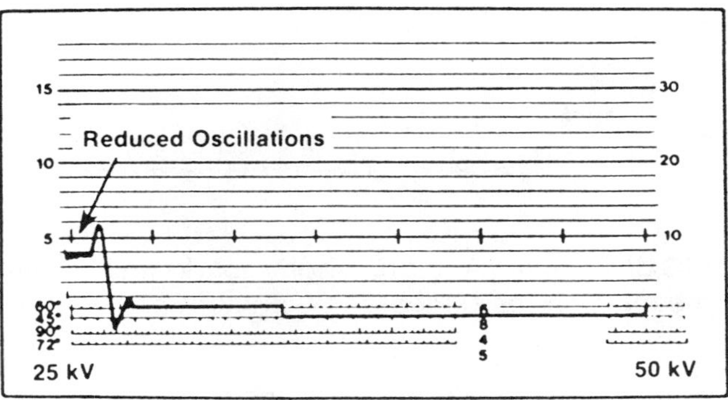

FIGURE 7-26 Shorted primary coil windings and a leaking condenser

- *Condenser Series Resistance.* Characterized by point arcing plus reduced condenser oscillations (Figure 7-24). Can be caused by either a defective condenser or by a bad condenser lead or ground connection.
- *Open Condenser.* Characterized by large oscillations, with little or no separation between the condenser and coil portions of the pattern (Figure 7-25). Sporadic point arcing may also be present. The cause is obvious: an intermittently open condenser.
- *Shorted Primary Coil Windings and Leaking Condenser.* Both problems are characterized by reduced coil and condenser oscillations (Figure 7-26). It is necessary to check the primary coil winding with an ohmmeter (as described later in this chapter) to see if it is the cause; if not, it is probably a leaking condenser.

NORMAL SECONDARY CIRCUIT PATTERNS

Secondary circuit voltage is divided into three categories: required, available, and reserve. Re-

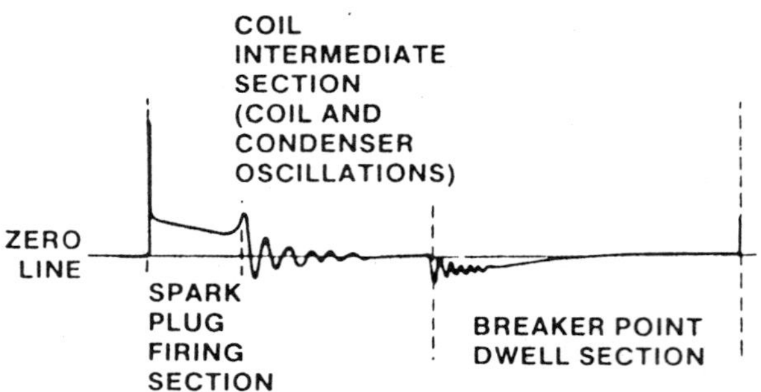

FIGURE 7-27 A normal secondary circuit superimposed pattern

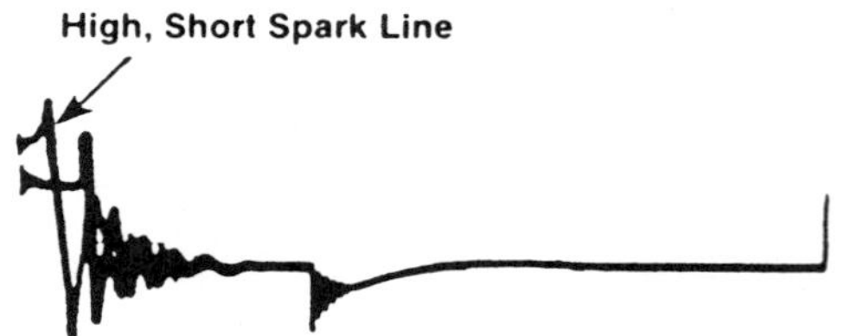

FIGURE 7-28 A high, short spark line usually means high resistance.

quired voltage is the voltage needed to fire the spark plugs and varies depending on operating conditions. Available voltage is the absolute most voltage the ignition system can deliver. In most cases, an engine with a breaker point ignition system has about 20,000 to 25,000 volts available for use. The reserve voltage is simply the difference between the required and the available voltage. Ideally, a reserve voltage of 60 percent is desirable.

Like its counterpart in the primary circuit, the secondary circuit pattern is broken into firing, intermediate, and dwell sections (Figure 7-27). Because these voltages are much higher than those in the primary circuit, the scope's kV scales are used for measurement. The secondary superimposed pattern shows its firing section beginning as a straight vertical line; this indicates how much voltage is needed to create an arc across the spark plug air gap. It is called the firing line, or voltage spike. Once the arc has been established, less voltage is required to maintain it. Continued current flow across the gap is indicated by the horizontal spark line, which is only about one-quarter the height of the voltage spike.

The intermediate section begins when the arc is extinguished, with the remaining voltage dissipated through oscillations between the coil and condenser. The oscillations start at the very beginning of this section, then gradually lessen. As in the primary circuit, the dwell section begins with a points-close signal. The series of small oscillations that follows is due to the buildup of current in the coil primary winding that induces low voltage in the coil secondary winding. Once the primary current flow reaches full strength, the dwell section continues as a straight line at zero voltage until the points open again at the far right portion of the screen. The pattern then starts all over again at the left.

ABNORMAL SECONDARY CIRCUIT PATTERNS

Variations in the secondary circuit pattern are an indication of faulty or improperly adjusted components, similar to the primary circuit. Following are examples of secondary circuit pattern abnormalities and their likely causes. The traces are shown in the superimposed pattern, unless they can be seen more clearly in the parade or raster pattern.

- *High Resistance.* Characterized by one of the spark lines being higher and shorter than the rest (Figure 7-28). This indicates high resistance between the distributor cap and spark plug and is usually caused by a loose or damaged cable or too wide a plug gap. High resistance may also be characterized by one of the spark lines starting higher than the rest, then angling more sharply downward (Figure 7-29). In this case, the likely cause is corrosion of the cable terminals and/or distributor cap.
- *Low Resistance.* Characterized by one of the spark lines being lower and longer than the rest (Figure 7-30). This indicates low resistance between the distributor cap and spark plug; the cause is either carbon tracks in the distributor, poorly insulated cable, or a fouled spark plug.
- *Open Plug Cable.* Characterized by an unusually large oscillation for one of the cylinders, with little or no spark line (Figure

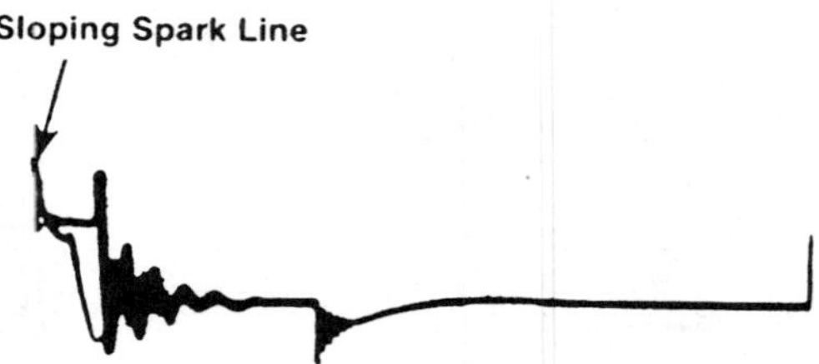

FIGURE 7-29 High resistance can also be characterized by one of the spark lines starting higher, then angling sharply downward.

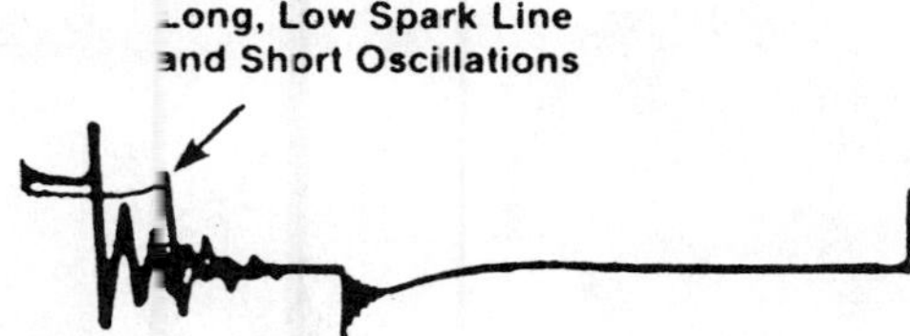

FIGURE 7-30 Low resistance is characterized by one of the spark lines being lower and longer than the rest.

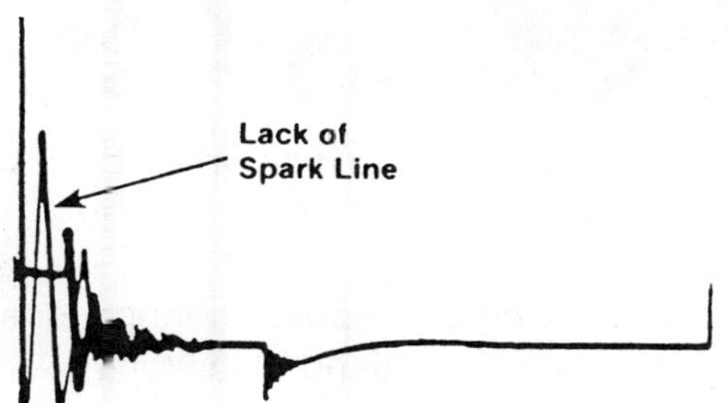

FIGURE 7-31 A large oscillation with little or no spark line is a sign of an open in one of the plug circuits.

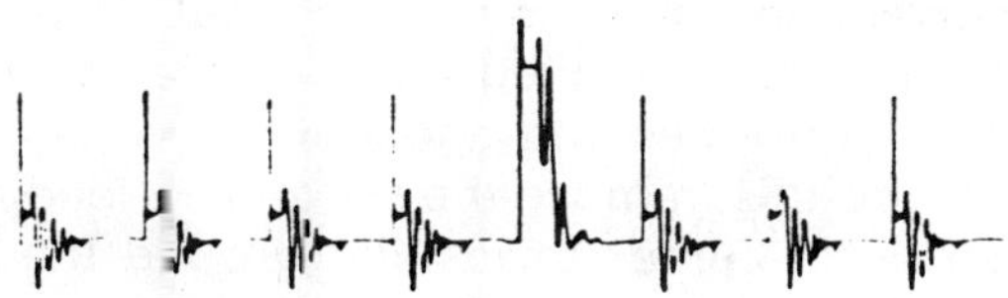

FIGURE 7-32 A leakage to a ground can be seen on the scope when the spark plug cable has been disconnected.

7-31). Caused by a disconnected or open spark plug cable. The raster pattern should be used to locate the faulty plug circuit.

- *Leakage to Ground.* Characterized by a short spark line in the parade pattern (Figure 7-32), despite a disconnected spark plug cable. This indicates that high voltage somewhere in the circuit is causing a current leak to a ground. Likely trouble spots are through the ignition cable insulation, the distributor cap or the rotor. Carbon tracks are often a telltale sign that leakage has occurred.
- *Reversed Coil Polarity.* Characterized by an upside-down pattern (Figure 7-33). Reversed coil polarity is, more often than not, caused by reversed primary connections at the coil.
- *Intermittent Open.* Characterized by a jumping, erratic pattern (Figure 7-34). The likely cause is an intermittent open somewhere in the coil secondary winding.

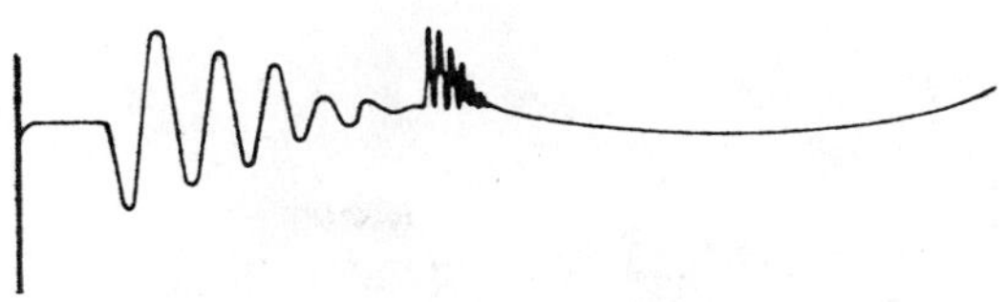

FIGURE 7-33 Reversed coil polarity is characterized by an upside-down pattern.

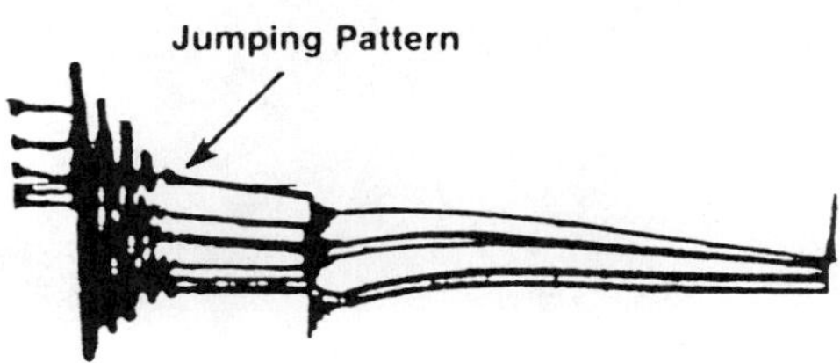

FIGURE 7-34 A jumping pattern is caused by an intermittent open in the coil secondary winding.

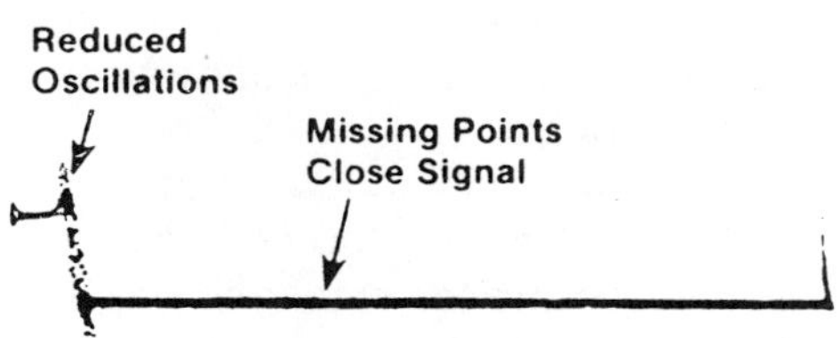

FIGURE 7-35 This pattern usually means a problem between the coil and distributor cap.

- *Reduced Oscillations.* Characterized by a missing points-close signal, as well as reduced coil oscillations (Figure 7-35). Usually caused by a problem between the coil and the distributor cap.

SPARK PLUG CABLE REPLACEMENT

Secondary circuit components fail for one of two reasons: physical damage or insulation failure. Close visual inspection can usually uncover physical damage, but insulation failure is not as easily detectable. It sometimes reveals itself as carbon tracks, but the most reliable method of detection is still the oscilloscope. Use the following procedure to replace a set of spark plug cables:

1. Starting at the front of the engine, remove one cable from the distributor cap and spark plug.
2. Be sure the cap tower is clean and free of corrosion, then insert the new cable into

FIGURE 7-36 The oscilloscope is a very important tool for checking the ignition system.

the cap so that the terminal is firmly seated and the rubber boot seals over the tower. Squeeze the boot to release any air trapped in the tower.

3. Install the spark plug end of the cable on the plug. Be sure the connector is firmly seated on the terminal and the boot seals the plug insulator.
4. Install the other cables in the same manner. Be sure all cables are routed as originally placed by the vehicle manufacturer to avoid crossfiring and cable damage .
5. Secure the cables in their holders. Make sure the cables do not touch the hot exhaust manifold.

PRIMARY CIRCUIT TESTING AND SERVICING

The equipment used for primary circuit testing ranges from simple meters (Figure 7-36) to the oscilloscope. Service frequency is usually specified by the engine or ignition system manufacturer and varies between 5,000 and 10,000 miles. Remember that the type of vehicle operation has a great influence on the service life of the points and other system components; short-term, stop-and-go driving requires more frequent inspection and servicing. As with all engine tests, always compare the results with the latest manufacturer's specifications. Two important precautions should be taken during *all* ignition system tests:

- Turn off the ignition switch with the key before disconnecting any system wiring (Figure 7-37).

FIGURE 7-37 Be sure the ignition switch is off before beginning any ignition test.

- Do not touch any exposed connections while the engine is cranking or running.

VOLTMETER TESTS

Use the following voltage tests to help pinpoint any excessive voltage drops in the primary circuit. Before beginning, make sure that the battery capacity is correct and that the battery is fully charged. Unless otherwise specified, voltmeter tests should always be performed with the ignition system disabled. To do this, remove the secondary lead from the distributor center tower and ground the lead (Figure 7-38).

Available Coil Voltage Test

1. Connect the voltmeter positive lead to the positive coil terminal, and the voltmeter negative lead to a ground (Figure 7-39).
2. Connect a remote starter switch and bump the engine with the starter motor until the breaker points close.
3. Turn on the ignition switch and note the voltmeter reading; it should read 5 to 7 volts, or approximately half the normal battery voltage.
4. Again, bump the engine with the starter motor until the breaker points open.
5. Turn on the ignition switch and note the voltmeter reading; it should now read approximately 12 volts, or full battery voltage.
6. If either reading was not within specifications, first check the primary resistor circuit for loose or damaged connections, and make any necessary repairs. Test the resistor and coil as described later in this chapter, and replace if needed. Then repeat the available coil voltage test.

Bypass Circuit Voltage Drop Test

1. Connect the voltmeter positive lead to the positive battery terminal, and the voltmeter negative lead to the positive coil terminal.
2. While cranking the engine, note the voltmeter reading. If it is less than 0.5 volt, the bypass circuit is in good shape. If not, proceed to the next step.
3. To locate the area of excessive resistance, check the primary resistor circuit for loose or damaged connections, and make any necessary repairs. Then repeat the bypass circuit voltage drop test.

Breaker Point Voltage Drop Test

1. Connect the voltmeter positive lead to the negative coil terminal, and the voltmeter negative lead to a ground.

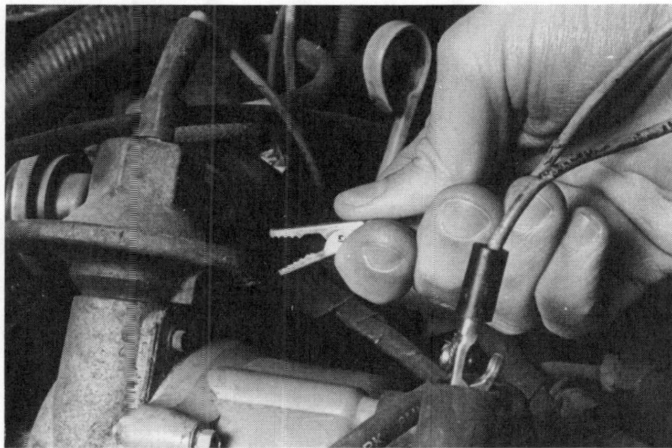

FIGURE 7-38 Disabling the ignition system

FIGURE 7-39 Testing available coil voltage

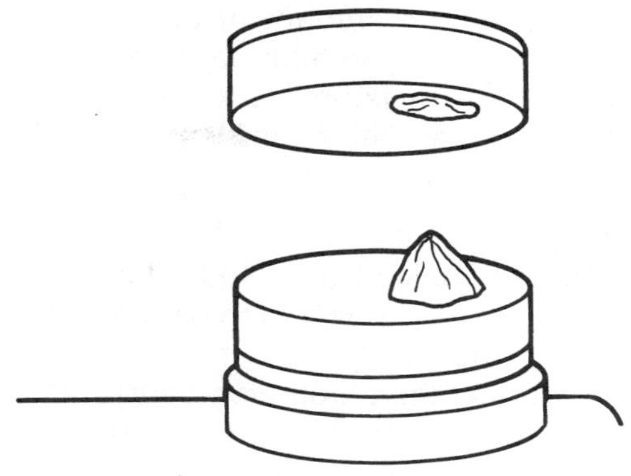

FIGURE 7-40 A faulty condenser can result in a pitted contact.

2. Connect a remote starter switch and bump the engine with the starter motor until the breaker points close.
3. Turn on the ignition switch and note the voltmeter reading. If it is less than 0.2 volt, the points are in good condition. If not, proceed to the next step.
4. Inspect the points for the following: loose wiring connections; faulty ground connections; poor alignment; burned, pitted, or oxidized contacts; and excessive wear. Note that a faulty condenser can cause material to transfer to one side of the contact, with a pit forming on the other side (Figure 7-40).
5. Adjust or replace the points as needed, then repeat the breaker point voltage drop test.

Primary Resistor Voltage Drop Test

1. Connect the voltmeter positive lead to the positive battery terminal, and the voltmeter negative lead to the positive coil terminal.
2. While running the engine at 1500 rpm, note the voltmeter reading. If it is within specifications (generally 1.5 to 3.5 volts), the resistor circuit is in good condition. If not, proceed to the next step.
3. Check for any loose or damaged connections, and make any necessary repairs. Test the resistor and coil as described later in this chapter, and replace if needed. Then repeat the primary resistor voltage drop test.

PRIMARY RESISTOR TESTING

The number one problem with primary resistors is high (or infinite) resistance due to an internal open. The maximum amount of acceptable resistance differs; consult the manufacturer's specifications. Ballast-type resistors are tested in the following manner using an ohmmeter:

1. Make sure the ignition switch is off. As an added precaution, the battery ground cable can be disconnected (Figure 7-41).
2. Disconnect the ignition wiring connectors from the resistor.
3. Connect the ohmmeter leads to the resistor terminals and note the meter reading. If it is within the manufacturer's specifications, the resistor is fine. If not, replace it.

If the resistor is a length of resistance wire, it may not be feasible to test it with an ohmmeter, especially if one end of the wire is connected to the ignition switch in the driver's compartment and the other end to the ignition coil in the engine compartment. Instead, test the resistor by performing the voltmeter tests explained earlier in this chapter.

FIGURE 7-41 Disconnect the battery ground cable before testing a ballast resistor.

COIL PRIMARY WINDING TESTS

Because coil resistance is directly affected by temperature, always test the coil at its normal operating temperature. Make the following checks before beginning the test:

- Make sure the coil is securely mounted, and all connections are tight and clean.
- Inspect the coil tower for signs of cracking or burning, and replace the entire coil if necessary.
- Inspect the housing for cracking or other signs of damage, and replace the entire coil if necessary.
- Check for signs of oil leakage.

Winding Resistance Test

1. Disconnect the wires from the coil primary terminals.
2. Set the ohmmeter on the lowest scale. Connect one of the ohmmeter leads to the positive coil primary terminal, and the other lead to the negative coil primary terminal (Figure 7-42).
3. Note the ohmmeter reading. If it is within the manufacturer's specifications, proceed to the next step. If not, replace the coil.
4. Disconnect the ohmmeter leads and set the ohmmeter on the highest scale.
5. Connect one of the ohmmeter leads to the coil tower secondary terminal. Touch the other lead to one of the coil primary terminals and note the reading. Then move this lead to the other coil primary terminal and again note the reading (Figure 7-43).
6. Compare the *lowest* reading from the previous step to the manufacturer's specifications. If it is within the acceptable range, proceed to the next step. If not, replace the coil.
7. Disconnect the ohmmeter leads and set the ohmmeter on the lowest scale.
8. Connect one of the ohmmeter leads to either coil primary terminal. Touch the other lead to the coil's metal casing, and note the reading. It should show infinite

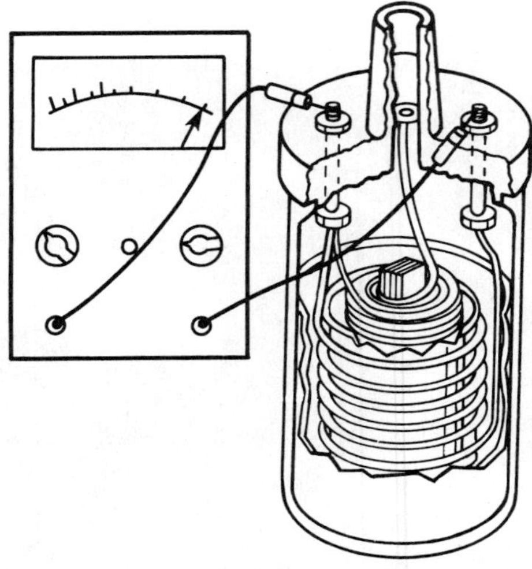

FIGURE 7-42 Using an ohmmeter to test the resistance of the coil primary windings

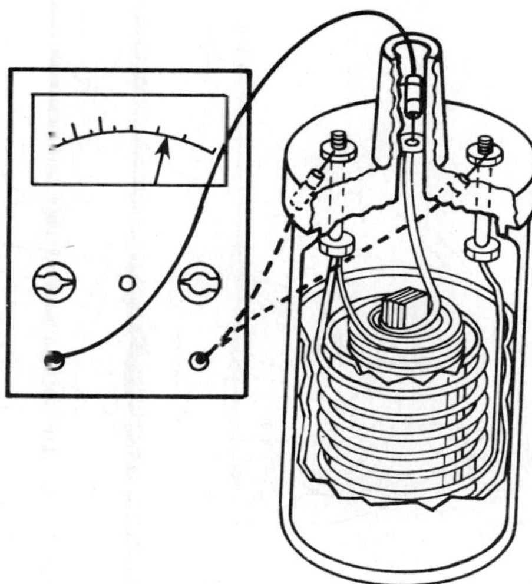

FIGURE 7-43 Using an ohmmeter to test the resistance of the coil secondary windings

FIGURE 7-44 Corroded battery connections is a prime cause of low current draw.

resistance; if not, the coil windings are shorted to the casing and the coil must be replaced.

Current Draw Test

1. Turn off the ignition switch. Disconnect the positive coil primary wire from the coil.
2. Connect the ammeter positive lead to the positive coil primary wire, and the ammeter negative lead to the positive coil primary terminal.
3. Turn on the ignition switch. Depending on the manufacturer's recommendations, either start the engine or close the breaker points.
4. Note the ammeter reading and compare it to the manufacturer's specifications. If it is less than what is specified, proceed to the next step; if it is more than what is specified, proceed to step 6. If no current draw registers, there is an open somewhere in the primary circuit; in this case, the ignition system would not operate at all.
5. If low current draw is the problem, check for a discharged battery; high resistance in the primary wiring or the coil primary winding; or loose or corroded battery connections (Figure 7-44).
6. If high current draw is the problem, check for a shorted coil or resistor. There is also a possibility that the wrong type of coil or resistor is being used in the vehicle.

Polarity Test

If the coil polarity has been reversed, the engine will run—but the voltage requirement will be increased by some 20 to 40 percent. The result is misfiring under some operating conditions. If the polarity cannot be determined by checking the markings at the coil primary terminals, it can be done using a voltmeter that has both a positive and negative scale.

1. Connect an adapter with an exposed terminal into one of the spark plug cables, either at the plug or at the distributor cap (Figure 7-45).

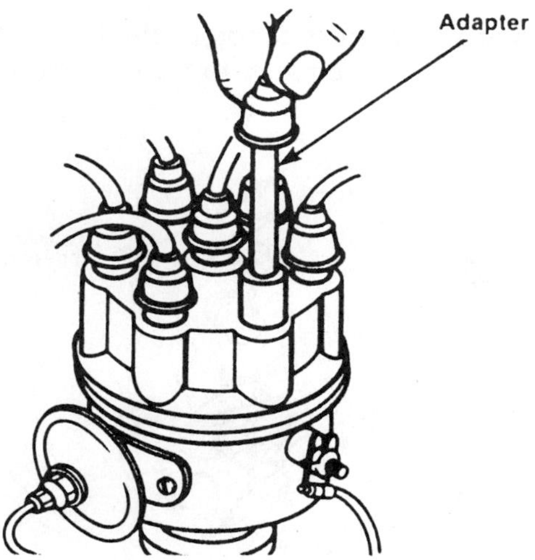

FIGURE 7-45 Connecting an adapter with an exposed terminal to perform a primary circuit coil polarity test

2. Start the engine and run it at idle speed.
3. Set the voltmeter on the highest scale.
4. Connect the voltmeter positive lead to an engine ground, and touch the voltmeter negative lead to the adapter terminal.
5. If the voltmeter needle moves toward the positive end of the scale, the coil polarity is fine. If it moves toward the negative end of the scale, the polarity is reversed; proceed to the next step.
6. Check for reversed primary circuit connections at the coil, or reversed cable connections at the battery. The latter condition would damage the alternator and any other solid-state system components.

BREAKER POINT REPLACEMENT

Breaker points must be replaced whenever they are found to be burned or pitted. While used points can be filed and cleaned for reuse, this practice is not recommended. Most breaker points are made with specially hardened contact surfaces; when these surfaces are filed away, the points wear out very quickly. It is much more efficient to replace old points with a new set.

Most replacement breaker points are sold as preassembled sets, with the alignment and spring tension set at the factory. However, replacement points that come as two-piece units require alignment, as well as spring tension and dwell adjustment. Special spring scales are available to check the tension; always compare the reading with the manufacturer's specifications. In most cases, the tension is adjusted by sliding the spring along its slot (Figure 7-46).

Point alignment is accomplished using a special tool (Figure 7-47). Be sure to bend only the stationary contact support, never the movable contact arm. Always clean and lubricate the distributor cam and breaker point rubbing block; special grease is available for this purpose.

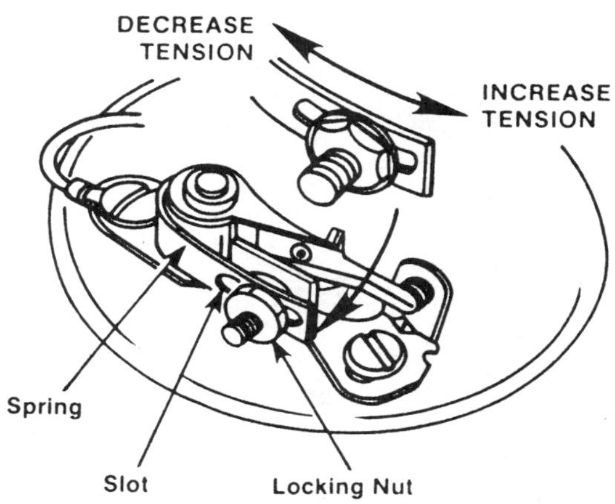

FIGURE 7-46 Adjusting breaker point spring tension

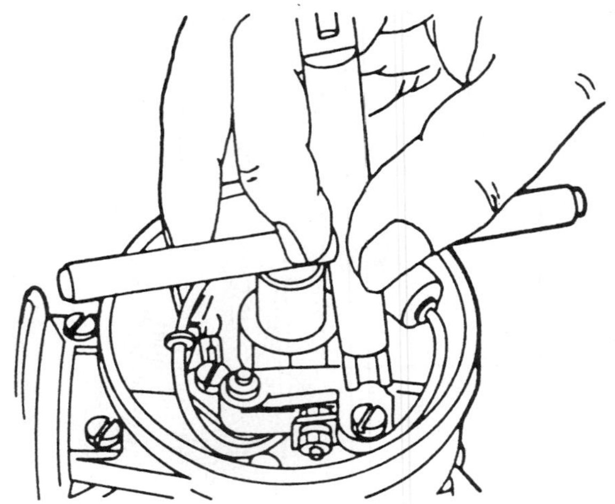

FIGURE 7-47 Bend only the stationary contact support when performing breaker point alignment.

Setting Breaker Points

Set newly aligned breaker points with a feeler gauge, using the following procedure:

1. Remove the distributor cap and rotor (Figure 7-48).
2. Connect a remote starter switch and bump the engine with the starter motor until the rubbing block is *exactly* on a high point of the cam.

FIGURE 7-48 The first step in setting newly aligned breaker points is to remove the distributor cap and rotor.

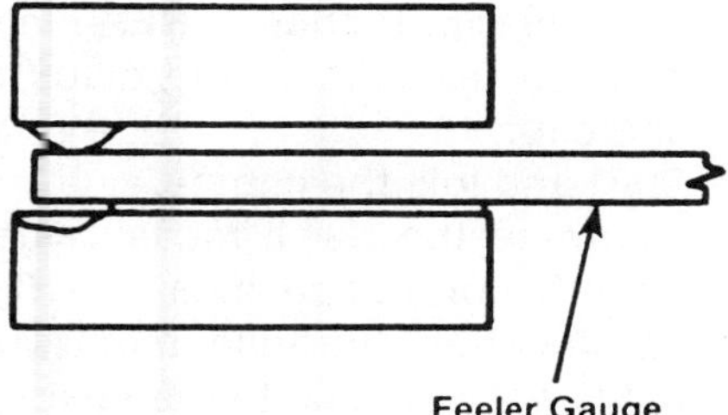

FIGURE 7-49 The uneven surfaces of used points will produce an inaccurate feeler gauge reading.

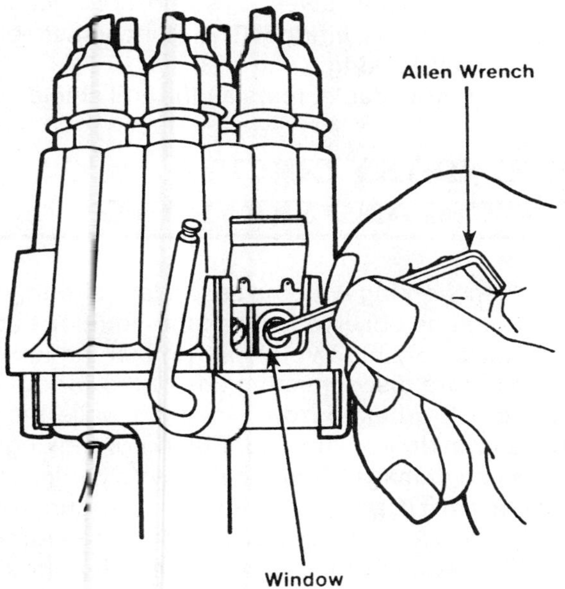

FIGURE 7-50 Some Delco-Remy distributors have points that can be adjusted through a window in the distributor cap.

3. Slide a clean feeler gauge of the proper thickness between the points. It should slide between the contacts with a slight drag. Consult the manufacturer's specifications for proper gap dimension, and adjust accordingly. Make the adjustment by shifting the position of the point assembly.
4. Tighten the holddown screw and recheck the gap. Make any further adjustments that may be necessary.

WARNING: Do not attempt to set used breaker points in this manner (Figure 7-49). The uneven contact surface will make any feeler gauge measurement inaccurate.

CHECKING THE DWELL

The most accurate method of measuring dwell—especially on used breaker points—is by using a dwellmeter or oscilloscope. A dwellmeter can measure both rpm and dwell angle, and usually has two test leads. The positive lead connects directly to one of the coil primary terminals (as specified by the tester manufacturer) and the negative lead to an engine ground. Dwellmeters with additional test functions built in frequently have two extra leads that connect to the battery terminals. It is important to keep in mind that most dwellmeters use the average of all the individual cylinder readings to arrive at the dwell angle. Also, keep in mind that internal distributor problems such as shaft or bushing wear, breaker plate bushing/pivot pin wear, and flyweight/pivot pin wear can cause the dwell readings to be out of specification.

SHOP TALK

Most Ford coils require a special "alligator clip" adapter in order to use a dwellmeter. Consult the vehicle service manual for details. Delco-Remy V-6 and V-8 distributors feature a distributor cap window through which the points can be adjusted while the engine is running (Figure 7-50). The primary or secondary trace pattern is displayed while the dwell is adjusted.

ADJUSTING THE DWELL

The dwell adjustment procedures that follow can be done equally well with a dwellmeter or an oscilloscope. Note that the procedures differ, depending on the type of distributor.

Internal Adjustment Distributors

1. For an oscilloscope test, select the primary superimposed pattern. For a dwellmeter test, select the dwell position and the correct number of cylinders.
2. Remove the distributor cap and rotor.
3. Connect a remote starter switch, then turn the ignition on.
4. While observing the scope screen or meter dial, crank the engine with the remote starter switch. If the dwell measurement is within the manufacturer's specifications, no adjustment is necessary. If it is not, proceed to the next step.
5. Adjust the point assembly, either by turning the adjustment screw (Figure 7-51) or

FIGURE 7-51 Adjusting a point assembly by turning the adjustment screw

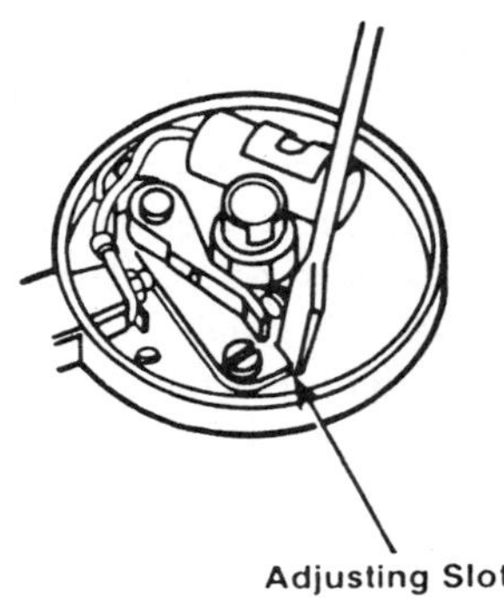

FIGURE 7-52 Adjusting a point assembly with an adjusting slot

inserting a screwdriver in the adjusting slot (Figure 7-52).

6. Tighten the point set lockscrew.
7. Recheck the dwell measurement. If it does not remain within specifications after the lockscrew is tightened, repeat the adjustment.
8. Disconnect the remote start switch, and replace the distributor cap and rotor.
9. Start and idle the engine. Compare the dwell measurement to specifications, and repeat the adjustment if necessary.

External Adjustment Distributors

1. For an oscilloscope test, select either the primary or the secondary superimposed pattern. For a dwellmeter test, select the dwell position and the proper number of cylinders.
2. If the distributor has an RFI shield, remove it. Then reassemble the distributor rotor and cap.
3. Start and idle the engine.
4. Compare the dwell measurement to the manufacturer's specifications. If it is within specifications, no adjustment is necessary. If it is not, proceed to the next step.
5. Adjust the points as needed, using an Allen wrench through the window in the distributor. Turn the screw clockwise to increase the dwell, and counterclockwise to decrease the dwell.
6. When the dwell is within specifications, close the window. The point adjustment is self-locking.
7. If applicable, reinstall the RFI shield.

SECONDARY CIRCUIT TESTING AND SERVICING

When working with the secondary circuit, caution should be observed; secondary ignition cables carry extremely high voltage. Turn off the ignition switch before disconnecting any of the wiring, and never touch an exposed connection while the engine is cranking or running. Pulling or kinking the cables can damage them. To remove a cable, grasp the boot, twist it, and pull it straight off (Figure 7-53).

Before beginning any of the tests, do a thorough visual inspection of the entire secondary circuit. If any damaged components are discovered, they should be replaced immediately.

FIGURE 7-53 Removing an ignition cable

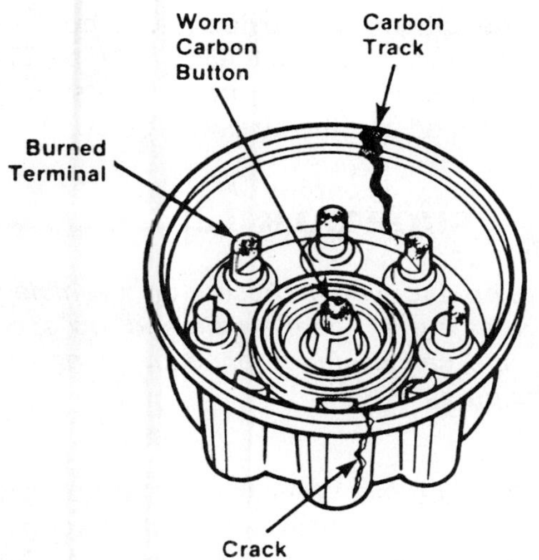

FIGURE 7-54 Potential distributor cap problems

1. Check all ignition cables and boots for cracked, burned, or brittle insulation.
2. Make sure all cable connectors are secure and well insulated.
3. Inspect the distributor cap for cracks; carbon tracks from the arcing current; a worn or sticking carbon button; burned or corroded inside terminals; and signs of corrosion inside the towers (Figure 7-54).
4. Inspect the rotor for cracks; a bent or broken contact strip; a burned or worn tip; a cracked or broken positioning lug; and carbon tracks.

TESTING CABLE RESISTANCE

Ignition cables used with breaker point ignition systems should measure approximately 4,000 ohms of resistance per foot. A completely open cable that reveals infinite resistance when tested is a common occurrence. Use an ohmmeter to perform the test as follows:

1. Remove the distributor cap from the housing.
2. Remove the spark plug cable being tested from the spark plug.
3. Connect one ohmmeter lead to the cable at the spark plug terminal, and the other to the corresponding terminal inside the distributor cap.
4. Note the ohmmeter reading. If it is within specifications, the cable is in good condition. If it is not within specifications, check the cable connections and repeat the test. Replace the cable if necessary.

SPARK PLUG SERVICE

Spark plug service is a vital part of a complete engine tune-up. Vehicle manufacturers' recommendations for service intervals can range anywhere between 5,000 and 50,000 miles, depending on a variety of factors. Among the most important are:

- The type of ignition system
- Engine design
- Spark plug design
- Vehicle operating conditions
- The type of fuel used
- Types of emission control devices used

Regardless of what other tools are used, a spark plug socket is essential for plug removal and installation. Spark plug sockets are available in two sizes: 3/16-inch (for 14-millimeter gasketed and 18-millimeter tapered-seat plugs) and 5/8-inch (for 14-millimeter tapered-seat plugs). They can be either 3/8- or 1/2-inch drive, and many feature an external hex so that they can be turned using an open end or box wrench.

To remove spark plugs, proceed as follows:

1. Remove the cables from each plug, being careful not to pull on the cables. Instead, grasp the boot and twist it off gently.

SHOP TALK

To save time and avoid confusion later, use masking tape to mark each of the cables with the number of the plug it attaches to.

2. Using a spark plug socket and ratchet, loosen each plug a couple of turns (Figure 7-55).
3. Use compressed air to blow any dirt away from the base of the plugs.
4. Remove the plugs, making sure to remove the entire gasket as well (if applicable).

Inspecting Spark Plugs

Once the spark plugs have been removed, it is important to "read" them (Figure 7-56). In other words, inspect them closely, noting in particular any deposits on the plugs and the degree of electrode erosion. A plug in good working condition can still have minimal deposits on it; they will usually be light tan or gray in color. However, there should be no evidence of electrode burning, and the air gap should be no more than 0.001 inch for every 10,000 miles of engine operation. A plug that fits this de-

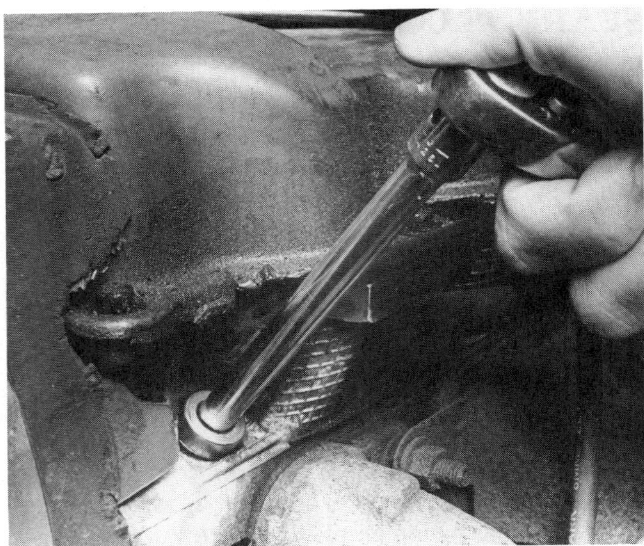

FIGURE 7–55 Using a spark plug socket and ratchet

scription can be reinstalled after filing the electrodes and resetting the air gap to specifications.

It is possible to diagnose a variety of engine conditions by examining the firing end of the spark plugs. If an engine is in good shape, they should all look alike. Whenever plugs from different cylinders look different, a problem exists somewhere in the engine. Following are examples of plug problems and how they should be dealt with.

- *Cold Fouling.* This condition is the result of an excessively rich air/fuel mixture. It is characterized by a layer of dry, fluffy black carbon deposits on the tip of the plug (Figure 7–57). Cold fouling can be caused by a faulty choke, a blocked air cleaner, incorrect idle adjustment, a dirty carburetor, a malfunctioning ignition system, or prolonged engine operation at idle speed. If only one or two of the plugs show evidence of cold fouling, sticking valves are the likely cause. The plug can be used again, provided its electrodes are filed and the air gap is reset.

SHOP TALK

If cold fouling is present on a vehicle that operates a great deal at idle and low speeds, plug life can be lengthened by using hotter spark plugs.

- *Wet Fouling.* When the tip of the plug is practically drowned in excess oil, this condition is known as wet fouling (Figure 7–58). In an overhead valve engine, the oil may be entering the combustion chamber past worn valve guides or valve guide seals. If the vehicle has an automatic transmission, the likely cause of wet-fouled plugs is a defective vacuum modulator that is allowing transmission fluid into the chamber. On older engines, check for worn rings or excessive cylinder wear. While switching to a hotter plug can relieve this condition, care should be taken, because too hot a plug can cause preignition. The best solution is to replace the vacuum modulator or overhaul the engine. Wet-fouled plugs can be reused if serviced properly.
- *Splash Fouling.* This condition occurs immediately following an overdue tune-up. Deposits in the combustion chamber, accumulated over a period of time due to misfiring, suddenly loosen when the temperature in the chamber returns to normal. During high-speed driving, these deposits can stick to the hot insulator and electrode surfaces of the plug (Figure 7–59). If the plug is shorted out,

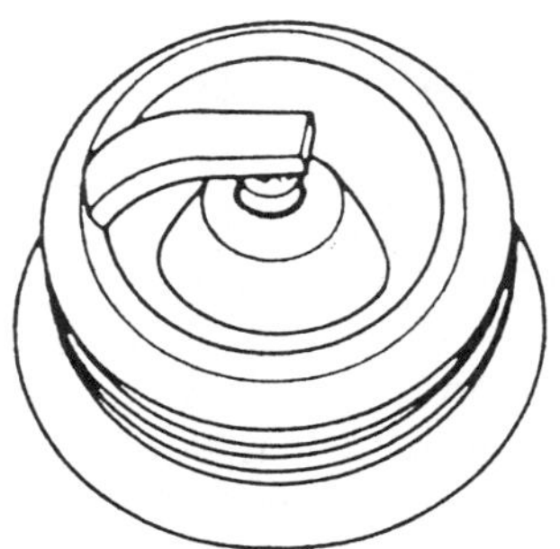

FIGURE 7–56 A spark plug in good working order

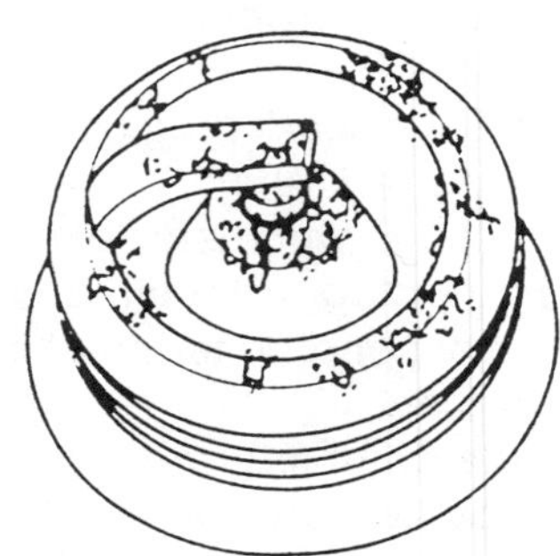

FIGURE 7–57 A cold-fouled spark plug

remove it for service; since the engine has essentially purged itself, the plug is acceptable for further use.

- *Gap Bridging.* A plug with a bridged gap is rarely seen in automobile engines. It occurs when flying carbon deposits within the combustion chamber accumulate over a long period of stop-and-go driving. When the engine is suddenly placed under a hard load, the deposits melt and bridge the gap, causing misfiring (Figure 7-60). This condition is also corrected by servicing the plug.
- *Glazing.* Under high-speed conditions, the combustion chamber deposits can form a shiny, yellow glaze over the insulator (Figure 7-61); when it gets hot enough, the glaze acts as an electrical conductor, causing the current to follow the deposits and short out the plug. Glazing can be prevented by avoiding sudden wide-open throttle acceleration after sustained periods of low-speed or idle operation. Because it is virtually impossible to remove glazed deposits, any glazed plugs should be replaced.
- *Overheating.* This condition is characterized by white or light gray blistering of the insulator; there may also be considerable electrode gap wear (Figure 7-62). Overheating can result from using too hot a plug, overadvanced ignition timing, detonation, a malfunction in the cooling system, an overly lean air/fuel mixture, using too-low octane fuel, an improperly installed plug, or a heat-riser valve that is stuck closed. Overheated plugs must be replaced.
- *Turbulence Burning.* When turbulence burning occurs, the insulator on one side of the plug wears away as the result of normal turbulence in the combustion chamber (Figure 7-63). As long as the plug life is normal, this condition is of little consequence; however,

FIGURE 7-58 A wet-fouled spark plug

FIGURE 7-59 A splash-fouled spark plug

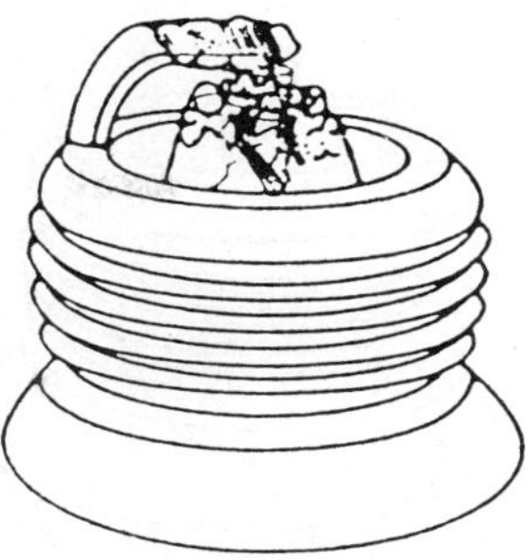

FIGURE 7-60 A plug with a bridged gap

FIGURE 7-61 A glazed spark plug

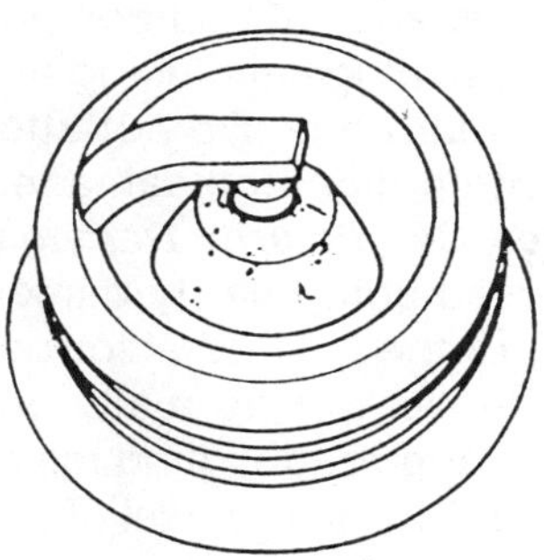

FIGURE 7-62 An overheated spark plug

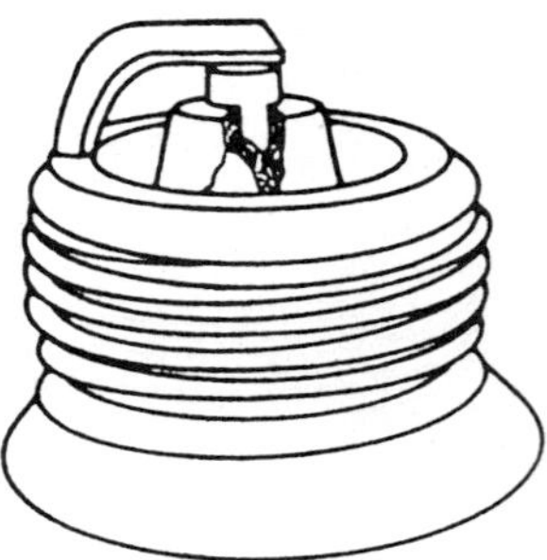

FIGURE 7-63 This plug has experienced turbulence burning.

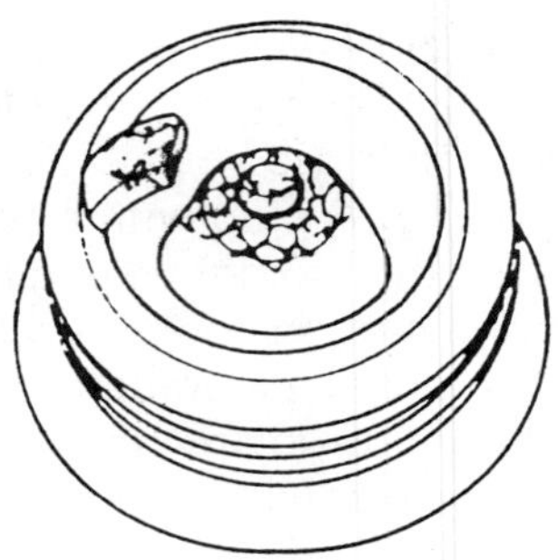

FIGURE 7-64 This plug shows evidence of preignition damage.

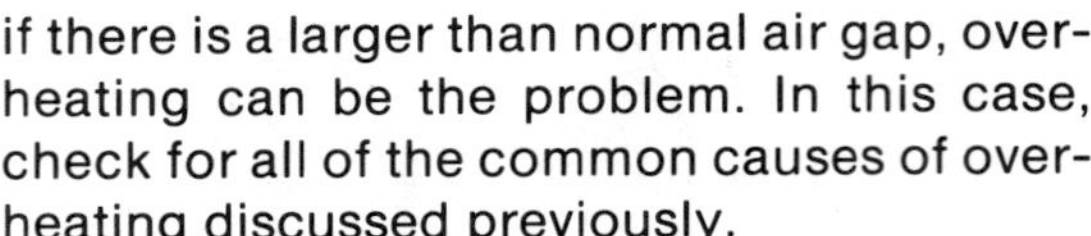

if there is a larger than normal air gap, overheating can be the problem. In this case, check for all of the common causes of overheating discussed previously.

- *Preignition Damage.* This condition is caused by excessive engine temperatures. Preignition damage is characterized by melting of the electrodes, or chipping of the electrode tips (Figure 7-64). When this problem occurs, look for the general causes of engine overheating, including overadvanced ignition timing, a burned head gasket, and using too-low octane fuel. Other possibilities include loose plugs and using plugs of the improper heat range. Do not attempt to reuse plugs with preignition damage.
- *Reversed Coil Polarity Damage.* A sure sign of reversed coil polarity damage is a slight dishing of the ground electrode; the center electrode will not normally be worn badly. Misfiring and rough idling may also be present. In addition to reversal of the coil primary leads, older vehicles can experience reverse coil polarity damage by reversing the battery polarity.

SHOP TALK

It is a good idea to brush used spark plug electrodes with several strokes of a flat distributor point file or spark plug gauge file (Figure 7-65). This will help reduce the required firing voltage of the plug.

Regapping Spark Plugs

Both new and used spark plugs must have their air gaps set to the engine manufacturer's specifications. Whenever possible, use an approved tool not only to measure the gap, but also to bend the side

FIGURE 7-65 It may be helpful to use a file on the electrodes to get them fully clean.

electrode to make the adjustment. The combination gauge and adjusting tool shown in Figure 7-66 is designed for use on new plugs only; this is because it utilizes flat gauges for the adjustment procedure. The gauges are mounted on the tool like spokes on a wheel; above the gauges is an anvil, which is used to apply pressure to the electrode. On the opposite end of the tool is a curved seat. This seat performs two functions: it supports the plug shell during the procedure; and it compresses the ground electrode against the gauge, thus setting the air gap.

The tapered regapping tool is simply a piece of tapered steel whose leading and trailing edges are different dimensions; between these two points, the gauge varies in thickness. A scale, located above the gauge, indicates the thickness at any given point. When the gauge is slid between electrodes, it stops when the air gap size reaches the thickness on the gauge. The scale reading is made in thousandths of an inch. Adjusting slots are available to bend the ground electrode as needed to perform the air gap adjustment.

In many instances, the electrodes on a used spark plug are no longer flat, thus rendering a flat gauge useless for regapping purposes. Instead, a round wire gauge is required. Combination round and flat feeler gauge sets are available; these multipurpose tools can be used to check air gaps, adjust contact points, bend electrodes, and file contact points and electrodes. When regapping a spark plug, keep the following points in mind:

- Always check the air gap of a new spark plug before installing it; never assume the gap is correct just because the plug is new.
- Do not try to reduce a plug's air gap by tapping the side electrode on a bench.
- Never attempt to set a wide gap, electronic ignition plug to a smaller gap specification. Likewise, never attempt to set a small gap, breaker point ignition plug to the wide gap necessary for electronic ignitions. In either case, damage to the electrodes results.
- Never try to bend the center electrode to adjust the air gap; this will crack the insulation.

Spark Plug Installation

Use the following procedure to install a spark plug:

1. Wipe dirt and grease from the plug seats with a clean cloth (Figure 7-67).
2. Be sure the gaskets on gasketed plugs are in good condition and properly placed on the plugs. If reusing a plug, install a new gasket on it. Be sure that there is only one gasket on each plug.
3. Adjust the electrode gap as needed.
4. Install the plugs and fingertighten. If the plugs cannot be installed easily by hand, the threads in the cylinder head may require cleaning with a thread-chasing tap. Be especially careful when working with aluminum heads.
5. Tighten the plugs with a torque wrench, following the vehicle manufacturer's specifications, or the values listed in Table 7-1.

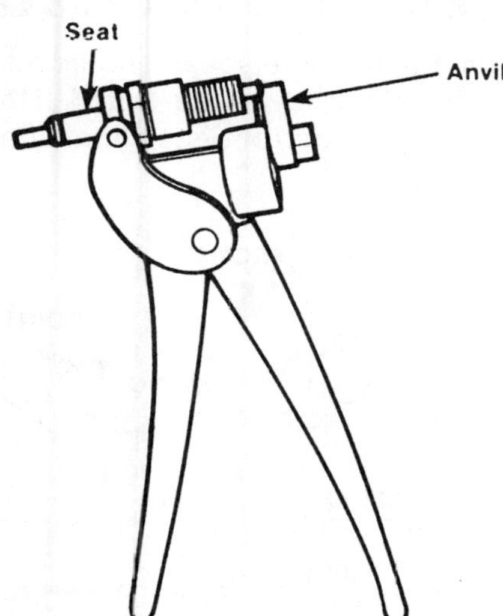

FIGURE 7-66 A combination spark plug gauge and adjusting tool

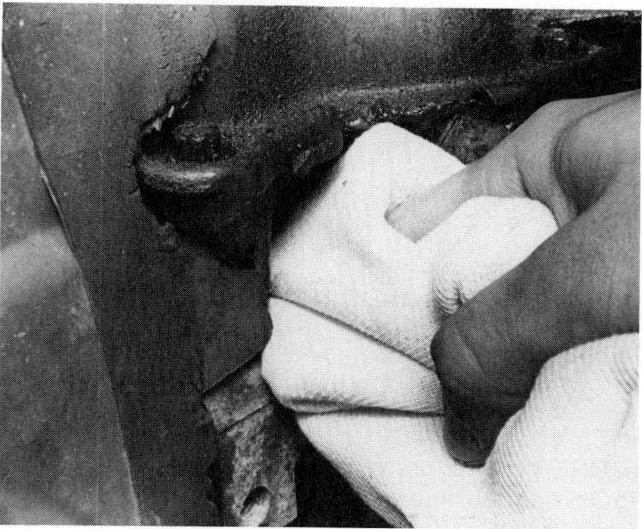

FIGURE 7-67 Use a clean cloth to wipe dirt from the spark plug seats.

CAUTION: If thread lubricant is used, reduce the torque setting. (Keep in mind that many spark plug manufacturers do not recommend the use of thread lubricant.)

TABLE 7-1: PLUG INSTALLATION TORQUE VALUES (IN FOOT-POUNDS)

Plug Type	Cast-Iron Head	Aluminum Head
14-mm Gasketed	25 to 30	15 to 22
14-mm Tapered Seat	7 to 15	7 to 15
18-mm Tapered Seat	15 to 20	15 to 20

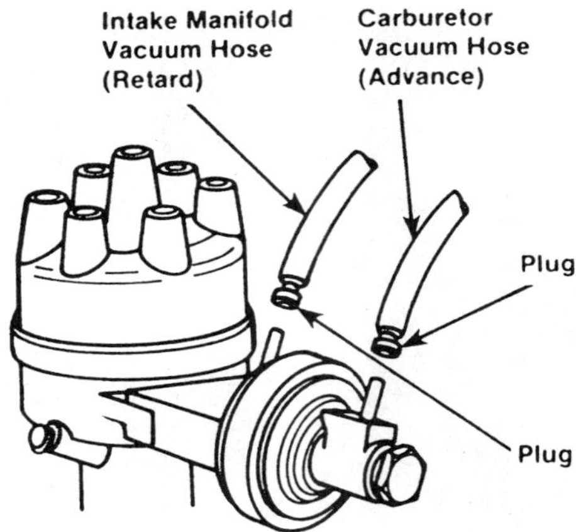

FIGURE 7-68 Most vehicles must have their distributor vacuum lines disconnected and plugged before they can be timed.

IGNITION TIMING TESTS AND ADJUSTMENTS

It is possible to set the ignition timing two ways: statically (with the engine turned off) and dynamically (with the engine running). Timing is usually set in relation to the number 1 cylinder. Remember that most vehicles must be timed with their distributor vacuum lines disconnected and plugged (Figure 7-68). Always consult the tune-up specifications for the vehicle being adjusted. For late-model engines, the timing setting is given on the VECI decal.

STATIC TIMING

Static timing is normally set only when the distributor is being reinstalled in the engine. Use the following procedure to set ignition timing statically:

1. Rotate the engine until the timing mark aligns with the proper timing specification.
2. Loosen the distributor holddown clamp and bolt.
3. Rotate the distributor housing in the direction of the rotor rotation so that the points are closed.
4. Connect one lead of the timing light to the distributor primary lead terminal on the coil. Connect the other lead to a ground.
5. Turn the ignition on. The timing light should not come on if the points are closed.
6. Slowly rotate the distributor housing in the opposite direction of the rotor rotation until the timing light comes on. This indicates that the points have opened.
7. Tighten the holddown clamp and bolt.

DYNAMIC TIMING

Two factors must be present before the timing can be set dynamically. First, the dwell must be correct. Second, the engine must be running at normal operating temperature. Consult the vehicle service manual for specific instructions regarding the disconnecting of the distributor vacuum lines, and be sure to note their position prior to disconnecting them. Use the following procedure to set ignition timing dynamically:

1. Connect the tachometer and timing light.
2. Start the engine and run it at the speed specified in the service manual.
3. Observe the position of the timing marks; the mark indicating proper degrees must be aligned with the pointer (Figure 7-69). If adjustment is necessary, loosen the distributor holddown clamp and bolt.
4. To advance the timing, rotate the distributor in the opposite direction of the rotor rotation. To set the timing back, rotate the distributor in the direction of the rotor rotation.
5. Tighten the holddown clamp and bolt. Recheck the timing.
6. When satisfied with the timing, reconnect the vacuum lines and readjust the idle (if necessary). Whenever idle speed is adjusted, the timing should be rechecked.

ADVANCE MECHANISM TESTING

Dwell, timing, and idle speed should all be adjusted before the spark advance is tested. If the vehicle contains a timing control system that affects the

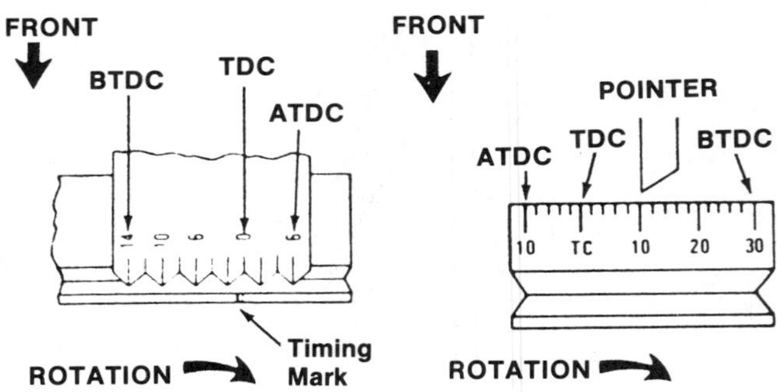

FIGURE 7-69 Observing the position of the timing marks

vacuum advance for emission control, this should also be checked; do it after the centrifugal and vacuum systems are tested. Distributor advance specifications are crucial in order to achieve accurate test results. Known as advance curves, these specs are listed as part of the manufacturer's part number.

Adjustable Timing Lights

Use the following procedure to perform advance mechanism tests with an adjustable timing light:

1. Connect a tachometer and an adjustable timing light to the engine. Set the timing advance knob on the light to zero.
2. Start the engine and run it at idle speed. Check the initial (base) timing setting.
3. Accelerate the engine to 2500 rpm, or whatever speed is specified in the service manual for checking total advance.
4. Turn the advance knob on the timing light while observing the timing marks. Continue turning until the marks are aligned at the initial setting established in step 2.
5. If the total advance reading on the timing light meter is not within specifications, the individual advance mechanisms must be tested to determine where the fault lies; proceed to the next step. In fact, it may be a good idea to perform the individual tests even if the reading is within specifications.
6. Reduce the engine speed to idle, then disconnect and plug the distributor vacuum lines. Timing advance will now be provided only by the centrifugal advance mechanism.
7. Accelerate the engine to the recommended test speed.

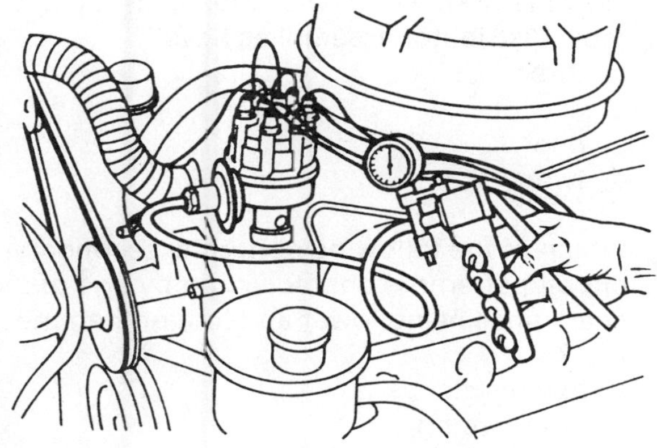

FIGURE 7-70 Using a hand-operated vacuum pump to apply vacuum to the advance diaphragm

8. Rotate the timing light advance meter while observing the timing marks, as in step 4. Continue turning until the marks are aligned at the initial timing setting.
9. The reading now recorded on the timing light meter is the centrifugal advance. If it is within specifications, proceed to the next step. If not, the centrifugal advance must be repaired or replaced.
10. While maintaining the test speed, unplug and reconnect the vacuum advance line to the distributor vacuum unit.

SHOP TALK

If working on a vehicle with a speed- or transmission-controlled spark system that prevents vacuum advance, vacuum must be applied to the advance diaphragm. This can be accomplished either by means of a hand-operated vacuum pump (Figure 7-70) or by connecting a manifold vacuum line to the diaphragm.

11. Observe the timing marks again, and adjust them until they are aligned at the initial (base) timing setting.
12. Subtract the centrifugal advance reading of step 9 from the new reading. The difference is the vacuum advance reading. If the vacuum advance is within specifications, the vacuum advance unit is working properly. If not, it requires adjustment or replacement.

If the distributor in question has a dual-diaphragm vacuum advance unit, the vacuum retard must be checked. Proceed as follows:

1. Disconnect and plug the distributor vacuum lines. Run the engine at idle speed.
2. While observing the timing marks with the adjustable timing light (Figure 7-71), con-

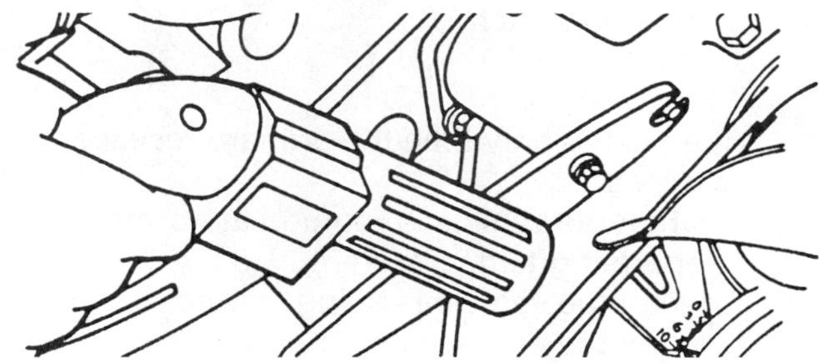

FIGURE 7-71 Using an adjustable timing light

nect the manifold vacuum hose to the inner vacuum chamber. The timing should immediately retard from the initial setting.
3. Observe the degree of retard on the timing light advance meter. If it is within specifications, the vacuum retard is operating properly. If not, the unit must be replaced.

Nonadjustable Timing Lights

Advance mechanism tests can be performed just as easily using nonadjustable timing lights. Use the following procedure:

1. Connect a tachometer and a nonadjustable timing light to the engine.
2. Start the engine and run it at idle speed.
3. Disconnect and plug the distributor vacuum lines.
4. Accelerate the engine to 2500 rpm while observing the timing marks. They should advance smoothly and steadily, indicating that the centrifugal advance is working.
5. While holding the engine speed above idle, unplug and connect the vacuum advance line to the distributor. The timing marks should advance an additional amount and engine speed should increase, indicating that the vacuum advance is working.
6. To check the vacuum retard on dual-diaphragm units, return the engine speed to idle.
7. Unplug and connect the manifold vacuum line to the retard vacuum chamber. Timing should retard as specified in the service manual (usually about 6 degrees to 12 degrees), and engine speed should decrease.

REVIEW QUESTIONS

1. Which of the following is not considered a component of the primary circuit?
   a. starting bypass
   b. battery
   c. ignition switch
   d. spark plugs

2. When is the only time the primary resistor can be bypassed?
   a. when the vehicle is running at normal operating temperature
   b. when the vehicle is being started
   c. when the engine is not running
   d. none of the above

3. Which of the following statements concerning primary circuit operation is incorrect?
   a. The action of the condenser prevents any arcing at the breaker points.
   b. Dwell is the number of degrees the distributor cam rotates while the breaker points are closed.
   c. Vehicles driven consistently at low speed experience more rapid point failure than those driven faster.
   d. Current flow has a more difficult time reaching full potential at high engine speeds than at low engine speeds.

4. If a pitted set of breaker points has a built-up metal mound on the positive joint, this is a sign that the condenser capacity is ______________ .
   a. too low
   b. too high
   c. erratic
   d. just right

5. The heart of the vacuum advance mechanism is the ______________ .
   a. distributor breaker plate
   b. spring-loaded diaphragm
   c. rubbing block of the movable point
   d. weight attached to the distributor shaft

6. When performing an available coil voltage test on a primary circuit, what should the initial voltmeter reading be when the breaker points close?
   a. 7 to 12 volts
   b. 5 to 7 volts
   c. 3 to 5 volts
   d. 12 to 15 volts

7. How many spark advance mechanisms do most distributor assemblies have?
   a. three
   b. five
   c. one
   d. two

8. Which of the following abnormal secondary circuit patterns is characterized by one of the spark lines being lower and longer than the rest?
   a. low resistance
   b. point arcing
   c. point bounce
   d. high resistance

9. During breaker point alignment, bend only the ______________.
   a. movable contact arm
   b. stationary positioning pin
   c. stationary contact support
   d. movable positioning pin

10. How much resistance per foot should the typical spark plug cable measure?
   a. 8,000 ohms
   b. 4,000 ohms
   c. 400 ohms
   d. zero

11. When replacing a set of spark plug cables, Technician A squeezes the boot to release trapped air from the tower. Technician B never squeezes the boot. Who is correct?
   a. Technician A
   b. Technician B
   c. Both A and B
   d. Neither A nor B

12. Upon inspection, a spark plug reveals a layer of dry, fluffy black carbon deposits on its tip. Technician A says the plug cannot be used again. Technician B says the plug can be cleaned and serviced, then reused. Who is correct?
   a. Technician A
   b. Technician B
   c. Both A and B
   d. Neither A nor B

13. Which of the following statements concerning regapping spark plugs is incorrect?
   a. New plugs should be checked for the correct air gap before being installed.
   b. In some cases, a flat gauge cannot be used for regapping.
   c. The air gap can be adjusted by bending the plug's center electrode.
   d. Never try to set a small gap, breaker point ignition plug to the wide gap needed for electronic ignitions.

14. Before timing can be set dynamically, ______________.
   a. the dwell must be correct
   b. the engine must be running at its normal operating temperature
   c. both a and b
   d. neither a nor b

15. What happens when the low-voltage current flow in the coil primary winding is interrupted by the switching device?
   a. The magnetic field collapses.
   b. A high-voltage surge is induced in the coil secondary winding.
   c. Both a and b.
   d. Neither a nor b.

16. Placing the ignition switch in the START or RUN position serves to ______________.
   a. start the low-voltage current flow
   b. route the low-voltage current flow to the accessory circuits
   c. both a and b
   d. neither a nor b

17. In maintaining the primary circuit voltage at 9 or 10 volts, the primary resistor protects the ______________.
   a. ignition switch
   b. relay
   c. contact points
   d. ballast resistor

18. The fixed contact, movable contact, movable arm, rubbing block, pivot, and spring make up the ______________.
   a. condenser assembly
   b. primary circuit
   c. secondary circuit
   d. breaker point assembly

19. Technician A says the typical automotive condenser capacity is 0.18 to 0.32 microfarads. Technician B says it is 0.32 to 0.50 microfarads. Who is right?
   a. Technician A
   b. Technician B
   c. Both A and B
   d. Neither A nor B

20. The condenser discharges its stored energy in the form of ______________.
   a. ions
   b. windings
   c. electrodes
   d. current

21. The insulated boots on the ends of the spark plug cables ______________.
   a. prevent voltage loss
   b. prevent dust and water infiltration

c. strengthen the connections to the coil, distributor cap, and plugs
d. all of the above

22. On an oscilloscope, which section ends with a points-open signal?
a. intermediate
b. dwell
c. firing
d. both a and b

23. When an open spark plug cable is suspected, which oscilloscope pattern should be used to locate the faulty plug circuit?
a. raster
b. stacked
c. display
d. superimposed

24. A breaker point voltage drop test produces a reading of 0.1 volt. Technician A says this means there is a problem. Technician B says it means that the points are in good condition. Who is right?
a. Technician A
b. Technician B
c. Both A and B
d. Neither A nor B

CHAPTER EIGHT

TROUBLESHOOTING ELECTRONIC IGNITION SYSTEMS

Objectives

Upon completion of this chapter, you should be able to:

- Identify the limitations of a conventional ignition system.
- Name the basic parts of an electronic ignition system and explain how they work.
- Explain the similarities and differences between different electronic switching devices.
- Describe the early electronic systems used on domestic cars.
- Explain the use of an oscilloscope to diagnose electronic ignition system performance.
- Implement electric troubleshooting techniques to test individual electronic components.

The conventional breaker point system described in the previous chapter satisfied the performance requirements placed on automotive ignition systems for many years. It performed the following three functions that every ignition system must do:

1. It provided sufficient voltage to the spark plug to force a spark across the air gap with sufficient intensity to ignite the air/fuel mixture in the combustion chamber.
2. It timed the spark to arrive at the correct cylinder at a specific moment in the compression stroke of that cylinder.
3. It varied the time the spark arrived at the cylinder in order to improve engine performance, fuel efficiency, and emission levels.

However, spark distribution and timing advance were mechanically controlled. As automotive engineers sought for ways to decrease emissions and increase fuel efficiency, the breaker point system was abandoned in favor of electronically controlled ignition systems.

VOLTAGE VERSUS BREAKER POINT LIFE

A breaker point ignition system has several disadvantages that are overcome by electronic systems. Foremost is the limited voltage that can be handled by the system. The high secondary system voltages needed to ignite lean air/fuel mixtures are unattainable with breaker points, due to excessive wear of the points. When the points are suddenly opened, the primary current momentarily arcs between the points, electrolyzing and corroding the metal surface of the contacts. The greater the current, the faster the points wear (Figure 8-1A).

A

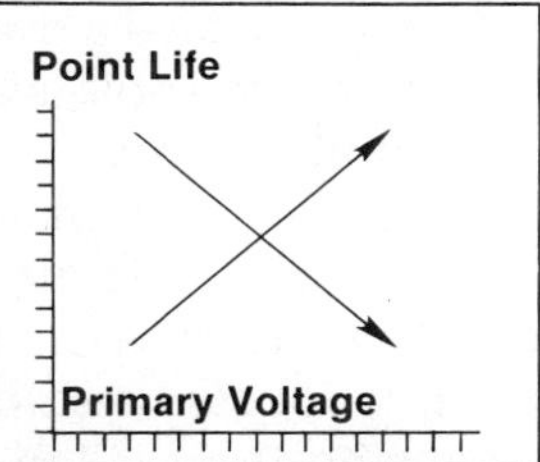

B

Dwell Period
Engine rpm

FIGURE 8-1 Disadvantages inherent in a breaker point ignition system

The primary circuit in most breaker point systems operates on 3.5 to 4.0 amperes (depending on temperature). This current value is a compromise between contact point life and system performance. Maximum point life is realized at approximately 1 ampere. As current increases from that level, point life decreases steadily until reaching a current level of just over 4 amperes. If the current increases above 4 amperes, point life begins to decrease at an increased rate, resulting in a very limited point life. Faster wearing points require extra maintenance and result in an ever decreasing dwell period, which in turn decreases the potential voltage induced in the secondary system.

With the electronic systems introduced in the early 1970s, primary voltages were significantly increased beyond 4 amperes, while actually decreasing the maintenance required on the system.

ENGINE SPEED VERSUS DWELL PERIOD

Another handicap of the mechanical breaker point system is that as engine speed increases, the dwell time of the system decreases: this, in turn, decreases the output of the coil (Figure 8–1B). For the ignition coil to generate maximum secondary voltage, the maximum primary current must be attained before the primary circuit is opened. This allows the magnetic field in the coil to reach full strength (saturation). In a breaker point system, the length of time the primary current is closed is controlled by the speed of the breaker cam. This period of time is called a *dwell angle* and is expressed in a number of degrees of distributor shaft rotation. For example, many V-8 engines have a dwell angle of 30 degrees during which time the points are closed and current builds in the coil primary. This dwell angle remains constant regardless of engine speed, but as engine rpm increases, the actual time the points are closed decreases. Any increase in engine speed above a specific rpm reduces the saturation time of the ignition coil, causing the available voltage and coil energy to decrease.

This phenomenon is due to the fact that the current in the coil does not instantaneously reach a value of 4 amperes when the contact points close. Voltage in the coil must build for several milliseconds for this value to be reached. At 1000 engine rpm, the distributor shaft rotates once every 0.12 seconds. Of this time, the points are closed for 0.10 seconds, or 10 milliseconds, for every cylinder of an 8-cylinder engine. This is sufficient time for the primary current to build up to its maximum current of just over 4 amperes. This time versus voltage relationship is shown in Figure 8–2.

When the engine speed increases to 2000 rpm, the time during which the points are closed for each plug firing is reduced to 5 milliseconds. A dwell period of 5 milliseconds allows the primary current to build to 3.8 amperes. At 3000 rpm, the dwell period drops to 3.3 milliseconds and the current to 3.2 amperes. The reduced saturation time weakens the secondary voltage, causing the spark plugs to misfire at high rpms. This increases exhaust emission and decreases fuel economy and engine performance.

ELECTRONICALLY CONTROLLED DWELL

An electronic ignition system, however, is not limited by a fixed dwell angle. The system's control unit can vary the "ON" time of the primary circuit based on engine speed, load, and temperature. As the chart in Figure 8–2 shows, GM's HEI system maintains a constant voltage level until reaching 3000 rpm. And because coil primary current levels are not limited by breaker points, low-resistance coils are used in the electronic ignition system. By decreasing the resistance in the primary circuit, the saturation time of the coil is greatly reduced. It takes 10 milliseconds for the current to reach maximum saturation in a coil with a resistance of 2.6 ohms. In a coil used in electronic ignition systems, the primary winding can have a resistance as low as 0.5 ohms. This allows full current to be reached in about 3.4 milliseconds. Because it takes less time to reach full current, coil saturation can be obtained at much higher engine speeds. For example, the High-Energy Ignition (HEI) system developed by General Motors in 1974 is able to generate 35,000 volts at engine speeds above 3000 rpm. A typical engine breaker point system, on the other hand, develops a peak of 20,000 volts at 1000 rpm; above this speed, the voltage drops off.

In summary, electronic ignition systems have many advantages over breaker point ignition systems. An electronic system costs more than a breaker point system, but the following advantages outweigh the increased cost:

- High secondary voltage, even at high engine speeds
- More reliable performance at all engine speeds
- Potential for more responsive and variable spark advance curves
- Decreased maintenance requirements
- Reduced exhaust emissions due to lack of change in ignition dwell angle

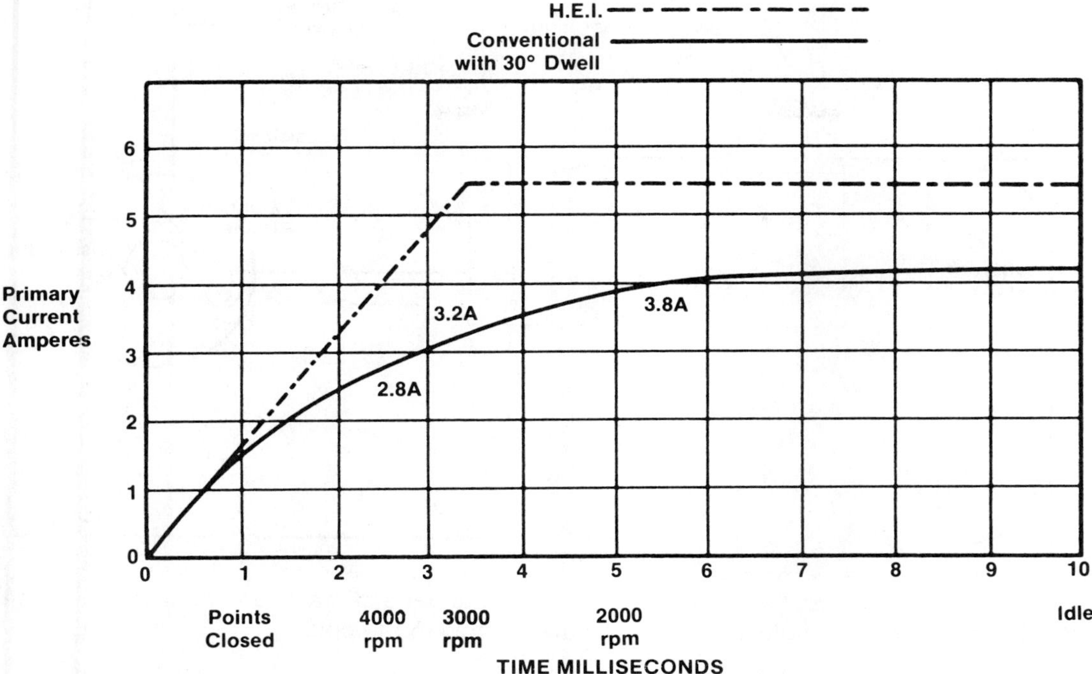

FIGURE 8-2 Current levels in GM's high-energy ignition system remain higher over a longer range of engine rpm as compared to the performance of a conventional system.

ELECTRONIC SYSTEM OVERVIEW

Early electronic ignition systems are very similar to contact breaker point systems. A coil is still used to generate secondary system voltage and a distributor is still used to direct secondary current to the spark plugs. Spark advance is provided by conventional centrifugal and vacuum advance mechanisms.

However, there are two basic differences between conventional and electronic ignition systems.

1. Mechanisms used to interrupt the primary system
2. Method used to time the spark delivery to the plugs

TRANSISTORIZED SWITCHES

Electronic ignition systems control the primary circuit, using an NPN type transistor. The NPN type transistor has three terminals: emitter, base, and collector. The emitter is connected to the negative (-) battery terminal and takes the place of the fixed contact point. The collector is connected to the negative (-) end of the coil primary circuit, taking the place of the movable contact point. When a small amount of current is supplied to the third transistor terminal (the center base), the collector and emitter act as if they are closed contact points (a conductor), allowing current to build up in the coil primary circuit. When the current to the base is interrupted, the collector and emitter act as an open contact (an insulator), interrupting the coil primary current. An example of how this works is shown in Figure 8-3, which is a simplified diagram of an early electronic ignition system using contact points.

Instead of running the entire coil primary voltage through the contact points, a small amount of voltage is bled off from the P section of an NPN transistor. As long as current flows out of the P section, the transistor is a conductor and current flows through the entire system. When the points open, current stops flowing out of the P section, the transistor becomes an insulator, and current stops flowing in the system. This causes the coil's magnetic field to collapse, firing the coil secondary. The advantage of this system is longer point life since the points carry very little current from the transistor base.

The transistor is normally part of a control module. The control module might be remotely mounted somewhere under the hood, or it might be integrated into the distributor or attached to it. In addition to the transistor switching device, the control module usually contains circuits for amplifying the ignition pulse and for regulating the dwell time.

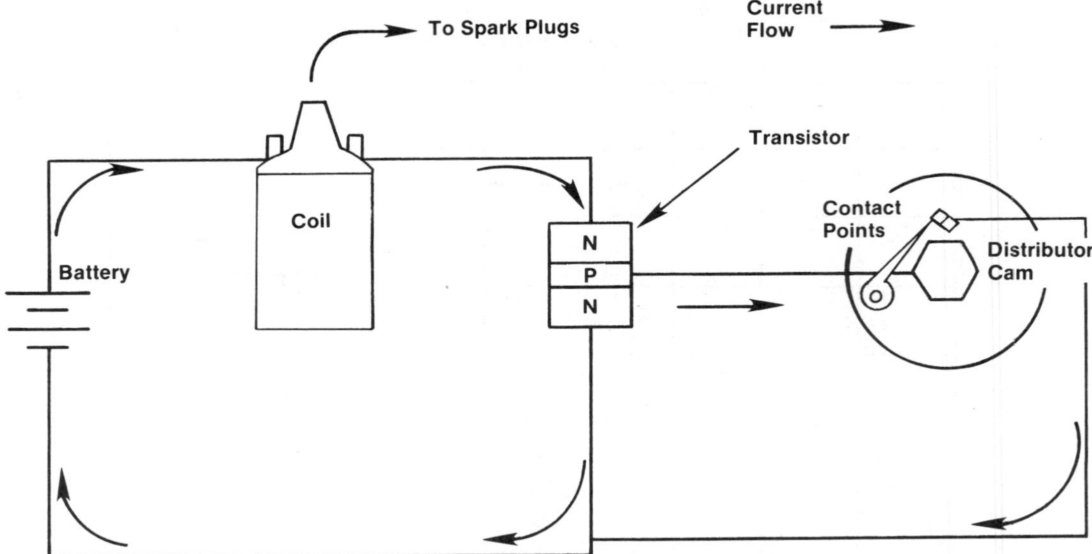

FIGURE 8-3 While current flows through the center base, the transistor acts as a conductor. When the points open, the transistor acts as an insulator, stopping current flow in the coil.

ENGINE POSITION SENSORS

The most significant difference between electronic ignition systems is in the method used to time the spark. On a conventional system, the pistons, crankshaft, and distributor shaft are joined by connecting rods and gears so that the up and down movement of the pistons is converted into the rotational movement of the distributor shaft. The cam is pressed onto the end of the distributor shaft and the breaker points are positioned so that the cam forces the points open each time a piston reaches approximately top dead center on its compression stroke. In this way, spark is generated in fixed relation to piston position.

An electronic ignition system must also generate a spark in relation to piston (or crankshaft) position. Because an electronic transistor has replaced the points as the primary system switching device, a device must be used to both sense crankshaft position and controlled current flow through the base of the transistor. The following paragraphs describe several popular engine position-controlled switching means.

BREAKER POINT SWITCHING

The first solid state ignition systems were an addition to the breaker point system, rather than a replacement for it. An electronic control circuit was added to the existing primary circuit (Figure 8-4). No change was made to the distributor or contact points, except that the condenser was no longer necessary.

However, the contact points in a breaker point system with electronic control do not carry the primary voltage. Battery voltage is provided directly to the points in a low current level (approximately 1 ampere). When the points are closed, current flows through the points to the emitter-to-base circuit of

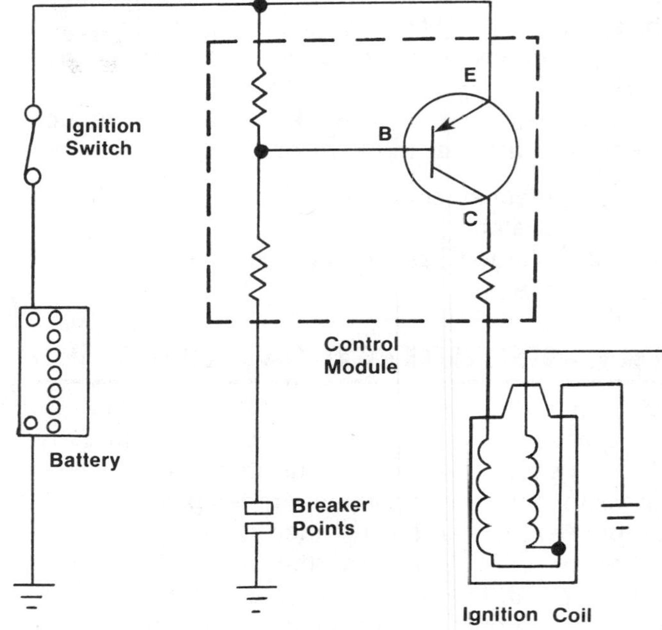

FIGURE 8-4 A transistorized breaker point system

the transistor. This activates the emitter-to-collector circuit, completing the ground of the primary coil windings. Just as in a conventional breaker point system, current builds in the primary ignition system while the points are closed.

When the revolving cam opens the points, the current to the emitter-to-base circuit is interrupted. This also interrupts the current flow through the emitter-to-collector circuit. The primary ignition circuit is thus abruptly opened, inducing a high level voltage surge through the secondary ignition circuit.

A breaker point system fitted with solid state control of the primary system has several advantages over a conventional breaker point system. Because the emitter-to-base circuit carries a low current level, the points are not subject to arcing each time they open and close. The service life of the points is thus greatly extended. Electronic switching of the primary system also improves performance. The transistor opens the primary circuit very rapidly, much quicker than the mechanical switching provided by breaker points. As a rule, the quicker the magnetic field in the coil collapses, the greater the induced voltage in the secondary system will be.

The electronic breaker point system has some of the same disadvantages as a conventional breaker point ignition. The rubbing block of the points is subject to wear by the cam. This eventually affects both the timing and dwell period. At high engine speeds, the dwell period is also reduced significantly, creating the same performance problems associated with breaker points as described earlier in this chapter.

MAGNETIC PULSE GENERATOR

The first electronic switching device to replace the breaker point system was the magnetic pulse generator. A magnetic pulse generator contains a trigger wheel (also called a *reluctor*) pressed over the distributor shaft in place of the conventional cam and a pick-up coil that has been substituted for contact points. Figure 8-5 shows the magnetic pulse generator used by Chrysler.

The trigger wheel performs much the same function of a cam. It is pressed onto the distributor shaft in relation to the crankshaft position. The position of the trigger teeth as they rotate past the pick-up coil correspond to each cylinder as individual positions approach top dead center.

The pick-up coil assembly is a permanent magnet with fine wire windings around its pole. The permanent magnet induces a voltage in the pick-up coil. As the teeth of the reluctor rotate past the coil, the magnetic field and induced voltage level changes (Figure 8-6). As a tooth moves into the magnetic field, the field strengthens and the voltage in the coil increases. The voltage level is greatest when the tooth is aligned with the pole. When the control unit senses the increase of voltage in the pickup, it turns on the primary system of the coil.

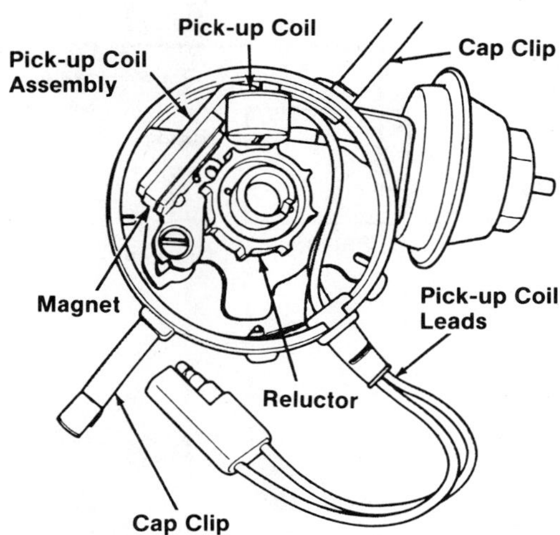

FIGURE 8-5 A magnetic pulse generator used on early Chrysler electronic ignition systems

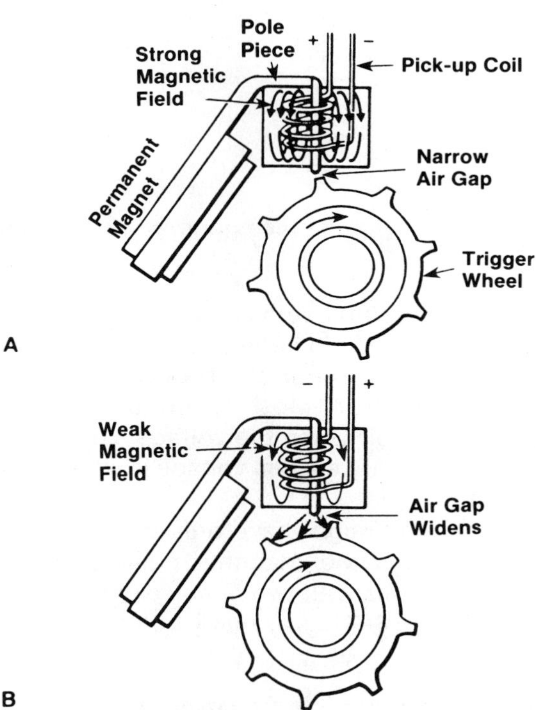

FIGURE 8-6 The magnetic field in the pole piece expands as the reluctor teeth approach. As the teeth move away, the field collapses. The changing strength of the magnetic field varies the voltage induced in the pick-up coil.

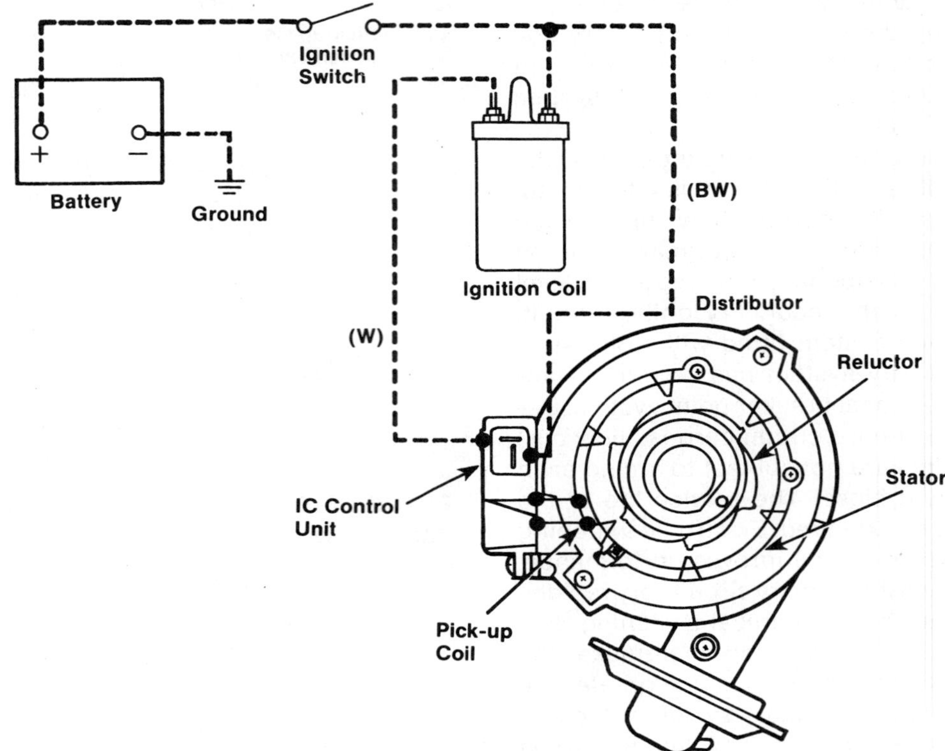

FIGURE 8-7 A magnetic pulse generator used in early Datsun electronic ignition systems

As the reluctor tooth begins to move away from the pole, the magnetic field begins to collapse. This induces a negative voltage in the pick-up coil and reverses the current flow. Most control units open the primary coil circuit when the voltage signal reverses. Thus, secondary system voltage is very suddenly induced.

The pick-up coil might have only one pole as shown in Figure 8-5. Other magnetic pulse generators have pick-up coils with two or more poles. The one shown in Figure 8-7 has as many poles as it has trigger teeth.

METAL DETECTION SENSOR

Another early electronic switching device is the metal detection sensor. It performs much like the magnetic pulse generator. A trigger wheel is pressed over the distributor shaft in place of the traditional cam and a pick-up coil detects the passing of the trigger teeth as the distributor rotates. However, the pick-up coil does not have a permanent magnet. Instead, the pick-up coil is an electromagnet; a low level of alternating current is supplied to the coil by an electronic control unit, inducing a weak magnetic field around the coil. As the reluctor on the distributor shaft rotates, the trigger teeth pass very close to the coil (Figure 8-8). As the teeth pass in and out of the coil's magnetic field, the magnetic field builds and collapses, producing a corresponding change in the coil's voltage. The voltage changes are monitored by the control unit to determine the crankshaft position.

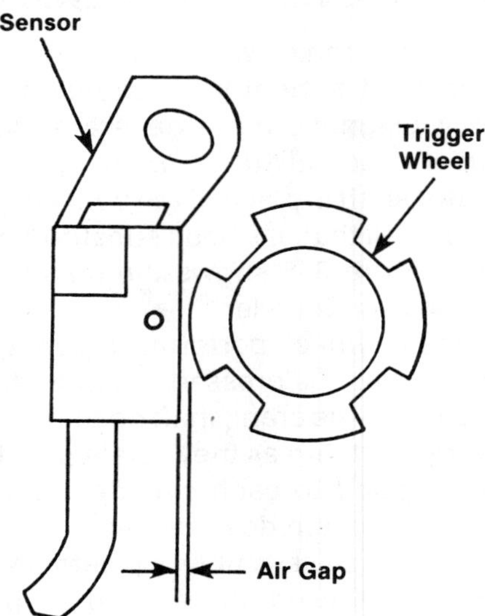

FIGURE 8-8 In a metal detecting sensor, the revolving trigger wheel teeth alter the magnetic field produced by the electromagnet in the pick-up coil.

HALL-EFFECT SENSOR

A fourth electronic engine position sensor used in early electronic ignition systems is the Hall-effect sensor. In a Hall-effect sensor, voltage signals are not magnetically induced. Instead, a sensor incorporating a semiconductor chip and a permanent magnet generates electrical pulses using the Hall-effect principle.

The Hall-effect principle states that when electrons flow through a conductor that is permeated by lines of magnetic force, the electrons are deflected perpendicular to the direction of the magnetic field and perpendicular to the direction of current flow (Figure 8-9). This concentrates electrons along the edge of the conductor. This effect is especially pronounced when using a semiconductor material. In the Hall-effect sensor, the concentration of electrons along the edge of the semiconductor chip sends a constant low-level current to the electronic control unit. As shown in Figure 8-10, a Hall-effect sensor has a reluctor with wide shutters rather than teeth. As the reluctor rotates with the distributor shaft, the shutters pass between the semiconductor chip and the permanent magnetic (Figure 8-11). The low reluctance shutters absorb the magnetic field, which allows the electrons concentrated along the edge of the semiconductor to disperse. Current flow to the control unit stops.

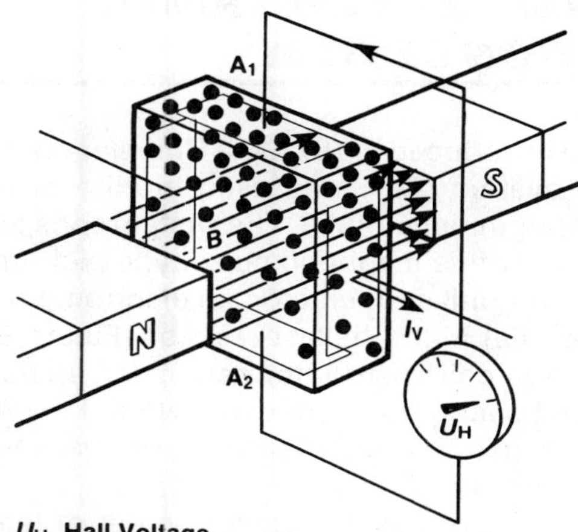

FIGURE 8-9 The Hall-effect: electrons move perpendicular to lines of magnetic flux and electrical current.

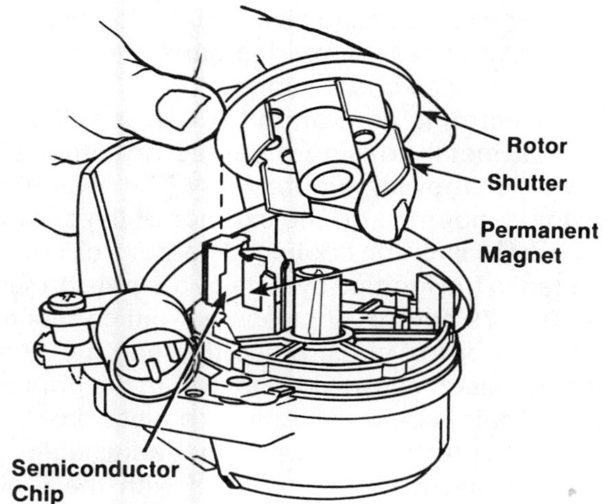

FIGURE 8-10 A Hall-effect triggering device used in early Chrysler electronic ignition systems

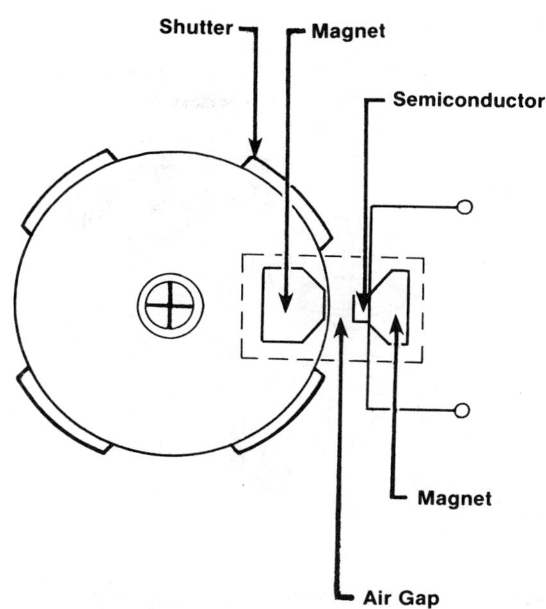

FIGURE 8-11 As the reluctor rotates with the distributor shaft, the shutters pass between the magnet and semiconductor, interrupting the Hall-effect voltage generation.

When the shutter interrupts the voltage signal to the control unit, the control unit turns on the primary ignition circuit. The width of the shutter determines the dwell period. After the shutter passes by the sensor, voltage is again created along the edge of the chip and a signal is sent to the control unit. The transistor then interrupts the primary ignition circuit and secondary system voltage is induced.

The Hall-effect sensor has several advantages over other crankshaft position sensors. Unlike magnetic pulse generators, Hall-effect voltage is not affected by engine speed. The strength of the voltage induced by magnetic lines of force varies as the speed of the lines increases or decreases. At high engine speeds, engine performance is reduced.

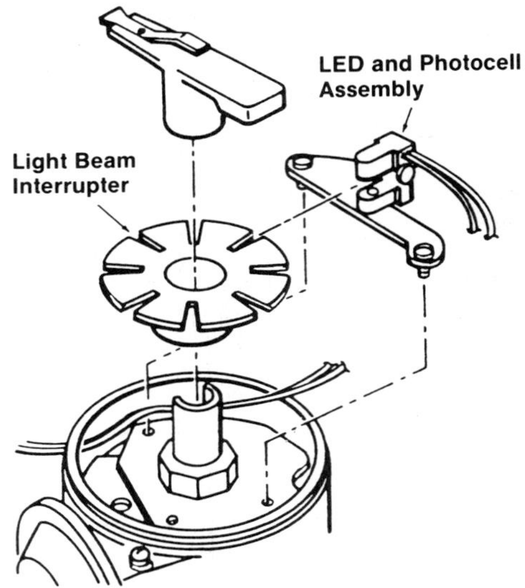

FIGURE 8–12 A photoelectric triggering device

The Hall-effect sensor on the other hand produces a constant strength signal. Thus its performance is more consistent over a wide range of engine operating conditions.

PHOTOELECTRIC SENSOR

A fifth type of crank angle sensor is the photoelectric sensor. The parts of this sensor include a light emitting diode (LED), a light sensitive phototransistor (photo cell), and a slotted disc called a light beam interrupter (Figure 8–12).

The slotted disc is attached to the distributor shaft. The LED and the photo cell are situated over and under the disc opposite of each other. As the slotted disc rotates between the LED and photo cell, light from the LED shines through the slots. The intermittent flashes of light are translated by the photo cell into voltage pulses. When the voltage signal occurs, the control unit turns the primary system on. When the disc interrupts the light and the voltage signal ceases, the control unit turns the primary system off, causing the magnetic field in the coil to collapse and sending a surge of voltage to a spark plug.

The photoelectric sensor sends a very reliable signal to the control unit, especially at low engine speeds. However, it has found very limited application on original equipment ignition systems.

ELECTRONIC IGNITION SYSTEMS

Early ignition systems used one of the triggering devices just described. Base ignition timing was determined by the rotating reluctor and the alignment of the trigger teeth with the pick-up coil as each piston approached top dead center. Timing advance was provided by the centrifugal and vacuum advance mechanisms found on conventional systems.

As engine electronics progressed, timing advance became a function of computers. Computer-controlled spark advance systems are described in the next chapter. The remainder of this chapter discusses early electronic ignition systems (those with mechanical advance components) and their diagnostic and servicing procedures.

CHRYSLER'S ELECTRONIC IGNITION SYSTEM

First introduced in 1971, the Chrysler electronic ignition system (EIS) was used in all production engines beginning in 1973. The system consists of a distributor with a magnetic pulse type pick-up unit and reluctor unit and reluctor, an electronic control unit (module) and a ballast resistor (Figure 8-13). The ignition coil, distributor cap, rotor, spark plug wires, and spark plugs are conventional in design. Spark advance is provided with centrifugal and vacuum advance mechanisms.

The reluctor has the same number of teeth as the engine has cylinders (Figure 8-5). The pick-up unit consists of a permanent magnet and coil wound around a pole piece. As the tooth of the reluctor passes the pick-up unit, an electrical impulse is sent to the electronic control unit. This pulse signals the transistor to open the primary circuit, firing the plug. Once the plug stops firing, the transistor closes the primary coil circuit.

Since the reluctor and pick-up unit do not touch each other and have no wearing parts, they require no periodic service or adjustment.

The electronic control unit (ECU) is self-contained in a metal housing mounted on either the fenderwell or firewall, Figure 8–14. The switching transistor is mounted on the exterior of the control unit in a heat sink for cooling. The control unit is connected to the ignition system by a 5-pin connector on 1971–79 control units and a 4-pin connector on 1980–82 control units. Since the length of time that the transistor allows current flow in the primary ignition circuit is determined by the electronic circuitry in the control unit, dwell is not adjustable.

A dual ballast resistor is used with the 5-pin control unit. The ceramic ballast resistor is mounted on the firewall and has a 0.5 ohm resistance that maintains a constant primary current with variation

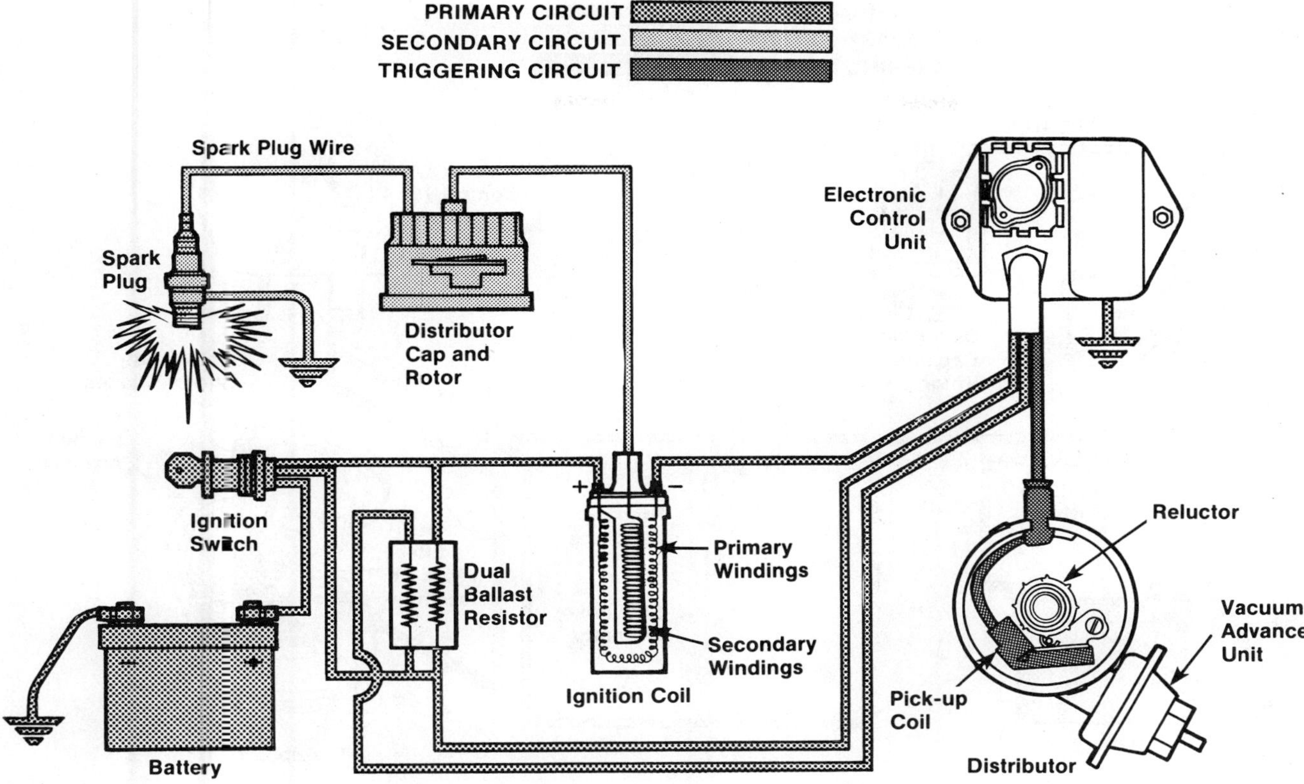

FIGURE 8–13 Chrysler's EIS ignition system

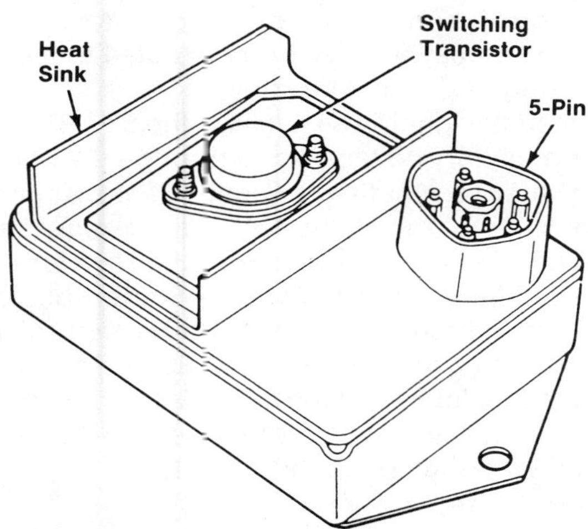

FIGURE 8–14 Chrysler electronic control unit used on 1971 to 1979 vehicles

in engine speed. The auxiliary ballast resistor uses a 5-ohm resistance to limit voltage to the control unit.

On the 4-pin control unit, a single element 1.2 ohm ignition resistor is used. During cranking the ignition resistor is bypassed, applying full battery voltage to the coil.

CHRYSLER EIS HALL-EFFECT

The 1980 Federal Omni and Horizon are both equipped with a Hall-effect electronic ignition system. The system consists of a Hall-effect distributor, electronic control unit (module), a conventional coil centrifugal and vacuum advance mechanisms, spark plug wires and spark plugs (Figure 8–15). The distributor rotor is of a special design. The distributor cap, advance mechanism, and spark plug wires are of a conventional design.

The distributor contains an integrated circuit mounted on the distributor switch plate, which includes a Hall-effect element (pick-up coil) and signal conditioner. A permanent magnet is also mounted on the distributor plate facing the module with an air gap between them (Figure 8–16).

The distributor rotor has four metal shutters (on a 4-cylinder engine), which rotates through the air gap when the distributor shaft rotates. The passing of the metal shutters (reluctor) between the magnet and pickup causes a voltage change in the sensor. As stated earlier in this chapter, with the Hall-effect distributor the signal voltage is not changed by the speed of the distributor as it is in an inductive magnetic signal generating system. Since the distributor

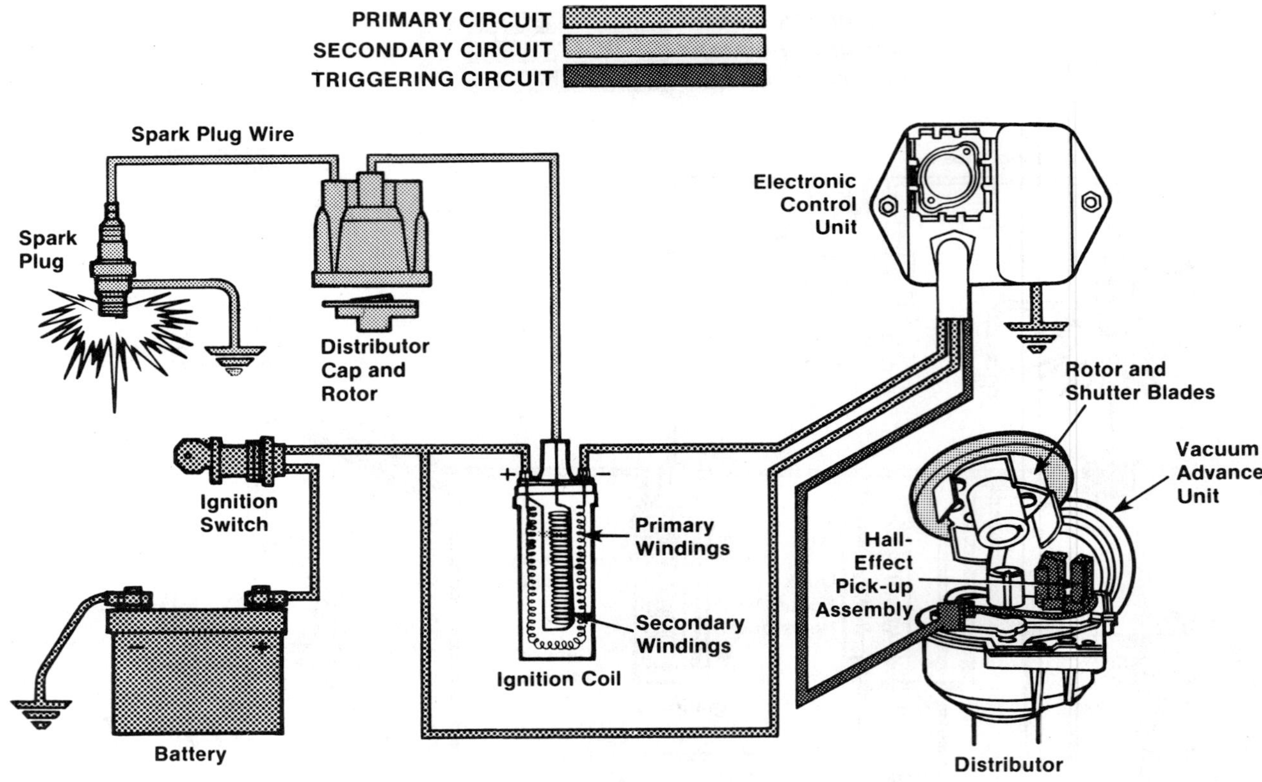

FIGURE 8–15 Chrysler Hall-effect EIS ignition system

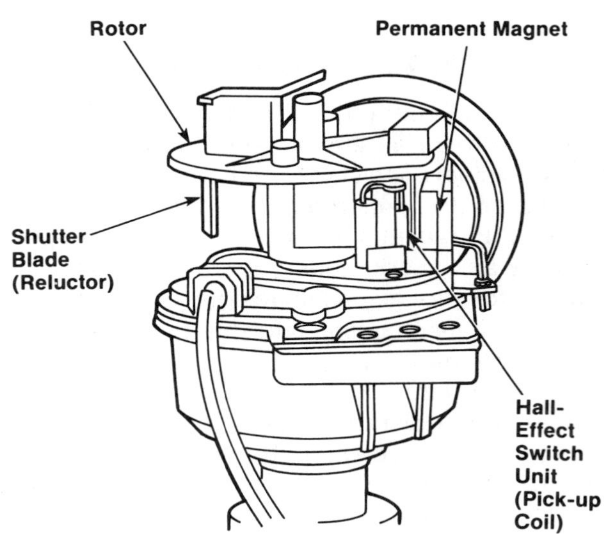

FIGURE 8–16 Chrysler Hall-effect distributor

has no contacting parts, adjustment is not necessary. The distributor rotor must be grounded through the distributor shaft for the system to operate.

The special Hall-effect electronic control unit is located in a metal housing on the firewall. Like the ECU used on EIS systems with magnetic pulse generators, it contains an exposed switching transistor and a heat sink, and is connected to the system through a 5-pin connector.

When the ignition switch is in the START or ON position, battery voltage flows through the coil primary windings, creating a magnetic field. As shown in Figure 8–15, the Hall-effect system requires no ignition resistor and uses no bypass circuit for starting. The voltage signal from the Hall-effect pickup is amplified and sent on to the electronic control unit, which contains the switching transistor. The switching transistor interrupts the flow of current in the coil primary circuit, inducing the high secondary voltage which fires the spark plugs.

AMC'S BREAKERLESS INDUCTIVE DISCHARGE SYSTEM

The Prestolite breakerless inductive discharge (BID) system (Figure 8–17) was used by AMC from 1975 through 1977. The BID ignition system consists of an electronic control unit, distributor, coil, secondary ignition wires and spark plugs. It was also used, with some modifications, on 1978 to 1981 Interna-

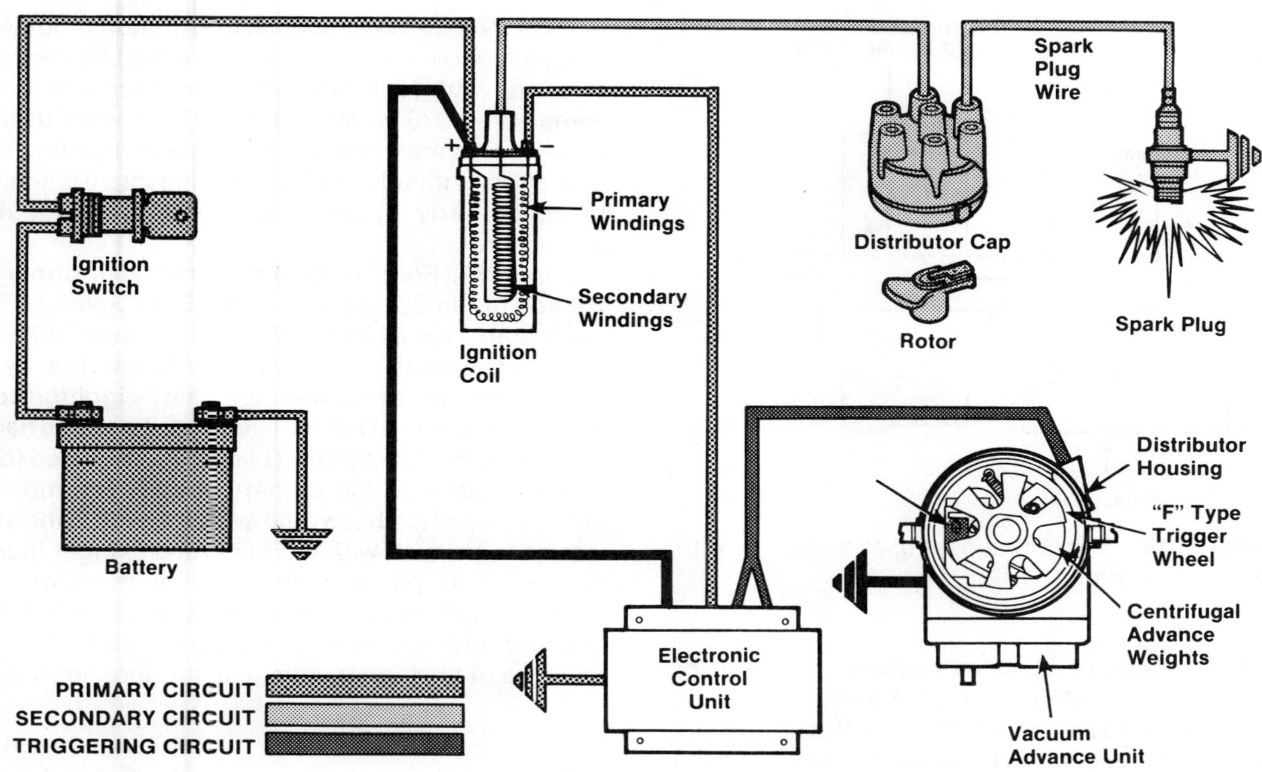

FIGURE 8–17 AMC Prestolite BID ignition system

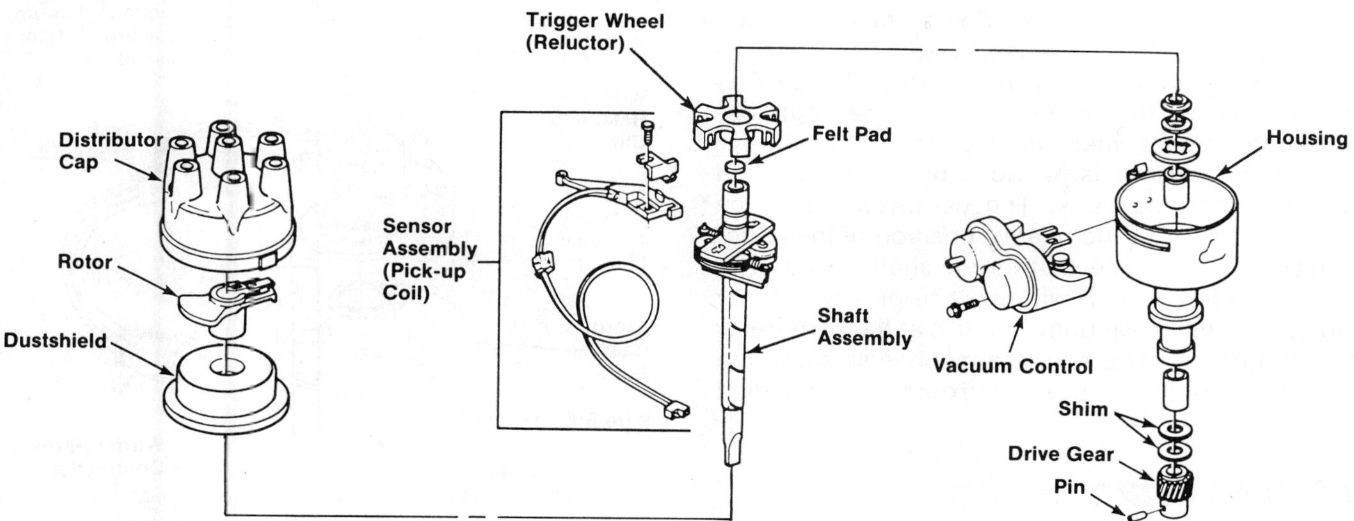

FIGURE 8–18 AMC distributor with a metal detection type impulse generator

tional Harvester vehicles. The distributor has conventional mechanical (centrifugal) and vacuum advance systems except that a sensor and trigger wheel replace the contact points, condenser, and distributor cam (Figure 8–18).

The electronic control unit (module) is a solid state, nonserviceable sealed unit. The control unit has a built-in current regulator and reverse polarity protection circuits within it. Because there is current regulation in the control unit, battery voltage is supplied to the positive (+) side of the coil whenever the ignition switch is on. Therefore, ignition bypass during cranking is not required. Current flow in the coil primary is controlled by the control unit.

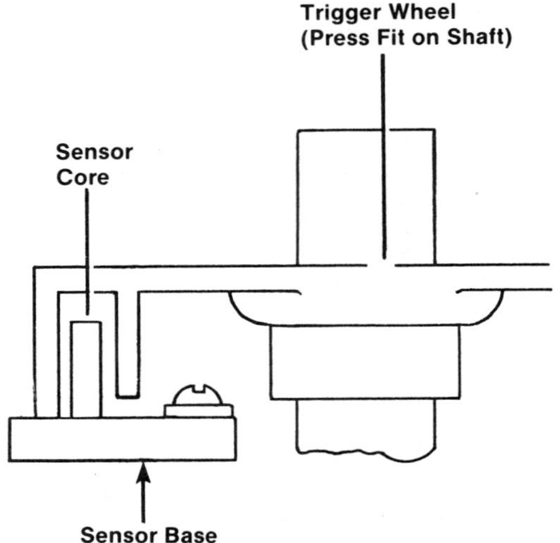

FIGURE 8-19 The BID sensor uses a reluctor with "F"-shaped shutters.

The distributor contains a pick-up coil and "F" type trigger wheel (Figure 8-19). Alternating current is supplied to the sensor from the control unit which builds up a magnetic field around the sensor. As the metal teeth of the trigger wheel (reluctor) pass through this electromagnetic field, a signal that opens the coil's primary circuit, is sent to the control unit. Since there are no wearing surfaces between the trigger wheel and sensor, dwell remains constant and no adjustments are required. The dwell is determined by the control unit and the angle between the trigger wheel and pickup.

Spark advance is provided by centrifugal and vacuum advance units. The centrifugal advance mechanism varies the relative position of the reluctor according to the distributor shaft speed. The vacuum advance unit will vary the pick-up sensor position with the vacuum changes at the carburetor spark port. The two advance mechanisms operate separately, but work together to obtain the specified spark advance.

FORD'S ELECTRONIC IGNITION SYSTEMS

Ford's solid state ignition (SSI) system was first introduced in 1974. As shown in Figure 8-20, the SSI system uses a distributor containing a pick-up coil (also called a stator), and a reluctor (called an armature by Ford). The system has an ignition module, a conventional coil, and a ballast resistor wire (Figure 8-21).

On 1974 to 1975 models, the ignition module is connected to the ignition system using two connectors, one with four wires and the other with three wires. The 1976 version of the SSI module uses a four-wire connector and a two-wire connector. The operation of this electronic ignition system is similar to other early systems with a magnetic pulse generator.

In 1977, Ford introduced a high-performance version of the SSI system, called Dura Spark I (Figure 8-22). This system was used through 1979 on certain California V-8 engine vehicles. The Dura Spark I system used a new high energy ignition coil, one with lower resistance. The ignition module had a current limiting circuit that eliminated the need for a ballast resistor in the primary circuit. The module also had a circuit that would switch off the transistor if the engine is off with the ignition on longer than 1 second. This prevents the ignition module from overheating if the engine stalls. A larger distributor cap and rotor are used on the Dura Spark I system because of the higher voltage in the secondary sys-

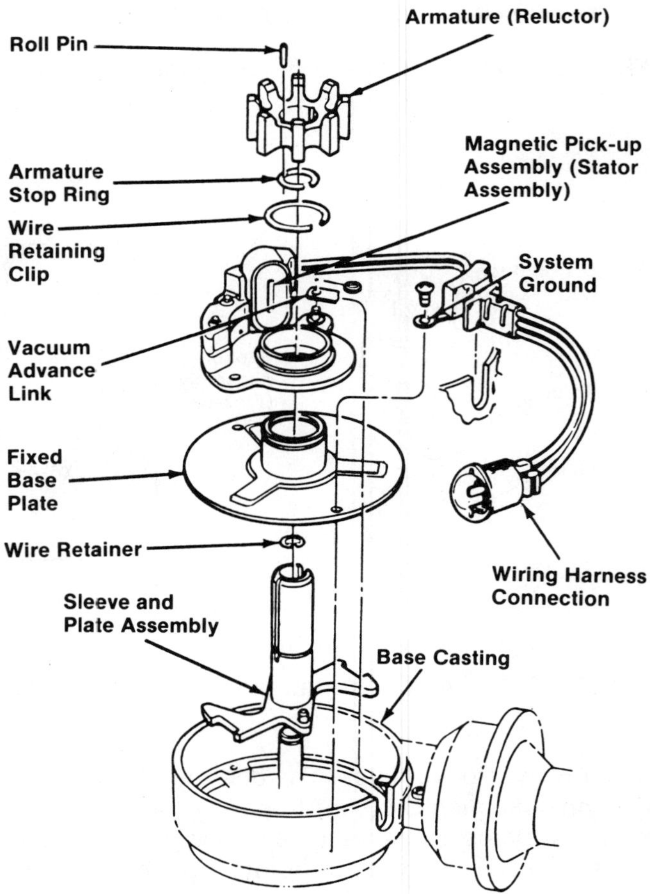

FIGURE 8-20 Ford breakerless ignition distributor

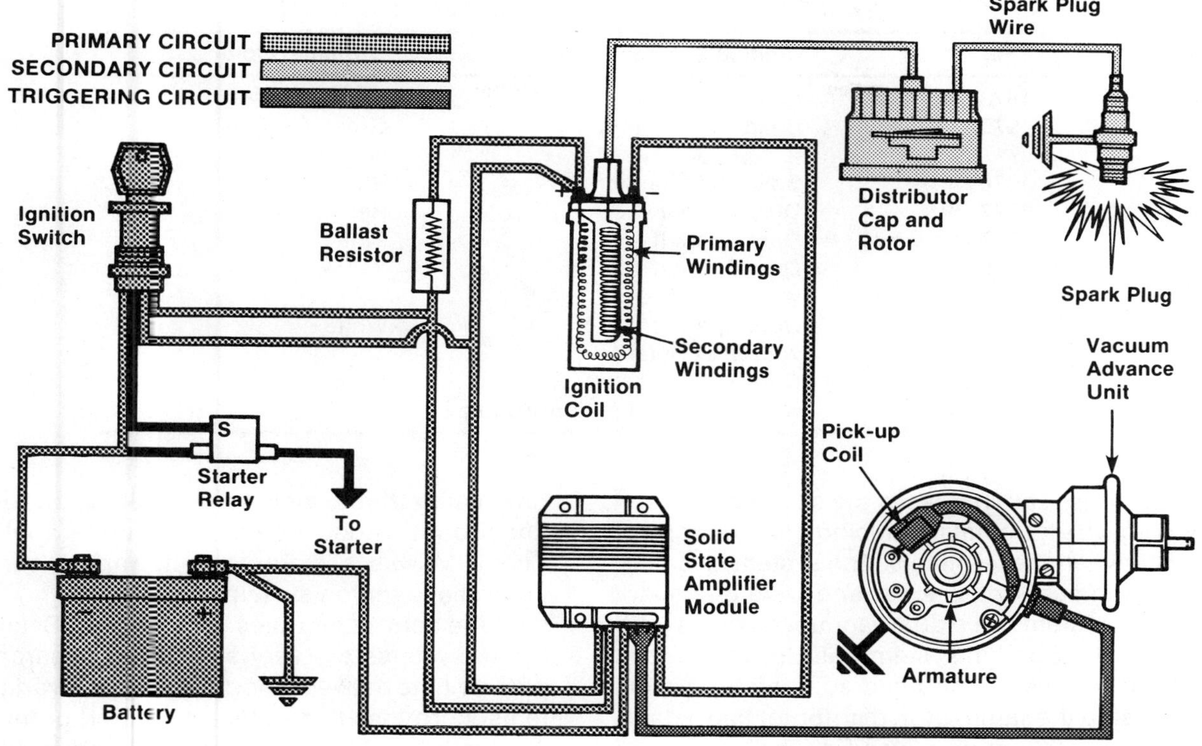

FIGURE 8–21 Ford SSI system

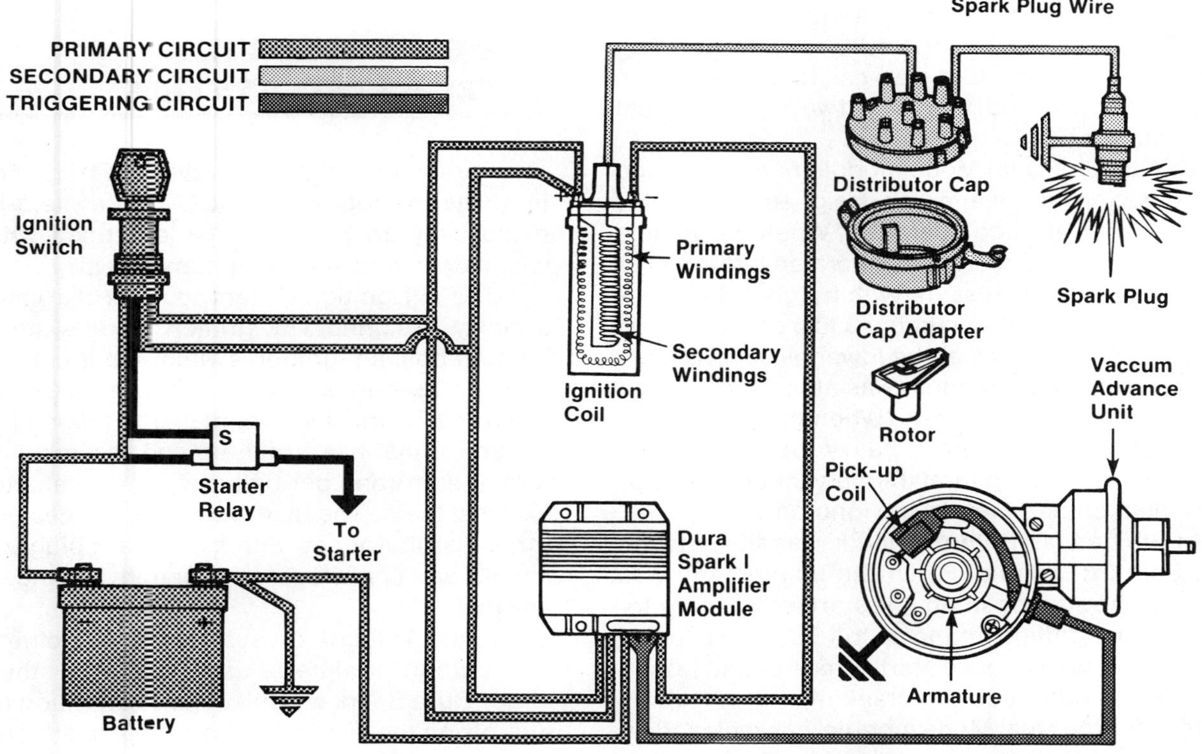

FIGURE 8–22 Ford Dura Spark I system

TABLE 8-1: IGNITION MODULE IDENTIFICATION

Year	Ignition System	Grommet Color
1973-74	Solid state ignition	Black (7 wires)
1975	Solid state ignition	Green (early—7 wires; late—6 wires)
1976	Solid state ignition	Blue
1977-79	Dura Spark I	Red
1977 and later	Dura Spark II	Blue
1979-80	Dura Spark II with Dual Mode	Yellow
1979 and later	Dura Spark II with Start Retard	White
1981 and later	Dura Spark with Universal Ignition Module	Yellow

tem. The larger cap increases the spacing between spark plug wire terminals, reducing the likelihood of arcing between the terminals. The system also uses high tension spark plug wires and wide gap spark plugs. The system has a dual vacuum advance unit.

Ford Dura Spark II system was also introduced in 1977. It too was a modified solid state ignition system, using the same rotor, distributor cap, adapter, and secondary wiring as used in the Dura Spark I system. The Dura Spark II system had an increased primary current flow, providing a stronger spark for firing lean fuel mixtures.

The Dura Spark II control module was modified in 1978 to retard the timing under certain operating conditions. This module was called a Dual Mode ignition module, and there are two types of Dual Mode modules.

One type of Dual Mode module had an altitude compensation circuit and was designed for vehicles made for high altitude operation. When the vehicle equipped with this module is operated a low elevations, a barometric pressure switch signals the module to retard the ignition timing 3 to 6 degrees. This prevented spark knock at the lower elevations.

The Dual Mode module was also used with an economy calibration system. When the engine was under hard acceleration or heavy load, a vacuum switch sensed a drop in manifold vacuum and signaled the module to retard the ignition timing.

In 1979, a Start Retard function was added to the Dura Spark II module. When the ignition key was turned to the start position, the start retard circuitry would retard ignition timing up to 18 degrees during startup. This would make starting easier and lessen the emission output during cranking.

In 1980, the Dual Mode module was replaced by the universal ignition module (UIM). This smaller, compact module could be programmed at the factory to provide ignition retard for either fuel economy calibrations, altitude compensation, or spark knock control, depending on the model application. The UIM was a major step toward a computer-controlled ignition system.

The control modules used on the SSI and Dura Spark systems are very similar in appearance. To differentiate between module use in the various systems, Ford installed different colored grommets in the modules. The grommets seal and insulate the wire entryway. Table 8-1 lists the grommet color for each ignition system.

FORD'S THICK FILM INTEGRATED SYSTEM

Ford's thick film integrated (TFI) ignition system was introduced in 1982. This system is used exclusively on Ford vehicles equipped with front wheel drive and automatic transaxles.

The TFI ignition system consists of a distributor, a newly designed TFI ignition module, an E-core ignition coil, an ignition switch, a battery, and primary and secondary wiring (Figure 8-23). The distributor (Figure 8-24) contains a magnetic pulse generator that has a pick-up coil assembly with internal teeth (one per cylinder) and a reluctor with external teeth. The distributor has a redesigned rotor, a distributor cap with male spark plug wire terminals, and centrifugal and vacuum advance mechanisms.

In the TFI ignition system, a small blue plastic TFI ignition module is used instead of the much larger Dura Spark module. The TFI ignition module mounts on the distributor housing and attaches to it with two retaining screws. The module connects electrically to the distributor's pick-up coil assembly, without the use of a wiring harness.

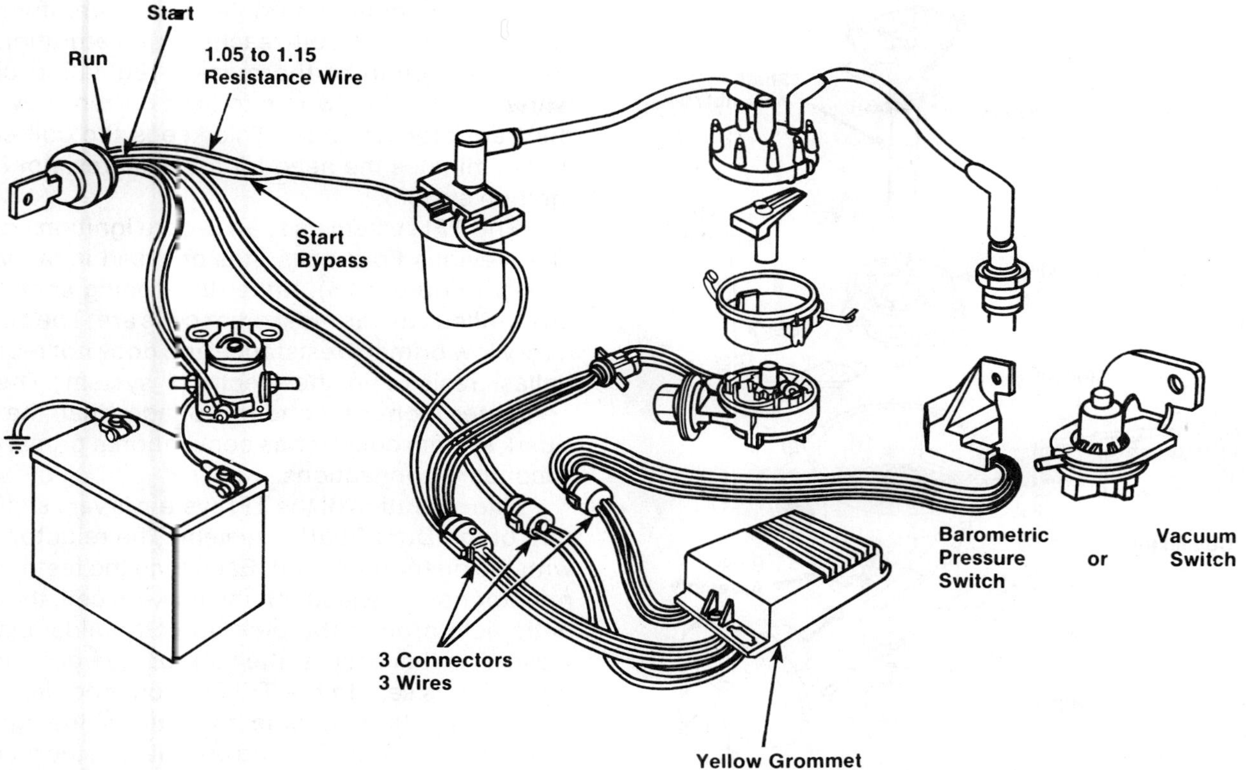

FIGURE 8–23 Ford Dura Spark II Dual Mode system

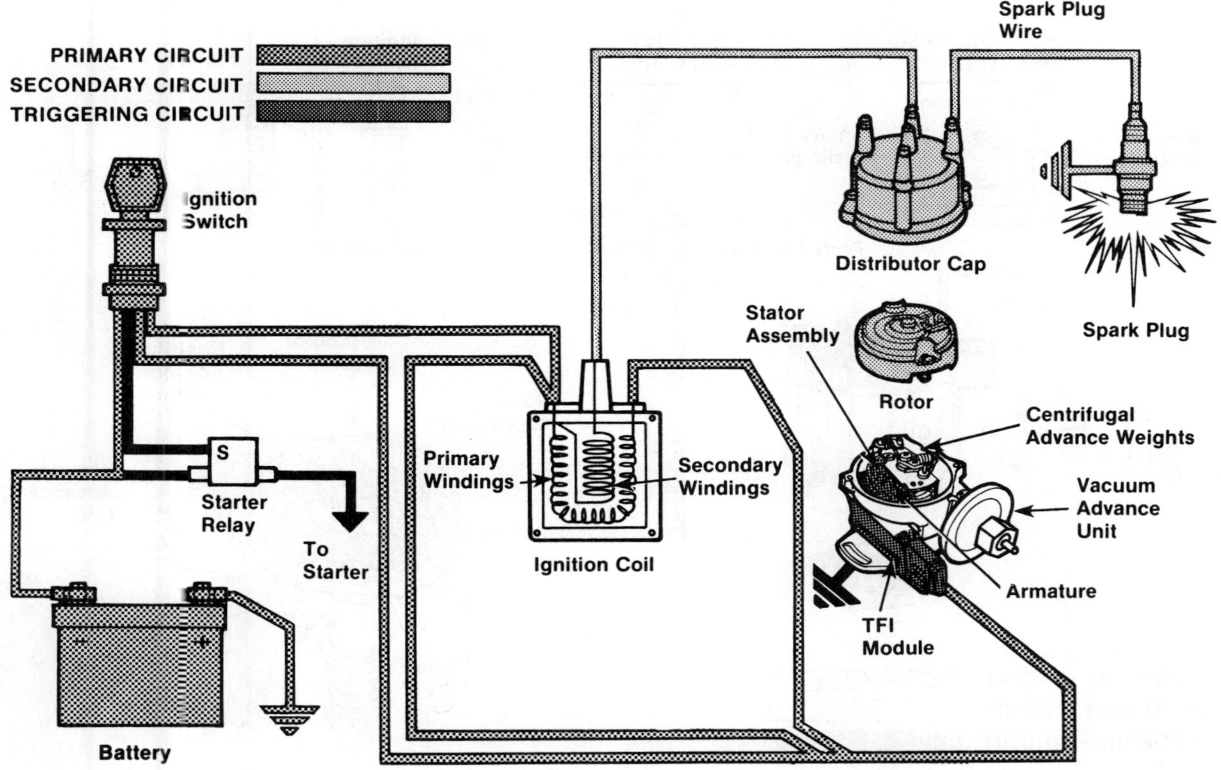

FIGURE 8–24 Ford thick film integrated system

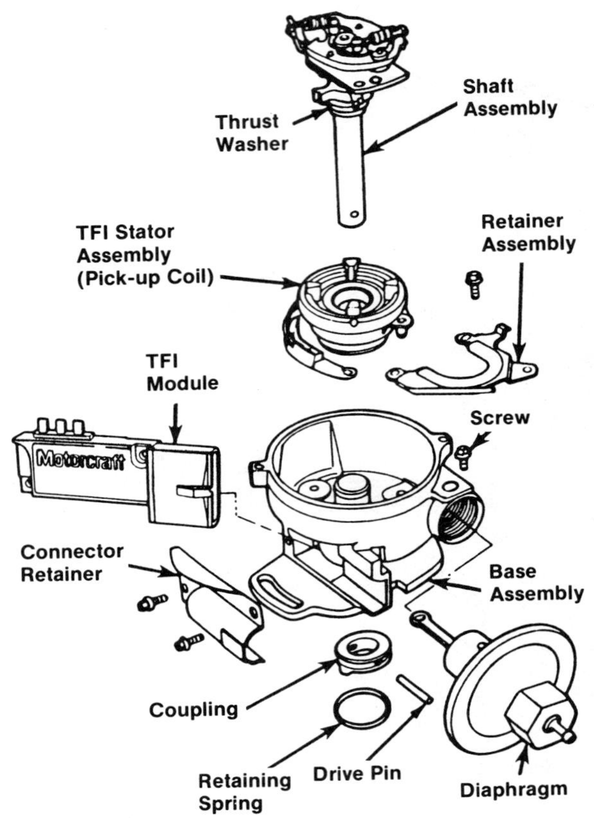

FIGURE 8–25 Ford TFI distributor

The TFI ignition module has a current limiting circuit. When the coil reaches full saturation, the switching action of the current regulating circuit within the module works to limit current flow until the transistor turns off. This keeps the coil cooler and eliminates the need for a ballast resistor in the ignition system.

The TFI system uses an E-core ignition coil unlike previous Ford coils. It is encased in laminated plastic (Figure 8–25), rather than being encased in an oil-filled case as most other coils are. The coil has a very low primary resistance and does not require a ballast resistor in the electrical system. The coil generates higher secondary voltages than the Dura Spark system coils. It has conventional primary and secondary connections.

The operation of the TFI system is very similar to that of the Dura Spark I system. The reluctor turns with the distributor shaft. Each time the teeth on the reluctor pass the teeth on the pick-up coil, the magnetic field around the pick-up coil builds and collapses. As this occurs, the pick-up coil generates a pulse that is sent to the TFI ignition module.

The ignition module then turns the ignition coil's primary circuit off and on, causing the magnetic field in the coil to build and collapse. This creates a high voltage surge in the secondary circuit, which fires the spark plugs.

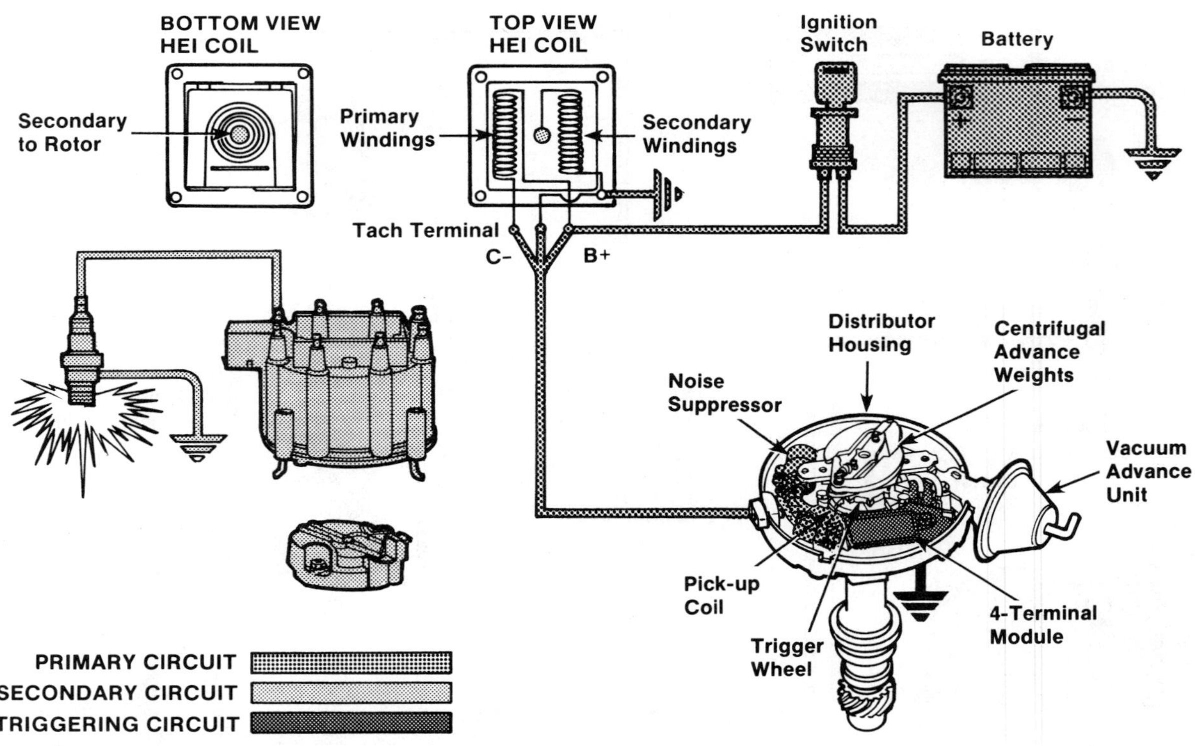

FIGURE 8–26 GM early HEI ignition system

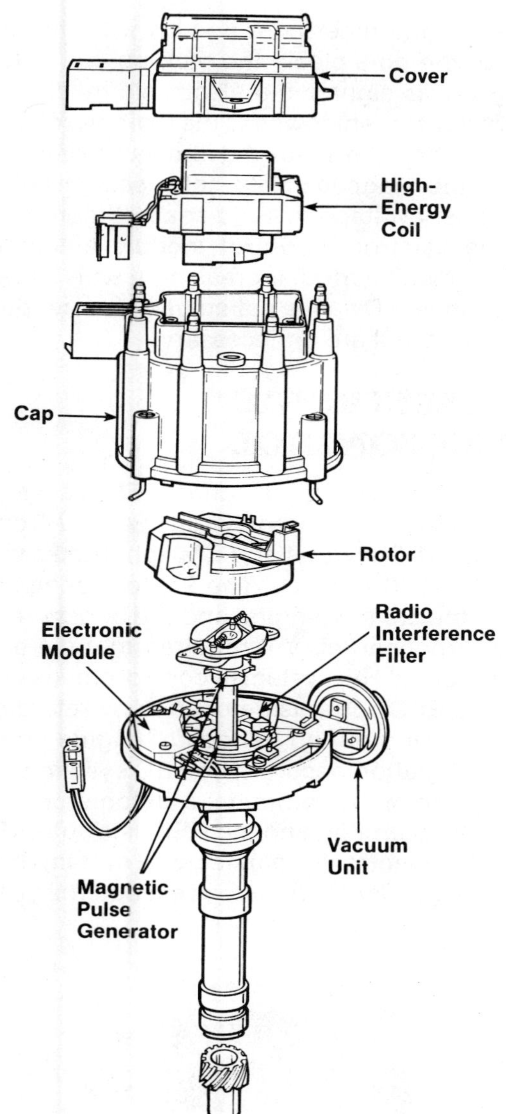

FIGURE 8–27 GM HEI distributor

GM'S HIGH-ENERGY IGNITION SYSTEM

General Motors introduced its high-energy ignition (HEI) system in 1974 as an option on GM V-8 engines. It became standard on all GM engines in 1975. This Delco-Remy system (Figure 8–26) is different from most of the other early ignition systems because a special distributor contains all of the components of the system. The HEI distributor (Figure 8–27) consists of the following components: an electronic control module, a pick-up coil assembly, a trigger wheel (reluctor), a rotor, a large distributor cap, vacuum and centrifugal advance mechanisms, a high-energy coil, and a capacitor for noise suppression. The spark plug wire in a HEI system is a carbon-impregnated conductor encased in an 8 millimeter diameter silicone rubber jacket. The high energy produced by the system allows the use of wide gap spark plugs.

The HEI system uses a special, high output transformer type ignition coil. Early inline 4- and 6-cylinder engines had a separate, externally mounted ignition coil. All engines since 1978, except 4-cylinder engines, have all ignition components integrated into the distribution cap.

Early model integral coils (1974–75) and all external coils have a direct wire connection between the primary and secondary sides, as a conventional coil does. Late model (1976 and later) integral coils have one side of the secondary windings grounded and no direct connection to the primary.

The magnetic pulse generator in the HEI distributor has a reluctor (timer core) with a tooth for each cylinder and a pick-up coil with an equal number of poles. The timer core, mounted on the distributor shaft, rotates with the shaft inside the pick-up coil assembly (Figure 8–28). When the external teeth

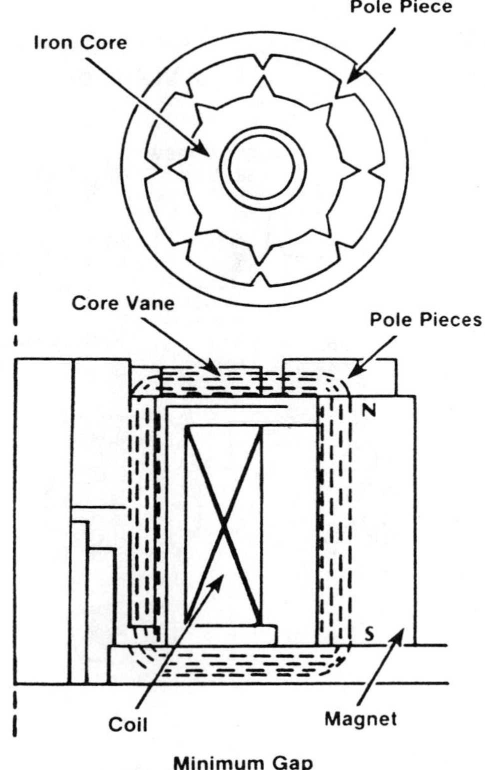

FIGURE 8–28 The magnetic field expands through the pick-up coil as the reluctor teeth align with the poles.

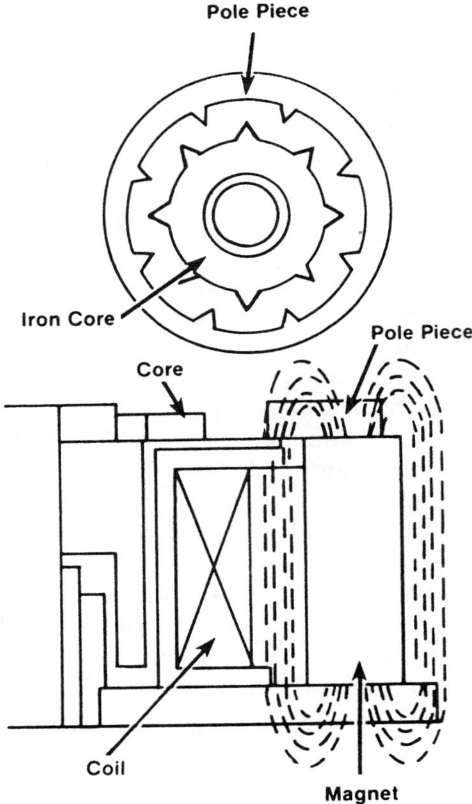

FIGURE 8-29 The magnetic field collapses as the reluctor teeth move away from the poles.

(vanes) of the timer core line up with the internal teeth of the pole piece, the pick-up coil induces a voltage. This signals the electronic module inside the distributor, which opens the ignition coil primary circuit. When the teeth move away from the pole pieces, the magnetic field collapses, allowing the coil primary voltage to build again (Figure 8-29).

The electronic control module automatically controls dwell period, increasing it with increasing engine speed. Dwell is not adjustable, and periodic checks of dwell are unnecessary.

HEI ELECTRONIC SPARK CONTROL

The electronic spark control (ESC) system was introduced on 1979 turbocharged Buick V-6 engines and on 1981 Chevrolet and GM light trucks with 5.0 liter (305 CID) engines. The 9.2 to 1 compression ratio of the 5.0 liter engine and the boost pressures of the turbocharged V-6 required that the engines use a modified HEI system to control engine detonation. The ESC controls detonation by retarding the ignition timing during periods of engine operation when detonation occurs. The ESC system consists of three major components: a detonation (knock) sensor, a controller, and the HEI distributor (Figure 8-30). The electronic spark control system operates just like a standard HEI system, except when detona-

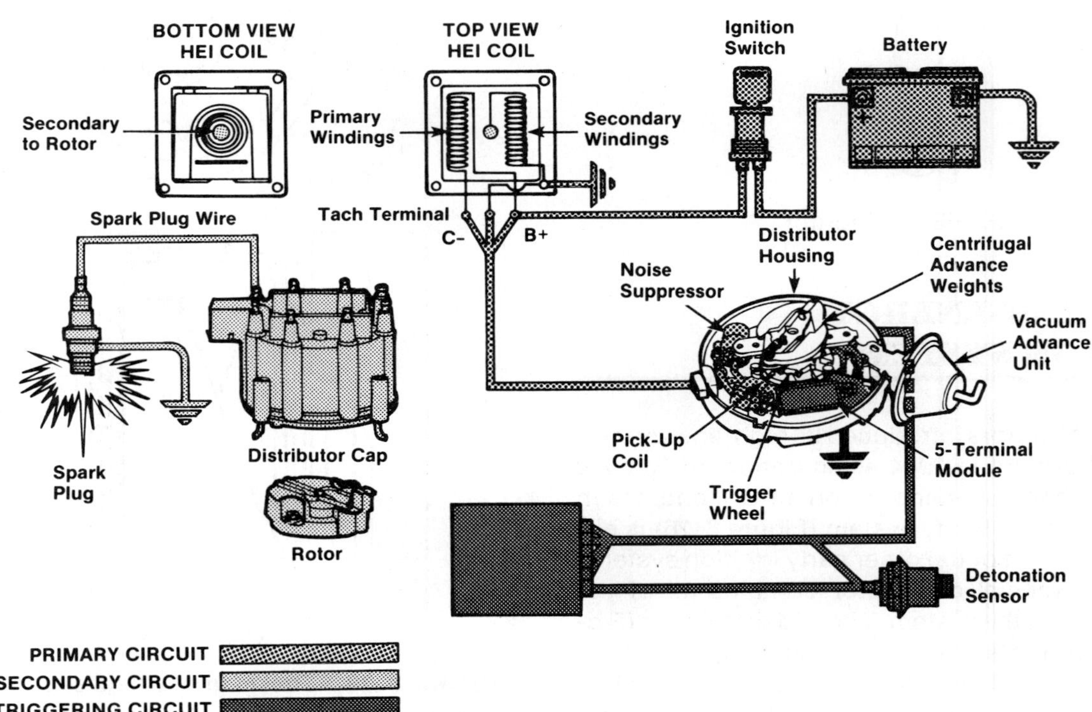

FIGURE 8-30 GM electronic control spark system

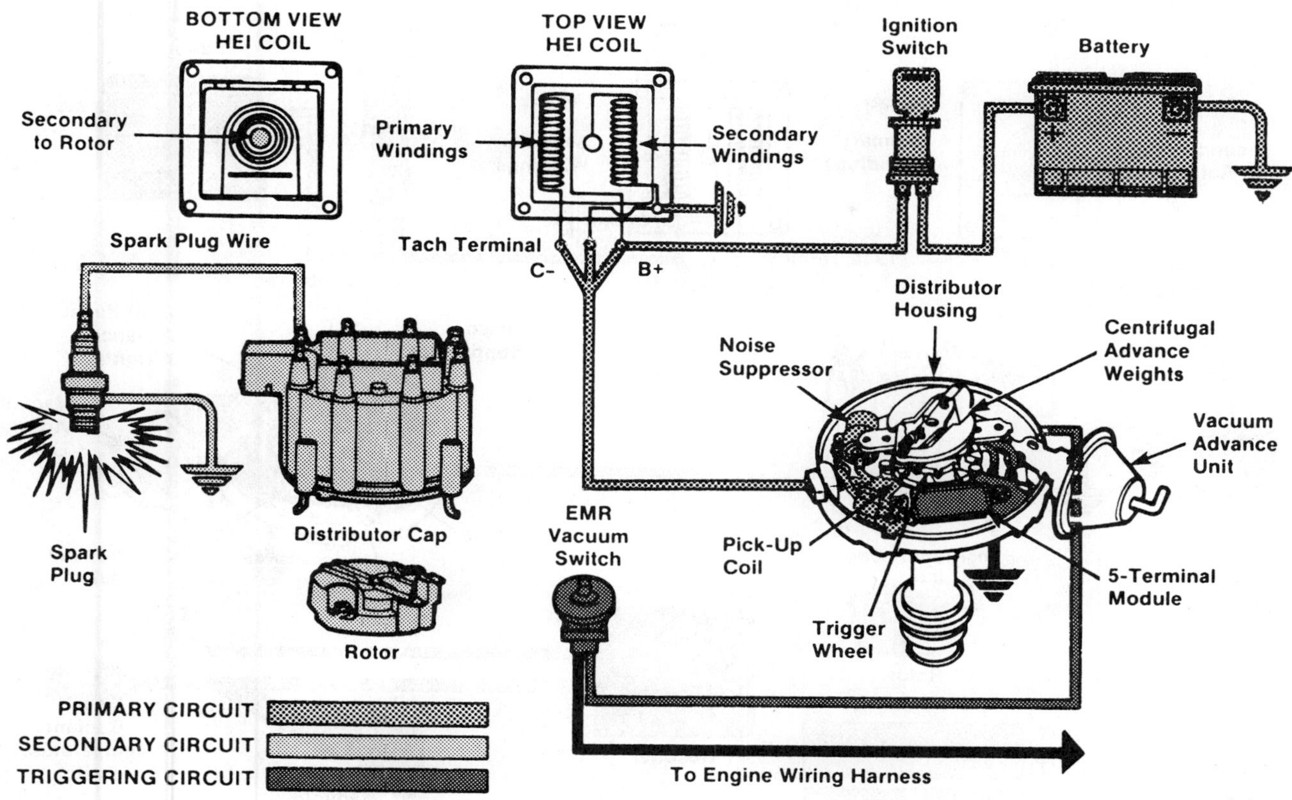

FIGURE 8-31 GM electronic module retard system

tion occurs. In the ESC system a knock sensor, which is mounted on the intake manifold, and an electronic controller detects engine detonation. The knock sensor produces a voltage signal when vibrated. The strength of the signal depends on the vibration frequency: the faster the vibration, the higher the voltage signal.

A controller monitors the voltage output of the knock sensor. The ESC controller is a metal box mounted in the passenger compartment. The box contains microelectronic circuits that perform two functions. The first is to filter out the background signal, or noise, from the knock signal (or from any signal that looks like that caused by detonation). The vibration caused by detonation is generally in a narrow frequency range near 6000 Hz. In that range, its amplitude is greater than that found when the engine is not detonating. These characteristics enable the knock control circuitry of ESC to detect and measure the increase in vibration caused by knock. The second function is to send a command to the distributor to retard the spark timing. This retard command is proportional to the magnitude and duration of the knock signal. The HEI distributor has a modified electronic module that responds to the signal from the controller. A maximum retard of 15 to 20 degrees is possible, depending on the model year. This process is a continual one, reducing or eliminating detonation in all driving ranges.

A burst detonation control feature was added to the ESC system in the 1982 model year. The burst detonation control retards spark for a very short duration when the throttle is opened quickly. This is done by sensing a rapid decrease in engine vacuum level. The short period of retard is generally sufficient to remove the quick rap of "burst" detonation that would otherwise normally occur. This keeps the spark from being retarded for a much longer period in the normal control mode, thus improving vehicle performance with the automatic transmission.

HEI ELECTRONIC MODULE RETARD

The electronic module retard (EMR) system, introduced in 1980, operates just like a standard HEI system, except during cold engine operation and below 2000 rpm. The components of the system (Figure 8-31) are an HEI control module with a timing retard circuit, a vacuum switch, and a tempera-

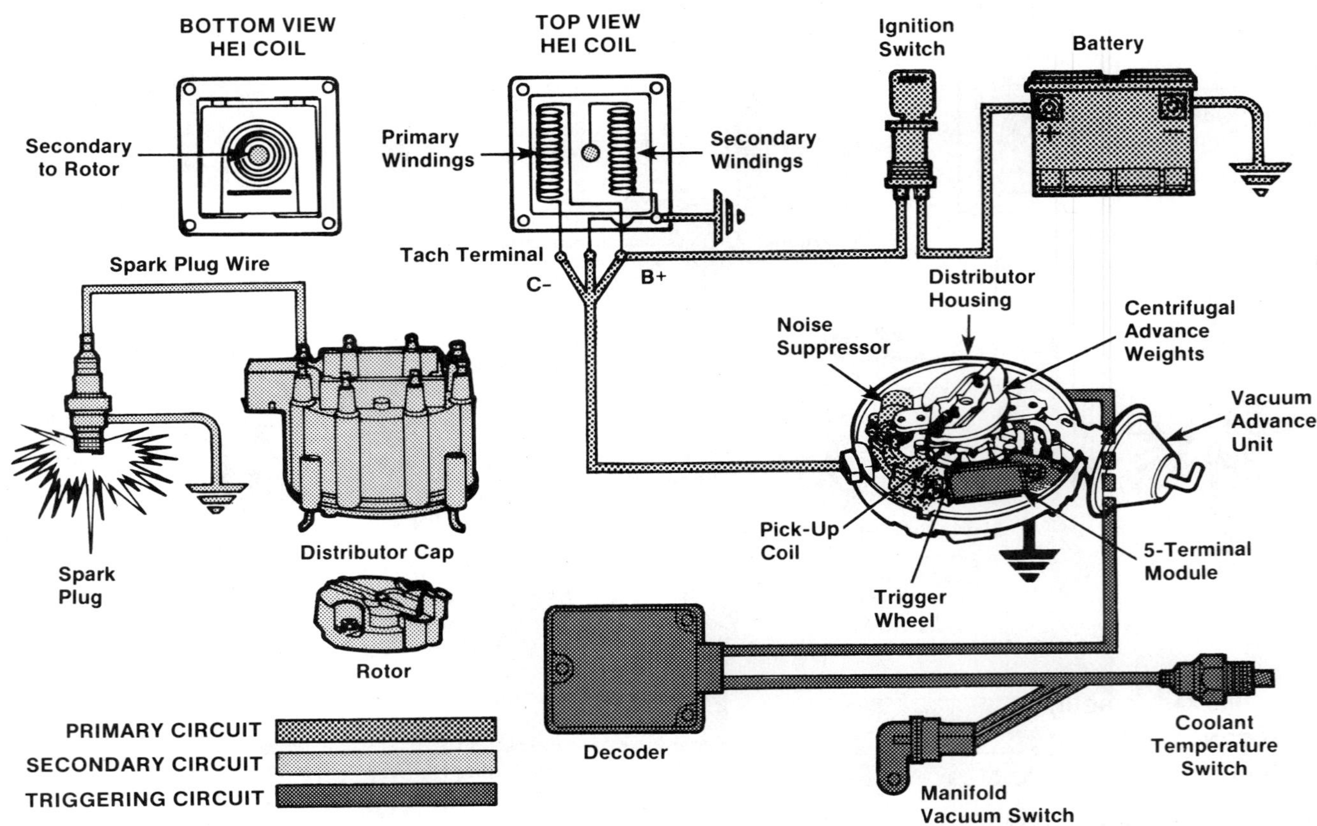

FIGURE 8–32 GM electronic spark selection system

ture vacuum switch (not shown). When the timing retard circuit is grounded, the firing of the spark plugs is delayed for a calibrated number of crankshaft degrees. A vacuum-operated electrical switch controls the grounding circuit. At approximately 2000 rpm, manifold vacuum opens the circuit, allowing the distributor's centrifugal and vacuum advance mechanisms to control ignition timing. However, the temperature vacuum switch shuts off vacuum to the EMR vacuum switch until the coolant temperature reaches 149 degrees Fahrenheit. This ensures a retard during cold engine operation, regardless of engine speed.

HEI ELECTRONIC SPARK SELECTION

General Motor's electronic spark selection (ESS) system (Figure 8–32) was a major step toward electronically controlled timing. Used between 1978 and 1980 on Cadillac vehicles with electronic fuel injection, the system is able to retard or advance the entire spark curve under certain operating conditions (cranking and high cruising). This improves hot engine restarting, increases fuel economy at cruising speeds, and reduces exhaust emissions.

Components of the system included an electronic decoder and high-energy ignition system distributor, which has been modified for the electronic spark system. The pick-up coil signal is sent to the electronic decoder to provide engine speed and ignition timing information. The ESS distributor has a specific 5-pin module in place of the 4-pin module used in standard HEI distributors. The electronic decoder output signal is sent to this new additional pin on the module. From this point, the decoder signal either delays or does not delay the shutting off of current in the primary winding of the ignition coil.

Existing components provide input to the electronic decoder to further identify engine operating conditions. Those components and the information they provide to the decoder are listed in Table 8–2.

During the cranking mode, an electronic decoder controls the spark timing. While the engine is being cranked, the decoder receives input signals from an engine coolant temperature switch, pick-up coil assembly, and manifold vacuum switch. The decoder then outputs a signal to the control module to retard the spark accordingly.

Spark retard improves hot restarting by delivering the spark to the cylinder when the piston is closer to top dead center (later in the compression stroke). By retarding the timing, ignition and the associated start of combustion pressure buildup in the cylinder occur closer to the point where the piston starts its downward motion at the beginning of the power stroke. This reduces demand on the starting system because the starter is not required to crank the engine as long against a high combustion pressure during starting.

On California cars, spark advance is also retarded from normal engine timing when coolant temperature is below approximately 130 degrees Fahrenheit. This shortens converter warmup time, thereby reducing exhaust gas hydrocarbon emissions. The converter is brought up to operating temperature faster than normal because of the higher exhaust gas temperatures produced when the spark is retarded. This is because ignition occurs later and therefore causes burning of the air/fuel mixture to begin closer to the start of the exhaust stroke. This results in higher exhaust gas temperatures at the end of the exhaust stroke and therefore also at the converter.

After the engine has been cranked and warmed up to operating conditions, the decoder will not modify the ignition timing until the vehicle reaches high cruising speeds. At idle or low cruising speeds, the centrifugal and vacuum advance mechanisms provide normal spark advance.

In addition to the input from the manifold vacuum switch and the coolant temperature switch, the decoder also receives signals from the pick-up coil. This allows the decoder to monitor ignition timing and engine speed. When the decoder detects that high cruising speeds have been reached ignition timing is advanced over normal engine timing. Ignition, therefore, takes place earlier in the compression stroke, giving the air/fuel mixture more time to burn. This increases performance efficiency and improves fuel economy.

TABLE 8-2: ESS DECODER INPUT

Component	Input Provided to Decoder
Fuel economy switch	Engine vacuum
EGR solenoid	Coolant temperature
Coolant temperature switch (3-way)	Coolant temperature
Ignition switch	Engine cranking

Note: The coolant temperature switch (3-way) is not used in 1978.

IMPORT ELECTRONIC IGNITION SYSTEMS

The technology and components that a technician will find on imported cars and light trucks is very similar to the domestic systems already discussed. In every case, some kind of crankshaft position sensor mounted on the distributor shaft signals a transistor, which in turn controls the timing of the secondary spark delivery. Figures 8-33 through 8-42 illustrate some of the import systems a technician will encounter when doing tuneup work.

TESTING ELECTRONIC IGNITION SYSTEMS

When servicing a vehicle with an electronic ignition system, many of the procedures discussed in

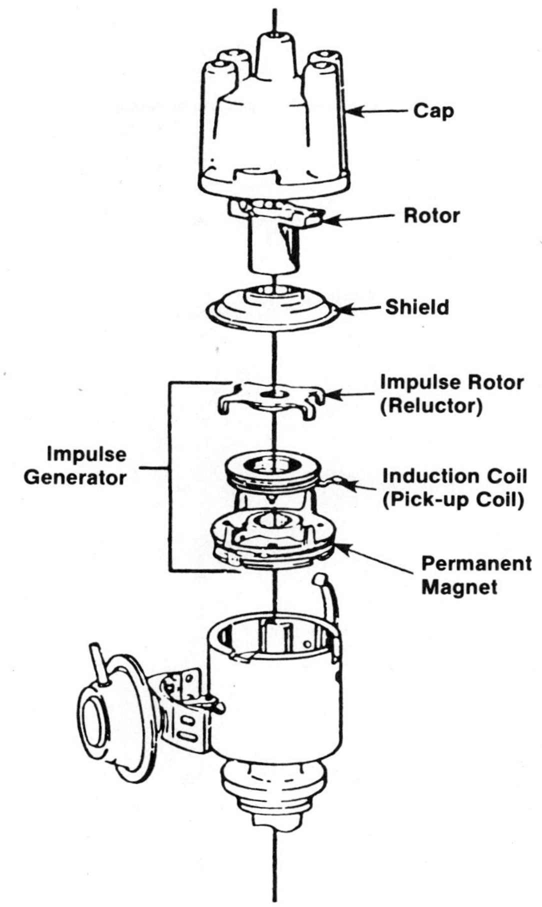

FIGURE 8-33 This Bosch distributor has a pulse generator and is found on many Fiat, Volvo, Volkswagen, and Audi vehicles.

FIGURE 8-34 Another Bosch system found on Volkswagen and Audi vehicles uses a Hall generator. This system also uses an idle stabilizer that advances ignition timing whenever the idle speed drops below 940 rpm.

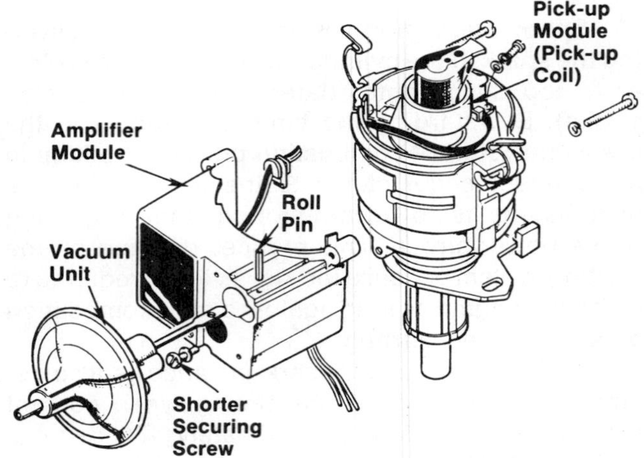

FIGURE 8-36 The Lucas electronic ignition system (used on British Leyland cars (such as the MG Midget and MGB, Triumph Spitfire and TR-7, and Jaguar XJ-6) has a plastic reluctor in which are embedded equally spaced ferrite rods. A voltage signal is generated when the rods pass through the magnetic field of the pickup coil. An amplifier and power transistor are connected to the distributor.

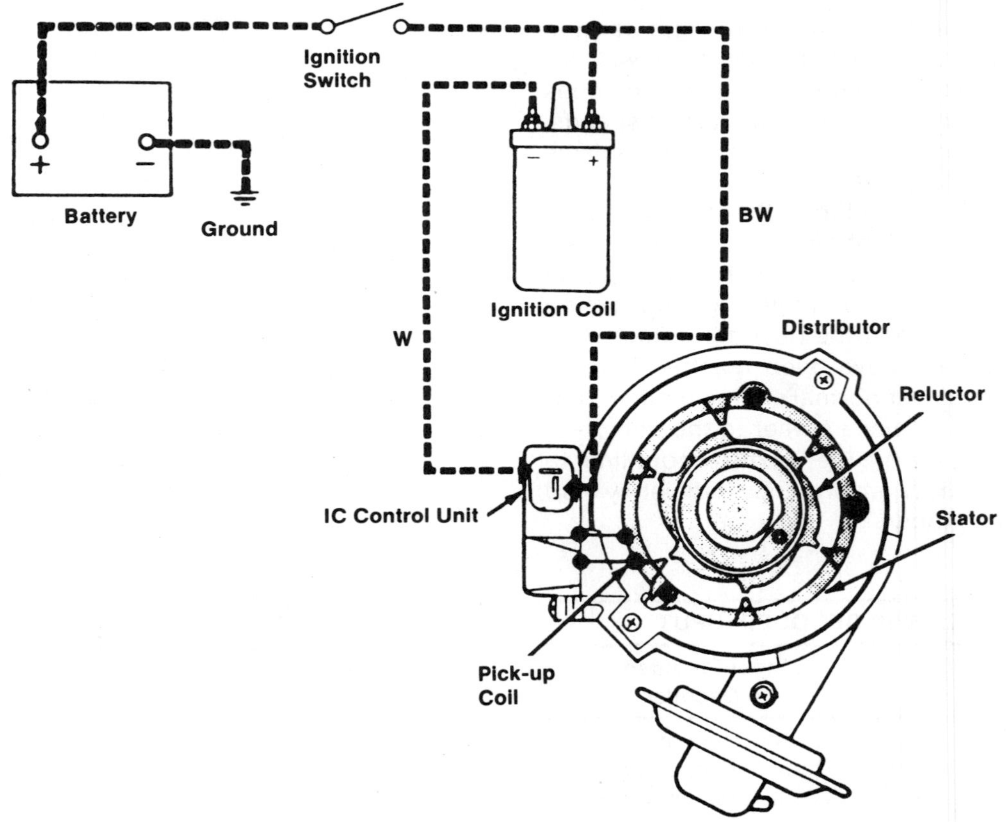

FIGURE 8-35 This Hitachi distributor has an integrated circuit (IC) control unit mounted to the exterior of the distributor.

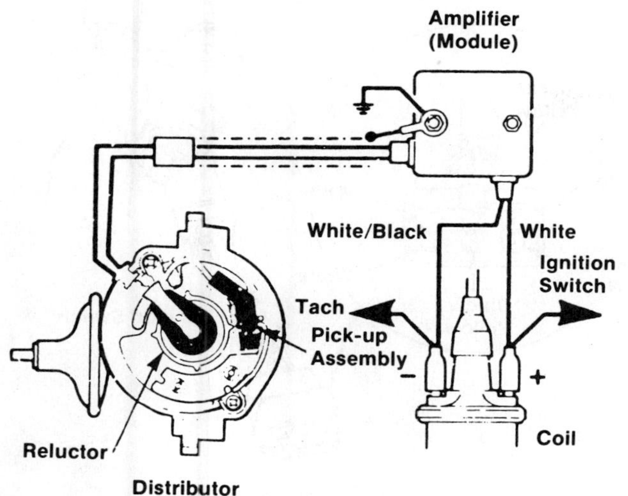

FIGURE 8-37 This Lucas electronic ignition system uses a toothed reluctor and an ignition module separate from the distributor. It replaced earlier Lucas system on most British Leyland vehicles in the early 1980s.

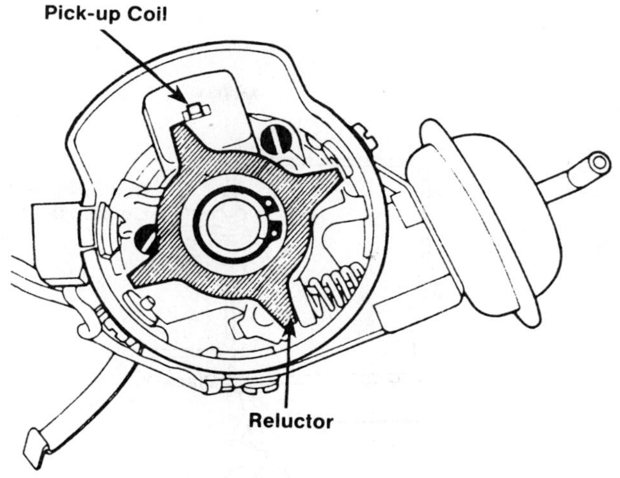

FIGURE 8-39 This pulse generator is found on early 1980s Renault vehicles and uses a General Motors HEI module. On the Renault 18i, the pulse generator has dual pickup coils. The second pickup is offset to provide spark advance when the engine is cold.

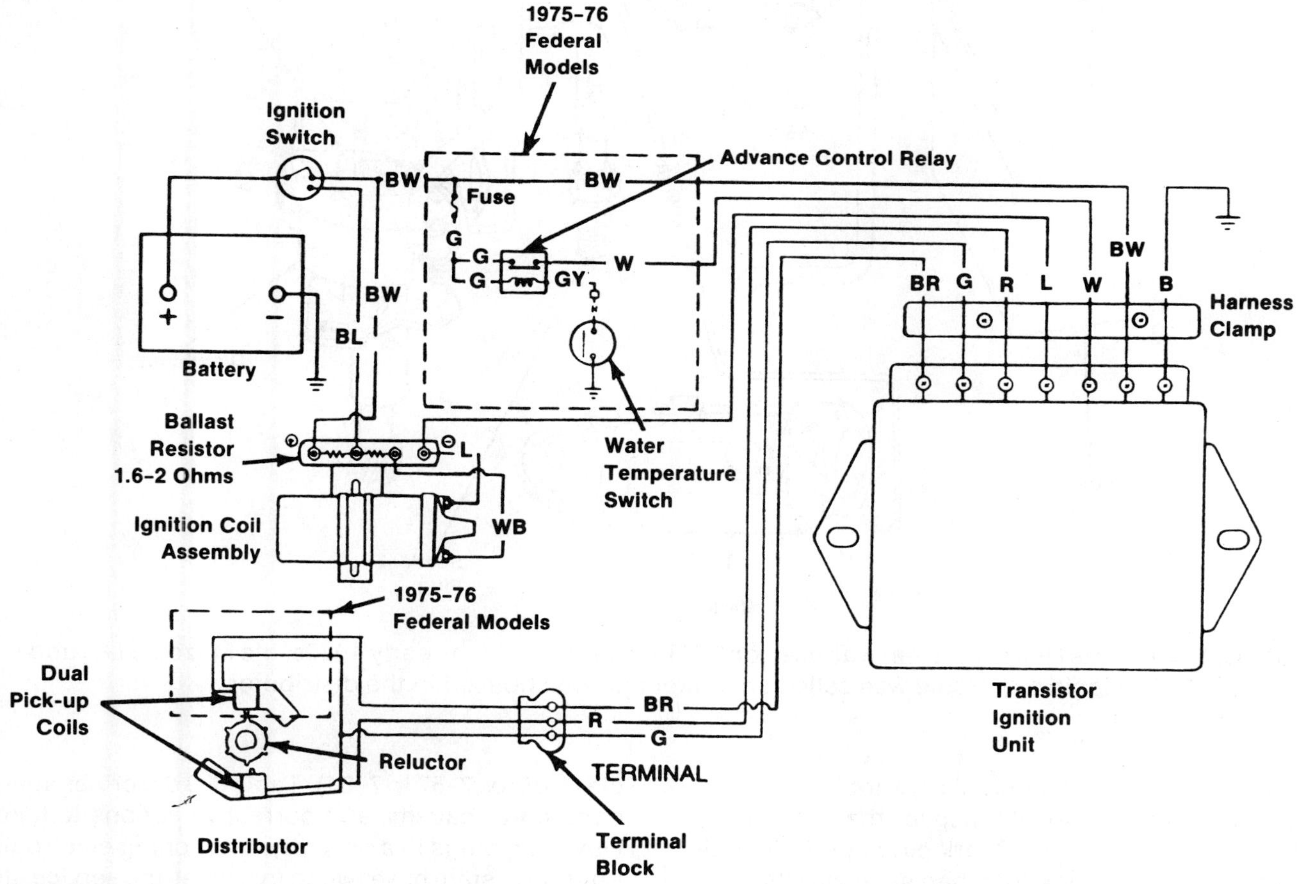

FIGURE 8-38 This Hitachi system, used on Datsun (Nissan) vehicles in the late 1970s, was called Trignition and had a dual pickup distributor that advanced spark timing during warmup.

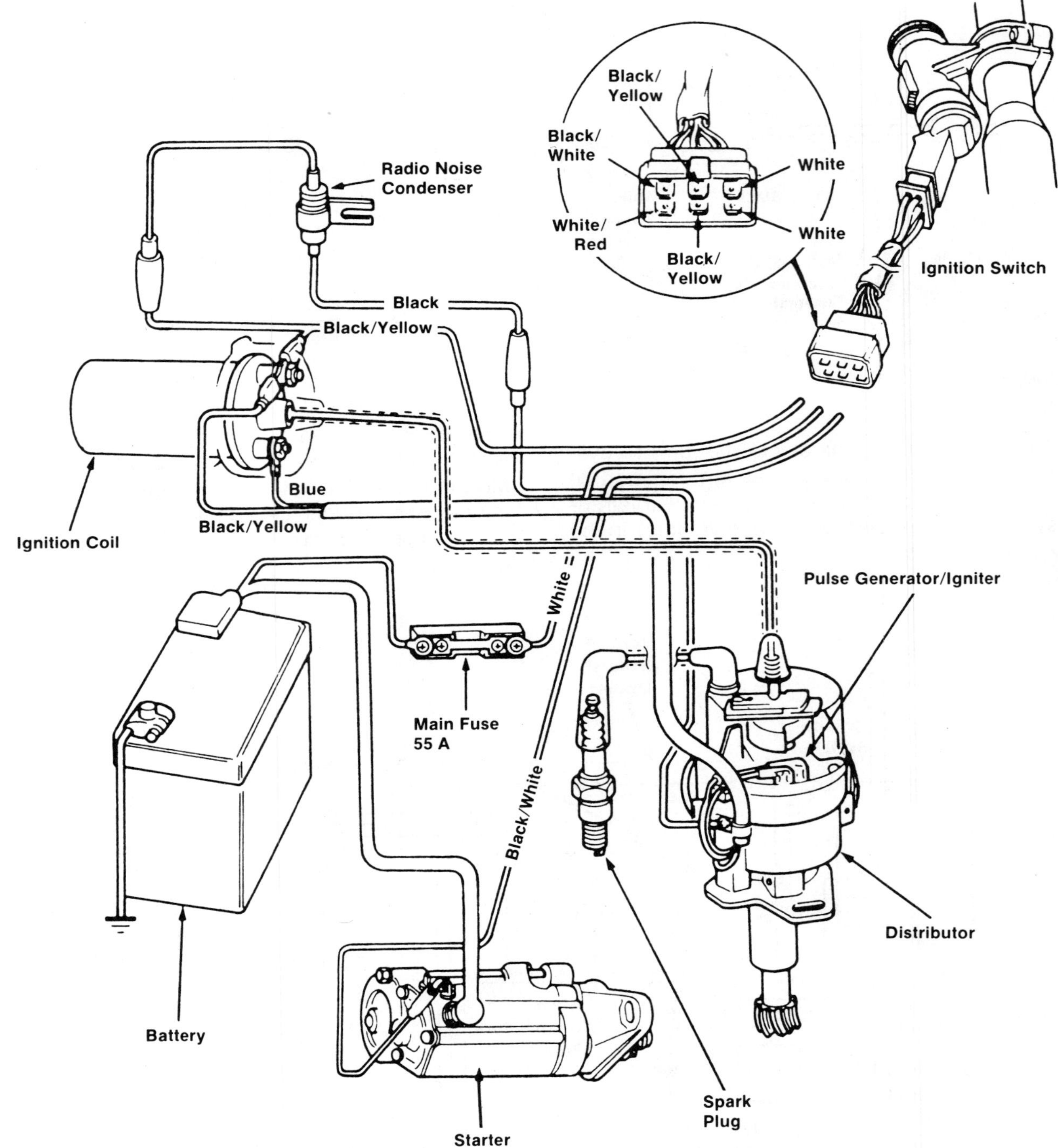

FIGURE 8-40 This Hitachi system was used on all Honda models in the early 1980s, the Mazda GLC, and the Datsun 310. The ignition module was called an igniter and was housed in the distributor.

the previous chapter on breaker point ignitions apply. For example, closely inspect the spark plugs when they are removed. Spark plug condition indicates problems to look for when servicing the vehicle (Figures 7-57 to 7-64). Table 8-3 describes spark conditions, causes, and corrective actions to take. However, plugs in a properly functioning electronic ignition system have two to five times the service life

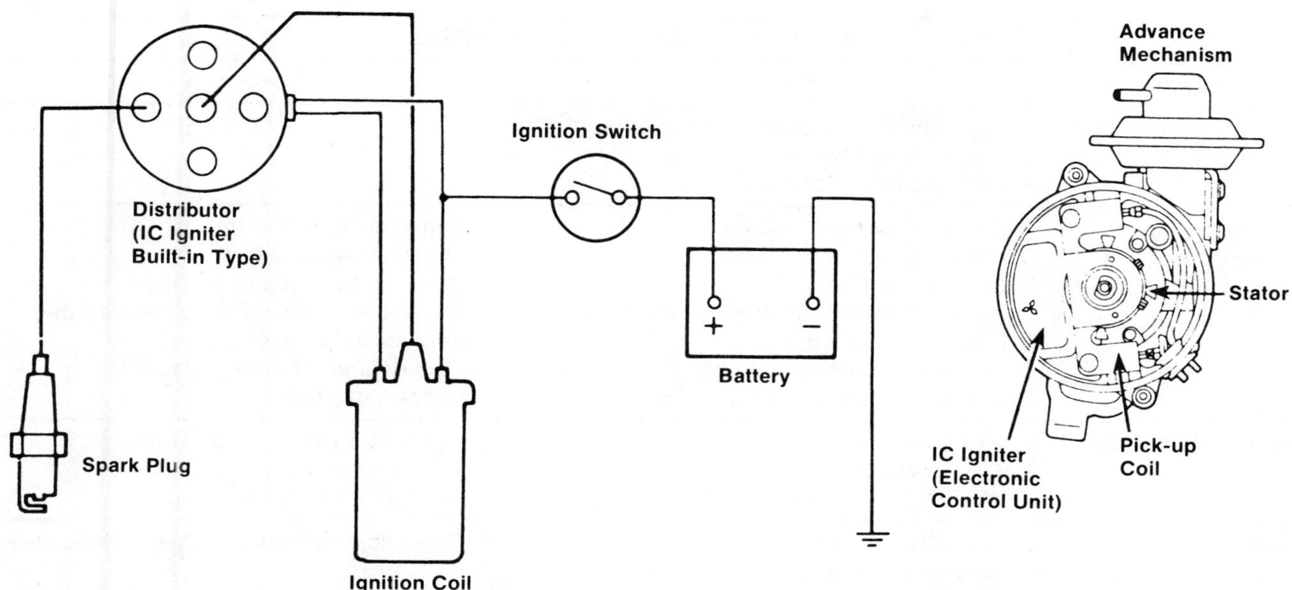

FIGURE 8-41 This Mitsubishi system was found on Ford courier trucks, various Mazda vehicles, Chrysler imports, and Chrysler domestic vehicles with 2.6L engines in the late 1970s and early 1980s. The ignition control unit might be mounted on the coil, on the distributor housing, or in the distributor, depending on the model and year.

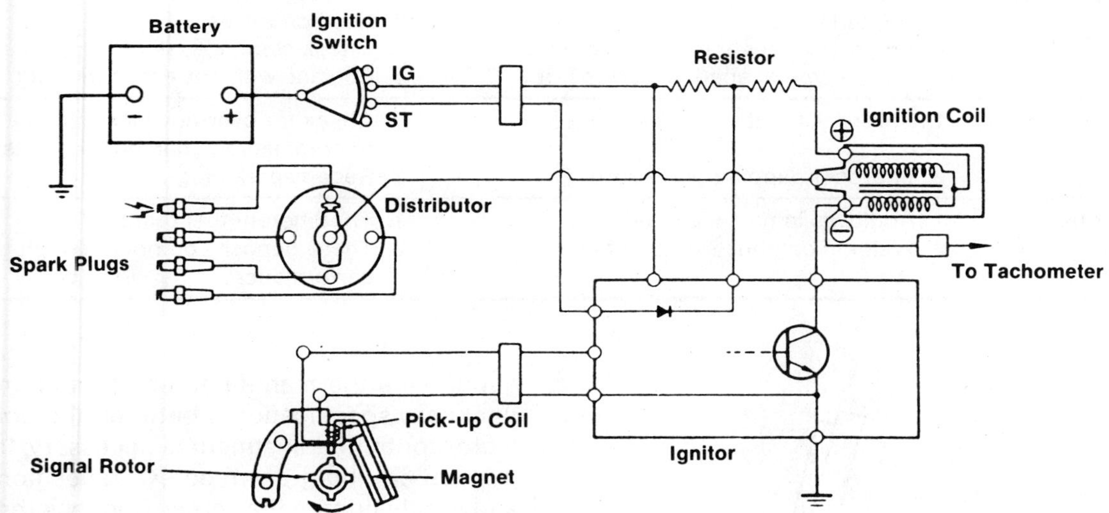

FIGURE 8-42 This Nippondenso system is used on all Toyota vehicles and the Chevrolet LUV in the late 1970s and early 1980s.

of plugs in conventional systems. If they appear to be in good condition, clean, regap, and replace them.

The major difference between tuning a vehicle with a conventional ignition system and one with an electronic system is that there are no points to gap and replace and no dwell to adjust. The ignition module controls dwell, and the magnetic pulse generator should seldom if ever need adjustment.

The air gap between the reluctor and pick-up coil (Figure 8-43) should never change because there are no wearing parts. The air gap can be checked by aligning the reluctor teeth with the pick-up coil pole(s) and by placing a nonferrous feeler

TABLE 8-3: SPARK PLUG DIAGNOSIS

Condition	Possible Cause	Correction
Normal Spark Plug Condition	1. Light tan or gray deposits on insulator 2. Electrode not burned or fouled 3. Gap tolerance not changed	
Cold Fouling or Carbon Deposits	1. Over-rich air/fuel mixture 2. Faulty choke 3. Clogged air filter 4. Incorrect idle speed or dirty carburetor 5. Faulty ignition wiring 6. Prolonged operation at idle 7. Sticking valves or worn valve guide seals	1. Adjust air/fuel mixture. 2. Replace choke assembly. 3. Clean and/or replace air filter. 4. Reset idle speed and/or clean carburetor. 5. Replace ignition wiring. 6. Shut engine off during long idle. 7. Check valve train.
Wet Fouling or Oil Deposits	1. Worn rings and pistons 2. Excessive cylinder wear 3. Worn or loose bearings	1. Install new rings and pistons. 2. Rebore or replace block. 3. Tighten or replace bearings.
Gap Bridged	1. Deposits in combustion chamber becoming fused to electrode	1. Clean combustion chamber of deposits.
Blistered Electrode	1. Engine overheating 2. Wrong type of fuel 3. Loose spark plugs 4. Over-advanced ignition timing	1. Check cooling system. 2. Replace with correct fuel. 3. Re-tighten spark plugs. 4. Reset ignition timing.
Preignition or Melted Electrodes	1. Incorrect type of fuel 2. Incorrect ignition timing 3. Burned valves 4. Engine overheating 5. Wrong type of spark plug, too hot	1. Replace with correct fuel. 2. Reset ignition timing. 3. Replace valves. 4. Check cooling system. 5. Replace with correct spark plug.
Chipped Insulators	1. Severe detonation 2. Improper gapping procedure	1. Check for over-advanced timing or combustion chamber deposits. 2. Regap spark plugs.
Rust-Colored Deposits	1. Additives in unleaded fuel 2. Water in combustion chamber	1. Try different fuel brand. 2. These deposits do not affect plug performance.

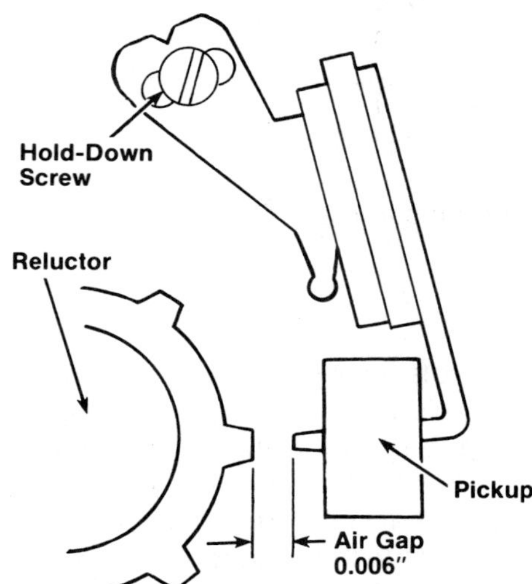

FIGURE 8-43 The air gap between reluctor and pick-up coil usually needs no adjustment.

gauge (see the manufacturer's service manual for thickness specifications) between the pole and reluctor tooth. If adjustment is necessary, loosen the pick-up coil hold-down screw, reposition the coil, and lightly tighten the screw. Recheck the gap with the feeler gauge and if it is correct, tighten the hold-down screw securely.

Like a conventional ignition system, electronic ignition systems can be tested with an oscilloscope. Individual components can also be tested for voltage, resistance, and/or current to pinpoint specific problems. The rest of this chapter will explain how to interpret oscilloscope patterns when testing a domestic electronic ignition system and will demonstrate how to troubleshoot individual components, using Ford's solid state ignition and Dura Spark systems as an example. However, always consult the manufacturer's service manual for specific instructions and specifications.

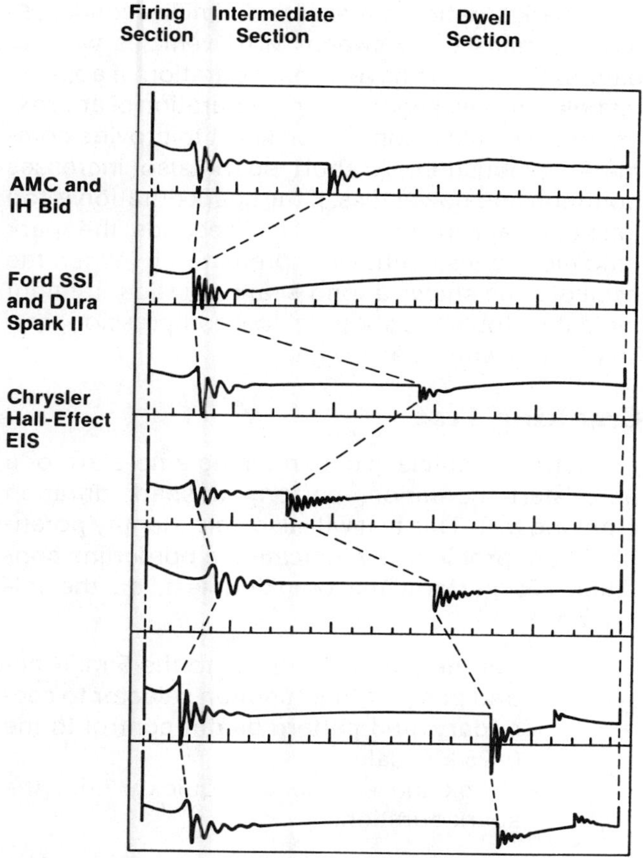

FIGURE 8-44 Primary scope patterns for electronic ignition systems

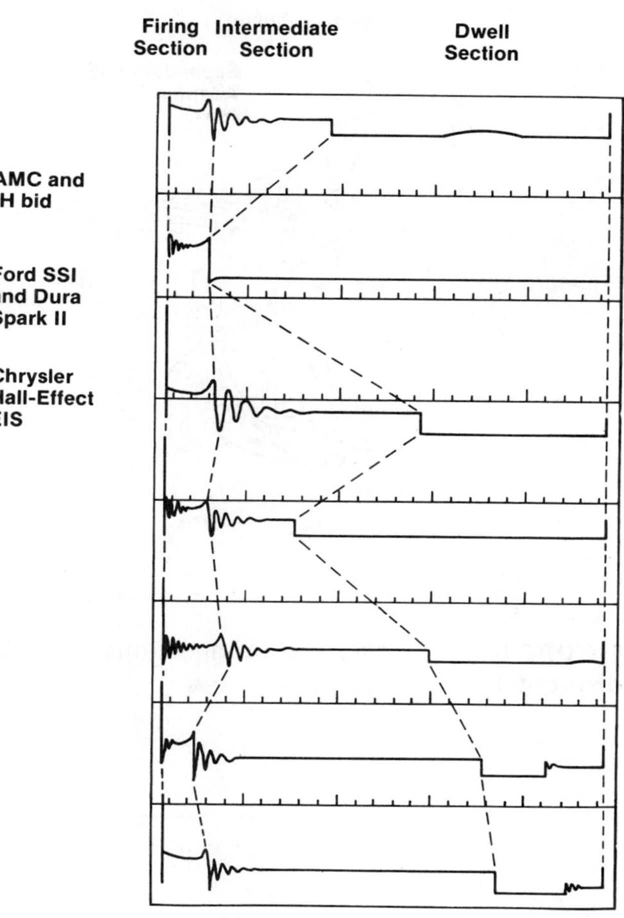

FIGURE 8-45 Secondary scope patterns for electronic ignition systems

SCOPING AN ELECTRONIC IGNITION SYSTEM

As explained in Chapter 7, one of the quickest methods of identifying a problem in an ignition system is to test the system with an oscilloscope. The waveform patterns produced by electronic system are very similar to those produced by a conventional system. Figures 8-44 and 8-45 show the primary and secondary patterns for the domestic electronic ignition systems covered in this chapter. The basic difference between the patterns is in the dwell section. The length of this section depends on when the control module turns on the transistor. On Chrysler's EIS pattern, the transistor turns on immediately after the spark extinguishes. In other systems, such as Ford's TFI system, the transistor does not turn on until much later.

The patterns produced by the electronic systems have fewer oscillations than conventional patterns. This is because most of the systems do not utilize a condenser to control dissipation of energy remaining in the coil. However, a small oscillation or hump might be noticeable near the end of the dwell section. This hump occurs when the current limiting circuit in the module is activated.

Connecting the oscilloscope to an electronic ignition system is usually identical to the connection described in Chapter 7 for a secondary system. The General Motor's HEI system requires a special adapter because the coil mounts in the distributor cap. The adapter and scope connections are shown in Figure 8-46. These connections must be made prior to performing the following tests on the oscilloscope.

SPARK DURATION TEST

The amount of energy available to ignite the air/fuel mixture can be determined by performing the spark duration cranking test and running test on the oscilloscope. These tests correspond to the cranking and maximum coil output test performed on conventional ignition systems.

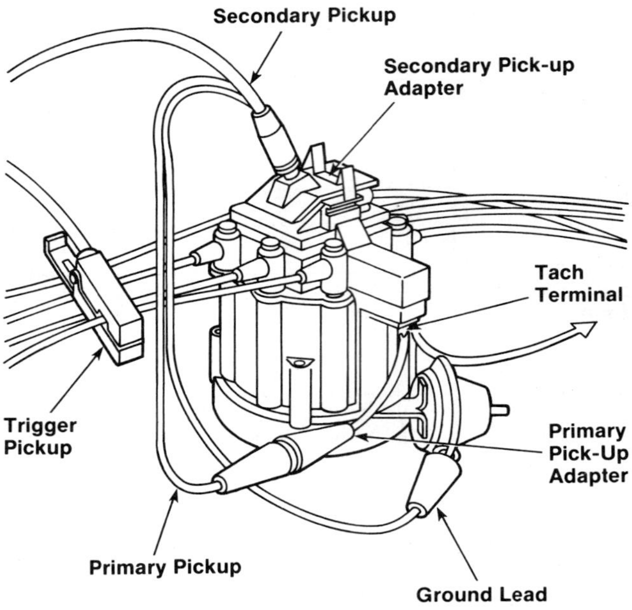

FIGURE 8–46 Oscilloscope connections to an HEI distributor

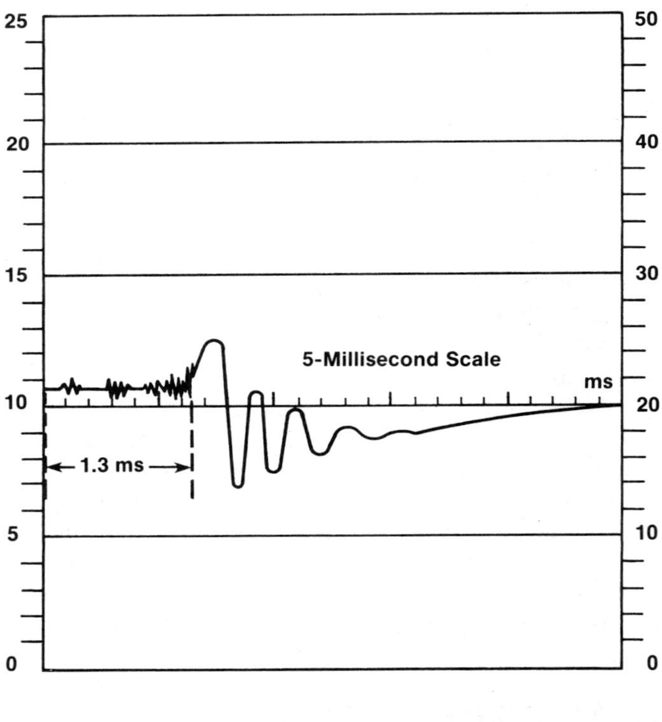

FIGURE 8–47 A 5-millisecond pattern showing a spark duration of 1.3 milliseconds during a cranking test

Spark duration is measured in milliseconds, using the millisecond sweep. Most vehicles with an electronic ignition have a spark duration of approximately 1.5 milliseconds. A spark duration of approximately 0.8 millisecond is too short to provide complete combustion. A short spark also increases pollution and power loss. If the spark duration is too long (over approximately 2.0 milliseconds) the spark plug electrodes might wear prematurely. When the oscilloscope shows a long spark duration, it might indicate a fouled spark plug, low compression, or a spark plug with a narrow gap.

Cranking Test

When a vehicle is experiencing a no-start, or a hard start, condition, perform a spark duration cranking test. This test will also help identify potential future problems by indicating a borderline condition. To perform the cranking test, do the following:

1. Set the pattern selector to the 5 millisecond position, the function selector to secondary, and pattern height control to the 0–25 kV scale.
2. Crank the engine and quickly note the spark duration.

SHOP TALK

Many oscilloscope manufacturers use different ways to prevent engine startup during a cranking test. Consult the scope manufacturer's instruction manual to properly disable an engine.

A secondary millisecond pattern in the 5 millisecond mode is shown in Figure 8–47. The spark duration shown in this pattern is normal, measuring approximately 1.3 milliseconds.

Running Test

The running test measures spark duration while the engine is running at a specific rpm. To perform a running test on the oscilloscope, do the following:

1. Set the pattern selector to the 5 millisecond position, the function selector to secondary, and pattern height control to the 0–25 kV scale.
2. Start the engine and adjust the speed to 1000 rpm.
3. Note the spark duration.

TABLE 8-4: CONVERTING PERCENT OF DWELL TO MILLISECONDS

	MS	600 RPM	1200 RPM	2400 RPM	Spark Duration
	.5	2%	4%	8%	too short
8	1.0	4%	8%	17%	minimum
cylinder	1.5	6%	13%	25%	average
	2.0	8%	17%	33%	too long
	.5	1.5%	3%	6%	too short
6	1.0	3%	6%	12%	minimum
cylinder	1.5	4.5%	9%	18%	average
	2.0	6%	12%	24%	too long
	.5	1%	2%	4%	too short
4	1.0	2%	4%	8%	minimum
cylinder	1.5	3%	6%	13%	average
	2.0	4%	8%	17%	too long

Some oscilloscopes do not have a millisecond sweep. Instead, they are equipped with a percent of dwell scale. In that case, the percent of dwell must be converted to milliseconds. A conversion chart is given in Table 8-4. The table lists percent of dwell at various rpm and the corresponding spark duration in milliseconds. On oscilloscopes without a millisecond sweep, the pattern selector should be set to superimposed or raster when performing the spark duration cranking and running tests.

The minimum acceptable spark duration (0.8 millisecond) can be computed in percent of dwell if the rpm is known, using the following formulas:

Multiply engine rpm x 0.007 for an 8-cylinder engine
Multiply engine rpm x 0.005 for a 6-cylinder engine
Multiply engine rpm x 0.003 for a 4-cylinder engine

The answer to these equations will be the minimum spark duration expressed as a percent of dwell.

For example, if a 4-cylinder engine is operating at 1000 rpm, multiply 1000 x 0.003. The answer, 3, is the minimum acceptable spark duration expressed as a percent of dwell. The raster pattern in Figure 8-48 is a secondary pattern of a 4-cylinder engine at 1000 rpm. The spark duration is 3 percent of dwell, or 0.08 millisecond.

COIL POLARITY

As explained in Chapter 7, less voltage is required to fire the spark plugs with proper coil polarity. If the coil polarity is reversed, 20 to 40 percent more voltage is needed to fire the spark plugs. To test coil polarity on the oscilloscope, do the following:

1. Set the pattern selector to the display position, the function selector to secondary, and the pattern height control to the 0–25 kV scale.
2. Start the engine.
3. Observe the pattern displayed on the scope.

If the coil polarity is correct, the firing lines of the display pattern will extend upward. If the polarity is reversed, the firing lines will extend downward. If reverse polarity is indicated by the test, check the coil primary connections: they are probably reversed.

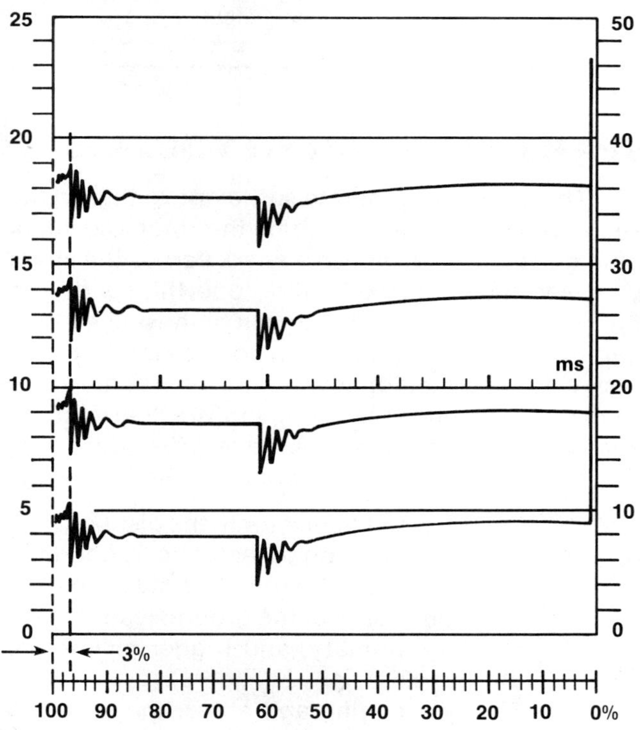

FIGURE 8-48 A secondary raster pattern showing firing duration as a percent of dwell

TABLE 8-5: FIRING LINE DIAGNOSIS

Condition	Possible Cause	Correction
Firing Voltage Lines the Same, but Abnormally High	1. Retarded ignition timing 2. Fuel mixture too lean 3. High resistance in coil wire 4. Corrosion in coil tower terminal 5. Corrosion in distributor coil terminal	1. Reset ignition timing. 2. Readjust carburetor. 3. Replace coil wire. 4. Clean and/or replace coil. 5. Clean or replace distributor cap.
Firing Voltage Lines the Same, but Abnormally Low	1. Fuel mixture too rich 2. Breaks in coil wire causing arcing 3. Cracked coil tower causing arcing 4. Low coil output 5. Low engine compression	1. Readjust carburetor. 2. Replace coil wire. 3. Replace coil. 4. Replace coil. 5. Determine cause and repair.
One or More, but Not All Firing Voltage Lines Higher Than the Others	1. Carburetor idle mixture not balanced 2. EGR valve stuck open 3. High resistance in spark plug wire 4. Cracked or broken spark plug insulator 5. Intake vacuum leak 6. Defective spark plugs 7. Corroded spark plug terminals	1. Readjust idle mixture. 2. Inspect and/or replace EGR valve. 3. Replace spark plug wires. 4. Replace spark plugs. 5. Repair leak. 6. Replace spark plugs. 7. Replace spark plugs.
One or More, but Not All Firing Voltage Lines Lower	1. Curb idle mixture not balanced 2. Breaks in plug wires causing arcing 3. Cracked coil tower causing arcing 4. Low compression 5. Defective or fouled spark plugs	1. Readjust idle mixture. 2. Replace spark plug wires. 3. Replace coil. 4. Determine cause and repair. 5. Replace spark plugs.
Cylinders Not Firing	1. Cracked distributor cap terminals 2. Shorted spark plug wire 3. Mechanical problem in engine 4. Defective spark plugs 5. Spark plugs fouled	1. Replace distributor cap. 2. Determine cause of short and replace wire. 3. Determine problem and correct. 4. Replace spark plugs. 5. Replace spark plugs.

SPARK PLUG FIRING VOLTAGE

The coil must generate sufficient voltage in the secondary system to overcome the rotor and spark plug gaps and to establish a spark across the spark plug electrodes. On the oscilloscope, this spark plug firing voltage level is seen as the highest line in the pattern. The spark plug firing voltage might be affected by the condition of the spark plugs and/or the secondary circuit, engine temperature, fuel mixture, and compression pressures. To test the spark plug firing voltage on an oscilloscope, do the following:

1. Set the pattern selector to the display position, the function selector to secondary, and the pattern height control to the 0–25 kV scale. Connect the ground lead, the secondary, primary, and trigger pickups to the vehicle.
2. Start the engine and adjust the speed to 1000 rpm.
3. Observe the firing voltage of all cylinders for height and uniformity. The normal height of the firing voltages should be between 7 and 18 kV with no more than a 3 kV variation between cylinders.
4. Record the highest and lowest firing voltage.

If during the test, one or more of the firing voltages are uneven, low, or high, consult Table 8-5 for possible causes and corrections.

ROTOR AIR GAP VOLTAGE DROP

When high firing voltages are present in one or more of the cylinders, perform a rotor air gap voltage drop test. The purpose of this test is to determine the amount of secondary voltage that is required to bridge the rotor gap.

To perform the test, do the following:

1. Set the pattern selector to the display position, the function selector to secondary, and the pattern height control to the 0–25 kV scale.

2. Start the engine and adjust the speed to 1000 rpm.
3. Observe the height of the firing voltages. Record the height and firing order number of any abnormal cylinder.
4. Shut the engine off.
5. Remove the spark plug wire of the abnormal cylinder from the distributor cap. Connect one end of the jumper lead to ground and the other end to the large portion of a grounding probe. Place the other end of the grounding probe in the distributor cap tower terminal.

SHOP TALK

Do not remove the spark plug wire from the distributor cap tower terminal while the engine is running. This will cause an open circuit and might damage the ignition system components.

6. Start the engine and adjust the speed to 1000 rpm.
7. Observe the firing voltage of the abnormal cylinder previously recorded. There should be a slight drop in the firing voltage when using the grounding probe.

The rotor air gap measurements should not exceed manufacturer's specifications. If during the test, the firing voltage remains high or drops to a level that does not meet manufacturer's specifications, the rotor or distributor cap might be defective. Visually inspect both and replace as necessary.

ROTOR REGISTER

The procedure for performing a rotor register test is very similar to that for performing a rotor air gap voltage drop test, except, in this test, vacuum is applied to the distributor vacuum advance unit with an external vacuum source.

SHOP TALK

If the vehicle is equipped with computer-controlled spark advance, consult the manufacturer's service manual for proper test procedures.

1. Set the pattern selector to the display position, the function selector to secondary, and the pattern height control to the 0–25 kV scale.
2. With the engine off, remove any spark plug wire from the distributor cap, except for the one connected to the trigger pickup. Connect one end of the jumper lead to ground and the other end to the large portion of a grounding probe. Place the other end of the grounding probe in the distributor cap tower terminal.
3. Start the engine and adjust the speed to 1000 rpm.
4. Note the height of the firing voltage of the circuit connected to the grounding probe.
5. Using the external vacuum source, apply 16 to 22 inches of mercury to the distributor vacuum advance unit. Readjust the engine speed to maintain 1000 rpm.
6. Again note the height of the firing voltage of the circuit connected to the grounding probe. There should be no more than a 3 kV difference between the firing voltage measurements obtained in steps 4 and 6. If at any time during the test, the difference between the firing voltages is over 3 kV, see Table 8–6.

SECONDARY CIRCUIT RESISTANCE

Analysis of the spark line of a secondary pattern will reveal the condition of the secondary circuit. The amount of resistance in the secondary circuit is indicated by the slope of the spark line. Excessive resistance in the secondary circuit will cause the spark line to have a steep slope with a shorter firing duration. The excessive resistance restricts the current flow necessary to generate a good spark.

To check secondary circuit resistance, do the following:

1. Set the pattern selector to the display position, the function selector to secondary, and the pattern height control to the 0–25 kV scale.
2. Start the engine and adjust the speed to 1000 rpm.
3. Observe the spark lines for height, length, angle, and oscillations.

A good spark line should be relatively even and measure 2 to 4 kV in height. High resistance in the secondary circuit produces a spark line that is higher in voltage with a shorter firing duration. If during testing high resistance is shown in the spark lines, see Table 8–6.

TABLE 8-6: TROUBLESHOOTING SPARK PLUGS		
Symptom	**Probable Cause**	**Remedy**
ROTOR REGISTER TEST		
Difference between firing voltages is over 3 kV	1. Defective rotor 2. Defective distributor cap 3. Wrong vacuum advance unit used	1. Replace rotor. 2. Replace distributor cap. 3. Check manufacturer's part number, or use distributor tester to verify advance curve.
SECONDARY CIRCUIT RESISTANCE TEST		
High resistance in spark lines	1. Rotor tip burned 2. High resistance in the spark plug wire 3. Distributor cap segments burned 4. Faulty spark plug	1. Visually inspect. 2. Perform ohmmeter test. 3. Visually inspect. 4. Perform secondary resistance test using the grounding probe.
After grounding spark plug, abnormal spark lines appear normal	1. Faulty spark plug	1. Substitute spark plug.
After grounding spark plug, abnormal spark lines still show high resistance	1. Corroded distributor cap towers 2. Distributor cap segments burned 3. High resistance in the spark plug wires	1. Visually inspect. 2. Visually inspect. 3. Perform ohmmeter test.
All the spark lines show high resistance	1. Defective coil wire 2. Rotor tip burned	1. Perform ohmmeter test. 2. Visually inspect.
SPARK PLUGS UNDER LOAD TEST		
One or more of cylinders show a voltage rise over 4 kV	1. Spark plug gap too wide 2. Worn spark plug electrodes 3. Open spark plug wire 4. Improper fuel mixture 5. Open spark plug resistor 6. Leakage in vacuum system	1. Check gap, then regap plug. 2. Replace the spark plug. 3. Perform ohmmeter test. 4. Adjust carburetor. 5. Substitute spark plug. 6. Check for leaky vacuum hoses and diaphragms, or gaskets.
One or more cylinders show a voltage rise less than 3 kV or no voltage rise at all	1. Shorted spark plug wire 2. Broken spark plug insulator 3. Fouled spark plug 4. Low compression	1. Perform secondary insulation test. 2. Replace spark plug. 3. Clean or replace spark plug. 4. Perform compression test.
CYLINDER TIMING ACCURACY TEST		
Transistor turn off signals exceed specs for engine being tested	1. Bent distributor shaft 2. Worn distributor bushings 3. Worn gear on distributor 4. Worn camshaft gear 5. Worn timing chain	1. Perform distributor test. 2. Perform distributor test 3. Visually inspect. 4. Visually inspect. 5. Visually inspect.

To pinpoint the cause of high resistance, perform the following steps:

1. Stop the engine and install spark plug connectors between the spark plugs and leads on all abnormal cylinders.
2. Connect one end of the jumper lead to ground and the other end to the large portion of a grounding probe.
3. Start the engine and adjust the speed to 1000 rpm.
4. Touch the spark plug connector with the point of the grounding probe and observe the spark lines.

If after grounding the spark plug, the abnormal spark lines appear normal, or the spark lines show high resistance, see Table 8-6.

SECONDARY INSULATION

A break in the secondary circuit insulation will result in reduced firing voltages. If sufficient voltages go to ground, the spark plug might not fire. Check for a break in the circuit by using a ground probe and the following procedure:

1. Set the pattern selector to the display position, the function selector to secondary, and the pattern height control to the 0–25 kV scale.
2. Connect one end of the jumper lead to ground and the other end to the large portion of the grounding probe.
3. Start the engine and adjust the speed to 1000 rpm.
4. Run the grounding probe around all the secondary circuit components, one by one. Most of the common places where a breakdown in secondary insulation may occur are listed here.
   - Coil tower
   - Coil high tension wire and boots
   - Distributor cap and terminals
   - Spark plug wires and boots
5. While scanning the secondary circuit with the grounding probe, watch the firing lines on the scope. If there is an insulation break in the circuit, the firing lines will decrease in height when the high secondary voltage arcs across to the probe.

SHOP TALK

When the voltage arcs across to the grounding probe, an electrical cracking sound might be heard and/or the spark might be visible.

SPARK PLUGS UNDER LOAD

The voltage required to fire the spark plugs increases when the engine is under load. The voltage increase will be moderate and uniform if the spark plugs are in good condition and properly gapped. However, if any unusual firing characteristics are displayed on the scope patterns when load is applied to the engine, the spark plugs are probably faulty. This condition is most evident in the firing voltages displayed on the scope. To test the spark plug operation under load, do the following:

1. Set the pattern selector to the display position, the function selector to secondary, and the pattern height control to the 0–25 kV scale.
2. Start the engine and adjust the speed to idle rpm.
3. Note the height of the firing voltages.
4. Quickly open and release the throttle (snap accelerate), and note the rise in the firing voltages while checking the firing voltages for uniformity. A normal rise would be between 3 and 4 kV upon snap acceleration.

If during testing, one or more of the cylinders show a voltage rise of over 4 kV, or one or more cylinders show a voltage rise less than 3 kV or no voltage rise at all, check Table 8–6.

COIL CONDITION

The energy remaining in the coil after the spark plugs fire gradually diminishes in a series of oscillations. These oscillations are observable in the intermediate sections of both the primary and secondary patterns. If the scope pattern shows an absence of normal oscillations in the intermediate section, check for a possible short in the coil by testing the resistance in the primary and secondary windings. Also perform a coil stress test for capacity and leakage. To observe the coil oscillations, do the following:

1. Set the pattern selector to the raster position and the function selector to either secondary or primary.
2. Start the engine and adjust the speed to 1000 rpm.
3. Observe the intermediate section for gradually diminishing oscillations.

SHOP TALK

This test can be performed on all electronic ignition systems, except Chrysler's EIS.

DWELL AND DWELL VARIATION

Although dwell is not adjustable on electronic ignition systems, it is an indication of how well the ignition module is functioning. On the oscilloscope, the dwell period is read as a percentage of the total duration of one cylinder's firing cycle. This must conform to the manufacturer's specifications not only at a set idle speed but also as the engine speed varies. Depending on the system, the dwell will either remain constant, decrease, or increase as engine

TABLE 8-7: DWELL VARIATION IN ELECTRONIC SYSTEMS

System	DWELL VARIATION		
	No Change	Decrease	Increase
Chrysler EIS		•	
Chrysler EIS (with Hall-Effect Switch)	•		
Chrysler ELB		•	
Chrysler ESC		•	
Chrysler Hall-Effect with ESC			•
Ford SSI	•		
Ford Dura Spark I			•
Ford Dura Spark II	•		
GM HEI			•
GM HEI/ESC			•
GM HEI/ESS			•
GM HEI/EMR			•

speed is increased. Table 8–7 lists the dwell variation characteristics of most domestic electronic ignition systems covered in this chapter.

To observe the dwell of an electronic ignition system on the oscilloscope, do the following:

1. Set the pattern selector to the superimposed position and the function selector to either secondary or primary.
2. Start the engine and while observing the dwell section of the pattern displayed on the scope, operate the engine through various rpm levels.
3. Compare the results shown on the scope with the dwell characteristics listed for the particular ignition system under test. If during testing, the dwell characteristics of the engine being tested are abnormal, check for a defective control module and replace it if necessary.

CYLINDER TIMING ACCURACY

The ignition timing should not vary cylinder to cylinder more than 2 degrees. On the oscilloscope the percent of dwell scale can be used to observe variations in timing between the cylinders. This is done by placing and selecting the superimposed pattern selector waveform and observing the variations between the transistor turn off signal. To perform the test, do the following:

1. Set the pattern selector to the superimposed position and the function selector to either secondary or primary.
2. Start the engine and adjust the engine speed to 1000 rpm.
3. Observe the transistor turn off signals.
4. Compare the timing variations observed on the scope with the timing variation specifications listed for the engine being tested.

If during testing the transistor turn off signals exceed the specifications for the engine being tested, check Table 8–6.

SHOP TALK

Two degrees of crankshaft rotation is equal to 2 percent of dwell on 4-cylinder engines, 4 percent on 6-cylinder engines, and 6 percent on 8-cylinder engines.

INDIVIDUAL COMPONENT TESTING

Unlike most conventional ignition systems, electronic ignition systems have many characteristics unique to the systems manufacturer. It would be impossible in any one textbook to explain all the

TABLE 8-8: IGNITION SECONDARY QUICK CHECK CHART

Visually inspect spark plug wires, coil wire, plug wire boots, rotor and distributor cap for damage.
OK
NO
Repair or replace damaged components as necessary.
Disconnect spark plug wire and insert spark plug tester. Ground plug, crank engine, and check for spark.
No Spark
Remove coil wire from distributor and attach spark plug tester. Ground plug, crank engine, and check for spark.
Plug Sparks
No Spark
Plug Sparks
If plug sparks, problem is most likely not in the ignition system. Check the fuel system or engine for mechanical problems.
Proceed to "Table 8-9: Ignition Primary Quick Check Chart."
If plug sparks, problem is in spark plug wires, distributor cap or rotor. Replace components as necessary.

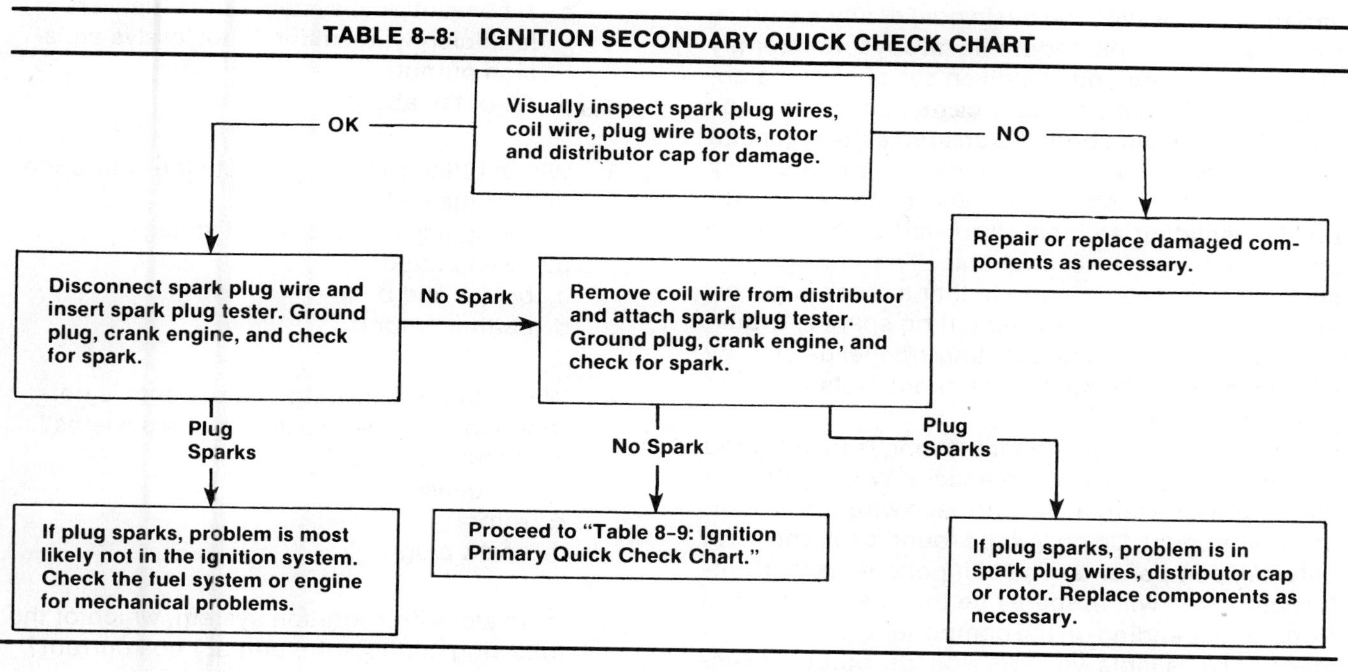

TABLE 8-9: IGNITION PRIMARY QUICK CHECK CHART

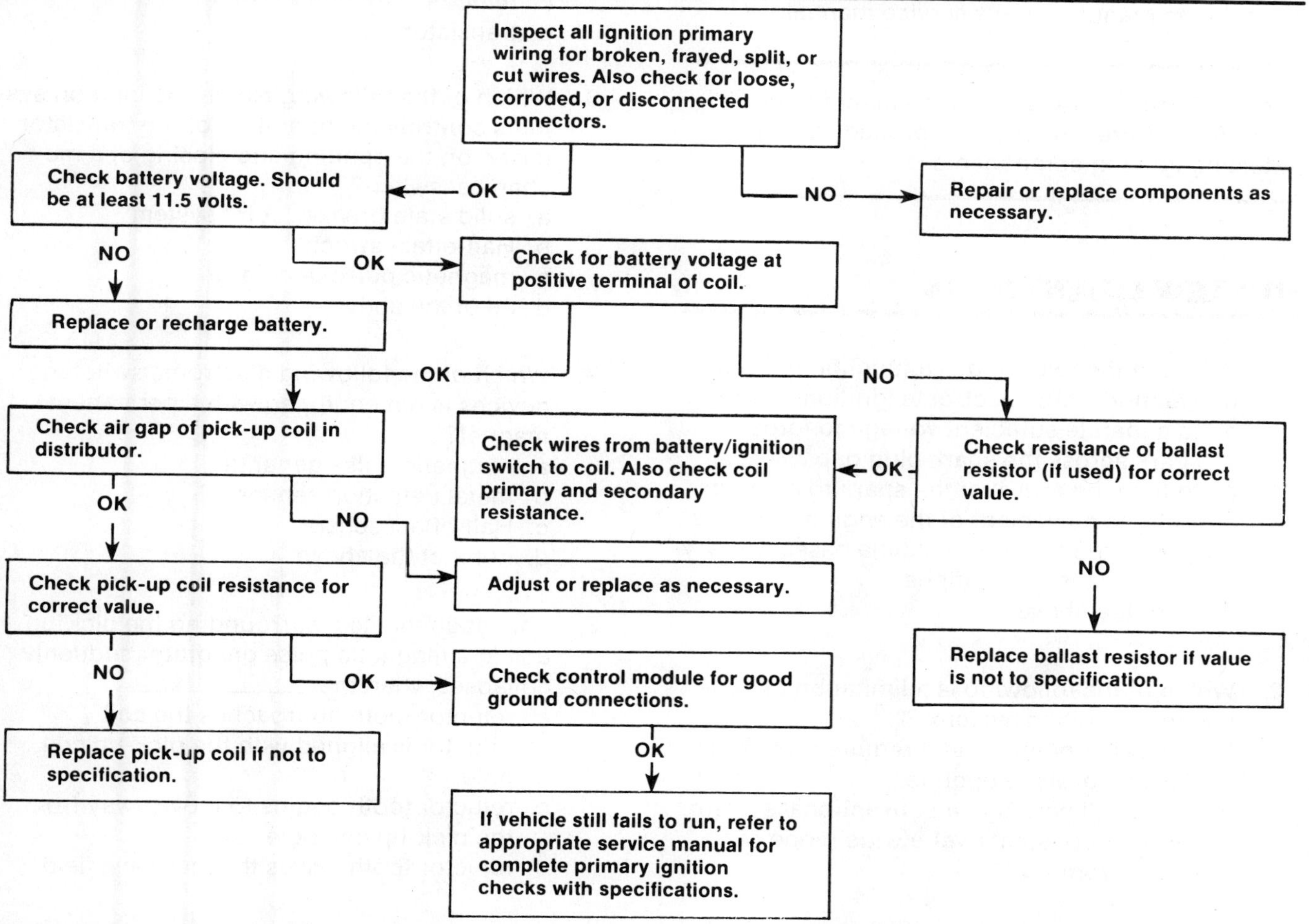

Inspect all ignition primary wiring for broken, frayed, split, or cut wires. Also check for loose, corroded, or disconnected connectors.
OK
NO
Repair or replace components as necessary.
Check battery voltage. Should be at least 11.5 volts.
NO
OK
Replace or recharge battery.
Check for battery voltage at positive terminal of coil.
OK
NO
Check air gap of pick-up coil in distributor.
Check wires from battery/ignition switch to coil. Also check coil primary and secondary resistance.
OK
Check resistance of ballast resistor (if used) for correct value.
OK
NO
Adjust or replace as necessary.
NO
Check pick-up coil resistance for correct value.
Replace ballast resistor if value is not to specification.
NO
OK
Check control module for good ground connections.
Replace pick-up coil if not to specification.
OK
If vehicle still fails to run, refer to appropriate service manual for complete primary ignition checks with specifications.

variations. However, the basic goal of any electrical troubleshooting procedure is always to identify electrical activity under a given set of circumstances. The manufacturer's service manual will provide the technician with both the desired activity (specification) and the set of circumstances (procedures).

Tables 8–8 and 8–9 outline a procedure for quickly isolating an ignition related problem when an oscilloscope is not available. The first troubleshooting tree determines whether a spark is generated in the secondary system. If no spark is available, the second troubleshooting tree leads step by step through individual component tests until the problem is located.

The secret to component testing is to use good electrical troubleshooting practices. Work systematically through a circuit, testing each wire, connector, and component. Do not jump around back and forth between components. The component inadvertently overlooked will probably be the one causing the trouble. Depending on the component, checks must made for available voltage, voltage output, resistance of wires and connectors, and available ground. Always compare the readings with specifications given in the manufacturer's service manual.

CAUTION: Always follow the manufacturer's suggested precautions and procedures when testing ignition components.

REVIEW QUESTIONS

1. Which of the following is a function of both conventional and electronic ignition systems?
   a. to generate sufficient voltage to force a spark across the spark plug gap
   b. to time the arrival of the spark to coincide with the movement of the engine's pistons
   c. to vary the spark arrival time based on varying operating conditions
   d. all of the above

2. Which of the following is a limitation of conventional ignition systems?
   a. A breaker point system requires very high voltage levels to operate.
   b. The dwell period in a conventional system remains constant over a wide range of engine rpm.
   c. A conventional system burns fuel very inefficiently, resulting in excessive emission output.
   d. All of the above

3. Which of the following affects the saturation time of the coil?
   a. resistance of the primary voltage
   b. dwell period
   c. both a and b
   d. neither a nor b

4. Which of the following components is not found in most electronic ignition systems?
   a. distributor cap
   b. condenser
   c. rotor
   d. spark plug

5. In an electronic ignition system, which of the following controls the primary coil current?
   a. breaker points
   b. pick-up coil
   c. reluctor
   d. transistor

6. Which of the following electronic ignition systems controls the operation of the transistor based on the opening and closing of a mechanical switch?
   a. solid state breaker point system
   b. Hall-effect switch
   c. magnetic pulse generator
   d. all of the above

7. Which of the following electronic switching devices is not equipped with a permanent magnet?
   a. magnetic pulse generator
   b. metal detection sensor
   c. Hall-effect sensor
   d. none of the above

8. The magnetic field surrounding the pick-up coil in a magnetic pulse generator suddenly collapses when the ____________.
   a. reluctor tooth approaches the coil
   b. reluctor is aligned with the pick-up coil pole
   c. reluctor tooth begins to move away from the pick-up coil pole
   d. reluctor tooth leaves the magnetic field

9. Which of the following electronic switching devices has a reluctor with wide shutters rather than teeth?
   a. magnetic pulse generator
   b. metal detection sensor
   c. Hall-effect sensor
   d. all of the above

10. In which of the following switching devices is the strength of the pick-up coil's signal affected by the speed of the engine?
    a. magnetic pulse generator
    b. photoelectric sensor
    c. Hall-effect sensor
    d. all of the above

11. Which is *not* a function of the dual ballast resistor used in Chrysler's early electronic ignition system?
    a. to limit the voltage available to the spark plugs
    b. to limit the voltage available to the control unit
    c. to maintain a constant primary current at all engine speeds
    d. to protect the control module from power surges

12. Which ignition system component is eliminated by the use of current limiting circuits in the electronic control module?
    a. transistor
    b. ballast resistor
    c. condenser
    d. none of the above

13. In which of the following systems would you find an "F"-shaped reluctor?
    a. Prestolite's breakerless inductive discharge system
    b. Chrysler's electronic ignition system
    c. Ford's solid state ignition system
    d. all of the above

14. Which of the following was the high performance version of Ford's early electronic ignition systems?
    a. solid state ignition (SSI)
    b. Dura Spark I
    c. Dura Spark II
    d. none of the above

15. Depending on the model year and application, the control module of Ford's Dura Spark II system would retard ignition timing ____________.
    a. during hard acceleration or high load conditions
    b. during high altitude operation
    c. after startup
    d. in stop-and-go traffic

16. The control module on Ford's thick film integrated system is located ____________.
    a. in the distributor
    b. attached to the distributor
    c. attached to the firewall
    d. inside the passenger compartment

17. Which of the following is not an integral component of GM's high-energy ignition system distributor?
    a. control module
    b. pick-up coil
    c. ballast resistor
    d. coil

18. Which of the following is *not* true of early electronic ignition systems?
    a. Dwell is not affected by parts wear.
    b. Timing advance is controlled by a computer.
    c. Spark plug life is much longer than in conventional systems.
    d. The air gap must be adjusted on a regular basis.

19. GM's electronic spark control system controls detonation by ____________.
    a. retarding the timing
    b. increasing the dwell period
    c. advancing the timing
    d. limiting the voltage available to the pick-up coil

20. Which of the following GM ignition system includes a knock sensor?
    a. HEI electronic module retard
    b. HEI electronic spark control
    c. HEI electronic spark selection
    d. all of the above

21. The average spark duration of an electronic ignition system is ____________.

a. over 2.0 milliseconds
b. approximately 1.5 milliseconds
c. under 0.8 milliseconds
d. none of the above

22. The spark line on an oscilloscope pattern extends below the waveform. Technician A says the polarity of the coil is reversed. Technician B says the problem is fouled spark plugs. Who is correct?
a. Technician A
b. Technician B
c. Both A and B
d. Neither A nor B

23. The firing lines on an oscilloscope pattern are all abnormally low. Technician A says the problem is probably low coil output. Technician B says the problem could be an overly rich air/fuel mixture. Who is correct?
a. Technician A
b. Technician B
c. Both A and B
d. Neither A nor B

24. If the spark line on an oscilloscope pattern has a steep slope and is shorter than normal, the problem is probably *not* ______________.
a. high resistance in the spark plug wire
b. a faulty spark plug
c. silicone grease on the rotor tip
d. burned distributor cap terminals

25. Which of the following would be used to troubleshoot an electronic ignition system?
a. spark plug tester
b. straight pin
c. volt-ohmmeter
d. all of the above

CHAPTER NINE

COMPUTER-CONTROLLED IGNITION SYSTEMS

Objectives

Upon completion of this chapter, you should be able to:

- Explain how a computer-controlled ignition system computes spark advance based on engine operation conditions.
- Describe the components and operation of most domestic computer-controlled ignition systems.
- Perform test procedure on most domestic computer-controlled ignition systems.
- Explain the operation of a distributorless ignition system and perform tests on GM's DIS systems.

The electronic ignition systems, discussed in the previous chapter, were an improvement over breaker point ignition systems. However, as exhaust emissions standards and governmental legislation became more and more stringent, manufacturers needed a more precise method of controlling the combustion process. The result was the introduction of electronic spark timing control and computerized engine control systems.

The main difference between the systems discussed in this chapter and those described in Chapter 8 is the elimination of any mechanical or vacuum advance devices from the distributor. Spark timing on these systems is controlled by a computer that continuously varies ignition timing to obtain optimum air/fuel combustion. In most systems discussed in this chapter, the distributor's sole purpose is to generate the ignition primary circuit switching signal and distribute the secondary voltage to the spark plugs; timing advance is controlled by a microprocessor, or computer. Late-model systems have even removed the primary switching function from the distributor; the distributor's only job is to distribute secondary voltage to the spark plugs. In some state-of-the-art systems, the distributor has been eliminated altogether.

Early computer-controlled ignition systems were limited to regulating the spark timing. As emission and fuel economy standards tightened, it was necessary to add air/fuel mixture controls to the vehicle's computer functions. By the late 1970s most vehicle manufacturers were combining fuel and emission controls with spark controls to form an integrated engine control system. During the 1980s, computerized controls have been applied to other vehicle systems as well, such as suspension, brakes, and climate control.

The capabilities and limitations of electronic spark advance systems vary from one manufacturer to another. Early spark control computers were analog-type computers and were very simple compared to the digital microprocessors used in today's engine control systems. All the systems, however, are similar in design or structure (Figure 9-1). Every system is made up of the following components

- Input sensors that monitor engine operating conditions
- A central computer that processes input information and issues commands to various motors, solenoids, and switches
- Output control devices that respond to commands from the computer

COMPUTERIZED SPARK ADVANCE SYSTEMS

Since electronic ignition has already been discussed in this textbook, discussions in this section

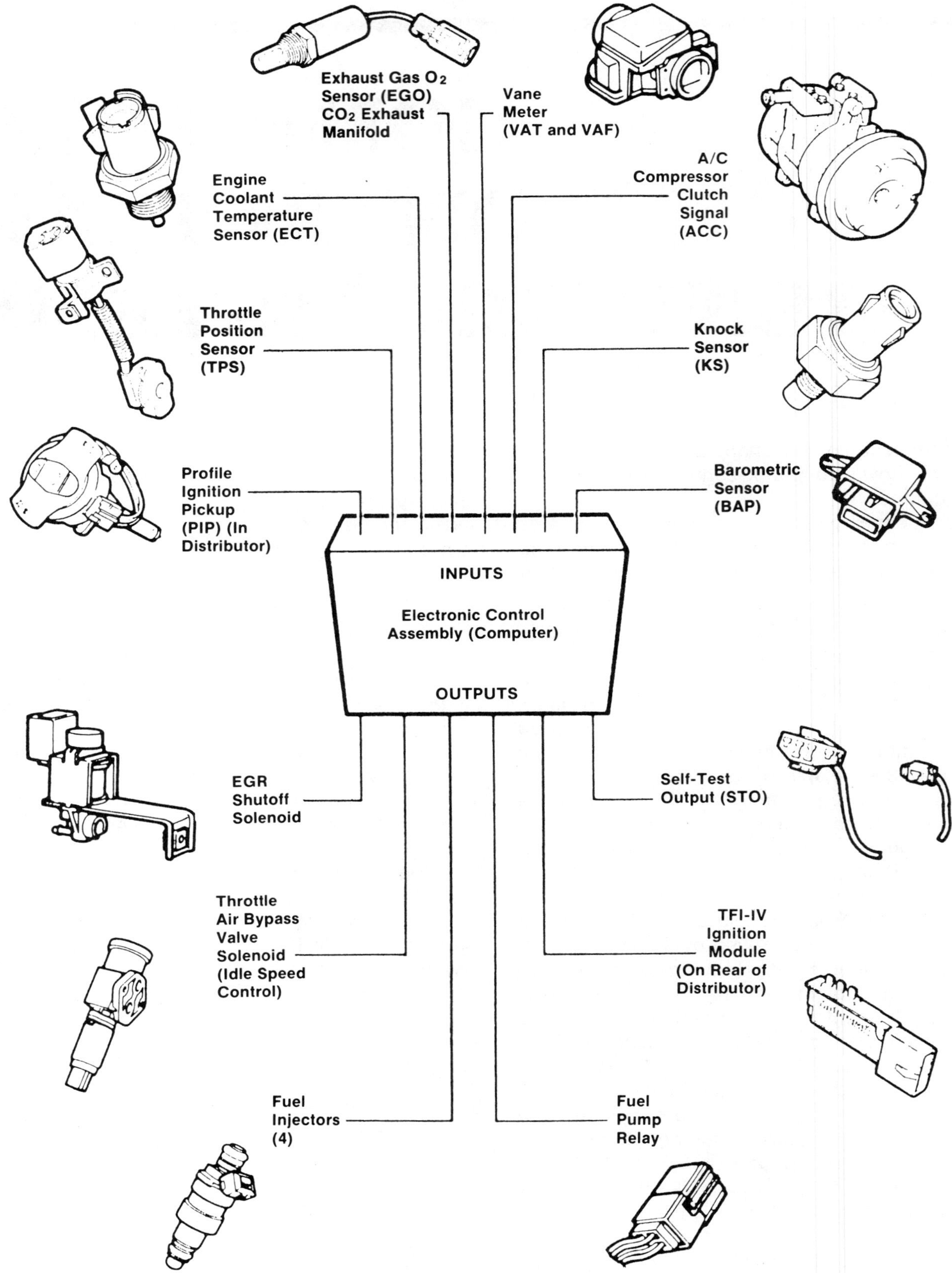

FIGURE 9-1 Typical components of a late-model Ford electronic engine control (EEC-IV) system

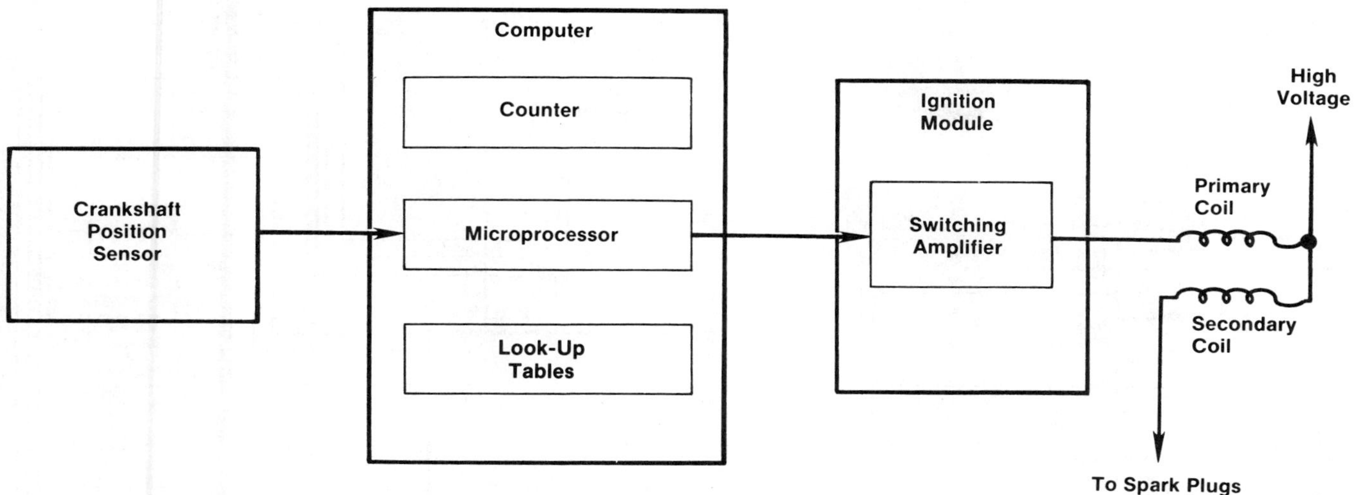

FIGURE 9-2 Typical engine timing control

will center on the electronic control of spark timing as it would occur in a computerized electronic control system. Most of the components and systems discussed in the previous chapter are also present in the early computer-controlled ignition systems.

A computer controls engine timing on vehicles equipped with an electronic engine control system. This eliminates the need for centrifugal and vacuum advance mechanisms that are required on early electronic ignition systems. The computer monitors the engine operating parameters with sensors and signals an ignition module when to collapse the primary switching circuit, allowing the secondary circuit to fire the spark plugs (Figure 9-2).

The logic (hardware and software) in a computerized system's program selects the method of spark timing control. During engine starting, the mechanical setting of the distributor controls spark timing. Once the engine is started and running, spark timing is under the control of the computer control system. This scheme ensures that the vehicle will start regardless of whether the electronic control system is functioning or not.

The goal of computerized spark timing is to produce maximum engine power, top fuel efficiency, and minimum emissions levels by adjusting the advance of the ignition firing in relation to TDC. The spark timing can be chosen to produce the best engine performance with input variables such as engine speed, engine coolant temperature, initial and operating manifold or barometric pressure. The total spark advance parameters are determined by computing the information received from the various engine sensors. The computer then adjusts the timing according to information that has been calibrated or programmed into it. The computer typically has programmed into it specific information on:

- *Warmup Spark Advance.* This is used when the engine is cold, since a greater amount of advance is required while the engine warms up.
- *Special Spark Advance.* This is used to improve fuel economy during steady driving conditions.
- *Spark Advance Due to Barometric Pressure.* This is used when barometric pressure exceeds a preset calibrated value.

All of this information is then added together and the initial mechanical advance is subtracted to determine the final spark advance. The calibrated or programmed information in the computer is contained in what is called software look-up tables. A block diagram of a typical computerized spark advance system is shown in Figure 9-3.

The computer receives a timing pulse from an input sensor that indicates crankshaft position for top dead center and engine rpm. The computer makes a decision based upon this information, the data from all the input sensors, and the information in look-up tables. At that time, the computer sends a pulse to the ignition actuator circuit, which opens the ignition coil primary circuit. This generates a secondary voltage pulse to fire the spark plugs. In some cases, the circuitry to open the primary of the coil may be in the computerized control unit. Spark selection and distribution are performed mechanically by the distributor and rotor contacts in the exact same manner as it is done in a nonelectronic-controlled system.

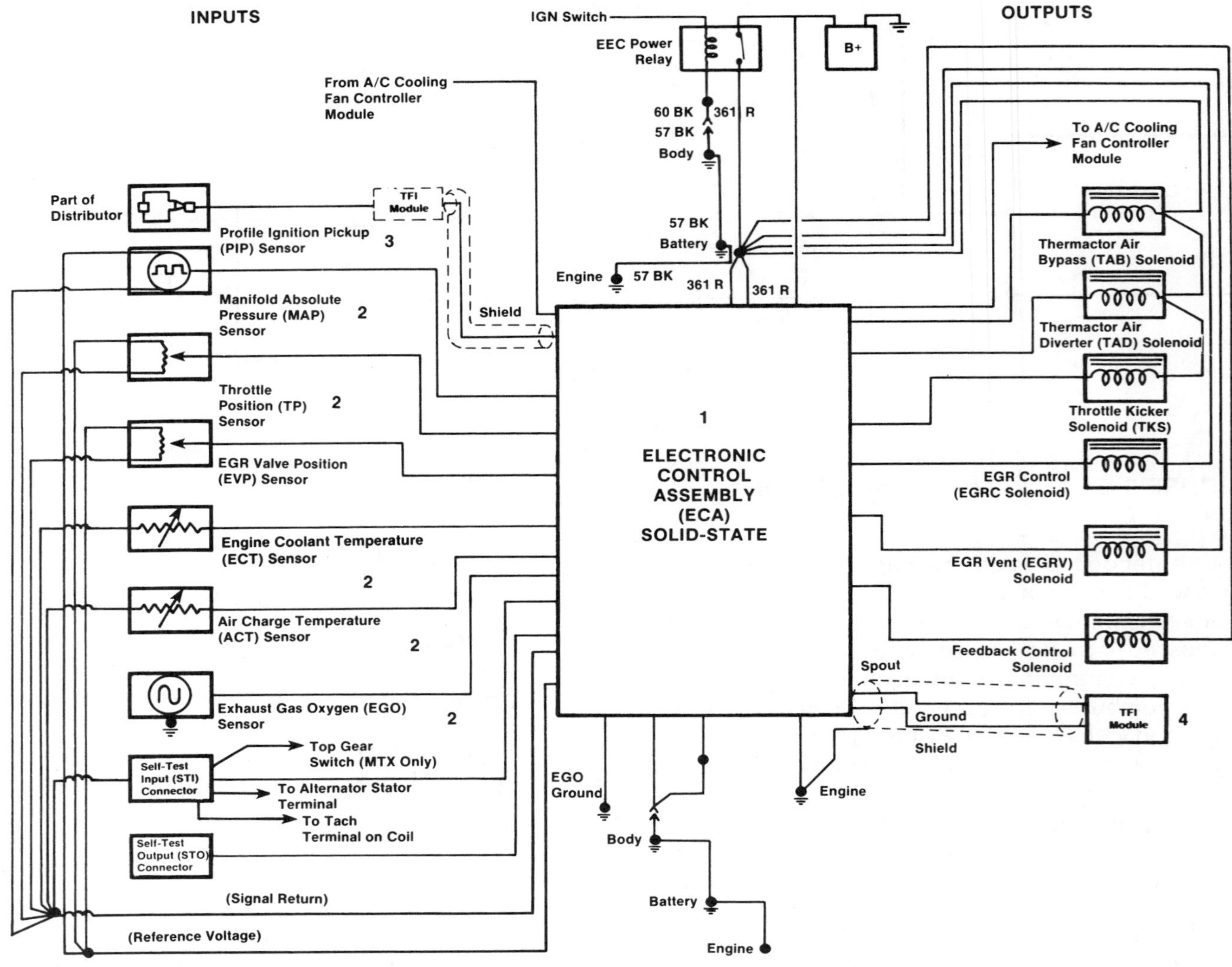

FIGURE 9–3 Input signals and output signals in an EEC system

The ignition timing also works in conjunction with the electronic fuel control system to provide emission control and optimum fuel economy and driveability because they are all dependent on spark advance of the engine timing.

This system operation just described operates in an open loop. In the open loop mode, the computer establishes spark timing according to fixed programmed values. Most computerized ignition systems receive an input from a knock sensor. These systems operate in a closed loop mode that allows the ignition system to monitor the engine for mechanical changes, such as engine knock, and makes spark advance adjustments accordingly. When the computer receives a signal from the knock sensor, it retards the spark advance from the programmed value until the knocking stops and then starts increasing the spark advance again. This cycle is repeated as long as engine knock occurs. Figure 9–4 illustrates a typical spark advance, closed loop mode of operation.

TROUBLESHOOTING COMPUTER-CONTROLLED ENGINES

When presented with a complaint by a vehicle owner, do not automatically assume the computer system is at fault. The key to making an accurate diagnosis of engine performance problems on a vehicle that is equipped with computer-controlled engine functions is to approach the vehicle as though it did not have the system. In other words, do not forget there is still an engine under all those sensors and actuators.

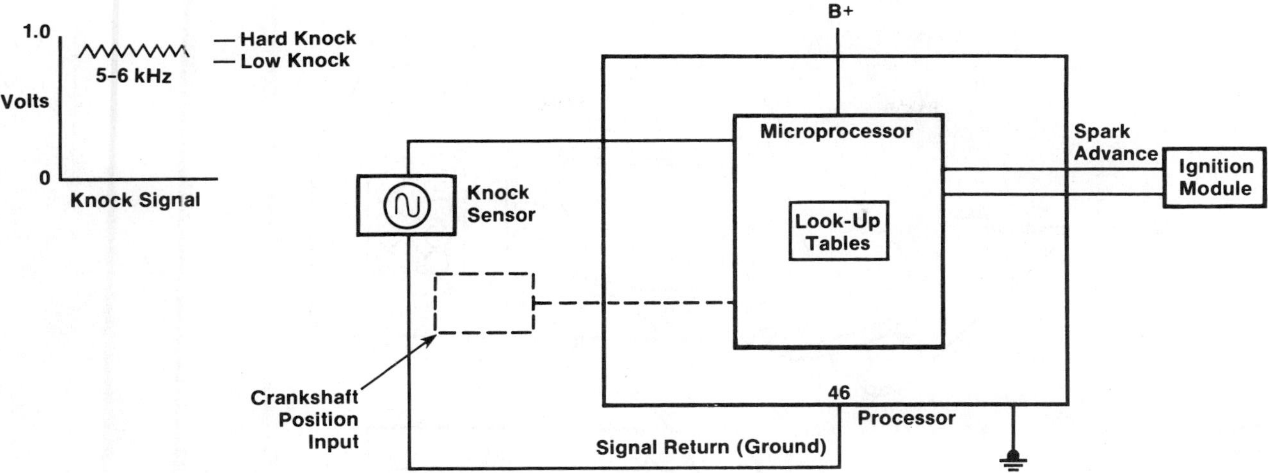

FIGURE 9-4 Schematic of spark control system designed to control detonation

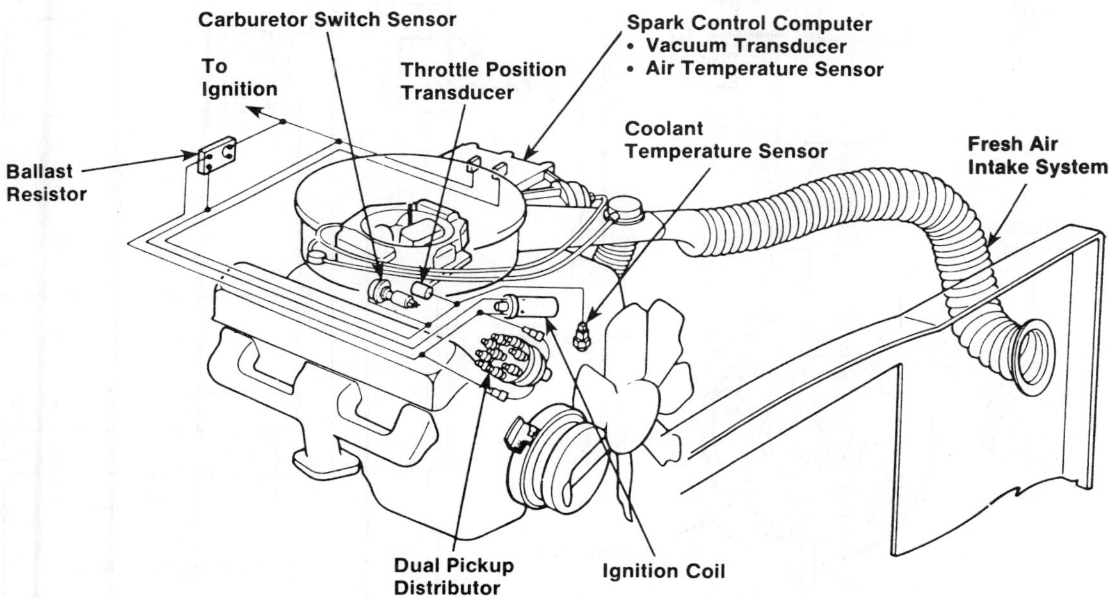

FIGURE 9-5 Chrysler electronic lean burn system

Perform standard diagnostic routines and visual inspections first. Start by looking for loose, corroded, or damaged electrical connections. Next, check all vacuum lines and fittings for general condition and proper routing, and take vacuum readings with a vacuum gauge (see Chapter 3). Any type of leak in the air induction system, especially on fuel-injected engines, will greatly affect engine performance. On turbo-equipped models, something as inconspicuous as a loose oil filler cap can cause the engine to run rough at idle.

Finally, check engine compression and fuel delivery. Connect the vehicle to an oscilloscope and verify that the ignition is performing as it should. No matter how sophisticated the electronic control system is, the computer cannot compensate for an open plug wire, burned valve, or clogged fuel filter.

CHRYSLER'S SPARK ADVANCE SYSTEMS

The electronic spark control system was first introduced in 1976 and was named the electronic lean burn system (ELB) (Figures 9-5 and 9-6). With the introduction of the four-cylinder Plymouth Ho-

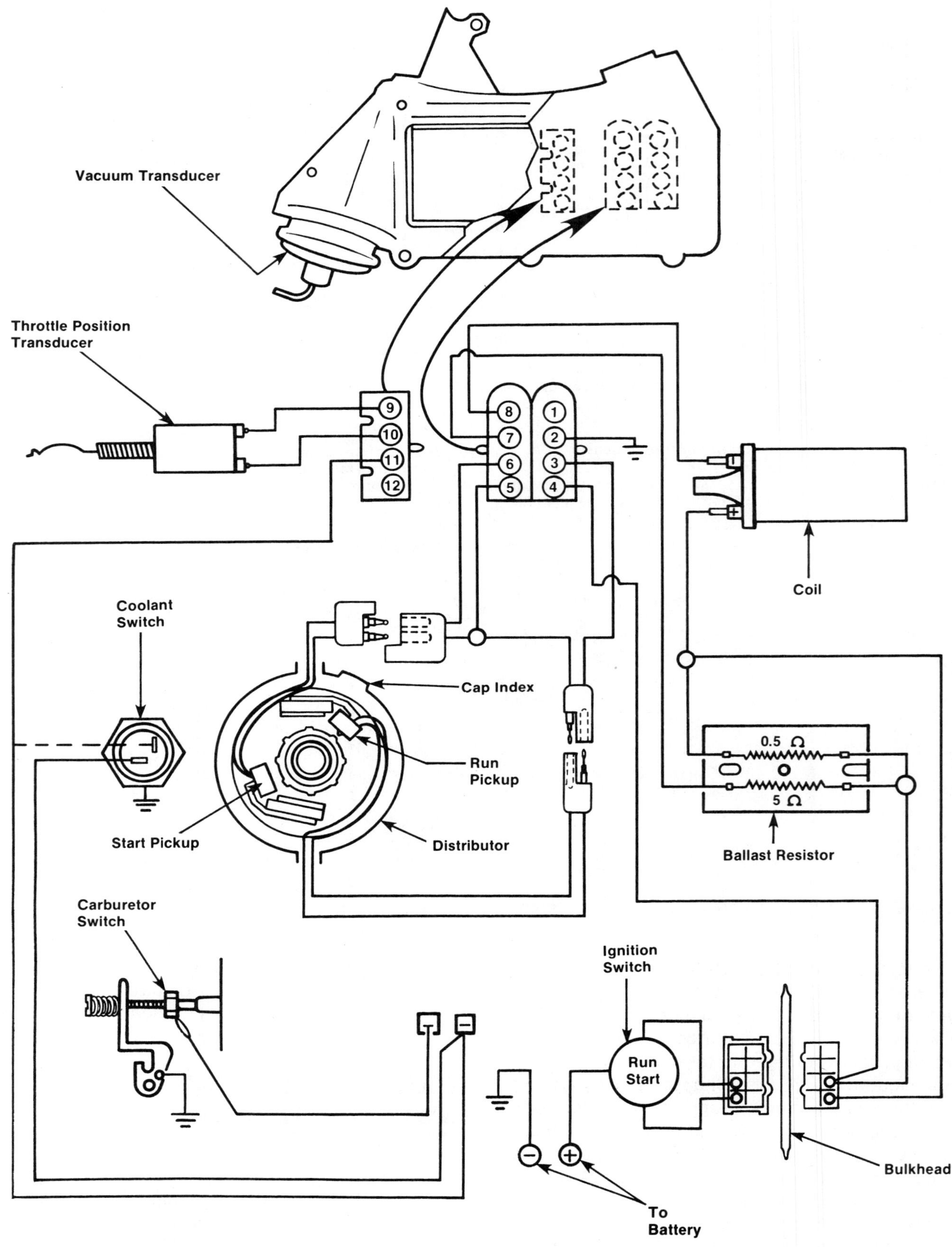

FIGURE 9–6 Electronic lean burn wiring schematic

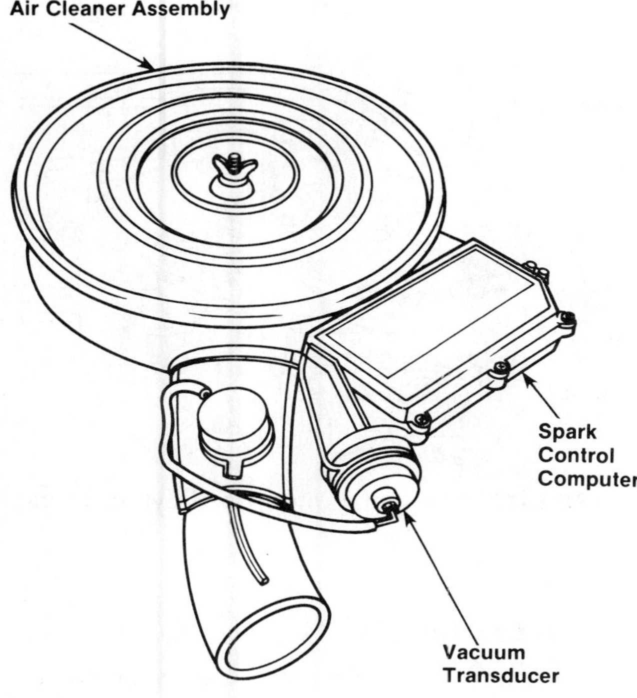

FIGURE 9-7 The spark control computer (SCC) is mounted on the air cleaner in a lean burn system.

rizon and Dodge Omni, an electronic lean burn system used a Hall-effect distributor. In 1979 the system was changed to the electronic spark control system (ESC). This system used an analog-type computer to process signals from the various engine sensors to accurately control engine timing. The new system provided improved engine performance, fuel economy, and emission control. In 1980 the system was changed to electronic spark advance (ESA); the change included a digital computer to replace the existing analog type. Some systems incorporate a feedback system to control the air/fuel mixture based on inputs from an oxygen sensor in the exhaust manifold.

SYSTEM COMPONENTS

Spark Control Computer (SCC)

The spark control computer is the brains of the system (Figure 9-7). It controls the ignition timing, based on the engine's vacuum and rpm, and in a feedback fuel system, the air/fuel mixture. The computer receives inputs from various engine sensors and within milliseconds determines the correct ignition timing and signals the ignition coil to produce the electrical impulses to fire the spark plugs. On six- and eight-cylinder engines it is mounted in a plastic housing on the air cleaner, and on four-

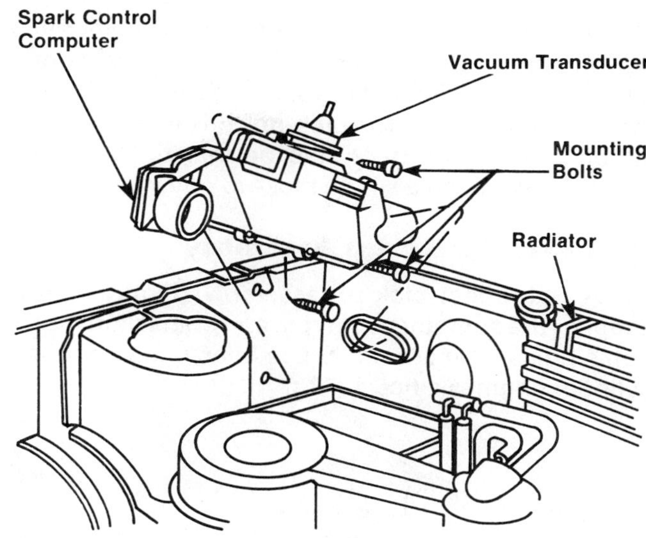

FIGURE 9-8 SCC mounting on 4-cylinder engines

cylinder engines, it is mounted on the fenderwell (Figure 9-8). The ignition module is contained within the assembly. The SCC assembly is serviced as an entire assembly.

Sensors

In the ESC system, a variety of sensors and switches supply the SCC with the necessary information to determine the ignition timing and the air/fuel ratio for the feedback system. The sensors include the distributor (start and run pickups, or Hall-effect distributor), carburetor switch, coolant switch, throttle position transducer, vacuum transducer, and the detonation (knock) sensor.

Magnetic Pick-up Assembly

Distributors on 6- and 8-cylinder carbureted engines have two pick-up coils (Figure 9-9). The

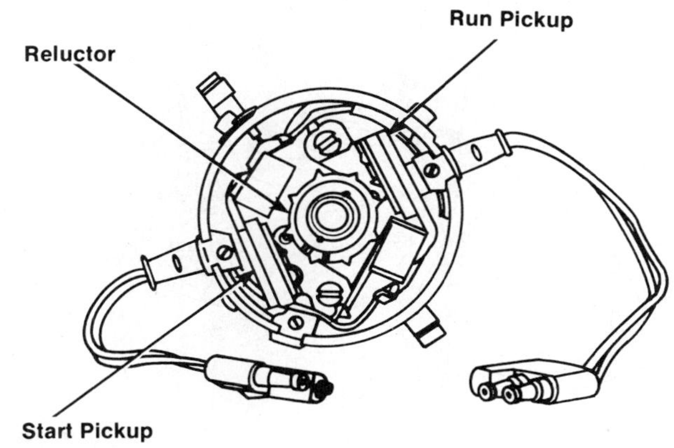

FIGURE 9-9 Dual pick-up distributor

start pick-up coil supplies a signal to the computer during cranking only. The timing advance is determined by the position of the start pick-up coil. When the engine starts, the run pick-up coil takes over and supplies the signal. Electronic fuel injection engines (1981 to 1982) use a single pick-up coil.

Hall-Effect Pick-up Assembly

A Hall-effect pick-up assembly (Figure 9-10) supplies the computer with rpm and crankshaft location signals. The assembly has been used on four-cylinder engines since 1978.

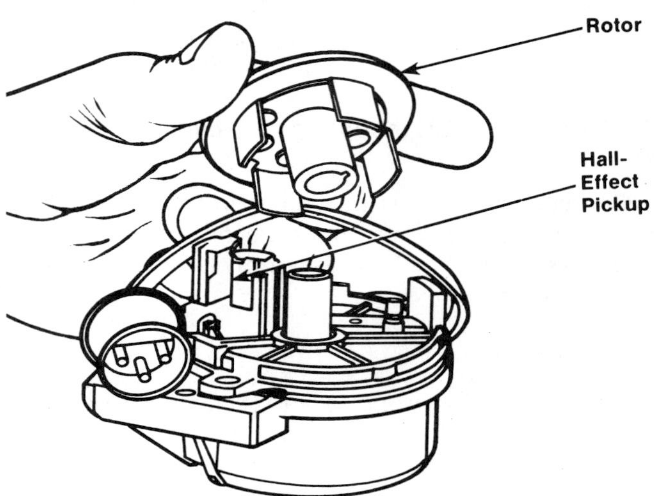

FIGURE 9-10 Hall-effect distributor

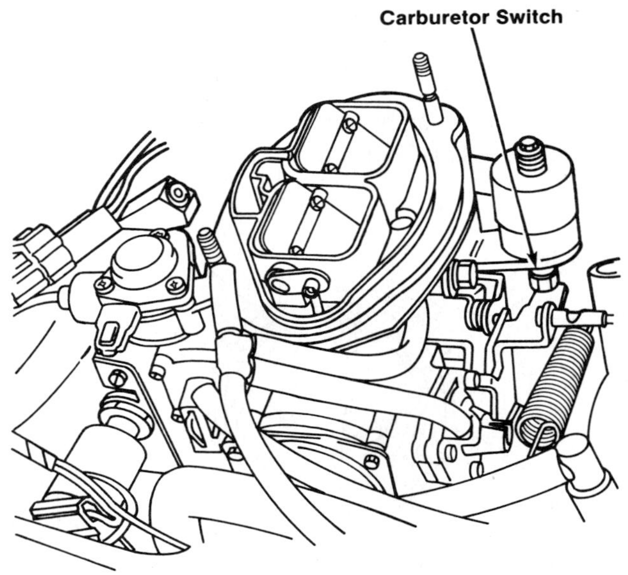

FIGURE 9-11 Carburetor switch on 4-cylinder engines

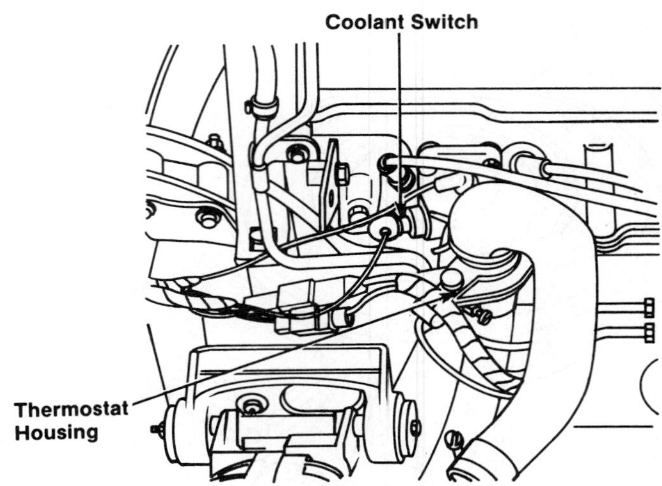

FIGURE 9-12 Coolant switch on 4-cylinder engines

Carburetor Switch

All systems use a carburetor switch located at the end of the idle stop (Figure 9-11) to signal the computer when the engine is idling. In the idle position the switch is grounded, causing the cancellation of spark advance and with a feedback carburetor a return to idle control of the air/fuel ratio.

Coolant Switch

All systems use a coolant switch (Figure 9-12), which supplies a signal to the computer when the coolant temperature reaches a predetermined temperature. This signal controls the amount of spark advance timing and on feedback fuel systems prevents changing of the air/fuel ratio until the engine reaches normal operating temperature.

Throttle Position Transducer

The throttle position transducer (Figure 9-13), which is used on 1976 to 1979 six- and eight-cylinder engines and in 1978 four-cylinder engines, is located on the carburetor. It signals the throttle position and rate of change to the computer. The computer advances the timing base of the throttle's position.

Vacuum Transducer

The vacuum transducer (Figures 9-7 and 9-8) is used on all systems. It is located on the SCC and signals the engine's vacuum to the computer. This is used to affect the ignition timing and to change the air/fuel ratio on feedback carburetor vehicles.

Detonation Sensor

Introduced in 1980, the detonation suppressor system (Figure 9–14) helps to control spark knock (detonation) on all eight-cylinder engines. The sensor mounts on the intake manifold and sends a low-voltage signal to the SCC to retard the timing when detonation occurs. The amount the timing that is retarded is proportional to the severity of the detonation, but limited to 11 degrees maximum. Once detonation is eliminated, the timing returns to its original value.

OPERATION

The basic design of the Chrysler spark control systems can be illustrated in a simple process flow diagram (Figure 9–15). The control loop process is basically the same as for any control system. Sensor inputs go to the computer, which processes them and generates a signal to advance or retard the spark timing as required. The resulting engine operation affects the sensors, which send new data to the computer, continuing the cycle of operation.

During cranking the computer fires the spark plugs at a fixed amount of advance. The start mode only functions during engine cranking and starting. The run mode operates after the engine starts. During engine operation, the run and start modes never operate together.

In dual pick-up distributors, the start pick-up coil inputs to the SCC in the start mode. After the engine starts, the run pick-up coil inputs the reference signal to the SCC. Based on the inputs from all the sensors, the SCC determines how much ignition advance is needed. The advance, based on vacuum and engine rpm, is given by the SCC when the carburetor switch is open (off idle). If there is a failure of the computer, the system will go into a "limp-in" mode. This will enable the driver to operate the vehicle until it can be repaired. During this period, very poor engine performance and fuel economy will result. If for some reason there is a failure of the pick-up coil (both pick-up coils on a dual pick-up distributor) or the start mode of the computer, the engine will not start or run.

The spark control systems can be seen in more detail using simplified schematics. Figure 9–16

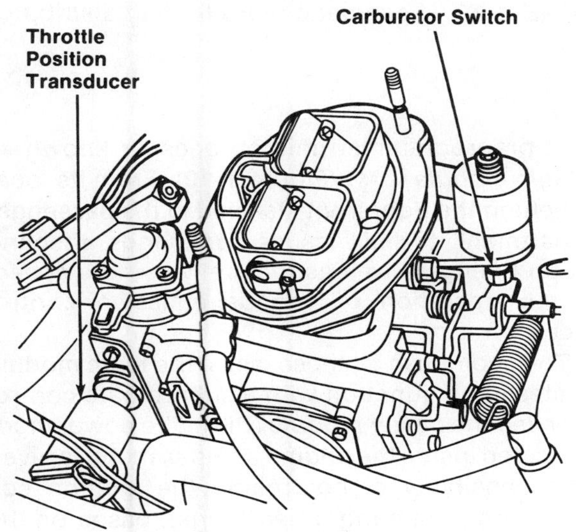

FIGURE 9–13 Throttle position transducer on 1978 4-cylinder engines

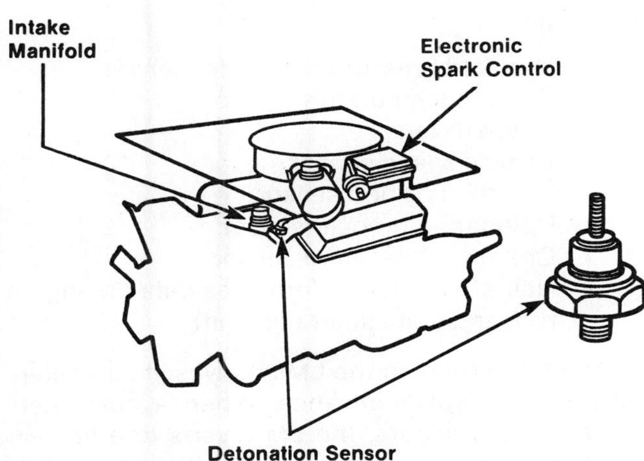

FIGURE 9–14 Detonation suppressor system

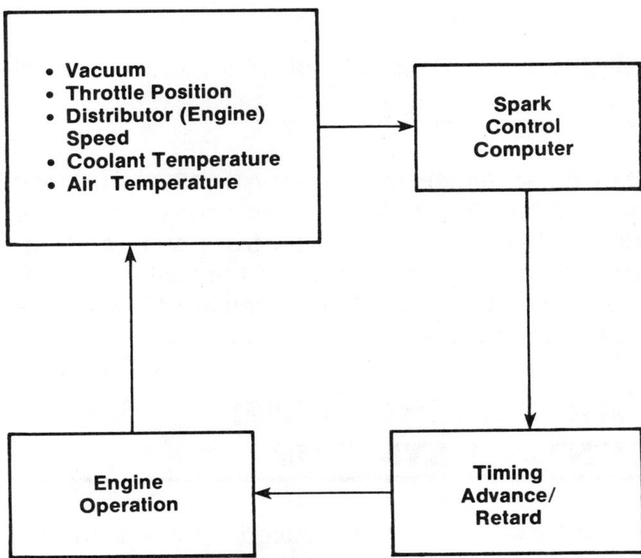

FIGURE 9–15 Process flow chart of spark control system

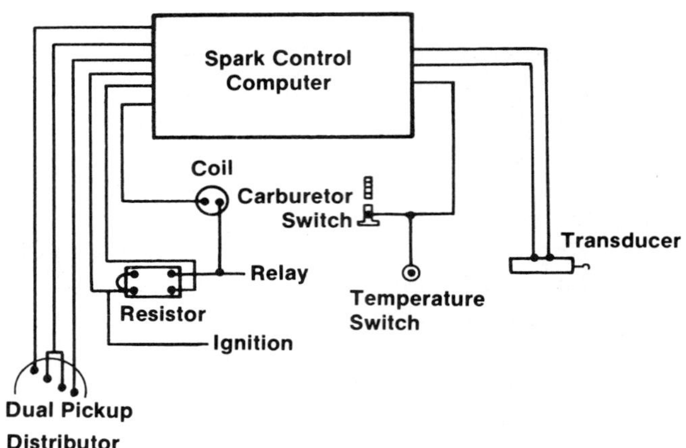

FIGURE 9–16 Schematic of 1976-77 lean burn system

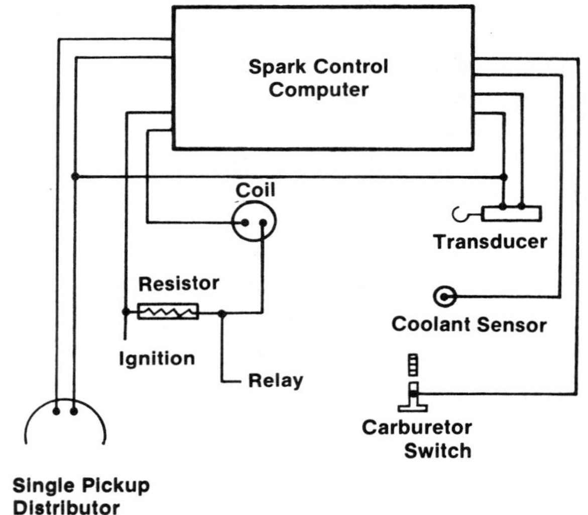

FIGURE 9–17 Schematic of 1978-79 spark control system

shows a schematic for the Chrysler 1976–1977 lean burn system; Figure 9–17 shows a schematic of the 1978–1979 electronic spark control system; and Figure 9–18 shows a schematic of the Hall-effect distributor spark control system used in 1978 and 1979 four-cylinder Omnis and Horizons.

CHRYSLER MODULAR CONTROL SYSTEMS

In 1986 Chrysler introduced a state-of-the-art electronic engine management system that was founded on their 2.2 liter EFI turbocharged engine. At the heart of the computer-controlled system is a

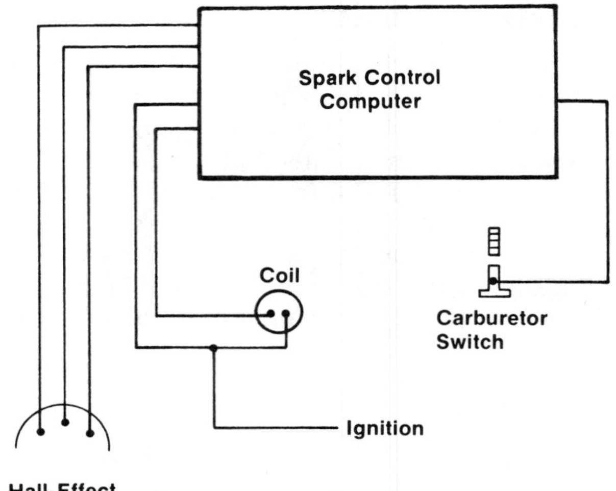

FIGURE 9–18 Schematic of Hall-effect distributor system

digital preprogrammed microprocessor known as the *logic module* (LM) (Figure 9–19). From its location behind the right front kick pad in the passenger compartment, the LM issues commands affecting ignition timing as well as fuel delivery, idle speed, and the operation of various emission control devices.

To accomplish all these tasks, the logic module operates in conjunction with a subordinate control unit called the *power module* (PM). The power module, located inside the engine compartment, is given the responsibility of controlling the ignition coil ground circuit (among other things), based on the LM's commands.

Under normal operating conditions, the LM bases its spark advance control decisions on information from several input devices. These include the following:

- Manifold pressure (absolute) sensor
- Coolant temperature sensor
- Oxygen sensor
- Vehicle speed sensor
- Throttle position sensor
- Detonation (knock) sensor
- Charge temperature sensor
- Hall-effect pickup (provides data on engine rpm and crankshaft position)

From the information the LM receives, it can determine precise spark advance. When a computer-related failure occurs, the LM reverts to a limp-in mode whereby the computer will either work with fixed values in place of failed sensor input, or depending on which input was lost, it can also operate a modified value by combining two or more related sensor inputs.

The modular control system also has the following capabilities when a problem occurs:

- Stores a fault code identifying the problem circuit in its memory.
- Turns on the POWER LOSS/POWER LIMIT lamp located in the vehicle instrument panel. This lamp is lit only for certain malfunctions—not all faults.

Once a fault code has been stored in the computer's (LM) memory, it will remain there until the battery is physically disconnected or the computer no longer sees a problem and erases the code automatically. In the later case, the LM is programmed to track fault conditions for 20 to 40 engine starts. After that the code will be erased—with or without a problem existing.

TESTING AND SERVICING CHRYSLER SPARK CONTROL SYSTEMS

Logic is the most important troubleshooting tool. As a basic principle, do not assume that any and all problems are caused by the electronic or computer control system. Only the ignition system primary side directly ties in with the electronic engine controls. Except for some later year four-cylinder models, Chrysler's ESC systems do not

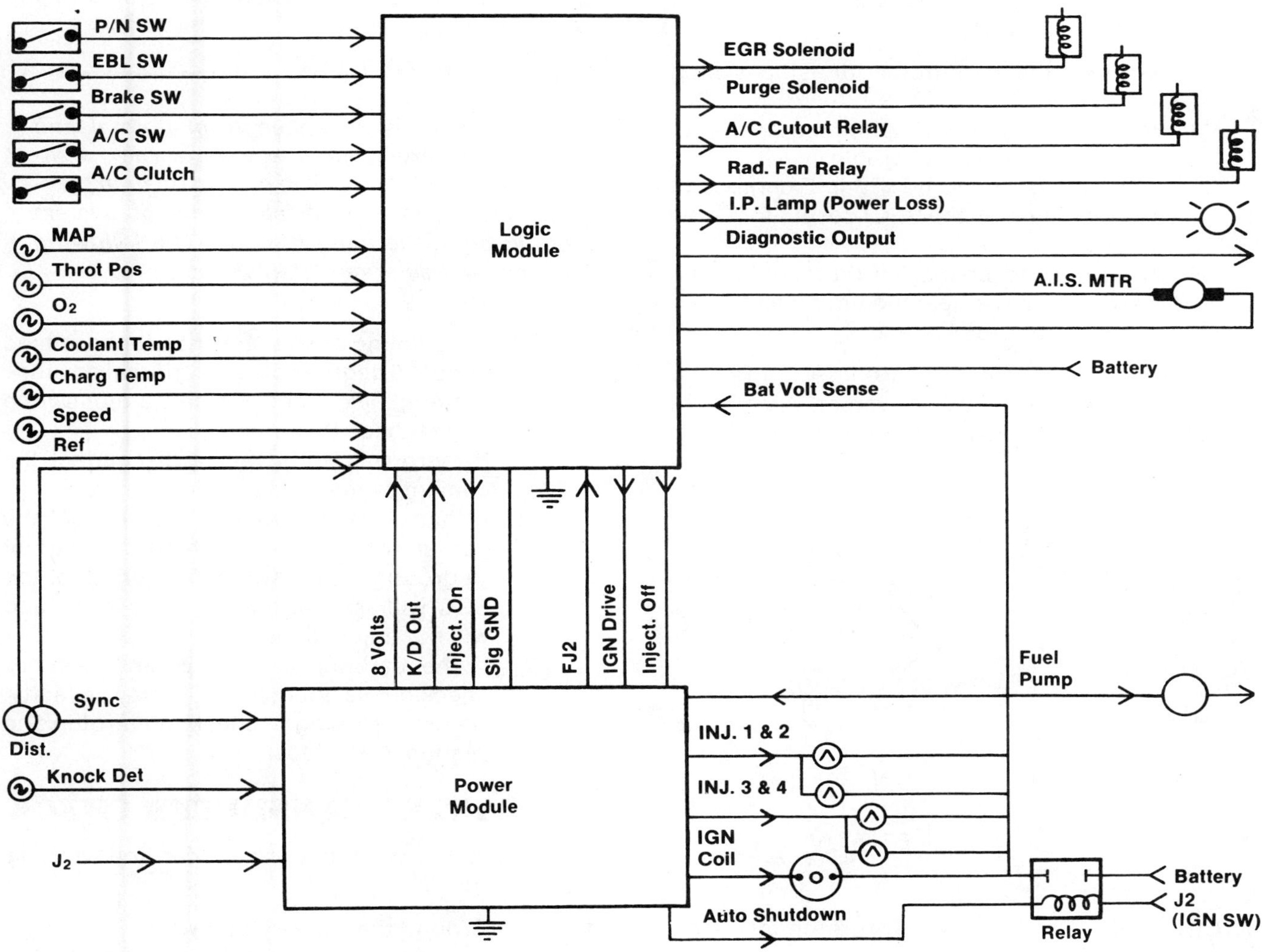

FIGURE 9-19 On-board diagnostics

have a self-diagnostic capability. Therefore, use a logical approach to troubleshooting and have a thorough knowledge of how Chrysler's systems work. The following preliminary precautions should be observed:

- Always turn the ignition off before disconnecting any electrical connectors to prevent damage to the computer.
- The carburetor switch is mounted with an idle-speed up solenoid on many engines. Do not confuse the two because voltage readings from the solenoid are meaningless.
- The computer cannot be bypassed by disconnecting it since the ignition primary uses the computer to complete its circuit. Unplugging the computer will prevent the engine from starting.

BASIC IGNITION TIMING TEST

To test the timing on most lean burn and ESC systems, do the following:

1. Ground the carburetor idle-stop switch (Figure 9-20).
2. Start the engine and allow the engine to warm-up to slow (curb)-idle.
3. The timing should be within 2 degrees of specification on the VECI decal (under the hood).
4. If it is not, loosen the distributor holddown hardware and adjust the timing as for any other distributor system.

Before checking the timing on 1983 and later cars with the modular control system, the coolant temperature sensor must be disconnected and reconnected to put the system into the limp-in mode. After checking and adjusting the timing, the computer power supply connector near the battery must be disconnected and reconnected to clear trouble codes from the computer's memory.

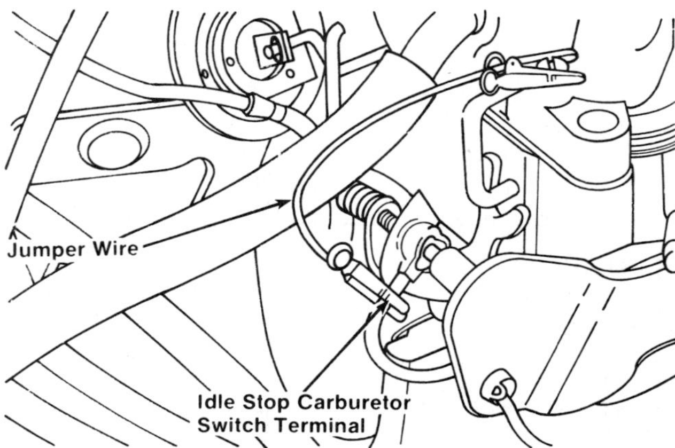

FIGURE 9-20 Ground the carburetor idle stop switch and run engine at warmed-up slow idle to test timing.

TIMING ADVANCE CHECKS

To check the timing advance, do the following:

1. Connect an advance timing light to the engine. Allow the engine to thoroughly warm up.
2. Start the engine and run at idle for 1 minute. Remove the vacuum hose from the SCC and check the position of the timing mark. The mark should be at the base timing position specified by the emission control sticker (VECI decal).
3. Accelerate the engine to 2000 rpm and again observe the timing mark. The mark should be advanced a significant amount.

SHOP TALK

If timing is advanced from normal and does not vary when the engine rpm is increased, the SCC (or run pickup on dual pickup models) may have failed and the system could be in the limp-in mode. The following checks will determine if this is so.

4. If the timing did not advance (or advanced about 15 degrees on a centrifugal advance system), disconnect the carburetor switch and recheck the timing.
5. If there was no advance, disconnect the coolant switch and recheck timing.
6. If there was still no advance, inspect the carburetor switch wiring for a short circuit to ground. If the switch is grounded, the SCC will assume the engine is idling and will not adjust timing.
7. If the carburetor and coolant switches check out as good, the SCC may be faulty. Before replacing it, check the throttle and vacuum transducers.

THROTTLE TRANSDUCER CHECK

To check the throttle transducer, follow these steps:

1. Ground the carburetor switch.
2. With the engine running and the timing light connected, move the throttle trans-

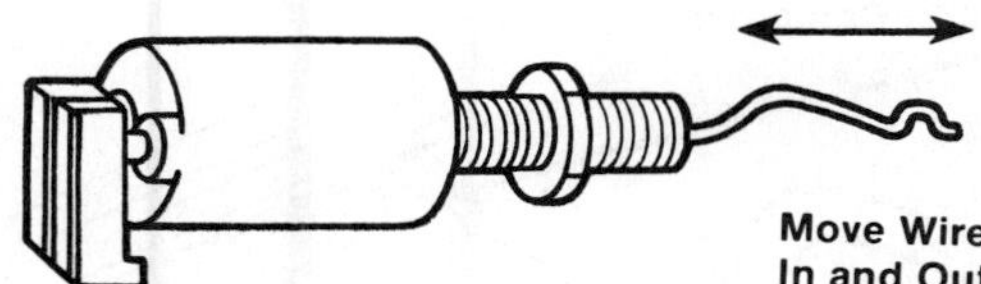

FIGURE 9-21 Move the wire in and out to check the transducer.

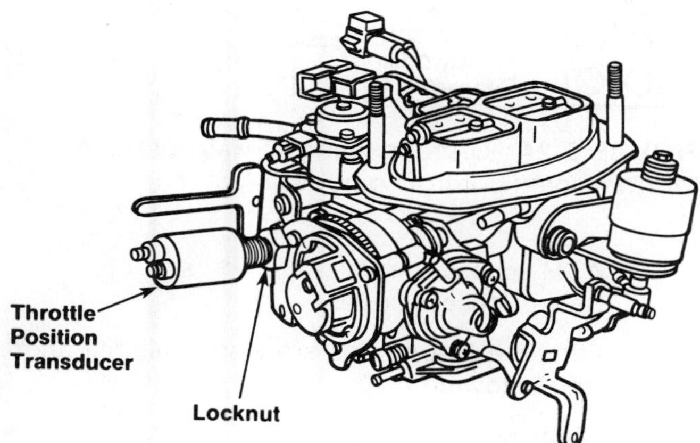

FIGURE 9-22 Using measurement gauge to adjust transducer

ducer wire in and out (Figure 9-21).

3. If there is no change in timing, either the throttle transducer or the SCC is bad. Either replace the transducer and recheck timing or test the transducer with a voltmeter.
4. There should be 6 or 8 volts at the transducer terminal. If not, and there are no wiring breaks, the SCC is bad.
5. If there is voltage available at the transducer terminal, it should vary about 1.5 volts as the transducer wire is moved in and out. If not, replace the transducer.

VACUUM TRANSDUCER CHECK

To check the vacuum transducer, follow these steps:

1. With the engine running at about 2000 rpm and the timing light connected, disconnect the vacuum transducer hose.

SHOP TALK

The carburetor switch should not be grounded for this test.

2. If timing retards about 20 degrees, the transducer is OK. If it does not, either the vacuum transducer or the SCC is bad. They must be replaced as a unit.

THROTTLE TRANSDUCER ADJUSTMENT

Adjust the throttle transducer as follows (Figure 9-22):

1. With the ignition switch off, disconnect the connector from the throttle transducer.
2. Loosen the transducer locknut.
3. Select the proper gauge block for the engine under test.

CARBURETOR	INCHES	MM
All 4-bbl and 5220 2-bbl	0.540	13.7
All other 2-bbl	0.685	17.4

4. Insert the gauge block between the outer portion of the transducer and the transducer mounting bracket.
5. Turn the transducer until the gap just admits the gauge block.
6. Tighten the locknut and reconnect the connector to the transducer.

DETONATION (KNOCK) DIAGNOSIS AND TESTING

Detonation or knock problems can be caused by overadvanced spark timing. If the timing advance is suspected, check the timing advance against the specification and adjust accordingly. If necessary, check and adjust the throttle transducer. If these checks are inconclusive, perform the following steps:

1. Disconnect the throttle transducer wiring harness connector.
2. Test drive the vehicle. If the detonation or knock disappears and all other checks are good, replace the throttle transducer.
3. If the engine still detonates, reconnect the throttle transducer.
4. Disconnect and plug the vacuum transducer hose.
5. Test drive the vehicle. If the detonation disappears, and all the other checks are good, replace the SCC.

SPARK PERFORMANCE TEST

To do a spark performance test, perform the following steps:

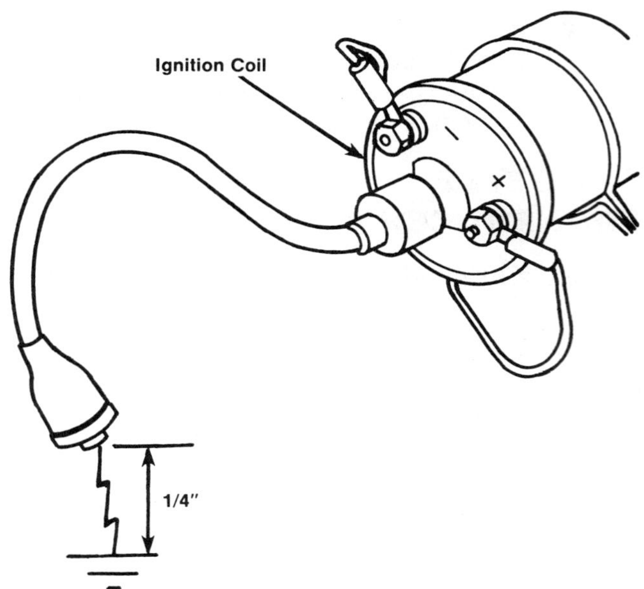

FIGURE 9-23 Checking for spark from coil

1. With the ignition on, use a jumper wire to momentarily short the negative (-) terminal of the coil to ground while holding the coil secondary wire 1/4 inch away from a good engine ground (Figure 9-23). A spark should be produced each time the wire is removed from ground.
   - If no spark is produced, turn the ignition off and disconnect the 10-way connector going into the spark control computer (SCC) (Figure 9-24).
3. With the ignition on again attempt to produce a spark from the coil wire by shorting the negative (-) terminal of the coil to ground.
   - If no spark is produced, proceed to troubleshoot the circuit.
   - If a spark is now obtained, replace the SCC.

ON-BOARD DIAGNOSTIC SYSTEM CHECK

For those Chrysler vehicles equipped with the latest modular control system (LM and PM), an on-board diagnostic capability is available to test the vehicle electronic control systems. Within the framework of Chrysler's system, a variety of service codes and test modes are used. Service codes are subdivided into

- Fault codes
- Indicator codes

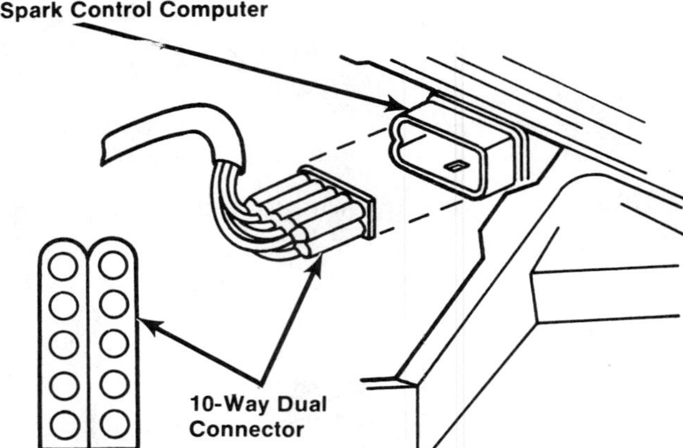

FIGURE 9-24 Disconnecting 10-way dual connector from spark control computer

- ATM access codes
- Sensor access codes

In addition, five test modes are also available and include the following:

- Diagnostic test
- Circuit actuation test
- Switch test
- Sensor test
- Engine running test

The diagnostic test and the circuit test are most important to the spark advance system. The diagnostic test can be performed using the POWER LOSS/POWER LIMIT lamp on the instrument panel or by using an aftermarket diagnostic readout box. The circuit test can only be performed by using a diagnostic readout box.

For the on-board diagnosis capability (diagnostic test) using the POWER LOSS/POWER LIMIT lamp, use the following procedure:

1. Start the engine. If the engine will not start, crank momentarily, then leave the ignition key on.
2. Move the transmission shift lever through all positions and back into PARK.
3. If the vehicle has air conditioning, turn the A/C switch on and then off.
4. Turn the ignition key to off.
5. Then turn the ignition key (in sequence): on, off, on, off, and then to on.
6. Read codes from the LM LED (for 1984 models) or the POWER LOSS/POWER LIMIT lamp (1985 and later models) on the instrument panel. Codes are indicated by a series of flashes. For example, code 14 will be a flash, a pause, followed by four

flashes. After a slightly longer pause all other codes are displayed in numerical order. The first code displayed will be 88 (start of test). The last code displayed will be 55 (end of message).

7. Once the lamps begin to display fault codes, they cannot be stopped. If count is lost, the test must be rerun. A specific fault code indicates a symptom failure, not a specific component. Therefore, a particular code denotes the area of the malfunction, not necessarily the component itself.

Typical Chrysler Fault Codes

88 Start of test
11 Engine not cranked since battery disconnected
12 Memory standby power lost
13 MAP sensor pneumatic circuit*
14 MAP sensor electrical circuit*
15 Vehicle speed/distance sensor circuit
16 Loss of battery voltage sense (if this signal drops below 4 volts, or between 7.5 and 8.5 after the engine has been running for one minute, the Logic Module will excite the alternator field circuit at a fixed rate)*
17 Engine running too cold
21 Oxygen sensor circuit
22 Coolant temperature sensor circuit*
23 Throttle body temperature sensor circuit
24 Throttle position sensor circuit*
25 Automatic idle speed motor driver circuit
26 Peak injector current has not been reached
27 Internal problem in Logic Module fuel circuit
31 Purge solenoid circuit
33 A/C cutout relay circuit
35 Cooling fan relay circuit
37 Shift fan relay circuit
41 Charging system excess or no field current
42 Automatic shutdown relay driver circuit
43 Spark interface circuit
44 Battery temperature out of range
46 Battery voltage too high*
47 Battery voltage too low
51 Oxygen sensor system stuck at lean position
52 Oxygen sensor system stuck at rich position
53 Logic Module has internal problem
55 End of message

***Power loss lamp lit**

FIGURE 9-25 Typical fault codes output by Chrysler's onboard diagnostic system

8. Refer to the Chrysler applicable service manual to interpret the fault codes received (Figure 9-25).
9. After the self-diagnostic codes have been read, proceed to the problem area indicated and begin troubleshooting.

FORD'S SPARK ADVANCE SYSTEMS

Ford introduced its first electronic engine control (EEC) system in 1978. This was the EEC-I. Subsequent systems included the EEC-II in 1979, the EEC-III in 1980, and the EEC-IV in 1983. Ford also introduced the microprocessor control unit (MCU) system in California in 1979 and nationwide in 1980. Ford's electronic engine controls work on the traditional principles of all computer/microprocessor systems. They control a majority of the engine's operating parameters, including spark timing. Each Ford system-controlled spark timing comes in slightly varying designs. This discussion will highlight the EEC systems, explaining similarities and differences and describing the spark timing designs used by the systems.

EEC-I, II, III, and IV

The EEC-I, II, and III systems each have a computer, called the electronic control assembly (ECA), mounted to the left of the steering column under the dash (Figure 9-26). The EEC-I ECA has three sepa-

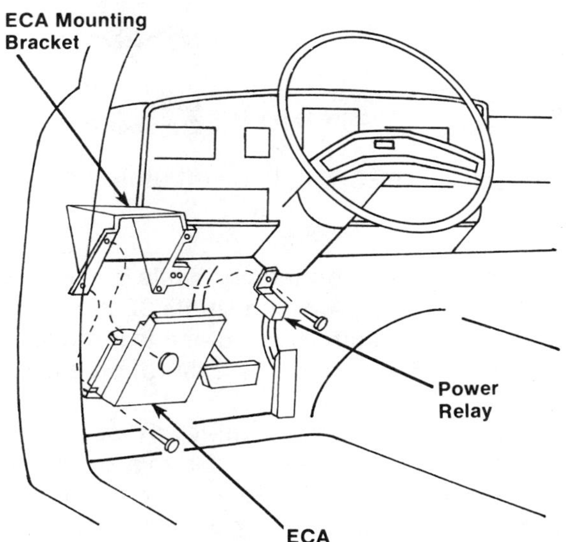

FIGURE 9-26 ECA mounting location of EEC-I, II, and III

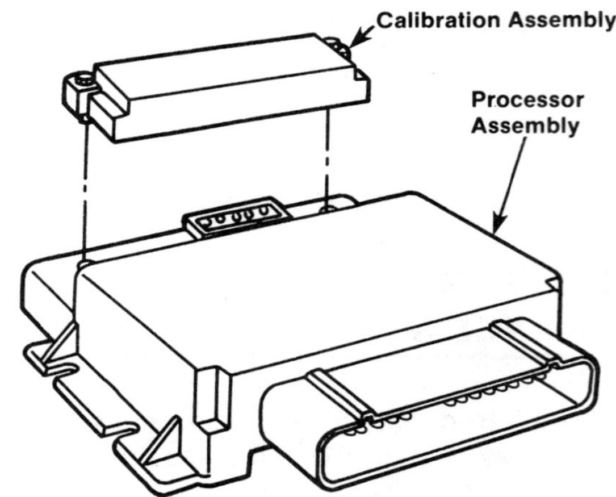

FIGURE 9–27 EEC-III system ECA

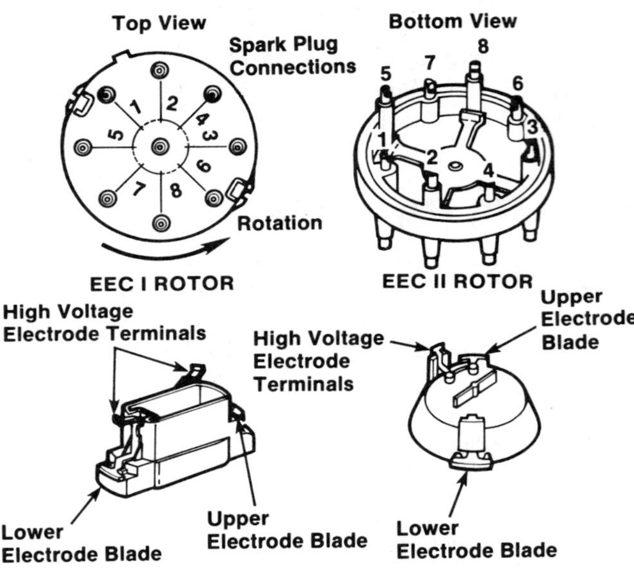

FIGURE 9–28 EEC-I, II, III distributor

rate wiring harness connectors, while EEC-II and III systems have an ECA with a single 32-pin connector. The ECA is a digital computer and is made up of a processor assembly and a calibration assembly (Figure 9–27).

The EEC systems have a bi-level distributor cap and rotor. This design is intended to prevent crossfire between adjacent distributor cap terminals. As shown in Figure 9–28, the cap and rotor have upper and lower terminals. Distributor cap terminals alternate between long and short. The rotor also has high and low electrodes. As the rotor spins, the longer cap terminals align with the lower rotor blades and the shorter cap terminals align with the upper rotor blades. A center electrode plate in the cap directs the high secondary voltage from the coil to the high voltage electrode terminals of the rotor. The voltage, in turn, is directed to the long (or lower) terminals of the cap by the lower electrode of the rotor and to the short (or upper) cap terminals by the upper electrodes of the rotor.

SHOP TALK

The numbers on the top of the distributor cap indicate the spark plug (or cylinder) number. They do not refer to firing order. The order of the numbers will vary, depending on the engine application.

The distributor cap is also larger than those used on a conventional distributor. To accommodate the larger caps, a spacer is mounted between the distributor and the cap (Figure 9–29). The large

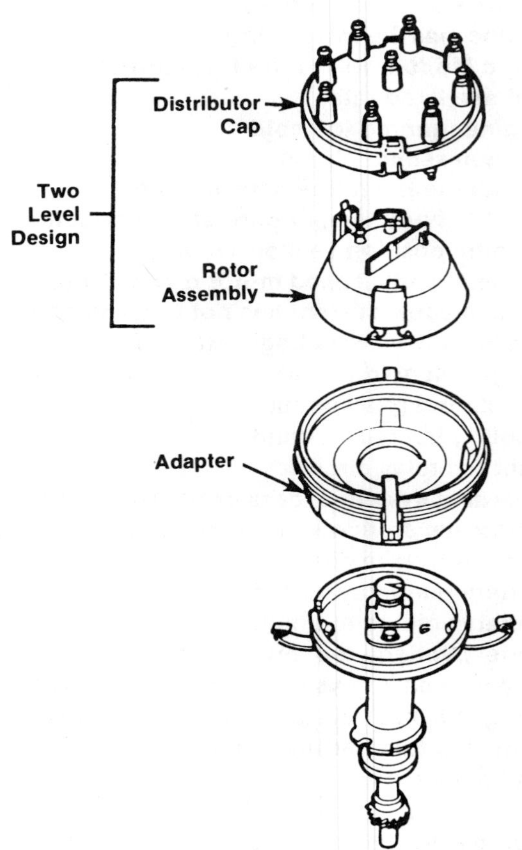

FIGURE 9–29 The EEC distributor has a large cap and adjuster.

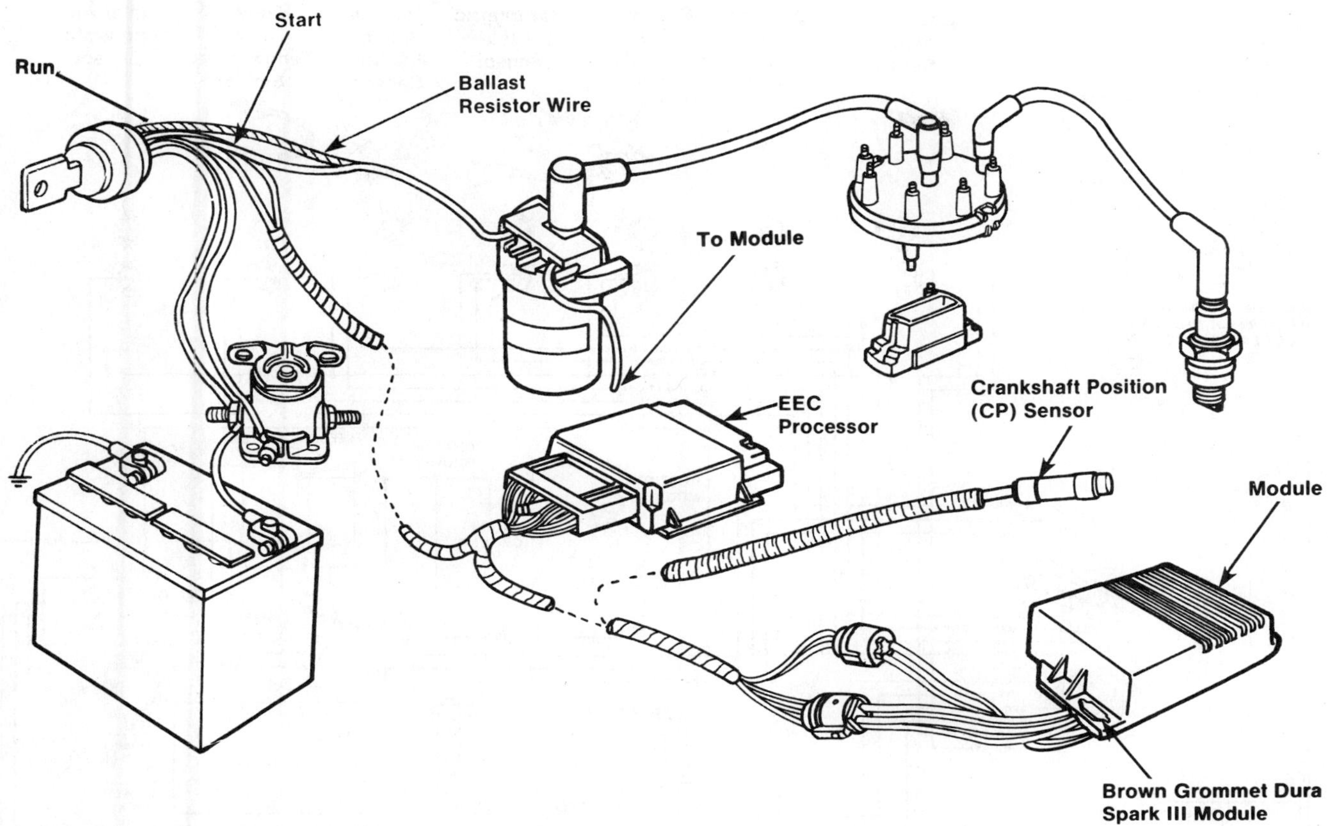

FIGURE 9-30 EEC-II and III, Dura Spark III electronic ignition system

cap allows the distributor to handle high voltages without arcing between cap terminals.

Since the ECA controls the ignition system, the distributor has no advance mechanisms or magnetic pickup assembly. Its only function is to distribute spark to the spark plugs. Crankshaft position and engine speed is monitored by a crankshaft position sensor (Figure 9–30).

Basically the EEC systems function alike in that the ECA receives inputs from the sensors, and then processes and determines what command signals are to be sent to various actuators to control engine operation. A signal from the ECA and the ignition module controls the ignition coil's primary current flow. The distributor only directs the secondary high voltage from the coil to the appropriate spark plug.

EEC-I

The EEC-I system was introduced on the 1978 Lincoln Versailles 5.0-liter engines and used through 1979 (Figure 9-31). The system uses 7-pin, 9-pin, and 10-pin connectors to attach the wiring harnesses to the ECA. The engine crankshaft position sensor mounts at the rear of the engine block. The

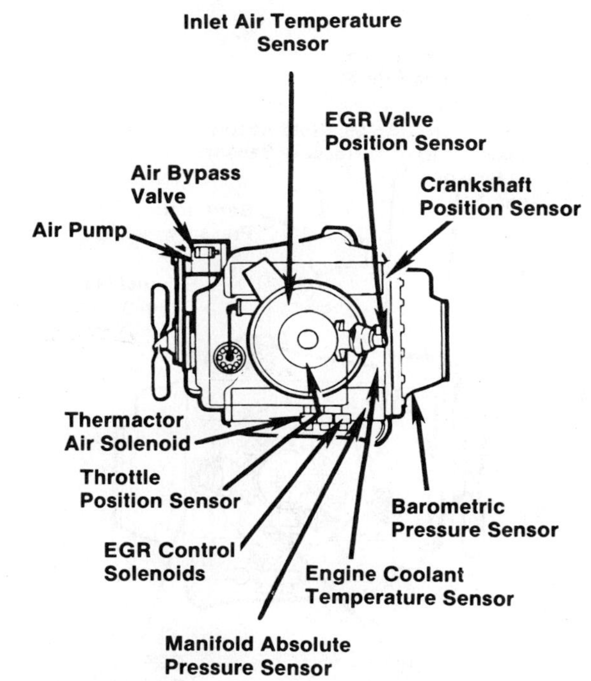

FIGURE 9-31 1978 EEC-I system sensors and actuators

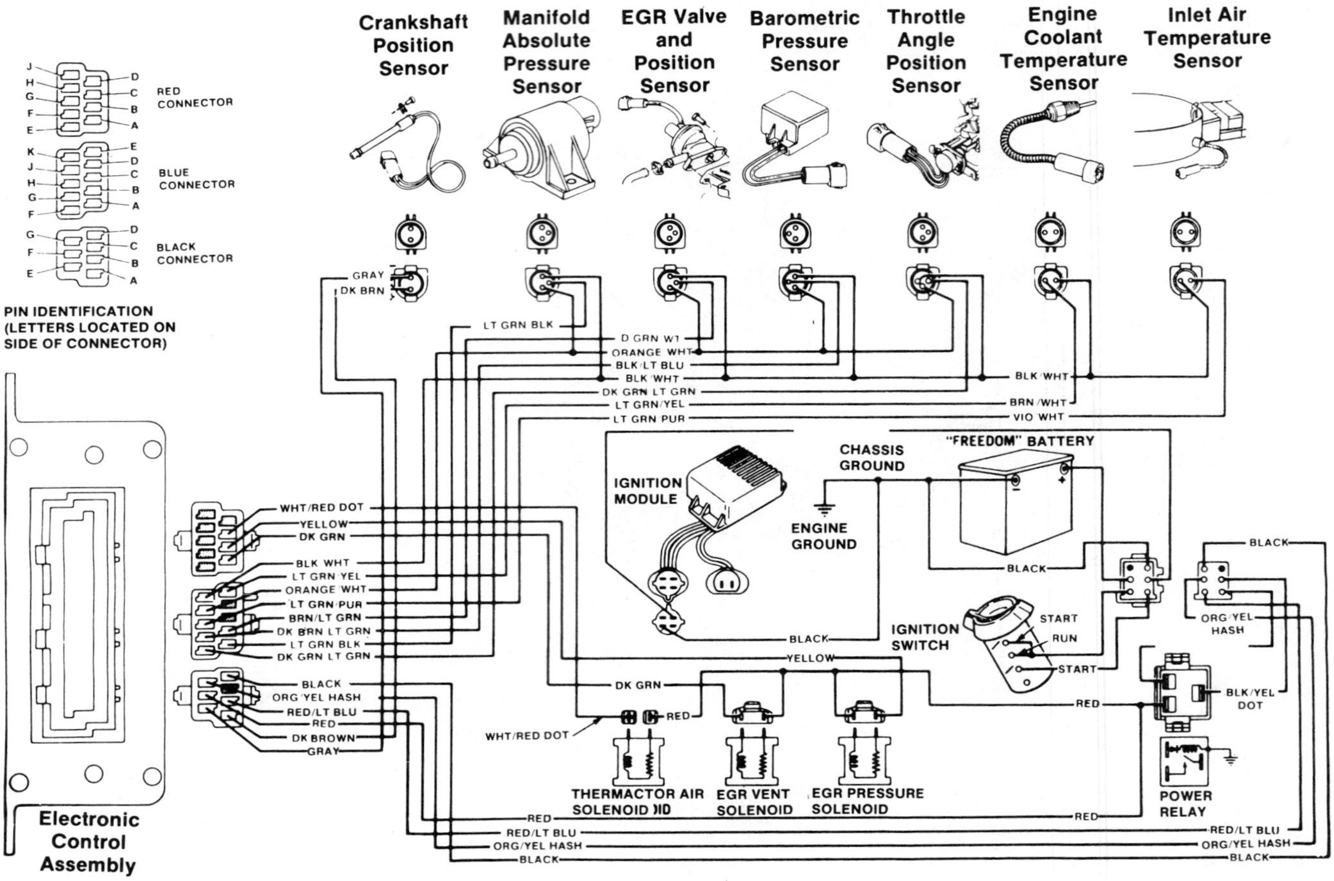

FIGURE 9–32 EEC-I schematic

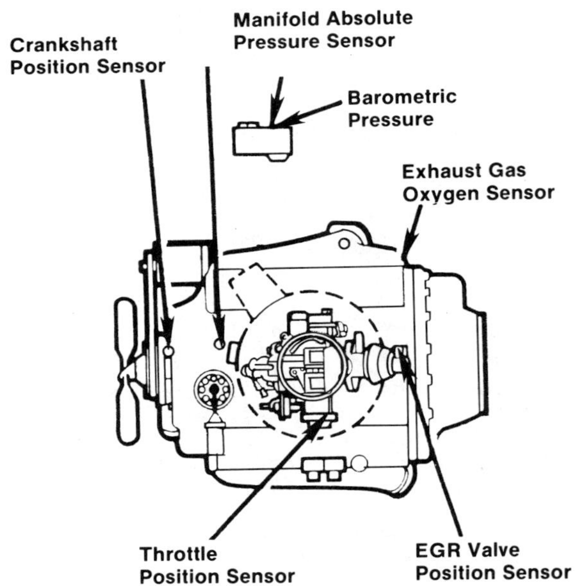

FIGURE 9–33 1979 EEC-II system sensors and actuators

ECA mounts under the dash to the left of the steering column. The ECA controls timing advance from TDC to 60 degrees BTDC. The EEC-I system uses the Dura Spark II ignition module. The module controls the primary circuit and dwell. A schematic diagram of the EEC-I system is shown in Figure 9–32.

EEC-II

The EEC-II system was used in 1979 and 1980. The crankshaft sensor is located at the front of the engine (Figure 9–33). The system uses a single 32-pin connector to attach the ECA to the system. The system controls timing advance up to 50 degrees BTDC. The EEC-II system uses the Dura Spark III ignition module. A schematic diagram of the EEC-II system is shown in Figure 9–34.

EEC-III

The EEC-III system was introduced in 1980. The ECA mounts under the seat. The system includes either a feedback carburetor or an electronic fuel injection system (Figure 9–35). There is essentially no difference between EEC-II and EEC-III, and the

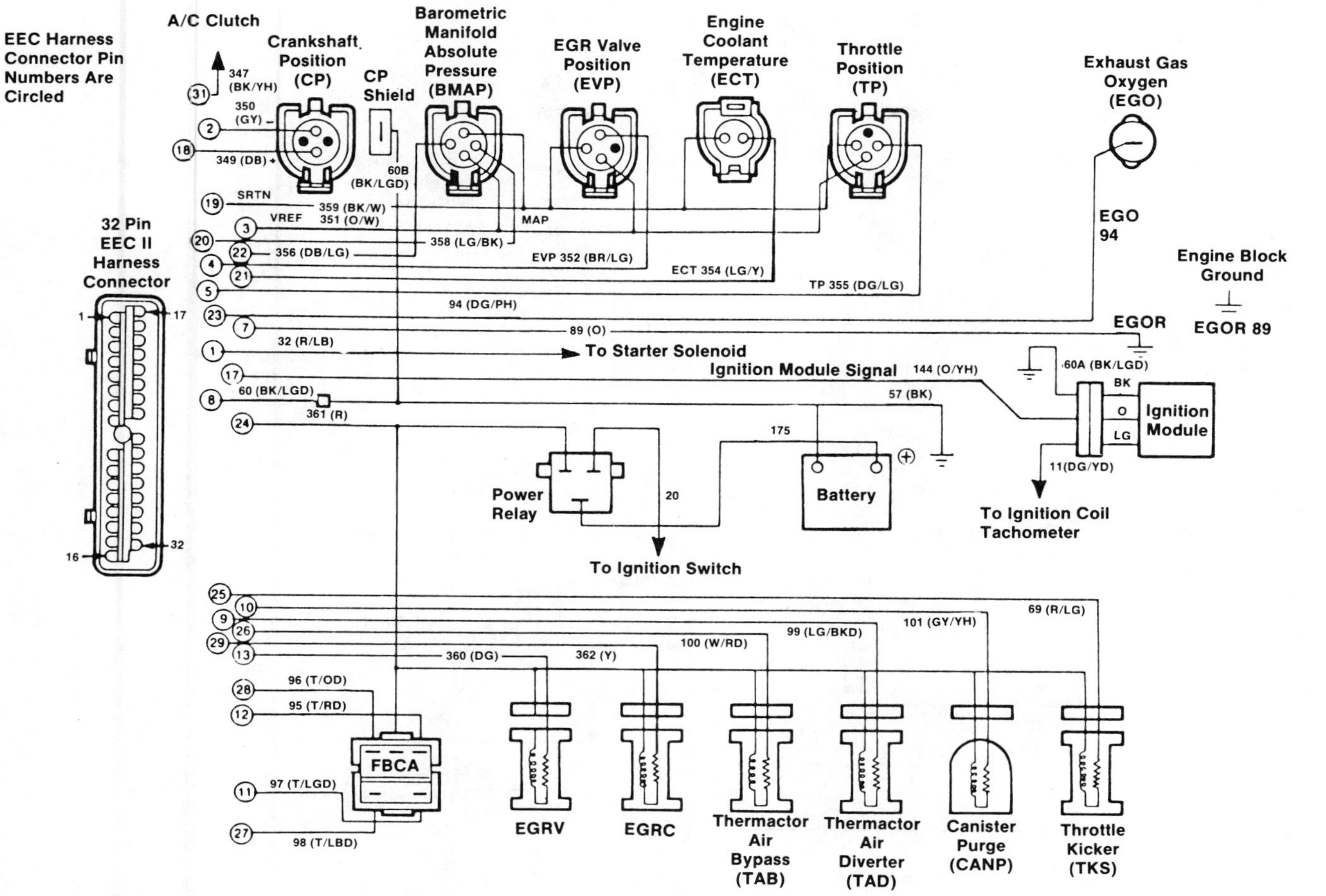

EEC Harness Connector Pin Numbers Are Circled

A/C Clutch

Crankshaft Position (CP)

CP Shield

Barometric Manifold Absolute Pressure (BMAP)

EGR Valve Position (EVP)

Engine Coolant Temperature (ECT)

Throttle Position (TP)

Exhaust Gas Oxygen (EGO)

347 (BK/YH)

350 (GY)

349 (DB)

60B (BK/LGD)

SRTN 359 (BK/W)

VREF 351 (O/W)

MAP

358 (LG/BK)

356 (DB/LG)

EVP 352 (BR/LG)

ECT 354 (LG/Y)

TP 355 (DG/LG)

EGO 94

32 Pin EEC II Harness Connector

Engine Block Ground

94 (DG/PH)

89 (O)

EGOR

EGOR 89

32 (R/LB)

To Starter Solenoid

Ignition Module Signal 144 (O/YH)

60A (BK/LGD)

BK

O

LG

Ignition Module

60 (BK/LGD)

57 (BK)

361 (R)

175

Power Relay

20

Battery

11(DG/YD)

To Ignition Coil Tachometer

To Ignition Switch

69 (R/LG)

101 (GY/YH)

99 (LG/BKD)

100 (W/RD)

360 (DG)

362 (Y)

96 (T/OD)

95 (T/RD)

FBCA

97 (T/LGD)

98 (T/LBD)

EGRV

EGRC

Thermactor Air Bypass (TAB)

Thermactor Air Diverter (TAD)

Canister Purge (CANP)

Throttle Kicker (TKS)

FIGURE 9–34 EEC-II and III schematic

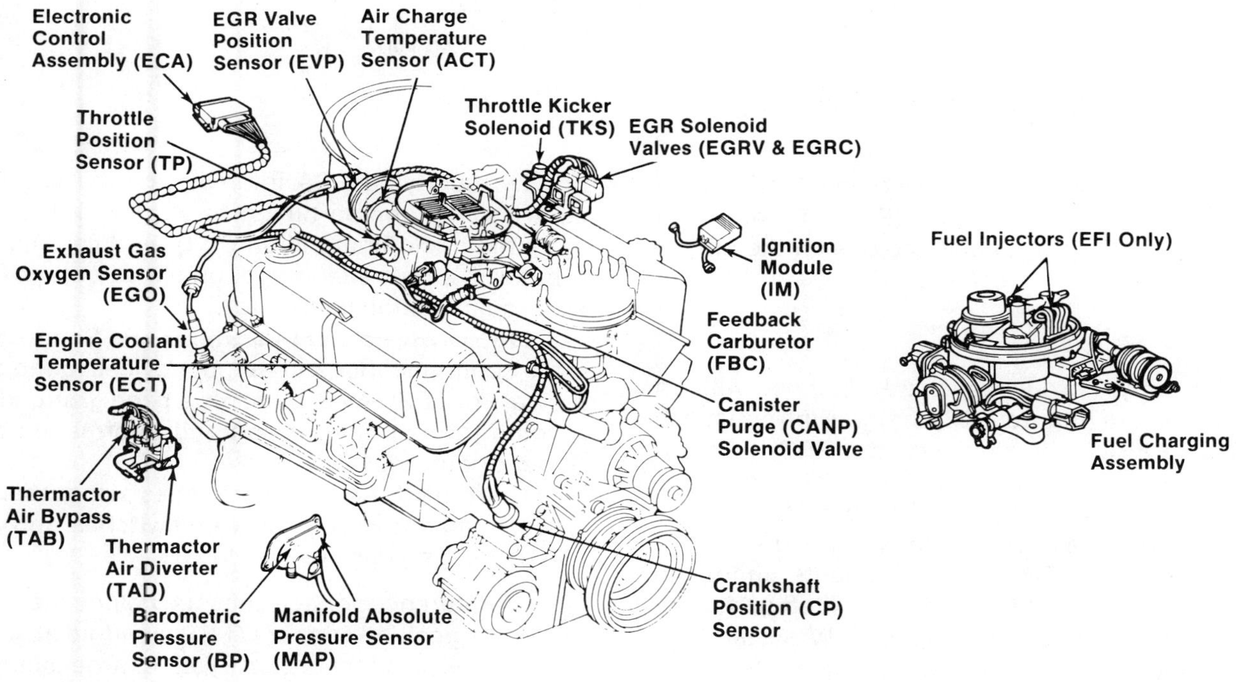

FIGURE 9–35 EEC-III system components

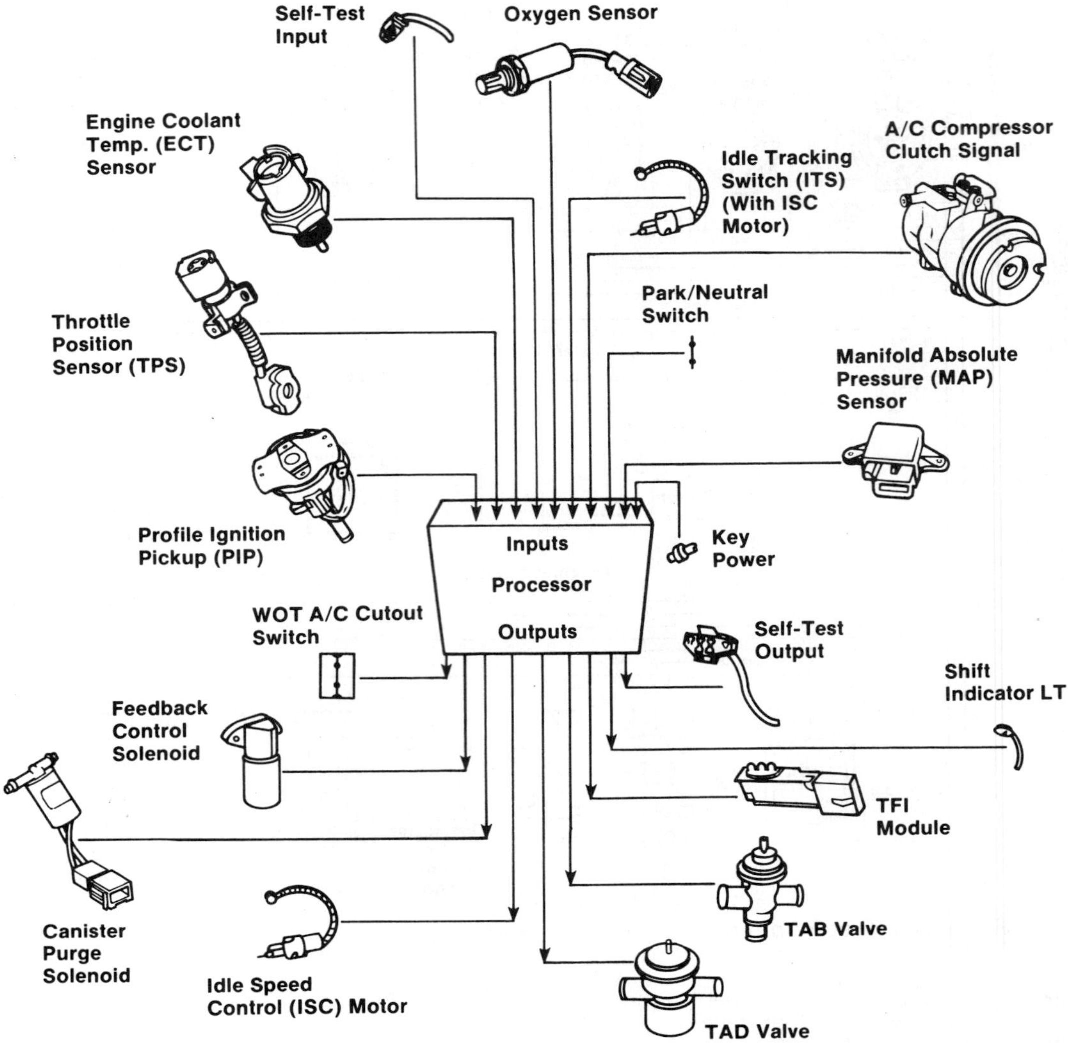

FIGURE 9–36 EEC-IV system components

EEC-III ECA is the service replacement for the EEC-II ECA. The schematic diagram (Figure 9–34) for the EEC-III system is identical to the one for the EEC-II system.

EEC-IV

Currently, Ford vehicles are equipped with an EEC-IV system (Figure 9–36) that was introduced in 1983. The EEC-IV is similar to the EEC-II and III systems but has no crankshaft position sensor. Instead, a distributor-mounted profile ignition pickup (PIP) provides the ECA with engine speed and position signals. The ECA usually mounts under the right-hand kickpanel but can be located under the front seat or dashboard. The EEC-IV system has self-diagnostic capabilities through a self-test connector. Some 1985–86 models have a learning mode that can compensate for changes in sensor calibration by modifying ECA internal circuits. The ECA controls the ignition timing through a distributor-mounted thick film ignition (TFI) module. The distributor is a Hall-effect type, unlike any other Ford electronic ignition distributor.

Electronic Control Components. Some of the more common input and output devices found on a typical Ford EEC-IV system have been grouped according to operating design and information type (input or output).

There are three basic categories of input sensor types: reference voltage sensors, switches, and voltage generating devices.

- Reference voltage sensors include the throttle position sensor (TPS), manifold absolute pressure sensor (MAP), air charge temperature sensor (ACT), EGR valve position sensor (EVP), and the engine coolant temperature sensor (ECT). On fuel-injected models, the EEC-IV system includes a vane airflow (VAF) and vane air temperature (VAT) sensor.

- Switches used in the EEC-IV system include the clutch engaged switch, idle tracking switch (ITS), power steering pressure switch (PSPS), neutral gear switch (NGS), and A/C clutch compressor signal switch (ACC).
- Voltage generating sensors capable of producing their own voltage signal include the exhaust gas oxygen sensor (EGO), profile ignition pickup (PIP), and knock sensor (KS).

Output devices can be divided into three categories also. They are: solenoids, electric motors, and controller modules.

- Solenoids include the EGR shut-off solenoid, canister purge solenoid (CANP), carburetor feedback solenoid (FBC), idle speed control solenoid (ISC), throttle kicker solenoid (TKS), and fuel injectors.
- Electric motors used in the EEC system include the idle speed control motor (ISC), fuel pump, and cooling fan.
- Controller modules are used to control more than one device. On the EEC-IV system two controller modules are used. One is the A/C and cooling fan controller module and the other is the integrated controller module (ICM).

The electronic control assembly (ECA) is the brains of the EEC-IV system. The ECA is essentially a two-chip microprocessor system in which all of the input/output and computational power resides on one chip, while the application's specific program memory resides on the other.

MCU SYSTEM

The MCU was introduced by Ford in 1980 on California vehicles equipped with 2.3-liter engines. Since then, it has been incorporated into vehicles nationwide. The MCU system (Figure 9–37) can be

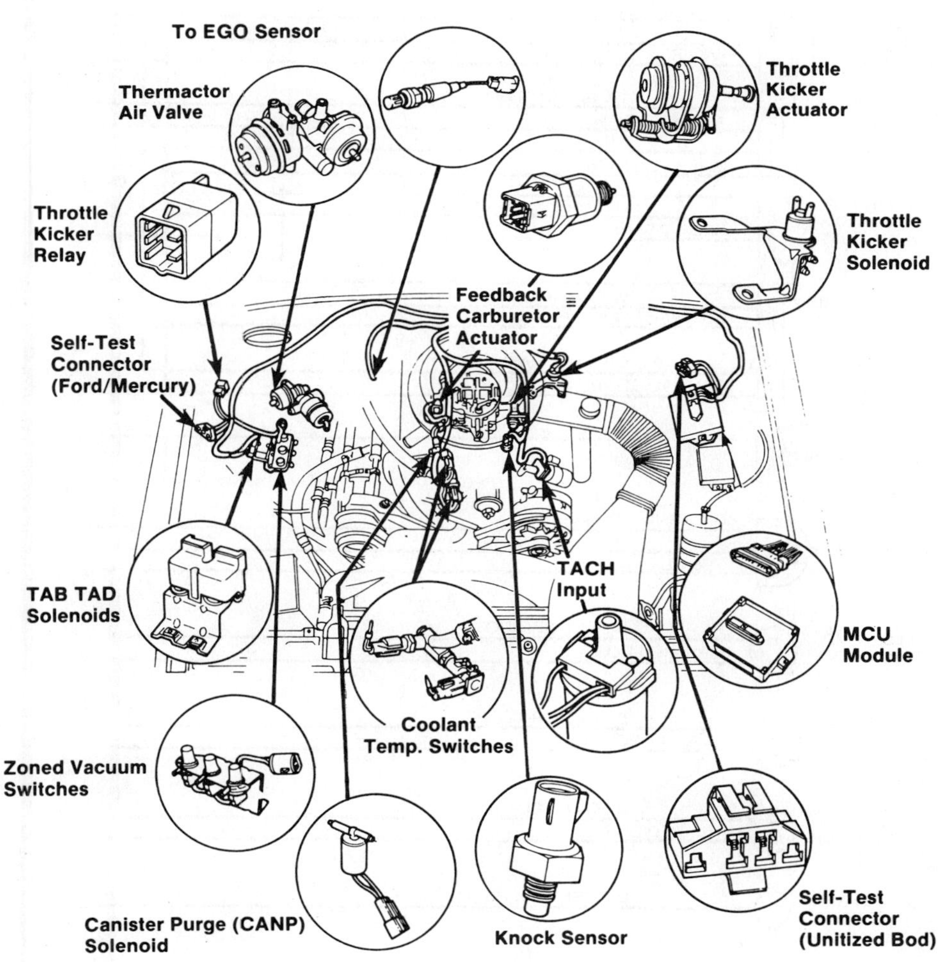

FIGURE 9–37 MCU components

identified by the control module called a microprocessor control unit, or MCU. This unit is usually located under the hood on the left fenderwell. The control unit is closely connected to a small two- or three-wire ignition module. The ignition system is a conventional electronic type, either Dura Spark II or TFI. Unlike other electronic systems discussed in this chapter, the distributor has conventional mechanical and centrifugal advance mechanisms.

The primary purpose of the MCU system is emission control. The MCU controls Thermactor airflow, carburetor fuel metering, and ignition timing. On certain engines, the MCU system can retard spark a maximum of 15 to 20 degrees in response to a signal from a knock sensor or from a zone vacuum switch. The zoned vacuum switch is a three-diaphragm vacuum sensor that provides inputs at low, intermediate, and high speeds. The MCU uses these inputs to control the amount of ignition timing retard. There is no other ignition control by the MCU.

The MCU system has self-diagnostic ability through a self-test connector located near the MCU module. Figure 9-38 shows a schematic diagram of the MCU wiring system.

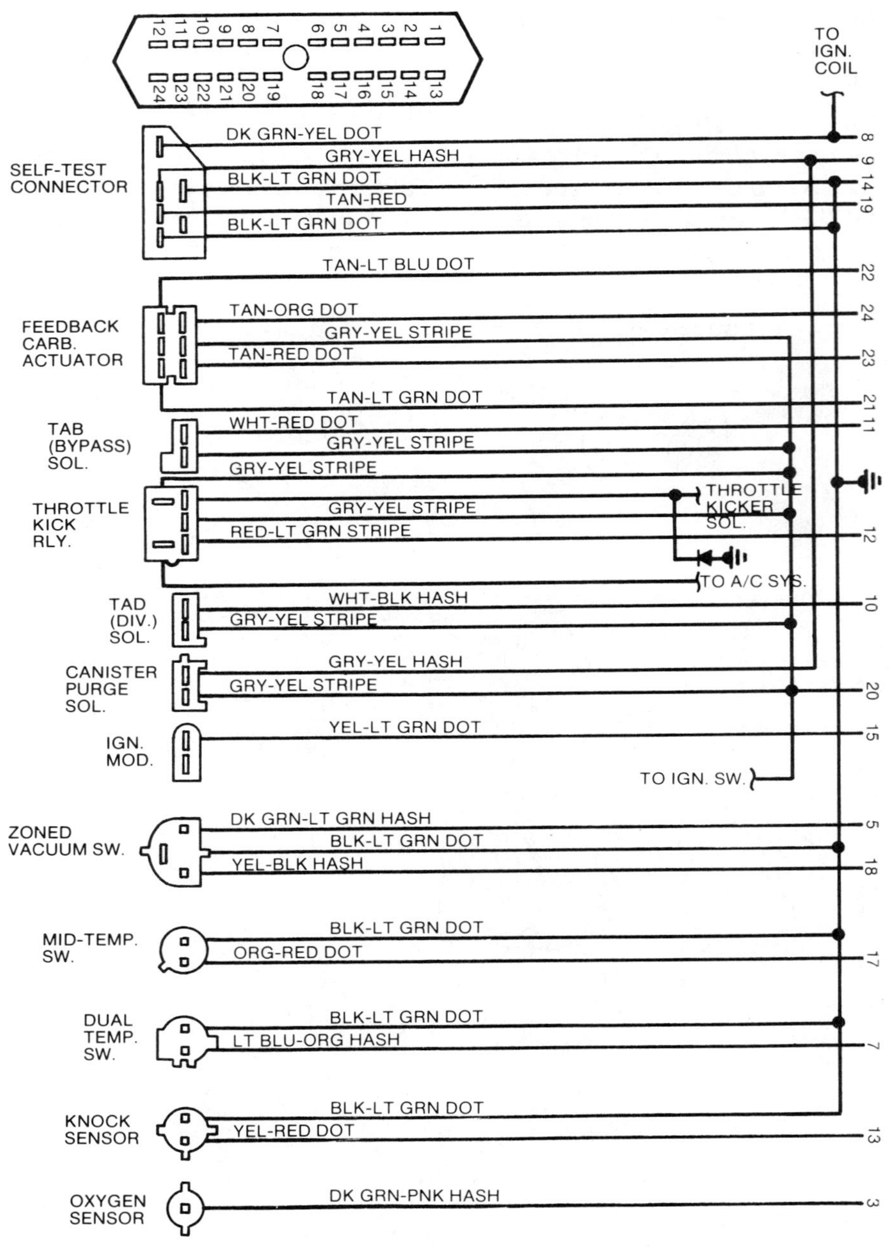

FIGURE 9-38 MCU wiring schematic

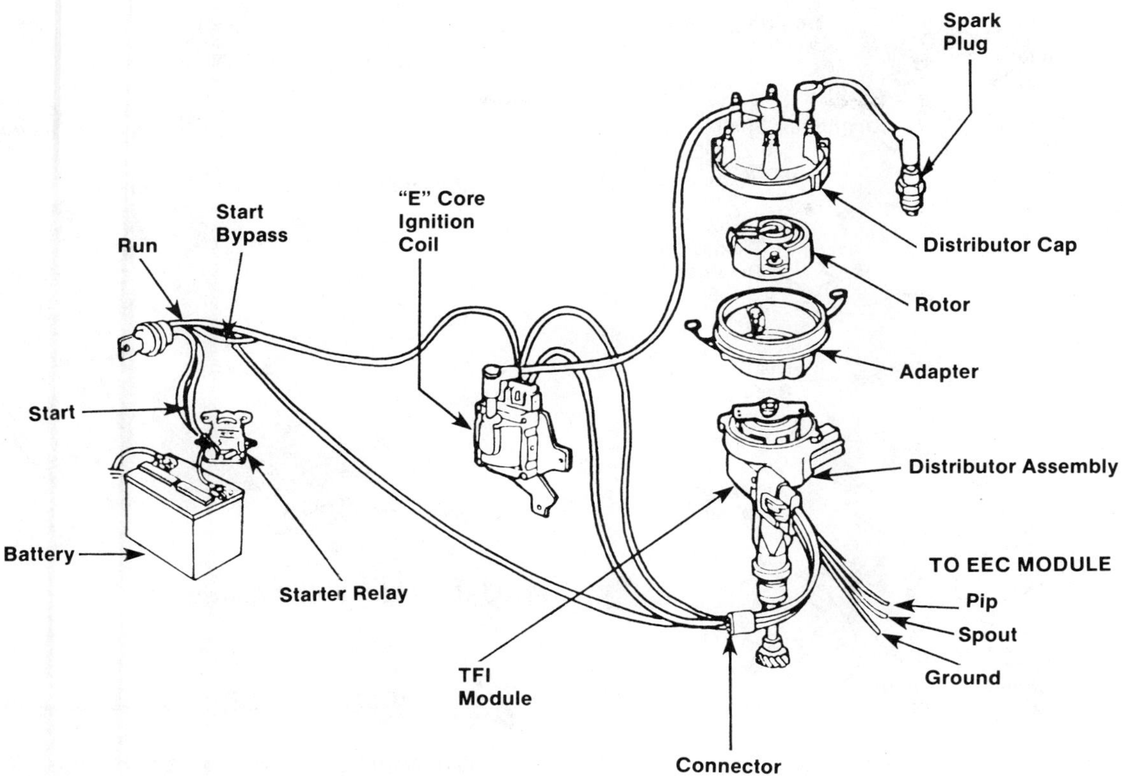

FIGURE 9–39 TFI-IV system

FORD'S TFI IGNITION SYSTEM

The thick film integrated (TFI) ignition system was first introduced in Ford vehicles in 1982. Since then, application of the system has greatly expanded and has largely replaced the Dura Spark system. TFI can be found in both EEC-IV and non-EEC Ford systems. TFI-I, as the original system is called, was introduced in non-EEC 1982 Ford models with carbureted 1.6-liter engines. From 1982 to 1985, all Ford carbureted models with 1.6-liter engines feature TFI-I systems. This version was last used in 1986 1.9-liter carbureted engines.

The most prevalent system today is TFI-IV, which is used in the EEC-IV ignition system. TFI-IV was first introduced in 1983 passenger cars with 1.6-liter EFI and 2.3-liter turbocharged engines and in Bronco II and Ranger truck models with 2.8-liter V-6 engines. Many Ford cars and light trucks today feature EEC-IV with TFI-IV (Figure 9–39).

Since its inception, the TFI system has experienced trouble. The module is subject to a lot of heat due to its location on the distributor housing. The heat often causes the module to malfunction. Ford relocated the TFI module on the firewall on some 1988 cars to alleviate excess heat conditions and designed the module to withstand more heat.

Components

The TFI module is easily recognizable. The small blue or gray plastic module is mounted on the outside of the distributor housing with two screws and replaces the larger and bulkier Dura Spark module. Inside the module, an integrated circuit is fused onto a plastic film, eliminating the conventional soldered circuit board. The distributor's pick-up mechanism (either pick-up coil assembly or a Hall-effect switch) connects directly to the three tabs on top of the TFI ignition module.

Besides the ignition module, TFI systems also include the distributor, ignition switch, battery, and primary and secondary wiring. A TFI system also features a special E-core ignition coil, which is housed in a laminated plastic instead of an oil-filled case. The E-core ignition coil has a very low primary resistance and is used without a ballast resistor in the electrical system. It has normal primary connections.

The differences between TFI-I and TFI-IV are most evident in the distributor. The TFI-I distributor contains a pick-up coil assembly (or stator) with internal teeth, a reluctor (or armature) with external teeth mounted on the distributor shaft, and centrifugal and vacuum advance mechanisms (Figure 9–40).

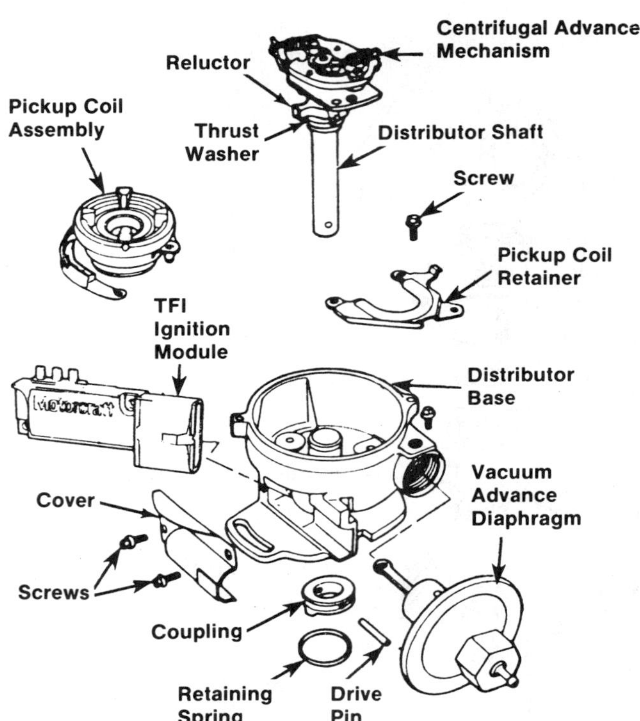

FIGURE 9-40 TFI-I distributor

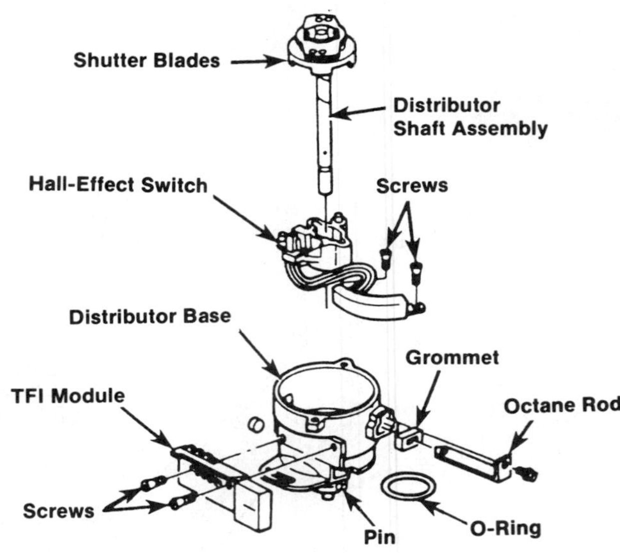

FIGURE 9-41 TFI-IV distributor

The TFI-IV distributor, on the other hand, uses a Hall-effect switch of stator known as the profile ignition pickup, or PIP, instead of the pick-up coil assembly used in TFI-I. No centrifugal or vacuum advance mechanisms are used. An octane rod can be installed if desired to vary the ignition advance (Figure 9-41).

Operation

The ignition process begins in the TFI-I distributor as the reluctor teeth rotate past the pick-up coil assembly. As each tooth approaches the pole of the pickup, a trigger signal is generated in the coil and is transmitted to the TFI ignition module. The ignition module then switches its power transistor on to allow current flow in the primary circuit of the ignition coil. As the reluctor tooth passes the pick-up coil assembly, it transmits another trigger signal to the TFI module to switch the power transistor off. This causes a high-voltage surge in the secondary, which fires the spark plug. The process is repeated for each cylinder. A wiring diagram of the TFI-I system is shown in Figure 9-42.

During the ignition process, the vacuum and centrifugal advance units will vary the spark advance with engine speed and load, as in other electronic ignition systems prior to computer-controlled spark advance.

SHOP TALK

After initial timing has been set on TFI-IV models, adjustments for octane concerns can be made by installing the appropriate octane rod in the distributor.

On TFI-IV-equipped engines, spark timing is controlled electronically with the aid of a Hall-effect switch (Figure 9-43). The Hall system creates an electrical reference signal that corresponds with crankshaft rotation. This reference pulse, or PIP signal, is transmitted to the TFI ignition module, which sends the signal further to the electronic control assembly (ECA) for modifications. A modified

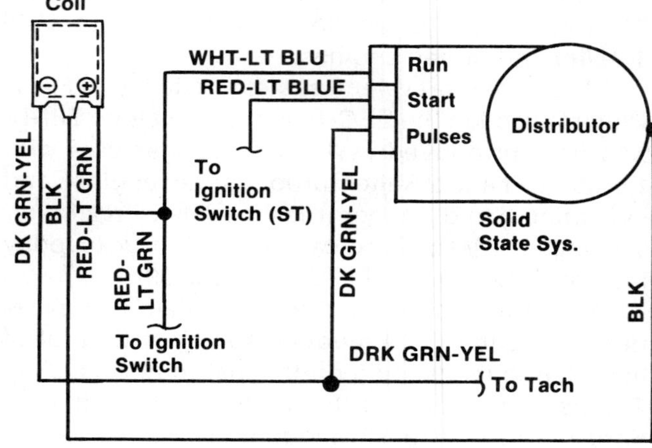

FIGURE 9-42 TFI-I wiring schematic

spark timing signal (SPOUT signal) is returned to the ignition module, which turns the ignition primary circuit on and off. Figure 9-44 shows a wiring diagram of a TFI-IV system. See "PIP and SPOUT (TFI-IV)."

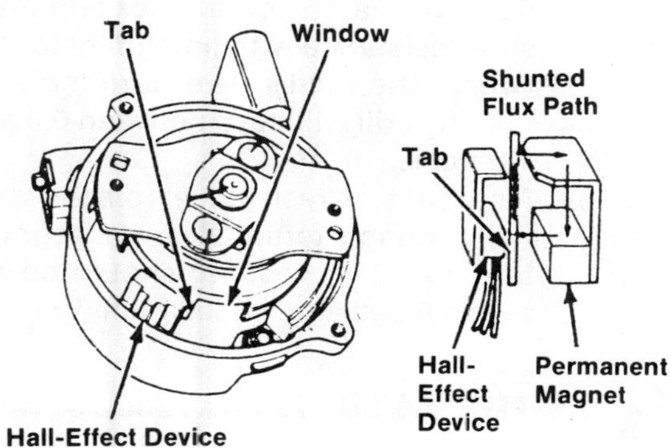

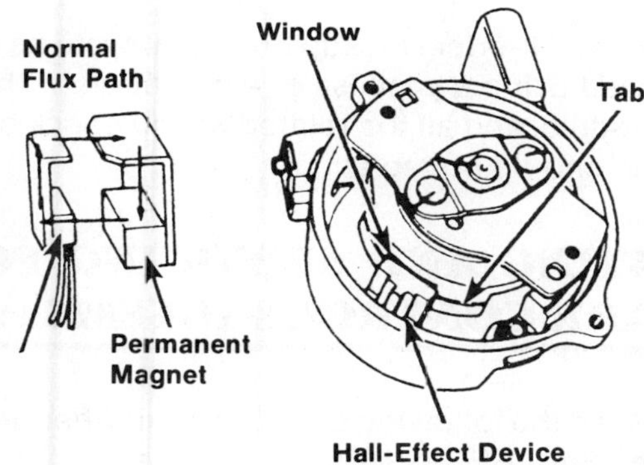

FIGURE 9-43 Operation of the TFI-IV PIP (Hall-effect) signal

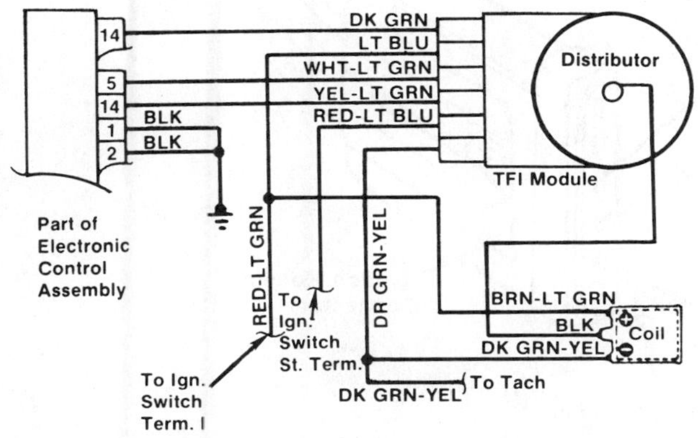

FIGURE 9-44 TFI-IV wiring schematic

PIP and SPOUT (TFI-IV)

The PIP signal is the signal generated by the Hall-effect stator or profile ignition pickup (PIP) assembly. The signal indicates crankshaft position (base timing) and engine rpm. The PIP signal is fed to both the TFI module and the ECA.

The PIP signal is processed by the ECA along with other input information from various engine sensors. After taking all the sensors' information (including the PIP signal), the ECA produces a new signal that represents the engine operating condition electronically. This signal is called the SPOUT signal. The SPOUT is sent back to the TFI module for comparison with the original PIP signal. TFI-IV uses both these signals to fire the ignition coil at the proper timing interval. A schematic of the electrical paths that these signals follow is given in Figure 9-45.

Basic Service

The faulty TFI module often results in a no-start condition because not only can the ignition module fail internally, its external connector can often deteriorate because of heat. For this reason, a good visual check of the module is important when diagnosing TFI problems.

A bad TFI module can also cause rough idling, hard starting, and poor fuel economy. However, more often than not, these symptoms are the result of fuel-related problems (for example, clogged injectors).

The following are three separate ohmmeter checks for the TFI ignition module. See Figure 9-46 for specifications and details.

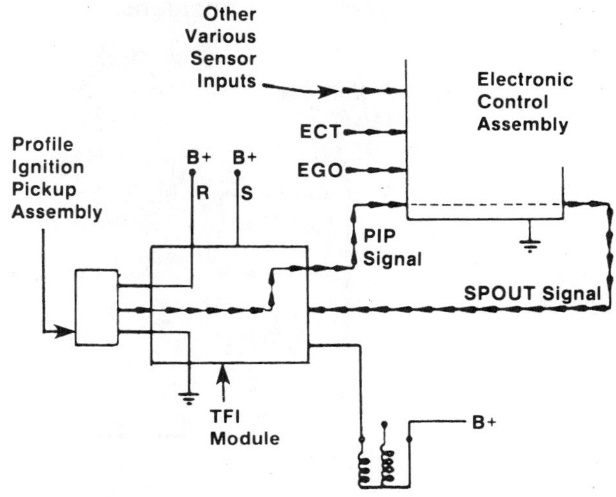

FIGURE 9-45 Schematic of PIP and SPOUT signal flow

MEASURE BETWEEN THESE TERMINALS	RESISTANCE SHOULD BE
1. GND—PIP IN	1. Greater than 500 Ohms
2. PIP PWR—PIP IN	2. Less than 2K Ohms
3. PIP PWR—TFI PWR	3. Less than 200 Ohms
4. GND—IGN GND	4. Less than 2 Ohms
5. PIP IN—PIP	5. Less than 200 Ohms

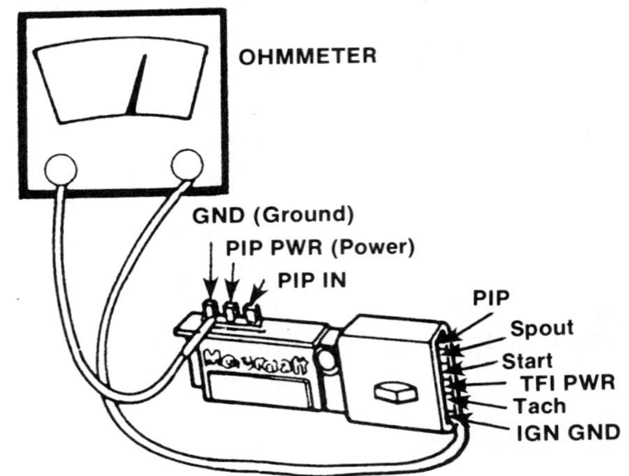

FIGURE 9–46 These resistance checks can be made to test TFI-IV ignition module.

1. Check for spark, using a spark tester. If the first plug wire gives no spark, try a second wire.
2. Check wiring harness by removing the ignition module connector and performing the following procedures using a voltmeter and small straight pin (Figure 9–47).
   a. Probe terminal 2 with the key in the RUN position.
   b. Probe terminal 3 with the key in both RUN and START positions.
   c. Probe terminal 4 with the key in the START position.

   The results should be that each of these terminals should produce at least 90 percent of battery voltage. Low readings at these terminals can be due to faulty wiring, failed ignition parts, or possibly a worn or damaged ignition switch.
3. Check the ignition coil/connector.
   a. Remove the coil connector and measure resistance with an ohmmeter between the coil's two terminals. A healthy coil will have between 0.3 and 1.0 ohms.
   b. Also measure resistance from the coil's high voltage terminal to its negative terminal. This reading should be between 8,000 and 11,500 ohms.

SHOP TALK

If a new ignition module is put in and fails shortly thereafter, the coil may be internally shorted, killing the modules.

The Hall-effect pickup on TFI-IV systems can be replaced using a process of elimination. If the coil, the module, and all the related wiring check out, the pickup should be replaced.

TESTING AND SERVICING FORD SPARK CONTROL SYSTEMS

Keep the following points in mind when working on any Ford EEC system:

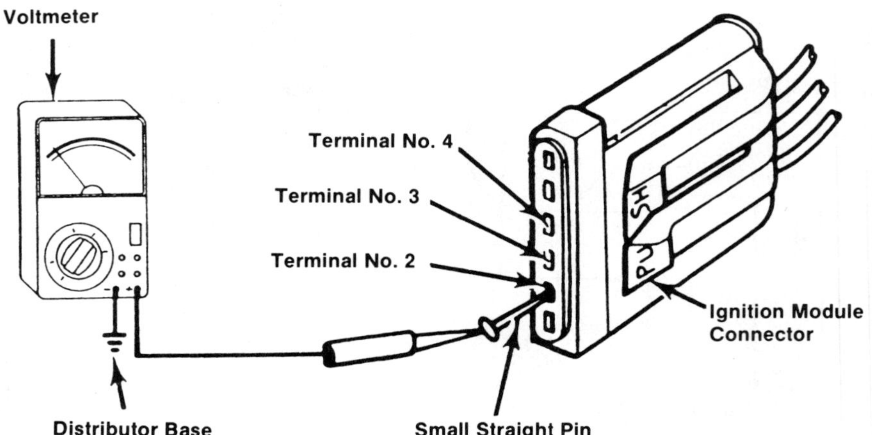

FIGURE 9–47 Method of testing the TFI–IV wiring harness

- Some sensors have a reference voltage fed into them. Always check for this voltage when testing component operation.
- EEC and MCU systems rarely have wiring problems unless they have been tampered with. However, do not discount a possible wiring problem without testing.
- Shorting across a solenoid can burn out internal circuits in the computer.
- Check the battery and the charging system for normal output, because high or low voltage can affect computer operation.

Ford EEC-III and EEC-IV and MCU systems have a self-diagnostic capability, as previously explained. The self-diagnostic tests require the use of aftermarket testers to locate problems within the systems. In addition, a service manual is needed to obtain specifications and specific procedures.

EEC-I, II, AND III PERFORMANCE TEST

SHOP TALK

Do not attempt to adjust the base ignition timing on any EEC-I, II, or III systems.

1. Attach an adjustable timing light to the engine, and check the timing at idle. Do not attempt to adjust base timing. Base timing is not adjustable on EEC systems.
2. Advance the engine speed and note the timing advance. Consult the service manual for the exact advance specifications, but a good rule of thumb is 30 to 40 degrees advance BTDC at 2000 rpm.

SHOP TALK

Some EEC-I, II, or III systems may not have timing marks. In this case, a timing mark can be fabricated by painting a spot on the vibration damper near a stationary engine part, with the number 1 piston at TDC, and the engine off. This can be used to check advance only, not initial timing, which is nonadjustable.

3. If timing advance is correct, go to step 4 of this procedure. If timing is not correct, go to step 11 of this procedure.

SHOP TALK

The following test is used on carbureted engines only. If the engine has fuel injection, go to step 11.

4. Disconnect the oxygen sensor at the harness. The engine should be warm and off.
5. Start the engine and run at 2000 rpm for at least 1 minute.
6. Connect a voltmeter between the sensor and a good ground.
7. Depress the carburetor control vacuum regulator rod (CVR) for 10 seconds. Voltage should be over 0.5 volts.
8. Disconnect the PCV valve and run the engine at fast idle. Voltage should be under 0.5 volts.
9. Repeat steps 7 and 8 at least once.
10. If voltage readings are incorrect, replace the oxygen sensor and retest. If readings are correct, go to step 11.
11. Check the oxygen sensor readings. When checking resistance, also check that the reference voltage is getting to the sensors that require it.
12. If all sensors are good, replace the ECA and recheck.

If the problem is still not found, recheck all wiring and conventional engine problems.

EEC-III SYSTEM SELF-TEST

The major difference between the EEC-III and the EEC-I and II systems is the self-test feature programmed into the computer memory. The ECA can be activated to run a series of checks to verify proper electrical connections and operation of the various sensors and actuators.

A scan tool can be used to activate the test and to display the test results. Be sure to connect the tester according to the manufacturer's instructions. The tester will either flash the results of the test on a blinking light or give the numerical equivalents of the codes. Consult a service manual for a list of applicable trouble codes for the vehicle being tested.

The self-test is activated by applying vacuum to the barometric/manifold absolute pressure (B/MAP) sensor located on the right fender apron (Figure 9-48). The vacuum tricks the computer into thinking that something is wrong and so runs the test. The results of the test are outputed as a series of electrical pulses to the thermactor air control solenoids

(Figure 9–49), which are also located on the right fender apron.

There are two ways to manually read the test results (trouble codes). Self-powered test lamps can be attached to the solenoids, or vacuum gauges can be tied into the solenoid vacuum lines. The electrical pulses will actuate both the solenoids and the test lamps, or if vacuum gauges are used, the gauges will momentarily register vacuum as the solenoids are energized: each sweep of the gauge needle corresponds to an electrical pulse from the computer.

To conduct the self-test, begin by installing either the test lamps or the vacuum gauges. If using the test lamps, connect one lead from each lamp to the orange wires at the solenoid and ground the other leads. If gauges are used, tie the gauges into the upper vacuum lines of each solenoid.

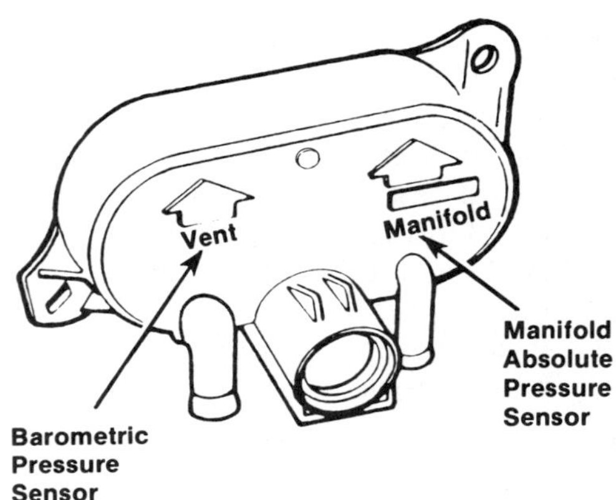

FIGURE 9–48 Barometric and manifold absolute pressure sensor

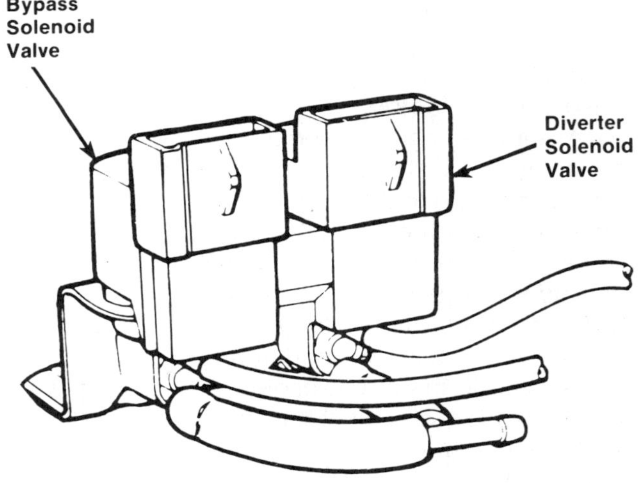

FIGURE 9–49 Thermactor air bypass (TAB) and diverter (TAD) solenoid valves

Then, start the engine and warm it to normal operating temperatures. With the engine idling, connect a hand vacuum pump to the B/MAP sensor nipple marked VENT.

Pump the sensor vacuum down to 20 inches of mercury. Hold the vacuum for at least 8 seconds. Because this is far below normal barometric pressure, the computer will assume that something is wrong with the system and go into the self-test mode.

During the self-test, the computer will check voltage points throughout the system. The test takes about 1-1/2 minutes.

If no problems are found in the system, a code 11 will be indicated on the test lamps or gauges. If a fault is discovered, the appropriate trouble code will be outputed. Consult the appropriate service manual for trouble code interpretations. If no trouble codes are displayed on the lights or gauges or if only one lamp or gauge is activated, one or both of the solenoids (or their associated wiring) are faulty. Locate and correct the problem; then, run the self-test again.

EEC-IV SYSTEM SELF-TEST

Like most computer-controlled systems, Ford's EEC-IV has self-diagnostic capabilities. By entering a self-test mode, the computer is able to evaluate the condition of the entire electronic system, including itself. If problems are found, they are identified as either hard faults (on-demand) or intermittent failures.

A hard fault means a problem has been found somewhere in the EEC system at the time of the self-test. An intermittent problem, on the other hand, indicates a malfunction occurred (for example, a poor connection causing an intermittent open or short), but is not present at the time of the self-test. A feature known as the keep-alive memory (KAM) allows intermittent faults to be stored for up to 20 ignition key on/off cycles. If the trouble does not reappear during that period, it is erased from the computer's memory.

The ECA communicates its problems by means of trouble codes that can only be read with the aid of special test equipment. Unlike GM C-3 or Chrysler computer-controlled systems, Ford does not display trouble codes on a flashing CHECK ENGINE LIGHT or POWER LOSS lamp. Ford does not use flashing lights or other warning devices to signal failures.

To read trouble codes on the EEC-IV system, one of the following is necessary: a STAR tester, scan tester, or analog (needle-type) voltmeter. The STAR (Self-Test Automatic Readout) tester is designed to read trouble codes on EEC-IV systems

FIGURE 9-50 Scan tools can be used to access EEC trouble codes. *(courtesy of All-Test, Inc.)*

only. Once it is plugged into the diagnostic connector, it can only do one of two things: display trouble codes (digitally) or activate and deactivate the self-test mode.

A hand-held scan tester is more versatile (Figure 9-50). In addition to displaying trouble codes on Ford's systems, most scan tools are programmed to work on GM and Chrysler computer-controlled systems as well.

An ordinary analog voltmeter can also be used to perform the self-test on EEC-IV systems. The voltmeter method is not nearly as convenient as the other testers, but once the self-test sequence is mastered, the voltmeter method will give the same results that the more expensive testers provide. The following paragraphs will explain how to perform the EEC-IV self-test using the analog voltmeter.

Self-Test Preliminaries

Before initiating the self-test procedure, some important preliminary steps must be taken.

Before entering the self-test mode, the engine has to be fully warmed up. If it is not, the computer will set a code 21 (ECT out of range) or code 41 (system always lean). Normal operating temperature has been reached when the electric cooling fan comes on (assuming there is one) and the upper radiator hose feels hot and pressurized. If during the test the engine cools off too much between steps, a code 43 may be shown, which means that the EGO (oxygen) sensor has cooled. To warm the sensor, start and run the engine for 2 minutes at 1500 rpm.

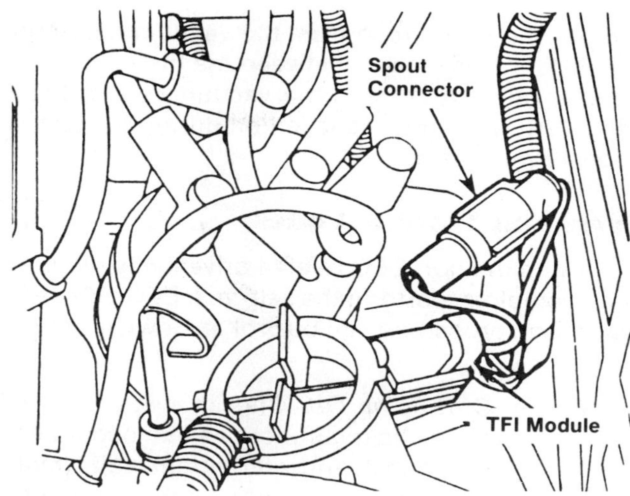

FIGURE 9-51 The SPOUT connector must be disconnected to check base timing.

Next, check the base ignition timing. To check base timing on EEC-IV systems, disconnect the single wire, in-line connector (SPOUT connector) from the TFI module on the distributor (Figure 9-51), hook up a timing light and check the base timing. Base timing is usually about 10 degrees BTDC, but check the emission decal to be sure. Adjust the timing if necessary, and then plug the in-line connector back in. Leave the light hooked up because it will be needed during one of the self-test procedures.

SHOP TALK

The self-test procedure given below applies to all EEC-IV-equipped vehicles, except the system used on 1983 1.6-liter EFI/EEC-IV engines. For these vehicles, follow the self-test procedures contained in the 1983 Engines/Emissions Diagnosis manual.

To avoid sending a false reference signal to the ECA while it is in the self-test mode, turn off all electrical loads and keep the doors closed. Any changes in current draw during the test can confuse the ECA, causing it to set an erroneous code.

On Tempo/Topaz models equipped with the 2.3-liter HSC engine and automatic transmission, remove the rubber cap from the vacuum restrictor (VREST) that is located between the thermactor air control valve (ACV) and the vacuum retard delay valve (VRDV). If the cap is left on during the test, a code 41 will be set.

Tests and Trouble Codes

When the non-EEC related driveability diagnosis is complete, perform the self-test. EEC-IV's self-diagnostic procedure can be broken down into four parts:

- *Key On/Engine Off.* Checks system inputs for hard faults (malfunctions that occur during the self-test) and intermittent faults (malfunctions that occurred sometime prior to the self-test and were stored in memory).
- *Computed Ignition Timing Check.* Checks the ECA's ability to advance or retard ignition timing. This check is made while the self-test is activated and the engine is running.
- *Engine Running Segment.* Checks system outputs for hard faults only.
- *Continuous Monitoring Test (Wiggle Test).* Allows the technician to look for and set intermittent faults while the engine is running.

Within these four tests, there are six types of service codes. There are:

1. **On-demand codes** are the same as hard faults. This means there is something wrong somewhere in the EEC-IV system at the time of the self-test. The term on-demand means asking the computer if a problem exists right now.
2. **Memory codes** mean a malfunction was noted sometime during the last twenty vehicle warmups, but is not present now (if it were it would be recorded as a hard fault).
3. **A separator code (10)** indicates that the on-demand codes are over and the memory codes are about to begin. The separator code occurs as part of the key-on/engine-off segment of the self-test only.
4. **A dynamic response code** is when a code 10 appears during the engine running segment of the self-test. The dynamic response code is a signal to the technician to "goose" the throttle momentarily so that the ECA can verify the operation of the throttle position (TP) and manifold absolute pressure (MAP) sensor. Failure to respond to the dynamic code within 15 seconds after it appears will set a code 77

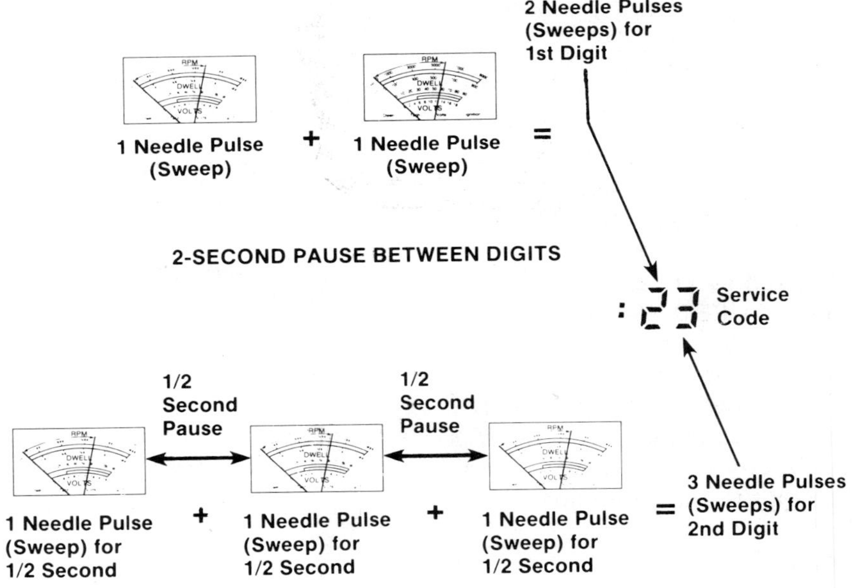

FIGURE 9–52 Reading EEC service codes with a voltmeter

(operator did not perform the "goose" test).

5. **Fast codes** are of no value to the service technician. They are for factory use only and are transmitted about 100 times faster than even a STAR tester can read. On the voltmeter, fast codes will cause the needle to rapidly pulse between 0 (zero) and 3 volts. On the STAR tester, the tester's LED light will flicker. Although fast codes have no practical use in the service bay, pay attention to when they occur. Fast codes will appear twice during the entire self-test sequence, once at the very beginning of the key-on/engine-off test (right before the on-demand codes) and again after the dynamic response code (prior to hard-fault transmission).
6. **Engine identification codes** are used to tell automated assembly line equipment how many cylinders the engine has. Two pulses indicate a four-cylinder; three pulses, a six-cylinder; and four pulses identify the engine as an eight-cylinder model. Engine I.D. codes will appear at the beginning of the engine-running segment only.

All trouble codes are represented by two digit numbers. Each two digit number is represented by wide sweeps of the voltmeter needle (Figure 9-52). For example, three sweeps, a 2-second pause, then four sweeps is a code 34. If the ECA has recorded more than one problem, there will be a 4-second pause between one code and the next. Separator codes and dynamic response codes (both 10) will be represented by one sweep (there is no sweep for zero) with a 6-second pause before and after.

Key-On/Engine-Off Test

To activate the self-test mode, two jumper wires (about 6 inches long each) with male spade terminals and a voltmeter are needed. Follow this procedure:

1. Locate the large gray-colored self-test output connector under the hood (Figures 9-53 and 9-54), and place one end of the first jumper lead in the pin 4 slot of the self-test connector and clamp the other end of to the negative voltmeter lead.
2. Connect the positive voltmeter lead to the positive (+) terminal of the battery.
3. Bridge the pin 2 slot of the self-test connector and the self-test input connector with the other jumper.
4. Set the voltmeter scale on the 0–15 scale.

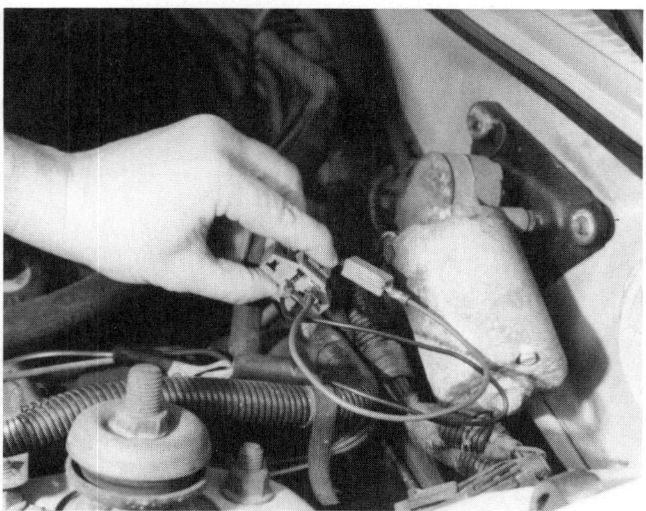

FIGURE 9-53 EEC self-test connectors

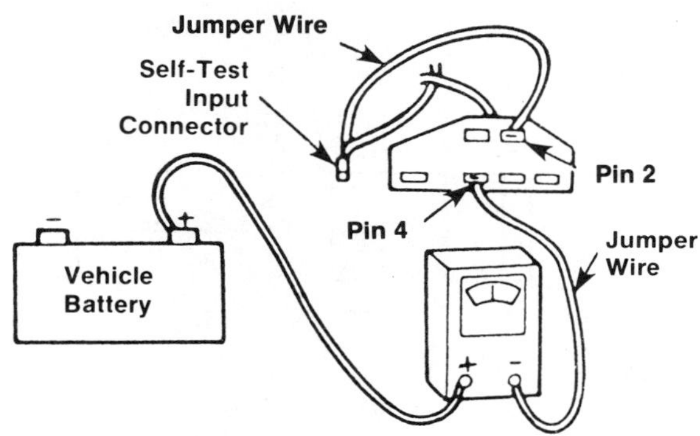

FIGURE 9-54 EEC-IV self-test connections for voltmeter and jumper wires

5. Place the ignition switch in the RUN position (do not start the vehicle yet) to begin the self-test process.

Immediately after the self-test is activated, the voltmeter needle will flutter rapidly between 0 (zero) and 3 volts. These are the fast codes, and can be ignored.

After the fast codes are finished, the first usable code (on-demand) will be displayed. Pay attention and write everything down. Once the codes start coming it is easy to misinterpret or lose track of codes.

If the system is okay, the needle will sweep once, pause 2 seconds, then sweep again, indicating a code 11. This means the system is okay. In 4 seconds, the code will be repeated just in case the technician missed it the first time.

If there is a problem that the computer has recognized, the code for that particular fault will be

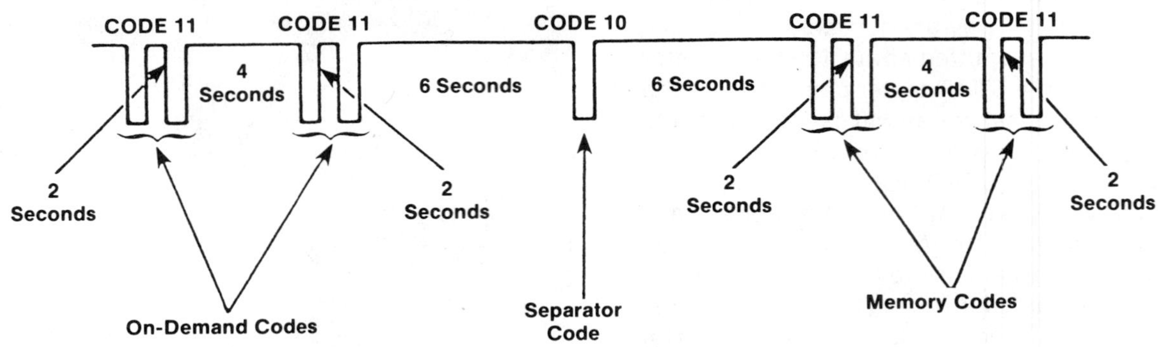

FIGURE 9–55 Key on/engine off self-test output codes

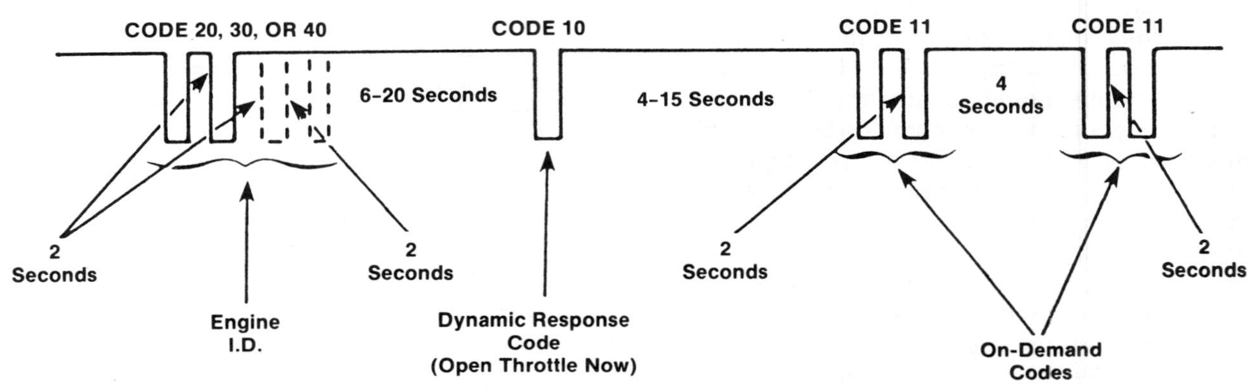

FIGURE 9–56 Engine running self-test output codes

counted out on the voltmeter. After each code is given once, the whole code sequence will be repeated. After a 6-second pause, the single sweep separator code (10) will be given (Figure 9–55), indicating it is time for memory codes. After another 6 seconds, the memory codes will start to count out and are read the same way as the on-demand codes. Once again, if the system passes, meaning no memory codes have been stored, a code 11 will be given and repeated once.

Computed Timing Advance Check

After the key-on/engine-off test sequence is finished, start the engine and perform the computed ignition timing check (this portion of the self-test must be performed within 2 minutes of self-test activation or the self-test will have to be repeated). Check the timing at this point: it should have advanced 20 degrees over base timing. If it has not advanced, refer to the service manual for diagnosis.

Engine Running Test

If the computer proves it is capable of controlling spark advance, shut the engine off, remove the #2 jumper wire and timing light, then start the engine again and let it run at 1500 rpm for 2 minutes to warm the EGO sensor. Shut the engine off, reattach the jumper, wait 10 seconds, then start the engine again (this procedure must be followed to the letter).

The first needle sweeps, set during the engine-running test, are the engine I.D. codes pulsing out the cylinder codes (Figure 9–56). Within 6 to 20 seconds, the single sweep dynamic response code will be given. This is the signal to goose the throttle (within 15 seconds of code). After another 4 to 15 seconds, the fast codes will be given, followed by the engine-running on-demand codes. As in the key-on/engine-off segment, a code 11 means the system is okay. This code will be repeated in 4-second signaling the end of the engine-running sequence.

As each self-test segment is completed, look up the codes to find out what specific problems they relate to. Some codes require immediate attention, meaning a specific problem must be repaired before proceeding to the next test segment. When tracing the cause of a specific code, follow the pinpoint test instructions as contained in *Ford's Engine/Emission Diagnosis* manual (or generic equivalent) before attempting to repair any faults.

The Wiggle Test

The final self-test procedure is called the wiggle test. This test derives its name from the action required to check for the set possible intermittent codes. In the continuous monitor test mode, an attempt is made to recreate intermittent fault conditions by wiggling connectors, manipulating movable sensors or actuators and heating and cooling thermistor type sensors with a heat gun.

This procedure can be performed while the engine is off, running, or both. Before looking for intermittent faults while the engine is off, repair any faults discovered during the first test segments; then deactivate the self-test and place the ignition switch in the RUN position. With the voltmeter still connected, begin wiggling the wiring of sensors and other computer system related components. If an intermittent problem is discovered, the voltmeter needle will deflect and a memory code will be set. After the wiring has been thoroughly shaken, tugged, and pulled, reactivate the self-test, and perform the key-on/engine-off segment. The intermittent faults that were recorded during the wiggle test will read out at this time.

To perform the wiggle test while the engine is running, the self-test must remain activated. Approximately 2 minutes after the last code in the engine-running segment has been displayed, the continuous monitor test mode will start. From this point on, the test is run the same as the key-on/engine-off procedure given on page 219.

After the self-test is complete, consult the service for an interpretation of the trouble codes. Troubleshooting trees in the manual will provide procedures for diagnosing and pinpointing the trouble indicated by the codes.

Erasing Memory Codes

Once intermittent problems have been identified and repaired, the trouble codes will be erased automatically from the KAM after twenty ignition key-on/key-off cycles. To erase the codes immediately, use the following procedure:

1. With the ignition key off, reconnect a voltmeter and jumper wires as shown in Figure 9-54.
2. Turn the ignition key on.
3. As soon as the voltmeter begins to read out trouble codes, even a code 11, disconnect the jumper wire from the single lead self-test input connector. This will erase the intermittent codes from the memory.
4. To confirm that the codes have been erased, turn the ignition off, reconnect the jumper wire, and turn the ignition switch back on again. Observe the voltmeter to see that it registers only code 11s (system okay).

MCU System Self-Test

The MCU also has a self-test mode used to diagnose malfunctions within the MCU system. A STAR tester, an aftermarket scan tool, or an analog voltmeter can be used to access and read the test results (service codes). A factory service manual is also needed to interpret the codes and give diagnostic procedures for pinpointing the trouble indicated by the self-test. To initiate and monitor the MCU self-test, follow these procedures:

1. Connect the positive lead of the voltmeter to the battery positive terminal, and the negative lead to the self-test output socket, as shown in Figure 9-57. Also ground the self-test trigger as shown.

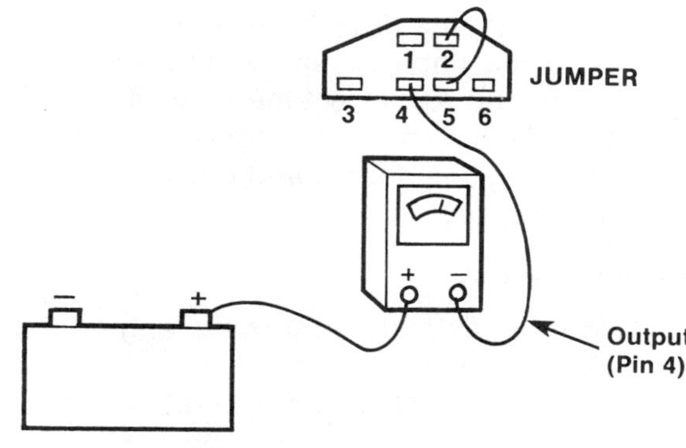

FIGURE 9-57 MCU self-test connections for voltmeter and jumper wires

2. Set the voltmeter on 0–15 volt scale. Battery voltage can be shown.
3. Remove the PCV valve from the valve cover.
4. Connect a vacuum gauge to the canister end of the canister purge solenoid valve hose.
5. Uncap the thermactor bypass vacuum restrictor while performing the tests.
6. Turn the key on (do not start the engine).
7. Watch the voltmeter for code pulses, which should appear within 5 to 30 seconds. Service codes are shown by voltage pulses. The first digit is indicated by a series of pulses. Then, the needle drops to 0 (zero) for 2 seconds. Finally, the second digit of the code is displayed. If more than one service code is given, a 5-second pause will separate the codes.

SHOP TALK

Ignore the vacuum gauge readings at this time. Ignore any initial surge of voltage when the ignition is turned on. Record code(s) as they are given.

8. Start the engine and run at 2000 rpm for 2 minutes or until warm.
9. Turn the ignition key off and immediately restart engine.
10. Observe the voltmeter and vacuum gauge for pulses after restarting engine. The throttle kicker should come on (and engine speed increase) at this time and remain on throughout the test.
11. Observe and record the service codes. When the self-test is complete (about 90 seconds), the throttle kicker will retract.
12. Stop the engine, remove the test equipment, and reconnect the PCV valve.
13. Check the engine systems indicated as faulty by the codes, and make the necessary repairs.

MCU Quick Checks

1. Connect a timing light to the engine (one with an advance meter).
2. Start the engine and run at idle for 2 minutes.
3. Check the timing marks with the timing light. If the timing is at or near the correct setting (check the underhood VECI decal for specifications), the spark advance system is functioning properly.

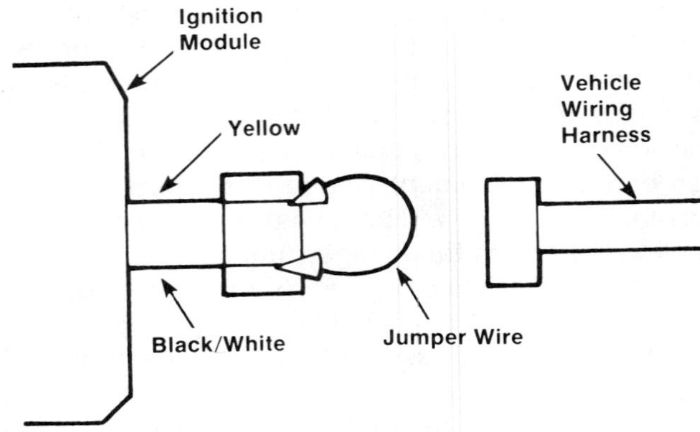

FIGURE 9–58 Jumping the ignition module connector terminals to retard timing

4. If the timing is retarded approximately 15 to 20 degrees, disconnect the two-wire connector at the ignition module. If the timing does not change, double check by connecting a jumper wire between the two wires (Figure 9–58). If the timing retards an additional 15 to 20 degrees, remove the jumper wire, and check for a jumped timing chain or belt. If the timing did not change when the wires were jumped, replace the ignition module and recheck the timing.
5. If the timing advanced when the ignition module was disconnected, check the wiring for grounds. If the wiring is good, reconnect the ignition module, disconnect the knock sensor, and recheck the timing.
6. If the timing advances when the knock sensor is disconnected, replace the knock sensor. If the timing does not change, test the zone vacuum switch. Verify that it is receiving vacuum and that the timing varies when the vacuum hose is disconnected. Figure 9–59 shows an example of checking this switch.
7. If there is no timing change, when the vacuum switch hose is removed, substitute a known good zone switch and recheck the timing. If the timing advances, the zone switch was bad.
8. If the timing does not advance with a new zone switch, replace the MCU and recheck the timing.

SHOP TALK

Before replacing the MCU, do the MCU self-test procedure to ensure that the problem is in the MCU.

MCU Detonation Checks

When a vehicle with an MCU system is brought in with a detonation problem, first check the engine cooling system and the EGR system. Also make sure that the fuel system is not delivering an overly lean fuel mixture. Then, perform the following procedure:

1. Check and set initial timing. Disconnect the knock sensor.
2. Disconnect the two-wire ignition module and use a jumper wire to connect the two wires together (Figure 9–58).
   a. If the timing did not retard 15 to 20 degrees, replace the ignition module.
   b. If the timing did retard 15 to 20 degrees, reconnect the two-wire connector and set the engine to 2000 rpm.
3. Disconnect the zone vacuum switch hose (Figure 9–59).
   a. If the timing does not retard at least 5 degrees, check the yellow wire between the MCU and the ignition module for shorts or opens. If the wire is okay, replace the MCU and recheck the timing.
   b. If the timing retards more than 5 degrees, reconnect the knock sensor and then tap on the intake manifold with a 3/4-inch wrench (with the engine at 2400 rpm). If the timing does not retard, replace the knock sensor and recheck the timing. If the timing still does not retard, replace the MCU.

GENERAL MOTORS' SPARK ADVANCE SYSTEMS

In the previous chapter, several GM ignition systems were discussed that could retard the ignition timing to compensate for detonation, cold starts, or high altitude. However, those systems employed centrifugal and vacuum advance mechanisms to achieve normal spark advance. In 1977, GM introduced a system that controlled spark advance by a computer. The system was called MISAR (microprocessor sensing and automatic regulation) and was installed on Oldsmobile Toronados. The system was changed in 1978 and was then called the electronic spark timing (EST) system. In 1979, GM introduced the computer-controlled catalytic converter (C-4) system. Although the C-4 system initially controlled only the fuel system, it was integrated with the EST ignition system in 1980. This combined system was expanded in 1981 into a complete computerized engine control system called computer command control (CCC) system.

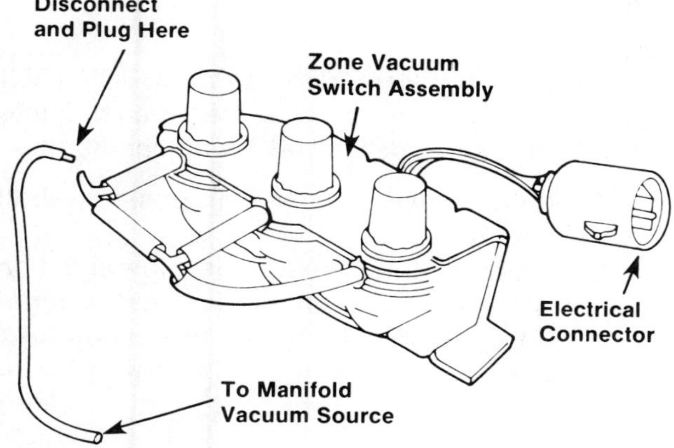

FIGURE 9–59 Checking zone vacuum switch

MISAR SYSTEM

Figure 9–60 shows a wiring diagram of the MISAR system. The MISAR system on 1977 Toronados does not use a standard HEI distributor. The distributor does not have a vacuum advance unit, a mechanical advance unit or a pick-up coil and pole piece. A different rotor is also used in the MISAR distributor, but the distributor cap, the ignition module, and the integral coil are the same as those found in other HEI distributors. Crankshaft position and engine speed are monitored by a crankshaft sensor located on the vibration damper. Spark timing is controlled by an electronic control module (called a controller). Another component in the system is the engine coolant sensor.

The crankshaft position sensor (Figure 9–61) consists of a pulse generator disc attached to the vibration damper at the front of the engine, and a sensor mounted to the front of the engine below the power steering pump. The operating principles of the crankshaft position sensor are similar to those involved in the magnetic pulse generator used in other HEI distributors. As the teeth of the disc rotate past the magnetic pole of the sensor, changes in the magnetic field are produced with a corresponding change in sensor voltage. The voltage fluctuations are monitored by the controller in such a way that it "knows" the crankshaft position and engine speed.

The engine coolant temperature sensor is different from the on-off switches used in other vehicles. It is part of the ignition system and resistance in the sensor changes with changes in coolant temperature. (Resistance lowers when temperature rises.) The controller monitors the change in voltage output to determine coolant temperature.

The controller assembly is an electronic unit and is mounted under the glove box. It receives

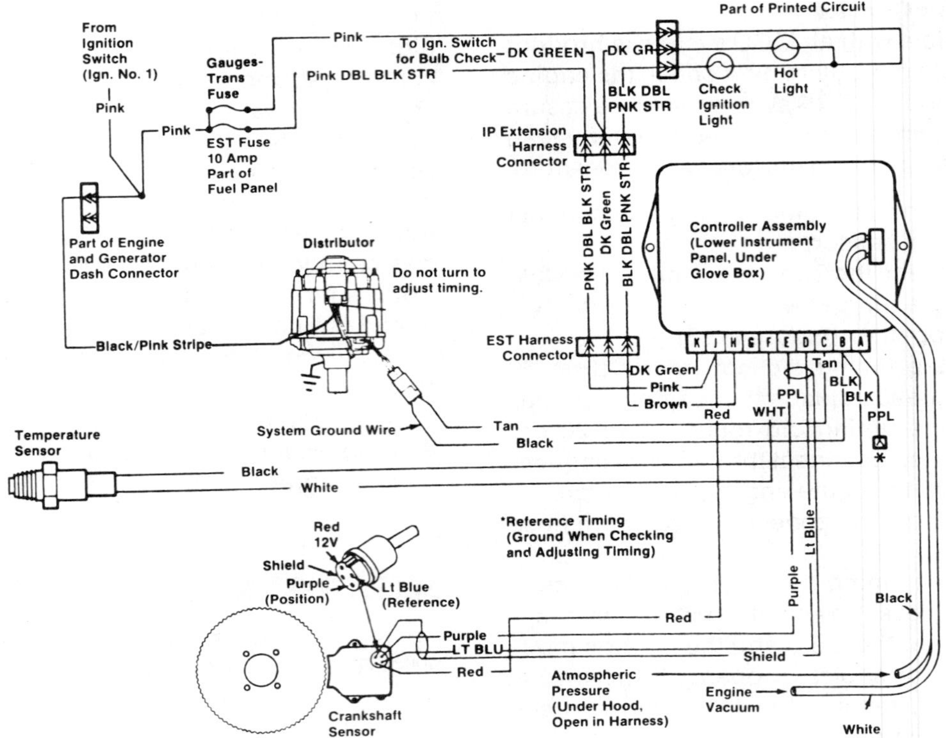

FIGURE 9-60 1977 MISAR schematic

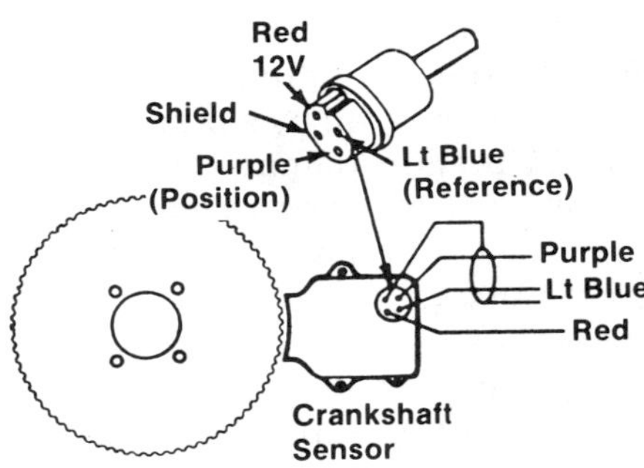

FIGURE 9-61 1977 MISAR crankshaft position sensor

signals from the crankshaft sensor (crankshaft position and engine rpm), engine coolant temperature sensor, engine vacuum, and atmospheric pressure. The controller assembly decides the most efficient spark advance based on the input signals and sends a command signal to the distributor module to fire the spark plugs.

The electrical harness connecting these units together and to the vehicle harness contains two vacuum tubes (Figure 9-60). They both connect to the controller assembly. The white one is also connected to manifold vacuum; the black one is not connected. The open end of the black one is in the engine compartment so that it will not be open to inside vehicle pressure but is left open to atmospheric pressure.

SHOP TALK

There are two different controller assemblies: one for use in California vehicles and another for use in other vehicles. The controller assembly used on California vehicles has different advance specifications. Be sure to consult the VIN for correct vehicle identification.

The MISAR system also has a CHECK IGNITION light located in the instrument panel cluster that will come on under the following conditions:

1. Ignition switch in the start position—bulb check
2. If electrical system voltage is low and there is a heavy electrical load such as the operation of power door lock, power windows, power seat, cigar lighter, rear window defogger, etc.

SHOP TALK

The CHECK IGNITION light will go off as soon as the electrical load is removed if the system voltage returns to normal.

3. When checking the reference timing and the reference timing connector is grounded
4. If there should be a controller failure so the spark timing would not advance

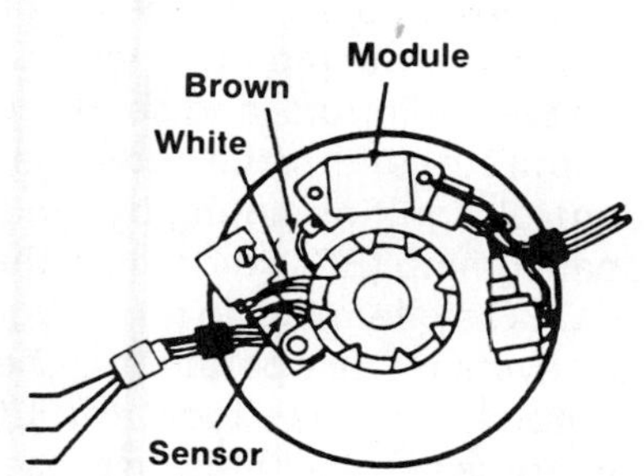

FIGURE 9-62 1978 EST distributor with magnetic pulse generator

ELECTRONIC SPARK TIMING

The EST system introduced in 1978 is very similar to the MISAR system. However, the crankshaft position sensor located on the vibration damper was replaced with a conventional pick-up coil and trigger wheel located in the HEI distributor (Figure 9-62). The standard four-terminal module was replaced with a three-terminal module. The pick-up coil and harness are also different. A spark shield was added under the rotor to protect the electronic circuits from false impulses. Figure 9-63 shows a wiring diagram of the EST system.

The EST controller performs the same function as the controller in the MISAR system. However, the EST controller has two harness connectors, rather than one and the black atmospheric pressure tube has been eliminated. Atmospheric pressure is sensed through a port on top of the controller.

Otherwise, the EST system operates like the MISAR system. Input from the pick-up coil, a coolant temperature sensor, and the vacuum tube and the air pressure port are used by the controller to compute the best timing. Low electrical system voltage and incorrect controller operation are indicated by a CHECK IGNITION light on the dash.

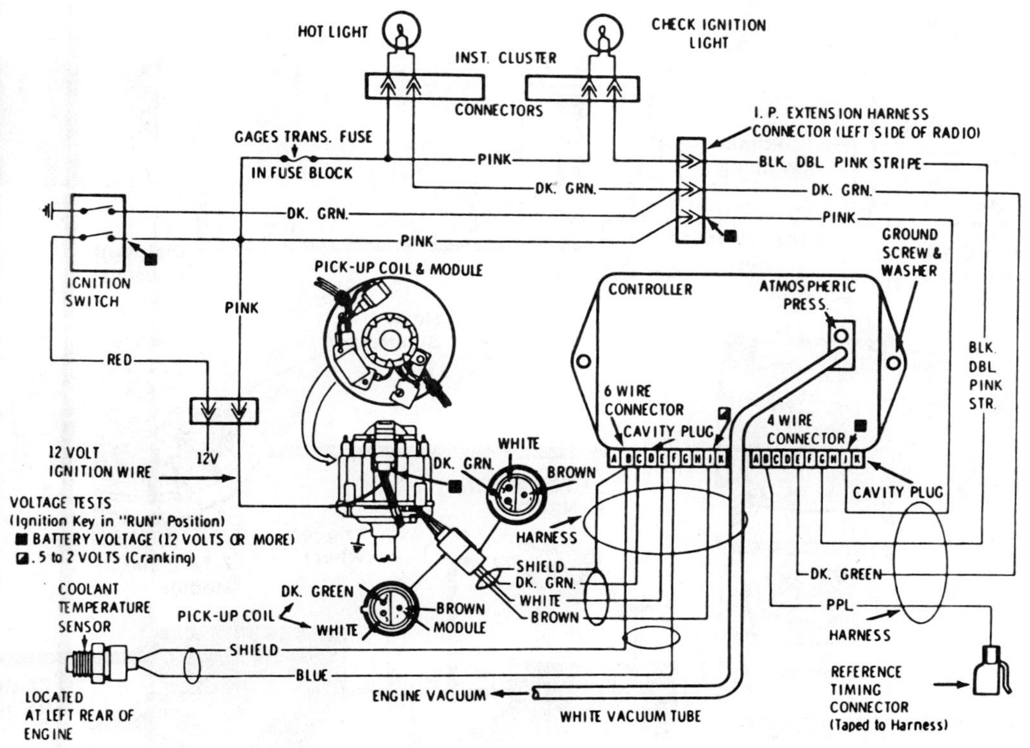

FIGURE 9-63 1978 EST schematic

EST/ESC

In 1981, electronic spark control was combined with EST on Buick turbocharged V-6 engines. The ESC (electronic spark control) system was discussed in the previous chapter. Figure 9-64 shows a schematic of the EST/ESC system. The EST/ESC system operates just like a standard EST system, except when detonation occurs. At all levels of engine operation, the electronic control module will retard the ignition timing for 20 seconds each time it receives a modified voltage signal from the ESC controller and the knock sensor. After 20 seconds, the spark timing will again be controlled by the normal parameters in the ECM memory until the next modified signal is received from the ESC controller. Then, the process is repeated. In this way, detonation is constantly monitored and corrected.

EST with a Hall-Effect Switch

Since 1982, 4- and 6-cylinder engines use a distributor with a Hall-effect switch (Figure 9-65). During cranking and up to 600 rpm, the pick-up coil and module in the distributor provide a fixed amount of spark advance independent of the electronic control module. The pick-up coil's signal is sent directly to the ignition coil. At engine speeds above 600 rpm, the ECM takes over and controls engine operation, including spark timing. At that point, the system operates like the previous EST systems, adjusting engine operation according to the input signals it receives from the various sensors.

COMPUTER COMMAND CONTROL

The CCC system uses an oxygen sensor, computer, and feedback carburetor or fuel injectors, as well as a computer-controlled engine timing. The computer is referred to as the electronic control module (ECM). The ECM monitors various sensors, processes the sensor inputs, and sends control signals to the carburetor or throttle body injectors. The ECM also controls ignition timing by regulating the electronic spark timing (EST) module in the distributor. An early CCC system is shown in Figure 9-66.

The CCC system can operate in either open or closed loop modes. In open loop, the ECM disregards sensor inputs and operates on preset information contained in the computer's memory circuits. In closed loop, the ECM considers both sensor read-

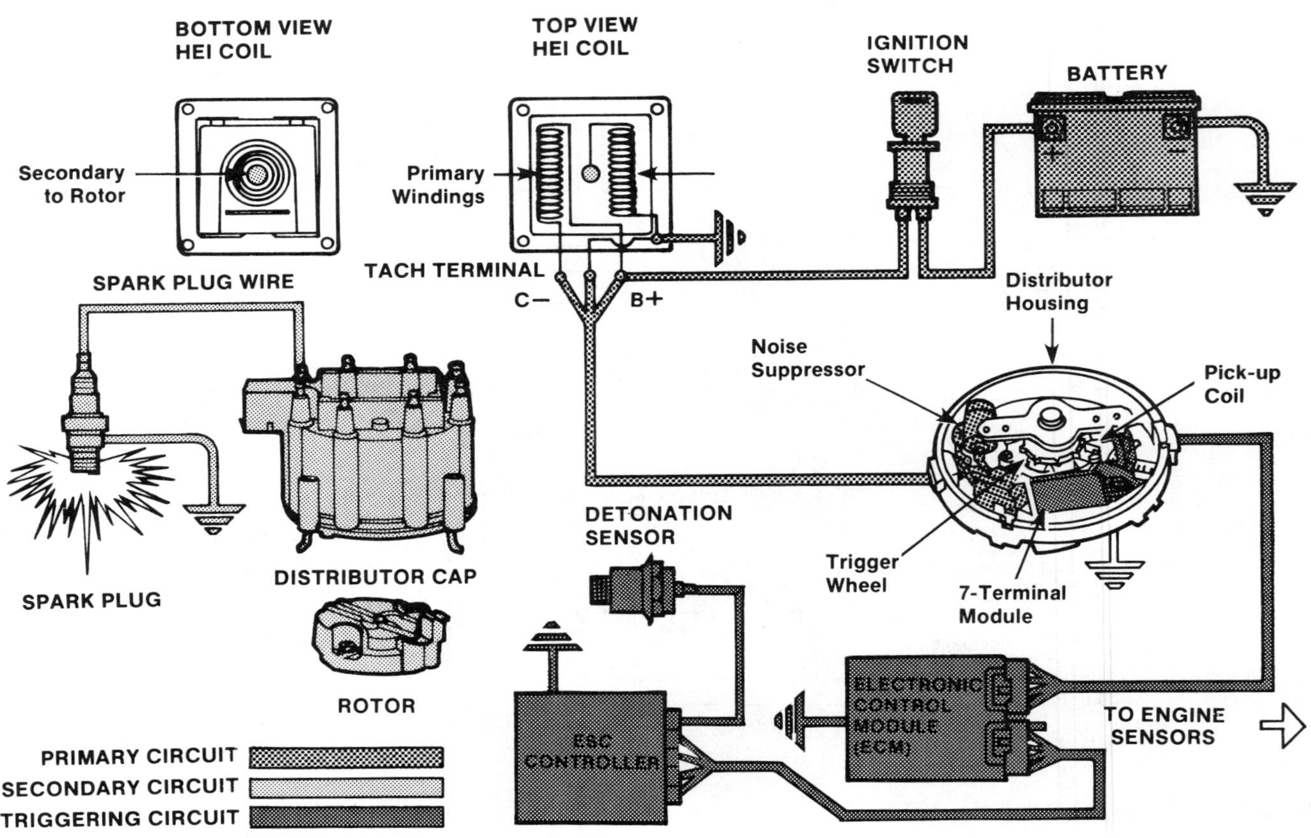

FIGURE 9-64 EST-ESC ignition system

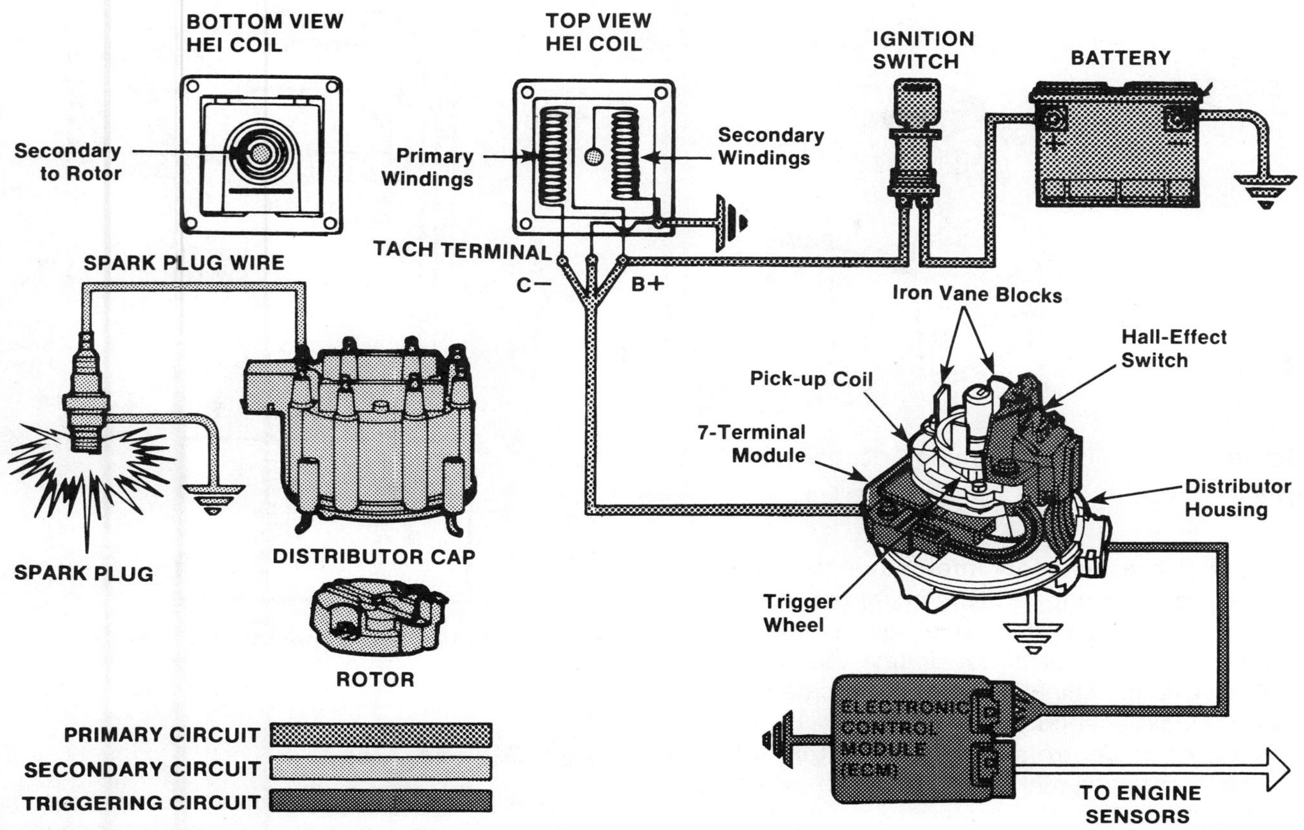

FIGURE 9-65 EST ignition system with a Hall-effect switch

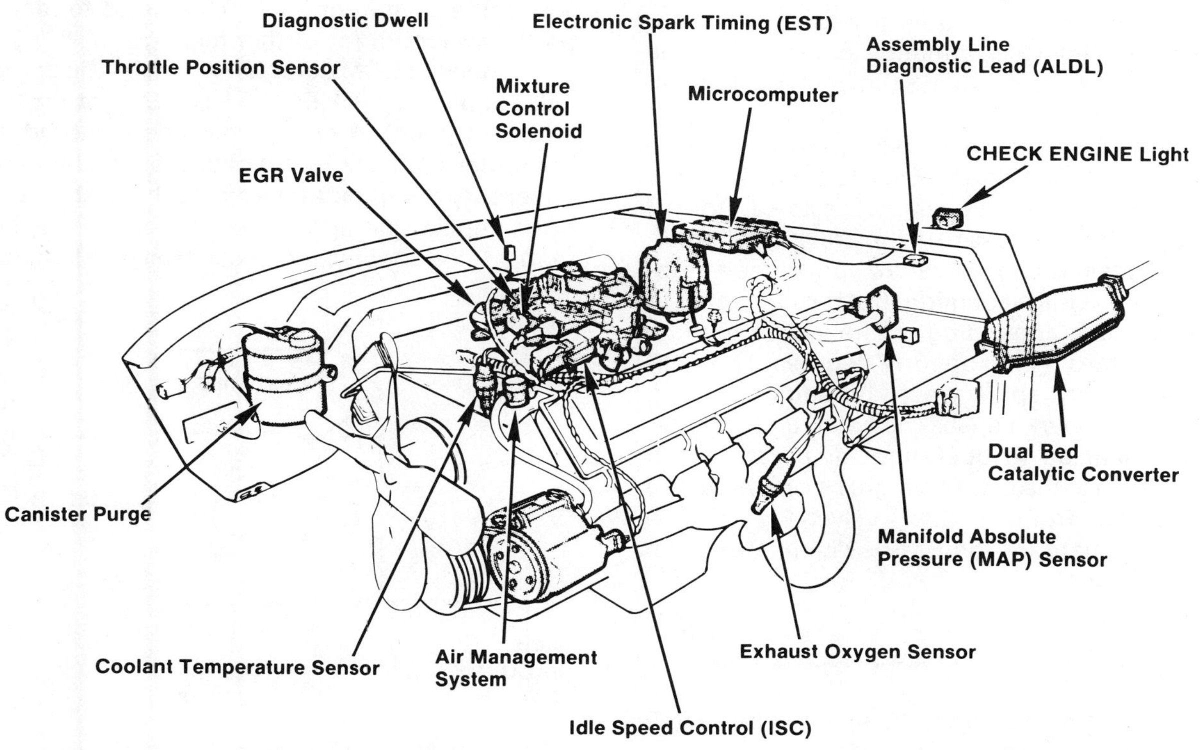

FIGURE 9-66 An early CCC system

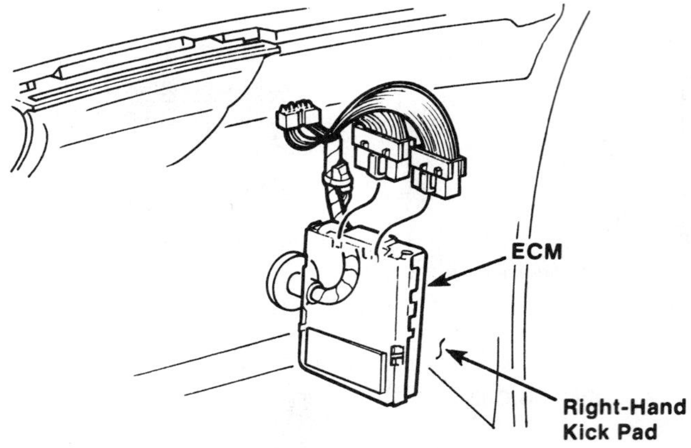

FIGURE 9-67 ECM mounting location for EST system

ings and internal memory information to determine proper engine settings and timing for maximum efficiency. The CCC system also has self-diagnostic capabilities to troubleshoot or test the electronic control systems. The ECM is usually located behind the right kickpanel (Figure 9-67).

The ECM controls spark timing through the electronic spark timing (EST) ignition system on most engines. The EST system consists of an HEI distributor with no vacuum or mechanical advance, and a 7-pin module. Ignition timing is determined by the ECM based upon signals from the ignition module and various other engine sensors, combined with the base timing information contained in the ECM's memory circuits. An EST schematic diagram is shown in Figure 9-68.

Some CCC systems also use a spark retard system, which is called either ESC (electronic spark control) or EMR (electronic module retard). Both of the systems were discussed in the previous chapter. The ESC system is used to retard spark when detonation occurs. It might or might not be connected to the CCC system, depending on the model application. It can retard timing up to 17 degrees. The EMR system can retard spark when the engine is cold. (Retarding the spark causes the catalytic converter to quickly heat up to operating temperatures.) It might be tied in to the ECM or might be operated by a vacuum switch from the EGR valve. Figure 9-69 shows a schematic of the EMR system. Maximum retard is 10 degrees.

SHOP TALK

Some Cadillacs with CCC also use the ESS (electronic spark selection) system.

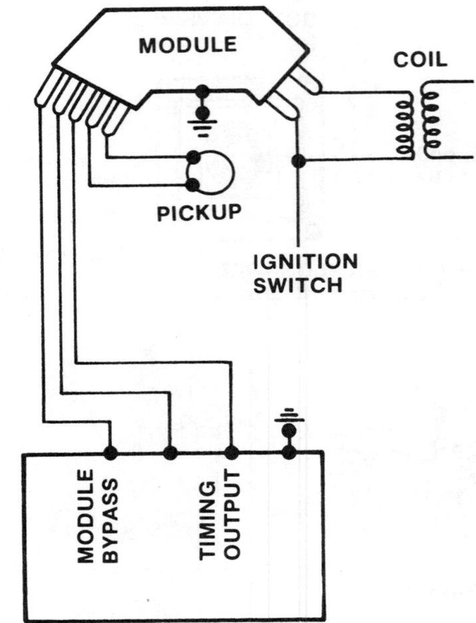

FIGURE 9-68 EST schematic

Self-Diagnostics

The CCC system also contains a self-diagnostic feature. If a problem in the CCC system occurs, the self-diagnostic system can detect and record failures in the electrical circuits and components and within the computer itself. This helps to identify a particular circuit for further testing.

When the ECM detects a problem, the computer will illuminate a CHECK ENGINE light on the instrument panel. A numerical code is stored in the computer memory to indicate the problem area. If necessary the computer adjusts the system to continue engine operation in what is called the limp-in mode. If the problem corrects itself, the light will go

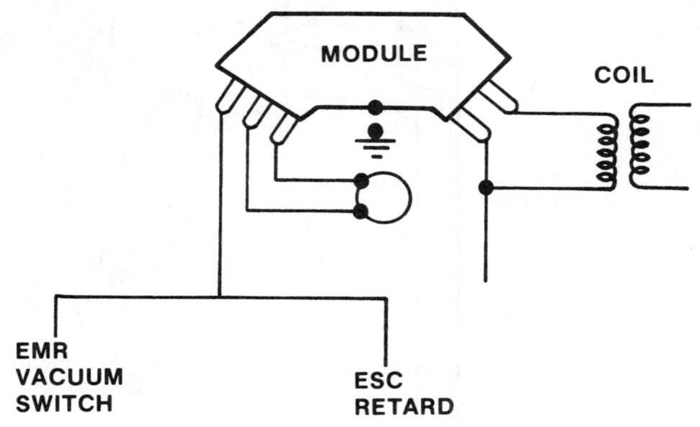

FIGURE 9-69 EMR schematic

out, but the trouble code will be preserved in the computer's memory until the battery is disconnected or until the engine has been started 50 times without the trouble reoccuring.

SHOP TALK

Trouble codes are erased from the ECM memory on Chevette and Pontiac J-1000 systems when the ignition switch is turned off.

The trouble codes can be easily retrieved from the computer's memory as an aid to troubleshooting problems in the system.

TESTING AND SERVICING GM'S SPARK CONTROL SYSTEMS

Early General Motors electronic engine controls do not have a self-diagnostic capability. The later systems (C-4 and CCC) have self-diagnostics that can greatly simplify troubleshooting. In both cases, use a logical approach to troubleshooting and make sure you have a thorough knowledge of how the entire system operates.

Remember that the computer can compensate for minor ignition and fuel system malfunctions. In many cases, this can hide problems that would otherwise be obvious.

Diagnosis can be accomplished with a test light, voltmeter, ohmmeter, jumper wire(s), and timing light. Special testers, such as the scan tester for CCC systems, are available for some systems and can greatly speed up the diagnosis procedure.

FIGURE 9-70 Adjusting rotor position on 1977 MISAR distributor

MISAR-ESC PERFORMANCE TEST

On 1977 MISAR models only, make the following adjustments before proceeding with the performance test procedures.

Distributor Rotor Alignment

Remove the distributor cap and crank the engine until the rotor points to the rear of the engine and the number 1 cylinder is at TDC (zero degrees on timing plate). The distributor is correctly positioned if the white stripe on the side of the rotor is aligned with the pointer in the distributor (Figure 9-70). If not, loosen the distributor bolt and rotate the distributor until it is aligned. If the original rotor has been replaced by one without a white stripe, align the pointer with the first flat edge past the hold-down screw. Tighten hold-down bolt.

Sensor Alignment

Check the crankshaft position sensor alignment and clearance at the crankshaft disc, as shown in Figure 9-71. Clearance should be 0.045 to 0.055 inch, measured at each end of the sensor.

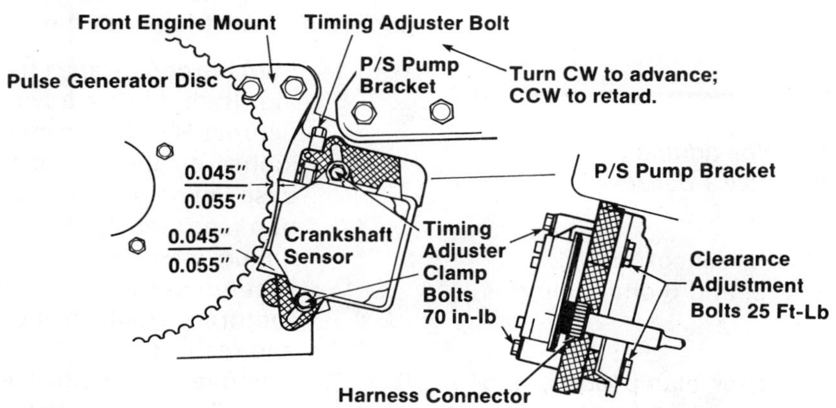

FIGURE 9-71 Crankshaft sensor adjustment bolts

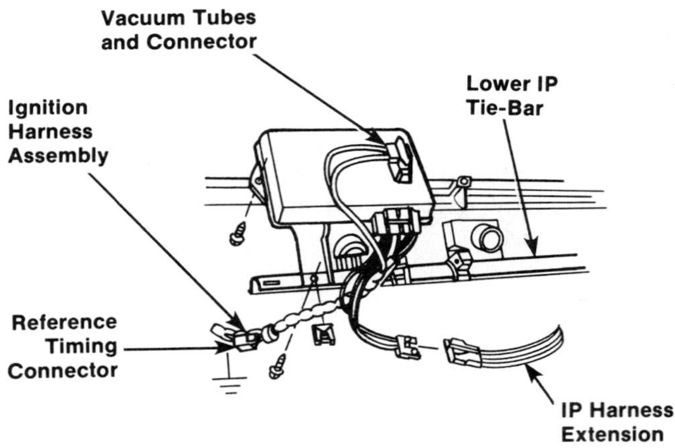

FIGURE 9–72 MISAR controller and timing connector

Adjust the air gap, if necessary. To do so, loosen the clearance adjuster bolts and insert a 0.050-inch carburetor gauge between the disc and sensor at two points. Retighten clearance adjuster bolts. Also, check to be sure that the sensor and disc are parallel to each other. Adjust as necessary.

After adjusting the air gap, check and adjust the timing. To do so follow this procedure:

1. Ground the reference timing connector as shown in Figure 9–72.
2. Start engine, and check timing. The timing should be as specified on the emissions VECI label. The CHECK IGNITION light should be on because of the ground wire on the reference timing connector.
3. Stop the engine.
4. Loosen timing adjuster clamp bolts (Figure 9–72).
5. Rotate the timing adjuster bolt clockwise to advance timing and counterclockwise to retard timing.

SHOP TALK

One complete turn of the adjuster bolt will change timing approximately 1 degree.

6. Start the engine and recheck timing.
7. Repeat steps 3, 4, 5, and 6 until timing is correct.
8. Stop the engine.
9. Retighten the adjuster clamp bolts, and disconnect the jumper wire from the reference timing connector.

Base Timing Test

On the 1978 MISAR-ESC model, check the base ignition timing before diagnosing performance problems. To do so, follow this procedure:

1. Connect the reference timing connector to ground with a jumper wire, as shown in Figure 9–72.

SHOP TALK

The reference timing connector is taped to the wiring harness near the controller assembly.

2. Start the engine and check the ignition timing. The CHECK IGNITION light on the dash will be illuminated.
3. If timing is incorrect, loosen the distributor hold-down bolt and turn the distributor to adjust the timing.
4. When timing is correct, tighten the distributor hold-down bolt, and remove the reference timing connector jumper wire.

SHOP TALK

If the CHECK IGNITION light is on during normal operation (when the reference timing connector is not grounded), the control assembly is faulty. Replace it and recheck the system. If the CHECK IGNITION light comes on at idle, check the charging system voltage and service as necessary.

Timing Advance Test

On MISAR-ESC systems perform the following performance test to check the timing advance.

1. Start the engine and check the ignition timing. It should be advanced to around 35 degrees BTDC with manifold vacuum at 18 inches or above. If it is not, recheck the base timing and the coolant sensor operation (see step 4). If both are okay, or if the timing remains at about 15 degrees, the vacuum sensor inside of the controller is ruptured. Replace the control assembly and recheck.
2. Remove the control assembly vacuum hose. This can be done at the controller or the intake manifold.

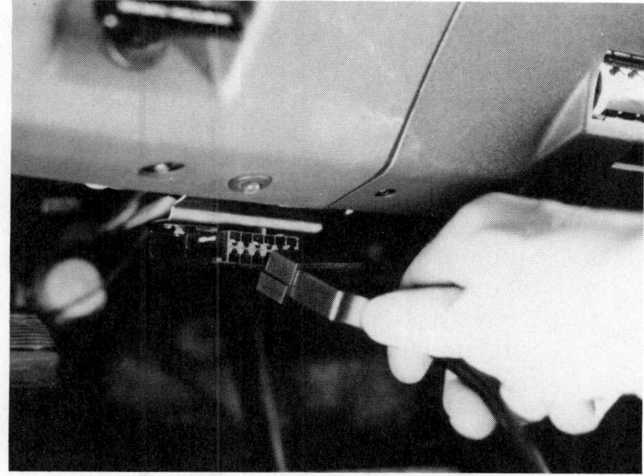

FIGURE 9-73 CCC system's ALCL connector

3. If ignition timing drops to 15 degrees when the hose is removed, the controller is okay. If timing does not drop, go to step 4.
4. Check the coolant temperature sensor resistance with an ohmmeter. At low temperatures the resistance across the sensor terminals should be 25,000 to 55,000 ohms. With the engine warmed up, the resistance should be 500 to 2,000 ohms. If resistance is not as specified, replace the sensor.
5. If the sensor resistance is within specifications, replace the controller and recheck the timing.

Detonation Diagnosis

If a detonation or engine knocking problems exists on a MISAR-ESC system, check and set the initial timing and check the timing advance as explained above. Also check for a plugged or kinked vacuum hose to the controller, which would prevent a rapid vacuum drop at the vacuum sensor. If the controller is overadvancing the spark, it should be replaced. Also check the EGR system for proper operation, and make sure the proper grade of gasoline is being used.

CCC SYSTEMS TESTING

The main components of the CCC on-board diagnostic system include: the electronic control module (ECM), a CHECK ENGINE or SERVICE ENGINE SOON warning light and an assembly line communications link (ALCL).

The CHECK ENGINE Light

The CHECK ENGINE warning light serves two purposes. First, it is a signal to the vehicle operator that a detectable CCC failure has occurred, and second, it can also be used by the service technician to aid in the location of system malfunctions.

To make sure that the CHECK ENGINE light is working, place the ignition switch in the RUN position and observe the light. With the ignition switch on, the light should be on also. If the light does not come on, connect a test light between terminal A and D in the ALCL connector. If the test light glows with the key on, check for a burned out CHECK ENGINE light bulb. If the bulb is okay, consult the service manual to find the cause of the inoperative light.

The ALCL Connector

The ALCL, located under the dash on most GM vehicles (Figures 9-73 and 9-74), is a multipin connector used to make certain diagnostic checks and read stored trouble codes. On 1981 CCC vehicles

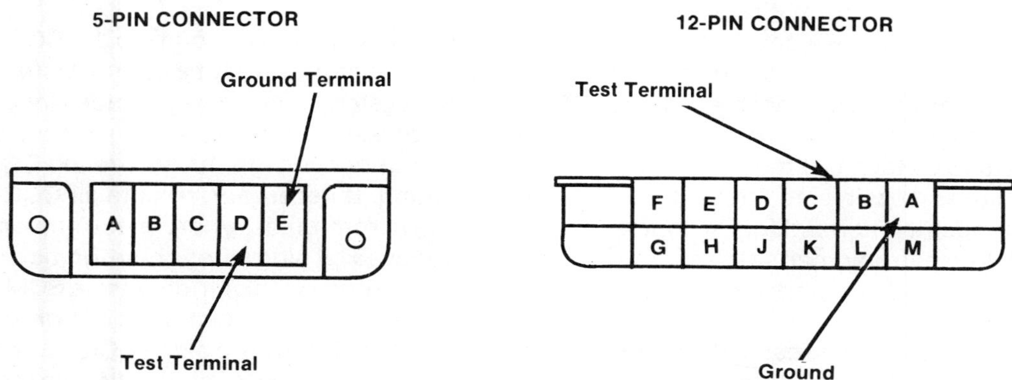

FIGURE 9-74 To initiate self-diagnostics, connect ground terminal and test terminal on ALCL connector.

the ALCL was a 5-pin connector (4 pin on Chevette and T-1000), but changed to a 12-pin design starting in 1982 (except Chevette and T-1000).

Each pin or terminal is identified alphabetically, starting with "A" at the upper right-hand position and ending with "M" at the lower left. However, due to normal variations in each system's design, all pins are rarely used at once. When they are, here is what they mean: terminal A is an ECM ground; terminal B is the diagnostic test terminal; terminal C is the air switching solenoid ground; terminal D is the CHECK ENGINE light ground; terminal E is used to access serial data on fuel-injected vehicles; terminal F is the torque converter clutch (TCC) solenoid ground; terminal G is a fuel pump test pin; terminals K and L are the ECM interface; and terminal M is used to test the level ride compressor.

By grounding or jumping different combinations of these test terminals, a variety of diagnostic checks can be performed from one central location.

Diagnostic Precautions

Some general service precautions need to be followed whenever performing CCC system diagnosis. Always keep in mind that the computer and related components are very sensitive to voltage changes (spikes/surges), heat and physical shock. Therefore, follow these precautions:

1. Never make or break electrical connections in any part of the CCC wiring harness with the ignition switched on.
2. Do not charge the vehicle's battery with the battery cables still connected.
3. Avoid using a booster battery to jump start a vehicle with a CCC system.
4. Before handling any electronic componentry, extreme care should be taken to protect against damage resulting from static electrical charges. Voltage spikes caused by the discharge of "normal" static electricity can result in severe and permanent damage to a variety of electronic components (for example, digital dash elements and components of the microcomputer). Because static voltage is difficult to detect and virtually impossible to prevent, the technican must avoid handling or touching sensitive electronic parts unless both the technician and the part are properly grounded.
5. To avoid potential damage caused by overheating, always remove the computer prior to welding or exposure to the high temperatures generated by items such as paint drying lamps or baking ovens.
6. If the computer must be removed for testing or service, protect it from sudden impact or physical shock while it is out of the vehicle.
7. To prolong the ECM's service life, make sure it is securely fastened into its special vibration resistant mounts during reinstallation.
8. Do not assume every driveability problem is related to the computerized engine controls. Perform normal diagnosis of the engine, electrical systems, fuel systems, and emission systems before suspecting the CCC system.

Self-Test Diagnosis Procedure

If the engine is mechanically sound, all vacuum hoses and wires are properly routed and the fuel delivery and ignition system check out, start to suspect the CCC system when one of the following conditions exist:

- The CHECK ENGINE light is glowing steadily, flashing intermittently or trouble codes are stored in the computer's memory.
- Ignition timing advance is incorrect.
- Idle problems such as stalling, wrong idle speed on engines equipped with idle speed control (ISC) circuits or rough idle exist.
- Rich or lean fuel condition, or symptoms that indicate carburetion problems such as surging, incorrect mixture control solenoid dwell, poor fuel economy, etc., exist.
- Failure or malfunction in the operation of computer-actuated devices such as the canister purge, EGR valve, transmission torque converter clutch (TCC), or air injector reactor (AIR) system are present.

If the initial diagnosis points to the CCC system, continue the diagnosis with a careful visual inspection of the system or systems affected (for example, ignition, emission, fuel). Look for loose, dirty, or corroded connections to make sure the ECM has a good ground. If necessary, reroute all CCC-related wires away from areas where electromagnetic induction caused by plug wires or the ignition coil may cause interference with or trigger false ECM signals. Only when everything appears to be in order should the self-diagnostic feature of the ECM be consulted.

When certain types of faults occur in the engine control system, the CHECK ENGINE warning light

FIGURE 9-75 Self-diagnosis is simplified, using a scan tool.

will come on and a trouble code will be stored in the computer's memory. As explained earlier, if the light remains on while the vehicle is running, it is an indication that a hard fault (problem present at the time of testing) exists. If the problem corrects itself or goes away (indicating an intermittent problem), a trouble code will still set, but the CHECK ENGINE light will go out 10 seconds after the problem is no longer detected. Trouble codes relating to intermittent problems will remain in the computer's memory for up to fifty engine starts or until power to the computer is interrupted.

Accessing Trouble Codes

Regardless of what causes the CHECK ENGINE light to come on, the first step to locating the problem involves checking the computer's memory to see if any trouble codes have been stored. This can be accomplished in one of two ways. The quickest method is to plug a hand-held scan tool into the ALCL connector and read the trouble codes directly (Figure 9-75). In lieu of a tester, the CHECK ENGINE light can be used to flash the codes.

To retrieve trouble codes from any CCC system using the CHECK ENGINE light, locate the ALCL connector, jumper terminal A and B together, and turn the ignition switch on. The light should come on and immediately start flashing. These flashes are trouble codes, so pay attention. A single flash followed by a pause and then two more flashes (flash-pause-flash-flash) indicates code 12. Each code will be repeated three times before the next code is displayed. With the key-on/engine-off, code 12 is considered normal. If there are any other trouble codes stored in memory, they will be flashed out in a similar manner and in consecutive numerical order. When all stored codes have displayed, code 12 will again reappear and flash three times signaling the end of code transmission.

Write down all trouble codes, but before proceeding, it is a good idea to clear the computer's memory first. This is accomplished by pulling the ECM fuse or disconnecting the ECM feed wire at the battery for 10 seconds. Once the memory is cleared, start and run the engine (road test if necessary) for at least 5 minutes to see if the codes reset. It is not unusual to find a vehicle with trouble codes that were left over from a previous repair. Therefore, it is always a good idea to start fresh and make sure all trouble codes are valid.

Troubleshooting the CCC System

After you have assessed the validity of all the trouble codes consult a factory service manual for applicable troubleshooting trees. The tree charts contain step-by-step instructions and procedures needed to determine the exact cause of any hard fault condition. Anytime multiple codes are given, priority should be given to codes in the 50's range since these codes indicate a problem in the ECM or PROM. Otherwise, check and repair codes in numerical order. If the problem is an intermittent one or if no codes are outputed, the factory manuals have detailed checklists that cover the most common driveability problems. Each symptom (for example, hesitation, surging, detonation) will have its own checklist; find the one that best describes the problem.

If the technician fails to find any trouble codes, the technician cannot assume the CCC system is okay. There are a lot of problems that can relate to CCC operation that will not set a trouble code. Problems of this type are often classified as "grey area" problems because they do not fit neatly into GM's diagnostic charts. For example, a misadjusted throttle position sensor (TPS) will still supply the computer with information, only it will not be the right information. The circuit will meet all the self-diagnostic criteria (meaning no code will set), but as long as the ECM's decisions are based on incorrect data, the engine will not perform right.

To help diagnose such situations, a hand-held scan tool comes in handy. With the scan tool, the actual voltage values generated by various sensors

FIGURE 9-76 A scan tool can also be used to test individual components.

can be read and monitored (Figure 9-76). Once the actual voltage values are known, compare the results to specifications to see if they are right.

Individual components can also be tested for voltage and resistance, using a digital multimeter. The service manual contains procedures for individual component testing and specifications.

CADILLAC'S ON-BOARD DIAGNOSTICS

The trend in self-diagnostics is toward on-board diagnostics. A good example of on-board diagnostics is the Cadillac DeVille's system. On this car, it is unnecessary to plug a scan tool into the ALCL connector. Instead, data stream from the ECM containing trouble codes, sensor voltages, and loop status is displayed on the instrument panel.

When the ECM detects a problem in the system, it illuminates a SERVICE SOON or SERVICE NOW light on the dash. To gain access to the trouble codes, switch on the ignition, and hold down the OFF and WARMER buttons on the electronic climate control panel (Figure 9-77) simultaneously, which will engage the self-diagnostic mode. The system will test itself, then begin displaying codes on the fuel data center panel (Figure 9-78). The numbers 8.8.8. will be displayed. Then codes will be displayed for any troubles the computers have recognized. Those in the ECM's memory (called "history codes") will be prefixed with "E," such as E52 for the MAT

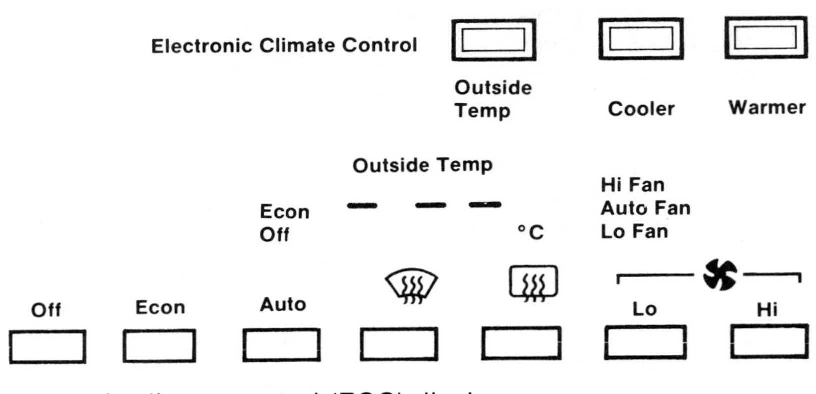

FIGURE 9-77 An electronic climate control (ECC) display

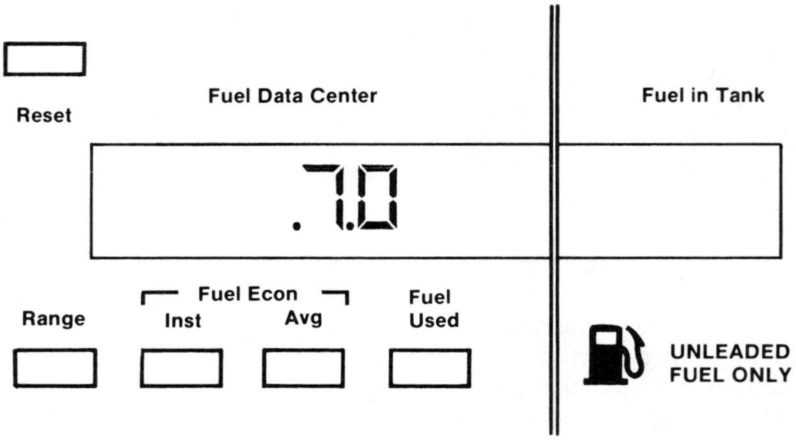

Reset
Fuel Data Center
Fuel in Tank
.7.0
Fuel Econ
Range
Inst
Avg
Fuel Used
UNLEADED FUEL ONLY

FIGURE 9-78 A fuel data center display

(manifold air temperature) sensor circuit. Current, or hard, trouble codes are prefixed by E.E. Next, BCM (body control module) history codes will appear with the prefix "F," then current ones with F.F.

With the code, check the circuit indicated for wiring problems, then the component itself if possible with a digital VOM. After making the repair, erase ECM codes by pressing the OFF and HI buttons until E.0.0. appears and BCM codes by holding down OFF and LOW until F.0.0. appears. The diagnostic mode can be terminated at anytime without erasing stored codes by pressing AUTO.

After the codes are displayed, "7.0" will appear automatically. This indicates a decision point—the system is ready to get into more involved investigations. To access engine and sensor parameters, start the motor, press LO, and get ECM data. The ECM will display voltage reading from various sensors, switches and actuators. E.9.0. appears, then each time the HI button is pressed, the system will advance one by one through the categories (LO will move the categories back). They are:

P.0.1.—Throttle angle in degrees
P.0.2.—Manifold absolute pressure in kilopascals
P.0.3.—Computed barometric pressure value in kilopascals
P.0.4.—Coolant temperature in degrees Celsius
P.0.5.—Manifold air temperature in degrees Celsius
P.0.6.—Injector pulse width in milliseconds
P.0.7.—Oxygen sensor voltage
P.0.8.—Spark advance in degrees
P.0.9.—Ignition cycle counter (0 to 50, the number of times the key has been turned on and off since a trouble that sets a code was last detected. After 50 cycles without the problem present, the code is cleared. A current or "hard" code will keep the reading at zero).
P.1.0.—Battery voltage (0 to 25.5 volts)
P.1.1.—Engine rpm (0 to 6,270—multiply the display by 10)
P.1.2.—Vehicle speed (0 to 255 mph)
P.1.3.—Oxygen sensor cross counts (0 to 255, indicating sensor activity—how often it switches between lean and rich signals in 1 second—and a low number means it is sluggish)
P.1.4.—Fuel integrator (0 to 255 counts, used to follow rich/lean trends. A reading of 128 means the integrator is not altering the amount of fuel supplied to the engine. If the oxygen sensor is rich most of the time, the integrator value will decrease, which will direct the ECM to provide less gas, and vice versa. A reading between 88 and 160 indicates that the ECM can control fuel delivery. A failed MAT sensor will give you a steady 60 and set code E37 or E38)
P.1.5.—Viscous torque converter clutch voltage (0 to 5.12 volts, indicating the voltage output of the converter clutch temperature sensor, which should be low when the sensor's cold).
P.1.6.—ECM PROM identification number (indicates a particular calibration; sometimes updated PROMs are specified to correct particular problems).

The last trouble code has been displayed, 7.0 will show again. ECM data is helpful by itself, but it can also be used in conjunction with another feature: In the diagnostic mode, the Electronic Climate Control (ECC) panel becomes a display of status lights (Figure 9-79). For example, to troubleshoot the oxygen sensor you start the engine, go to ECM data parameter P.0.4. (coolant temp), warm the engine up until a reading of 85 degrees Celsius is displayed, then move on to category P.0.7. (oxygen sensor voltage). Hold engine rpm at 1200-2000 for 2 minutes, then note the reading. If voltage is swinging from below .30 to above .60, look at the AUTO light on the ECC panel. If it is on, the sensor is okay. If it is off, a new part is needed. Also, the ECON light should flash on and off—if it glows constantly, the oxygen sensor signal is always rich, and if it never lights, the signal is always lean.

The Cadillac DeVille also has a body control module (BCM) which monitors such things as outside temperature, inside temperature, blower voltage, coolant temperature, and fuel level. To access trouble codes from the BCM, turn the ignition switch to ON and press OUTSIDE TEMP button. A series of twelve tests will be performed and results displayed of the fuel data center panel. Pressing the HI button will advance the BCM through the tests. Again, this information is supplemented by input from the ECC panel status lights.

SWITCH TESTING

Part of the on-board diagnostics is a series of tests for various switches. With the engine idling and 7.0 displayed, depress the brake pedal once and E.0.0. will appear, then E.7.1. through E.7.8. Each switch must be cycled within 10 seconds of when its number flashes on or the appropriate trouble code will be set even though the switch may be fine.

When E.7.1. is displayed, depress the brake again and the cruise control/VCC circuit will be tested. When E.7.2. is displayed, floor the accelerator once to check the throttle switch. With the brake on and E.7.4. showing, shift the transmission into RE-

ECM STATUS LIGHT DISPLAY						
	Light On	In 4th Gear	VCC Enabled	Closed Throttle	Rich	Closed Loop
	Light Off	Not in 4th Gear	VCC Disabled	Open Throttle	Lean	Open Loop
	Indicator			Off	Econ	Auto
	Function	4th Gear Input	VCC Output	Throttle Switch Input	Oxygen Sensor Input	ECM Operating Mode

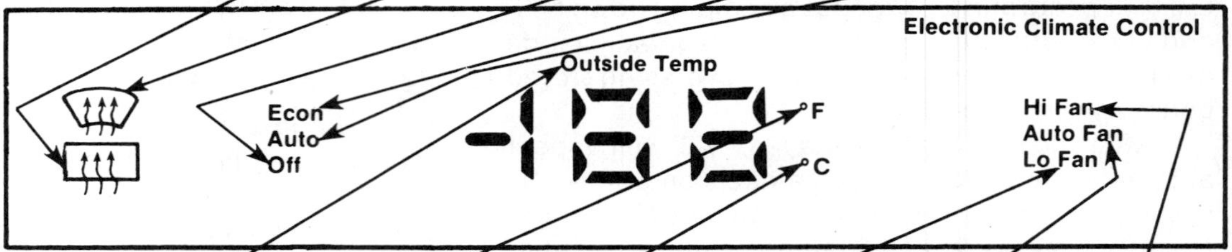

Electronic Climate Control

Outside Temp

Econ
Auto
Off

°F
°C

Hi Fan
Auto Fan
Lo Fan

ECM STATUS LIGHT DISPLAY							
	Function	A/C Clutch Output	Compressor Low Pressure Switch Input	Heater Water Valve Output	A/C-Def Mode Door Output	Cooling Fans Status	Up/Down Mode Door Output
	Indicator	Outside Temp	°F	°C	LO Fan	Auto Fan	HI Fan
	Light On	Energized	Open (Low Pressure)	Closed (No Water Flow)	A/C	Fans Running	Up
	Light Off	De-Energized	Closed	Open	Def	Fans Off	Down

FIGURE 9–79 Explanation of status lights on the ECC display

VERSE, then NEUTRAL, to test the PARK/NEUTRAL switch. Turn the cruise control switch off and on when E.7.5. is displayed; press and release the SET/ COAST button at E.7.6.; and operate the RESUME/ ACCEL switch with E.7.7. displayed. When E.7.8. appears, turn the wheel from straight ahead to full left or right to check the power steering pressure switch. With this sequence completed, the ECM will display the trouble codes for any switch that failed its test.

COMPONENT ACTUATION TEST

Another interesting feature is ECM output cycling. Here, the computer energizes the air switch and divert solenoids; the ISC (idle speed acontrol) motor; the cruise vacuum, cruise power, canister purge, EGR, and VCC solenoids; and the EFE relay every 3 seconds for 2 minutes, listen to and feel these components to find out if they are actually working.

To engage ECM output cycling, start the engine, shut it off, then turn the key back to on within 2 seconds (do not start the engine). Enter the diagnostics, and press HI at the 7.0 decision point. When E.9.5. (ECM output cycling ready) appears, floor the gas pedal and cycling will begin.

DISTRIBUTORLESS IGNITION SYSTEMS (DIS)

In 1984, Buick became the first major vehicle manufacturer to design and implement a distributorless ignition system (DIS). The system was called computer-controlled coil ignition (C3I) and was offered on select Rivieras and Regals equipped with a 3.8-liter turbocharged V-6.

Since then, Buick has expanded its C3I applications and others, including Chevrolet, Oldsmobile, Pontiac, Saab, Nissan, and Ford, have developed distributorless ignitions, too (Figure 9–80).

The development and spreading popularity of DIS is due to the same government and market pressures that encouraged the development of other electronic ignition systems. Automobile manufacturer's are constantly in search of new ways to reduce emissions, improve fuel economy, and boost component reliability.

DIS offers advantages in production costs and maintenance considerations. By removing the distributor, the vehicle manufacturers realize a substantial savings in ignition parts and related machining costs. Also, by eliminating the distributor, vehicle manufacturers also did away with cracked caps, eroded carbon buttons, burned through rotors, moisture misfire, base timing adjustments, etc.

GM is currently the largest supplier of vehicles equipped with distributorless ignition systems. Therefore, the discussion of DIS in this textbook will be limited to their systems.

GM DIS SYSTEMS

GM currently offers three types of DIS designs. The C3I system (Figure 9-81) previously mentioned

FIGURE 9-80 Ford's distributorless ignition system found on supercharged 3.8L engines

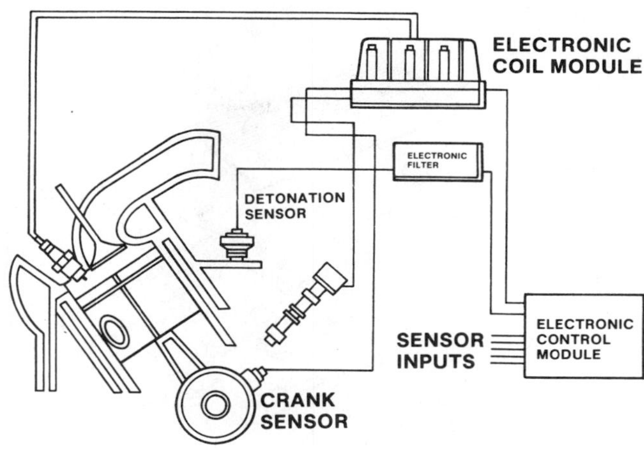

FIGURE 9-81 Schematic of a C3I system

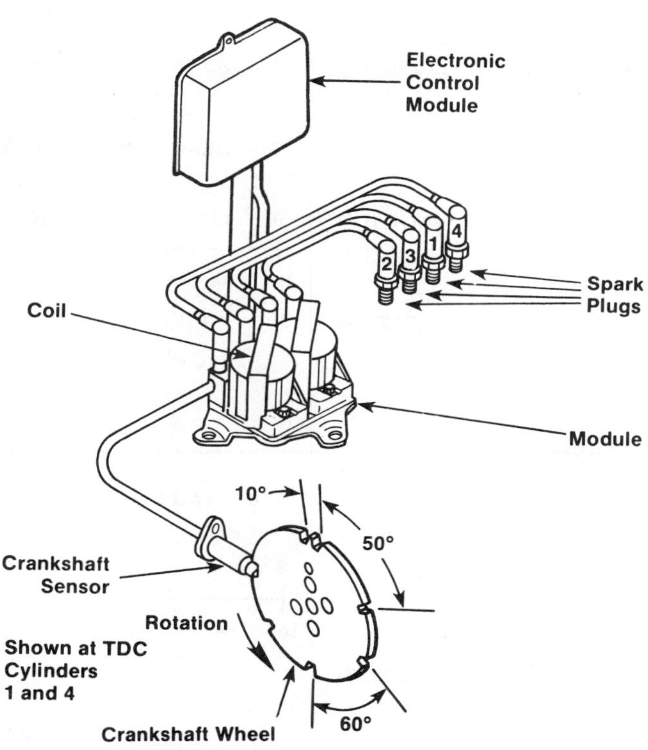

FIGURE 9-82 Schematic of a DIS system

is exclusive to the Buick 3.8 liter, 3.0 liter, and late model 3800 V-6 engines. The Chevy 2.8 liter and 2.0 liter ("Gen II" models only) and the Pontiac 2.5 liter Tech 4 engines are equipped with a system called DIS, which stands for direct ignition system (within GM, DIS carries dual meanings) (Figure 9-82).

The newest GM distributorless system is called integrated distributorless ignition (IDI for short). It was introduced on the Oldsmobile 2.3-liter quad 4 engine. Although similar to DIS in operation, IDI is

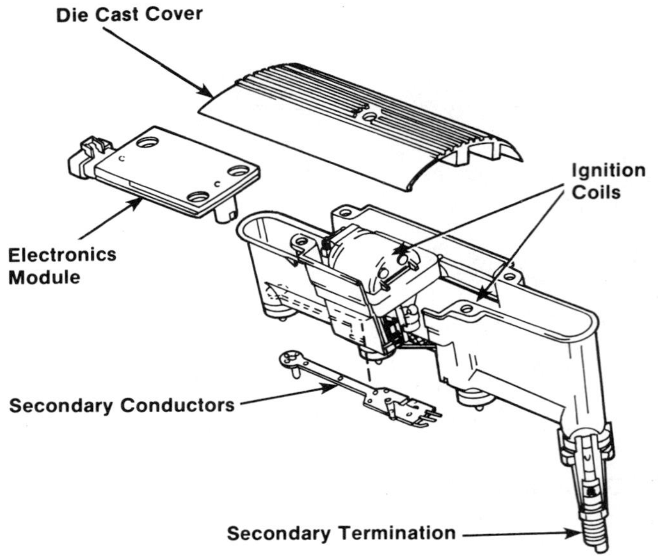

FIGURE 9-83 Quad 4 ignition system components

FIGURE 9-84 Type 3 coil/module used on the Buick 3800 V-6

TOP DEAD CENTER ON CYLINDER NUMBERS

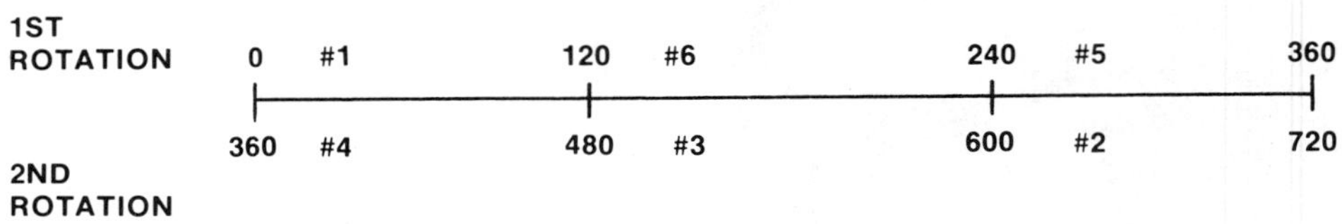

FIRING ORDER 1-6-5-4-3-2
BUICK 3.0L and 3.8L

TOP DEAD CENTER ON CYLINDER NUMBERS

1ST ROTATION: 0 #1 — 120 #2 — 240 #3 — 360

2ND ROTATION: 360 #4 — 480 #5 — 600 #6 — 720

CHEVROLET 2.8L
FIRING ORDER 1-2-3-4-5-6

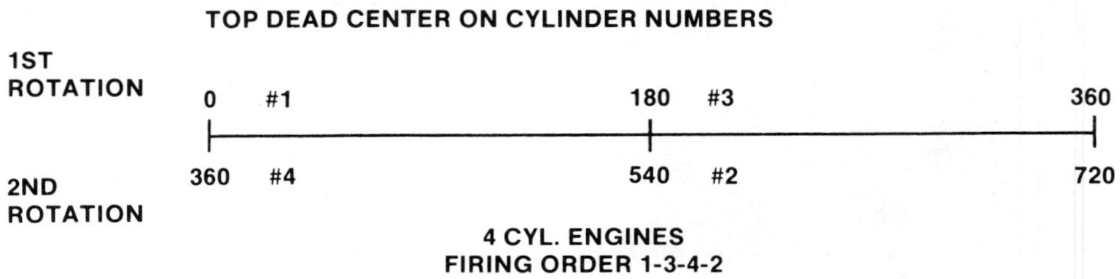

FIGURE 9-85 1984–1988 GM cylinder firing order/top dead center points for distributorless ignition system

so named because most of its ignition parts are "integrated" into a specially designed housing (Figure 9-83).

OPERATION

From a general operating standpoint, all of GM distributorless ignition systems are based on the "waste spark" method of spark distribution. The "waste spark" system uses one coil (Figure 9-84) for every two cylinders. Each of the coil secondary terminals are attached to a spark plug. Each coil fires the plugs of a pair of cylinders whose pistons rise and fall together. In all V-6s, the paired cylinders are 1 and 4, 2 and 5, and 3 and 6 (or 4 and 1 and 3 and 2 on 4-cylinder engines) (Figure 9-85). With this arrangement, one cylinder of each pair will be on its compression stroke while the other is on the exhaust stroke. Both cylinders get spark simultaneously, but only one spark generates power while the other is wasted out the exhaust. During the next revolution, the roles are reversed.

Due to the way the secondary coils are wired, when the induced voltage cuts across the primary and secondary windings of the coil, one plug will fire in the "normal" direction—positive center electrode to negative side electrode—and the other plug will fire just the reverse (side to center electrode). As shown in Figure 9-86, both plugs fire simultaneously, completing the series circuit. Each plug will always fire the same way on both the exhaust and compression strokes.

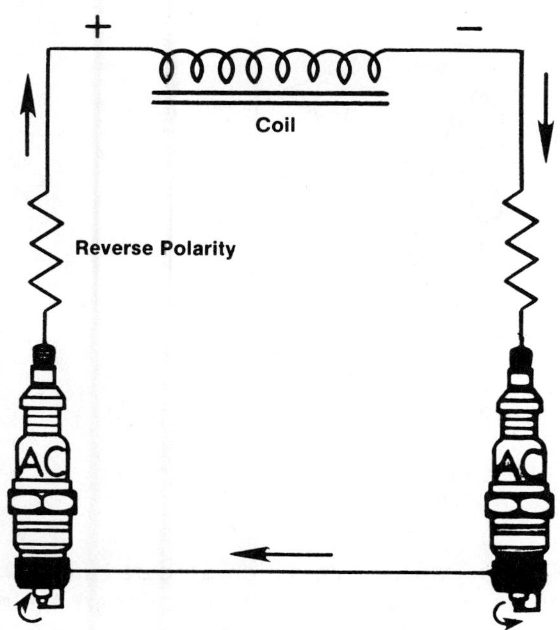

FIGURE 9-86 DIS current flow

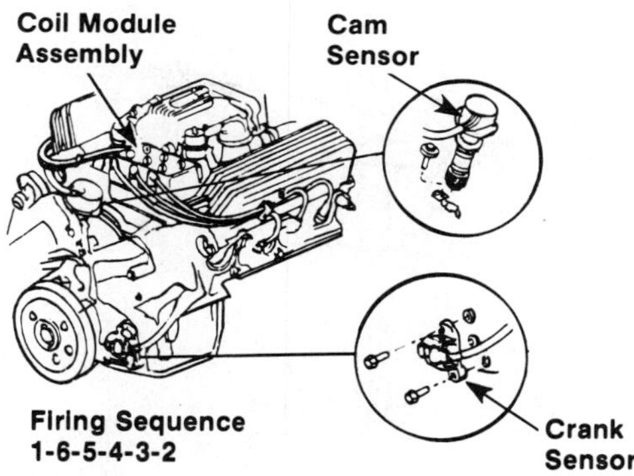

FIGURE 9-87 Hall-effect crank and cam sensors on a 3.8-liter Buick SFI turbocharged engine

The coil is able to overcome the increased voltage requirements caused by reversed polarity and still fire two plugs simultaneously because each coil is capable of producing up to 60,000 volts. Second, there is very little resistance across the plug gap on exhaust, so the plug requires very little voltage to fire, thereby providing its mate (the plug that is on compression) with plenty of available voltage.

Beyond waste spark, however, each GM system has its own personality. For example, the input devices that communicate camshaft and crankshaft position to the ECM are unique from engine to engine. Buick's 3.8-liter SFI (including turbo versions) engine uses separate Hall-effect crankshaft and camshaft sensors to control the ignition cycle and fuel injection firing order (Figure 9-87). The 3.0-liter engine determines its relative camshaft (sync pulse) and crankshaft positions from a single, combined Hall-effect switch that is mounted beyond the harmonic balancer (Figure 9-88).

The DIS system used on the 2.8-liter, 2.5-liter, and 2.0-liter engine and the IDI design used on the quad 4 receive reference pulses from a single electromagnetic crankshaft sensor that is triggered by a crankshaft mounted reluctor (Figure 9-89).

Other notable design differences are related to firing sequences, coil/module design (utilized or individual design), number and type of connections, and component location.

The ECM, ignition module, and position sensor(s) combine to control spark timing and advance. The ECM, as usual, is basically a master data interpreter. Information is collected from a variety of input sensors, including the manifold absolute pressure (MAP) sensor, manifold air temperature (MAT) sensor, mass airflow (MAF) sensor, coolant temper-

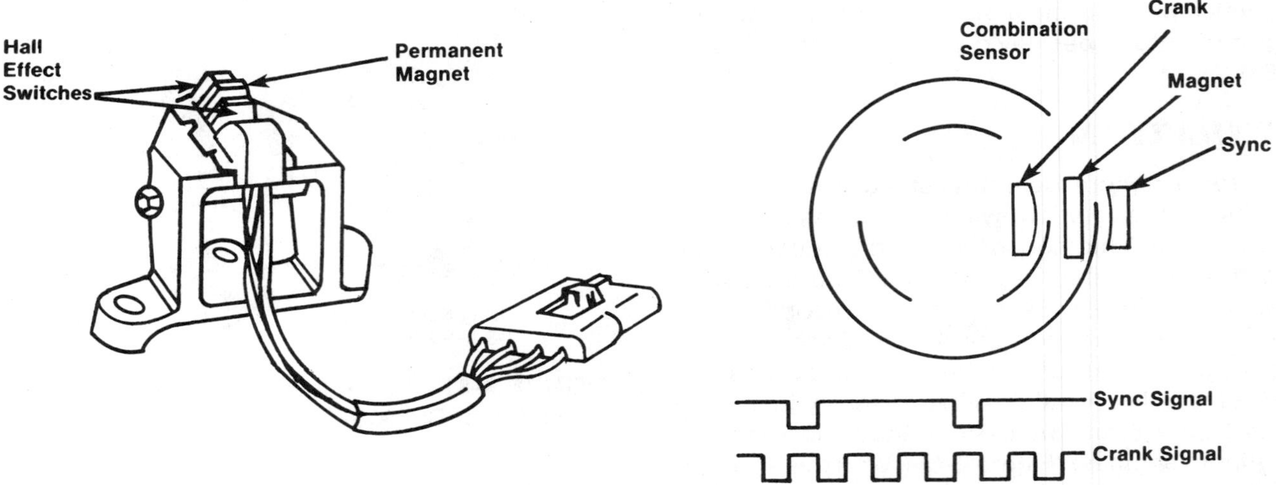

FIGURE 9–88 Buick's 3.0-liter V-6 engine uses a crankshaft sensor with two Hall-effect switches. Two rings on the back of the harmonic balancer rotate between the switches and the permanent magnet to provide the crank position signal and synchronization signals to the control module.

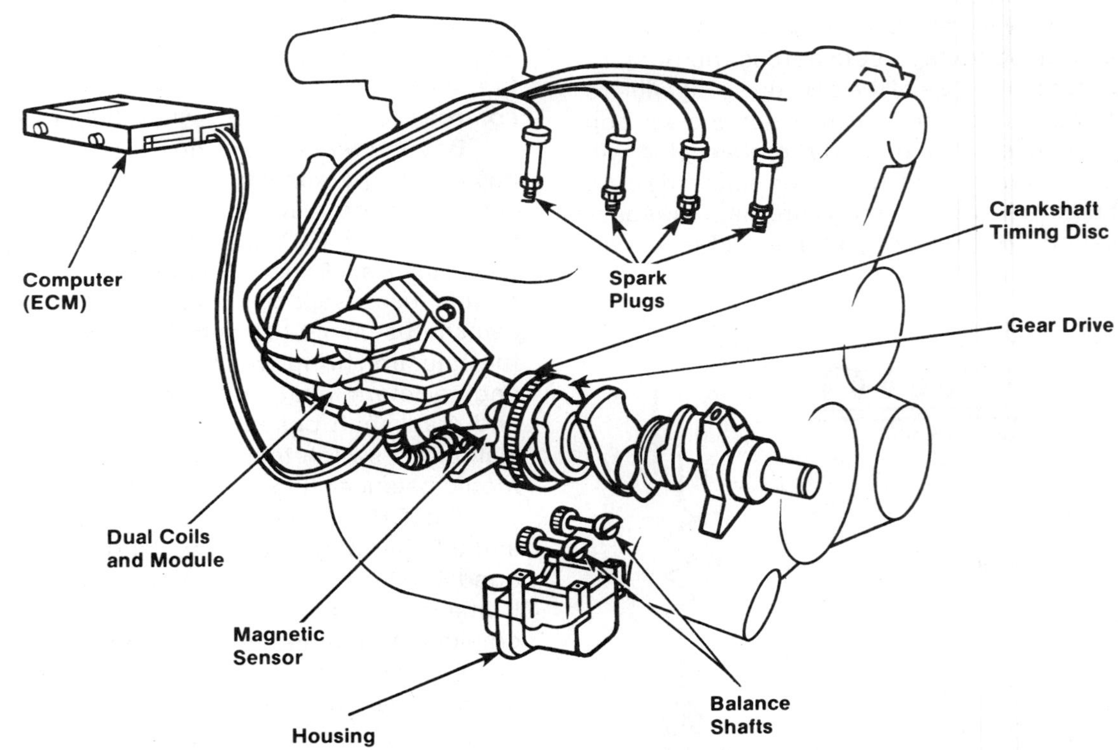

FIGURE 9–89 C-3 distributorless ignition system on GM 2.5-liter engine with a crankshaft sensor

ature switch (CTS), and the throttle position sensor (TPS). This input is used to selectively alter spark timing. The ignition module uses crank/cam sensor data to control the primary circuit of the coils.

The ignition module, which is located under the coil assembly, is also responsible for spark timing below 400 rpm (for starting) and when the ECM bypass circuit is grounded. The coils themselves are

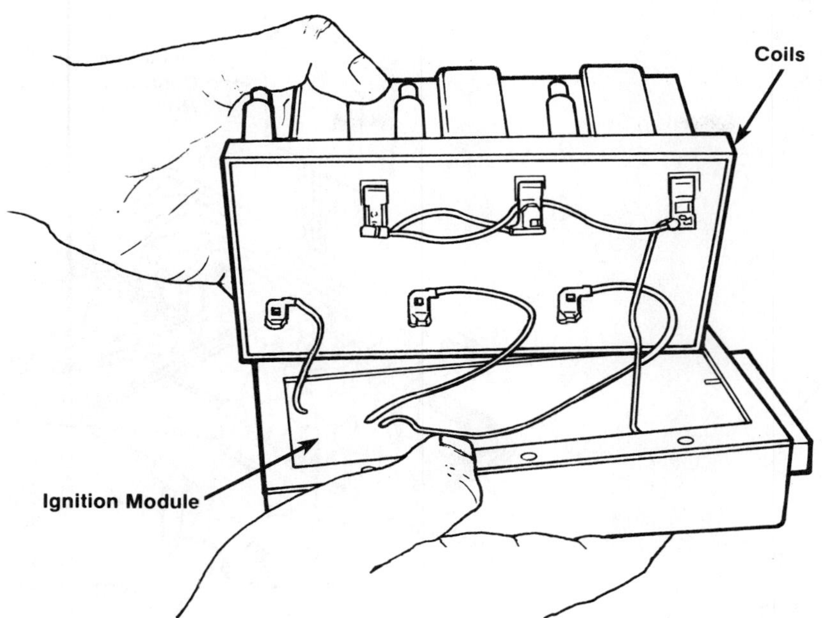

FIGURE 9-90 Type 1 C3I coil/module

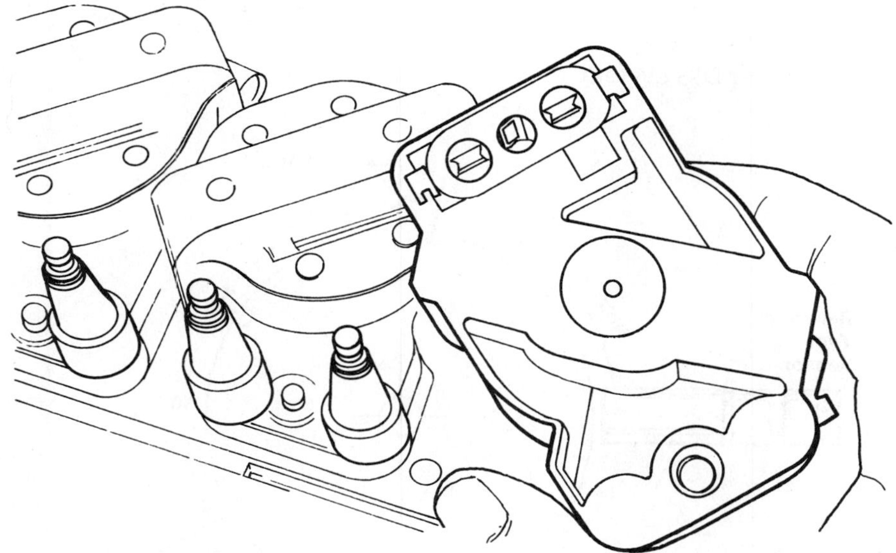

FIGURE 9-91 Type 2 C3I coil and module

either serviced as a complete unit (Type 1) (Figure 9-90) or may be serviced separately (Type 2) (Figure 9-91). When Buick introduced the 3800 engine in 1988, its C3I system was given a Type 3 designation because of internal module changes.

To compensate for the timing reference that was lost by the removal of the distributor, a new method of monitoring valve and piston position had to be devised.

The principal parts of the Chevrolet 2.8-liter system (Figure 9-92), for example, include a magnetic sensor and slotted timing disc called a reluctor (Figure 9-93). The reluctor is a special wheel cast on the crankshaft that has seven machined slots on it, six of which are spaced exactly 60 degrees apart. The seventh notch is located 10 degrees from the number 6 notch and is used to synchronize the coil firing sequence in relation to crankshaft position.

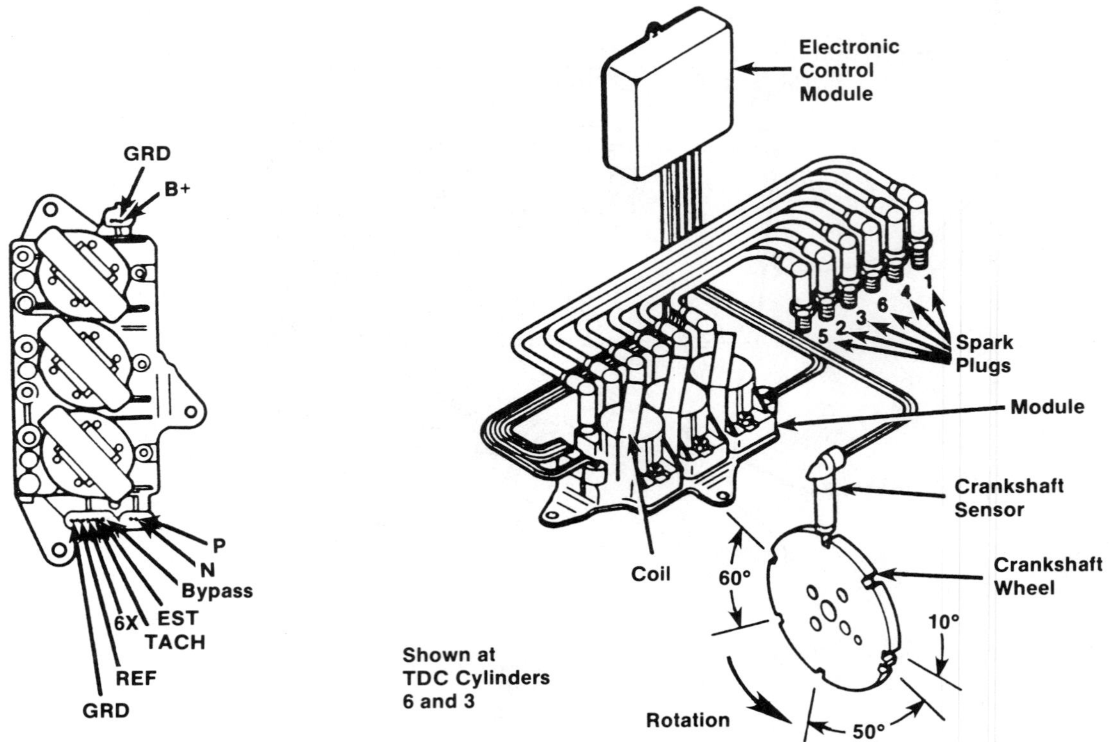

FIGURE 9-92 GM's 2.8-liter DIS system

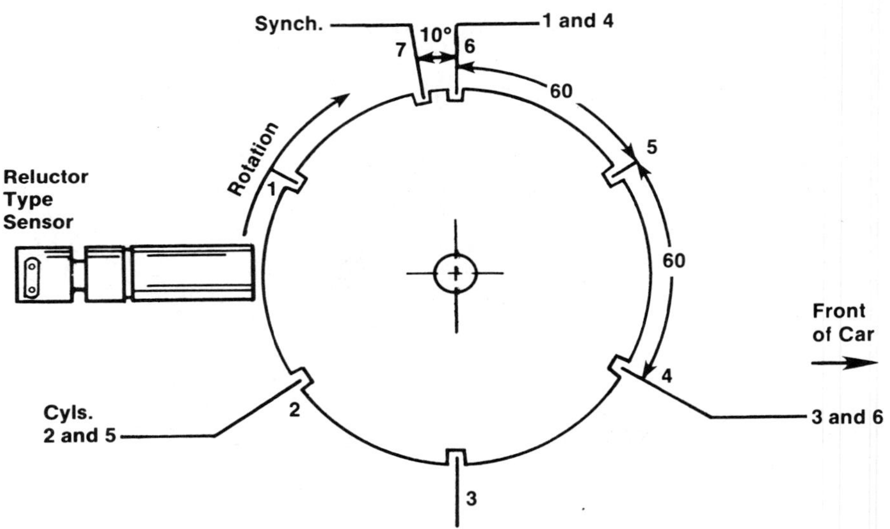

FIGURE 9-93 Crankshaft sensor found on GM's 2.8-liter V-6 engine with DIS

The magnetic sensor, which protrudes into the side of the block to within 0.050 inch (± 0.020 inch) of the crankshaft reluctor, generates a small AC voltage each time one of the machined slots passes by.

By counting the time between pulses, the ignition module picks out the unevenly spaced seventh slot (rotating in a clockwise direction), which starts the calculation of the ignition coil sequencing. Once it has its bearings, the module is programmed to accept the AC voltage signals of select notches for firing purposes.

For example, when the module "sees" the sixth, second and fourth notches—in that order—it triggers the coils that serve cylinder 1 and 4, 2 and 5, and

3 and 6. As a result, cylinders 1, 2, and 3 will fire on the first crankshaft revolution and cylinders 4, 5, and 6 will ignite the next time around (1-2-3-4-5-6 is the firing order of this engine). Remember that since the coil fires the plugs in pairs, the signals for notches number 1, 3, and 5 are there for reference only.

When the system is working properly, there is no base timing to adjust and there are no moving parts to wear. Of course, the lack of moving parts provides no guarantee that the system will be trouble-free.

DIS DIAGNOSTIC AND SERVICE CONSIDERATIONS

A distributorless ignition system will give a normal scope pattern provided the analyzer being used is compatible with DIS operation (special connectors are needed to get a normal-looking scope pattern). On the analyzer, firing voltages in the range of 3 kV on the exhaust plug and 6 to 12 kV on the power side are considered normal.

If it is decided to perform a cylinder balance test, handle the plug wires with care. Also, keep in mind that when one plug is pulled off and reconnected, raw fuel that is accumulated in the shorted cylinder could ignite on the exhaust stroke and create a potentially harmful manifold explosion.

To minimize these hazards, check the cylinder's power contribution by disconnecting individual fuel injectors instead of spark plugs. An injector balance test is just as accurate.

On applications where the injectors are inaccessible, there is an easy way to monitor rpm drop by shorting each plug with a 12-volt test light. To perform this monitoring function, do the following:

First, cut a length of 1/4-inch diameter vacuum hose into six 1-inch sections. For this test to work, the vacuum hose must be the type that contains carbon so that it conducts electricity (use an ohmmeter to check for hose conductivity). Next, remove the plug wires from each coil and insert the precut sections of hose over the exposed terminals. Reconnect the coil wires to each hose end and start the engine. If the vacuum hose is conductive and tightly attached, the engine will start and run. To short the cylinders, touch each vacuum hose with the tip of a 12-volt test light connected to ground and note the rpm drop.

As far as plugs and wires are concerned, they are serviced in the usual manner, meaning check and replace as necessary. When dealing with a misfire complaint, closely inspect all plug connections and use a spark checker to test for available voltage to the plug. If there are any loose plug wires, look for burned terminals, check their resistance, and always remove and inspect the underside of the coil that serves the misfiring plug(s). A loose or damaged wire or bad plug can lead to carbon tracking of the coil. If this condition exists, replace the coil.

NO-START DIAGNOSIS

If the engine cranks but will not run, here is a general procedure to follow that can verify the operation of the ignition system. To simplify things assume that the fuel system has been checked and is in working order.

Start the diagnosis by plugging a scan tool into the ALDL connector and turning the ignition switch on. The SERVICE ENGINE SOON dash light should flicker and a check of TPS voltage should indicate less than 2.5 volts at closed throttle. A reading higher than 2.5 volts could mean the TPS is directing the ECM to enter the "clear flood" mode.

If the TPS voltage is within acceptable limits, select a test mode that will allow monitoring of the engine rpm and use a spark checker to test for spark. Always check for spark on two wires from different coil packs (for example, 2–4 or 4–6), but test only one wire at a time, leaving all other wires connected. Also, check for a cranking rpm valve.

If a healthy spark is noted on both wires and an rpm signal is observed, the ignition system is working fine, so go back and check for correct fuel pressure and injector operation.

If rpm is indicated but there is no spark from the coils, check the connections at the ignition module. If the connections look clean and tight and no fault codes are present, replace the module.

If spark appears on only one wire, check both wires of the coil that did not fire for proper resistance. Each wire should measure less than 30,000 ohms. If the wires are okay, try switching the coils and test for spark again. If the problem moved when the coils were switched, replace the faulty coil. If the problem did not move and the substitute coil fails to produce a spark, replace the module.

It was mentioned earlier that a check for rpm should be made during cranking. In a situation where there is no spark and no cranking rpm value, begin the diagnosis by disconnecting the ignition module's 2-pin connector and connect a 12-volt test light between the harness terminals (black/white wire is ground, the other is battery positive) (Figure 9–94). With the ignition on, the test light should light. If the light is off, verify the supplied ground by con-

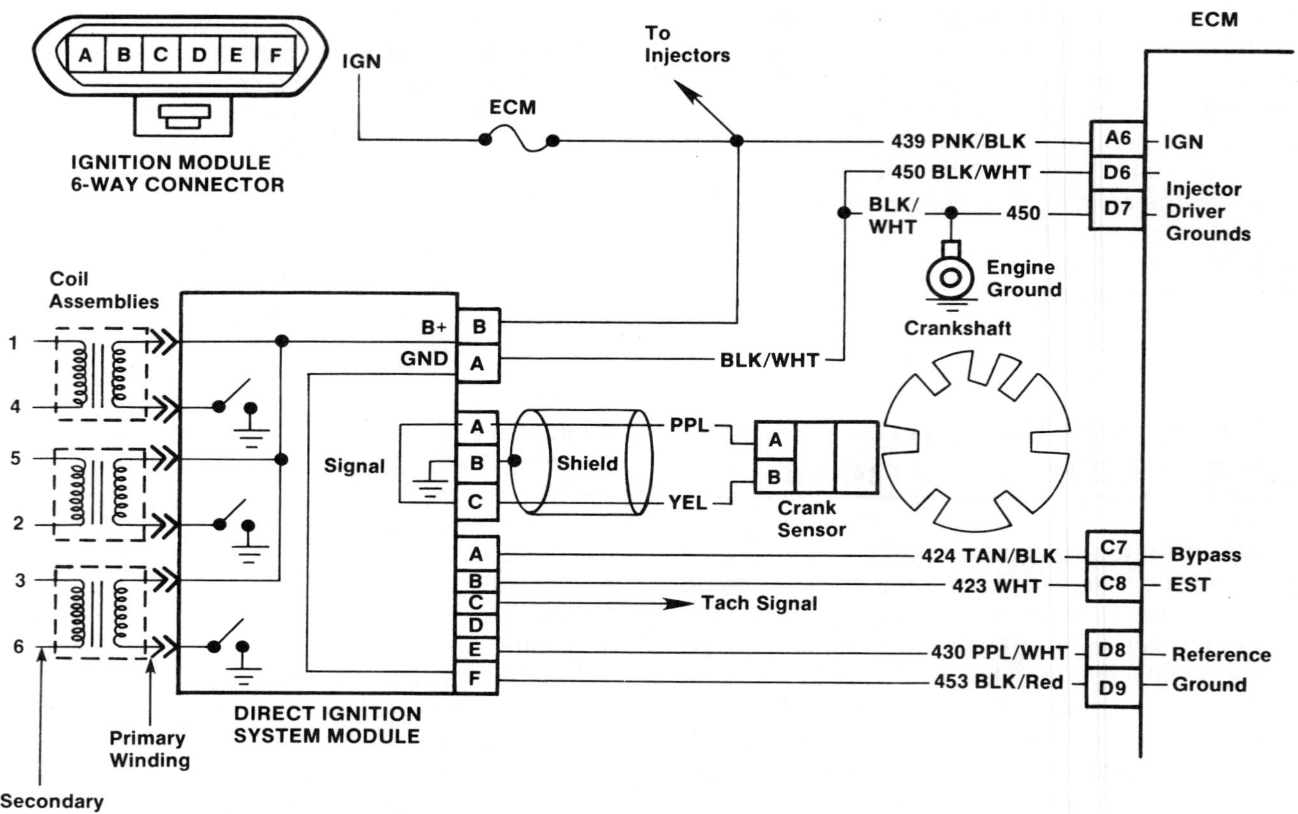

FIGURE 9–94 Electrical schematic of the DIS system on 2.8-liter engines

necting the test light between the "B+" of the connector terminal and a known good engine ground. If the light now glows, the module's ground circuit is open. If the alternate ground does not make the light glow, look for an open circuit on the ignition feed wire.

If battery voltage is at the 2-pin harness connector, check the crank sensor next. To determine the sensor's status, disconnect the sensor's three-way connector at the ignition module. With an ohmmeter set in the 2K ohms position, probe harness terminals A and C (the two outside terminals). The reading should be between 900 and 1200 ohms. A reading that is out of specs indicates one of three things: a short in the sensor leads (low resistance), an open sensor lead (high resistance), or a faulty sensor.

If the resistance is okay, set the VOM to the 2-volt AC position and observe the voltage reading across terminals A and C. A reading above 100 mV (0.1 V) means the sensor is okay and condemns the ignition module as the cause. A reading of less than 100 mV indicates a problem with either the sensor or harness connector.

The final problem that a technician could run into is a no-start situation where there is spark at both test locations, but no rpm reading. When this situation occurs, disconnect the six-way connector from the ignition module, turn the ignition on, and momentarily touch CKT 430 (purple/white wire) with a test light connected to 12 volts while the engine is being cranked. If an rpm reading is observed, the problem is either a bad six-way connection or a faulty ignition module. If the test light connection fails to produce an rpm reference, test CKT 430 for continuity. The presence of continuity points to a faulty ECM.

Because DIS is also part of the electronic engine control system, there are certain trouble codes the technician should become familiar with. For example, a code 42 means that the ECM has seen an open or short to ground in the electronic spark timing (EST) or bypass circuits. A code 43, on the other hand, relates to the electronic spark control (ESC) circuit. A code 43 logs anytime the signal wire from the knock sensor (CKT 496—dark blue/white wire) develops an open or short to ground. Since the diagnostic sequences that pertain to these codes are explained in detail in the service manual, it is suggested that the technician continue the diagnosis from there.

REVIEW QUESTIONS

1. Which of the following components is most likely to be found in a computer-controlled ignition system?
   a. centrifugal advance mechanism
   b. vacuum advance mechanism
   c. crank angle sensor
   d. all of the above

2. Which of the following would not be considered an input into a computer-controlled ignition system?
   a. carburetor idle switch
   b. air injection pump
   c. coolant temperature switch
   d. all of the above

3. Which of the following components provide the computer with information about the air/fuel mixture?
   a. oxygen sensor
   b. catalytic converter
   c. EGR valve
   d. all of the above

4. Which of the following components provides the computer with information about detonation?
   a. Hall-effect distributor
   b. coolant temperature switch
   c. knock sensor
   d. none of the above

5. Which of the following is not an input provided by the distributor in most computer-controlled ignition systems?
   a. engine speed
   b. manifold pressure
   c. crankshaft position
   d. all of the above

6. When a component of a computer-controlled ignition system necessary for operation fails, the computer goes into a ____________ mode of operation.
   a. limp-in
   b. open loop
   c. closed loop
   d. none of the above

7. The computer is operating in closed loop when it is receiving information from which of the following components?
   a. knock sensor
   b. EGR valve
   c. barometric pressure switch
   d. none of the above

8. In which of the following systems would you find a dual pick-up distributor?
   a. Chrysler electronic spark control system
   b. Ford electronic engine control system
   c. GM computer-controlled combustion system
   d. none of the above

9. The self-diagnostic feature of computer-controlled ignition systems displays ____________ to identify trouble area.
   a. trouble codes
   b. engine rpm
   c. torque specifications
   d. all of the above

10. Which of the following cannot be adjusted on most computer-controlled ignition systems?
    a. fast idle speed
    b. base timing
    c. advance timing
    d. all of the above

11. Memory codes are indications of ____________.
    a. hard failures
    b. intermittent failures
    c. manufacturer's specifications
    d. none of the above

12. Where is the best place to look for base timing and advance timing specifications?
    a. manufacturer's service manual
    b. aftermarket specification manuals
    c. vehicle emission control information decal
    d. none of the above

13. When a trouble code registers in the electronic control unit, what is the vehicle response?
    a. CHECK ENGINE light is lit.
    b. Problem is recorded in the ECU memory.
    c. The computer generates false signals to compensate for the failing component.
    d. All of the above.

14. Which of the following systems does not have a distributor?
    a. Ford's MCU system
    b. Chrysler's ESS system

c. GM's C3I system
d. All of the above

15. Which of the following is a characteristic of a distributorless ignition system?
a. waste spark
b. reversed polarity
c. high secondary voltage output
d. all of the above

16. An engine with Chrysler's ESC ignition system has a knocking problem. Technician A says the problem could be with the vacuum transducer. Technician B says the problem could be with the ignition timing. Who is correct?
a. Technician A
b. Technician B
c. Both A and B
d. Neither A nor B

17. A vehicle with an EEC ignition system is idling rough. Technician A says that the base timing should be adjusted. Technician B says that the TFI module may be defective. Who is correct?
a. Technician A
b. Technician B
c. Both A and B
d. Neither A nor B

18. A GM vehicle with a CCC ignition system is experiencing a driveability problem. Technician A thinks it is possible to use a scan tool to diagnose the problem. Technician B says that trouble codes can be accessed with just a jumper wire. Who is correct?
a. Technician A
b. Technician B
c. Both A and B
d. Neither A nor B

CHAPTER TEN

SERVICING THE FUEL SUPPLY SYSTEM

Objectives

Upon completion of this chapter, you should be able to

- Define the fuel delivery system components and their functions, including the fuel tank, fuel lines, filters, and pump.
- Conduct inspecting and servicing procedures for a fuel system.
- Define the components and their functions in an electronic engine control (EEC) fuel delivery system.

The automobile's fuel system is both simple and complicated: simple in the systems that transfer fuel to the engine and complex in the carburetor or fuel injector system that mixes that fuel with air in the correct amounts and proportion to meet all needs of the engine.

This chapter will focus on the simple part of the fuel system. It will cover the basic parts of the fuel delivery or transport system, including the fuel tank, lines, filters, and pump (Figure 10-1). Chapters 12 and 13 examine the more complex principles and functions of the fuel metering and atomization portion of the system. Remember, however, that the fuel transport portion of the system is very important in the operation of a gasoline driven vehicle. But to fully understand the automotive fuel system, a knowledge of the performance of petroleum-based material used to power the engine is necessary. Although there are several different types of fuels, gasoline is the most commonly used motor fuel in modern vehicles and is the only type covered in this chapter.

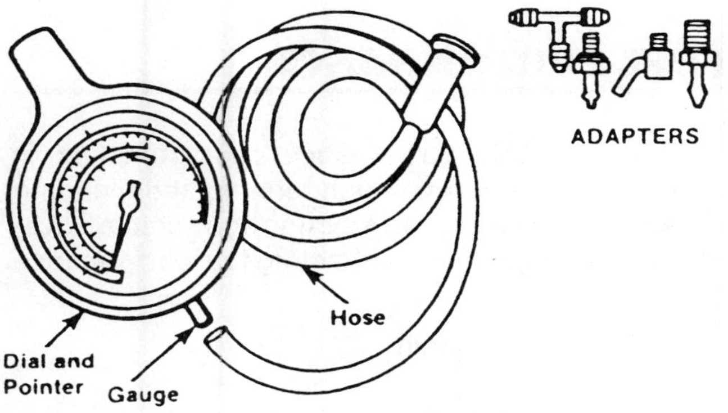

ADAPTERS

Hose

Dial and Pointer

Gauge

FIGURE 10-1 Typical components of a fuel system

WARNING: Extreme caution should be used while working with any of the components of the fuel system. Gasoline is a very volatile and flammable substance. Do not expose it to an open flame, spark, or high heat. Disconnect the negative terminal of the battery before doing any task that will release gasoline from any part of the system. Use containers to catch the gasoline and cloths to wipe up the minor spills. Use a flashlight rather than a trouble light or droplight. Gasoline spilled on a hot bulb could cause the bulb to explode and ignite the gasoline. Never attempt to weld a fuel tank. Even when a tank is empty of fuel, it can still contain enough gas fumes to cause an explosion if they come in contact with a flame or excessive heat. Always keep a Class B fire extinguisher nearby. It is specially intended for use on gasoline fires.

GASOLINE

Crude oil, as removed from the earth, is a mixture of hydrocarbon compounds ranging from gasses to heavy tars and waxes. The crude oil can be refined into products such as lubricating oils,

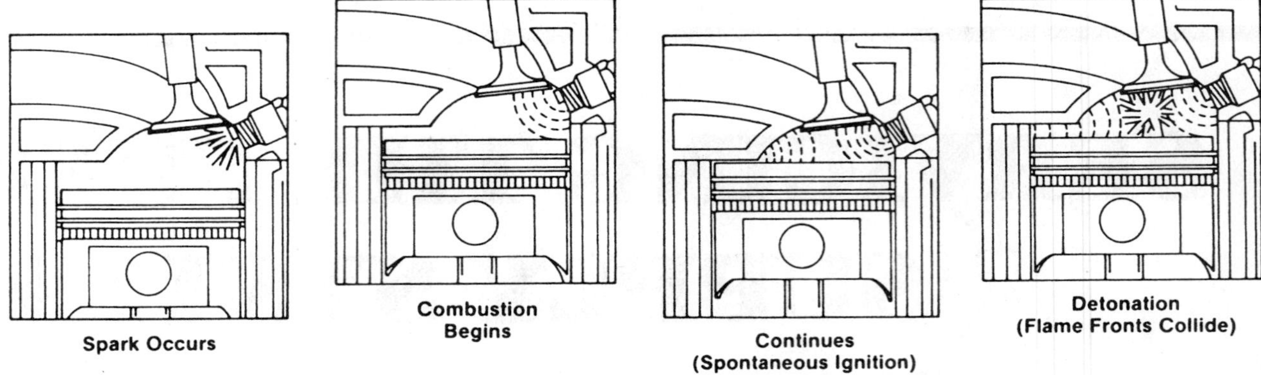

FIGURE 10-2 Normal combustion and detonation in a spark ignition engine

greases, asphalts, kerosene, diesel fuel, gasoline, and natural gas. Before its widespread use in the internal combustion engine of automobiles, gasoline was an unwanted byproduct of refining for oils and kerosene.

The automotive technician must remember that a good motor fuel must provide the following:

- Quick starts
- Fast warmup
- Rapid acceleration
- Smooth performance
- Minimum engine maintenance
- Good mileage under various driving conditions

Two important factors affect the power and efficiency of a gasoline engine.

1. **Compression Ratio.** The higher the compression ratio, the greater the power output and efficiency. The better the efficiency, the less fuel consumed to produce a given power output. To have a high compression ratio requires an engine of greater structural integrity in pistons, crankshafts, connecting rods, and heads. The 1915 Ford Model T had a low compression ratio of 3.6 to 1. As improvements were made to the engines, compression ratios slowly increased. A 1927 Model A was 4.5 to 1; and in the late 1960s it had approached 12.5 to 1 on some optional high-performance engines. Due to the use of low-octane unleaded gasoline in post-1975 models, the compression ratio generally ranges from 8.2:1 to 10.2:0.
2. **Detonation (Abnormal Combustion).** Normal combustion occurs gradually in each cylinder. The flame front (the edge of the burning area) advances smoothly across the combustion chamber until all the air/fuel mixture has been burned (Figure 10-2). Detonation occurs when the flame front fails to reach a pocket of mixture before the temperature in that area reaches the point of self-ignition. Normal burning at the start of the combustion cycle raises the temperature and pressure of everything in the cylinder. The last part of the air/fuel mixture will be both heated and pressurized, and the combination of those two factors can raise it to the self-ignition point. At that moment, the remaining mixture burns almost instantaneously. The two flame fronts create a destructive pressure wave between them that can destroy cylinder head gaskets, break piston rings, and burn pistons and exhaust valves. When detonating occurs, a hammering, pinging, or knocking sound is heard. But, when the engine is operating at high speed, these sounds cannot be heard because of motor and road noise.

FUEL PERFORMANCE

Many of the performance characteristics of gasoline can be controlled in refining and blending to provide proper engine function and vehicle driveability. The major factors affecting fuel performance are:

- Antiknock quality
- Volatility
- Sulfur content
- Deposit control

FIGURE 10-3 The octane rating or number as it appears on the gasoline pump

ANTIKNOCK QUALITY

An octane number or rating (Figure 10-3) was developed by the petroleum industry so the antiknock quality of a gasoline could be rated. It is a measure of burning quality; the tendency not to produce knock in an engine. The higher the octane number, the lesser the tendency to knock. By itself, the antiknock rating has nothing to do with the fuel economy or efficiency of an engine.

Two commonly used methods for determining the octane number of motor gasoline are the motor octane number (MON) method and the research octane number (RON) method. Both use a laboratory single-cylinder engine equipped with a variable head and knock meter to indicate knock intensity. The test sample is used as fuel, and the engine compression ratio and air/fuel mixture are adjusted to develop a specified knock intensity. There are two primary standard reference fuels, isooctane and heptane. Isooctane will not knock in an engine, but is not used in gasoline because of its expense. Heptane knocks severely in an engine. Isooctane has an octane number of 100; heptane has an octane number of zero.

A fuel of unknown octane value is run in the special test engine and the severity of knock is measured. Then various proportions of heptane and isooctane are run in the test engine to duplicate the severity of the knock of the fuel being tested. When the knock caused by the heptane/isooctane mix is identical to the test fuel, the octane number is established by the percentage of isooctane in the mixture. For example, if 85 percent isooctane and 15 percent heptane produce the same severity of knock as the fuel in question, the fuel is assigned an octane number of 85.

- *Lean Fuel Mixture.* A lean mixture burns slower than a rich mixture. The heat of combustion is higher, which promotes the tendency for unburned fuel in front of the spark-ignited flame to detonate.
- *Ignition Timing Overadvance.* Advancing the ignition timing induces knock. Slowing ignition timing suppresses knock because lower cylinder pressures result when the piston is on its way down as combustion nears completion.
- *Compression Ratio.* Compression ratio affects knock because cylinder pressures are increased with the increase in compression ratio. The higher the compression ratio, the more sensitive an engine is to knock.
- *Valve Timing.* The timing of intake and exhaust valves affect the cylinder pressure and knock. A valve timing that fills the cylinder with more air/fuel mixture promotes higher cylinder pressures, increasing the chances for detonation.
- *Turbocharging.* Turbocharging or supercharging forces additional fuel and air into the cylinder. This induces higher cylinder pressures and promotes knock. It also raises the temperature of the air coming in.
- *Coolant Temperature.* Hotspots in the cylinder or combustion chamber because of inefficient cooling or a damaged cooling system raise combustion chamber temperatures and promote knock.
- *Cylinder-to-Cylinder Distribution.* If an engine has poor distribution of the air/fuel mixture from cylinder to cylinder, the leaner cylinders could promote knock. The distributor spark advance curve must be set for the advance required for the leanest.
- *Excessive Carbon Deposits.* The accumulation of carbon deposits on the piston, valves, and combustion chamber causes poor heat transfer from the combustion chamber. Carbon accumulation also artificially increases the compression ratio. Both conditions cause knock.
- *Air Inlet Temperature.* The higher the air temperature when it enters the cylinder, the greater the tendency to knock. The higher air temperature increases the temperature of combustion.
- *Combustion Chamber Shape.* The optimum combustion chamber shape for reduced

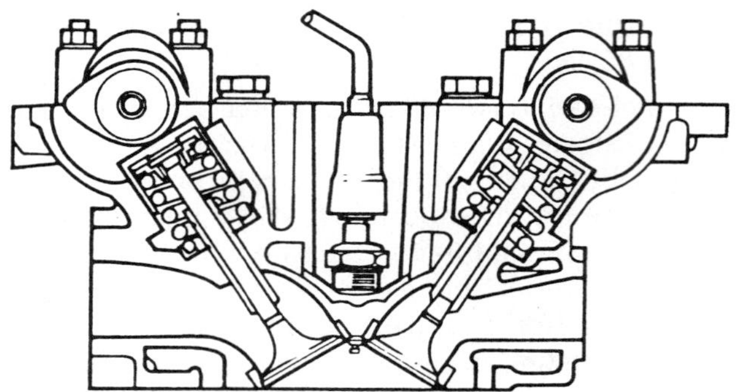

FIGURE 10-4 Cross section of a hemispherical combustion chamber

knocking is hemispherical with a spark plug located in the center (Figure 10-4). This hemi-head allows for faster combustion, allowing less time for detonation to occur ahead of the flame front.

- *Octane Number.* Recommendations about the best octane fuel to use in any given engine can only be approximate. Many factors determine octane ratings of fuel and octane requirements of engines. It is important to keep in mind that little is gained or lost by using gasoline of higher octane quality than required by the engine under a complete range of operating conditions. Only when an engine is designed and adjusted to take advantage of the higher octane gasoline can the fuel value be obtained. Most modern engines are designed to operate efficiently with regular grade gasoline and do not require high-octane premium grade (Table 10-1). Those designs will provide better fuel economy and performance when converted to a higher-octane gasoline.

VOLATILITY

As stated earlier, gasoline is very volatile. Actually it is the ability of gasoline to evaporate so its vapor will adequately mix with air for combustion. Only vaporized fuel will support combustion. To insure complete combustion, complete vaporization must occur.

The volatility of gasoline affects the following performance characteristics or driving conditions:

- *Cold Starting and Warmup.* Vaporization characteristics affect cold starting and warmup. A fuel can cause hard starting, hesitation, and stumbling during warmup if it does not vaporize readily. A fuel that vaporizes too easily in hot weather can form vapor bubbles in the fuel line and fuel pump, resulting in vapor lock or loss of performance. The volatility of a fuel must be seasonally blended at the refinery to insure proper operation of the engine.
- *Temperature.* Because a highly volatile fuel vaporizes at a lower temperature than a less volatile fuel, winter grade gasoline is more volatile than summer grade gasoline. Winter grade fuels, however, give poorer fuel economy than low volatility fuels. Seasonal blending provides a more volatile fuel for the northern states during the winter months than a fuel sold in southern states.

TABLE 10-1: PROPERTIES OF EXXON® GASOLINES

	EXXON 2000 (UNLEADED)	EXXON (LEADED)	EXXON EXTRA UNLEADED
Research octane number[a]	93	94	97
Motor octane number[a]	84	84	87
TEL content	Nil[b]	Approx. 2 grams/gal	Nil[b]
Approximate compression ratio served	8.2:1	8.2:1	8.2:1 and higher
Carburetor icing control		Adequate	
Volatility adjustments		Continually with season for each market area	
Detergent		For required carburetor deposit control	

[a] Lower for higher-altitude market areas.
[b] Trace lead contents are due to pickup in the distribution system.

- *Altitude.* Gasoline vaporizes more easily at high altitudes, so volatility is controlled in blending according to the elevation of the place where fuel is sold.
- *Carburetor Icing Protection.* Although carburetor icing is not as common in modern engines as in those of the past, it is most likely to occur at ambient temperatures between 28 to 55 degrees Fahrenheit when the relative humidity rises above 65 percent. The humid air enters the carburetor and mixes with drops of fuel. When the fuel evaporates, it removes heat from the air and surrounding metal parts. When this occurs, the throttle temperature is rapidly lowered to below 32 degrees Fahrenheit (if the ambient temperature is within the range indicated), and condensing water vapor forms ice. The ice will cause the engine to stall if it is idling during this phase. Gasoline with higher mid-volatility (that is, a higher percentage of evaporation at 212 degrees Fahrenheit) creates greater evaporative cooling. Therefore, the percentage of gasoline evaporated at 212 degrees Fahrenheit is the significant factor in minimizing carburetor icing.
- *Crankcase Oil Dilution.* A fuel must vaporize well to prevent diluting the crankcase oil with liquid fuel. If parts of the gasoline do not vaporize, droplets of liquid break down the oil film on the cylinder wall, causing scuffing or scoring. The liquid eventually enters the crankcase oil and results in the formation of sludge, gum, and varnish accumulation as well as decreasing the lubrication properties of the oil.
- *Driveability.* Poor vaporization can also affect the distribution of fuel from cylinder to cylinder since vaporized fuel will travel farther and faster in the manifold. This results in driveability problems and increased exhaust emissions.

SULFUR CONTENT

Gasoline can contain some of the sulfur present in the crude oil. Sulfur content is reduced at the refinery to limit the amount of corrosion of the engine and exhaust system.

When the hydrogen in the hydrocarbon of the fuel is burned with air, one of the products of combustion is water. Water leaves the combustion chamber as steam but can condense back to water when passing through a cool exhaust system. When the engine is shut off and cools, steam condenses back to a liquid and forms water droplets in the exhaust system and on engine parts. Steam present in crankcase blowby also condenses to water.

When the sulfur in the fuel is burned, it combines with oxygen to form sulfur dioxide. This sulfur dioxide can then combine with water to form highly corrosive sulfuric acid. This corrosion is the leading cause of exhaust valve pitting and exhaust system deterioration. With catalysts, the sulfur dioxide can cause the obnoxious odor of rotten eggs during vehicle warmup. To reduce corrosion caused by sulfuric acid, the sulfur content in gasoline is limited to less than 0.01 percent.

DEPOSIT CONTROL

Several additives are put in gasoline to control harmful deposits, including gum or oxidation inhibitors, detergents, metal deactivators, and rust inhibitors.

BASIC FUELS

Two basic motor fuels are used in modern vehicles: leaded and unleaded gasolines.

LEADED GASOLINE

The two most commonly used additives to regular and premium lead fuels are tetraethyl lead (TEL) and tetramethyl lead (TML). These additives are used to improve the antiknock quality. For example, if 3 grams of TEL are added per gallon of gasoline, the octane rating will increase by up to 15 numbers. Two factors determine the amount of TEL used: the response of the gasoline base stock to TEL and the economics of obtaining octane by refining processes versus the cost of TEL. Currently, there are federal restrictions that limit TEL or TML content to 3.5 grams of lead per gallon. Scavengers are often used with the TEL and TML additives to promote the removal of lead salts formed in the combustion chamber after combustion.

UNLEADED GASOLINE

Unleaded gasolines are the key to emission control. Since 1975, as previously stated, most domestic cars have used catalytic converters to treat the exhaust of the engine. Because of the deactivating or poisoning effect lead has on the catalyst, these gasolines are limited to a lead content of 0.06 grams per gallon. Since TEL or TML is not added to unleaded gasolines, the required octane number is obtained

FIGURE 10-5 Warning that appears on fuel gauge and fuel filter

by blending compounds of the required octane quality. Methylcylopentadienyl manganese tricarbonyl (MMT) is a catalyst-compatible octane improver. It may also be used to further improve the octane number of unleaded gasolines.

Vehicles with catalytic converters are labeled at both the fuel gauge and fuel filler UNLEADED FUEL ONLY (Figure 10-5).

FUEL DELIVERY SYSTEM

The components of a typical fuel delivery system (Figure 10-6) are as follows:

- Fuel tank
- Fuel lines
- Fuel filter
- Fuel pump

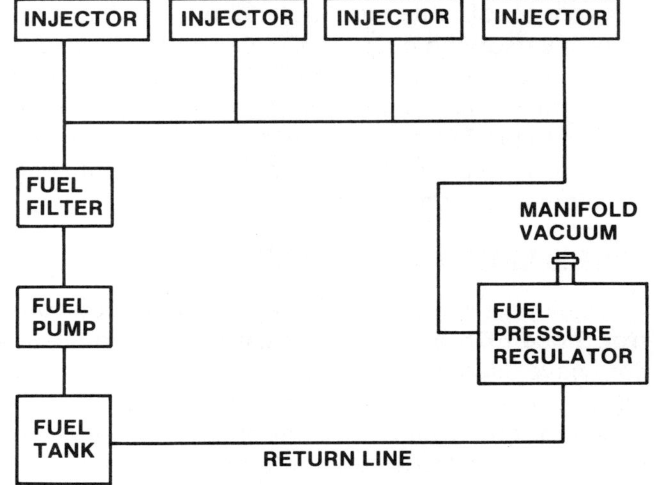

FIGURE 10-6 Typical fuel delivery system

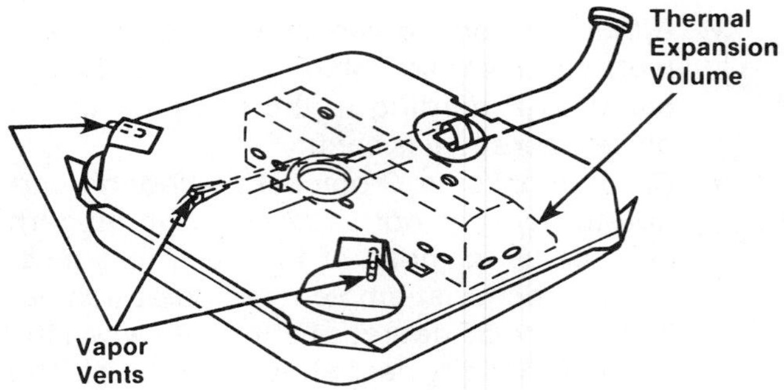

FIGURE 10-7 Fuel tank with an internal expansion tank to allow for changes in fuel volume due to changes in temperature.

FUEL TANKS

Modern fuel tanks have several features that differ from those of tanks in older cars. On early model cars, gas tanks were vented, meaning vapors escaped into the atmosphere, leading to both fire dangers and pollution. Modern fuel tanks include devices that prevent vapors from leaving the tank. For example, to contain vapors and allow for expansion, contraction, and overflow that results from changes in the temperature, the fuel tank has a separate air chamber dome at the top. Another way to contain vapors is to use a separate internal expansion tank within the main tank (Figure 10-7). All fuel tank designs provide some control of fuel height when the tank is filled. Frequently, this is achieved by using vent lines within the filter tube or tank (Figure 10-8). With tank designs such as this, only 90 percent of the tank is ever filled, leaving 10 percent for expansion in hot weather. Some vehicles have an overfill limiting valve to prevent overfilling of the tank. If a tank is filled to capacity, it will overflow whenever the temperature of the fuel increases.

From the fuel tank, the vapors travel through a line, or continuous tube, to a charcoal-filled canis-

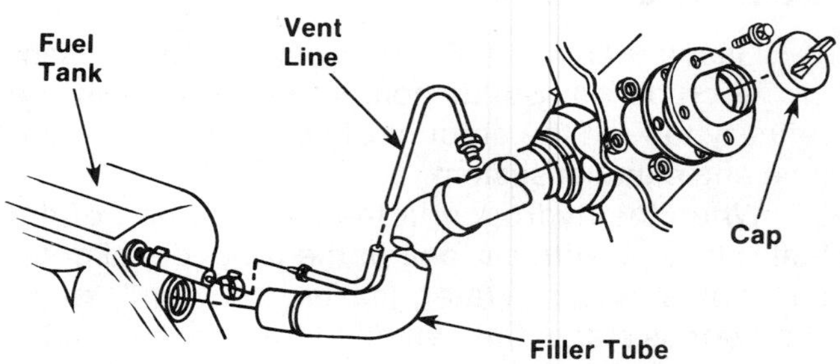

FIGURE 10-8 Ford EEC fill control vent system

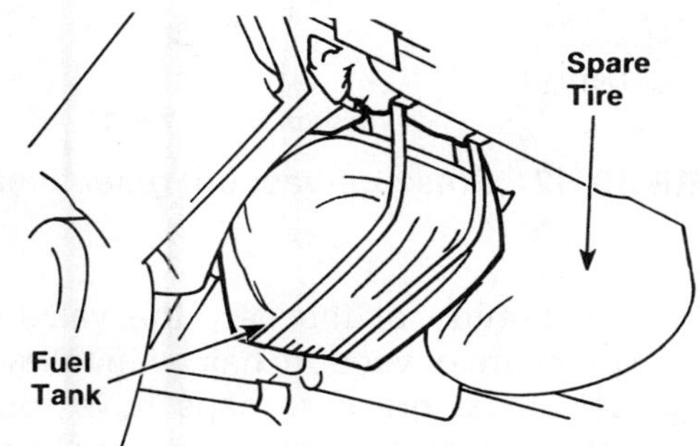

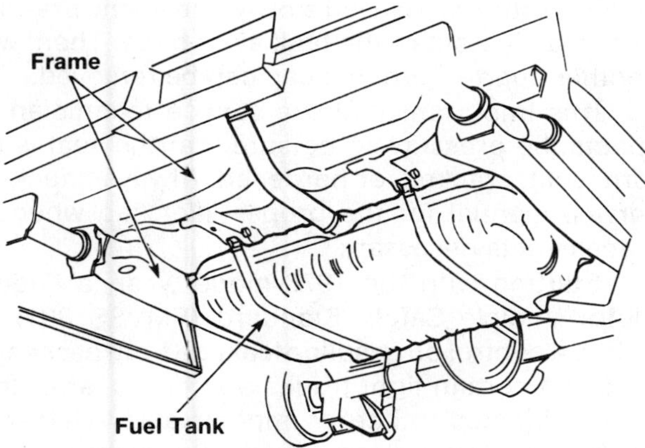

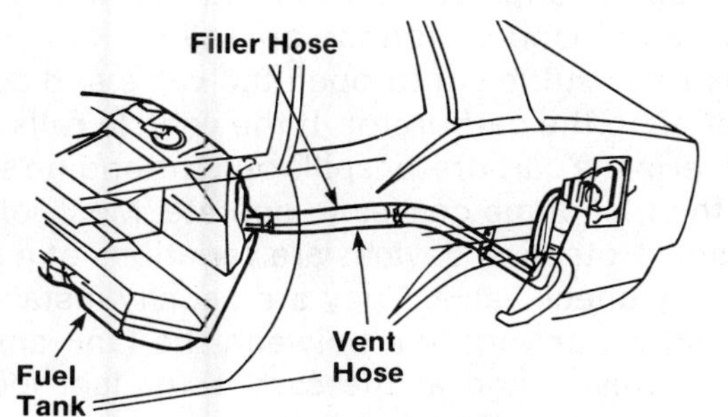

FIGURE 10-9 Gasoline tanks can be located almost anywhere that is safe from damage.

ter. When the engine starts, fresh air is drawn through the charcoal canister where it mixes with the vapors from the fuel tank. From there, these vapors are drawn through the carburetor or injectors. The vapors then pass through the intake manifold and into the cylinders, where they are burned along with the normal air/fuel mixture. For more information on evaporative emission control, see Chapter 14.

The fuel tank is usually made of pressed corrosion-resistant steel halves, which are ribbed for additional strength. Often, exposed portions of the tank are made of heavy-gauge steel for protection from road damage. The seams of the tank are welded. In an attempt by vehicle manufacturers to reduce overall car weight, some are now using aluminum or molded reinforced polyethylene fuel tanks.

Most metal tanks have slosh baffles or surge plates to prevent the fuel from splashing around inside the unit. In addition to slowing down fuel movement, the plates tend to keep the fuel pickup or sending assembly immersed in the fuel during hard braking and acceleration. The sheet metal plates or baffles are usually welded in place inside the tank and have holes or slots in them to permit the fuel to move from one end of the tank to the other.

Except for rear engine vehicles, the fuel tank in a passenger car is located in the rear of the vehicle for improved safety. Figure 10-9 illustrates the more common locations. To increase the driving range of a vehicle, some cars install an auxiliary or saddle tank, generally in the trunk. But when making such an installation, extreme care must be taken to be sure it is tightly secured in a safe location and that all emission regulations are met. Some states have laws against the use of auxiliary fuel tanks.

The fuel tank is provided with an inlet filler tube and cap. The location of the fuel inlet filler tube depends on the tank design and tube placement. It is usually positioned behind the filler cap or a hinged door in the center of the rear panel or in the outer side of either rear fender panel. Vehicles designed for unleaded fuel use have a restrictor in the filler tube that prevents the entry of the larger leaded fuel delivery nozzle at the gas pumps (Figure 10-10). The filler pipe can be a rigid one-piece tube soldered to the tank (Figure 10-11A) or a three-piece unit (Figure 10-11B). The three-piece unit has a lower neck soldered to the tank and an upper neck fastened to the inside of the body sheet metal panel.

From 1971 on, filler tube caps are the nonventing type and usually have some type of pressure-vacuum relief valve arrangement (Figure 10-12). Under normal operating conditions the valve is closed, but whenever pressure or vacuum is more

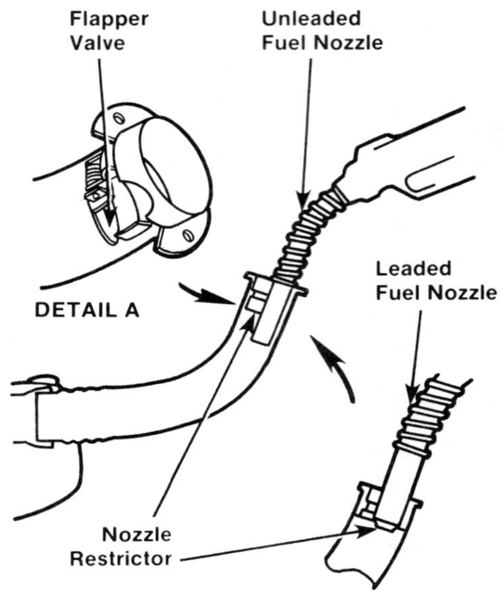

FIGURE 10-10 Cars that require unleaded fuel have restrictors in the filler tubes to allow only the entry of smaller unleaded fuel pump nozzles

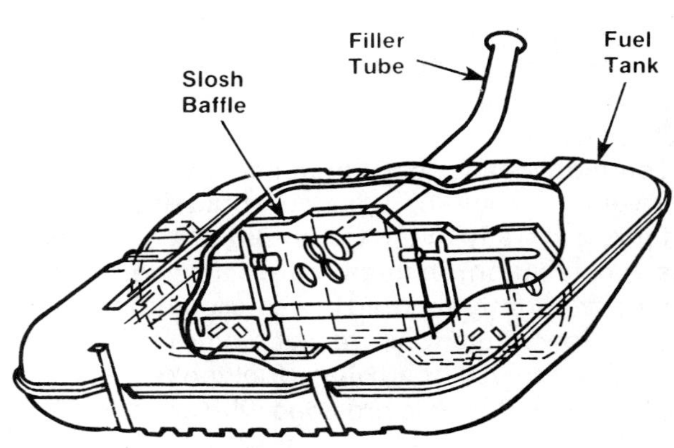

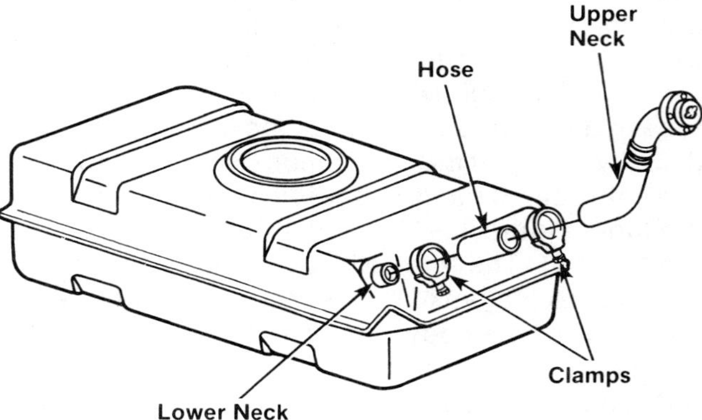

FIGURE 10-11 Filler tubes: (A) one piece; (B) three piece

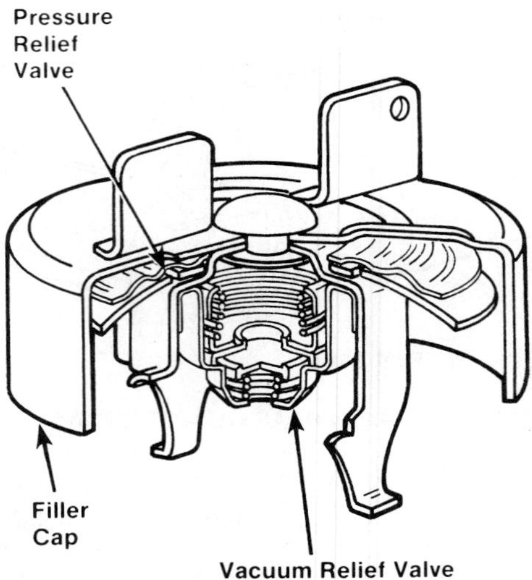

FIGURE 10-12 Pressure-vacuum relief filler cap

than the calibration of the cap, the valve opens. Once the pressure or vacuum has been relieved, the valve closes. Most pressure caps have four anti-surge tangs that lock onto the filler neck to prevent the delivery system's pressure from pushing fuel out of the tank. By turning such a cap one-half turn, the tank pressure is released slowly and keeps pressure from blasting out of the tank all at once. Then, with another quarter turn, the cap can be removed.

It is important that the service technician inspects the pressure to be sure that the seal is not torn, split, cracked, or hardened. Always check the service manual for the proper filler cap when replacement is necessary.

Starting with the 1976 model year, a Federal Motor Vehicle Safety Standard (FMVSS 301) required a control on gasoline leakage from passenger cars and certain light trucks and buses, after they were subjected to barrier impacts and rolled over. Tests conducted under these severe conditions showed the most common gasoline leak path to be the gasoline supply line from the fuel tank to the carburetor. Under normal operation, fuel pump pressure is sufficient to open the valve and supply gasoline to the carburetor. If the vehicle rolls over, the fuel in the carburetor spills out, the engine stops, and the fuel pump ceases to operate. Most rollover leakage protection devices are variations of a basic one-way check valve. They are usually installed in the fuel vapor vent line between the tank and the vapor canister and at the carburetor fuel feed or return line fitting (Figure 10-13). In some protection systems the check valve is incorporated into the carburetor inlet fuel filter (Figure 10-14). A check valve might also be fitted in the fuel tank filler cap,

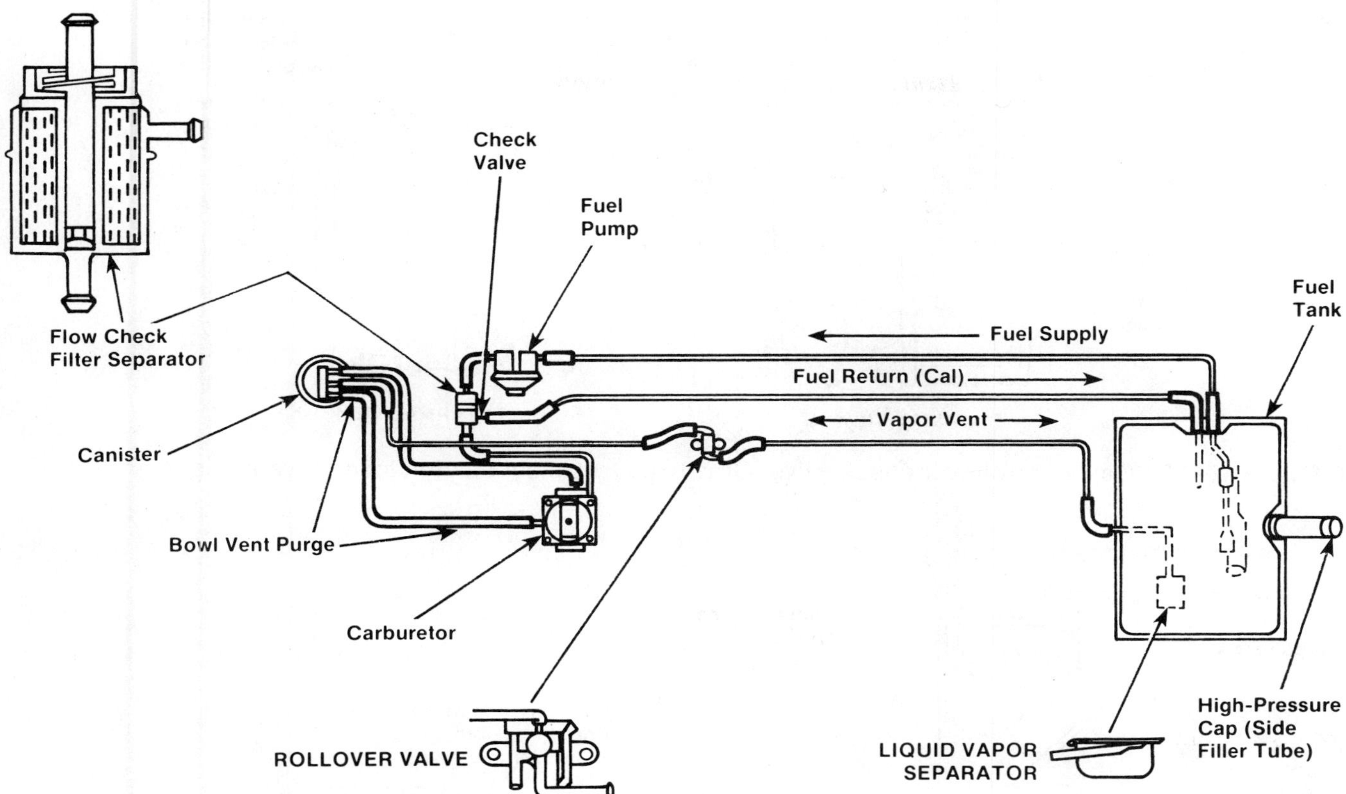

FIGURE 10-13 Rollover leakage protection devices

and most caps' pressure-vacuum relief valve settings have been increased so that fuel pressure cannot open them in a rollover.

Electric fuel pumps found on Ford vehicles and vehicles with a Bosch fuel injection system have an inertia switch that will shut off the pump if the vehicle is involved in a collision or rolls over. The Ford inertia switch (Figure 10-15) consists of a permanent magnet, a steel ball inside a conical ramp, a target plate, and a set of electrical contacts. The magnet holds the steel ball in the bottom of the conical ramp. In the event of a collision, the inertia of the ball will cause it to break away from the magnet, rollup the conical ramp, and strike the target plate. The force of the ball striking the plate will cause the electrical contacts in the inertia switch to open, cutting off power to the fuel pump. The switch has a reset button that must be depressed to close the contacts before the pump will operate again.

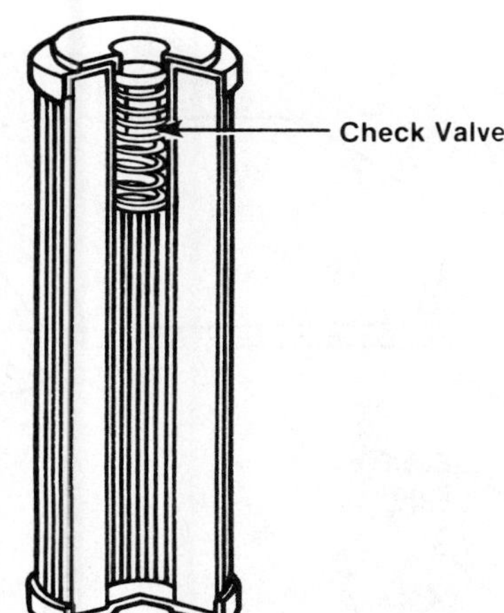

FIGURE 10-14 Gasoline filter with check valve

Some fuel line systems contain a fuel return arrangement which aids in keeping the gasoline cool, thus reducing chances of vapor lock. The return system (Figure 10-16) consists of a special fuel pump equipped with an extra outlet fitting and necessary fuel line. The fuel return line generally runs next to the conventional fuel line (Figure 10-17), except that flow is in the opposite direction. That is, the flow is through the return line toward the fuel tank, not toward the engine. The fuel return system allows a metered amount of cool fuel to circulate through the tank and fuel pump, thus reducing vapor bubbles caused by overheated fuel from upsetting the engine's operation by changing its fuel mixture.

Some form of liquid vapor separator is incorporated into most modern vehicles to stop liquid fuel or

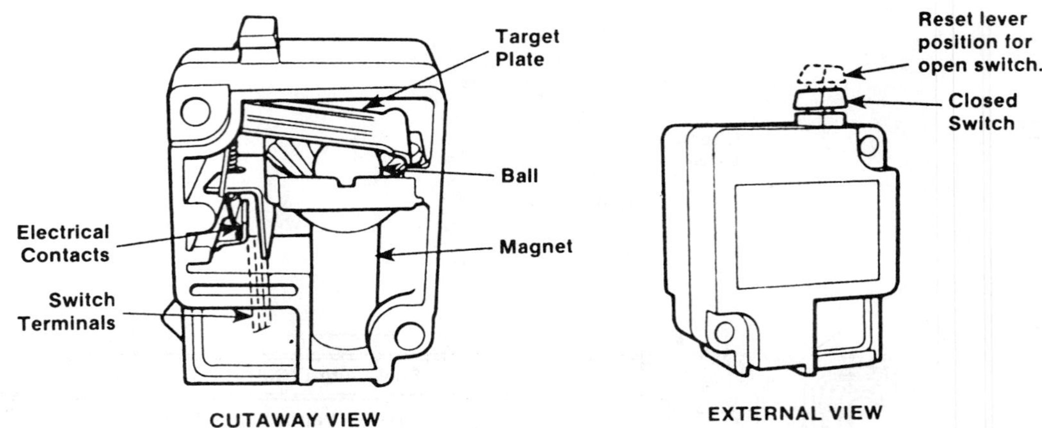

FIGURE 10-15 Cap with rollover check valve. Gravity causes steel ball to close off vent in cap.

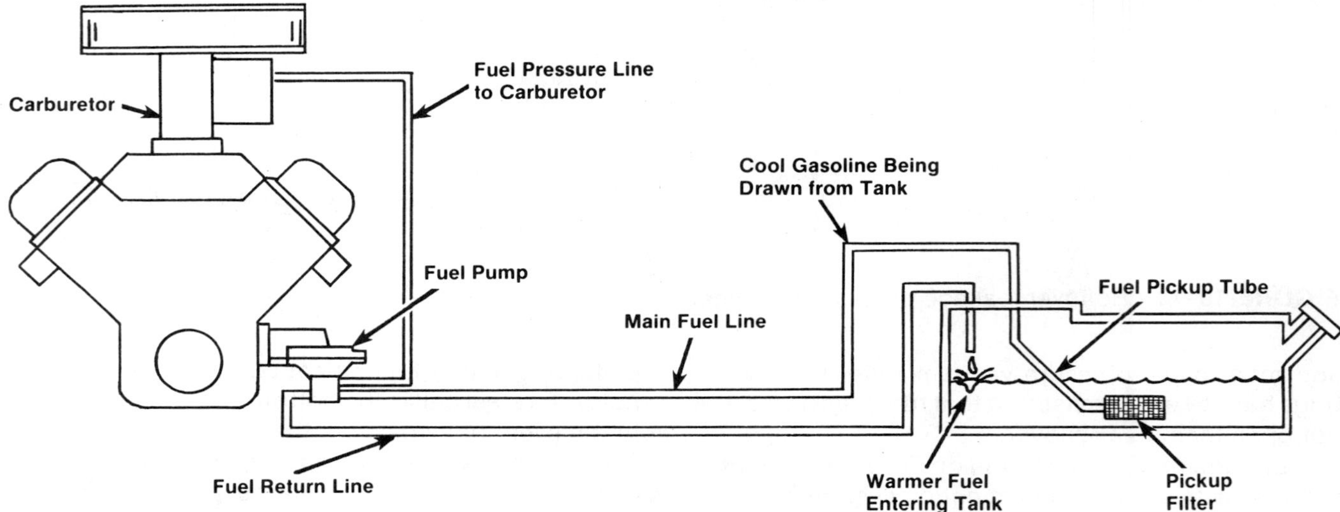

FIGURE 10-16 Gasoline is drawn from the tank into the fuel pump. From the pump, gasoline flows to the carburetor as well as back to the fuel tank. This tends to cool the gasoline in the pump and reduces the change of vapor lock.

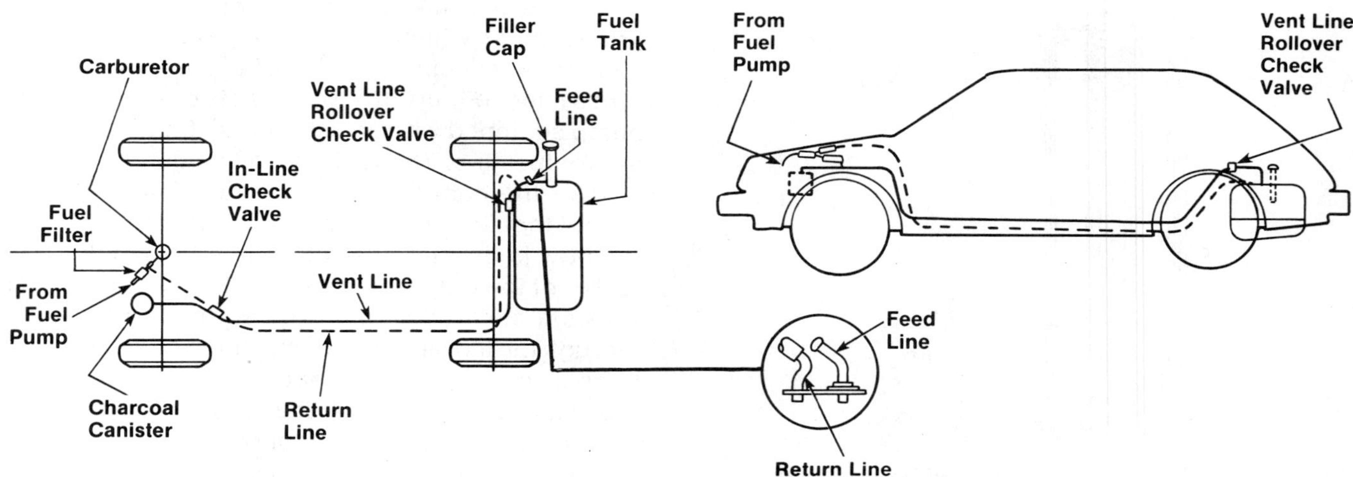

FIGURE 10-17 Parallel routing of fuel return line and vapor line

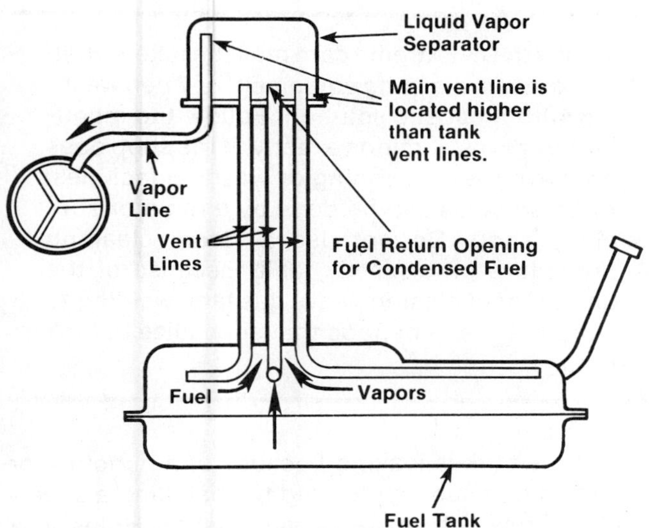

FIGURE 10-18 Fuel tank vapor separator allows some of the fuel vapors to condense back into liquid and return to the tank. Only vapors can normally enter the higher main vent tube opening.

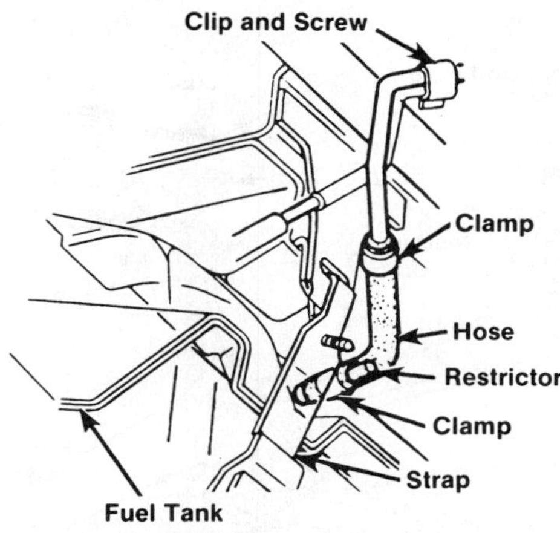

FIGURE 10-20 Vapor separator in fuel vent lines

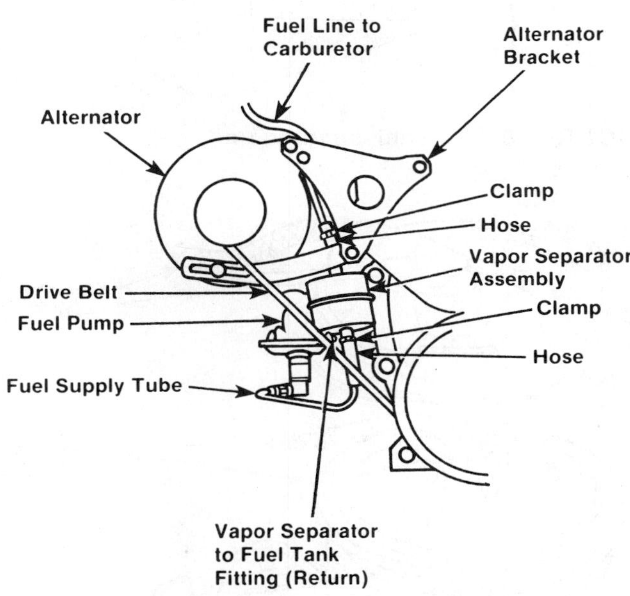

FIGURE 10-21 Vapor separator located near the fuel pump

bubbles from reaching the vapor storage canister or the engine crankcase (Figure 10-18). It can be located inside the tank, on the tank (Figure 10-19), or in fuel vent lines (Figure 10-20). It can also be located near the fuel pump (Figure 10-21). Check the service manual for the exact location of the liquid vapor separator and line routing.

Inside the fuel there is also a sending unit which includes a pickup tube and float-operated fuel gauge (Figure 10-22). The fuel tank pickup tube is connected to the fuel pump by the fuel line. Some electric fuel pumps are combined with the sending unit (Figure 10-23). The pickup tube extends nearly, but not completely, all the way to the bottom of the tank. Rust, dirt, sediment, and water cannot be drawn up into the fuel tank filter, which can cause clogging. The ground wire is often attached to the fuel tank unit.

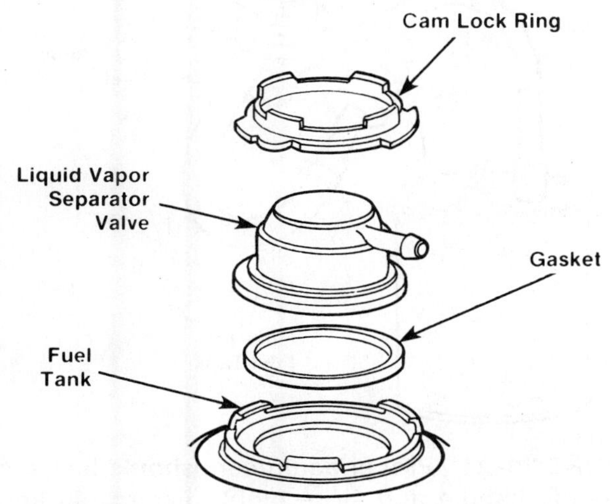

FIGURE 10-19 Vapor separator attached to the fuel tank

Servicing the Fuel Tank

Leaks in the metal fuel tank can be caused by a weak seam, rust, or road damage. The best method of permanently solving this problem is to replace the tank. Another method is to remove the tank and steam clean or boil in a caustic solution to remove the gasoline residue. After this has been done, the leak can be soldered or brazed.

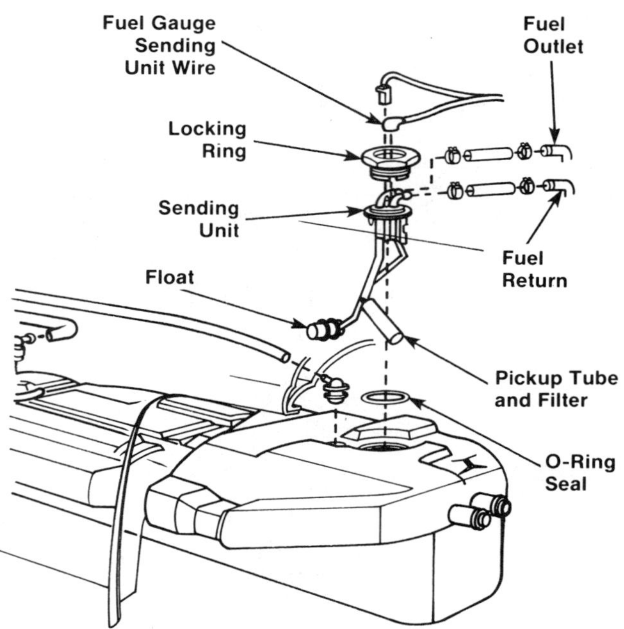

FIGURE 10-22 Tank sending unit

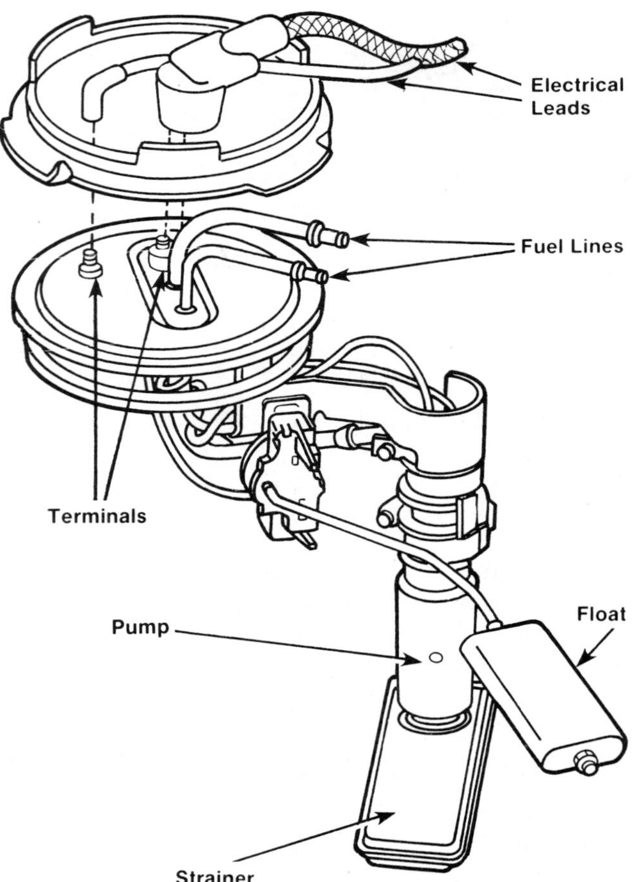

FIGURE 10-23 Electric fuel pump located inside a fuel tank

WARNING: Extreme care must be taken when using steam cleaning equipment or when washing with a caustic solution. Follow the manufacturer's instructions exactly. If the shop does not have steam cleaning or washing facilities, these services can be done by a radiator specialty shop. Do not use a steam cleaning procedure on a plastic tank. Because of the danger of explosion, leave gas tank welding to experts in repair shops that specialize in tank repair.

If the tank is leaking because of a puncture or small hole, it can be plugged by installing a sheet-metal screw with a neoprene washer. Holes in a plastic tank can sometimes be repaired by using a special tank repair kit. Be sure to follow manufacturer's instructions when doing the repair.

Replacing the Fuel Tank. When a fuel tank is leaking dirty water or has water in it, the tank must be cleaned, repaired, or replaced. To do this, disconnect the negative terminal from the battery, remove the tank filler cap, and drain the tank of all fuel. Then proceed as follows:

1. Disconnect the fuel line at the tank that runs to the fuel pump. Then connect a siphon hose or similar device to the tank and draw the fuel into a clean, approved safety can (Figure 10-24). Operate the siphon device as directed by its manufacturer.

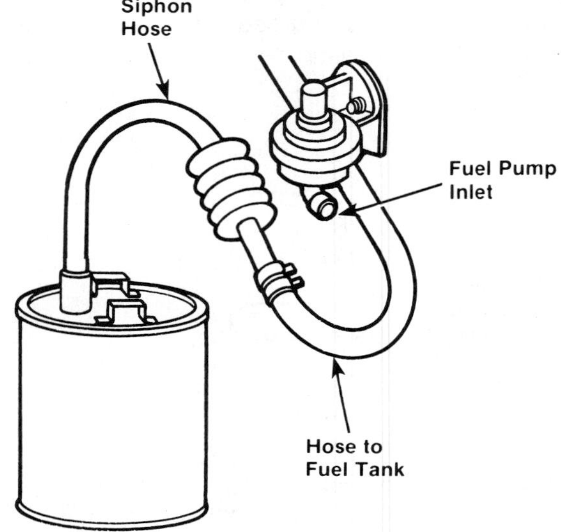

FIGURE 10-24 Proper equipment should be used to safely handle and store highly flammable and explosive gasoline.

CAUTION: Abide by local laws for the disposal of contaminated fuels. Also be sure to wear eye protection when working under the vehicle.

2. Attach a piece of masking tape identifying tag to each tank line to insure correct reinstallation. Disconnect the vent lines and the plugged wire connected to the sending unit. This wire is usually mounted on the upper front or top section of the fuel tank.
3. Unfasten the filler from the tank. If it is a rigid one-piece tube, remove the screws around the outside of the filler neck near the filler cap. If it is a three-piece unit, remove the neoprene hoses after the clamp has been loosened.
4. Loosen the bolts holding the fuel tank straps to the vehicle (Figure 10–25) until they are about two threads from the end. Holding the tank securely against the underchassis with one hand, remove the strap bolts and lower the tank to the ground. When lowering the tank, make sure all wires and tubes are unhooked. Also keep in mind that small amounts of fuel might still remain in the tank.

CAUTION: Use a drain pan to catch any spilled fuel or be sure to clean it up immediately.

5. To reinstall the new or repaired fuel tank, reverse the removal procedure. Be sure that all the rubber or felt tank insulators are in place. Then, with the tank straps in place, position the tank. Loosely fit the tank straps around the tank, but do not tighten them. Make sure that the hoses, wires, and vent tubes are connected properly. Also check the filler neck for alignment and for insertion into the tank. Tighten the strap bolts and secure the tank to the car. Install all of the tank accessories (vent line, sending unit wires, ground wire, filler tube, and so on). Fill the tank with fuel and check it for leaks, especially around the filler neck and the pick-up assembly. Reconnect the battery and check the fuel gauge for proper operation.

Removing and Replacing the Fuel Gauge Sending Unit. The sending unit is held in the tank by either a retaining ring or screws. The easiest way

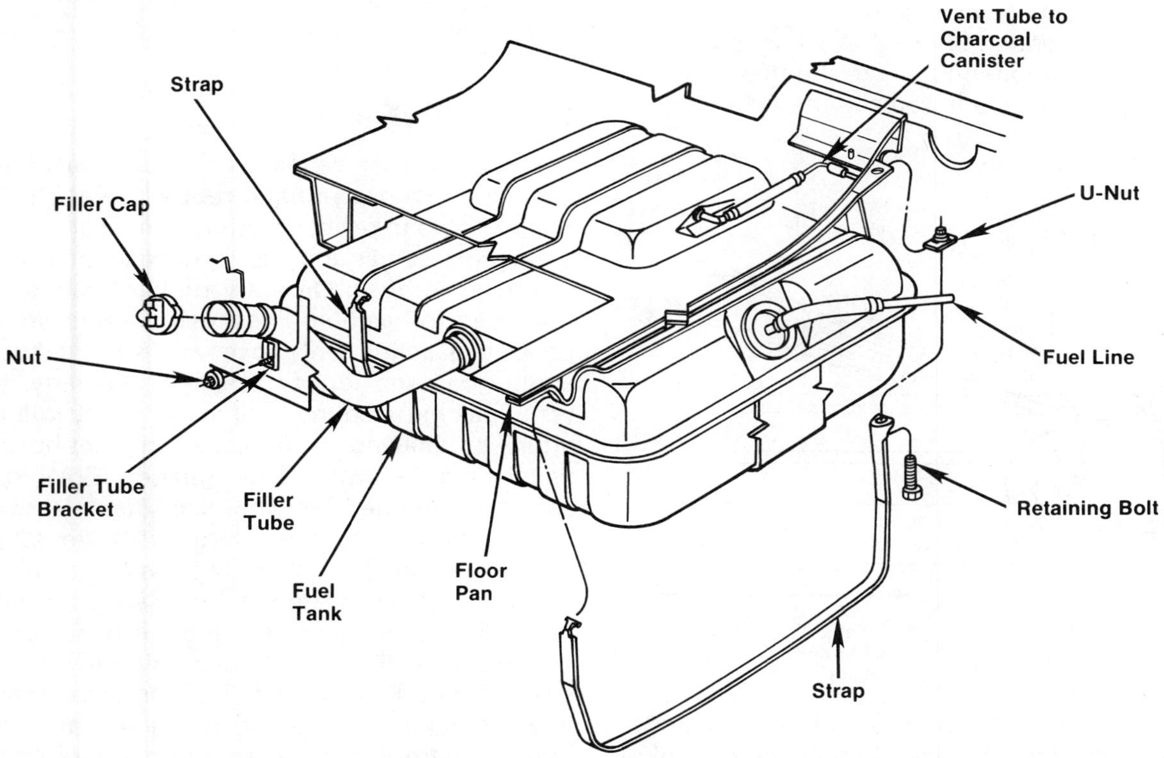

FIGURE 10–25 Fuel tank and mounting components

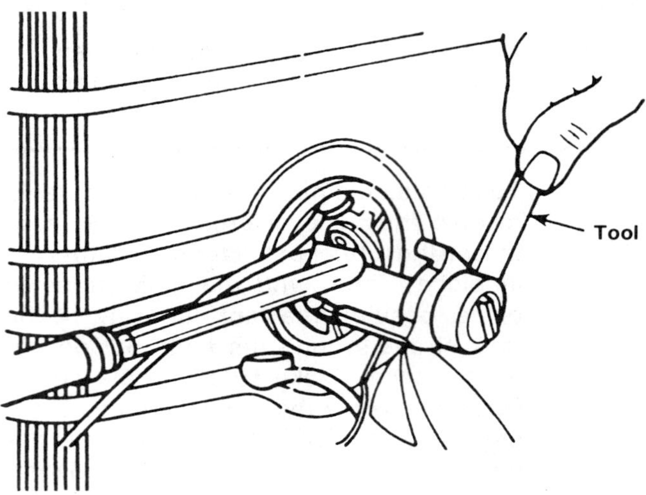

FIGURE 10-26 A special tool will help to remove a sending unit retaining ring.

to remove a sending unit retaining ring is to use a special tool designed for this purpose (Figure 10-26). This tool fits over the metal tabs on the retaining ring, and after about a quarter turn, the ring will come loose and the sender unit can be returned. If the special tool is not available, a drift punch or screwdriver and ball peen hammer will usually do the job (Figure 10-27).

When removing the sending unit from the tank, be very careful not to damage the float arm, the float, or the fuel gauge sender. Check the unit carefully for any damaged components. Shake the float, and if fuel can be heard inside it, replace the float. Make

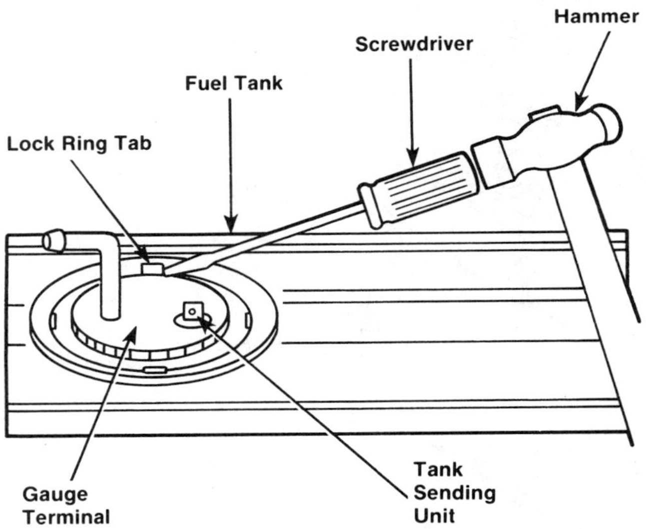

FIGURE 10-27 A drift punch or screwdriver and ball peen hammer can be used to remove the tank unit.

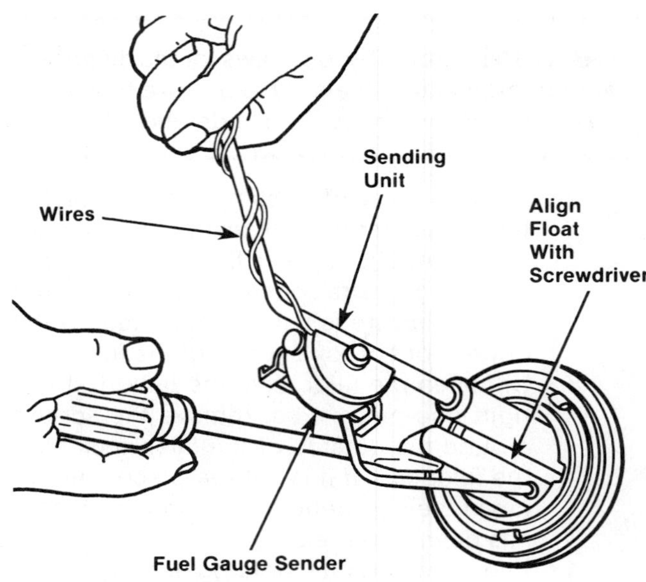

FIGURE 10-28 Be extremely careful not to damage the float, float arm, or screen when installing a tank sending unit.

sure the float arm is not bent. It is usually wise to replace the filter and O-ring before replacing the unit (Figure 10-28). Check the fuel gauge as described in the service manual. When reinstalling the pick-up pipe-sending unit, be very careful not to damage any of the components.

FUEL LINES

Fuel lines can be made of either metal tubing or flexible nylon or synthetic rubber hose. The latter must be able to resist gasoline. It must also be nonpermeable, so gas and gas vapors cannot evaporate through the hose. Ordinary rubber hose, such as that used for vacuum lines, deteriorates when exposed to gasoline. Only hoses made for fuel systems should be used for replacement. Similarly, vapor vent lines must be made of material that will resist attack by fuel vapors. Replacement vent hoses are usually marked with the designation EVAP to indicate their intended use. The inside diameter of a fuel delivery hose is generally larger (5/16 to 3/8 of an inch) than that of a fuel return hose (1/4 of an inch).

The fuel line system actually starts at the fuel tank. Here a fuel filler neck tube or hose connects with the tank. This short connection is made by a straight or curved tubing or hose that is roughly 1 1/2 to 2 1/2 inches in diameter. Where the latter is employed, there is usually an internal wire coil that resists collapsing.

Many fuel tanks have vent hoses to allow air in the fuel tank to escape when the tank is being filled with fuel. Vent hoses are usually installed alongside the filler neck hose.

The fuel lines carry fuel from the fuel tank to the fuel pump, the fuel filter, and to the carburetor or fuel injection metering pump. These lines are usually made of rigid metal, although some sections are constructed of rubber hose to allow for car vibrations. This fuel line, unlike filler neck or vent hoses, must work under pressure or vacuum. Because of this, the flexible synthetic hoses must be stronger. This is especially true for the hoses on fuel injection systems, where pressures reach 50 psi or more. For this reason flexible fuel line hose must also have special resistance properties. In fuel injection systems, unused gas is returned to the fuel tank where it becomes "sour gas," which ages when hydroperoxide molecules form. Hydroperoxides can cause some fuel line hoses to crack and disintegrate. Many auto manufacturers recommend that flexible hose should only be used as a delivery hose to the fuel metering unit in fuel injection systems. It should *not* be used on the pressure side of the injector systems. This application requires a special high-pressure hose.

All fuel lines should occasionally be inspected for holes, cracks, leaks, kinks, or dents (Figure 10–29). Many fuel system troubles that occur in the lines are blamed on the fuel pump or carburetor. For instance, a small hole in the fuel line will admit air but will not necessarily show any drip marks under the car. This usually occurs if the hole is high up in the line, especially where it curves up over the rear axle. Air can then enter the fuel line, allowing the fuel to gravitate back into the tank. Then, instead of drawing fuel from the tank, the fuel pump sucks only air

FIGURE 10–29 Checking fuel lines for leaks

through the hole in the fuel line. When this condition exists, the fuel pump is frequently tested, and if there is insufficient fuel, it is considered faulty, when in fact there is nothing wrong with it. If the hole is suspected, remove the coupling at the tank and the pump and pressurize the line with air. The leaking air is easily spotted.

Since the fuel is under pressure, leaks in the line between the pump and carburetor and injectors are relatively easy to recognize. When a damaged fuel line is found, replace it with one of similar construction—steel with steel, and the flexible with nylon or synthetic rubber. When installing flexible tubing, always use new clamps. The old ones lose some of their tension when they are removed and will not provide an effective seal when used on the new line.

CAUTION: Do not substitute aluminum or copper tubing for steel tubing. Never use hose within 4 inches of any hot engine or exhaust system component. A metal line must be installed.

Fuel supply lines from the tank to the carburetor or injectors are routed to follow the frame along the underchassis of vehicles. Generally, rigid lines are used extending from near the tank to a point near the fuel pump. To absorb engine vibrations, the gaps between the frame and tank or fuel pump are joined by short lengths of flexible hose (Figure 10–30).

Any damaged or leaking fuel line—either a portion or the entire length—must be replaced. To fabricate a new fuel line, select the correct tube and fitting dimension and start with a standard length that is slightly longer than the old line. With the old line as a reference, use a tubing bender to form the same bends in the new line as those that exist in the old. Although steel tubing can be bent by hand to obtain a gentle curve, any attempt to bend a tight curve by hand will usually kink the tubing. Since a kink in a brake line will weaken the line, a kinked line should never be used. To avoid kinking, always use a bending tool like the one shown in Figure 10–31.

The two most-used tubing fittings are either the compression type or the double-flare type (Figure 10–32). The double flare—which is the most common—is made with a special tool that has an anvil and a cone (Figure 10–33). The double flaring process is performed in two steps.

1. First, the anvil begins to fold over the end of the tubing.
2. The cone is used to finish the flare by folding the tubing back on itself, doubling the

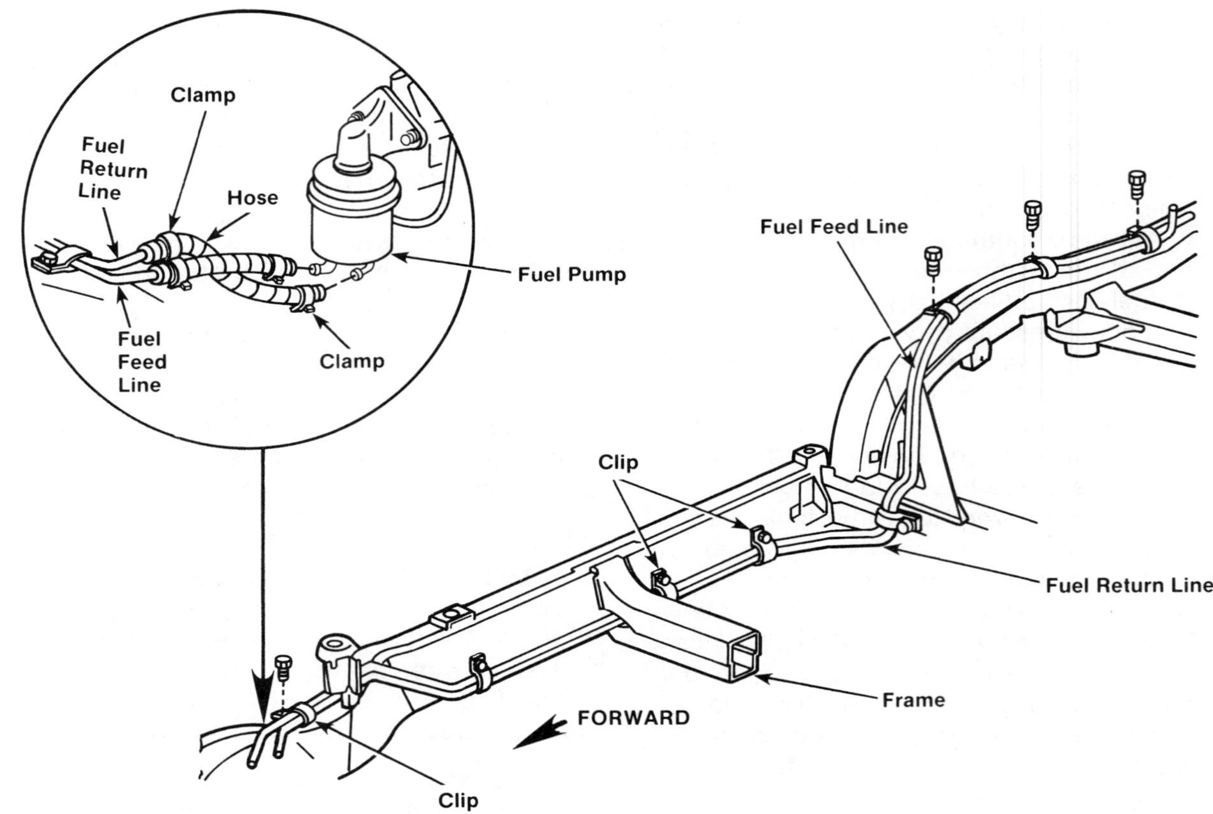

FIGURE 10–30 Gaps between the frame and tank or fuel pump are joined by short lengths of flexible hose.

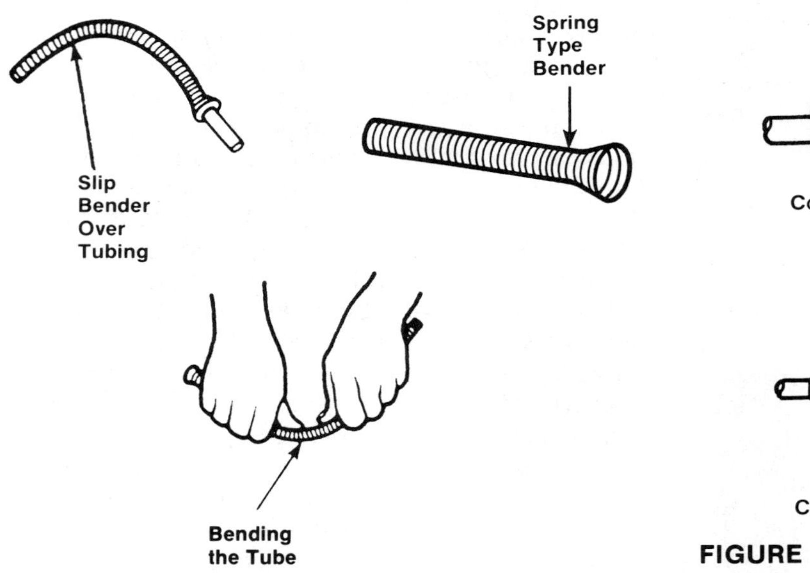

FIGURE 10–31 A bending tool avoids kinking.

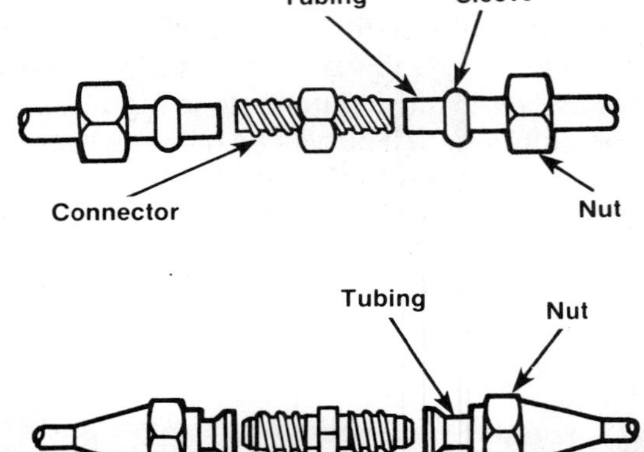

FIGURE 10–32 Tubing fittings

thickness and creating two sealing surfaces.

The angle and size of the flare are determined by the tool. Careful use of the double flaring will help to produce strong, leakproof connections. Figure 10–34 shows other metal fuel connections that are used by vehicle manufacturers.

The flare tool can also be used to make sure that nylon and synthetic rubber hoses will stay in place. That is, to make sure the connection is secure, put a

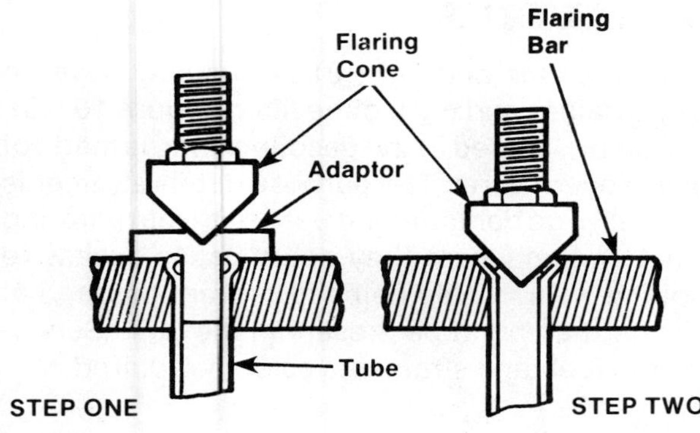

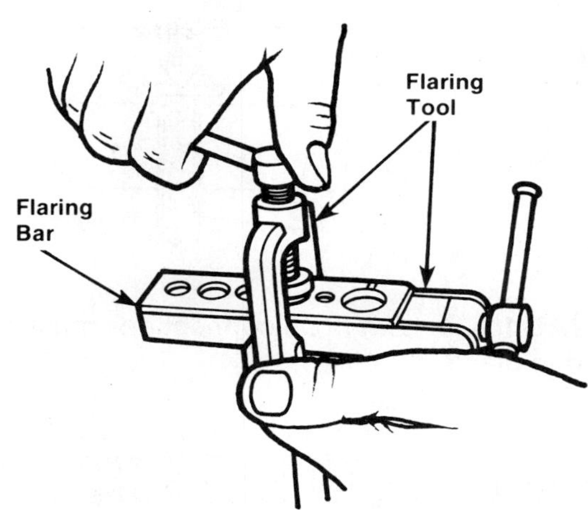

FIGURE 10-33 Double flare fit is made with a special tool.

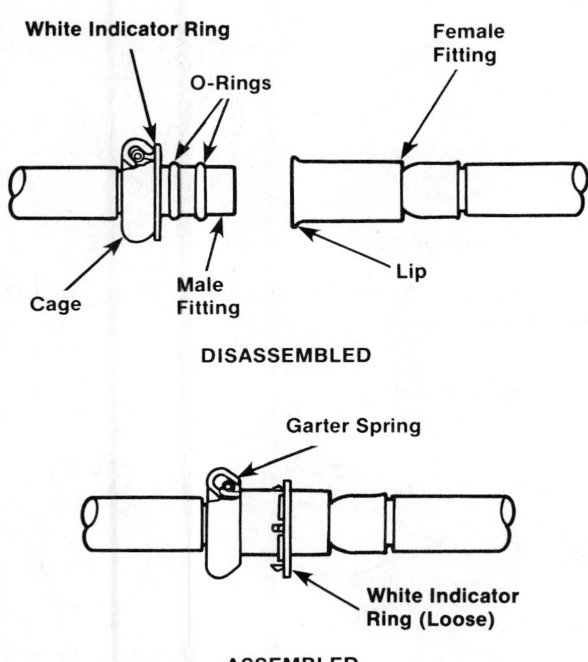

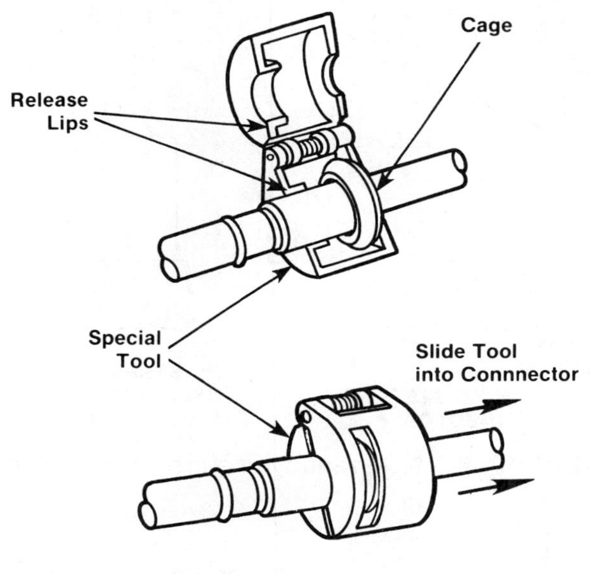

FIGURE 10-34 Metal fuel line connections

partial double-lip flare on the end of the tubing over which the hose is installed. This can be done quickly, with the proper flaring tool, by starting out as if it was going to be a double flare, but stopping halfway through the procedure (Figure 10-35). This provides an excellent sealing ridge that will not cut into the hose. The sealing ridge also holds tighter than a straight pipe, especially if a clamp is placed directly behind the ridge on the hose.

SHOP TALK

To insure a flexible fuel replacement hose is the right length, lay the old line alongside the new one. Then, with a sharp knife, cut the new line to the same length as the old one.

Nylon and synthetic rubber fuel line connectors are illustrated in Figure 10-36. Note that there are

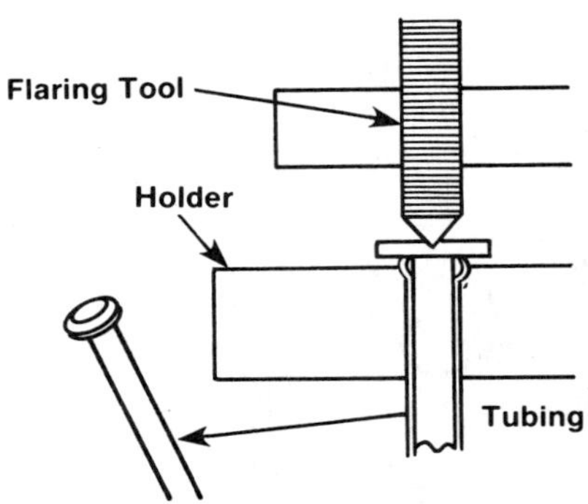

FIGURE 10-35 Cut hose connections must be secured in place.

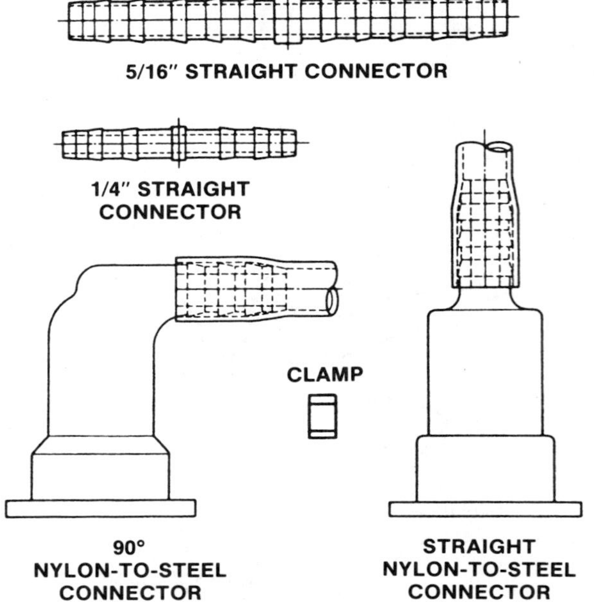

FIGURE 10-36 Nylon and synthetic rubber fuel lines

connectors available to combine rigid and flexible fuel lines. But if nylon or synthetic rubber fuel line is damaged, replace the entire line; do not attempt to patch it or use a patched section.

There are a variety of clamps used on fuel system lines, including the screw types (Figure 10-37). The crimp-type clamps shown in Figure 10-38 are used most for metal tubing, but they require a special tool to install.

To control the rate of vapor flow, a plastic or metal restrictor is placed in either the end of the vent pipe or in the vapor-vent hose itself (Figure 10-39). When the latter hose must be replaced, the restrictor must be removed from the old vent hose and installed in the new one.

FUEL FILTERS

Automobiles and light trucks usually have an in-tank strainer and a gasoline filter (Figure 10-40). The strainer, located in the gasoline tank, is made of a finely woven fabric. The purpose of this strainer is to prevent large contaminant particles from entering the fuel system where they could cause excessive fuel pump wear. It also helps to prevent passage of any water that might be present in the tank. Servicing of the fuel tank strainer is seldom required.

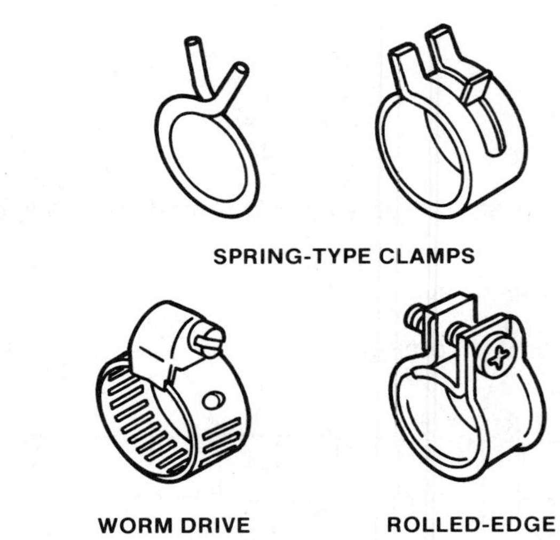

FIGURE 10-37 A variety of clamps used on fuel system hoses

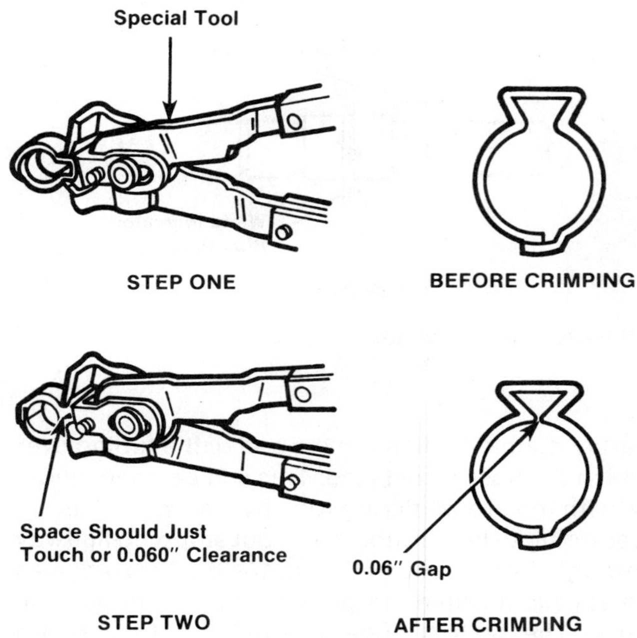

FIGURE 10-38 Crimp-type clamps

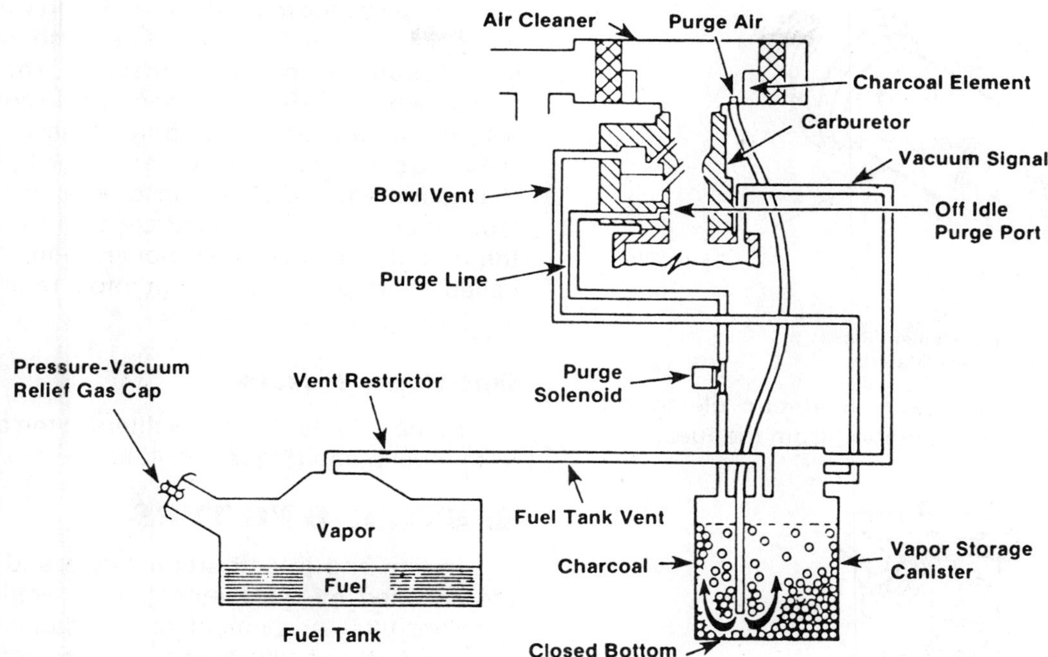

FIGURE 10-39 Open canister and closed canister evaporative emission control system schematic

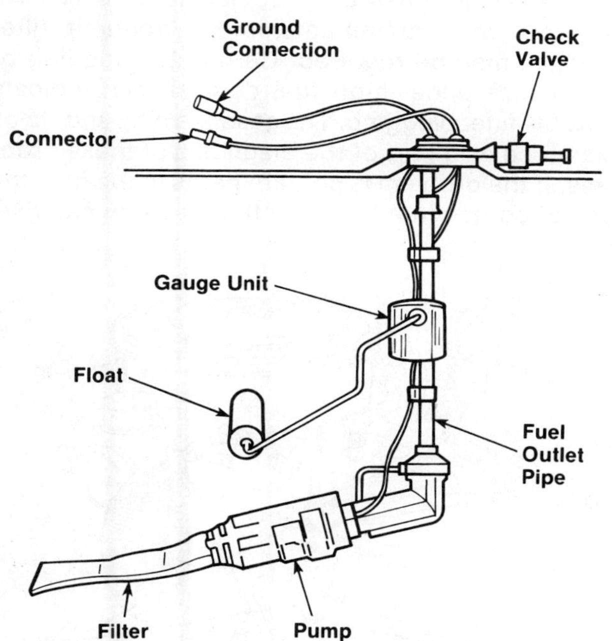

FIGURE 10-40 Cars and light trucks usually have an in-tank strainer and a gasoline filter.

The second gasoline filter is usually located in the engine compartment and is the one this section will examine because it is replaceable and might require service on a regular basis. Most automobile applications are covered by the following types of gasoline filters:

- In-carburetor filters
- In-line carburetor filters
- Out-pump filters

In-Carburetor Filters

There are three basic types of in-carburetor gasoline filters:

1. **Pleated Paper Filter.** This type uses pleated paper as the filtering medium. Paper elements are more efficient than screen-type elements, such as nylon or wire mesh, in removing and trapping small particles, as well as larger size contaminants. This type of filter is available in 1- and 2-inch lengths.
2. **Sintered Bronze Gasoline Filter.** Often referred to as a "stone" or "ceramic" filter. This filter is also an in-carburetor type.
3. **Screw-in Filter.** This type is designed to screw into the carburetor fuel inlet. The fuel line attaches to a fitting on the filter. One of the newest filters of this type has a magnetic element to remove metallic contamination before it reaches the carburetor (Figure 10-41).

Some imported vehicles as well as older domestic cars have a sediment bowl between the fuel pump

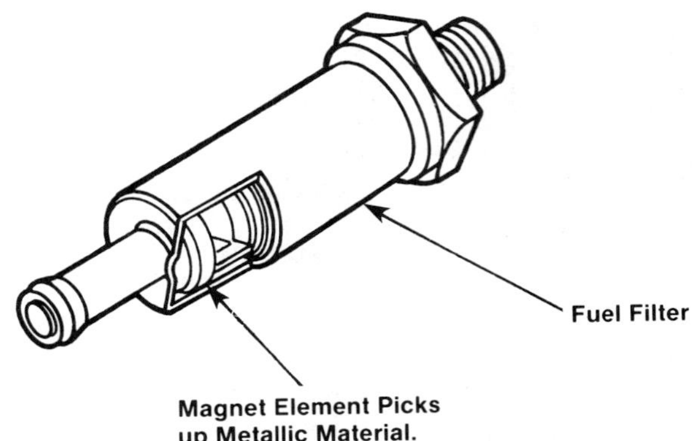

FIGURE 10-41 Filter with magnetic element to remove metallic contamination from the fuel

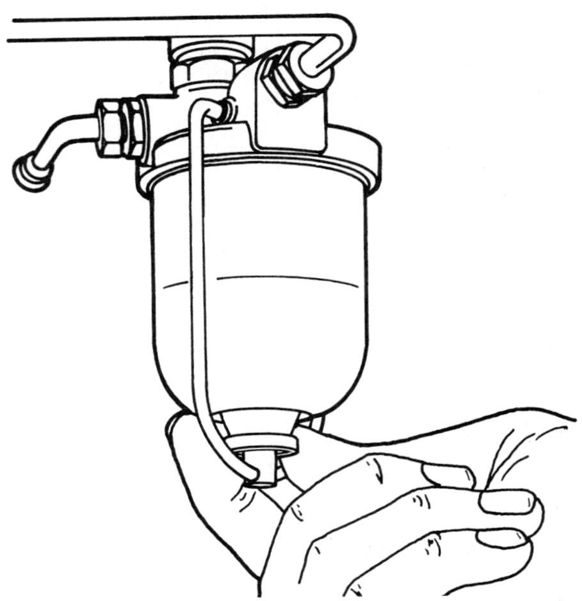

FIGURE 10-42 Sediment bowls can contain a paper, fiber, ceramic, or metal filter element.

and carburetor. The bowl contains a pleated paper and ceramic filter element. The bowl cover is held in place by a wire bail and clamp screw (Figure 10-42). It can be removed for filter cleaning or replacement. Ceramic can be cleaned and reused if necessary, but paper filter elements must be replaced when they are dirty.

In-Line Filters

This type of gasoline filter is installed in the fuel line (Figure 10-43). In carbureted engines, the in-line gasoline filter is usually installed between the fuel pump and the carburetor; in vehicles with a fuel injection system, the location of the fuel filter is determined by the manufacturer. Fitted with a pleated paper element, the in-line filter is sometimes installed as an extra protective measure. The optionally installed in-line filter will then work in conjunction with the in-tank and in-carburetor gasoline filters. Because of its large capacity, an in-line filter is often the most economical solution to a fuel system's contamination problems. The arrow on the filter shows the direction of fuel flow. Some in-line filters are called cartridge-type. They fit into a retainer housing (Figure 10-44).

Out-Pump Filters

Some vehicles have fuel filters in the outlet side of the fuel pump (Figure 10-45).

SERVICING FILTERS

In-line and in-carburetor filters and elements are serviced by replacement only. Replacing the gasoline filter or element at the intervals recommended by the vehicle or engine manufacturer is the most effective method of minimizing fuel starvation and other carburetor problems. On occasions when the fuel system has been subjected to excessive amounts of contaminants, more frequent filter changes may be required. Carburetor flooding or power loss under high fuel demand can indicate possible filter plugging. The gasoline filter should be examined as a part of the diagnosis of these problems. If the problem is contaminants in the fuel, the type of contaminant held by the fuel filter can give

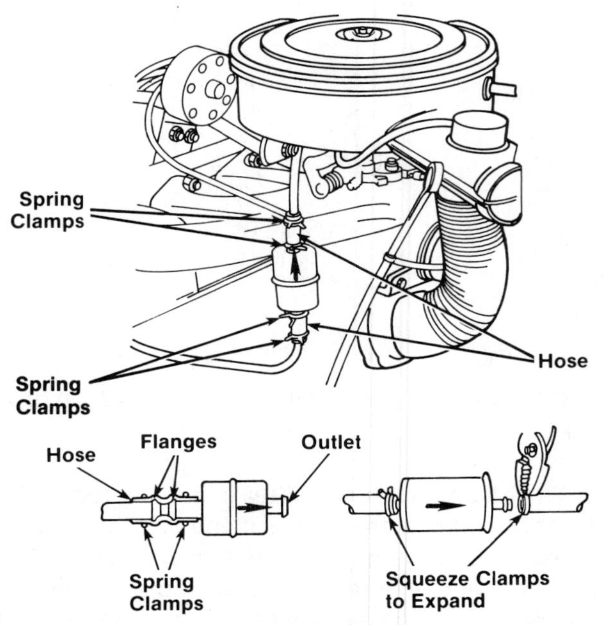

FIGURE 10-43 In-line fuel filter replacement

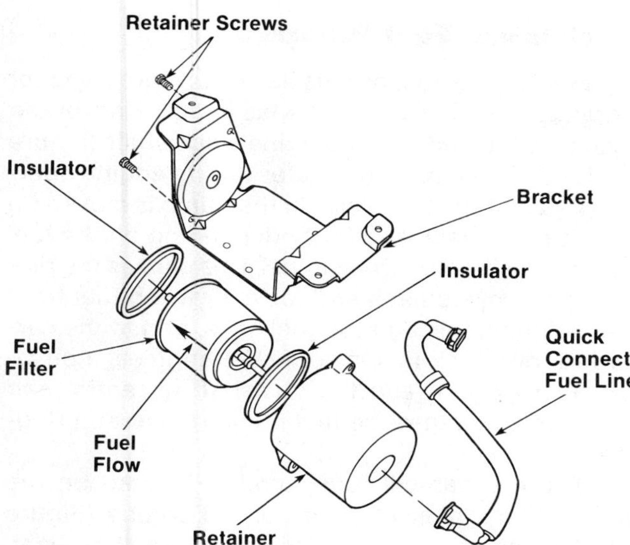

FIGURE 10-44 Cartridge-type in-line filter fits into a retainer housing.

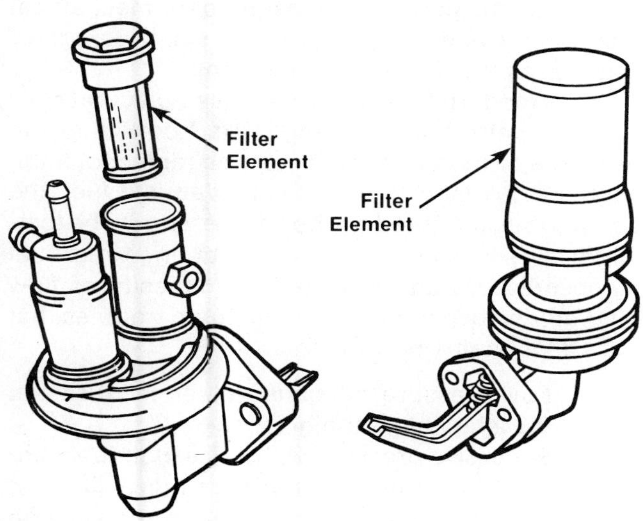

FIGURE 10-45 Fuel pump with out-pump filter

the experienced technician an indication of the source and severity of the problem.

Replacing an In-Carburetor Filter

To replace an in-carburetor gasoline filter, proceed as follows:

1. To remove an in-carburetor gasoline filter, remove the fuel line from the carburetor fuel inlet nut using two wrenches, one to hold the inlet nut, the other to turn the fuel line fitting (Figure 10–46).
2. After disconnecting the fuel line, remove the inlet nut from the carburetor and discard the used filter. The inlet nut will have a gasket that is loose or permanently attached. This gasket can usually be reinstalled.

SHOP TALK

Take care not to lose the spring located behind the filter. The spring is required to seal the filter to the inlet nut.

3. The correct size and type of gasoline filter for the engine or vehicle being serviced is listed in the service manual. Installation of the filter is accomplished by simply reversing the steps used to remove the used gasoline filter.
4. Check fuel line connections for leaks with the engine running.

Replacing a Sintered Bronze Filter

Replacement is similar to replacing the in-carburetor gasoline filter. Take care to reinstall the gasket at the fuel inlet to the filter.

Most in-line filters have tubes or threaded fittings. Tube fittings are used with rubber hoses and clamps. Threaded fittings may be inverted flare, or male or female pipe threads. When replacing an in-line filter, take care to match the end of the gasoline

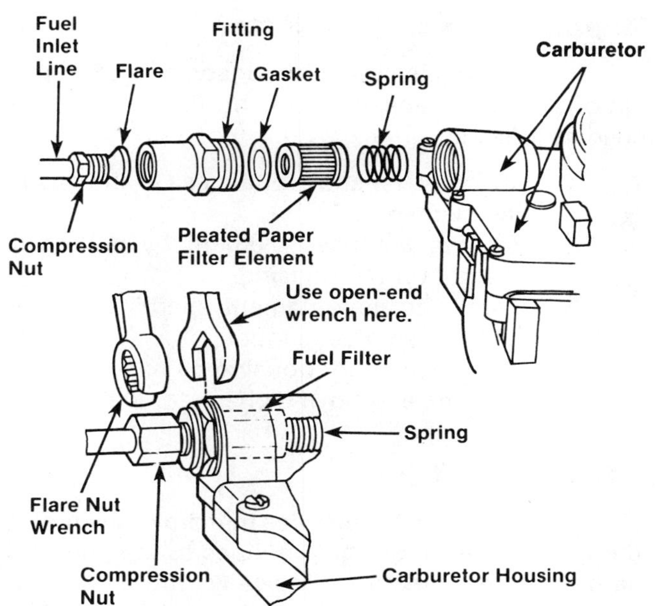

FIGURE 10-46 Removing parts of a carburetor fuel filter

line with the proper fitting on the filter housing. Most in-line gasoline filters have an arrow on the housing to show direction of fuel flow.

SHOP TALK

Most late-model automobiles have steel gasoline lines. Non-OE in-line gasoline filters can be installed using short rubber tubes and screw-type, tangential hose clamps. There should be no bends in the rubber tubes after installation. After installing an in-line gasoline filter, start the engine and check for leaks.

Replacing a Screw-in Filter

To replace a screw-in gasoline filter, proceed as follows:

1. Detach the fuel line from the gasoline filter.
2. Unscrew the filter from the carburetor using an open-end wrench. Discard the used filter.
3. Screw a new filter with gasket into the carburetor.
4. Reconnect the fuel line, start the engine, and check for leaks.

Most out-pump or pump outlet filters are the screw-in type and are replaced/installed as described above.

Replacing an In-Line Filter

The in-line gasoline filter for some truck applications has a replaceable element. To replace, proceed as follows:

1. Unscrew the shell for access to the gasoline element.
2. Remove the used element by pulling it straight off the housing.
3. Install the new element specified in the vehicle's service manual.
4. Tighten the filter housing by hand.
5. Start the engine and check for leaks.

FUEL PUMPS

The fuel pump is the device that draws the fuel in the fuel tank through the fuel lines to the engine's carburetor or injectors. Basically, there are two types of fuel pumps: mechanical and electrical. The latter is the most commonly used today.

Mechanical Fuel Pumps

Mechanical fuel pumps have a synthetic rubber diaphragm inside the unit that is actuated by an eccentric located on the engine's camshaft (Figure 10–47). As the camshaft rotates during engine operation, a shaft or rocker arm in the pump is moved up and down or back and forth, depending on the fuel pump's position on the engine. This causes the diaphragm to move back and forth, drawing fuel from the fuel tank, through the fuel lines, and to the carburetor or injectors. On some V-8 engines, a pushrod, located between the camshaft eccentric and fuel pump, actuates the fuel pump rocker arm (Figure 10–48).

The mechanical fuel pump is located on the engine block near the front of the engine (Figure 10–49). This, of course, subjects it to engine heat. Further, during fuel intake, a low-pressure area is developed in the fuel line. High temperatures and low pressure can lead to a vapor lock condition. To overcome this condition, most modern mechanical fuel pump systems contain a fuel vapor separator and vapor return line to the fuel tank.

The modern fuel pump is a sealed unit that cannot be repaired. If the pump leaks from either the vent hole or from a seam, it must be replaced. If the engine performance indicates inadequate fuel, the pump, while mounted on the engine, should be tested for pressure and volume. In fact, incorrect fuel pump pressure and low volume (capacity of flow rate) are the two most likely fuel pump troubles that will affect engine performance.

- Low pressure will cause a lean mixture and fuel starvation at high speeds.
- Excessive pressure will cause high fuel consumption and carburetor flooding.
- Low volume will cause fuel starvation at high speeds.

To determine that the fuel pump is in satisfactory operating condition, tests for both fuel pump pressure and fuel pump capacity (volume) should be performed. These tests are performed with the fuel pump installed on the engine and the engine at normal operating temperature and at idle speed.

SHOP TALK

Before making any tests, be sure the replaceable fuel filter has been changed within the recommended maintenance mileage interval. When in doubt, install a new filter.

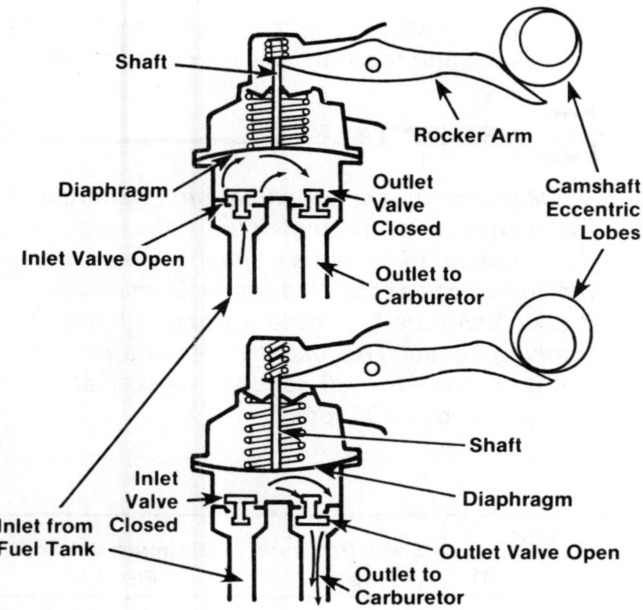

FIGURE 10–47 Mechanical fuel pump assembly

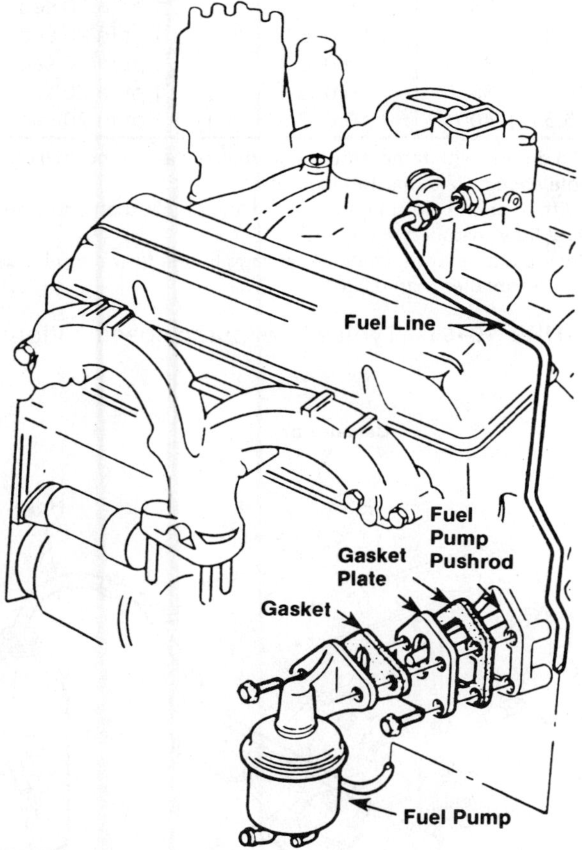

FIGURE 10–48 V-8 engines usually have a push-rod between the camshaft eccentric and the fuel pump.

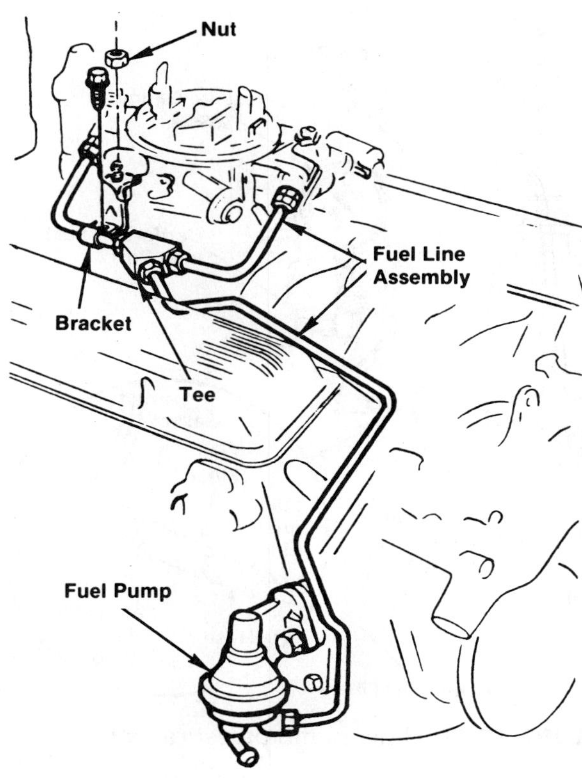

FIGURE 10–49 Typical fuel pump and line installation on a V-8 engine

To make a pressure test, proceed as follows:

1. Remove the air cleaner assembly. Disconnect the fuel inlet line or the fuel filter at the carburetor. **Use care to prevent combustion from fuel spillage.**
2. Connect a pressure gauge, a restrictor, and a flexible hose (Figure 10–50) between the fuel filter and the carburetor.

WARNING: The fuel supply lines will remain pressurized for long periods of time after the engine is shut down. This pressure must be relieved before servicing of the fuel system is begun. A valve is provided on the throttle body for this purpose.

---

3. Position the flexible hose and the restrictor so the fuel can be discharged into a suitable, graduated container.
4. Before taking a pressure reading, operate the engine at the specified idle rpm and vent the system into the container by opening the hose restrictor momentarily.

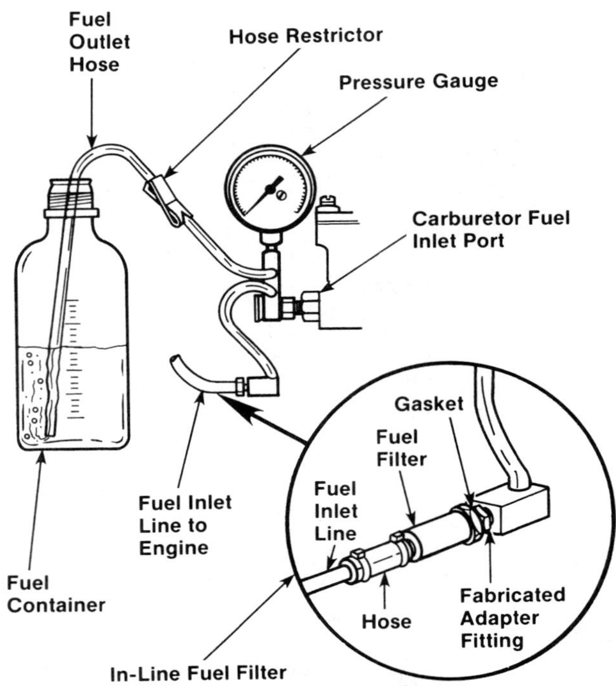

FIGURE 10–50 Fuel pump pressure test

5. Close the hose restrictor. Allow the pressure to stabilize and note the reading. If the pump pressure is not within specifications (Figure 10–51) and the fuel lines and filter are in satisfactory condition, the pump should be replaced.

SHOP TALK

Fuel pump pressure specifications are given by the manufacturer in pounds per square inch (psi) at a certain engine speed.

If the fuel pump pressure is within specifications, test the capacity (volume) as follows:

1. Operate the engine at the specified idle rpm.
2. Open the hose restrictor and allow the fuel to discharge into the graduated container (Figure 10–52) for the specified time. Then, close the restrictor. At least 1 pint of fuel should have been discharged within the specified time limit.
3. If the pump volume is below specifications, repeat the test using an auxiliary fuel supply and a new fuel filter.
4. If the pump volume meets specifications while using the auxiliary fuel supply, check for a restriction in the fuel supply from the tank and for improper tank ventilation.

SHOP TALK

Manufacturer's specifications generally are given in terms of 1 pint of fuel delivered in a given number of seconds while the engine is at idle and also at a specified speed. Some manual specifications just state a good volume at cranking speed. This usually means a pint of fuel is delivered in 30 seconds or less at an engine speed of 500 rpm.

Engine		Static Pressure	Minimum Volume
Liter	CID	(psi)[1]	Flow[1 3]
2.3		5.0–7.0[2]	1 pt in 25 sec
3.3	200	5.0–7.0	1 pt in 30 sec
4.1	250	5.0–7.0	1 pt in 20 sec
4.2	255	6.0–8.0	1 pt in 20 sec
5.0	302	6.0–8.0	1 pt in 20 sec
5.8	351W	6.0–8.0	1 pt in 20 sec
5.8	351M	6.0–8.0	1 pt in 20 sec
6.6	400	6.0–8.0	1 pt in 20 sec

[1]On engine with temperatures normalized and at normal curb idle speed, transmission in neutral
[2]With the pump-to-tank fuel return line pinched off and a new fuel filter installed in fuel line
[3]Inside diameter of smallest passage in test flow circuit must not be smaller than 0.220"

FIGURE 10–51 Typical fuel pump specifications

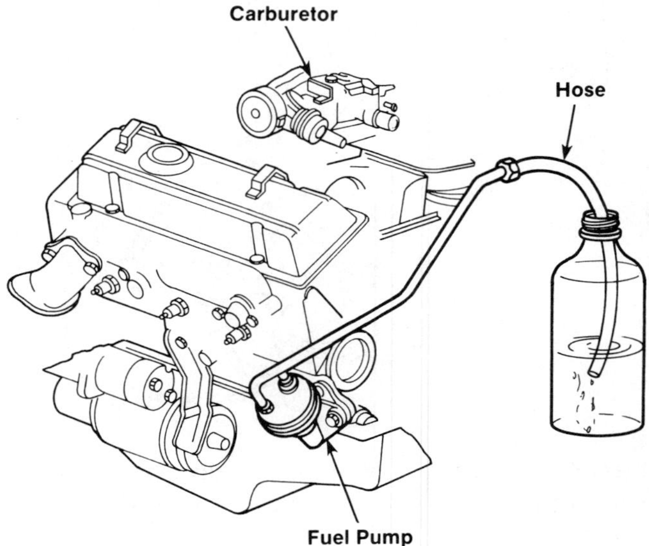

FIGURE 10–52 Fuel pump output test

5. If the pump fails the pressure and volume test, it generally requires replacement. But before replacing the unit make sure that there are no restricted or leaking fuel lines or clogged filter hoses and that there is no dirt or sludge at the fuel pickup in the tank. Blow out any restriction with compressed air or replace any damaged parts. Figure 10-53 shows the common problems with a mechanical pump.

To replace the fuel pump (Figure 10-54), remove the inlet and outlet lines at the pump. Use a plug to stop the flow of fuel from the tank. With the correct size socket wrench, remove the mounting bolts, then remove the pump from the engine. Clean the old gasket material from the engine block. Apply gasket sealer to the mounting surface on the engine and to the threads of the mounting bolts. Install the new gasket, then position the pump by tilting it either toward or away from the block to correctly place the lever against the cam. If the pump is driven by a pushrod, the rod must be held up to permit the rocker arm to go under it. When the pump is properly positioned there should be an internal squeaking noise with each movement of the pump. Tighten the mounting bolts firmly. Attach the inlet and outlet lines, start the engine, and check for leaks.

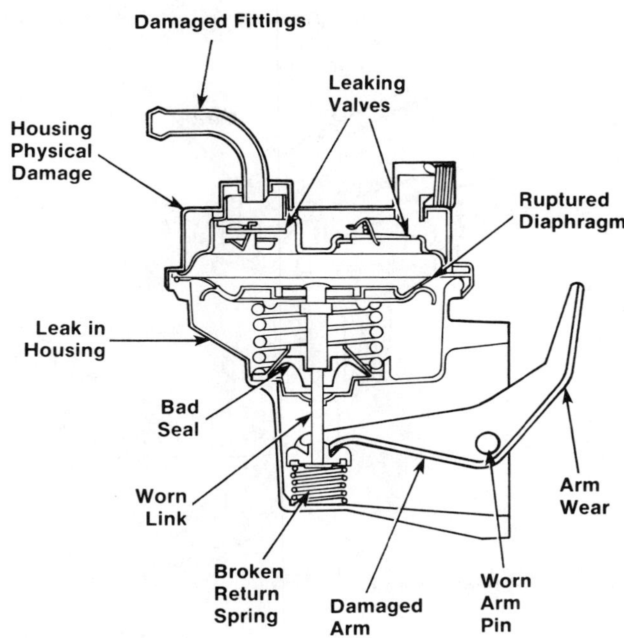

FIGURE 10-53 Common problems with a mechanical pump

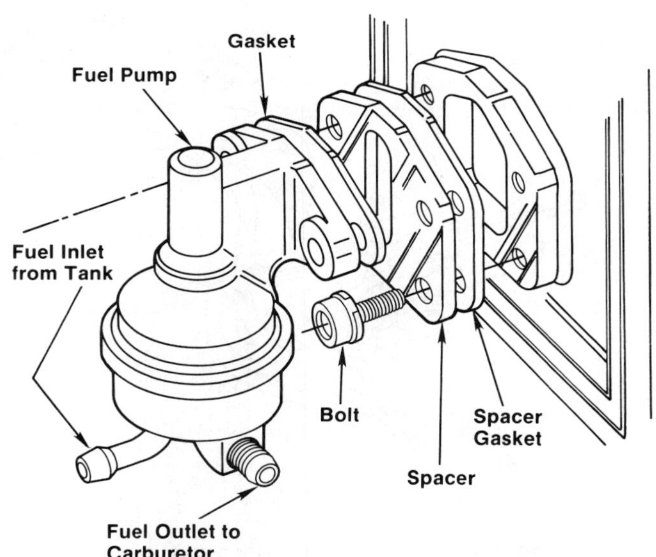

FIGURE 10-54 Before removing a fuel pump, disconnect the fuel lines at the pump. Using a proper-size wrench, remove the bolts holding the fuel pump to the engine block.

ELECTRIC FUEL PUMPS

Electric fuel pumps offer important advantages over mechanical fuel pumps. Because electric fuel pumps maintain constant fuel pressure, they aid starting and reduce vapor lock problems. Since they are not mechanically attached to the engine, fit problems are eliminated, making them especially useful when an exact replacement mechanical fuel pump is unavailable. Unlike mechanical fuel pumps, the operation of electric pumps is not affected by worn cams. It is also easy to install a hidden on/off switch for the electric pump as an antitheft precaution.

The electric fuel pump can be located inside (Figure 10-55) or outside (Figure 10-56) the fuel tank. There are four basic types of electric fuel pumps: the diaphragm, plunger, bellows, and impeller or rotary. The in-tank electric pump is usually a rotary type (Figure 10-57). The diaphragm, plunger, and bellows type (Figure 10-58) are usually the demand style. That is, when the ignition is turned on, the pump begins operation. It shuts off automatically when the carburetor bowl is full and the fuel line is pressurized. When the carburetor demands more fuel, the electric pump pumps more. When demand is lower, it pumps less, so proper fuel flow and pressure are constantly maintained. In most installations, the rotary electric fuel pump is considered to be a continuous operated type.

A typical wiring diagram for an electric pump is shown in Figure 10-59. Some major problem areas in an electric pump system are the following:

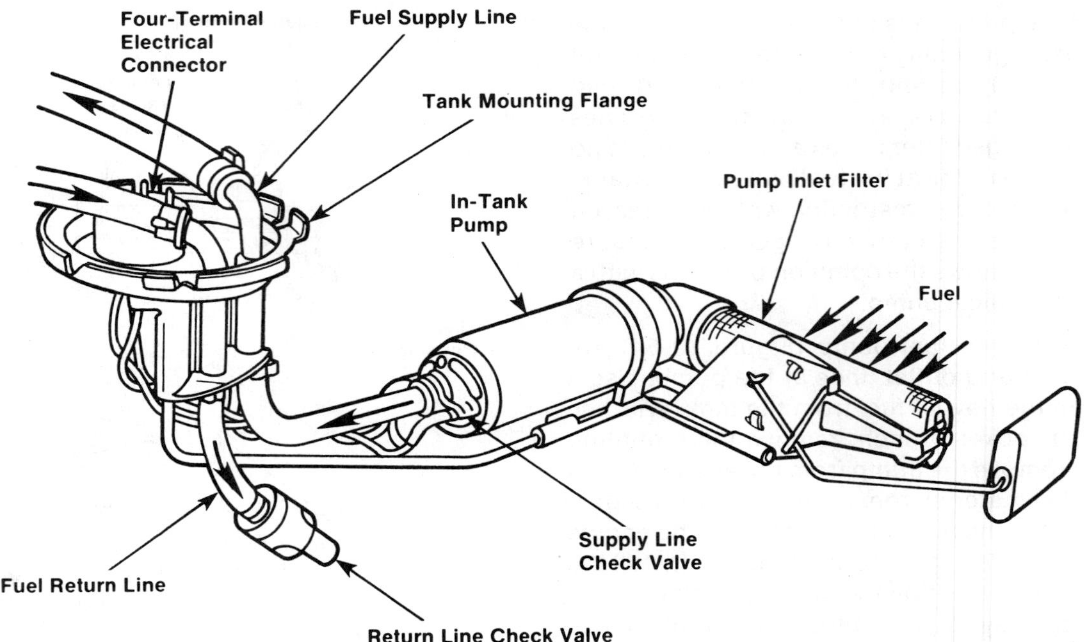

FIGURE 10–55 Electric fuel pump system located inside the fuel tank

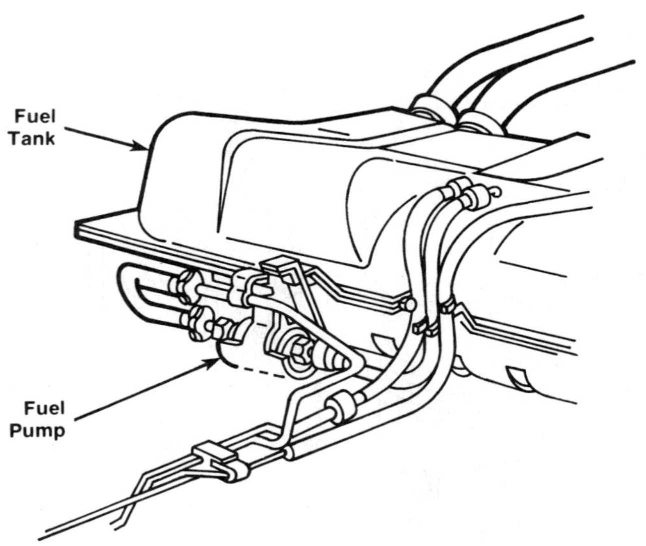

FIGURE 10–56 Electric fuel pump system located outside the fuel tank

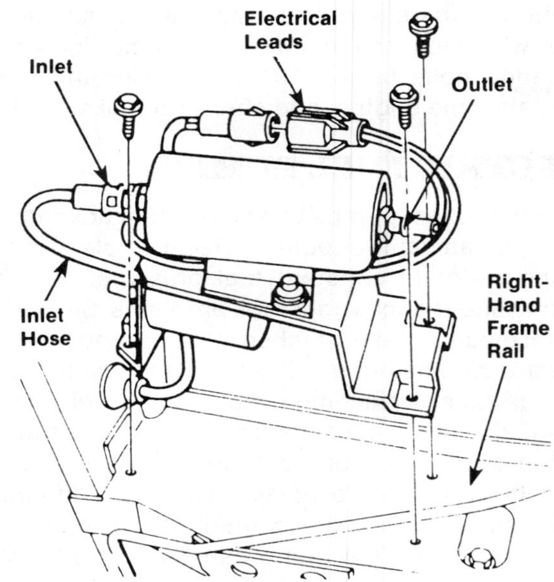

FIGURE 10–57 Rotary-type fuel pump

- Dead or inoperative pump
- Too much fuel pressure
- Too little fuel pressure

A high pressure reading usually indicates either a faulty pressure regulator or an obstructed return line. Disconnect the fuel return line at the tank. Use a length of hose to route the returning fuel into an appropriate container. Start the engine and note the pressure reading at the engine. If fuel pressure is now within specs, check for an obstruction in the in-tank return plumbing. The fuel reservoir check valve or aspirator jet might be clogged.

If the fuel pressure still reads high with the return line disconnected from the tank, note the volume of fuel flowing through the line (Figure 10–60). Little or no fuel flow can indicate a plugged return line. Shut off the engine and connect a length of hose directly to the fuel pressure regulator return

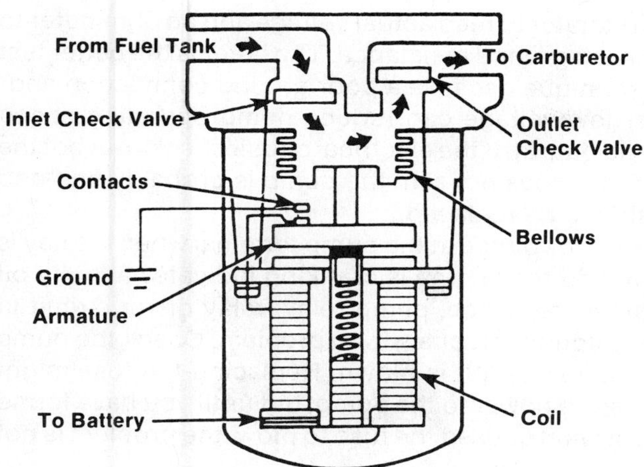

FIGURE 10-58 Bellows-type fuel pump

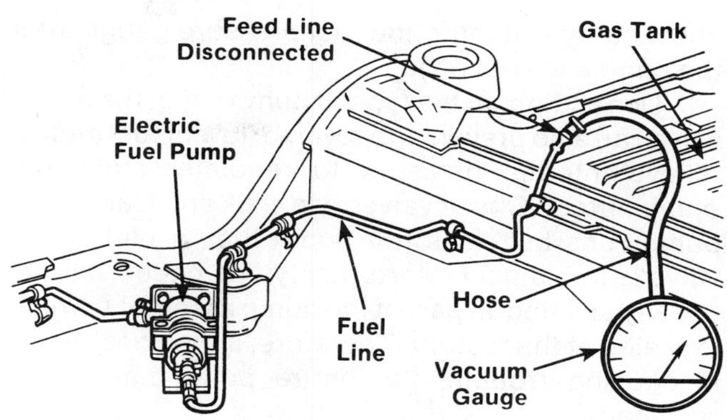

FIGURE 10-60 Vacuum test of electric pump

port to bypass the return hose. Restart the engine and again check the pressure reading. If bypassing the return line brings the readings back within specs, a plugged return line is the culprit. If pressure is still high, apply vacuum to the pressure regulator to see if that makes a difference (it should). If there is still no change, replace the faulty pressure regulator. If applying vacuum directly to the regulator lowers fuel pressure, the vacuum hose that controls the operation of the regulator might be plugged, leaking, or misrouted.

Low pressure, on the other hand, can be due to a clogged fuel filter, restricted fuel line, weak pump, leaky pump check valve, defective fuel pressure regulator, or dirty filter sock in the tank. It is possible to rule out filter and line restrictions as a cause of the problem by making a pressure check at the pump outlet. A higher reading at the pump outlet (at least 5 psi) means there is a restriction in the filter or line. If the reading at the pump outlet is unchanged, then either the pump is weak or is having trouble picking up fuel (clogged filter sock in the tank). Either way it will be necessary to get inside the fuel tank. If the filter sock is gummed up with dirt or debris, it is also wise to clean out the tank when the filter sock is cleaned or replaced.

Another source of trouble is the pump check valve. Some pumps have one, others have two (positive displacement roller vane pumps). The purpose of the check valve is to prevent fuel movement through the pump when the pump is off so residual pressure will remain in the injectors (which can be

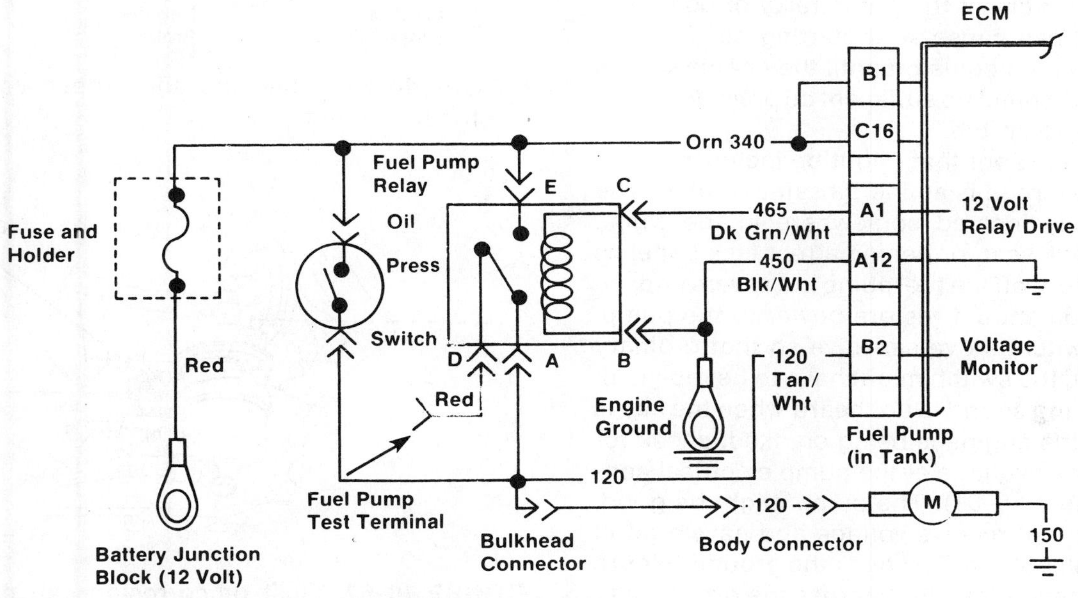

FIGURE 10-59 Typical wiring diagram for an electric pump

checked by watching the fuel pressure gauge after the engine is shut off).

Depending on the type of pump used, the check valve can also prevent the reverse flow of fuel and/or relieve internal pressure to regulater maximum pump output. Check valves can stick and leak, so if a pump runs but does not pump fuel, a bad check valve is to blame. Unfortunately, the check valve is usually an integral part of the pump assembly, which is sealed at the factory. Therefore, if the check valve is causing trouble, the entire pump has to be replaced.

When an engine fails to start because there is no fuel delivery, the first step is to check the fuel gauge. A gauge that reads higher than a half tank probably means there is fuel in the tank, but not always. A defective sending unit or miscalibrated gauge might be giving a false indication. Sticking a wire or dowel rod down the fuel tank filler pipe will tell whether or not there is really fuel in the tank. If the gauge is faulty, repair or replace as discussed later in this chapter.

Listen for pump noise. When the key is turned on, the pump should buzz for a couple of seconds to build system pressure. The pump is usually energized through an oil pressure switch (the purpose of which is to shut off the flow of fuel in case of an accident that stalls the engine). But on most late-model cars with computerized engine controls, the computer energizes a pump relay when it receives a cranking signal from the distributor pickup or crankshaft sensor. An oil pressure switch might still be included in the circuitry for safety purposes and to serve as a backup in case the relay or computer signal fails. Failure of the pump relay or computer driver signal can cause slow starting because the fuel pump will not come on until the engine cranks long enough to build up sufficient oil pressure to trip the oil pressure switch.

Another element that might be included in the pump wiring circuit is an inertia safety switch. The switch usually located somewhere in the trunk, under the back seat, or behind a rear kick panel, is designed to turn off the fuel pump in a severe impact. A reset button should restore power to the pump. But if the switch is oversensitive so that ordinary bumps trip it, the switch might have to be replaced.

If a buzzing sound is not heard when the key is on or while the engine is being cranked, check for the presence of voltage at the pump electrical connectors (Figure 10-61). The pump might be good, but if it does not receive voltage and have a good ground, it will not run. To check the ground (Figure 10-62), connect a test light across the ground and feed wires at the pump to check for voltage, or use a voltmeter to read actual voltage and an ohmmeter to check ground resistance. The latter is the better test technique because a poor ground connection and/or low voltage can reduce pump operating speed and output. If the electrical circuit checks out but the pump does not run, the pump is probably bad and should be replaced.

No voltage at the pump terminal when the key is on and the engine is cranking indicates a faulty oil pressure switch, pump relay, relay driver circuit in the computer, or a wiring problem. Check the pump fuse to see if it is blown. Replacing the fuse might restore power to the pump, but until you have found out what caused the fuse to blow, the problem is not

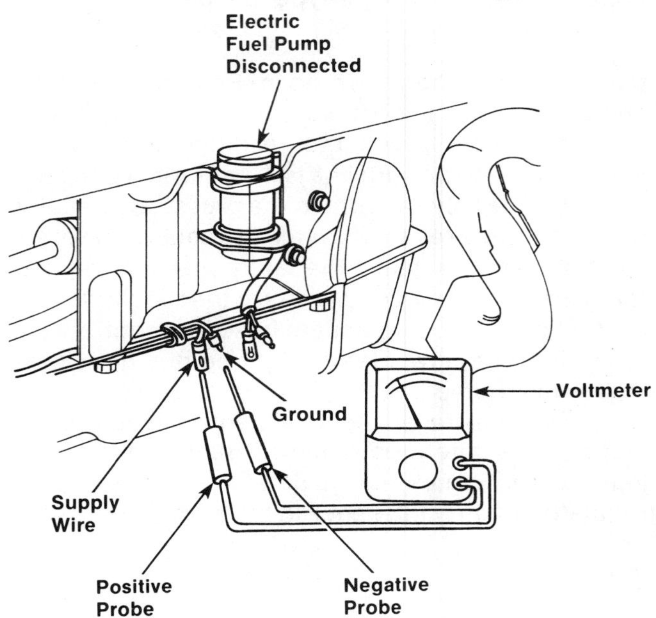

FIGURE 10-61 Checking for proper voltage to electric fuel pump

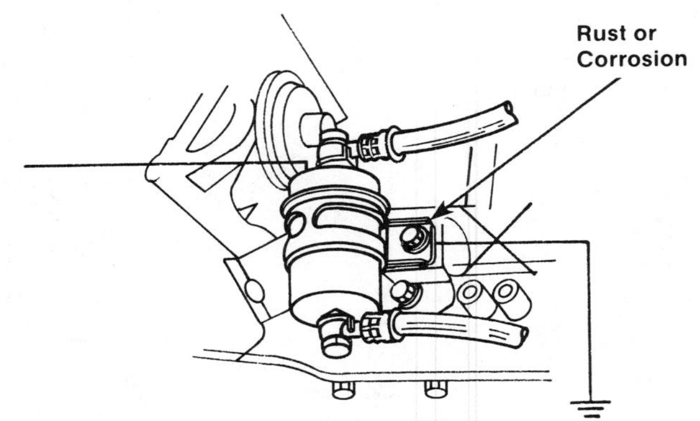

FIGURE 10-62 Rust or corrosion can prevent a good ground.

solved. The most likely cause of a blown fuse would be a short in the wiring between the relay and pump, or a hot short inside the oil pressure switch or relay.

A faulty oil pressure switch can be checked by bypassing it with a jumper wire. If that restores power to the pump and the engine starts, replace the switch. If an oil pressure switch or relay sticks in the *closed* position, the pump can run continuously whether the key is on or off depending on how the circuit is wired.

To check a pump relay, use a test light to check across the relay's hot and ground terminals. This will tell if the relay is getting battery voltage and ground. Next, turn off the ignition, wait about 10 seconds, then turn it on. The relay should click and you should see battery voltage at the relay's pump terminal. If nothing happens, repeat the test checking for voltage at the relay terminal that is wired to the computer. The presence of a voltage signal here means the computer is doing its job but the relay is failing to close and should be replaced. No voltage signal from the computer indicates an opening in that wiring circuit or a fault in the computer itself.

WARNING: When testing an electric fuel pump, do not let fuel contact any electrical wiring. The smallest spark (electric arc) could ignite the fuel.

When replacing an electric pump, be sure that the new or rebuilt replacement unit meets the minimum requirements of pressure and volume for that particular vehicle. This information can be found in the service manual.

To replace an electric fuel pump, proceed as follows:

1. Disconnect the ground terminal at the battery.
2. Disconnect the electrical connector(s) on the electric fuel pump. Label the wire(s) to aid in connecting it to the new pump. Reversing polarity on most pumps will destroy the unit.
3. Fuel and vapor lines should be removed from the pump. Label the lines to air in connecting it to the new pump.
4. The inside tank type pump can usually be taken out of the tank by removing the fuel sending unit retaining ring. Usually, however, it may be necessary to remove the fuel tank to get at the retaining ring. Removal of the fuel tank can be accomplished as described earlier in this chapter.

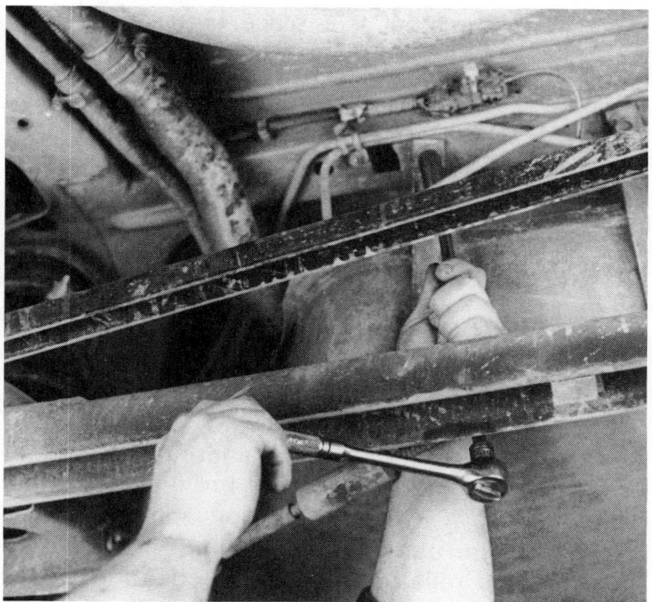

FIGURE 10–63 Remove the bolts holding the fuel tank in place.

5. On a fuel pump that is outside of the tank remove the bolts holding it in place (Figure 10–63). On intake models, loosen the retaining ring by rotating it counterclockwise with a brass drift and hammer. Pull the pump and sending unit out of the tank (if they are combined in one unit) and discard the tank O-ring seal.
6. To remove the pump, twist off the filter sock, then push the pump up until the bottom is clear of the bracket. Swing the pump out to the side and pull it down to free it from the rubber fuel line coupler. The rubber sound insulator between the bottom of the pump and bracket and the rubber coupler on the fuel line are normally discarded because new ones are included with the replacement pump. Some pumps have a rubber jacket around them to quiet the pump. If this is the case, slip off the jacket and put it on the new pump.
7. Compare the new or rebuilt electric fuel pump with the old one. If necessary, transfer any fuel line fittings from the old pump to the new one. Note the position of the filter sock on the pump so you can install a new one in the same relative position.
8. When inserting the new pump back into the sending unit bracket, be careful not to bend the bracket. Also make sure the rubber sound insulator under the bottom

of the pump is in place. Install a new filter sock on the pump inlet and reconnect the pump wires. Be absolutely certain you have correct polarity. Replace the O-ring seal on the fuel tank opening, then put the pump and sender assembly back in the tank and tighten the locking ring by rotating it clockwise with a brass drift and hammer. Some pump/sender assembly units are secured by bolts.

9. If the fuel was removed from the vehicle, replace it.

CAUTION: Avoid the temptation to test the new pump before replacing the fuel tank by energizing it with a couple of jumper wires. Running the pump dry can damage it because the pump relies on fuel for lubrication and cooling.

---

10. Reconnect the electrical connector(s).
11. Reconnect the ground terminal at the battery.
12. Start the engine and check all connections for fuel leaks.

REVIEW QUESTIONS

1. What percent of a modern fuel tank can ever be filled with fuel?
   a. 40 percent
   b. 60 percent
   c. 90 percent
   d. 100 percent

2. Technician A replaces any leaking metal fuel tank. Technician B steam cleans a damaged fuel tank and then solders or brazes it. Who uses the best method?
   a. Technician A
   b. Technician B
   c. Both A and B
   d. Neither A nor B

3. When installing new flexible tubing, Technician A replaces damaged or worn clamps. Technician B replaces all old clamps. Who is right?
   a. Technician A
   b. Technician B
   c. Both A and B
   d. Neither A nor B

4. What component, located in the gasoline tank, prevents large contaminant particles from entering the fuel system?
   a. fuel filter
   b. diesel filter
   c. air filter
   d. strainer

5. A vehicle's engine performance indicates inadequate fuel, and the fuel pump is suspected. Technician A tests the fuel pump for pressure and volume while mounted on the engine. Technician B removes the fuel pump to inspect it. Who is right?
   a. Technician A
   b. Technician B
   c. Both A and B
   d. Neither A nor B

6. In a gasoline engine, ______________ occurs when the flame front fails to reach a pocket of air/fuel mixture before the temperature in that area reaches the point of self-ignition.
   a. volatility
   b. carbon depositing
   c. detonation
   d. flaring

7. After a fuel pump is installed, an internal squeaking noise is heard with each movement of the pump. Technician A considers the noise to be normal. Technician B examines the pump for a defective or loose component. Who is right?
   a. Technician A
   b. Technician B
   c. Both A and B
   d. Neither A nor B

8. Which of the following components exists in both a carbureted fuel delivery system and a fuel injector system?
   a. inertia switch
   b. electronic control assembly (ECA)
   c. power relay
   d. fuel pump

9. Which of the following statements is incorrect?
   a. Winter grade fuels give poorer fuel economy than low volatility fuels.
   b. The sulfur content in gasoline is limited to less than 0.01 percent.
   c. Gasoline vaporizes more easily at low altitudes.
   d. Poor vaporization can affect the distribution of fuel from cylinder to cylinder.

CHAPTER ELEVEN

SERVICING THE VACUUM/AIR SUPPLY SYSTEMS

Objectives

Upon completion of this chapter, you should be able to:

- Define vacuum and explain how it affects and assists vehicle operation and driveability.
- Define the air intake system components and their functions, including air intake ductwork, air cleaner/filter assembly, and intake air temperature controls.
- Conduct inspection and servicing procedures for a vacuum system, air intake system, and exhaust system.
- Explain the operation of a turbocharger.

The vacuum/air supply systems play a very important part in engine diagnostic procedures.

VACUUM SYSTEMS

The term *vacuum* refers to a pressure level that is lower than the earth's atmospheric pressure at any given altitude. The higher the altitude, the lower the atmospheric pressure.

Vacuum pressure can be measured in relation to atmospheric pressure. Atmospheric pressure is the pressure exerted on every object on earth and is caused by the weight of the surrounding air. At sea level, the pressure exerted by the atmosphere is 14.7 psi (absolute). Most pressure gauges ignore atmospheric pressure and read zero under normal conditions. For service purposes, atmospheric pressure is zero psi gauge. But, the usual measure of vacuum is in inches of mercury instead of psi. Other units of vacuum measure seen on automotive service gauges are kilopascals and the manometer, which measures pressure in inches of water.

Vacuum in any four-stroke engine is created by the downward movement of the piston during the intake stroke. With the intake valve open and the piston moving downward, a partial vacuum is created within the cylinder and intake manifold (Figure 11-1). The automotive engine creates a partial vacuum that is relatively continuous. In a typical four-cylinder engine, one cylinder is always at some stage of the intake cycle. The amount of vacuum created is partially related to the positioning of the choke (on carbureted vehicles) and to the throttle plate(s). The throttle plate not only admits air or air/fuel into the intake manifold, it also helps to control the amount of vacuum available during engine operation. At closed throttle idle, the vacuum available is usually between 15 to 22 inches. At a wide-open throttle acceleration, the vacuum can drop to zero.

SHOP TALK

A perfect vacuum—29.92 inches mercury—is the complete absence of pressure (zero psi absolute). It is interesting to note that at 32 degrees Fahrenheit (0 degrees Celsius), water will boil in a "perfect vacuum." Remember that a perfect vacuum would prevent the engine from running. If the engine were airtight, air and fuel would not be able to enter the engine for combustion.

TYPICAL VACUUM-CONTROLLED SYSTEMS

During the first fifty years of automotive vehicle production, the controls for gasoline engines were

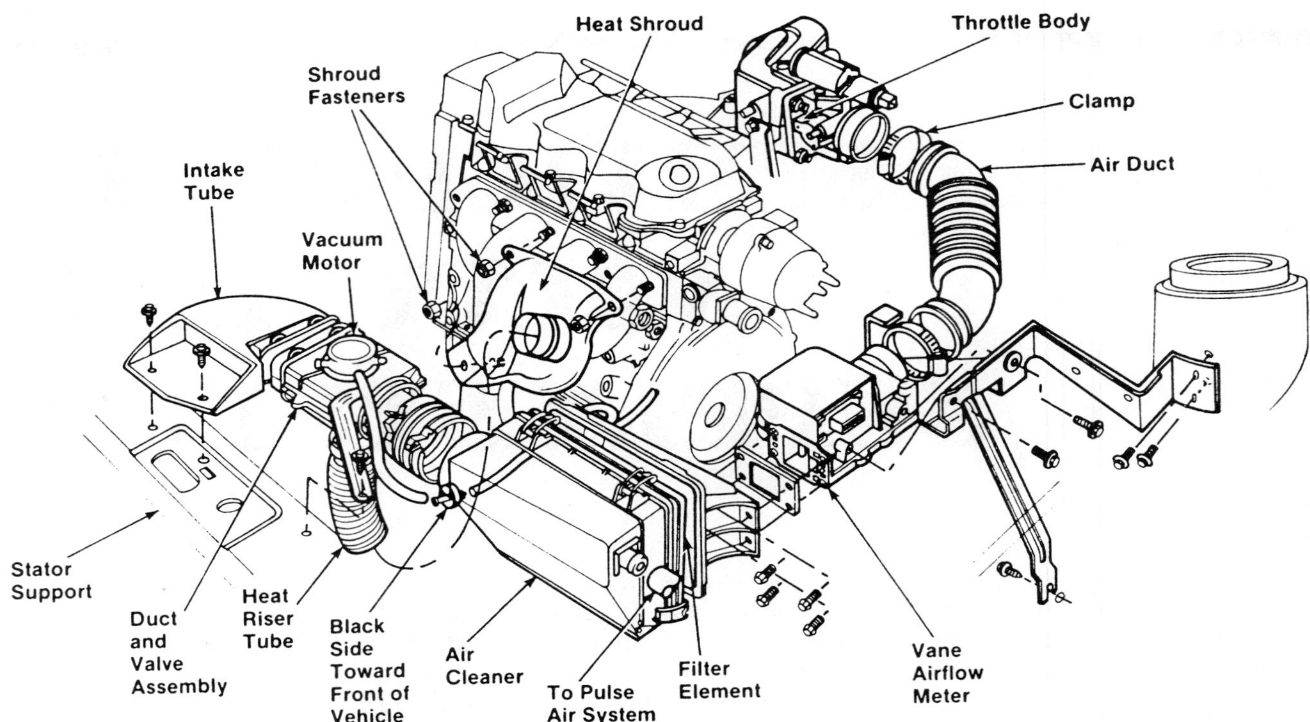

FIGURE 11–1 The automotive engine creates a partial vacuum

relatively distinct. Early electrical, ignition, fuel, and vacuum systems were easy to service because they had a limited operational interrelationship. For example, the effect and application of the vacuum system were quite apparent when operating windshield wipers; under heavy acceleration, the wipers would stop. Spark advance was vacuum controlled and as technology progressed to meet improved fuel economy and lower emissions, vacuum applications were increased.

Engine vacuum is used for emission controls, improved fuel economy, driveability, and to operate accessories. Electronic controls and vacuum applications have now made previously distinct systems interdependent. The following describe the engine systems that have vacuum inputs and controls:

- *Fuel Induction System.* Certain vacuum-operated devices are added to carburetors and some central fuel-injected throttle bodies to ease engine start-up, warm-and-cold engine driveaway, and to compensate for an air conditioner load on the engine. Chapters 8 and 9 describe the use of vacuum in the fuel induction system operation and servicing.
- *Emission Control System.* While some emission control output devices are solenoid or linkage controlled, many operate on vacuum. This vacuum is usually controlled by solenoids that are opened or closed, depending on electrical signals received from the electronic control assembly (ECA). The PCV valve is an important factor in the air induction system; it delivers gasses at normal cruise speeds when the engine can best handle them. Other systems use switches that are controlled by engine coolant temperature, such as a ported vacuum switch (PVS), or by ambient air such as a temperature vacuum switch (TVS). The complete use of engine vacuum in emission control operation and servicing is covered in Chapter 14.
- *Accessory Controls.* Engine vacuum is used to control operation of certain accessories, such as air conditioner/heater systems, power brake boosters, speed-control components, automatic transmission vacuum modulators, and so on.
- *Air Intake and Fuel System.* Vacuum is used to draw filtered air into the engine and to control the system devices, which, in turn, are used to control the temperature of the air allowed into the engine. This improves fuel vaporization so that fuel will burn more effi-

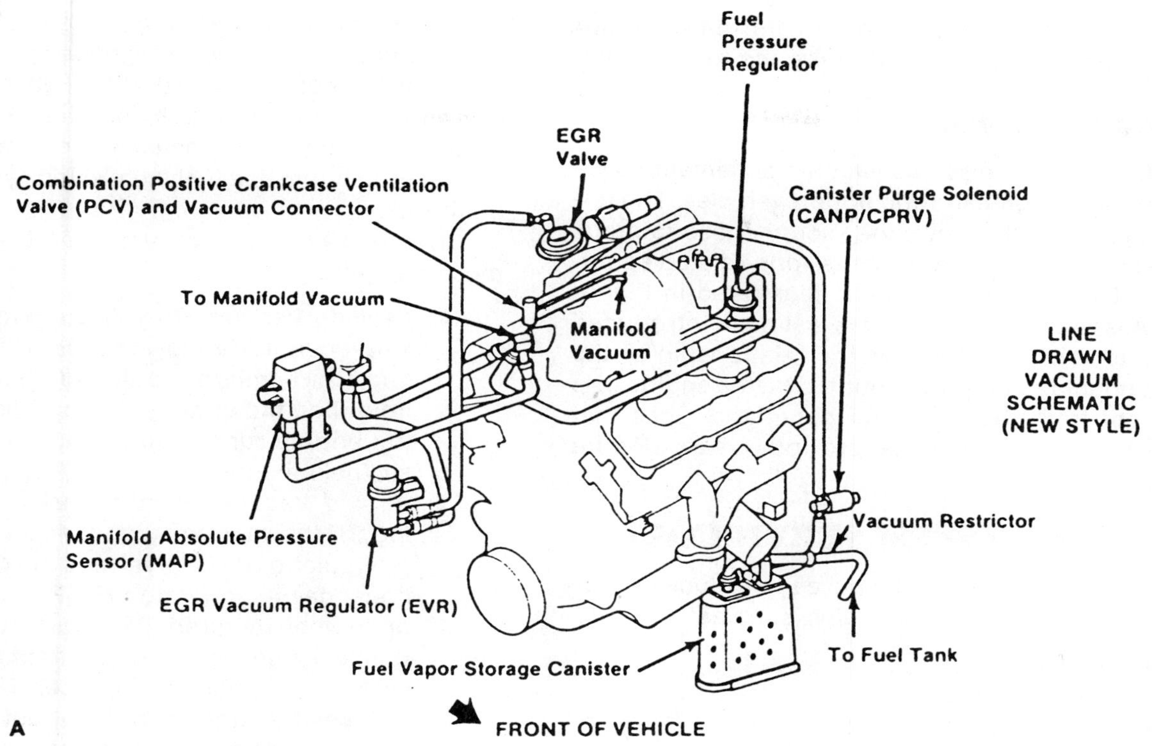

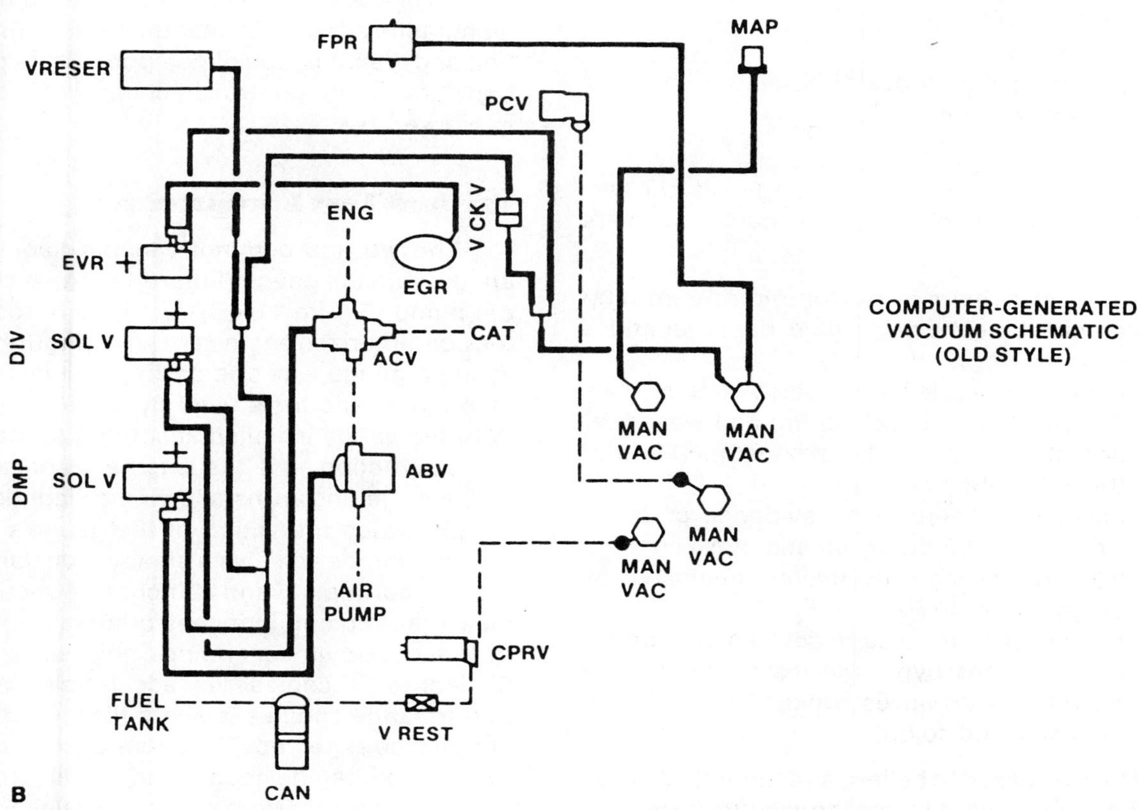

FIGURE 11-2 (A) New style line-drawn vacuum schematic; (B) old style computer-generated vacuum schematic

ciently. This chapter covers the operation and servicing of air intake and fuel systems.

Vacuum Schematic

An engine emissions vacuum schematic is required on all domestic and import vehicles. The following schematic shows the vacuum hose routings and vacuum source for all emissions-related equipment. The vacuum schematic illustrated in Figure 11-2A is the new type that has just been introduced in some vehicles. It shows the relationship of the components as they are mounted on the engine. It is different from the computer-generated types that have been used over the previous years (Figure 11-2B).

VACUUM SYSTEM SERVICING

Vacuum system problems can produce or contribute to driveability symptoms, such as:

- Stalls
- No start (cold)
- Hard start (floods hot)
- Backfire (deceleration)
- Rough idle
- Poor acceleration
- Rich or lean stumble
- Overheating
- Detonation, or knock or pinging
- Rotten eggs odor
- Poor fuel economy

As a routine part of problem diagnosis, the service technician suspecting a vacuum problem should first

- Inspect vacuum hoses for improper routing or disconnections (engine decal identifies hose routing).
- Look for kinks, tears, or cuts in vacuum lines.
- Check vacuum hose routing and wear near hot spots, such as the exhaust manifold or the EGR tubes.
- Make sure there is no evidence of oil or transmission fluid on vacuum hose connections (valves can become contaminated by oil getting inside).
- Inspect vacuum system devices for damage (dents in cans; bypass valves; broken nipples on VCV or PVS valves; broken "tees" in vacuum lines, and so on).

If there is reason to believe a vacuum leak exists, this schematic helps to locate the vacuum hose routing for emission-related vacuum components only. If there is a vacuum leak somewhere in the vacuum system, it could be in a nonemission-related component. Vacuum is used to regulate speed control, heater or A/C controls, and the like. If there is a crack or broken vacuum line in these areas, the vacuum system is affected in the same manner as a cracked or broken vacuum line in the emission system vacuum hoses.

The following vacuum system problems are the most common:

- *Leaking Vacuum Hose (Rough Idle).* A broken vacuum line allows a vacuum leak. A vacuum leak admits more air into the intake manifold than the engine is calibrated for. Then the engine runs roughly due to the leaner air/fuel mixture.
- *Kinked Vacuum Hose (Spark Knock/Pinging).* If the vacuum hose to the exhaust gas recirculation (EGR) valve is kinked, the exhaust gas recirculation (EGR) valve will not open when required. The engine is calibrated to allow a certain amount of exhaust gas to enter the combustion chamber. This not only reduces the amount of NO_x, but also cools down the combustion chamber to prevent spark knock and pinging.

To check vacuum controls, refer to the vehicle manufacturer's service manual for the correct location and identification of components. Typical locations of vacuum-controlled components are shown in Figure 11-3.

Vacuum Test Equipment

The two most common vacuum diagnostic tools are the vacuum gauge (Figure 11-4A) and the vacuum pump (Figure 11-4B). Until the introduction of the computerized engine analyzer (Figure 11-5), the vacuum gauge was one of the most important engine diagnostic tools used by service technicians. With the gauge installed according to manufacturer's instruction and the engine warm and idling properly, getting an instant engine diagnosis is easy. Simply watch the action of the gauge's needle. A healthy engine will give a steady, constant vacuum reading between 17 and 22 inches. (A vacuum gauge measures vacuum in inches of mercury.) On some four-and six-cylinder engines, however, a reading of 15 inches is considered acceptable. With high-performance engines, a slight flicker of the needle can also be expected. The operator should discount for the fact that the gauge reading will drop about 1 inch for each 1,000 feet above sea level. Figure 11-6 shows some of the common readings and what engine malfunctions they indicate.

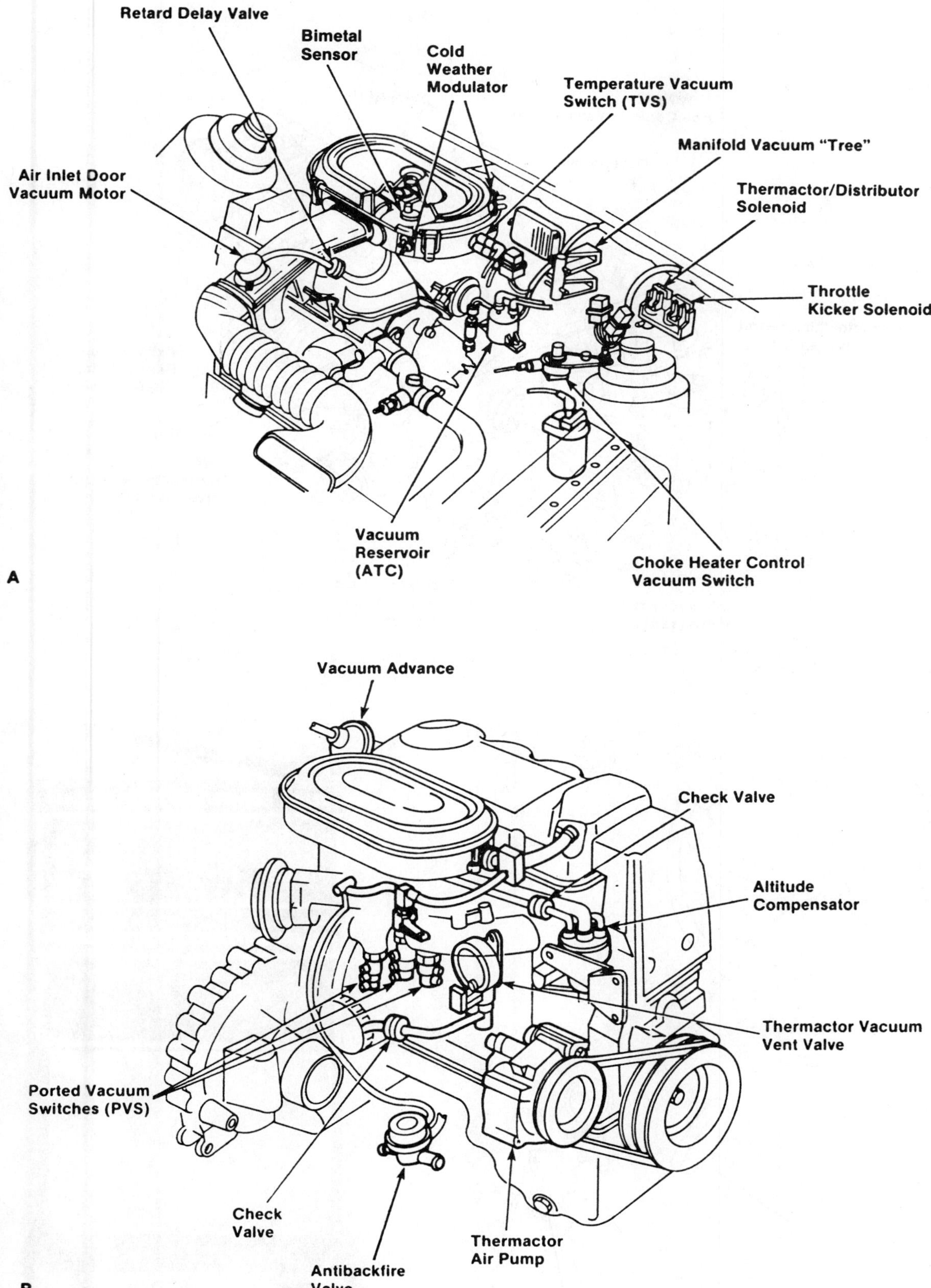

FIGURE 11-3 Vacuum devices and controls: (A) in dash panel inside engine compartment; (B) in a 1.6-liter engine with PVS; (C) in a 1.6-liter engine with an EGR valve

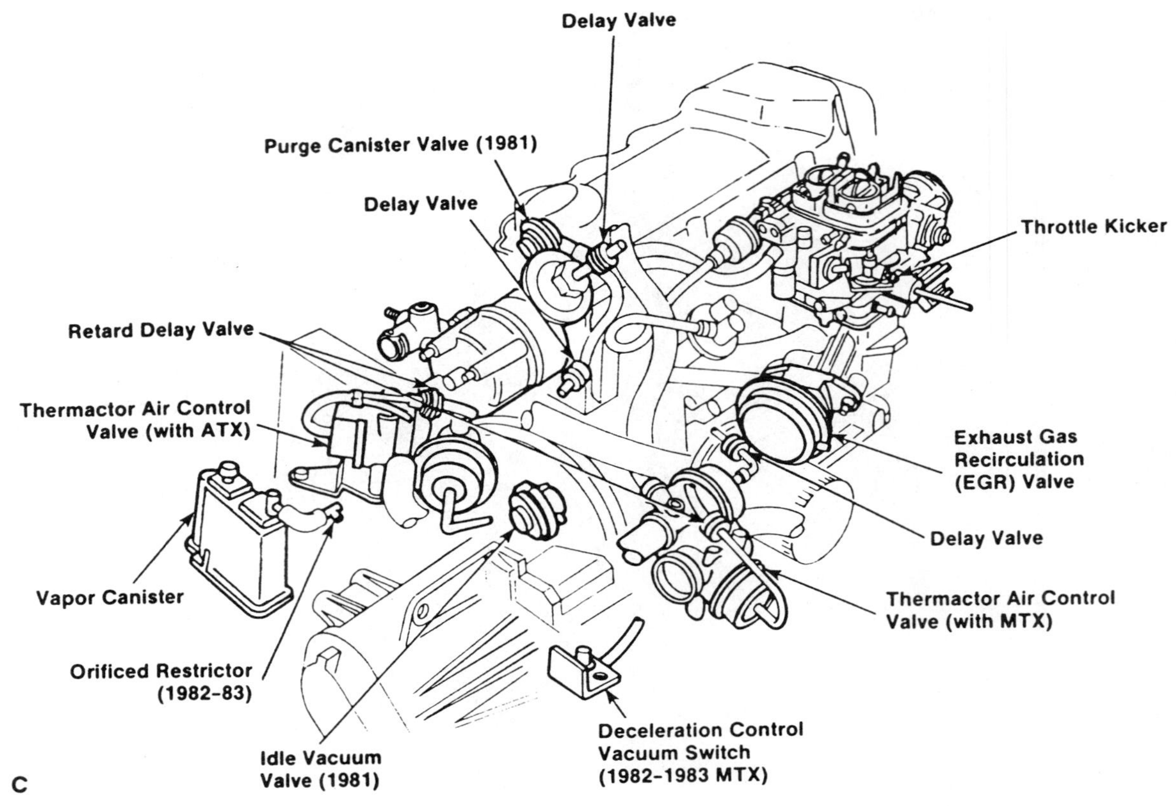

FIGURE 11-3 Continued

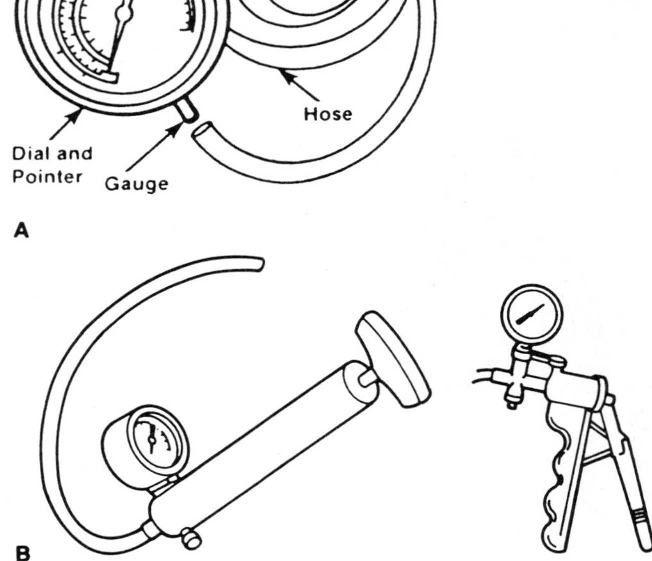

FIGURE 11-4 (A) Vacuum gauge and adapters; (B) vacuum pumps

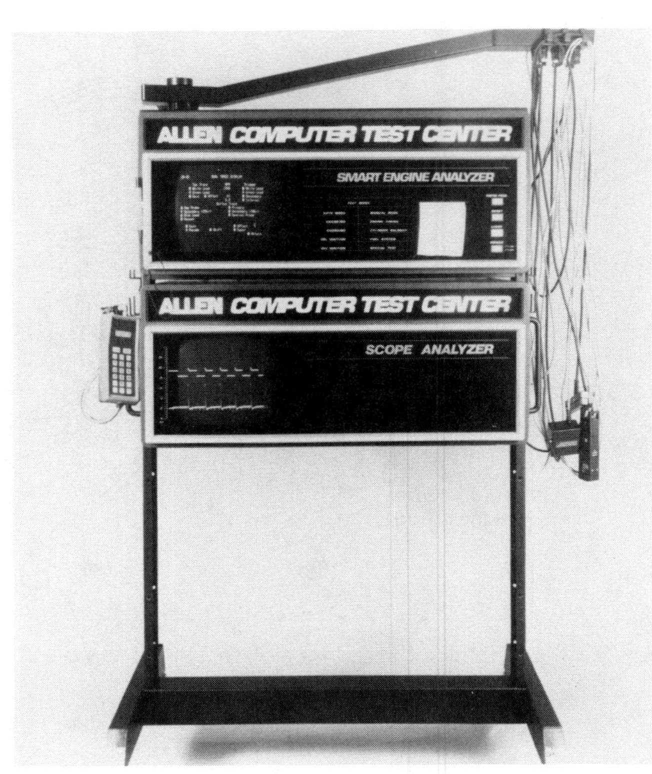

FIGURE 11-5 Computerized engine analyzer *(courtesy of Allen Testproducts Division)*

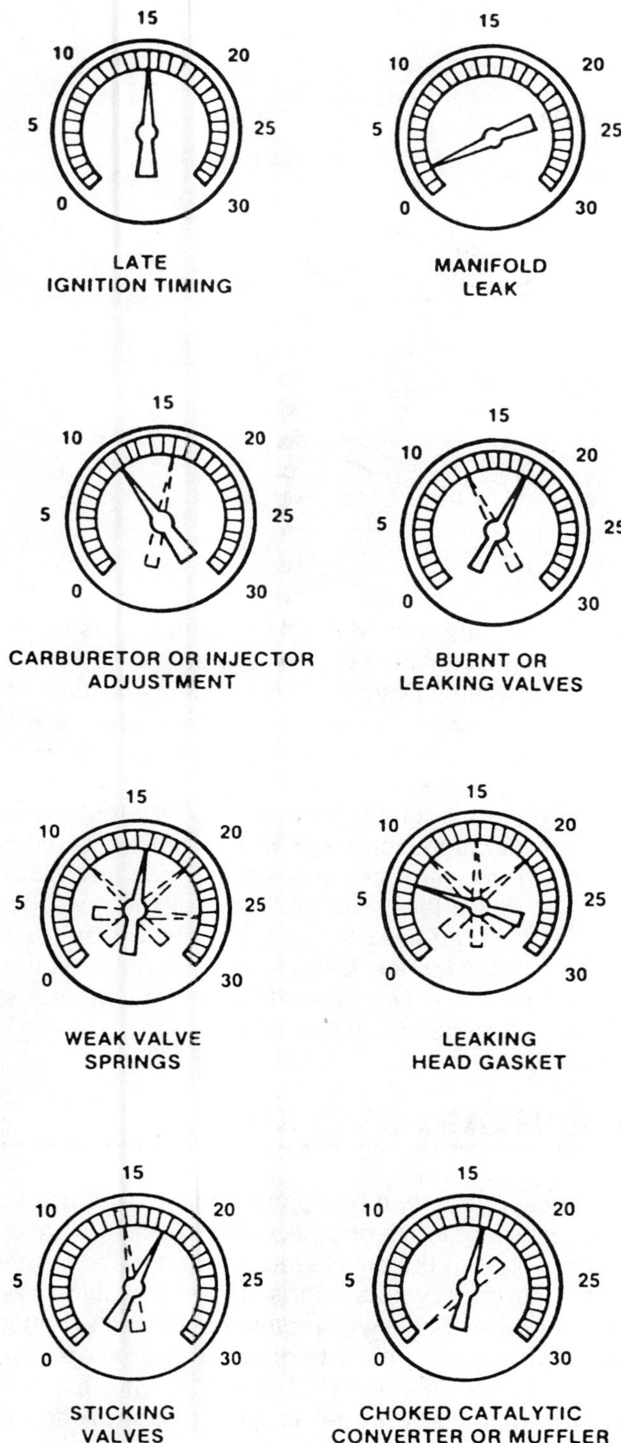

FIGURE 11-6 Common vacuum gauge readings

A vacuum gauge and hand-held vacuum pump are used to check many of the vehicle's vacuum emission valves (Figure 11-7). Testing procedures for vacuum emission valves are in Chapter 14.

SHOP TALK

When a vacuum valve is closed, vacuum will be blocked from passing through the valve. When a valve opens, vacuum will pass through. Refer to the vehicle's service manual for correct valve operating conditions and test procedures.

Vacuum Hose Lines

The major cause of problems in a vacuum system occurs in its hose lines. Check all vacuum hoses carefully for cracks, breaks, kinks, hardening, and loose or broken connections (Figure 11-8). Any defective hoses should be replaced one at a time to avoid misrouting. Refer to the vacuum schematic located inside the engine compartment or check the vehicle's service manual for correct hose routing if more than one hose is disconnected.

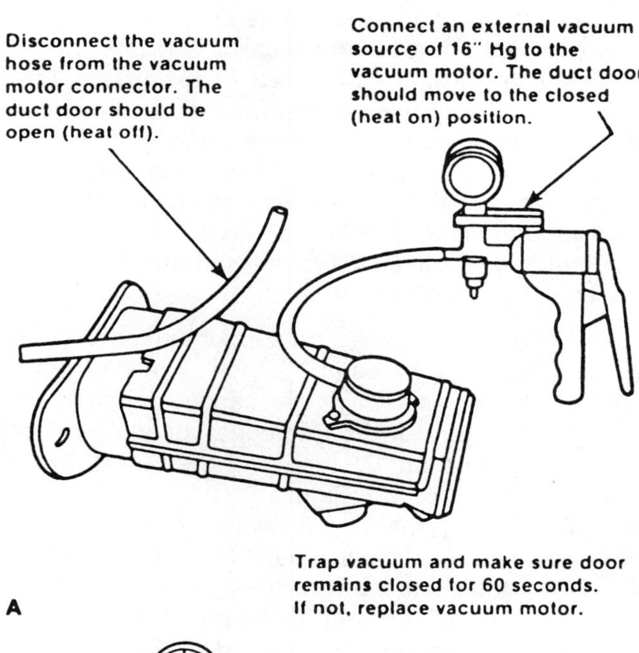

FIGURE 11-7 (A) Hand-operated vacuum pump being used to test an air cleaner duct vacuum motor; (B) two-port/four-port VCV or PVS test

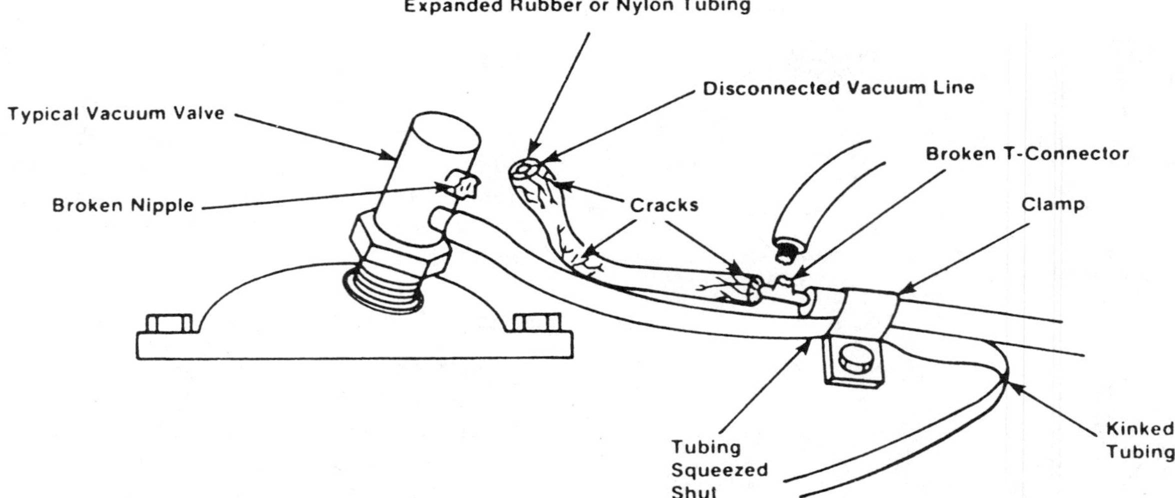

FIGURE 11-8 Vacuum line defects

OEM equipped vacuum lines are installed in a harness consisting of 1/8-inch or larger outer diameter and 1/16-inch inner diameter nylon hose with bonded nylon or rubber connectors. Occasionally, a rubber hose might be connected to the harness. The nylon connectors have rubber inserts to provide a seal between the nylon connector and the component connection (nipple). In recent years, many domestic car manufacturers have been using ganged steel vacuum lines.

The following is a typical vacuum hose replacement procedure:

1. If a nylon hose is broken or kinked and the damaged area is 1/2 inch or more from a connector, the hose can be repaired by cutting out the damaged section but not more than 1/2 inch and then installing a rubber union (Figure 11-9A).
2. If the remaining hose is too short or the damaged portion is more than 1/2 inch, replace the entire hose and connectors with rubber vacuum hoses and a tee. Use existing service stock of 5/32-inch hose, 7/32-inch hose, and tees (Figure 11-9B). The circled number in these illustrations identifies the same connection points on both the original and the repaired harnesses.

CAUTION: Care must be exercised to keep the nylon parts routed away from hot components. In addition, holes might be worn into the nylon hoses if allowed to rub against rough surfaces.

3. If only part of a nylon connector is damaged or broken, cut the connector apart (as illustrated in Figure 11-9C) and discard the damaged half of the harness. Replace it with rubber vacuum hoses and a tee from service stock.
4. Figure 11-9D shows how to replace a damaged harness with available service stock—rubber hoses (5/32-inch and 7/32-inch), tees, and elbows. Cut the hoses to the required lengths but never less than 1-1/2 inches. Be sure to properly route the hoses, using the identification numbers provided for this purpose.

AIR INTAKE SYSTEMS

This chapter will be concerned with the use of air in the automotive engine. It will cover the intake of fresh air into the air cleaner system to its expulsion of harmful pollutants through the vehicle's exhaust system. It also will explain how these exhaust gasses (air) can be used to drive a turbocharger for better engine efficiency. Chapter 14 will describe how air, as well as vacuum, is used to control engine emissions.

The air intake or induction system has the following main functions:

- Provide the air that the engine needs to operate
- Filter the air to protect the engine from wear
- Monitor air temperature for more efficient combustion and reduction of hydrocarbon (HC) and carbon monoxide (CO) emissions

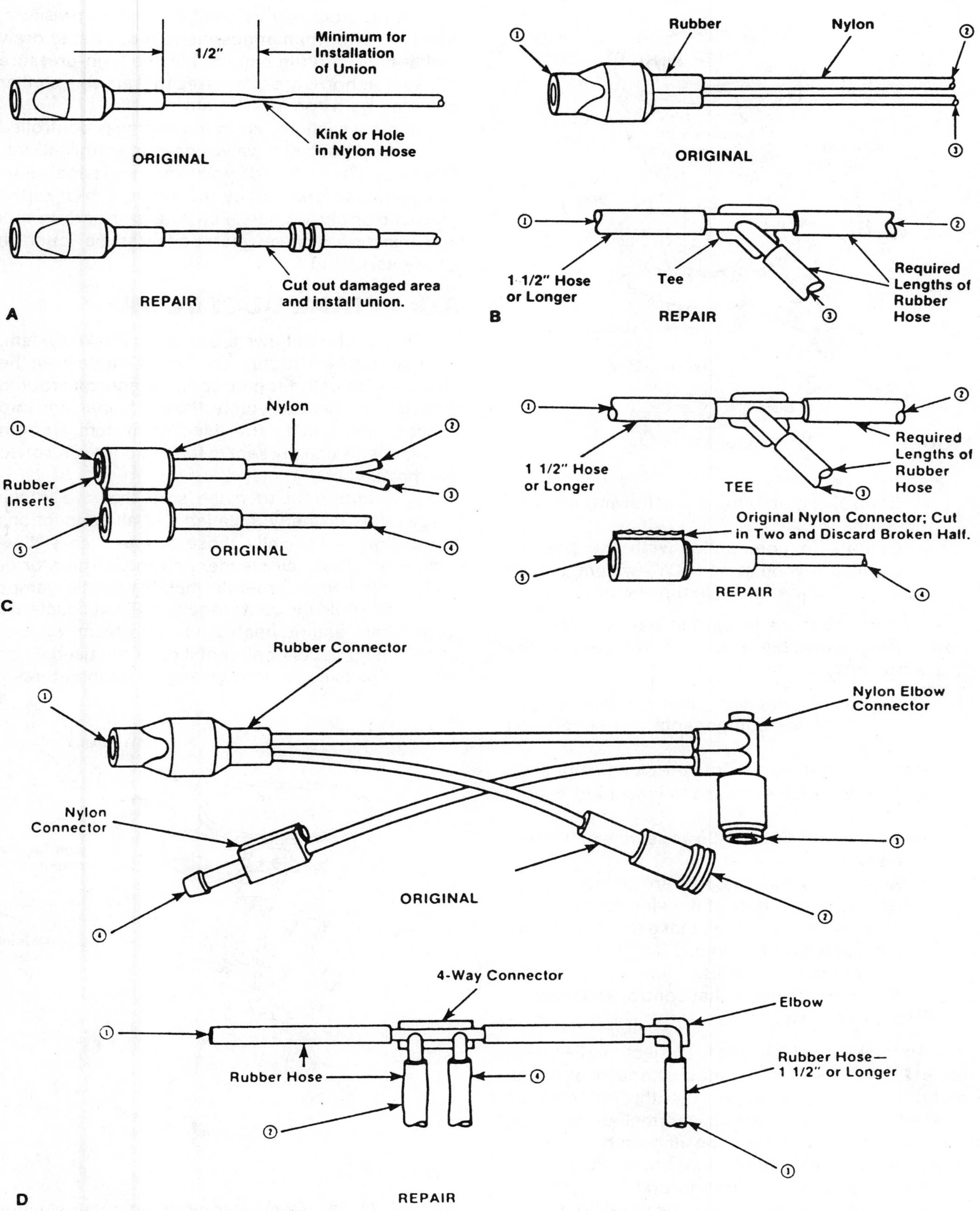

FIGURE 11-9 (A) Rubber union installation; (B) hose replacement procedure; (C) damaged nylon connector repair; (D) replacement of damaged harness

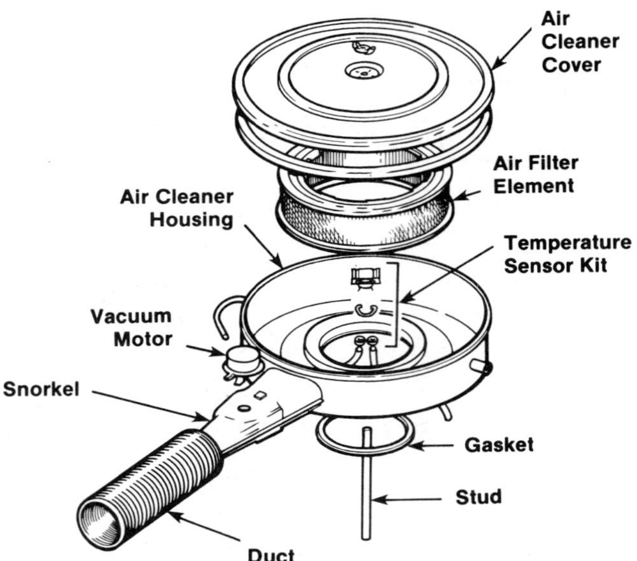

FIGURE 11-10 Air cleaner assembly

- Control the amount of air flowing into the engine
- Operate in conjunction with the positive crankcase ventilation (PCV) system to burn the crankcase fumes in the engine

To perform these tasks, the typical intake air cleaner/filter assembly (Figure 11-10) contains the following parts:

- *Air Cleaner Housing.* (metal or plastic container). Holds components and supporting parts.
- *Air Ducts.* Carry air into the cleaner housing.
- *Filtering Element.* Cleans inkeeping engine air.
- *Air Cleaner Cover.* Allows service of filtering element.
- *Air Cleaner Gaskets.* Prevent air leakage between components of the air cleaner.
- *Snorkel.* Rigid air inlet that extends out from the air cleaner housing.
- *Temperature Controls.*
- *Other components* that control and monitor the intake air.

Air intake for electronic fuel injection (EFI) systems (Figure 11-11) is very much the same as on the carbureted systems. When the throttle plate is opened, the outside air under normal atmospheric pressure is forced through the air horn because the pressure in the intake manifold is lower. The primary difference between carburetion and fuel injection systems is that the EFI models use injectors, controlled by a computer, to spray fuel into the engine.

Carburetors rely on intake manifold pressure, which is lower than atmospheric pressure, to draw fuel and air into the engine. For the high-pressure EFI, two air horns are often used that are very similar to the air horns on a carburetor.

In most systems, air to the engine is controlled by a butterfly valve, or valves on some applications. The valve(s) is actuated by a linkage and pedal cable arrangement, operated by the driver, similar to the one used on carburetors. Full details on carburetors and injection system operations can be found in Chapters 12 and 13.

AIR INTAKE DUCTWORK

Figure 11-12 shows a typical air intake system. Cool air is drawn through the fresh air tube from the leading edge of the engine compartment and routed through a preheat assembly, the air cleaner, and into the carburetor. In the fuel injection system, air from the cleaner is usually sent to the air sensor and on to the throttle body.

It is important to make sure that the intake ductwork is properly installed and all connections are airtight—especially those between an airflow sensor or remote air cleaner and the carburetor or fuel injection fold. Generally metal or plastic clamps are used to hold the ducts together. Plastic ducts are used where engine heat is not a problem. Special paper-metal ducts or all-metal ones are used when they will be exposed to high engine temperatures.

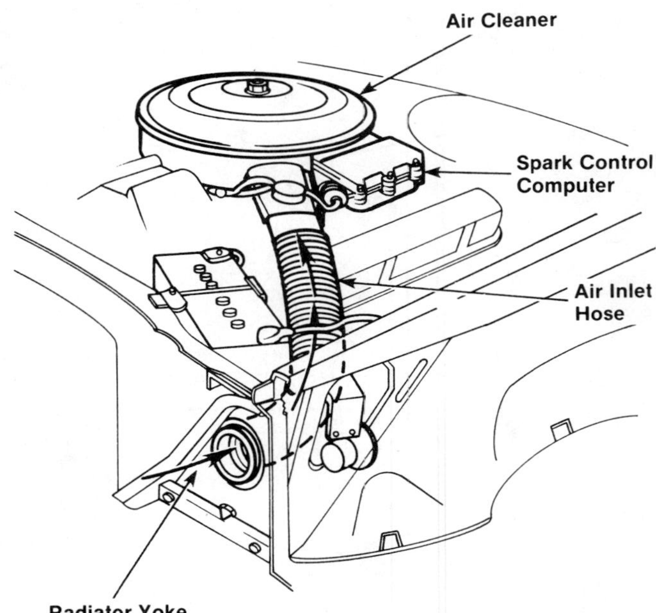

FIGURE 11-11 Computer mounted on air cleaner housing controls ignition spark.

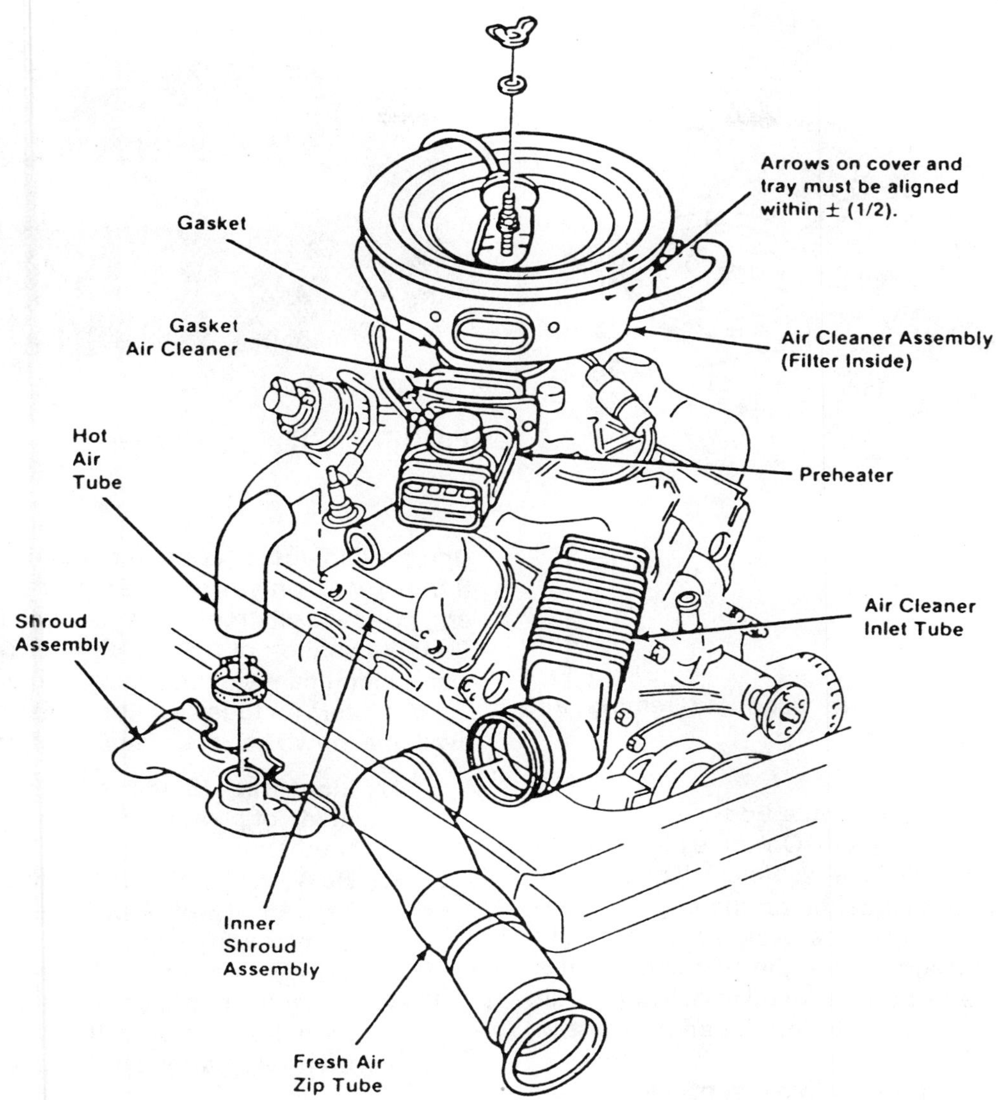

FIGURE 11–12 Typical air intake system

AIR CLEANER/FILTER

The primary function of the air filter is to prevent airborne contaminants and abrasives from entering the engine through the carburetor with the air/fuel mixture. That is, the air portion of the air/fuel mixture can carry minute particles of foreign matter such as dust, dirt, and exhaust carbons. Without proper filtration, these contaminants can cause serious damage and appreciably shorten engine life. Remember that all incoming air must pass through the filter element before entering the engine.

The air filter unit used on most gasoline driven cars and light-duty trucks is located inside the air cleaner housing (Figure 11–13), which can be mounted directly to a flange on the carburetor air

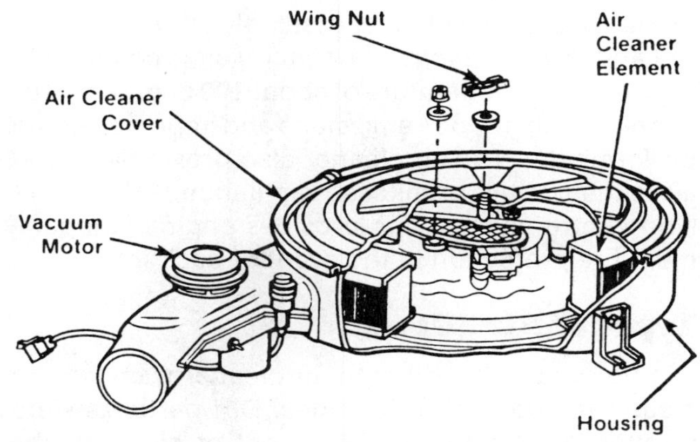

FIGURE 11–13 Carbureted engine air cleaner design

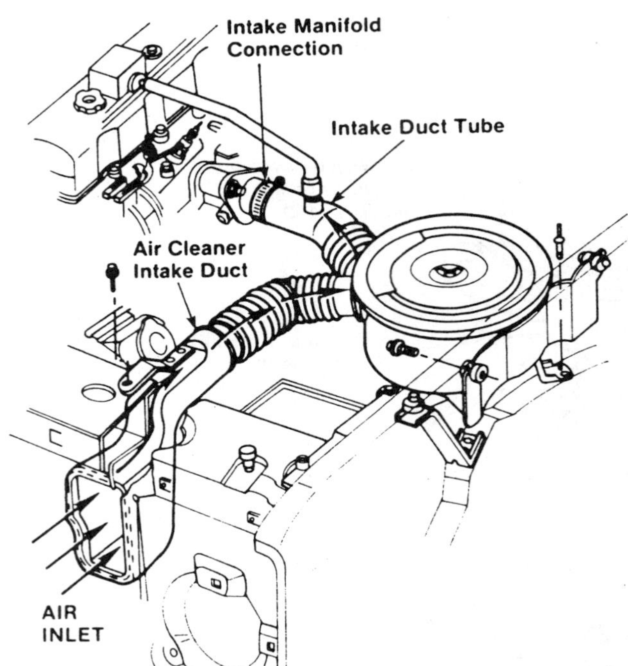

FIGURE 11-14 Ducts are used for remote air cleaner on a diesel engine

horn, the fuel injection throttle body, or the intake manifold (on diesel engines). On some automobiles and trucks (especially diesel vehicles), it can also be mounted in a remote location on the engine or engine compartment to reduce overall engine height (Figure 11-14). In such cases, the air cleaner unit is connected to the carburetor (or injector system), air horn, throttle body, or manifold by an air transfer tube or duct.

The air cleaner provides other services to the air intake system. For example, airflow through the carburetor and intake valve operation creates "noise" from the intake manifold. The air cleaner assembly helps to eliminate or reduce this noise. On late-model cars, the air cleaner assembly provides a control and mixing function. Incoming air is maintained at a constant temperature of about 100 degrees Fahrenheit to help reduce emissions and improve engine performance. The air cleaner also provides filtered air to the positive crankcase ventilation (PCV) emission control system and provides engine compartment fire protection in the event of backfire.

Types of Air Filters

For years, the oil bath air cleaner was used to clean the intake air of engines. Dirt particles were trapped on the surface of a pool of oil when the airstream was forced to make an abrupt 180 degree turn. However, the efficiency of such an air cleaner

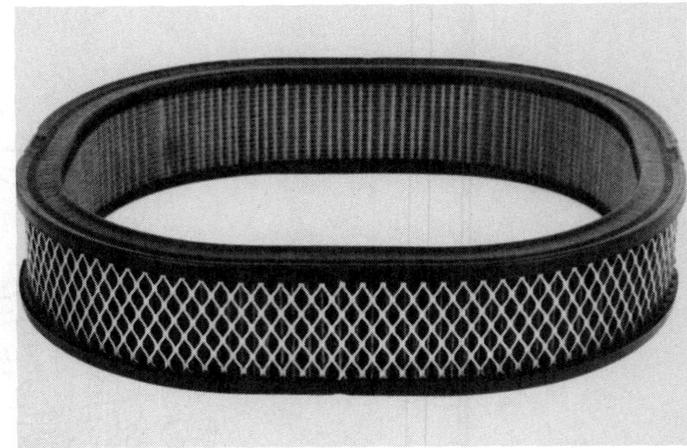

FIGURE 11-15 Dry air filter

(the percentage of contaminants it removed from the air) was only around 95 percent, and at low engine speeds it dropped by as much as 25 percent. So the invention of the dry air filter (Figure 11-15) was a significant step in prolonging engine life.

Air filters for today's automobiles and trucks are available in two basic air filter classifications:

1. **Light-Duty (Standard).** These are usually made of pleated paper or oil-wetted polyurethane.
2. **Heavy-Duty.** Better heavy-duty air filters have an oil-wetted polyurethane outer cover over a dry-type paper element (Figure 11-16). The polyurethane effectively traps larger dirt particles on its surface. The finer particles that pass through the polyurethane are trapped by the paper media. The

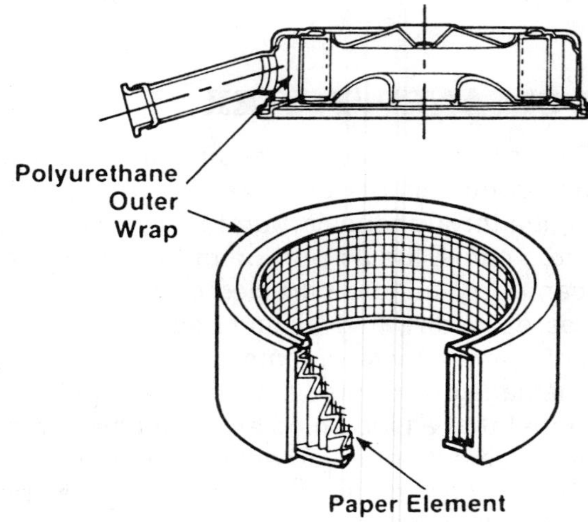

FIGURE 11-16 Polyurethane outer wrap and paper element

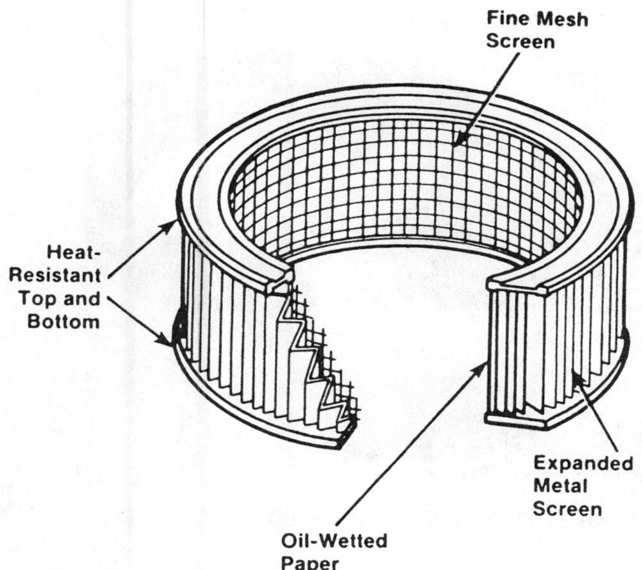

FIGURE 11–17 Automotive air filter design features

polyurethane outer cover can be removed, cleaned, and reinstalled, thus extending filter life.

When selecting a new air filter keep these design features in mind (Figure 11–17):

- Fine mesh screen on the inside reduces the possibility of fire hazards caused by engine backfire.
- Heat-resistant plastisol on the top and bottom with special sealing beads provides a positive dust seal over the wide range of automotive operating conditions.
- Oil-wetted resin-impregnated paper provides long-lasting filter efficiency.
- Wire or expanded metal outer screen adds construction strength and protects against accidental paper damage during handling and shipping.

Air Filter Servicing

The engine manufacturer's recommended air filter replacement interval is usually specified in the vehicle owner's service manual. However, replacement of the air filter might be required on a more frequent schedule if the vehicle is subjected to continuous operation in an extremely dusty or severe off-the-road environment.

The air filter should be replaced on a regular basis because a damaged air filter can accelerate wear of engine cylinder walls, pistons, and piston

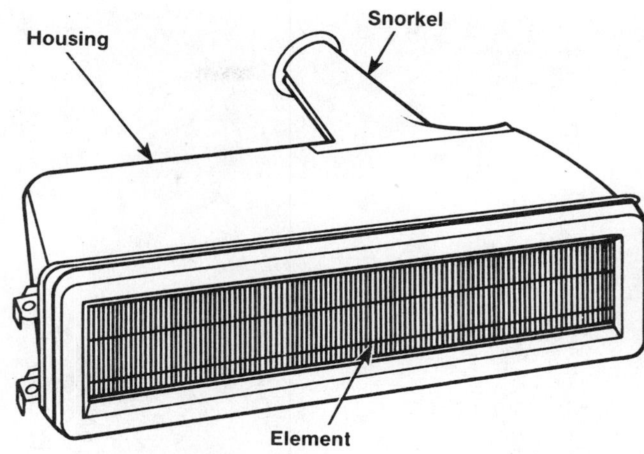

FIGURE 11–18 Paper element or cartridge

rings; an extremely dirty air filter can act as a choke, affecting engine performance and fuel consumption. To replace an air filter, proceed as follows:

1. Identify the correct replacement air filter for the engine.
2. Remove the wing nut, or other fasteners, from the engine air cleaner assembly; then remove the cover (Figure 11–18). Some unitized air cleaners with nonremovable covers must be replaced as a total assembly.
3. The air filter should easily lift out of the air cleaner assembly.
4. Visually inspect the air filter for damage and dirt accumulation. An air filter that shows signs of damage or heavy dirt accumulation should be replaced with a new air filter of the specified type.
5. Reassemble the air cleaner cover and wing nut or other fasteners. Do not overtighten the wing nut.

Certain types of heavy-duty air filters can be cleaned and reused. These filters have a heavy paper media encased in an element with metal end caps. Generally, reusable heavy-duty air filter elements should be replaced with a new element after six cleanings, or once each year.

WARNING: When working on pressurized gasoline systems, be sure to wear safety glasses. Shop rags should be used to wipe up any fuel leaks from the filter during service.

Before attempting to clean a reusable element, always examine it carefully. Look for a torn or punc-

FIGURE 11-19 Low-pressure shop air can be used to blow dirt from the element. Air pressure must be applied on the inner surface only since dirt has collected on the outer surface.

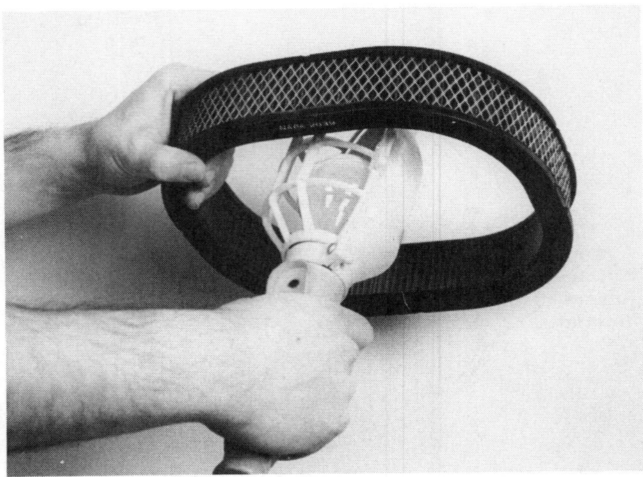

FIGURE 11-20 Inspect element with light.

tured paper media, bent end covers, or pinholes. If end gaskets are damaged or missing, replace the element. Air filter elements that have precleaner fins or tubes should be inspected for dirt accumulations. A stiff nylon or fiber bristle brush can be used to remove light coatings of dust.

If dust is the main contaminant, compressed air can often clean the element to near "new service" capability. Heavy-duty air filter elements that are lightly coated with dust can be cleaned by directing a stream of air up and down the pleats on the clean air side of the element (Figure 11-19).

CAUTION: Do not allow air pressure at the nozzle to exceed 30 psi and maintain a 6-inch minimum distance from the nozzle to the paper. Air pressure must be applied only on the inner surface since dirt has collected on the outer surface.

Heavy-duty elements that are extremely dirty or have soot or oil vapor deposits require a more thorough cleaning. Clean these elements as follows:

1. Remove all loose dust with a flow of water or compressed air.
2. Immerse the element in a solution of approved filter cleaner for 15 minutes or more. Do not allow the temperature of the solution to exceed 140 degrees Fahrenheit. Occasionally agitate the element to loosen particles.
3. Remove the element and rinse it thoroughly by running clean water through it from the clean air side to the dirty side until the water comes out clean. Use a hose without a nozzle and do not permit water pressure to exceed 40 psi.
4. Allow the element to air dry completely before reinstalling. A flow of warm air (temperature not to exceed 140 degrees Fahrenheit) can be used to shorten drying time.

Inspect the element with a light bulb after cleaning with compressed air or washing (Figure 11-20). Pinholes or slight ruptures will admit enough airborne dirt to render the element unfit for further service. Use of a damaged filter can cause rapid failure of piston rings.

After cleaning, check the gaskets on the ends for condition and proper location. If they are damaged or missing, replace the element.

INTAKE AIR TEMPERATURE CONTROLS

Air temperature is controlled by the heated or thermostatic air cleaner intake system. The heated air intake system aids vaporization of the fuel on cold start-up by warming the intake air at the exhaust manifold.

The air valve in the snorkel (cold air intake) should close off outside (ambient) air when a cold engine starts. At 75 to 105 degrees Fahrenheit in the air cleaner (depending on temperature setting of the bimetal sensor), the bimetal sensor causes the air valve to open to outside air and shut off heated air (Figure 11-21).

When a vacuum signal is applied to the motor, it closes the door to cold air and allows warm air to enter the air cleaner. When a vacuum signal is not applied, the door opens to cold air.

A bimetal sensor is installed in the air cleaner housing tray. When the sensor is below a specified temperature, it allows a vacuum signal to flow through it to the air cleaner vacuum door motor. At a given increase in temperature, the sensor bleeds off vacuum, permitting the vacuum motor to open the duct door to allow fresh air to enter the engine. The air cleaner sensor can be checked with a magnetic base thermometer (Figure 11-22). As a precaution, the thermometer can be taped in place to keep it from being drawn into the air horn.

FIGURE 11-22 Checking the air cleaner sensor with a magnetic base thermometer

The cold weather modulator is clipped to holes in the air cleaner housing assembly and senses the temperature of the inlet air. A bimetal disc actuates a check valve inside the modulator, depending on the air temperature.

- *Open.* The modulator passes vacuum freely either way.
- *Closed.* The modulator can trap vacuum at an actuator or block vacuum from being applied, depending on how the hoses are connected.

When the modulator is used to trap vacuum, it will trap the vacuum signal to the air cleaner vacuum door motor. This prevents the door from switching to cold inlet air when the vacuum drops during acceleration. After the engine warms up, there is no need to trap the vacuum to the door. Some intake air temperature control systems use a vacuum retard delay valve (VRDV) instead of the cold-weather modulator. Further details on the operation and servicing of air intake systems can be found in Chapter 14.

If the air filter becomes very dirty, the dirt will partially block the flow of air into the fuel charging assembly (Figure 11-23). Without enough air, the engine will constantly burn a rich air/fuel mixture, using up more fuel than it normally would. The use of extra fuel means poor fuel economy, and the reduced airflow can cause lack of power. Also, if the heated air inlet door sticks open, heated air cannot reach the engine, even if it is cold. The cold air that does enter the engine will not allow the fuel to vaporize completely. Liquid fuel will not ignite as easily as fuel that is vaporized. Typical vacuum/air intake problems are given in Table 11-1. Typical thermostatic air cleaner troubles are illustrated in Figure 11-24.

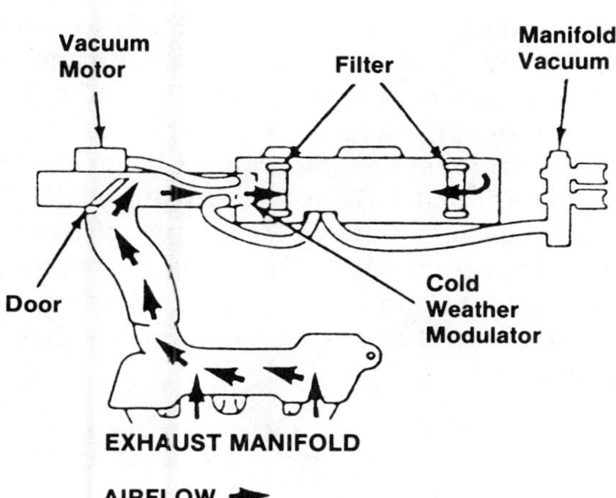

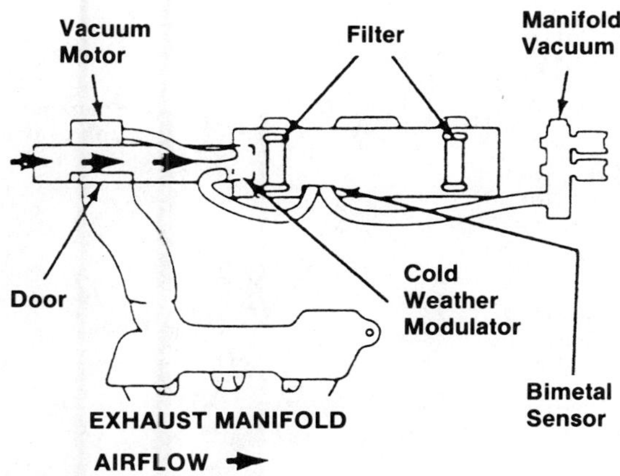

FIGURE 11-21 Heated air intake system operation

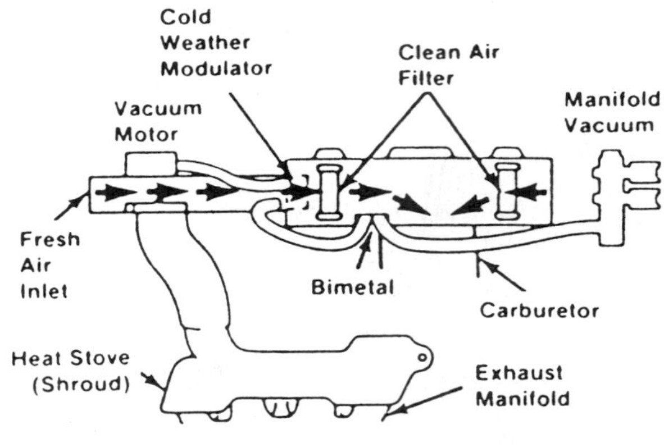

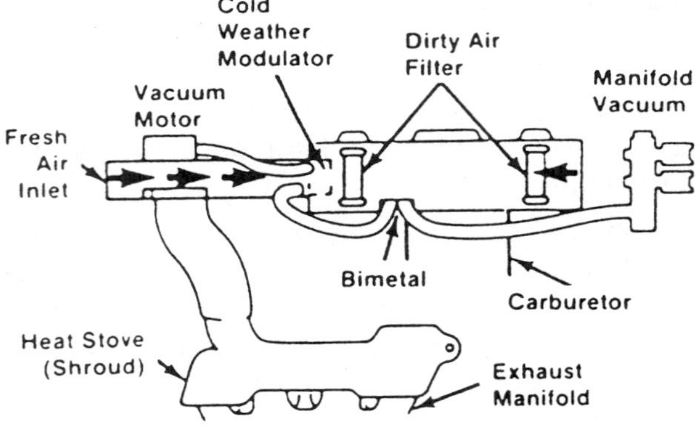

FIGURE 11-23 Restricted airflow—dirty air filter

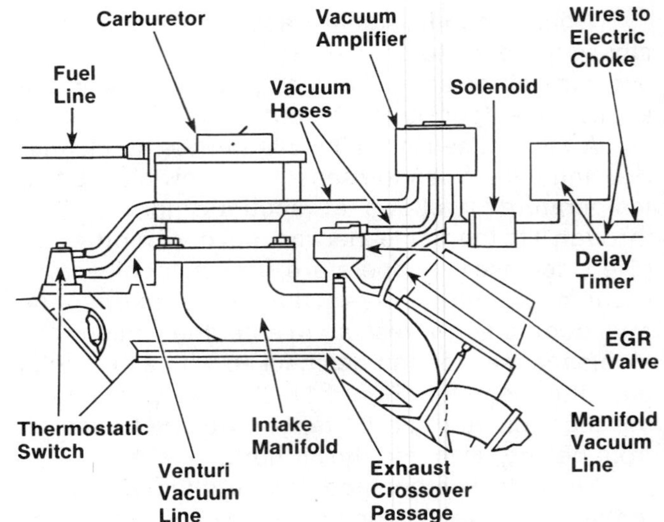

FIGURE 11-25 Exhaust gas recirculation (EGR) system

The exhaust gas recirculation (EGR) control reduces NO_X by diluting the air/fuel mixture and lowering combustion temperatures. Recirculating a small amount of exhaust back into the intake manifold keeps combustion temperatures below the 2500 degree Fahrenheit NO_X formation threshold. The heart of the system is the EGR valve, which opens to allow engine vacuum to siphon exhaust into the intake manifold (Figure 11-25). Complete operational details of the EGR as well as servicing are given in Chapter 14.

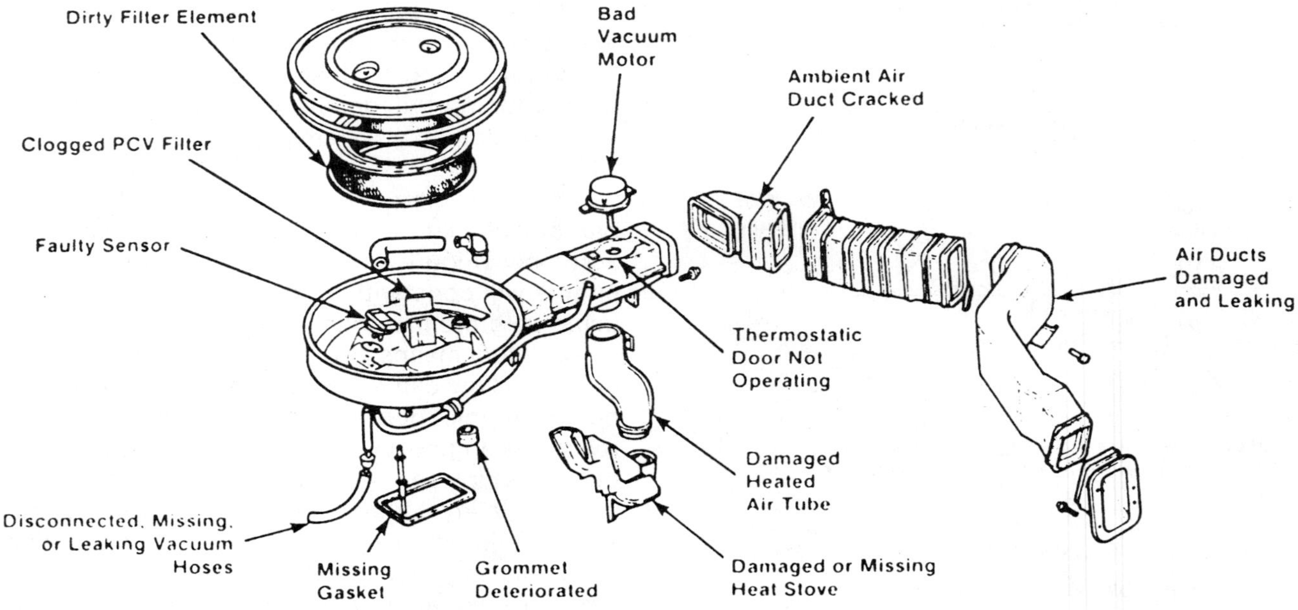

FIGURE 11-24 Any of these air cleaner problems can affect engine performance.

TABLE 11-1: VACUUM/AIR DELIVERY SYSTEMS DIAGNOSIS

Condition	Possible Cause	Remedy
Engine cranks normally, but will not start.	Improper starting procedure	Check with customer to determine if proper starting procedure is being used.
	Engine flooded	Remove the air horn. If foreign material is present, clean the fuel system and replace fuel filters as necessary. If excessive foreign material is found, completely disassemble and clean.
Poor gas mileage.	Customer driving habits	Run mileage test with customer driving if possible. Make sure car has 2,000 to 3,000 miles for the break-in period.
	Loose, broken, or improperly routed vacuum hoses	Check condition of all vacuum hose routings; correct as necessary.
Engine starts but will not keep running.	Vacuum hose routing loose, broken, or incorrect	Check condition and routing of all vacuum hoses; correct as necessary.
	Choke vacuum break units not adjusted to specification or are defective	Adjust both vacuum break assemblies to specification. If adjusted properly, check the vacuum break units for correct operation. Always check the fast-idle cam adjustment when adjusting vacuum break units.
Engine starts hard (cranks normally).	Loose, broken, or incorrect vacuum hose routing	Check condition and routing of all vacuum hoses; correct as necessary.
	Improper starting procedures	Check to be sure customer is using proper starting procedure.
Engine hesitates on acceleration.	Loose, broken, or incorrect vacuum hose routing	Check condition and routing of all vacuum hoses; correct or replace.
	Malfunctioning air valve	Check operation of secondary air valve and spring tension adjustment.
	Inoperative air cleaner heated air control	Check operation of thermostatic air cleaner system.
	Malfunctioning distributor vacuum or mechanical advance	Check for proper operation.
	EGR valve stuck open	Inspect and clean EGR valve.
Engine has less than normal power at normal accelerations.	Loose, broken, or incorrect vacuum hose routing	Check condition and routing of all vacuum hoses.
	PCV system clogged or defective	Clean or replace as necessary.
	Sticking choke	Check connections and operation of choke hot-air system.
	Air cleaner temperature improperly regulated.	Check regulation and operation of air cleaner system.
Less than normal power on heavy acceleration or at high speed.	Plugged air cleaner element	Replace element.
	Air valve malfunction (where applicable)	Check for free operation of air valve. Check spring tension adjustment. Make necessary adjustments and corrections.
Fuel starvation.	Air cleaner element plugged	Replace element.

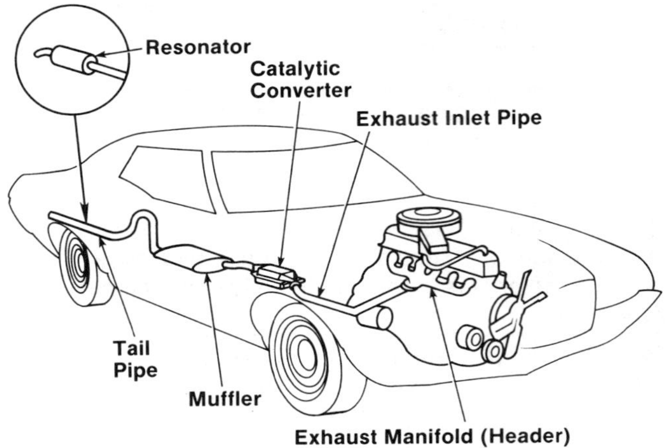

FIGURE 11-26 Exhaust system components

THE EXHAUST SYSTEM

The exhaust system is a relatively simple part of a vehicle, and because of this, there is a tendency on the part of many car owners to ignore it until there is an obvious problem. However, remember that exhaust performs the following important functions:

1. Allows the engine's spent combustion gasses to escape
2. Carries these gasses safely away from the passenger compartment and bodywork
3. Muffles engine noise
4. Cleans up the exhaust before it is released into the atmosphere

Each of these functions of the exhaust system is critically important. If the air output (or exhaust) system fails, the results can range from mere inconvenience to disaster.

The various components of the typical exhaust system are shown in Figure 11-26: They include the following:

- Exhaust manifold
- Exhaust connector pipe
- Catalytic converter
- Intermediate pipes
- Muffler
- Resonator
- Tail pipe
- Heat shields
- Clamps, gaskets, and hangers

EXHAUST MANIFOLD

The exhaust manifold (Figure 11-27) is a bank of pipes that collects the burned gasses as they are expelled from the engine cylinders and directs them to the exhaust pipe. Exhaust manifolds for most vehicles are made of cast iron, but some newer, smaller passenger cars have stamped, heavy-gauge sheet

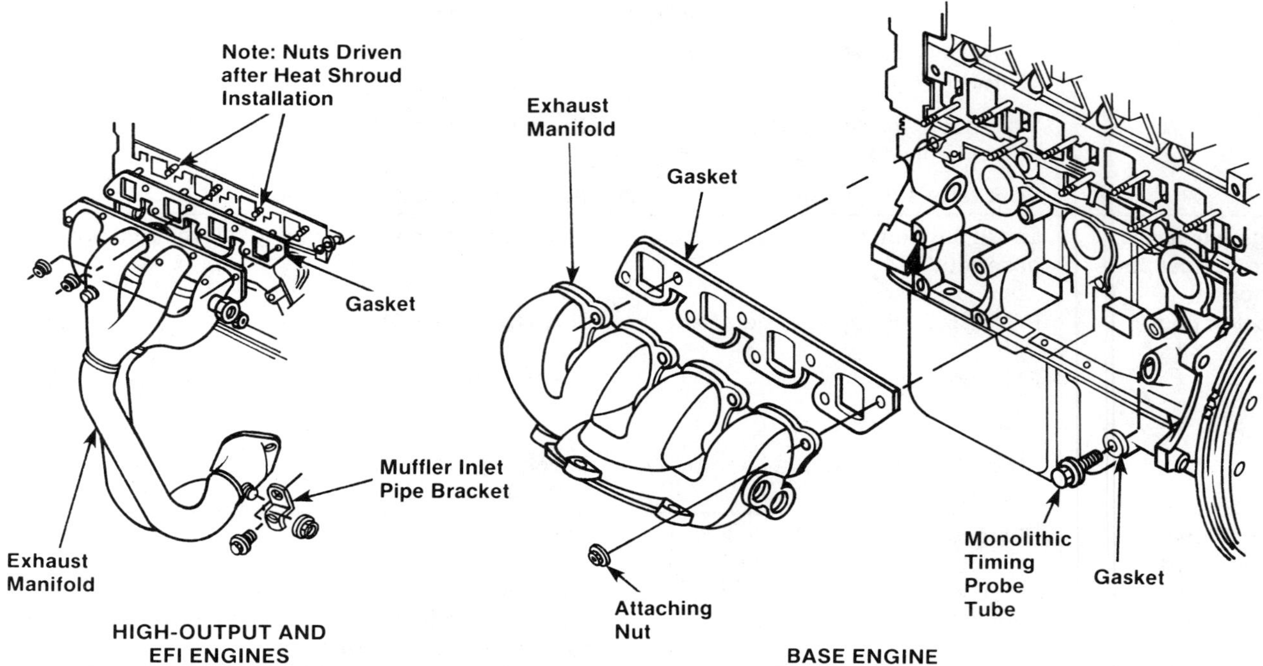

FIGURE 11-27 Exhaust manifold

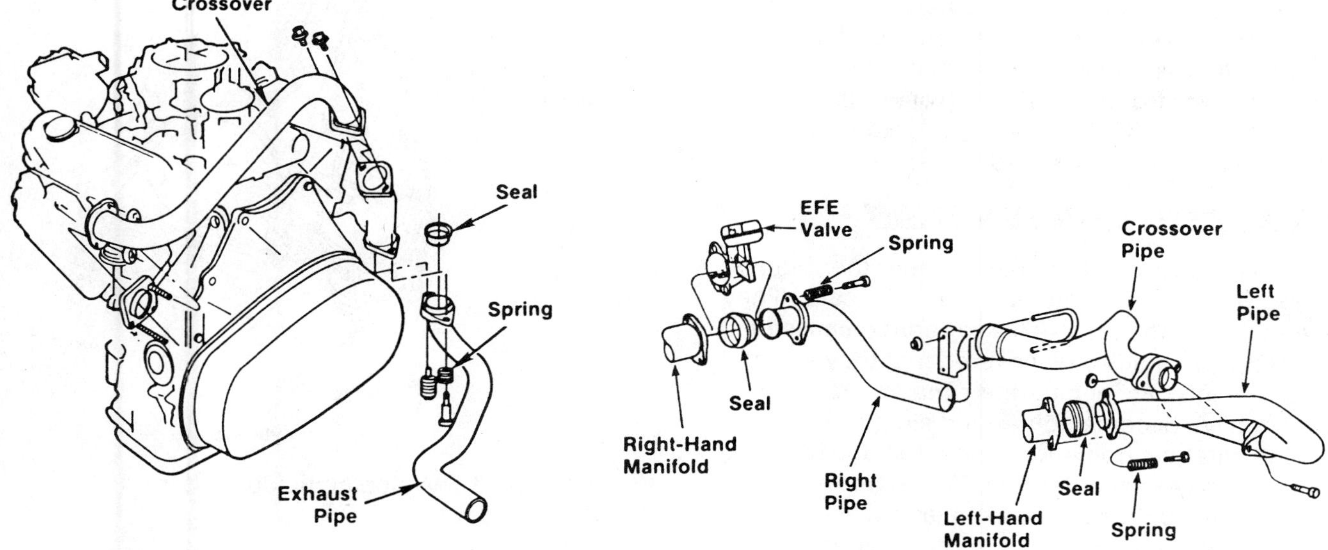

FIGURE 11-28 Crossover pipe arrangement

metal exhaust units. The manifold unit has one port for each cylinder.

The operation of the exhaust manifold is simple, but its design might not be. Near the end of the power stroke, the exhaust valve or port opens and exhaust gasses begin to leave the cylinder. The piston continues on its exhaust stroke, pushing the burned gasses into the exhaust manifold. But each time the exhaust valve opens, waves of high-pressure gasses are forced into the exhaust manifold. Sound, or noise, is produced by pressure waves.

Exhaust noises are reduced by several design factors. For example, one important factor is the length of passages within the manifold. If an exhaust manifold is designed correctly, pulses from the cylinders do not interfere with one another. Thus, exhaust gasses can flow easily through the manifold.

To assure a good flow of gasses with minimum back pressure, the exhaust manifold has large tubular sections. In addition, automotive design engineers try to avoid sharp bends that can slow the passages of gasses. However, the exhaust manifold often has to fit into narrow, cramped spaces between the engine and body of the vehicle. Access to spark plugs and other parts also helps determine the shape of the exhaust manifold. Thus, exhaust manifold shapes or designs in smaller cars are often compromises between good gas flow and limited mounting space.

There is a heat riser valve (HRV) in most exhaust manifolds to help cold starting. This valve, in conjunction with the EFE system, provides extra heat under the carburetor as the engine is warming up to improve performance. It is located at the outlet on one side of the exhaust manifold, usually on the right.

The manifold itself rarely causes any problems. The gasket that seals the manifold to the engine block and the gasket that seals the exhaust pipe to the manifold occasionally fail. A leak in these areas causes a ticking sound when the engine is running. Visually inspect the gaskets for signs of leakage, looking at the gasket and its surrounding area for a grayish-white deposit.

MANIFOLD EXHAUST CONNECTOR PIPE

The exhaust pipe carries collected gasses and vapor from the exhaust manifold to the next component in the exhaust system. A "Y" pipe or a crossover pipe arrangement (Figure 11-28) is often used to connect both exhaust manifolds of a V-type engine to form a single exhaust system. The crossover pipe can be rotated either below or behind the engine.

In a dual or V-type engine setup with a dual exhaust system, an "H" pipe is often used. This consists of right and left exhaust pipes connected by a balance pipe. In other words, each bank of engine valves has its own separate exhaust system. Incidentally, some car buffs believe that a dual exhaust system, which consists of two complete exhaust systems, is a worthwhile way to cut down back pressure and increase gasoline mileage. Dual exhaust systems are excellent when designed into the vehicle by the manufacturer, who usually specifies them only for large high-performance engines. The dual system is seldom very useful on smaller engines. The single system is capable of handling the car's exhaust at speeds from idle to the top of the speedometer and has no difficulty coping with today's *maximum* 65 miles per hour speed limits.

CATALYTIC CONVERTERS

Catalytic converters have been the biggest change in exhaust systems over the years. They were added in the mid 1970s to reduce emissions, but carmakers quickly discovered that they also did part of the job of silencing the exhaust. As a result, mufflers have become lighter and simpler.

The catalytic converter is located ahead of the muffler in the exhaust system. The extreme heat in the converter oxidizes the exhaust emissions that flow out of the engine. There are four types of catalytic converters on most passenger vehicles. They are:

- *Pellet Catalytic Converter.* Uses hundreds of small beads that act as the catalyst agent (Figure 11-29A).
- *Monolithic Catalytic Converter.* Uses a ceramic block shaped like a honeycomb or beehive that is coated with a special chemical, which helps the converter act on the exhaust gasses (Figure 11-29B).
- *Minicatalytic Converter.* Provides a close coupled converter that is either built in the engine exhaust manifold or located next to it (Figure 11-30). It is primarily used on four-cylinder vehicles, or on larger vehicles as a second converter.
- *Three-Way Converter.* The dual bed type treats all three controlled emission gasses: it oxidizes HC and CO by adding oxygen and reduces NO_x by removing oxygen from the nitrogen oxides.

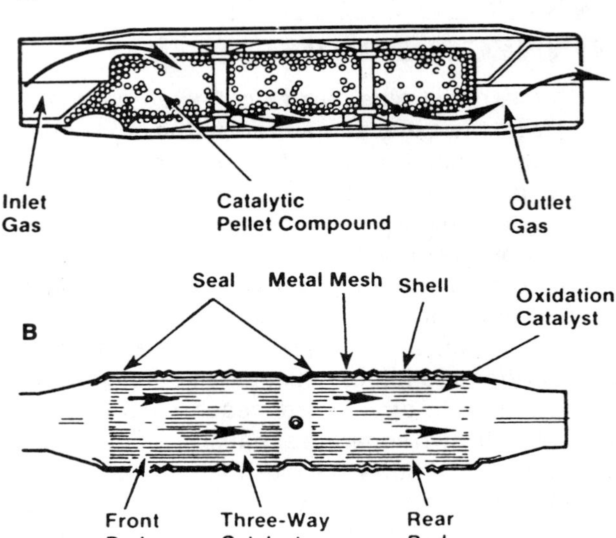

FIGURE 11-29 (A) Pellet catalytic converter; (B) monolithic catalytic converter

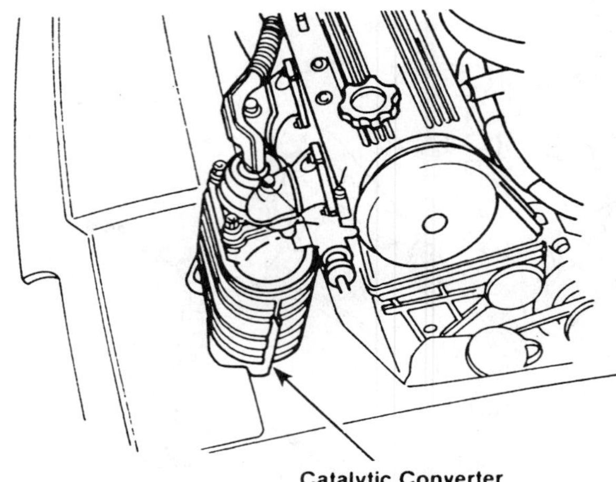

FIGURE 11-30 Minicatalytic converter

The converter is normally a trouble-free emission and exhaust control device, but two things can damage it. One is leaded gasoline. Lead coats the catalyst and renders it useless. The other is overheating. If raw fuel enters the exhaust because of a fouled spark plug or leaky exhaust valve, the temperature of the converter will soar. This can melt the ceramic honeycomb or pellets inside, causing a severe or complete exhaust blockage.

SHOP TALK

Under no circumstances should the converter be removed and replaced with a straight piece of pipe (a test pipe). Do-it-yourselfers can still get away with it (until the vehicle undergoes an emission test, then the converter will have to be put on), but this practice is ILLEGAL for the professional auto mechanic.

MUFFLERS

The muffler (Figure 11-31) is a cylindrical or oval-shaped component, generally about 2 feet long, mounted in the exhaust system about midway or toward the rear of the car. It consists of a series of baffles, chambers, tubes, and holes to break up, cancel out, or silence the pressure pulsations that occur each time an exhaust valve opens. There are two types of mufflers commonly installed on passenger vehicles (Figure 11-32):

1. **Reverse-Flow Mufflers.** These mufflers change the direction of exhaust gas flow through the inside of the unit. This is the most common type of muffler found on passenger cars.
2. **Straight-Through Mufflers.** These mufflers permit exhaust gasses to pass through a single tube. The tube has perforations that tend to break up pressure pulsations, but they are not as quiet as the reverse flow type.

While mufflers still deaden noise as the exhaust gasses pulse through them, there have been several important changes in recent years in their design. Most of these changes have been centered at reducing weight and emissions, improving fuel economy, and simplifying assembly. Many also affect the technician who repairs or replaces the system. These changes include:

- *New Materials.* More and more mufflers are being made of aluminized and stainless steel. Car designers, trying to make exhaust systems lighter, sometimes call for thinner metal for some components. Be sure the muffler is of high-quality steel.

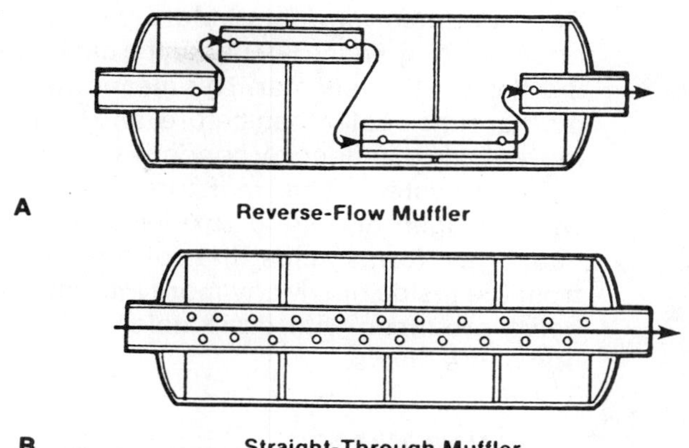

FIGURE 11-32 (A) Reverse-flow muffler; (B) straight-through muffler

SHOP TALK

Remember that stainless steel pipe tends to break rather than rust out, and it might take some tact to explain to the customers why a cracked pipe that does not look rusted is actually worn out. Also, if low-quality components are used where the car had stainless or aluminized steel, you are likely to end up with a premature rustout and an unhappy customer.

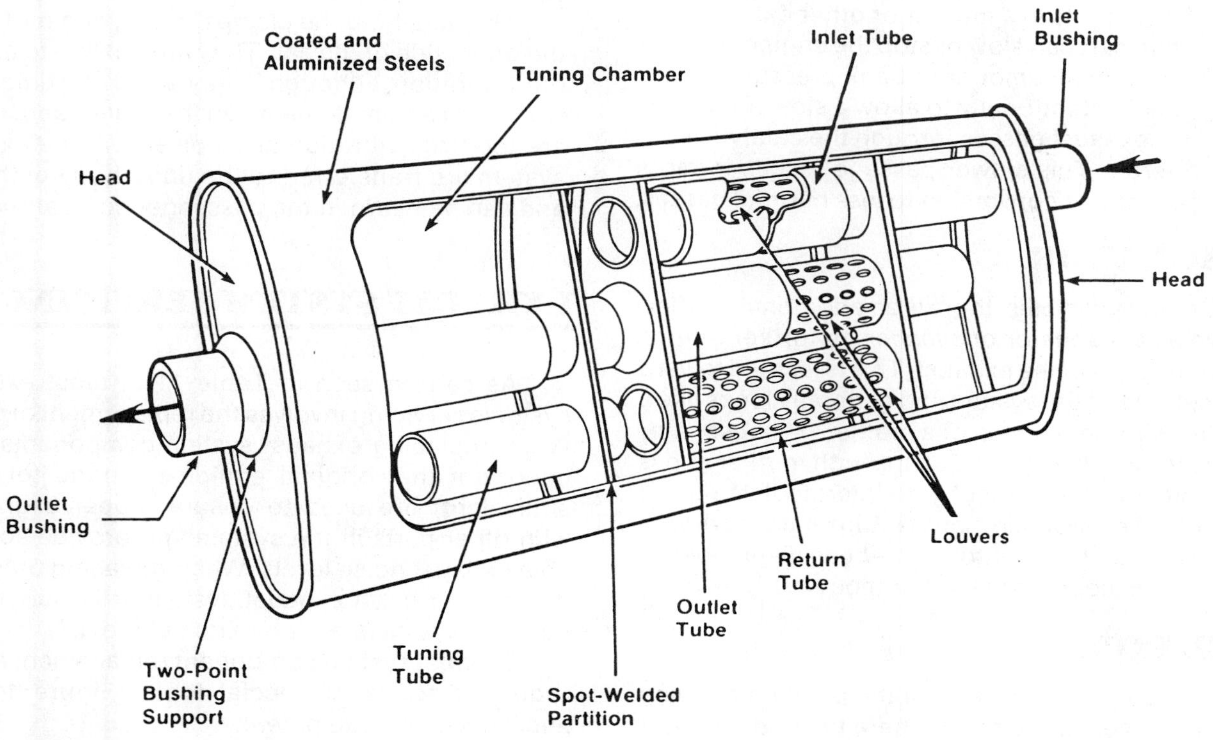

FIGURE 11-31 Typical muffler

- *Rear-Mounted Mufflers.* More and more often, the only space left under the car for the muffler is at the very rear. This means that the muffler runs cooler than before, and is much more easily damaged by condensation in the exhaust system. This moisture, combined with nitrogen and sulfur oxides in the exhaust gas, forms acids that rot the muffler from the inside out. Many more mufflers are being produced with drain holes drilled in one of the heads.

SHOP TALK

Many customers do not always like the holes, because they allow a small amount of gas and noise to escape, but they are one of the cheapest and most effective ways to fight corrosion. Do not plug the drain holes with a sheet metal screw.

- *Back Pressure.* Even a well-designed muffler will produce some back pressure in the system. In fact, a certain amount of back pressure is designed into every exhaust system, either in the form of calculated restrictions or valve overlap. Back pressure reduces an engine's volumetric efficiency, or ability to "breathe." Excessive back pressure caused by defects in a muffler or other exhaust system part can slow or stop the engine. However, a small amount of back pressure can be used intentionally to allow a slower passage of exhaust gasses through the catalytic converter. This slower passage results in more complete conversion to less harmful gasses.

RESONATOR

On some vehicles, there is an additional muffler, known as a *resonator* or *silencer,* to further reduce the sound level of the exhaust. This is located toward the back end of the system and generally looks like a smaller, rounder version of a muffler. The resonator is constructed like a straight-through muffler and is connected to the muffler by an intermediate pipe.

Because of vehicle manufacturers' concern with weight and cost, resonators have been eliminated on most cars except for the larger models.

TAIL PIPE

The tail pipe is the end of the pipeline carrying exhaust fumes to the atmosphere beyond the back end of the car. On cars equipped with a resonator, the resonator can be located in the middle of the tail pipe. In fact, the resonator on some cars is an integral part of the tail pipe, forming a one-piece unit. If one or the other is damaged, the entire unit must be replaced.

In most cases, the tail pipe opens at the rear of the vehicle below the rear bumper. In some cases, it opens at the side of the vehicle just ahead of or just behind the rear wheel. Decorative tail pipe exhaust extensions in either chrome or black are available. Most extensions are fastened to the tail pipe with double-lock screws. The downward exhaust spout deflects gasses down and away from the vehicle.

HEAT SHIELDS

Heat shields are used to protect vehicle parts from the heat of the exhaust system and catalytic converter (Figure 11–33). They are usually made of pressed or perforated sheet metal.

CLAMPS, GASKETS, AND HANGERS

Clamps provide leak-free connections at joints in the muffler system (Figure 11–34). Gaskets, usually made from asbestos (decreasing in use), pressed steel, or sintered iron, are used to ensure tighter connections between the exhaust manifold and the exhaust pipe.

Hangers hold the clamps, pipes, and muffler to the underside of the car. They are flexible to absorb road vibration, although if they are too flexible, road vibration can break them. On the other hand, if they are too firm, vibration and noise from the exhaust system are transferred to the underbody of the car and can be heard in the passenger compartment.

EXHAUST SYSTEM SERVICING

As can be seen in Table 11–2, most exhaust systems servicing involves the replacement of parts. When replacing exhaust system components, it is important that original equipment parts (or their equivalent) are used to ensure proper alignment with other parts in the system and provide acceptable exhaust noise levels. When replacing only one component in an exhaust system, it is not always necessary to take off the parts behind it.

Exhaust system component replacement might require the use of special tools (Figure 11–35) and/or welding equipment.

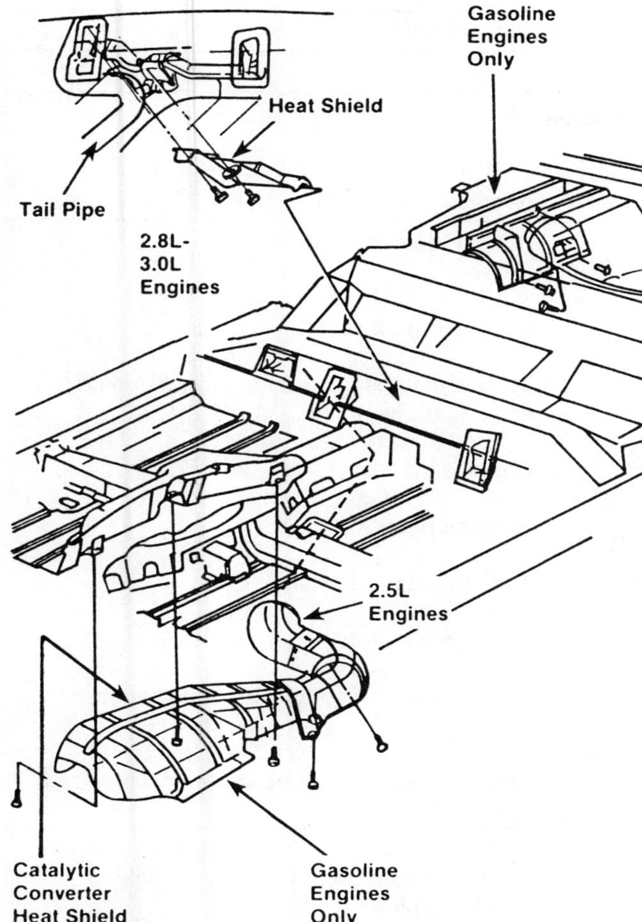

FIGURE 11-33 Typical heat shield

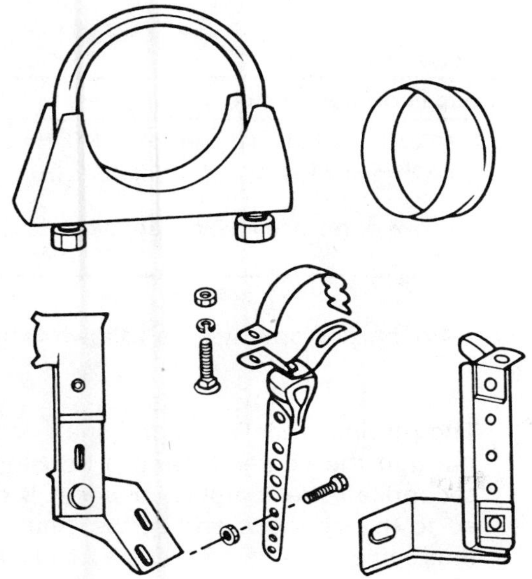

FIGURE 11-34 Clamps, gaskets, and hangers

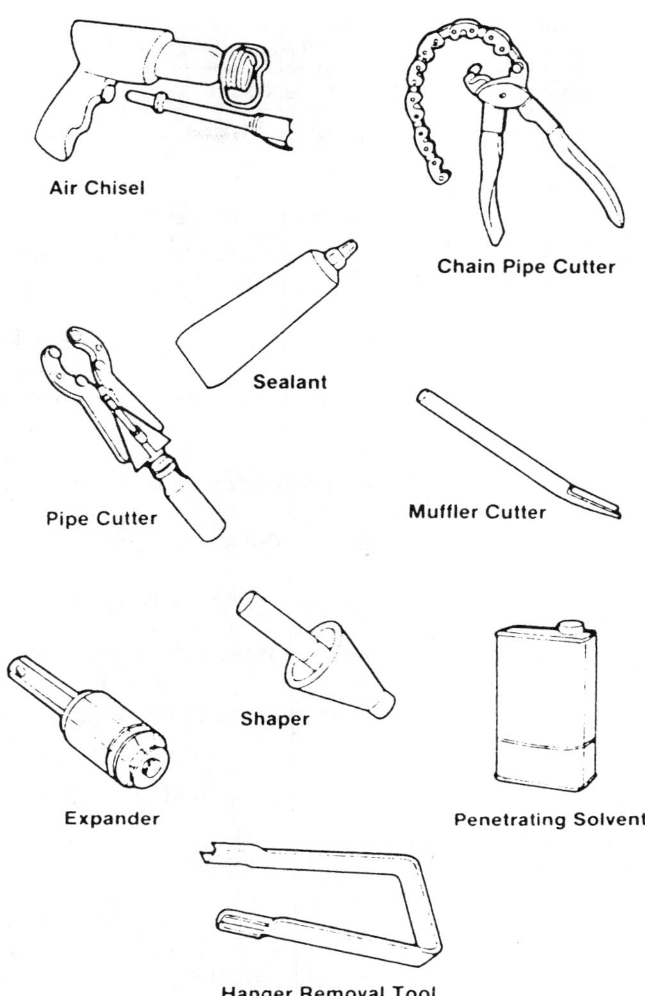

FIGURE 11-35 Exhaust system tools

TURBOCHARGER OPERATION AND SERVICING

On some vehicles, the turbocharger is considered part of the exhaust system since it is driven by the high velocity of the engine exhaust gasses. A typical turbocharger consists of the following components (Figure 11-36):

- Turbine or hot wheel
- Shaft
- Compressor or cold wheel
- Turbine housing
- Compressor housing
- Center housing (contains numerous bearings, turbine seal assembly, and compressor seal assembly)

TABLE 11-2: EXHAUST SYSTEM PROBLEM DIAGNOSIS

Condition	Possible Cause	Remedy
Heat control valve noisy	Loose, weak, or broken antirattle spring or shaft at heat control valve	Replace antirattle spring or heat control valve.
	Thermostat broken	Replace thermostat.
Excessive exhaust system noise	System components striking body or chassis	Reposition components.
	Broken or loose clamps or brackets	Repair or replace.
	Leaks at manifold or pipe connections	Torque clamps or leaking connections to specifications.
	Burned-out or blown-out muffler	Replace muffler.
	Burned-out or rusted-out pipe	Replace pipe.
	Exhaust pipe leaking at manifold flange	Torque nuts to specifications.
	Exhaust manifold cracked or broken	Replace manifold.
	Leak between manifold and cylinder head	Torque studs to specifications or replace gaskets.
	Restriction in muffler or tail pipe	Remove restriction if possible or replace as necessary.
Leaking exhaust	Leaks at pipe joints	Reseal joints with exhaust system sealer. Tighten clamp bolts securely. If leaks persist, replace pipes.
	Rusted-out pipes	Replace.
	Damaged or improperly placed gasket at exhaust pipe/exhaust manifold joint	Replace.
	Rusted-out muffler	Replace.
Restricted exhaust	Kinked or plugged pipe.	Repair or replace pipe.
Engine hard to warmup or will not enter to normal idle	Heat control valve frozen in OPEN position	Free up manifold heat control valve using a suitable solvent or penetrating oil.
	Blocked crossover passage in intake manifold	Remove or replace intake manifold.

Exhaust gasses are routed from the exhaust manifold to the turbocharger, mounted near the carburetor (Figure 11-37). A spring-loaded diaphragm sensitive to intake manifold pressure regulates the waste gate. The waste gate causes exhaust gasses to bypass the turbine when the predetermined manifold pressure is reached, thereby regulating manifold pressure at a safe level. Excessive turbocharging causes detonation and engine damage. After leaving the turbocharger, the gasses enter a normal exhaust system.

The turbocharger operates as follows (Figure 11-38):

1. The turbine wheel's integral shaft passes through the center (bearing) housing. A second fan, the compressor wheel, is bolted to the opposite end of the shaft, and thus spins at the same speed as the turbine wheel. Compressor housing design causes air from the air filter to be drawn into the housing, compressed, then expelled at

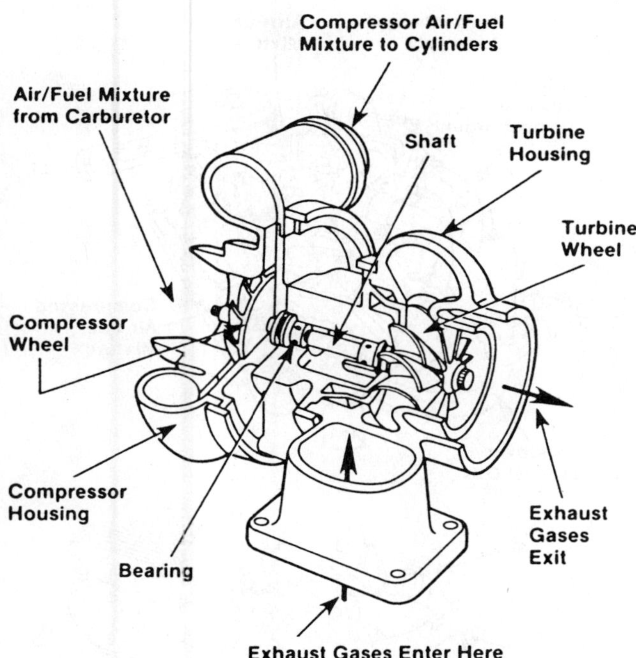

FIGURE 11-36 Typical turbocharger

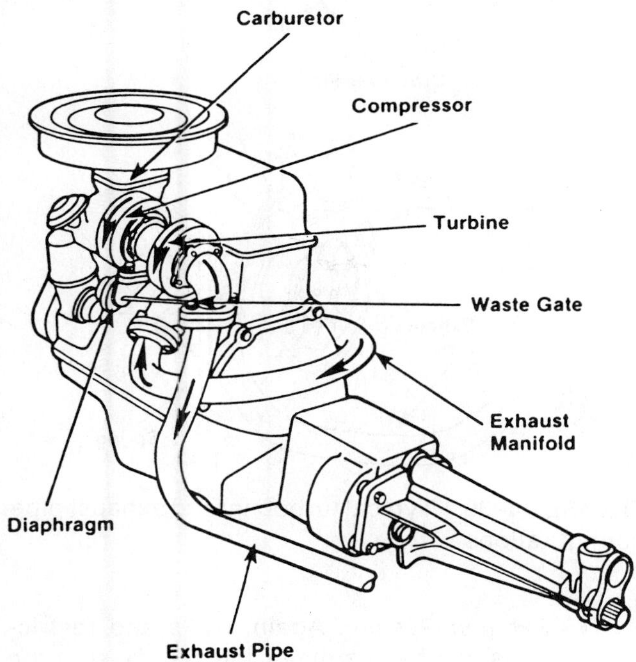

FIGURE 11-37 Exhaust gasses are routed from the manifold to the turbocharger.

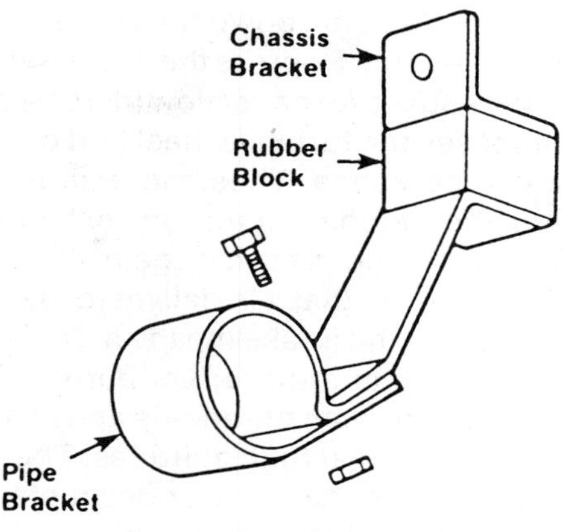

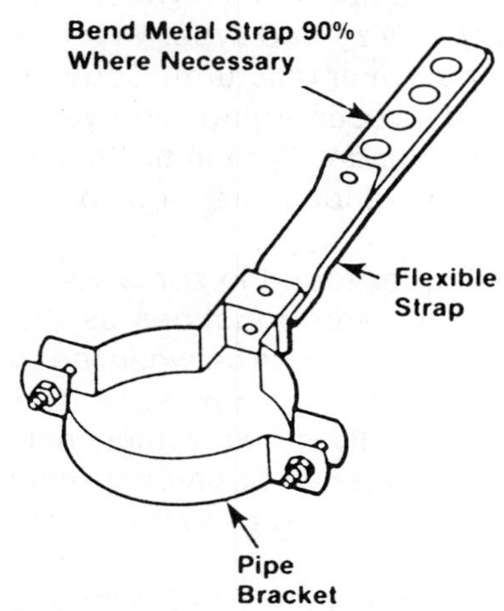

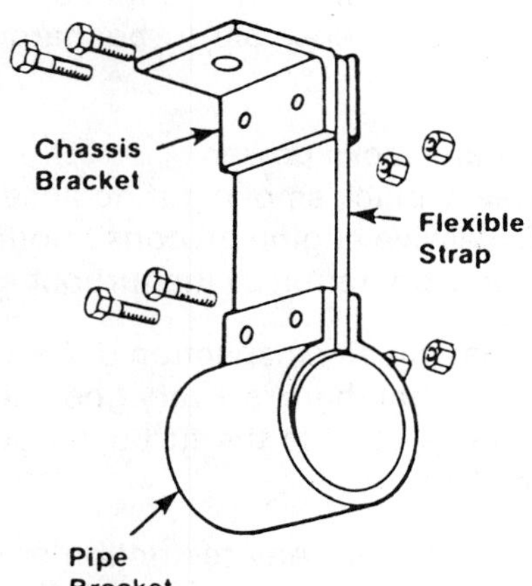

FIGURE 11-38 Turbocharger operation

elevated pressure and temperature into the engine's induction system (carburetor or fuel injectors). That is, as the compressor turns, it increases the amount of air/fuel mixture (air only on diesel engines) delivered to the engine's cylinders.

2. By greatly exceeding atmospheric pressure, the turbocharger force-feeds an en-

gine for dramatic power increases, but only on demand. Depress the accelerator and engine speed/exhaust flow increases. This increases the turbine wheel and compressor wheel speed. Thus, more air is forced into the engine, increasing exhaust flow and beginning the cycle again. It should be remembered that air delivered by turbocharger to the intake means more fuel can be burned, and more fuel burned means more power from the engine and maintains power output at high altitudes. The turbocharger also reduces undesirable emissions, and reduces combustion noise.

The design capacity of the turbocharger is dependent on engine size and type, which is determined by the vehicle manufacturer. Control devices limit the degree of turbocharging to prevent detonation and engine damage. Typical turbocharger exhaust pipe configurations are shown in Figure 11-39).

The supercharger found in some vehicles performs basically the same functions as the turbocharger. The major difference between the two is the method used to drive the compressor. The supercharger is mechanically driven rather than by the velocity of exhaust gasses. Superchargers are seldom used in automobile engine systems today; they are found mainly in trucks.

The purpose of system troubleshooting is to identify the reason for failure so repair can be made before installing a new unit. Common symptoms that might indicate possible turbocharger trouble are:

1. Engine lacks power
2. Heavy black smoke during acceleration
3. Excessive engine oil consumption
4. Noisy operation of the turbocharger

But before making any inspection of the turbocharger, remember that there is a very good chance that the problem lies outside the turbocharger. Here is what to look for:

- *Intake System.* Any restriction or leak in the system limits the volume of air the turbocharger can deliver. Check for air filter blockage or dirt buildup, then check all intake plumbing from the air filter to the turbocharger, including the clamps. Next, check the plumbing from the turbocharger to the intake manifold and be sure the manifold is properly torqued to the cylinder head.

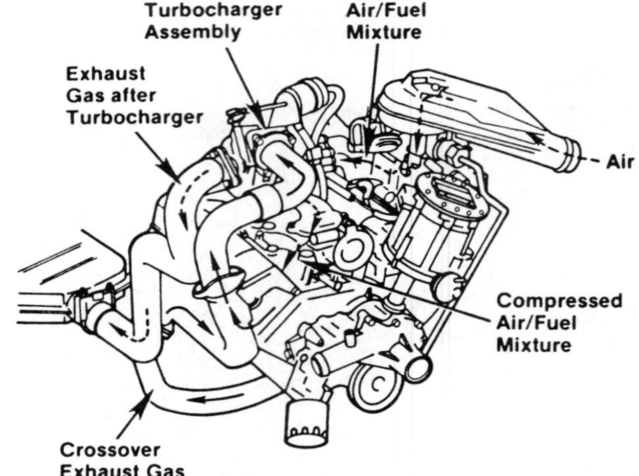

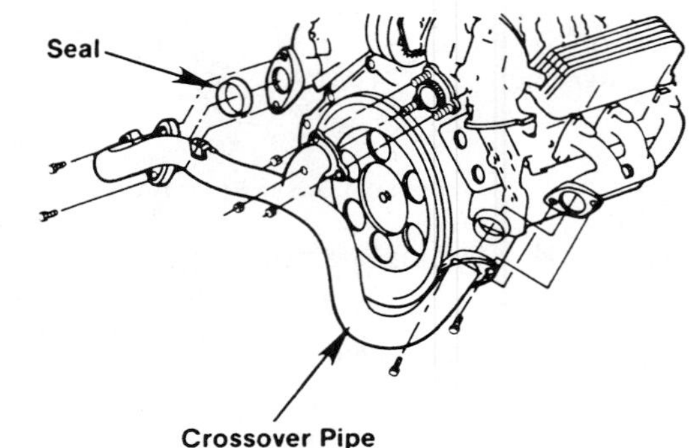

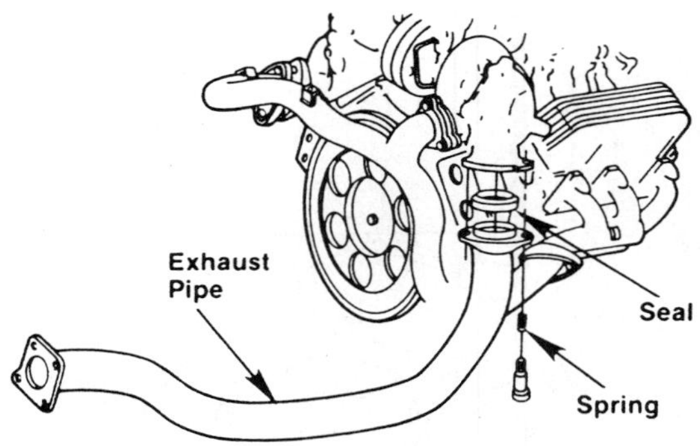

FIGURE 11-39 Typical turbocharged exhaust pipe configurations

- *Exhaust System.* Again, leaks and restrictions can have a major impact. Be sure the exhaust manifolds are properly torqued to the heads and that the turbocharger is properly torqued to the manifold. A leaking or improperly installed turbocharger/manifold gasket could be the culprit. Finally, check the entire exhaust system downstream from the turbocharger, including muffler and catalytic converter, for restrictions. Exhaust blockage causes sluggish and reduced engine performance.

- *Carburetion/Fuel Injection.* The air delivered by the turbocharger will be of no help if it is not mixed with the right amount of fuel. Check the function of carburetion, fuel injectors, and fuel pump. Make certain that all fuel lines and distribution blocks are unrestricted and that the entire fuel system is free of air-entry leaks.
- *Waste Gate.* This device, sometimes called an actuator, regulates the air pressure available from the turbocharger. It is mounted to the side of the turborcharger housing and a linkage rod connects to the waste gate mechanism (Figure 11-40). Vacuum from the intake manifold controls the movement of the waste gate control diaphragm. Remember that any attempt to increase boost pressure by modifying or removing the waste gate can easily allow an engine and turbocharger to rev to destruction. Most waste gates have anti-tampering seals to prevent unauthorized changing of the boost setting.

If all these checks fail to isolate the problem, it is time to inspect the turbocharger.

Inspecting the Turbocharger

To inspect a turbocharger, start the engine and listen to the sound the turbocharger system makes. As a technician becomes more familiar with this characteristic sound, he or she will be able to identify an air leak between the compressor outlet and engine or an exhaust leak between engine and turbocharger by a higher pitched sound. If the turbocharger sound cycles or changes in intensity, the likely causes are a plugged air cleaner or loose material in the compressor inlet ducts or dirt buildup on the compressor wheel and housing.

After listening, check the air cleaner for a dirty element. If in doubt, measure for restrictions per engine manufacturer's shop manual. Next, with the engine stopped remove the ducting from the air cleaner to the turbocharger and look for dirt buildup, oil contamination, or turbine or compressor wheel or damage (Figure 11-41). Even the slightest imperfection could throw the wheels out of balance and cause severe vibration or disintegration.

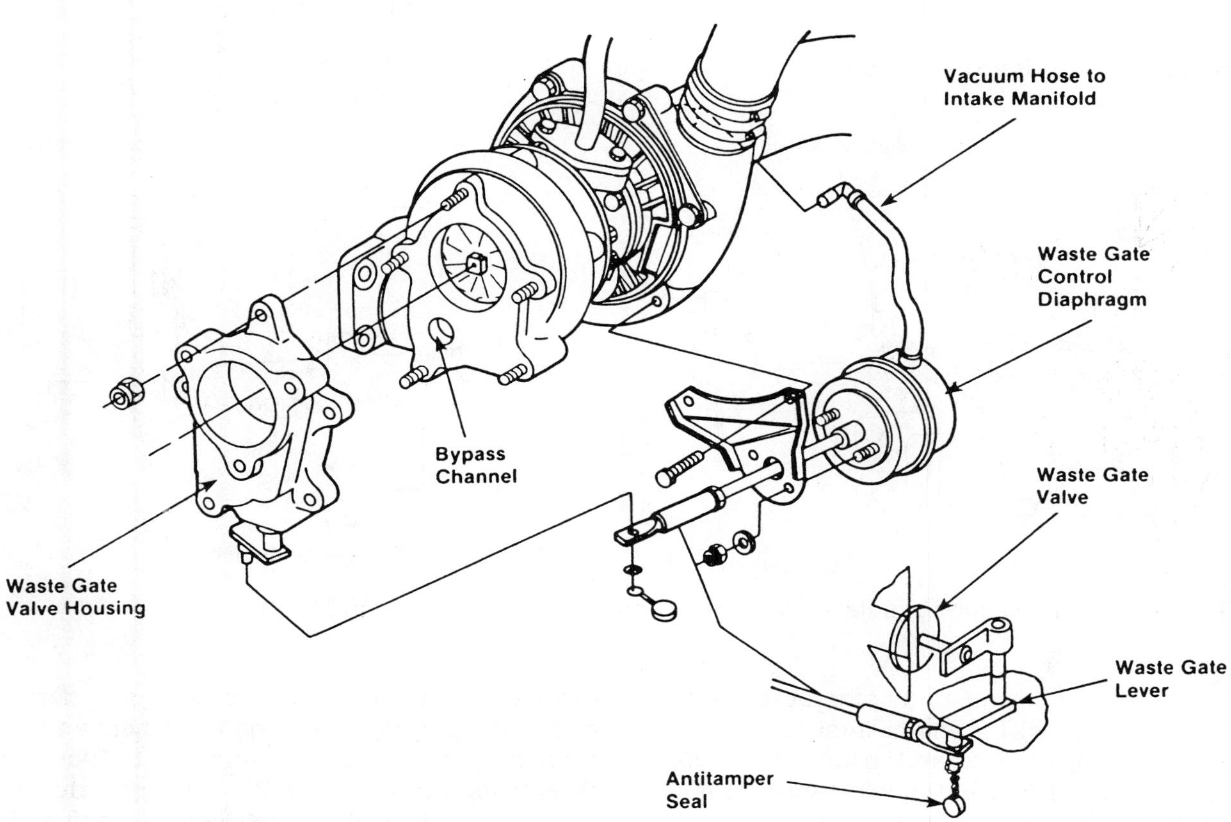

FIGURE 11-40 Waste gate control

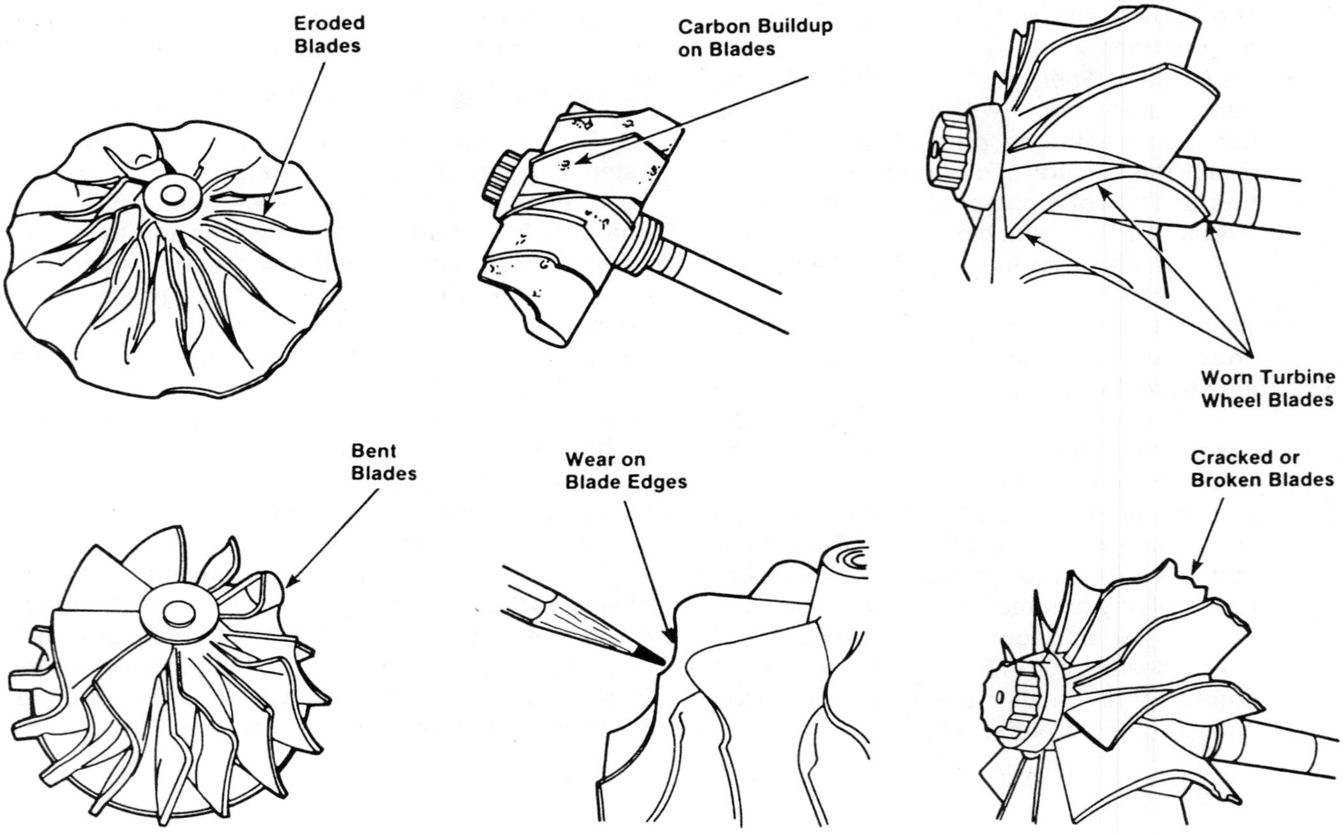

FIGURE 11–41 Turbine and compressor wheel damage and wear

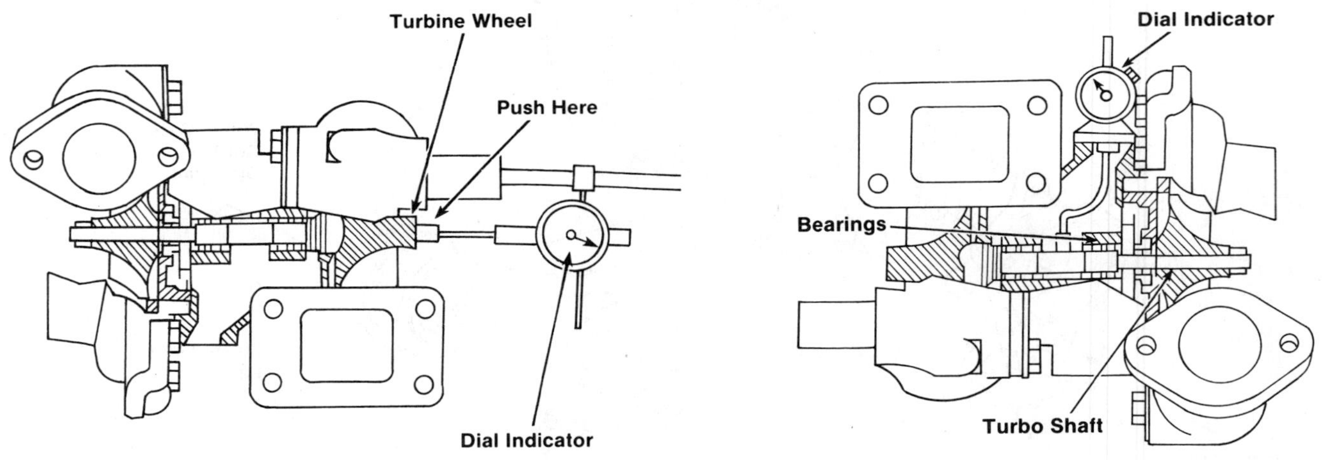

FIGURE 11–42 (A) Measuring radial clearance; (B) checking end-play on the shaft

Check for loose clamps on compressor outlet connections and check the engine intake system for loose bolts, leaking gaskets, and so forth. Then, disconnect the exhaust pipe and look for restrictions or loose material. Examine the engine exhaust system for cracks, loose nuts, or blown gaskets. Rotate the turbocharger shaft assembly. Does it rotate freely? Are there signs of rubbing or wheel impact damage? Axial shaft play is end-to-end movement and radial shaft play is side-to-side movement. These measurements can be made with a dial indicator. Measure the radial clearance between the shaft and the bearing (Figure 11–42A), as well as make a check on the shaft for end-play (Figure 11–42). Keep in mind that there is normally side-to-side play; however, if this play is sufficient to permit either of the wheels to touch the housing when the shaft is rotated by hand, then there is excessive wear. If none of these symptoms are present, the low power complaint is not being caused by the turbocharger. Consult the engine manufacturer's troubleshooting procedures for a basic engine.

Black or Blue Smoke

If the engine is giving low power and emitting black smoke, check for the following:

- Dirty air cleaner
- Loose ducting connections
- Engine over-fueled

If the engine is emitting blue smoke and oil consumption is high, check the air cleaner for restrictions per the engine manufacturer's shop manual. Higher than normal air cleaner restriction can cause compressor oil seal leakage. With the engine stopped, remove the turbocharger ducts and check the shaft assembly for free rotation, damage to wheels, or rubbing against housing walls. Next, check the oil drain line for restriction or damage, which can cause seal flooding and leakage. Also check for high crankcase pressure. If in doubt, measure the crankcase pressure, which must be within the engine manufacturer's specifications. Finally, loosen the exhaust manifold duct and check for oil in the engine exhaust. If oil is present, see the engine manual for appropriate repairs.

TURBOCHARGER PROBLEMS

The turbocharger, with proper care and servicing, will provide years of reliable service. Most turbocharger failures are caused by one of the following reasons:

- Lack of lubricant
- Ingestion of foreign objects
- Contamination of lubricant

As can be noted proper lubrication—clean oil and a fresh oil filter—is the most important factor determining service life of a turbocharger. The main shaft of a turbocharger rotates within its twin bearings at up to sixty times the crankshaft speed under maximum load, greater than 2,000 revolutions per second. At those speeds, contaminated lubricant or insufficient lubrication can destroy a turbocharger in less time than it takes to describe it.

To ensure maximum life for a turbocharger the oil and oil filter must be changed at least as often as recommended by the vehicle or engine manufacturer; more often when the vehicle is operated off-road or in similar, dirty environments. And, as with a nonturbocharged engine, the oil filter should be changed at each oil change.

FIGURE 11-43 On the left, an overheated bearing. Compare with new bearing on the right.

Proper oil level is equally important. Low level increases oil temperature and can lead to momentary interruption of oil flow to the turbocharger's bearings. At speeds up to 150,000 rpm, a moment is all it takes. The bearing surface area in any turbocharger is relatively small, especially compared to the large bearings on the slower-spinning crankshaft (Figure 11-43). The heat that turbocharger bearings must dissipate is much greater. Oil lubricating the turbocharger is contaminated by high temperatures and must be changed more often than the oil in a conventional engine.

Although preventive maintenance is not required for the turbocharger itself, oil lines leading to and from the turbocharger should be checked periodically for leaks or seepage and tightened as required (Figure 11-44).

Excessive Engine Oil Consumption

If there is no black or blue smoke, check the air cleaner for restrictions per engine manufacturer's shop manual. Check the compressor discharge duct for loose connections, and check crankcase pressure, which must be within the manufacturer's specifications. Check the turbocharger shaft assembly for free rotation. Also check for evidence of wheel rubbing on housing walls. If there is rubbing, it is possible to feel it when rotating the shaft while pulling or pushing on it. Check the oil supply line for damage or restriction. If no fault is found, consult

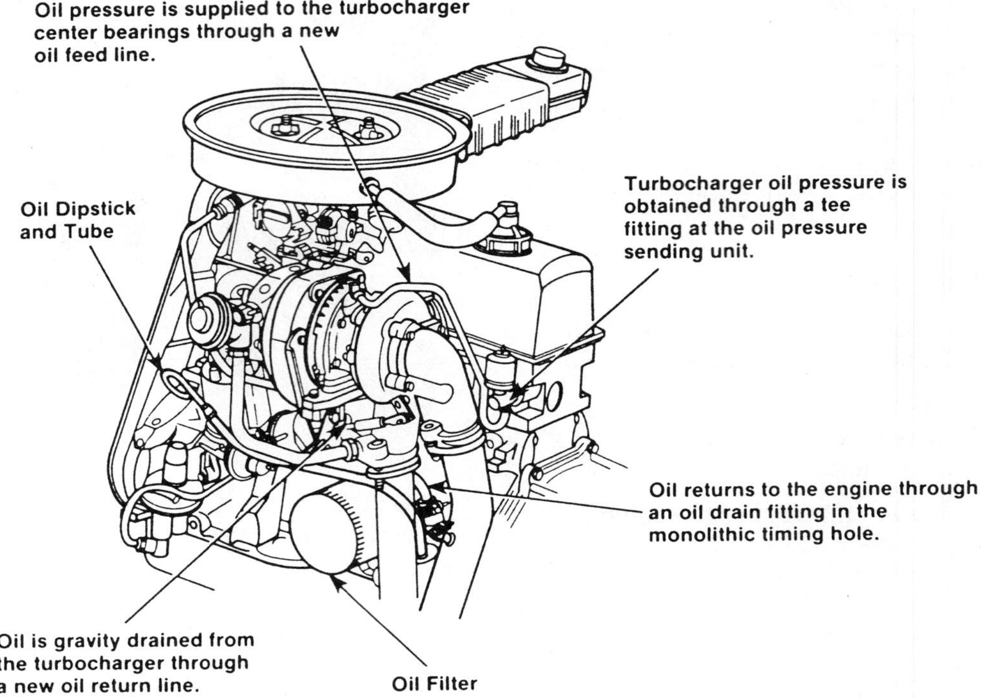

FIGURE 11-44 External components of a turbocharger lubrication system.

the engine manual for further troubleshooting procedures.

Noisy Turbocharger

Every engine has its own unique sound or noise level. Any change in the sound/noise level can indicate a turbocharger or engine problem and should be investigated immediately. If the noise changes to a higher pitch, look for an air leak between the air cleaner and engine or a gas leak in the exhaust system between the turbocharger and engine. Noise-level cycling can indicate a plugged air cleaner, restricted air inlet before the turbocharger, or dirt buildup in the compressor housing or on the compressor wheel.

With the engine running, uneven noise and vibration can indicate a malfunction in the shaft wheel assembly. If a problem is indicated, shut the engine down immediately and repair it. If the turbocharger is functional, check the air system for the following:

- Air cleaner restriction
- Hose clamps tight
- Intake manifold gaskets
- Cracked or deteriorated hoses

With the engine running at idle the following can be performed:

- Air tube and air cleaner/turbocharger connections can be checked by spraying exterior surfaces lightly with starting fluid; an engine speed increase indicates a leak.
- Air leaks between turbocharger and engine can be checked by feel and by application of lightweight oil or soap suds on the crossover tube, connections, and hoses. Look for bubbles.

Exhaust gas leaks between the engine block and inlet to turbocharger will create a noise-level change and reduced performance. Check the exhaust system for:

- Manifold gasket leakage
- Manifold retaining bolt tightness
- Cracked or porous manifold
- Compressor discharge ducting
- Turbocharger inlet gasket leakage
- Turbocharger inlet flange bolt tightness

Remember, exhaust gas leakage can be detected by carbon deposits or heat discoloration in the area of the leak.

It is also important to check the turbocharger shaft for looseness and look for wheel rubbing or impact damage to blades from foreign material. If rubbing or impact damage is found, remove and replace the turbocharger.

REPLACING A TURBOCHARGER

If the turbocharger proves faulty, it can be replaced with a new unit or rebuilt unit. While it is best to always follow the exact replacement procedure given in the service manual, the following is a good general guide.

1. Remove the old turbocharger by loosening bolts, connections, and gaskets.
2. Make sure that the unit is the proper one for the engine. The part can be checked against the service manual.
3. Install the new gaskets and seals, as needed, and be sure that they are properly positioned.
4. Install the new or rebuilt unit and torque fasteners to the service manual specifications and recommended sequence.
5. Spin the impeller-turbine to check for binding before installing unit into the engine.
6. Change the engine oil and flush the oil lines. If oil problems were the cause of the failure of the original unit, check the oil supply pressure in the feed line to the turbocharger.

Once the new or rebuilt unit is installed, the turbocharger should be started up using the following procedure:

Turbo Startup and Shutdown

Oil supply to a turbocharger's shaft bearings is critical to long life for this high-speed component. After replacement of a turbocharger, or after an engine has been unused or stored, there can be a considerable lag after engine startup before the oil pressure is sufficient to deliver oil to the turbocharger's bearings. To prevent the problem, which can lead to premature turbocharger failure, follow these simple steps:

1. When installing a new or remanufactured turbocharger, make certain that the oil inlet and drain lines are clean before connecting.
2. Be sure the engine oil is clean and at the proper level.
3. Fill the oil filter with clean oil to minimize cranking time.
4. Leave the oil drain line disconnected at the turbocharger and crank the engine without starting until oil flows out of the turbocharger drain port.
5. Connect the drain line, start the engine, and operate at low idle for a few minutes before operating at higher speeds.

After an oil and filter change, crank the engine without starting until the oil pressure gauge indicates normal oil pressure. Alternate: run the engine at low idle until a steady oil pressure is obtained. This same procedure should be followed if the engine has not been operated for some time.

Instruct the driver as to the proper shutdown procedure. If he or she shuts down from high speed, the turbocharger will continue to rotate after the engine oil pressure has dropped to zero, which can cause bearing damage. Also, advise the driver about oil lag. Allow 30 seconds for the oil flow to become established before running up a high rpm. Ask the driver if engine oil and oil filter are changed at recommended intervals, review the proper lube oil and filter change interval with the operator. Contaminated oil can cause sludge buildups within the turbocharger. Check the oil drain outlet for sludge buildup with the oil drain line removed. Failure to follow these steps can result in bearing failure on the turbocharger's main shaft. Remember, shaft rotation speed on modern turbochargers can easily exceed 2,000 revolutions per second, so momentary oil flow interruption is likely to lead to turbocharger failure.

REVIEW QUESTIONS

1. Which of the following statements about vacuum is true?
   a. Most vacuum pressure gauges read 14.7 psi under normal conditions.
   b. Vacuum pressure is greater than atmospheric pressure.
   c. At closed throttle idle, manifold vacuum is typically 15 to 22 inches of mercury.
   d. At wide open throttle, manifold vacuum is higher than at idle.

2. Which of the following is a symptom of a vacuum system problem?
   a. knock or pinging
   b. poor acceleration
   c. hard start
   d. all of the above

3. Vacuum is measured in ____________.
   a. inches of mercury
   b. pounds per square foot
   c. meters
   d. all of the above

4. Which of the following types of air filters is used in modern passenger car applications?

a. oil bath air filter
b. paper air filter
c. oil-wetted polyurethane filter
d. none of the above

5. If a paper element air filter is dirty, ______________.
a. clean it with compressed air and reinstall
b. wash it with filter cleaner solution and reinstall
c. replace it with a new one
d. none of the above

6. Which of the following is not a component of the thermostatic control air cleaner?
a. bimetal sensor
b. pressure gauge
c. cold weather modulator
d. air valve

7. Which of the following components depends on exhaust gasses to improve cold start?
a. heat riser valve
b. catalytic converter
c. vacuum retard-delay valve
d. all of the above

8. Which of the following components in the exhaust system is used to silence the exhaust system?
a. catalytic converters
b. mufflers
c. resonators
d. all of the above

9. Which of the following components of a turbocharger is responsible for limiting boost pressures?
a. turbine
b. waste gate
c. compressor
d. none of the above

10. An engine equipped with a turbocharger is emitting blue smoke. Technician A says the problem is probably a leaky compressor oil seal. Technician B says the problem is probably leaky turbine oil seals. Who is correct?
a. Technician A
b. Technician B
c. Both A and B
d. Neither A nor B

11. Turbocharger shaft wear can be checked using a ______________.
a. pressure gauge
b. dial indicator
c. vernier calipers
d. torque wrench

12. A cause of turbocharger noise might be ______________.
a. shaft bearing wear
b. exhaust back pressure
c. rich air/fuel mixture
d. low engine oil level

13. To check for air leaks between the turbocharger and the engine ______________.
a. check for loose connections in crossover pipe, waste gate hose, etc.
b. check for leaks using an ultrasonic leak detector
c. spray soap water on the crossover pipe, waste gate hoses, etc., and look for bubbling
d. all of the above

14. Which of the following could lead to premature turbocharger failure?
a. insufficient oil pressure
b. shutting off the engine without giving the turbocharger shaft time to stop spinning
c. dirty engine oil
d. all of the above

15. A customer complains of a lack of power when his or her vehicle is accelerating. Technician A says the problem is a leak in the PCV system. Technician B says the problem is a clogged air cleaner element. Who is correct?
a. Technician A
b. Technician B
c. Both A and B
d. Neither A nor B

CHAPTER TWELVE

CARBURETOR TESTING AND SERVICING

Objectives

Upon completion of this chapter, you should be able to:

- Explain the purpose and functions of a carburetor.
- Identify the main systems and components of a carburetor.
- Recognize carburetor-related performance problems.
- Troubleshoot and adjust a carburetor for optimum performance and efficiency.

The carburetor is a device used to mix, or meter, fuel with air in proportions that satisfy the energy demands of the engine in all phases of operation. Today's carburetor is a very complex mechanism (Figure 12-1). Some of the larger two- and four-barrel carburetors can have over 200 parts. These parts make up the metering systems and subsystems that are necessary for matching air and fuel delivery with engine performance demands. Each of these systems must be functional and properly adjusted if the engine is to operate efficiently.

Although fuel injection systems have replaced carburetion in many passenger cars and light trucks, there are many carbureted engines still on the road, and all technicians must understand the principles of carburetion and how carburetors are constructed and operate before they can successfully tune carbureted engines.

BASIC CARBURETOR CIRCUITS

Variations in engine speed and load demand different amounts of air and fuel (often in differing proportions) for optimum performance—and present complex problems to the carburetor. At engine idle speeds, for example, there is insufficient air velocity to cause fuel to be drawn from the discharge nozzle and into the airstream. Also, with a sudden change in engine speed, such as rapid acceleration, the venturi effect (pressure differential) is momentarily lost. Therefore, the carburetor must have special circuits or systems to cope with these situations. There are six basic circuits used on a typical carburetor.

1. Float
2. Idle
3. Main metering
4. Full power (or power enrichment)
5. Accelerator pump
6. Choke

FLOAT CIRCUIT

The float circuit (also called the *fuel inlet system*) (Figure 12-2) of a typical carburetor consists of the following:

- Fuel bowl
- Fuel inlet fitting
- Fuel inlet needle valve and seat
- Float

A fuel screen or filter is usually installed at the fuel inlet to prevent dirty fuel from mixing in the carburetor and causing a malfunction.

The float system stores fuel and holds it at a precise level as a starting point for uniform fuel flow. Fuel enters the carburetor through the inlet line and passes through an inlet filter to the inlet needle valve and seat. The incoming fuel is captured and stored

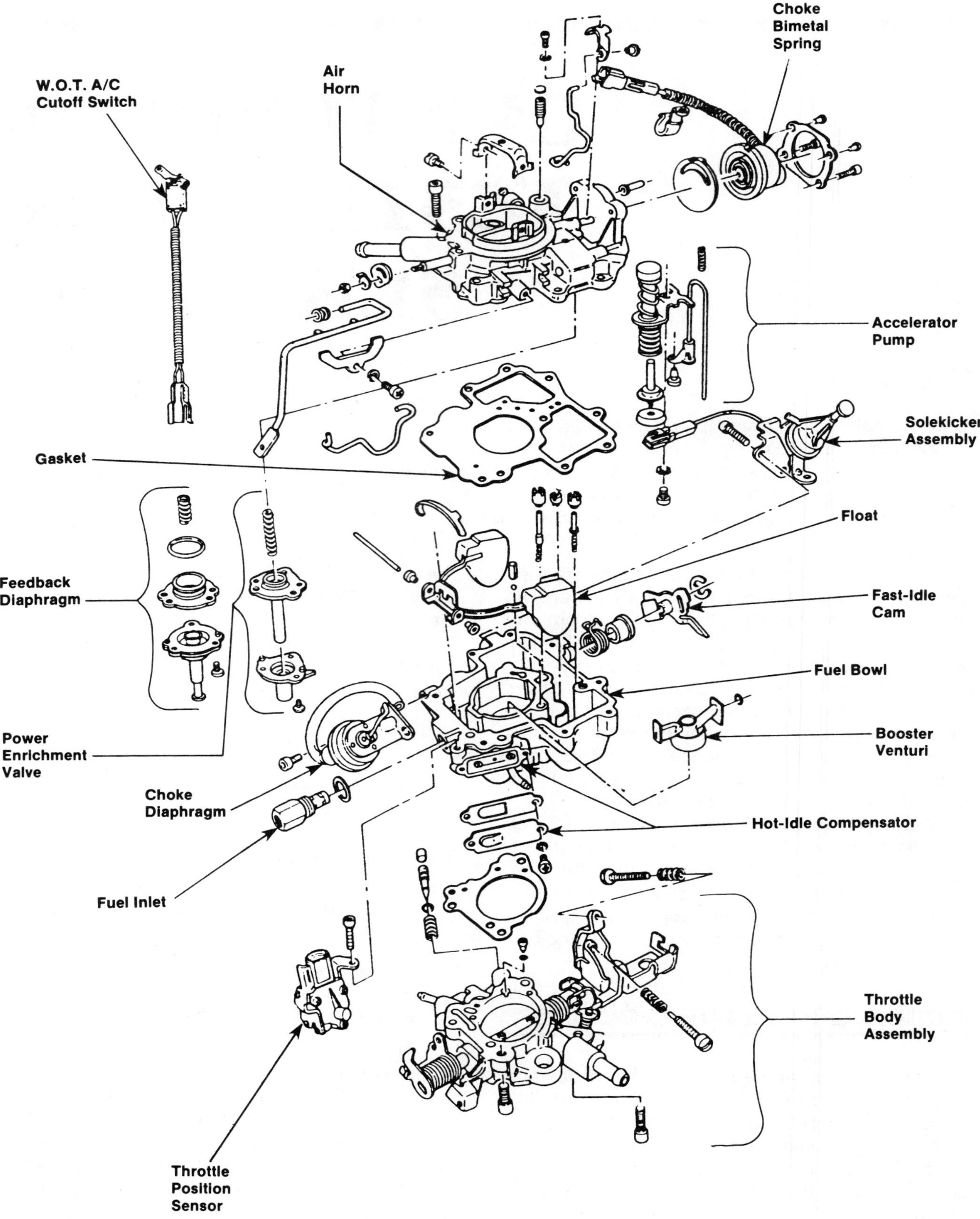

FIGURE 12-1 Typical carburetor

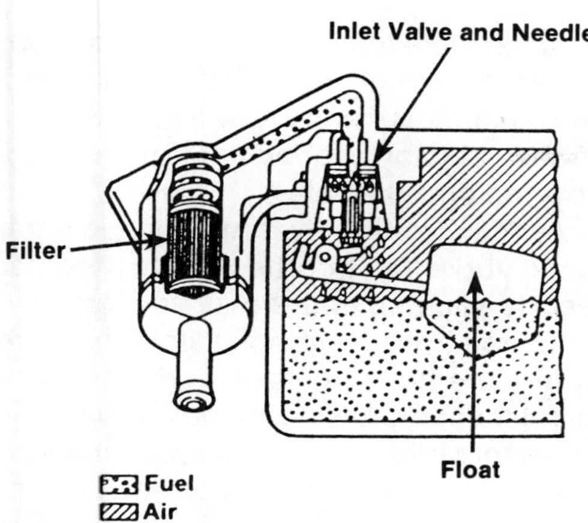

FIGURE 12-2 Fuel inlet system

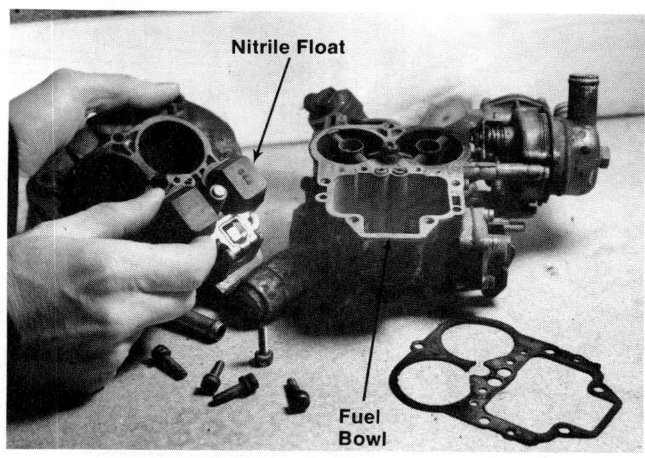

FIGURE 12-3 Nitrile rubber float

in the reservoir or fuel bowl. The fuel bowl can be an integral part of the main casting or it can be a separate casting attached to the carburetor body with screws. Carburetors with primary and secondary venturis might have two separate fuel bowls.

The level of the fuel in the bowl is maintained at a specified height by the rising and falling of the float in the fuel bowl. Early floats were made of brass stampings soldered into an airtight "lung." In the 1960s, floats made of nitrile rubber (Figure 12-3), a closed cell material made of thousands of tiny hollow spheres, were introduced. By the early 1980s, nitrile rubber floats were used almost exclusively on domestic cars. Today, brass floats, nitrile rubber floats, and hollow plastic floats are used in carburetors.

As fuel enters the bowl, the float, which is connected to a hinged lever, rises and closes the inlet needle valve. With the needle valve closed, fuel is prevented from entering the carburetor. Fuel pressure against the inlet needle valve tends to force it open while the buoyancy of the float in the bowl tends to force it closed. This action establishes the precise fuel level for the carburetor. To prevent the float from bouncing and vibrating, a bumper spring is usually installed under the float or to a tang connected to the float.

The metering systems of a carburetor are designed to function properly only when the fuel level in the bowl is at a specific level. The specific level can be adjusted externally on some carburetors by turning a threaded inlet valve assembly. A sight plug on the side of the fuel bowl can be removed to observe the fuel level. Carburetors without externally adjustable inlet valves must be removed from the engine and partially disassembled to adjust the float.

The fuel bowl is vented internally to the air horn by a vent tube in the carburetor body. Prior to the introduction of emission controls, most primary fuel bowls were also vented to the atmosphere when the engine was at idle or turned off. Since the introduction of evaporative control systems, all fuel bowls are vented by a valve to a charcoal canister (Figure 12-4) and the vapors are returned to the engine when it is restarted. Further details on the vapor venting system can be found in Chapter 14.

IDLE CIRCUIT

At idle, the engine requires a richer air/fuel mixture than during normal cruising conditions. This is because residual exhaust gasses remaining in the combustion chambers during low-engine rpm dilute the air/fuel charge.

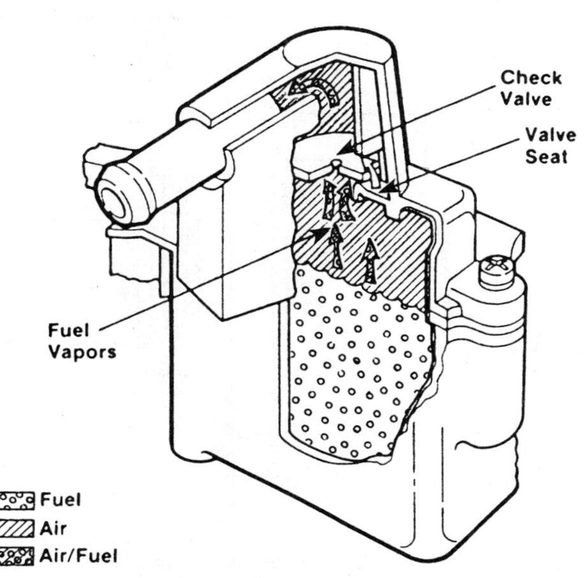

FIGURE 12-4 External vent system

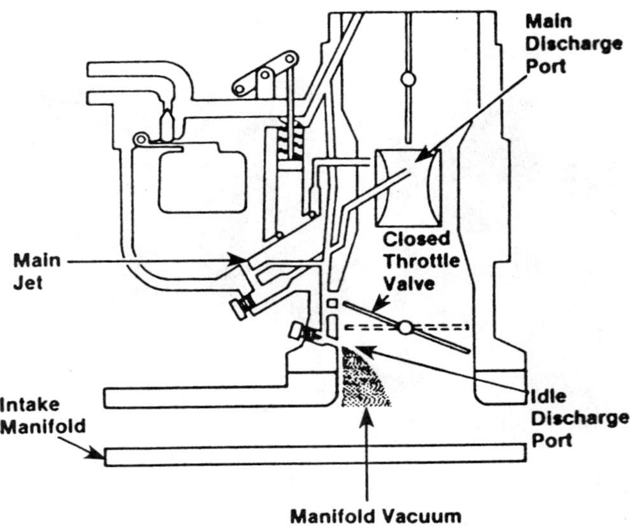

FIGURE 12-5 Off-idle system

The idle circuit (Figure 12-5) supplies the richer air/fuel mixture to operate the engine at idle and low speeds. Fuel is drawn into the idle system from the main well. Some applications have an idle tube in the main well to meter the fuel. Other applications use a horizontal passage from the main well to the idle well and an idle channel restriction for metering purposes.

In either type, air enters the idle air bleed and mixes with the fuel after the fuel flows through the idle tube or restriction. The mixture of air and fuel is called an *emulsion*.

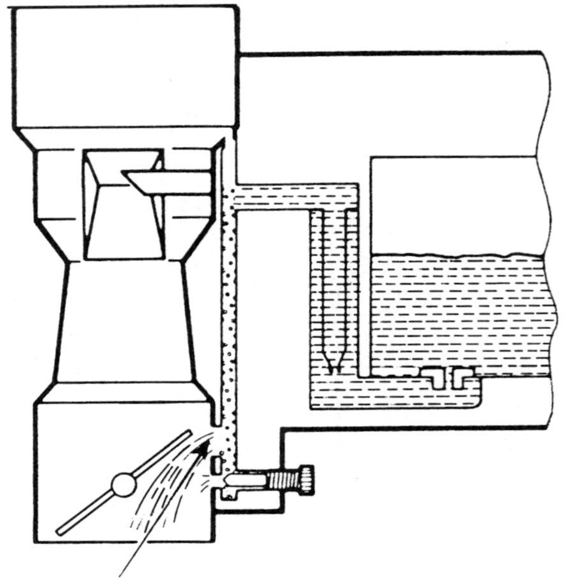

FIGURE 12-6 Off-idle operation

At curb idle the throttle valves are almost closed. This creates a high vacuum below the throttle valve with near atmospheric pressure above the valve. The air/fuel emulsion is drawn out of the curb idle port below the throttle plate.

As the throttle valves are opened a transfer slot located above the throttle plate is progressively exposed to vacuum and the air/fuel emulsion is also discharged from the transfer slot. The increased air/fuel mixture flow provides a smooth transition between idle and cruising modes of operation. Some carburetors have a series of holes called off-idle air passages, instead of a transfer slot. Like the transfer slot, the holes permit increased fuel delivery as the throttle opens (Figure 12-6). This is sometimes called an off-idle system.

To improve idle quality when meeting emission standards, a variable air bleed idle system is used on some carburetors (Figure 12-7). In this system there is the normal fixed idle fuel restriction. However, there are two idle air bleeds. One idle air bleed is installed normally in the air horn. An auxiliary idle air bleed is drilled into the lower skirt of the venturi. The air entering through the auxiliary passage is adjusted by an idle air adjusting screw. The screw is turned clockwise to enrich the idle mixture and counterclockwise to lean out the idle air/fuel mixture.

SHOP TALK

On most modern carburetors, the idle mixture is factory set and the adjustment screw is covered with a plate or plug. This ensures that the CO percentage in the exhaust meets emissions standards and prevents anyone from tampering with the factory setting. Earlier carburetors have an idle limiter cap attached to the idle mixture screw (Figure 12-8). The limiter cap also prevents the factory setting from being tampered with, but allows the idle mixture to be adjusted within a narrow range (about 3/4 of a turn).

MAIN METERING CIRCUIT

The main metering circuit (Figure 12-9) comes into operation when the engine speed reaches about 20 mph or higher. Opening the throttle plate past the idle position increases the air flowing through the carburetor venturi and creates enough vacuum to allow atmospheric pressure to force fuel through the main metering system and out the main fuel discharge nozzle, located in the center of the booster

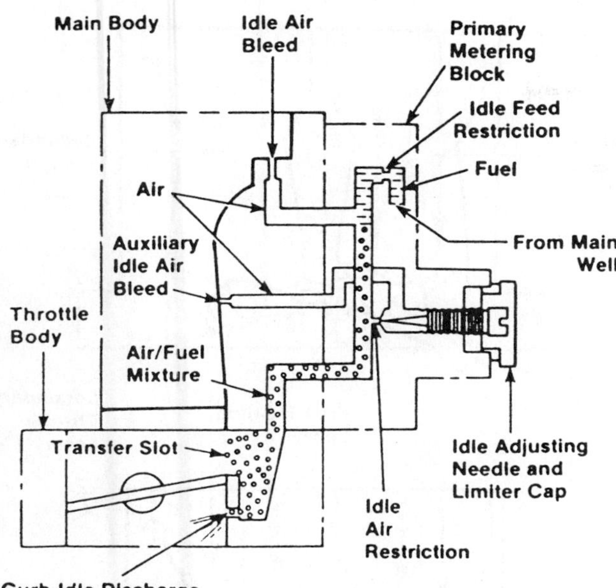

FIGURE 12-7 Variable air bleed idle system

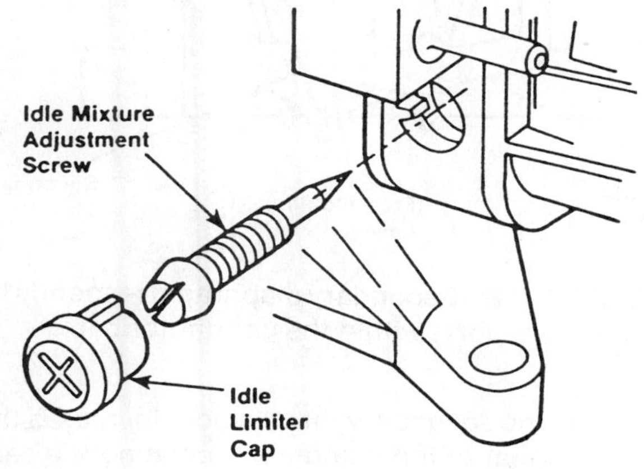

FIGURE 12-8 Idle limiter caps restrict the amount of adjustment allowed for the idle mixture.

venturi. As engine speed is increased, the vacuum at the discharge nozzle increases (Figure 12-10).

This vacuum or pressure differential causes fuel to flow out of the fuel bowl, through the main metering jet, and into the main well. On most carburetors the main well is vented through a precisely sized opening called the *main well air bleed.* The main well air bleed allows air to enter at the top of the main well. Fuel then flows from the main well up the main well tube to the discharge nozzle. When this happens, the pressure in the main well drops off. Air entering the calibrated main air bleed prevents a vacuum from developing in the main well. The air also allows for aeration of the fuel as it leaves the main well and travels up the well tube. Air enters the main well tube through small holes in it. This will allow the fuel to be partially atomized as it travels toward the discharge nozzle.

The main air bleeds draw air from the high-pressure areas in the carburetor barrel above the venturi. Air flows through the bleeds because the high-pressure areas are at atmospheric pressure levels while the main metering discharge nozzle is in the low-pressure venturi level. Air and fuel quantities, as well as bleed airflow, are controlled by the throttle opening.

As the air speed increases through the venturi, more fuel is drawn from the main well. This will lower the fuel level in the main well and expose more air bleed openings in the main well tube. This will cause

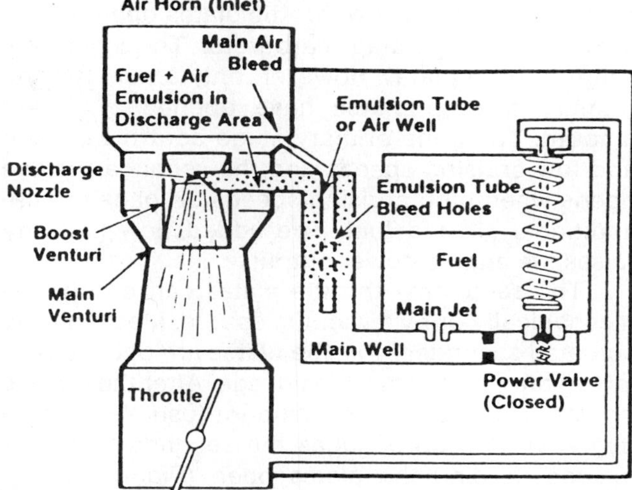

FIGURE 12-9 Air/fuel mixture routed to the booster venturi via the discharge nozzle.

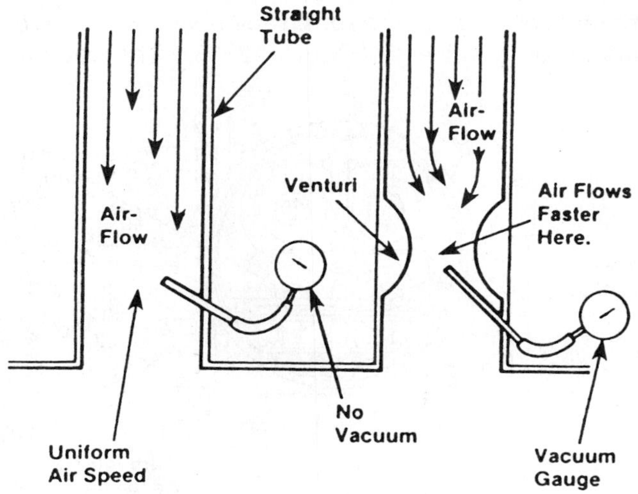

FIGURE 12-10 How a venturi works

extra air to enter the well tube, mix with the fuel, and, in essence, dilute the fuel. This action circumvents the richening effect caused by the increased carburetor airflow. If the fuel were not diluted as such, the air/fuel mixture would richen at high speed as the venturi vacuum increased faster than the engine's need for additional fuel.

Secondary Metering Systems

Some carburetors have more than one barrel or air horn. Each barrel has a throttle valve and main metering system (as well as other circuits). When all throttle valves open and close simultaneously, the carburetor is called a single stage carburetor. Some carburetors have two stages, called primary and secondary stages or venturis (Figure 12–11). In the primary stage, one or two throttle plates operate normally as in a single stage carburetor. The secondary stage throttle plates, however, only open after the primary throttle plates have opened a certain amount. Thus, the primary stage controls off-idle and low cruising speeds and the secondary stage opens when high cruising speeds or loads require additional air and fuel. The added flow capacity raises the engine power output.

The secondary throttle plates can be opened mechanically or by a vacuum source. Mechanically actuated secondary throttle plates are opened by a tab on the primary throttle linkage. After the throttle primary plates open a set amount (usually 40 to 45 degrees), the tab engages the secondary throttle linkage, forcing the plates open (Figure 12–12). Vacuum-actuated secondaries have a spring-loaded diaphragm. Vacuum is supplied to the diaphragm from ports in the primary and secondary throttle bores. When the vacuum in the primary bore reaches a specific level, the vacuum supplied to the diaphragm overcomes the spring and opens the secondary throttle valve(s) (Figure 12–13). The vacuum created in the secondary throttle bore increases the vacuum signal to the diaphragm, opening the secondary throttle valve(s) still farther.

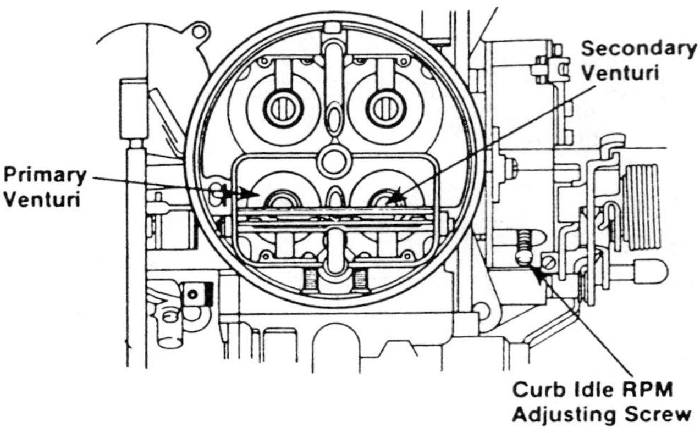

FIGURE 12–11 A two-stage carburetor; secondary throttle opens when more power is needed.

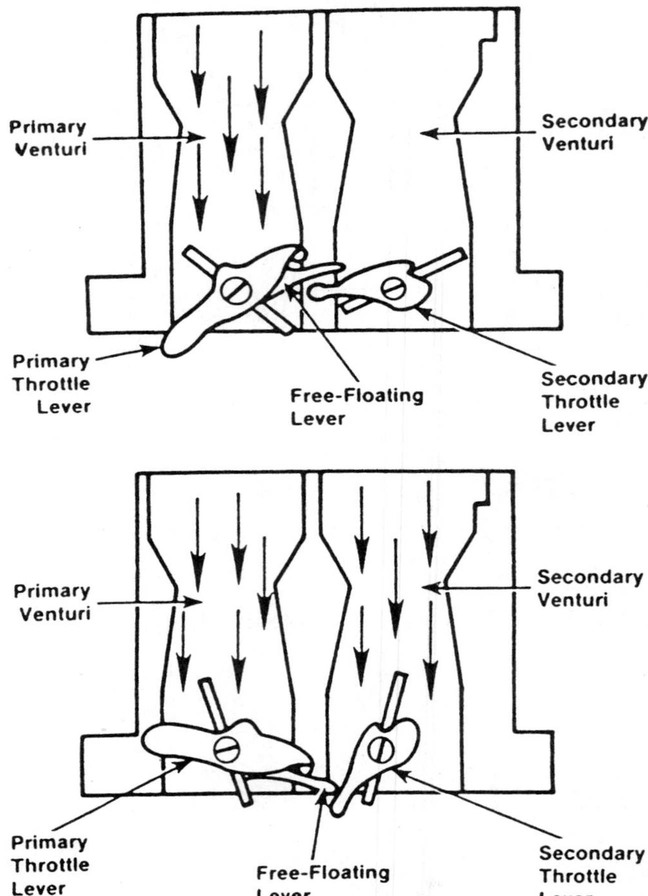

FIGURE 12–12 Secondary diaphragm responds to vacuum from ports within the carburetor barrels.

POWER ENRICHMENT CIRCUIT

At wide open throttle, the engine needs a richer than normal air/fuel mixture. This mixture cannot be supplied by the main metering system; so an additional fuel enrichment or full-power system is provided on most carburetors (Figure 12–14). The power enrichment system meters additional fuel into the mixture. This can be accomplished in several ways.

Metering Rods

In some carburetors, power enrichment is provided by metering rods placed in the main jets (Figure 12–15). The metering rods are actuated mechanically or by vacuum. When the throttle is not wide open (or nearly so), the throttle linkage keeps the rods in the jets, providing normal fuel flow. When the throttle is opened wide, either a mechanical link in

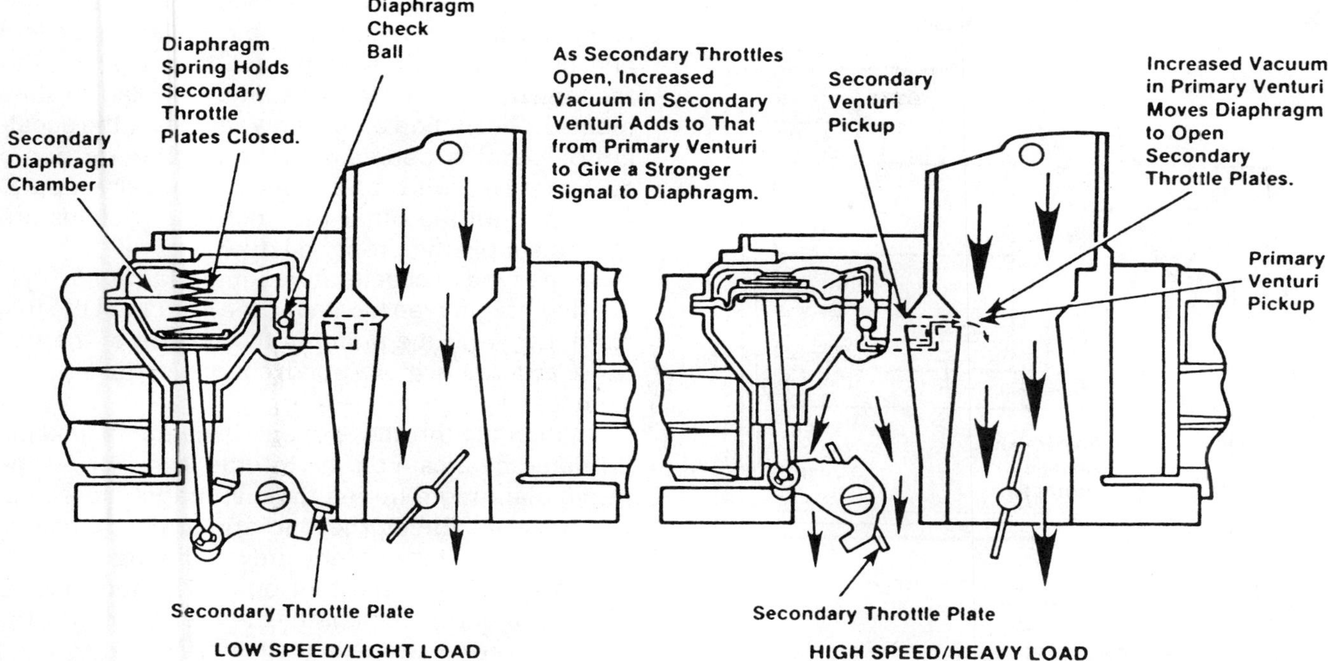

FIGURE 12-13 Vacuum-controlled secondary system

the throttle linkage or vacuum-actuated lever lifts the rods out of the jets, enabling more fuel to be forced into the main well. The additional fuel flow richens the air/fuel mixture.

Power Valves

The power valve (Figure 12-16) replaced mechanical metering rods as the carburetor developed.

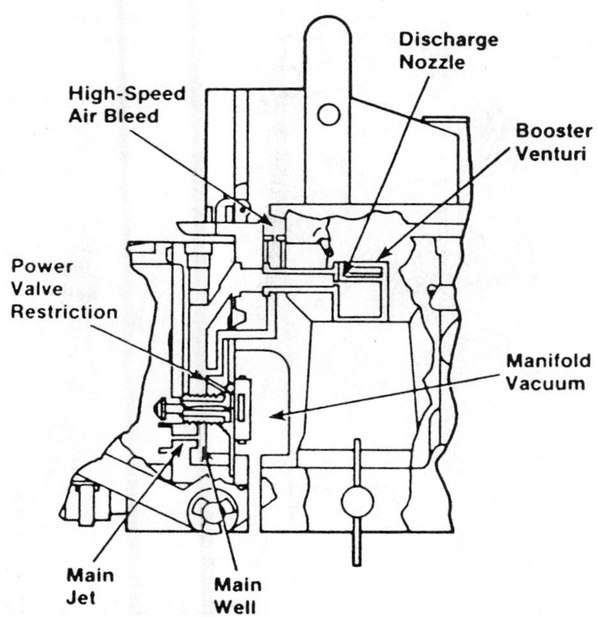

FIGURE 12-14 Power enrichment system

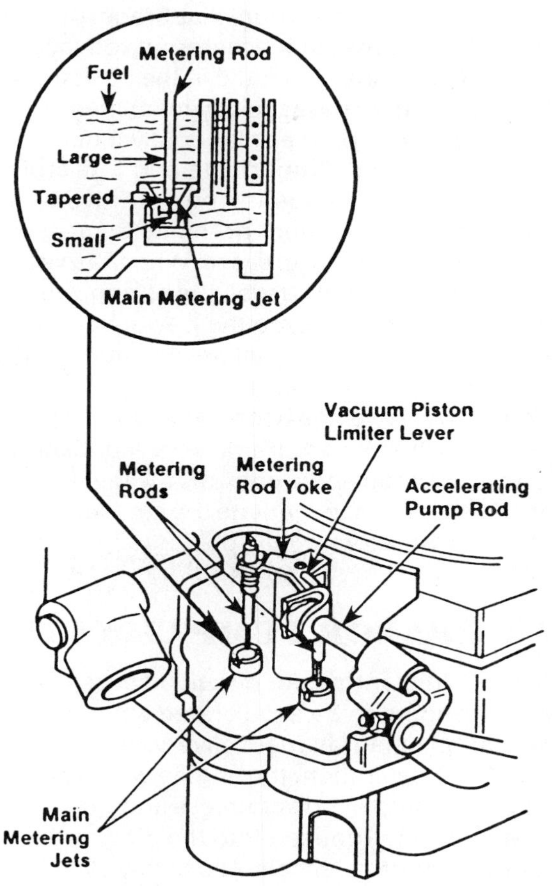

FIGURE 12-15 Some power systems consist of metering rods placed in the main jets.

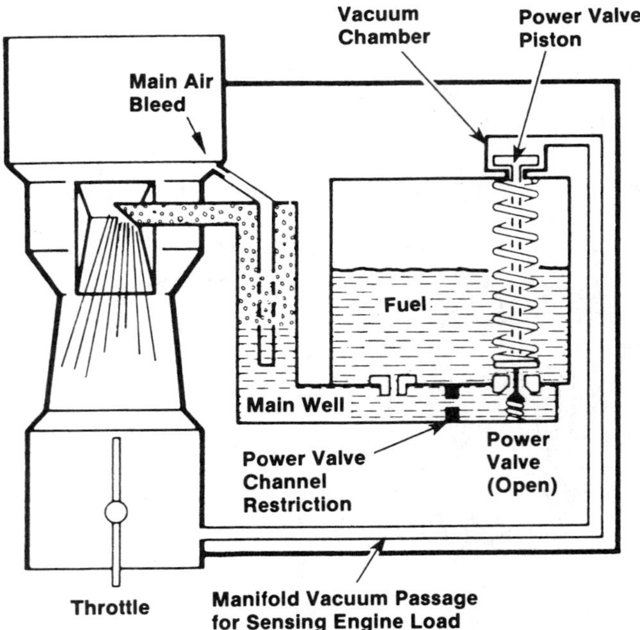

FIGURE 12–16 Power system

A power valve is basically a vacuum-operated metering rod. It consists of a vacuum diaphragm or piston, a spring-loaded valve, and a metering rod inside an auxiliary fuel jet usually located in the bottom of the fuel bowl. A vacuum passageway machined into the main body casting supplies manifold vacuum to the diaphragm or piston. During idle and low cruising speeds, the vacuum holds the power valve closed. However, manifold vacuum normally decreases as engine speed and load increases. When the vacuum signal drops to a specific level, the spring will overcome the vacuum and force the power valve out of the jet. This increases the fuel flowing into the main well.

The power valve has been replaced or modified in today's feedback carburetor. In a feedback system, an electrical solenoid controls the metering fuel jets or idle air bleeds to regulate the air/fuel mixture. Feedback carburetors will be discussed later in this chapter.

ACCELERATOR PUMP CIRCUIT

The off-idle or transfer circuit discussed earlier allows the engine to be accelerated smoothly without hesitation or lags in fuel delivery. However, during sudden acceleration, the engine will experience a momentary drop in power unless additional fuel is simultaneously introduced into the air charge.

During sudden acceleration the air flowing through the carburetor barrel reacts almost immediately to each change in the throttle valve opening. However, since fuel is heavier than air, it has a slower response time. Fuel in the main metering system and/or idle system takes a fraction of a second to respond to the throttle opening. This lag in time creates a hesitation of fuel flow whenever the accelerator pedal is depressed. The accelerator pump system solves this problem by mechanically supplying fuel until the other fuel metering systems are able to supply the proper mixture.

One type of accelerator pump (Figure 12–17) is the diaphragm type located in the bottom of the fuel bowl. Locating the pump in the bottom of the fuel bowl assures a more solid charge of fuel (fewer bubbles).

When the throttle is opened, the pump linkage, activated by a cam on the throttle lever, forces the pump diaphragm up. As the diaphragm moves up, the pressure forces the pump inlet check ball or valve onto its seat, thereby preventing the fuel from flowing back into the fuel bowl. The fuel passes through a short passage in the fuel bowl into the long diagonal passage in the metering body. It next goes into the main body passage and then the pump discharge chamber. The pressure of the fuel causes the discharge jet or valve to raise and fuel is then discharged into the venturi.

As the throttle returns toward the *closed* position, the linkage returns to its original position and the diaphragm return spring forces the diaphragm down. The pump inlet check valve is moved off its seat and the diaphragm chamber is refilled with fuel from the fuel bowl.

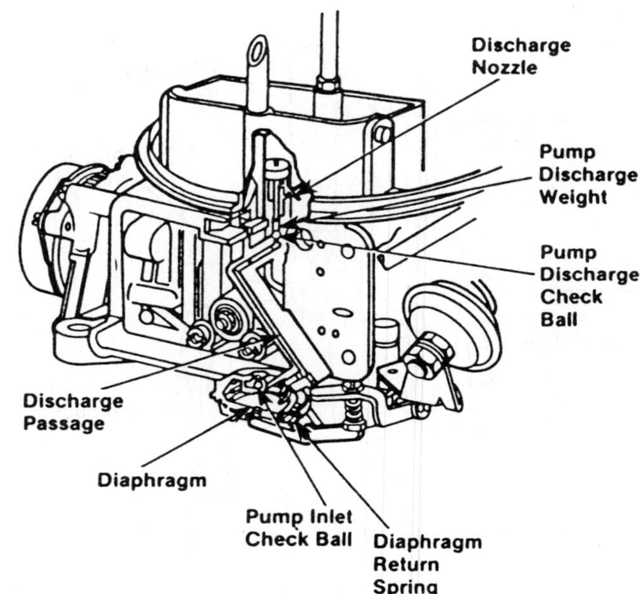

FIGURE 12–17 An arrow on a decal indicates lean direction.

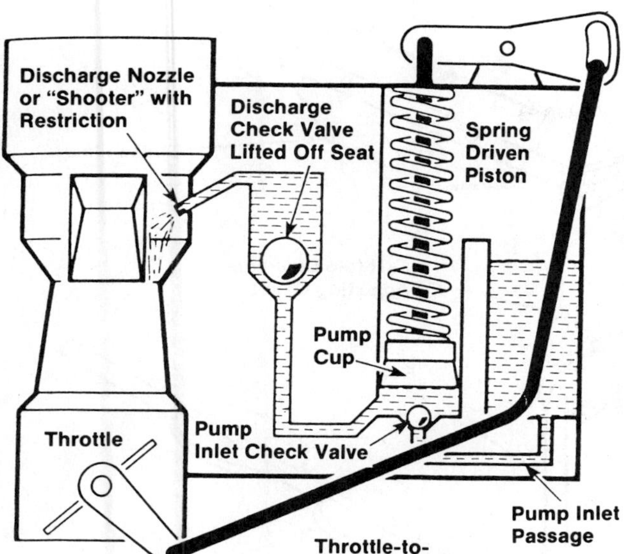

FIGURE 12-18 Partially open throttle of power system

Another common type of accelerator pump is shown in Figure 12-18. This type of pump uses a plunger rather than a diaphragm. As the throttle moves toward the *open* position, pressure from the plunger forces an inlet ball onto a check valve, sealing the inlet valve. At the same time, a ball is forced off the outlet check valve and fuel is discharged from the shooter nozzle. As the throttle moves back toward the *closed* position, the plunger retracts, allowing the inlet check valve to open and fuel to refill the pump. Some carburetors have a plunger cup that serves as an inlet valve when the plunger moves upward (Figure 12-19).

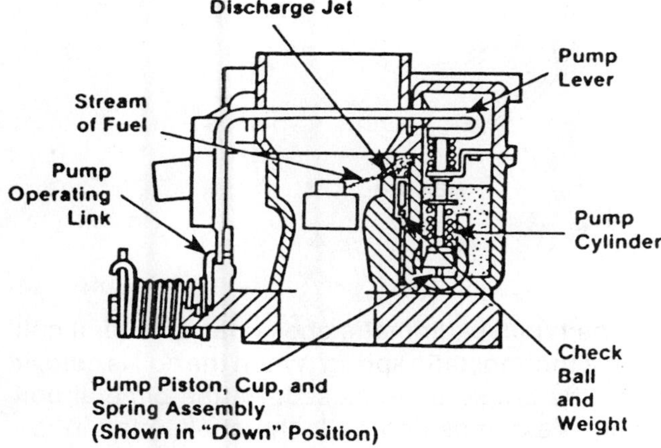

FIGURE 12-19 The accelerator pump forces a stream of fuel into the carburetor throat every time the gas pedal is pressed.

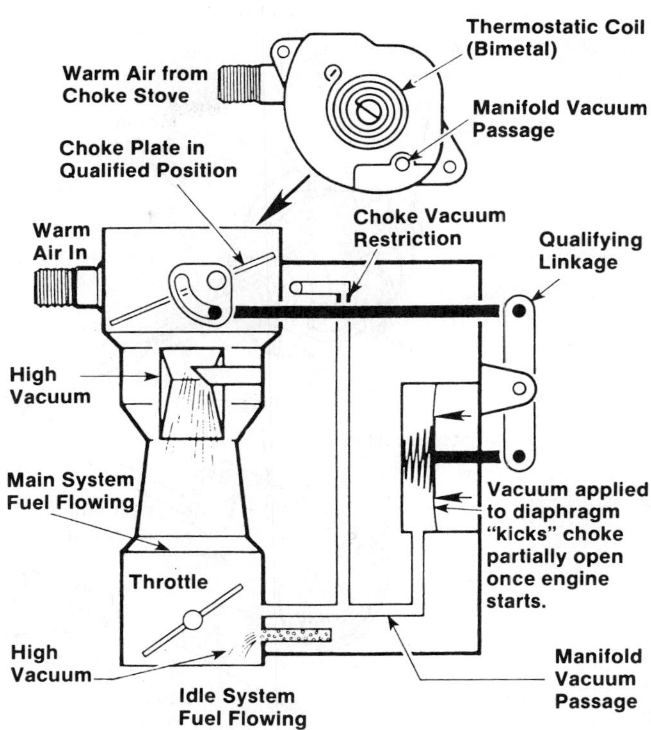

FIGURE 12-20 Choke system

CHOKE CIRCUIT

A cold engine needs a very rich air/fuel mixture during cranking and start-up. Providing the rich mixture is the job of the choke circuit (Figure 12-20).

During a cold startup, the choke should be closed. This creates a very high vacuum level in the air horn below the choke plate. As the air pressure outside the carburetor forces its way into the low-pressure areas, it draws with it a rich air/fuel mixture. When the throttle plate is closed, the mixture is forced out through the idle port or ports below the throttle valve. If the throttle valve is opened to assist in starting the engine, additional ports are exposed to the low-pressure manifold pressure and additional fuel is forced into the air horn. After the engine starts, a leaner mixture can be used to keep the engine running. Therefore, the choke should be opened some to allow increased airflow. After the engine has warmed to normal operating temperatures, the choke should be opened completely to allow the throttle to control airflow and fuel metering.

Before the introduction of automatic chokes, the opening and closing of the choke plate was manually controlled by the driver. A choke cable was connected to a knob inside the passenger compartment on the dash. To close the choke, the driver

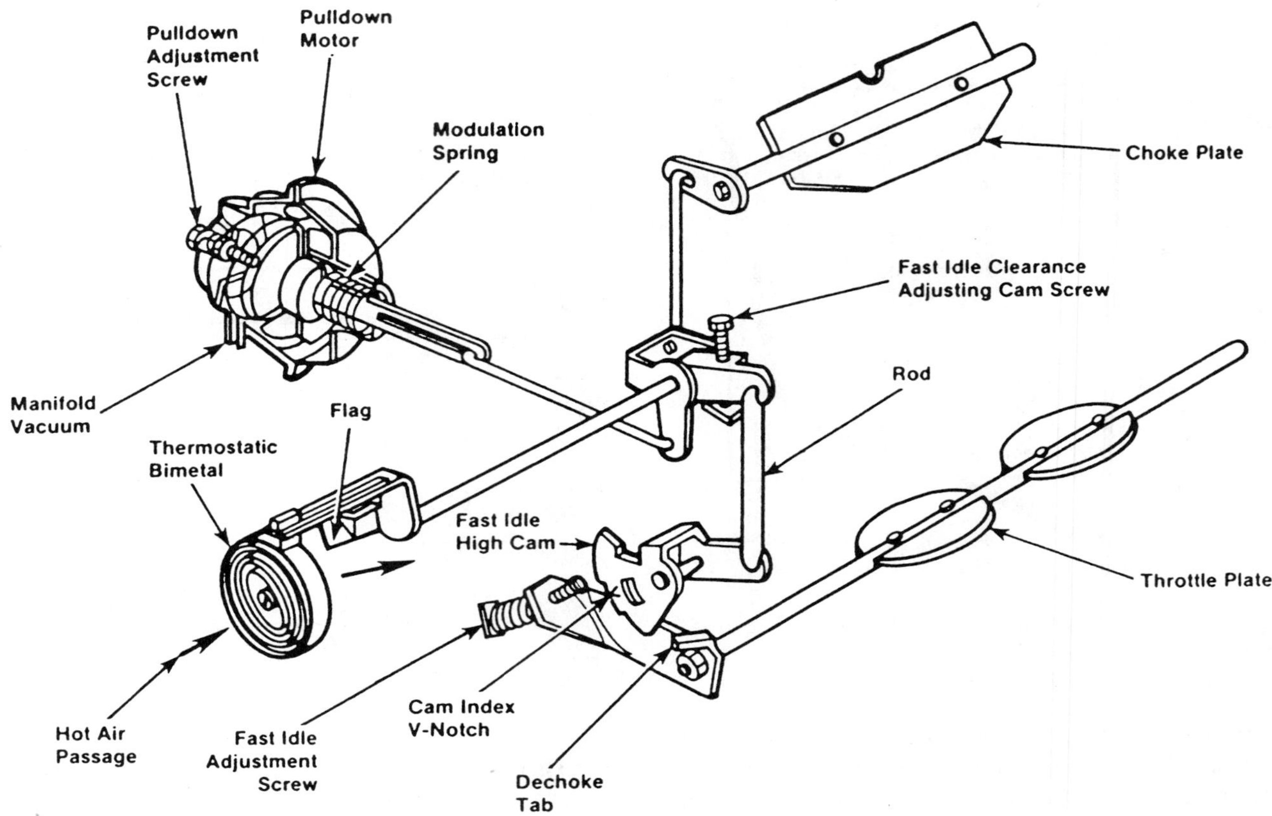

FIGURE 12–21 Typical automatic choke system

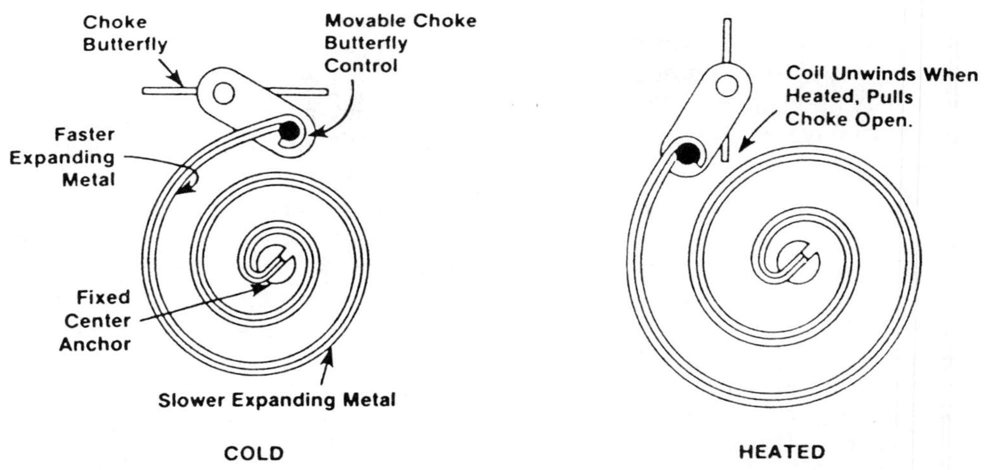

FIGURE 12–22 Principle of a bimetallic spring

simply pulled the knob out. As the engine warmed, the choke knob would be gradually pushed in to open the choke.

Most modern vehicles have an automatic choke (Figure 12–21) that operates without any driver assistance. Being more sensitive to engine temperature, an automatic choke is more efficient.

The typical automatic choke has a bimetal coil called a thermostatic spring. When the coil is cold, it forces the choke plate closed. As the bimetal coil warms, it expands and pulls the choke plate open (Figure 12–22).

The bimetal coil can be mounted directly on the carburetor; this type is called an integral choke (Figure 12-23). The bimetal coil may also be mounted on the intake manifold or in a heat well in the exhaust heat passage of the intake manifold; this type is called a divorced or remote choke (Figure 12-24).

The integral choke normally has a heat source to warm the bimetal. The heat source might be hot air (Figure 12-25) or coolant. Many integral chokes also have an electrical heater to assist in warming the coil during warm weather or hot start-up. The bimetal coil on many feedback carburetors is heated solely by a solid state heating element. Battery voltage is usually provided to the heating element from the alternator circuit so that the element does not affect the coil until the engine starts.

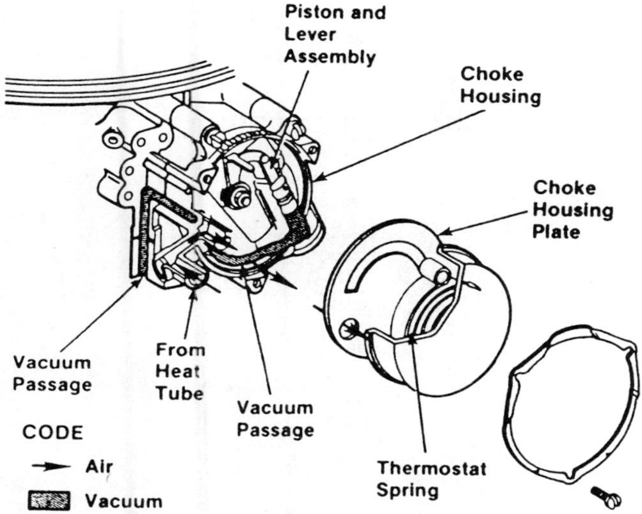

FIGURE 12-23 Integral automatic choke

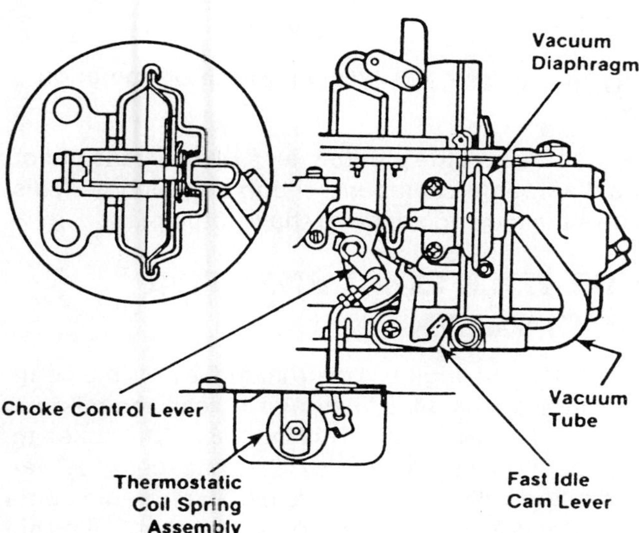

FIGURE 12-24 Divorced choke thermostatic element mounted in heat well

ADDITIONAL CARBURETOR CONTROL

To meet complex fuel economy and emission requirements, carburetors require the help of auxiliary controls. The following describes some of the more common assist devices you are likely to encounter when servicing a carburetor.

CHOKE QUALIFIER

Once the engine has started, a leaner mixture is needed. If the choke stays shut, the rich mixture will flood the engine and cause stalling. Therefore, the automatic choke has a choke qualifying mechanism to open the choke plate slightly after the engine has started.

Many integral chokes have a vacuum piston in the choke housing (see Figure 12-23) that opens the choke slightly when manifold vacuum reaches a certain level (immediately after the engine starts). A vacuum passage in the carburetor supplies vacuum to the piston. When the piston retracts in its cylinder, a passage is opened in the housing, allowing warm air from the heat tube to circulate through the choke housing. This delays warming the bimetal coil until the engine is started.

Some integral chokes and all divorced chokes have a qualifying diaphragm (also called a choke

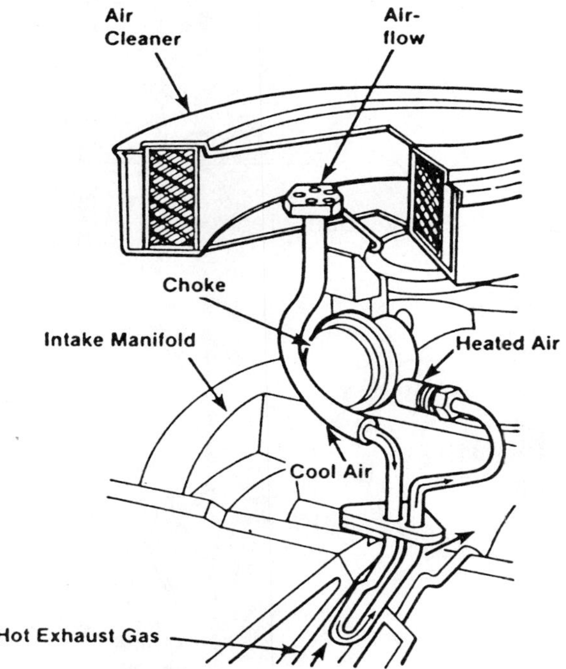

FIGURE 12-25 Late-model hot air choke system

pull-off diaphragm or vacuum break) instead of a vacuum piston (Figure 12-26). The diaphragm is connected to the manifold vacuum, and when the engine starts, the vacuum diaphragm retracts, pulling the choke open. The amount of opening, or the distance between the upper edge of the choke plate and the side of the air horn, is called the qualifying dimension or setting. In some carburetors, the pull-off diaphragm has a modulator spring that varies the qualifying setting based on ambient temperatures.

DASHPOT

The dashpot (Figure 12-27) is used during rapid deceleration to retard the closing of the throttle in the last few degrees. This allows a smooth transition from the main metering system to the idle system and prevents stalling due to an overly rich air/fuel mixture. It also controls the level of HC in the exhaust during deceleration.

The dashpot consists of a small chamber with a spring-loaded diaphragm and a plunger. A link from the throttle comes in contact with the dashpot plunger as the throttle closes. As the throttle linkage exerts force on the plunger, air slowly bleeds out of the diaphragm chamber through a small hole. This slows the closing action of the throttle plate.

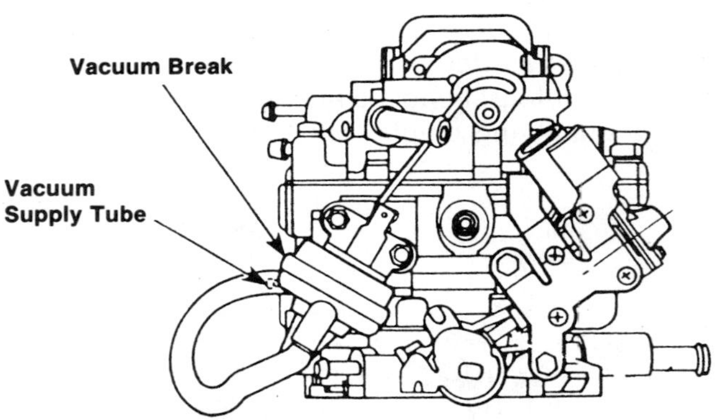

FIGURE 12-26 Vacuum break

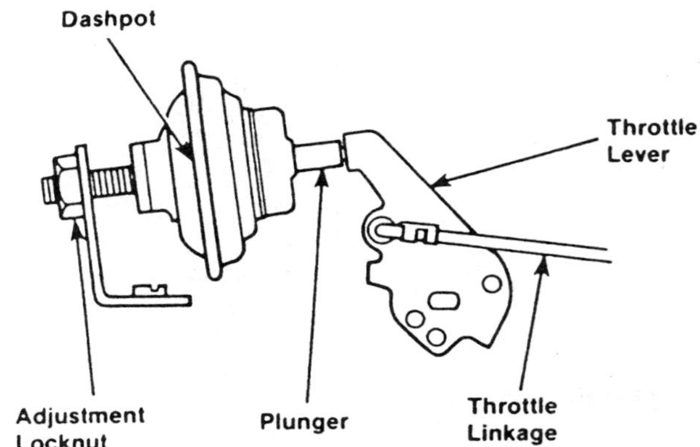

FIGURE 12-27 Typical carburetor dashpot installation

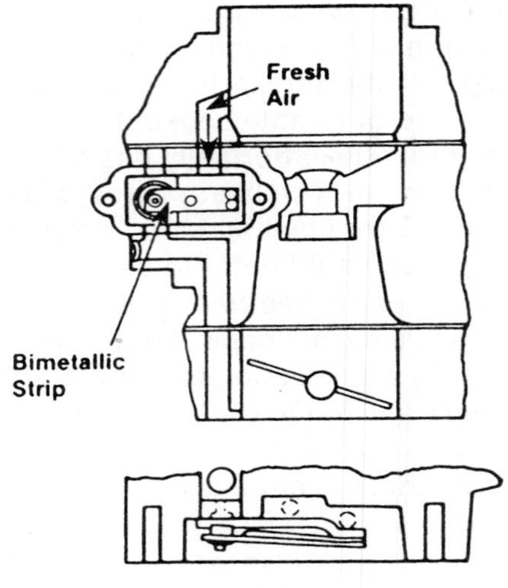

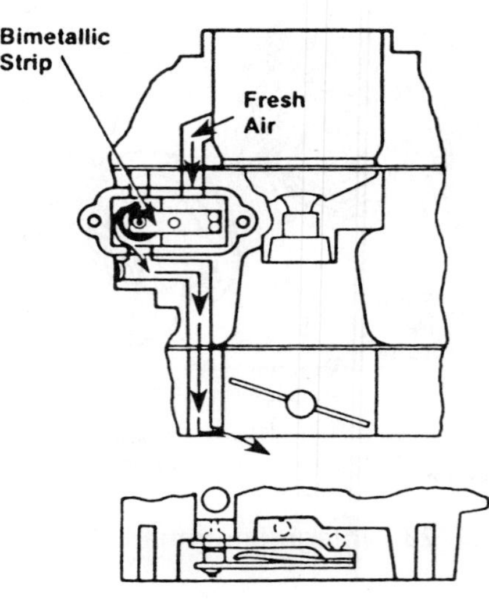

FIGURE 12-28 Hot-idle compensator operation

HOT-IDLE COMPENSATOR (HIC) VALVE

When the engine is overheated, a hot-idle compensator opens an air passage to lean the mixture slightly (Figure 12-28). This increases idle speed to help cool the engine and to prevent excess fuel vaporization within the carburetor. The hot-idle compensator is a bimetal, thermostatically controlled air bleed valve. By admitting additional air into the idle system, the mixture is leaned and the idle rpm raised, helping to cool the engine (by increased coolant flow) during prolonged idle.

DUAL VACUUM BREAK

Most modern carburetors are equipped with a dual vacuum break system (Figure 12–29)—a primary diaphragm and a secondary diaphragm. The primary vacuum diaphragm opens the choke valve slightly as soon as the engine starts to keep the engine from overchoking and stalling. The secondary vacuum diaphragm, which is also attached to the choke lever, opens the choke valve slightly wider in warm weather or when a warm engine is being started. Vacuum to the secondary diaphragm is controlled by a thermal vacuum switch or valve. The TVV releases vacuum to the secondary vacuum break when the engine reaches a certain temperature. This prevents a rich fuel mixture and the high emissions that result when a cold engine is started in warm weather or when a warm engine is started.

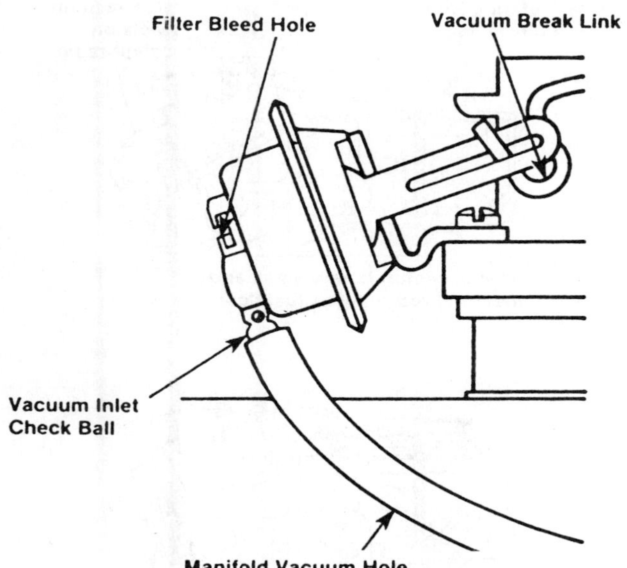

FIGURE 12–29 The vacuum break opens the choke slightly as soon as the engine starts.

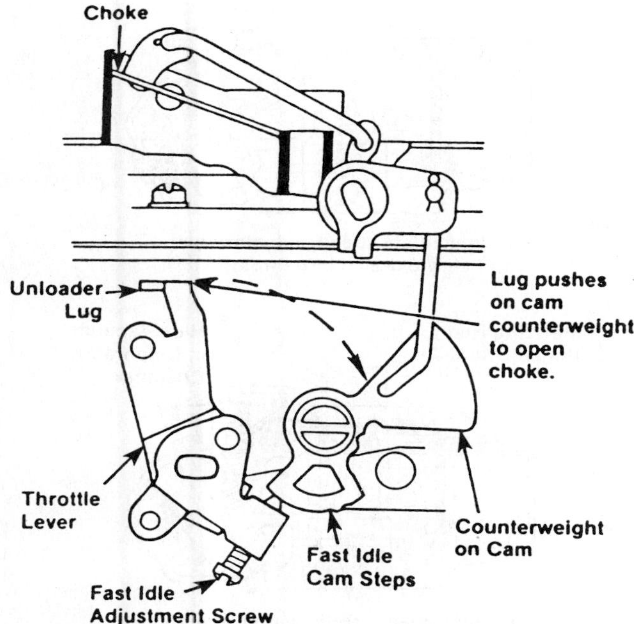

FIGURE 12–30 The choke unloader opens the choke any time the gas pedal is "floored."

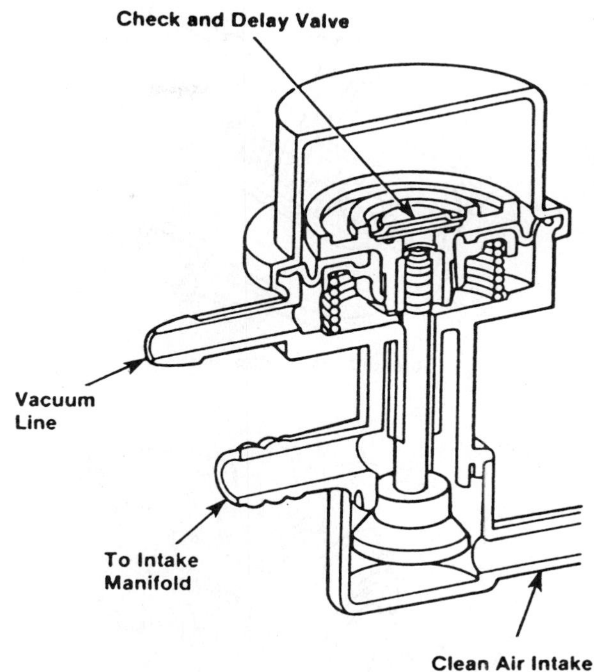

FIGURE 12–31 Deceleration valve

CHOKE UNLOADER

To be able to start a cold engine that has been flooded with gasoline, a choke unloader is required. The choke unloader (Figure 12–30) is throttle linkage actuated and opens the choke whenever the accelerator paddle is "floored." At wide-open throttle the partially opened choke allows additional air to lean out the mixture and reduce fuel flow.

DECELERATION VALVE

This valve (Figure 12–31) is designed to prevent backfire during deceleration as the fuel mixture becomes richer. The valve, which operates only during deceleration, is usually located between the intake manifold and the air cleaner. A typical valve has a cam-shaped diaphragm housing on one end. A control manifold-vacuum line is attached to a port under the diaphragm housing. The other end of the valve is connected by hoses to the intake manifold and air cleaner. When deceleration causes an increase in

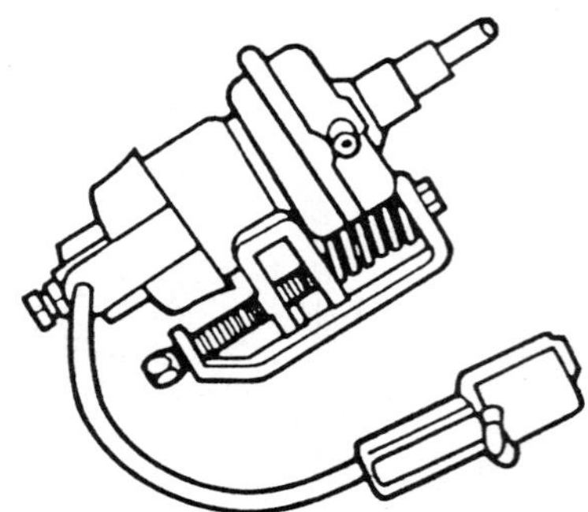
Solenoid Diaphragm

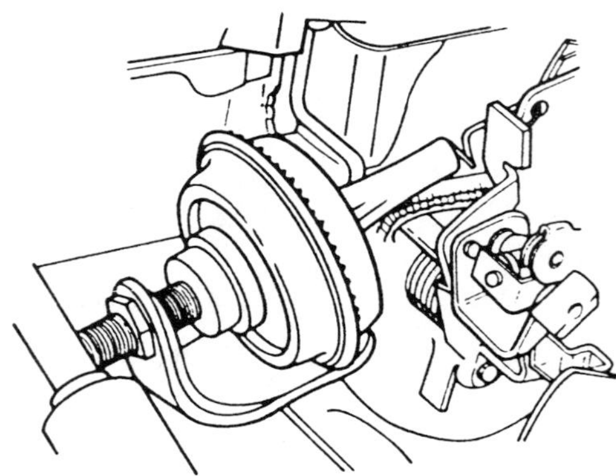
Vacuum Operated Throttle Modulator (VOTM)

FIGURE 12-32 Throttle positioners

manifold vacuum, the diaphragm opens the deceleration valve and allows air to pass from the air cleaner into the intake manifold, leaning the fuel mixture and preventing exhaust system backfire.

THROTTLE POSITION SOLENOID

The throttle position solenoid (TPS) is an electrical device used to control the position of the throttle plate (Figure 12-32). It can have several functions, depending on its application. When the basic function is to prevent dieseling, the solenoid is called a throttle stop solenoid or an idle stop solenoid. When the engine is started, the solenoid is energized and the plunger extends, pushing against the throttle linkage. This forces the throttle plate(s) open slightly to the curb idle position. When the ignition switch is turned off, the solenoid is de-energized and the plunger retracts, allowing the throttle plate to close completely and shutting off the air/fuel supply.

The throttle position solenoid is also used to increase the curb idle speed to compensate for extra loads on the engine. When this is its primary function, the solenoid might also be called an idle speed-up solenoid or a throttle kicker. This feature is most often used on cars with air conditioners; when the air conditioning is turned on, a relay energizes the solenoid so that the plunger extends farther, raising the idle rpm. This keeps the engine running at the high rpm necessary to ensure adequate emission control.

The throttle position solenoid is also used to control idle speed when the transmission is en-

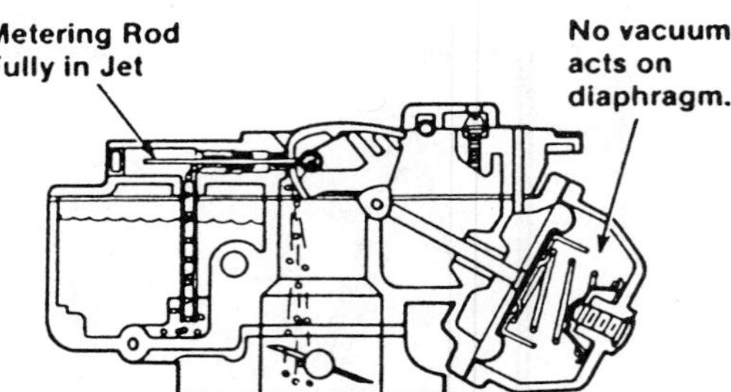

At idle, venturi is very small and metering rod reduces fuel flow to air horn.

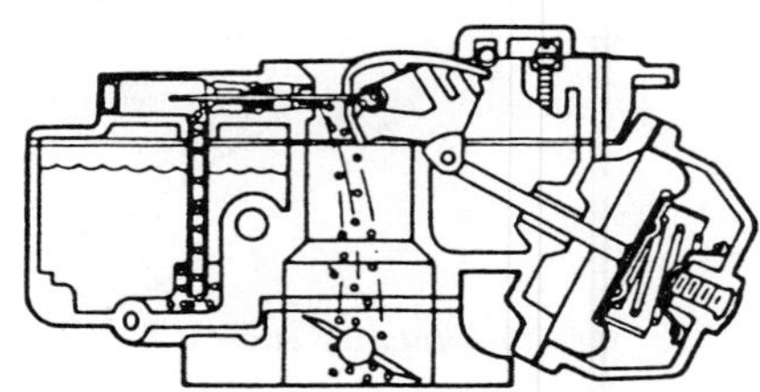
At idle, air horn is slightly larger. More fuel is drawn into it.

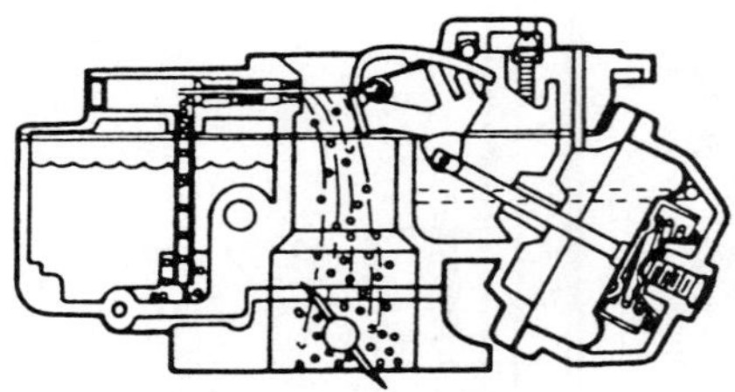
At immediate speed, still more fuel and air are drawn into throat.

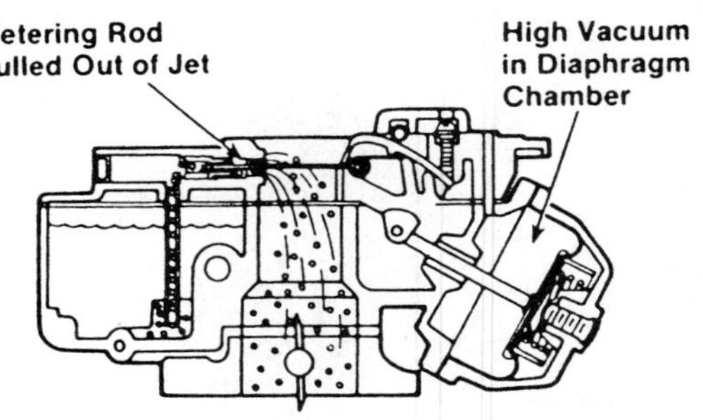

With wide-open throttle, variable venturi is fully withdrawn, as is metering rod.

FIGURE 12-33 Diaphragm controls variable venturis in the carburetor design

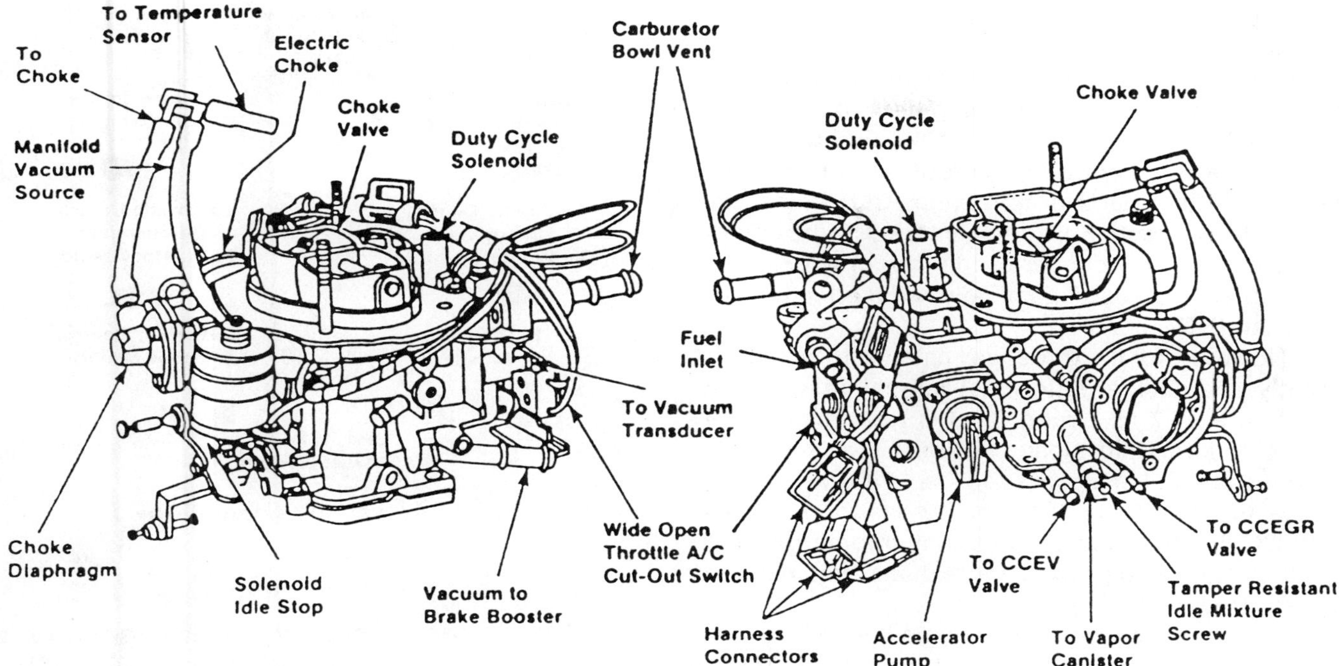

FIGURE 12–34 A 2 V two stage electronic feedback carburetor

gaged. A relay in the PARK/NEUTRAL switch signals the solenoid to extend when the transmission is shifted into gear. This opens the throttle slightly to compensate for the increased load on the engine.

VARIABLE VENTURI CARBURETORS

The carburetors discussed in this chapter have so far referred only to those that have a fixed venturi. A fixed venturi does not change shape and size to accommodate changing engine performance demands. Therefore, the speed of the air flowing through the venturi varies according to engine rpm and load. Because the vacuum in the venturi is the result of moving air, the amount of fuel drawn from the discharge nozzle varies as air velocity (and vacuum) in the venturi fluctuates. In some engine operation modes, the air speed, vacuum level, and fuel discharge are matched to the needs of the engine. At other times, the fuel discharge might be too little or too much. To compensate for the inadequacies of a fixed venturi, idle systems, power systems, and choke systems are needed to supplement the main metering system.

These assist systems are not necessary when a variable venturi is used. A variable venturi increases in size as engine demands increase. In this way, airflow speed through the venturi and the resulting pressure differential remains fairly constant. Thus, a variable venturi carburetor is also known as a *constant velocity* carburetor or a *constant depression* (vacuum) carburetor.

An example of a variable venturi carburetor is shown in Figure 12–33. This carburetor, built by Motorcraft, was introduced on Ford and Mercury models in 1977. Variable venturi carburetors were installed on Ford and Mercury V-6 and V-8 engines from 1977 to 1982, when they were replaced with electronic fuel injection systems. The two venturi valves are controlled by a single vacuum diaphragm that receives vacuum from ports in the throttle bores between the venturi valves and the throttle plates. As the throttle plates open, vacuum in the throttle bore increases and the adjustable venturi valves open farther. As the valves open, tapered metering rods attached to the valves retract from metering jets in the sides of the throttle bores. This increases the size of the jet openings, allowing additional fuel to be drawn into the airstream so that the air/fuel ratio remains constant. By metering both the fuel and airflow simultaneously, better fuel economy and lower emissions are possible than with a fixed venturi carburetor.

FEEDBACK CARBURETOR SYSTEMS

The latest type of carburetor system is the electronic feedback controlled design (Figure 12–34), which provides better combustion by control of the air/fuel mixture.

The feedback carburetor was introduced in the late 1970s, following the development of the three-way catalytic converter. Prior to 1978, two-way catalytic converters controlled the output of hydrocarbons (HC) and carbon monoxide (CO) in the exhaust. In 1978, both Ford and General Motors began using a three-way converter that not only oxidized HC and CO but also chemically reduced oxides of nitrogen (NO_x).

However, for the three-way catalyst to work efficiently, the air/fuel ratio must be maintained very close to a 14.7 to 1 ratio. In this ratio, called the stoichiometric ratio, air and fuel burn most efficiently. If the air/fuel mixture is too lean, NO_x will not be converted efficiently. If the mixture is too rich, HC and CO will not oxidize efficiently. Monitoring the air/fuel ratio is the job of the exhaust gas oxygen sensor (Figure 12-35).

An oxygen sensor (as its name implies) senses the amount of oxygen present in the exhaust stream. (For a full explanation of how an oxygen sensor is constructed and operates, refer to Chapter 13). A lean mixture will produce a high level of unburned oxygen in the exhaust. A rich mixture will produce little oxygen in the exhaust. The oxygen sensor, placed in the exhaust upstream from the catalytic converter, produces a voltage signal that varies in intensity in direct proportion to the amount of oxygen the sensor detects in the exhaust. If the oxygen level is high (a lean mixture), the voltage output is low. If the oxygen level is low (a rich mixture), the voltage output is high.

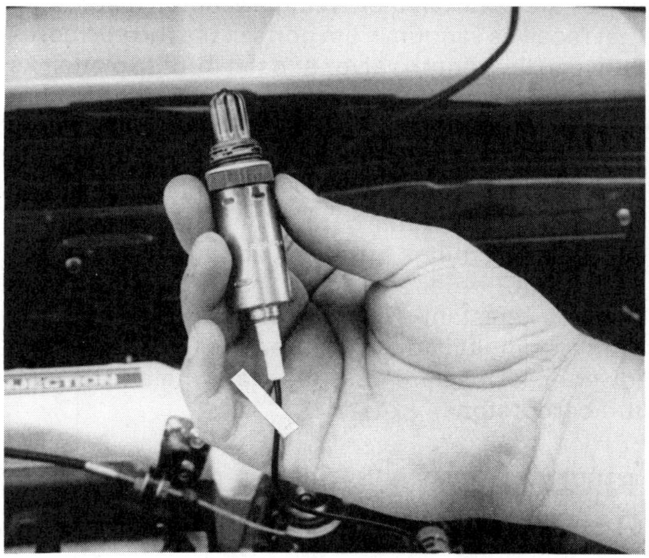

FIGURE 12-35 Exhaust gas oxygen sensor

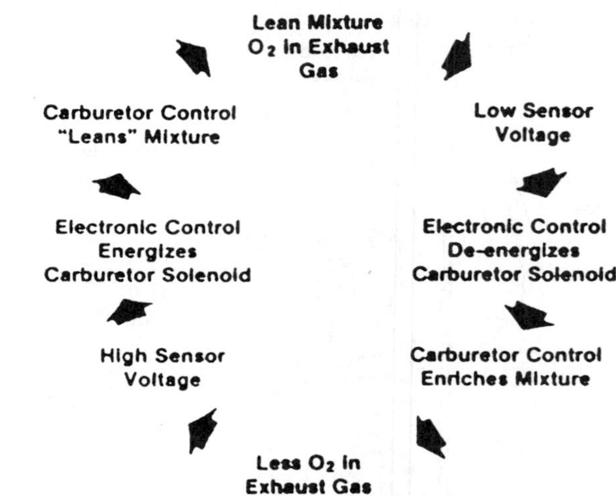

FIGURE 12-36 Closed loop operation

The electrical output of the oxygen sensor is monitored by an electronic control module (ECM). This microprocessor is programmed to interpret the input signals from the sensor and in turn generate output signals to a mixture control device that will meter more or less fuel into the air charge as it is needed to maintain the 14.7 to 1 ratio.

Whenever these components are working to control the air/fuel ratio, the carburetor is said to be operating in closed loop. Closed loop is illustrated in the schematic shown in Figure 12-36. The oxygen is constantly monitoring the residual oxygen in the exhaust, and the control module is constantly making adjustments to the air/fuel mixture based on the fluctuations in the sensor's voltage output. However, there are certain conditions under which the control module ignores the signals from the oxygen sensor and does not regulate the ratio of fuel to air. During these times, the carburetor is functioning in a conventional manner and is said to be operating in open loop (the control cycle has been broken).

The carburetor must operate in open loop until the oxygen sensor reaches a certain temperature (approximately 600 degrees Fahrenheit). The carburetor also goes into open loop when a richer than normal air/fuel mixture is required—such as during warmup and heavy throttle application. Several other sensors are needed to alert the electronic control module of these conditions. A coolant sensor provides input relating to engine temperature. A vacuum sensor and a throttle position sensor indicate wide open throttle.

Early feedback systems used a vacuum switch to control metering devices on the carburetor. Figure 12-37 shows a feedback carburetor introduced on Ford vehicles in 1978. Closed loop signals from the electronic control module are sent to a vacuum solenoid regulator (Figure 12–38), which in turn controls vacuum to a piston and diaphragm assembly in the carburetor. The vacuum diaphragm and a spring above the diaphragm work together to lift and lower a tapered fuel metering rod that moves a tapered fuel metering rod in and out of an auxiliary fuel jet in the bottom of the fuel bowl. The position of the metering rod in the jet controls the amount of fuel allowed to flow into the main fuel well.

The more advanced feedback systems use electrical solenoids on the carburetor to control the metering rods (Figure 12–39). These solenoids are generally referred to as duty-cycle solenoids or mixture

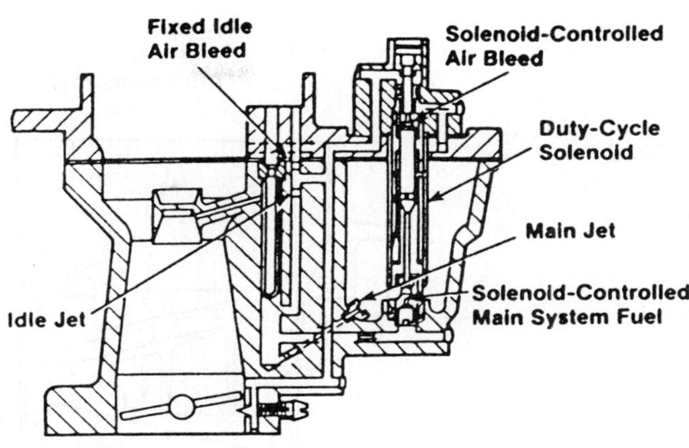

FIGURE 12–39 Electronic feedback carburetor

control (M/C) solenoids. The solenoid is normally wired through the ignition switch and grounded through the electronic control module. The solenoid is energized when the electronic control module completes the ground. The control module is programmed to cycle (turn on and off) the solenoid ten times per second. Each cycle lasts 100 milliseconds. The amount of fuel metered into the main fuel well is determined by how many milliseconds the solenoid is on during each cycle. The solenoid can be on almost 100 percent of the cycle or it can be off nearly 100 percent of the time. The M/C solenoid can control a fuel metering rod, an air bleed, or both.

In the Carter thermo-quad carburetor shown in Figure 12–40, variable air bleeds control the air/fuel ratio. This carburetor contains two fuel supply subsystems: the high-speed system and the low-speed system. The high-speed system meters fuel with a tapered metering rod positioned in the jet by the throttle. Fuel is metered into the main nozzle well where air from the feedback controlled variable air bleed is introduced. Since this air is delivered above the fuel level, it reduces the vacuum signal on the

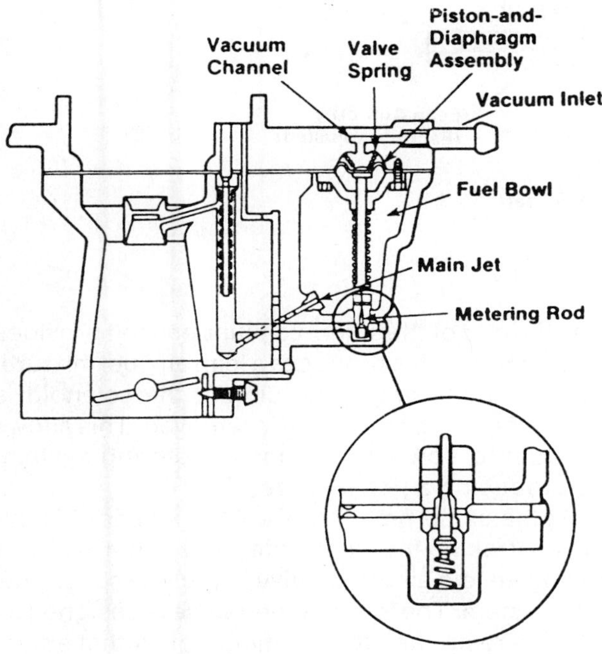

FIGURE 12–37 Vacuum feedback carburetor

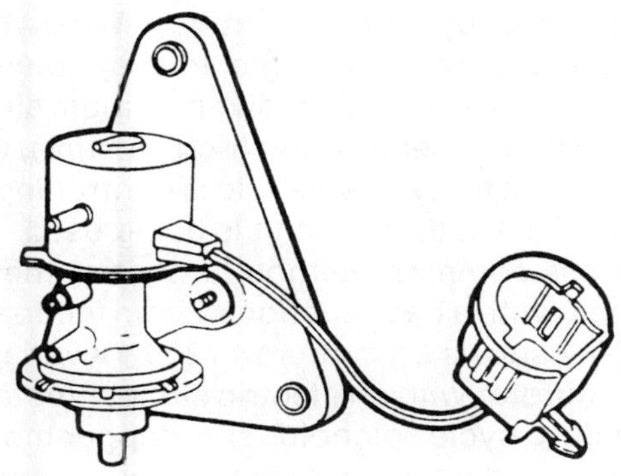

FIGURE 12–38 Remote-mounted fuel control solenoid

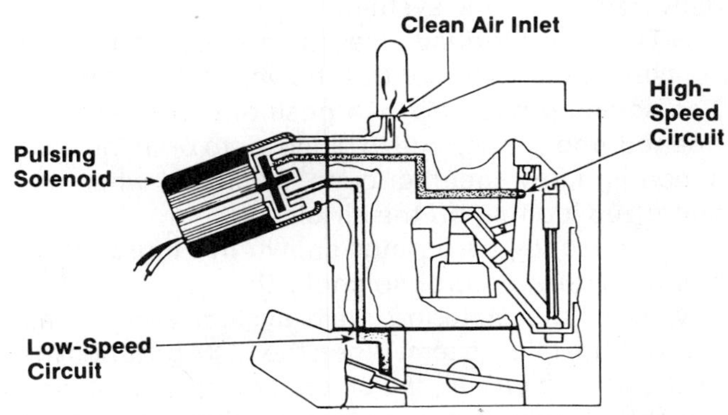

FIGURE 12–40 Thermo-quad with O_2 feedback

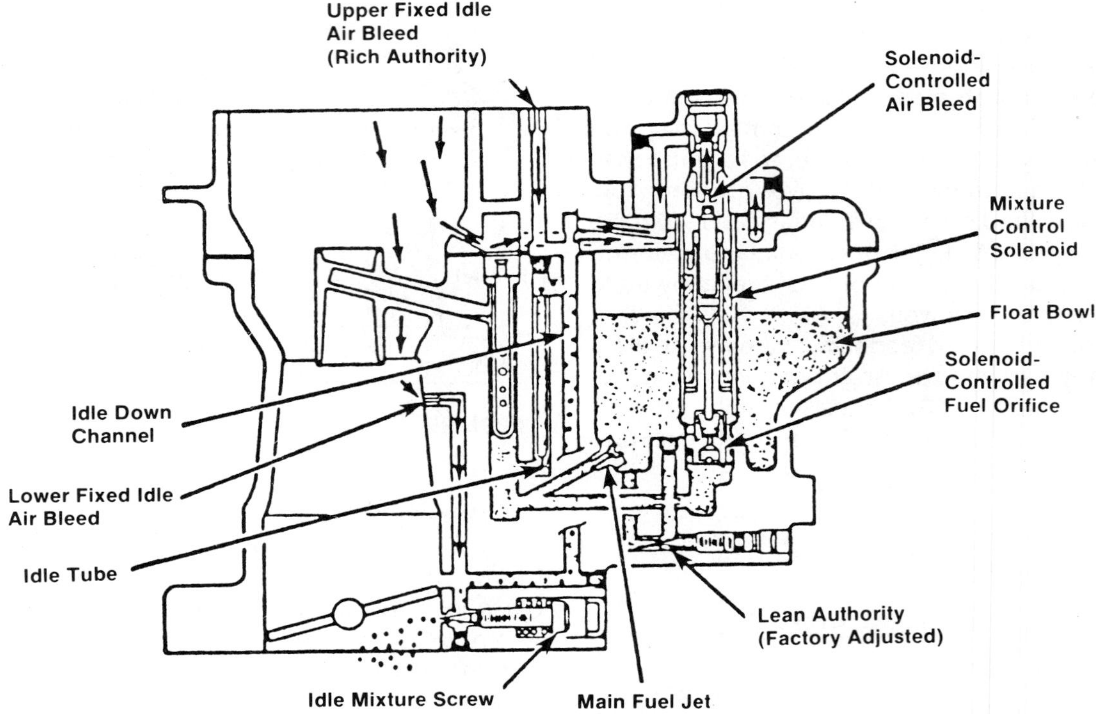

FIGURE 12–41 M/C solenoid

fuel consequently reducing the amount of fuel delivered from the nozzle.

The idle system is needed at low airflow through the venturi because there is insufficient vacuum at the nozzle to draw fuel into the air stream. After leaving the main jet, fuel is supplied to the idle system by the low-speed jet. It is then mixed with air from the idle bypass, then accelerated through the economizer and mixed with additional air from the idle bleed before being discharged from the idle ports below the throttle. Air from the variable air bleed is introduced between the idle air bleed and idle port. This air reduces the vacuum signal on the low speed jet and consequently the amount of fuel delivered to the idle system.

The thermo-quad uses a mixture control or "pulse" solenoid to control the variable air bleeds. The solenoid has only two positions of operation; opened when energized to bleed air to both the high speed and low speed circuits, or closed when de-energized, cutting off the air bleeds.

The Holley carburetor shown in Figure 12–41 has a mixture control solenoid that regulates fuel flow from the metering main system and the air bleed in the idle system. When the ECM grounds the M/C solenoid circuit, the solenoid is turned on and drives the solenoid plunger down. When the plunger is driven down, it drives a metering rod(s) down into the main metering jet(s) to restrict fuel flow. In some cases the end of the solenoid plunger itself provides the restriction. This produces a lean air/fuel mixture. When the ECM opens the circuit, the solenoid is turned off and the restriction is removed. This allows more fuel to flow into the main metering system, thus producing a rich mixture.

At the same time that the M/C solenoid is cycling or stroking the main metering system, it is also stroking an idle air bleed valve in the idle speed-low speed system. The idle air bleed valve is designed as a variable restriction to the amount of air that enters the idle speed-low speed circuit. When the M/C solenoid is down, the idle air bleed valve plunger drops and allows maximum air into the idle speed-low speed circuit (lean mixture). When the M/C solenoid plunger is up, the idle air bleed valve plunger is driven up and allows minimum air into the idle speed-low speed circuit (rich mixture). With this arrangement the M/C solenoid is controlling the air/fuel ratio in whatever circuit is being used.

A less common method to control the air/fuel mixture is with a back suction system feedback. Figure 12–42 shows a Motorcraft 7200 VV variable venturi carburetor with an electric stepper motor rather than a duty-cycle solenoid. The stepper motor controls the opening and closing of a vacuum passage to the fuel bowl. When the vacuum passage is open, a partial vacuum is created in the fuel bowl, reducing the fuel flow into the main metering system. The back suction system consists of an electric stepper motor, a metering pintle valve, an internal vent restrictor, and a metering orifice. The stepper motor regulates the pintle movement in the metering orifice, thereby varying the area of the opening communicating control vacuum to the fuel bowl. The larger this area is, the leaner the air/fuel mixture will be. Some of the control vacuum is bled off through the internal vent restrictor. The internal vent restrictor also serves to vent the fuel bowl when the back suction control pintle is in the closed position.

The 7200 VV carburetor was also produced with a feedback stepper motor that controls the main air bleed (Figure 12-43). The stepper motor controls the pintle movement in the air metering orifice thereby varying the amount of air being metered into the main system discharge area. The greater the amount of air, the leaner the air/fuel mixture will be. A hole in the upper body casting of the carburetor allows air from beneath the air cleaner to be channeled into the main system discharge area. The metered air lowers the metering signal at the main fuel metering jets.

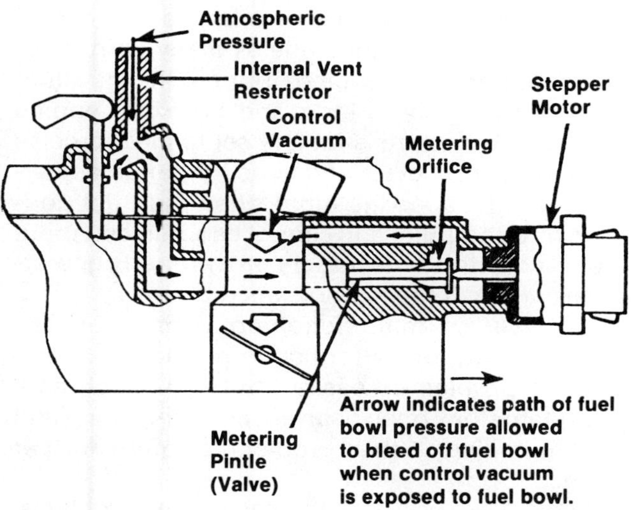

FIGURE 12-42 Back suction feedback system of the 7200 VV feedback carburetor

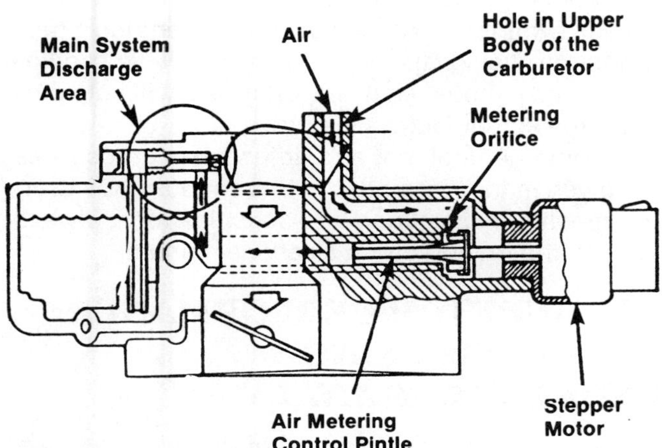

FIGURE 12-43 Air bleed feedback system of the 7200 VV feedback carburetor

ELECTRONIC IDLE-SPEED CONTROL

In order to maintain federally mandated emission levels, it is also necessary to control the idle speed. Most feedback systems operate in open loop when the engine is idling. To reduce emissions during idle, most feedback carburetors idle faster and leaner than nonfeedback carburetors.

To adjust idle speed, many feedback carburetors have an idle speed control (ISC) motor that is controlled by an electronic control module. The ISC motor is a small, reversible, electric motor. It is part of an assembly that includes the motor, a gear drive, and a plunger (Figure 12-44). When the motor turns in one direction, the gear drive extends the plunger; when the motor turns in the opposite direction, the gear drive retracts the plunger. The ISC motor is mounted so that the plunger can contact the throttle lever. The ECA controls the ISC motor and can change the polarity applied to the motor's armature in order to control the direction it turns (Figure 12-45). When the idle tracking switch is open (throttle closed), the ECA commands the ISC motor to control idle speed. The ISC provides the correct throttle opening for cold or warm engine idle.

The electronic control module receives input from various switches and sensors to determine the best idle speed. Some of the possible applications are:

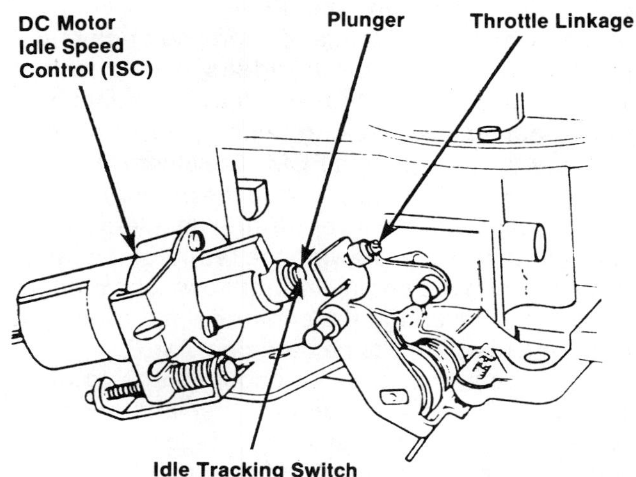

FIGURE 12-44 DC motor idle speed control (ISC)

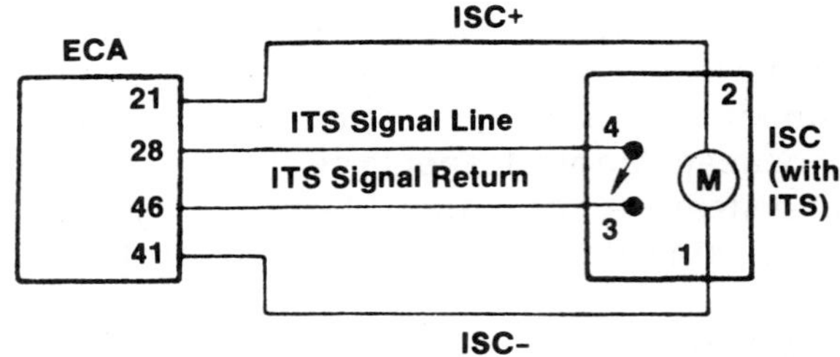

FIGURE 12-45 ISC motor circuit schematic

- Engine coolant temperature sensor
- Air charge temperature (ACT) sensor
- Manifold absolute pressure (MAP) sensor
- Barometric pressure (BP) sensor
- PARK/NEUTRAL or NEUTRAL gear switch
- Clutch engaged switch
- Power steering pressure switch
- A/C clutch compressor switch
- Idle tracking switch (ITS)

Based on the input signals from the system's sensors, the ECM will increase the curb idle speed if the coolant is below a specific temperature, if a load (such as air conditioning, transmission, power steering) is placed on the engine, or when the vehicle is operated above a specific altitude.

During closed choke idle, the fast-idle cam holds the throttle blade open enough to lift the throttle linkage off of the ISC plunger. This allows the ISC switch to open so that the ECM does not monitor idle speed. As the choke spring allows the fast-idle cam to fall away and the throttle returns to the warm idle position, the ECM notes the still low coolant temperature and commands a slightly higher idle speed.

As the engine warms up, the plunger is retracted by the electronic control module. If the A/C compressor is turned on, the ECM will extend the plunger a certain distance to increase engine idle speed to compensate for the added load. When the throttle is opened and the lever leaves contact with the plunger, an idle tracking switch (ITS) in the end of the plunger signals the ECM. The electronic control module will then fully extend the plunger where, upon contact with the lever (during deceleration), it will act as a dashpot, slowing the return of the throttle lever. When the engine is shut down, the plunger will retract, preventing the engine from dieseling. It will then extend for the next engine startup.

In some systems, if the engine starts to overheat, the ECM commands a higher idle speed to increase coolant flow. If system voltage falls below a predetermined value, the ECM commands a higher idle speed in order to increase generator speed and output.

Normally, idle speed adjustments are not possible on carburetors with electronic idle speed control. Attempting to adjust idle speed by adjusting the ISC plunger screw results in the ECM moving the plunger to compensate for the adjustment. Idle speed does not change until the ISC motor uses up all of its plunger travel trying to compensate for the adjustment, at which point the system is completely out of calibration. When idle speed driveability problems occur, the ISC system is usually responding to or being affected by the problem, not causing it.

CARBURETOR DIAGNOSIS AND ADJUSTMENT

The tuneup procedure of a late-model carbureted engine does not include as many carburetor adjustments as were required, prior to the introduction of feedback carburetors and electronic engine controls. Idle mixture is factory set to meet precise emission control levels and is no longer an adjustable item. Idle speed is more often than not controlled by a computer and cannot be adjusted. About the only adjustment necessary on a properly operating carburetor is the fast-idle speed.

However, a malfunctioning carburetor can cause a variety of performance problems. Sometimes the problem is easily observed—such as a choke plate stuck open or an accelerating pump that is not pumping. Other problems require further testing.

Table 12-1 lists specific carburetor problems, their possible causes, and possible remedies. Use of the chart assumes that the engine is in good mechanical condition and properly tuned. Keep in mind that many ignition and carburetor problems have similar symptoms. An analysis of the engine's performance, using an engine analyzer with oscilloscope and exhaust analysis functions, will help pinpoint the actual fault.

Some general test and adjustment procedures are given in the remainder of this chapter. However, for specific instructions and specifications, always refer to the manufacturer's service manual.

CARBURETOR PERFORMANCE DIAGNOSIS USING EXHAUST GAS ANALYSIS

The most precise method of diagnosing a carburetor is with an infrared exhaust analyzer (Figure

TABLE 12-1: CARBURETOR TROUBLESHOOTING GUIDELINES

Condition	Possible Cause	Remedy
Engine cranks but will not start or starts hard when cold.	Choke valve not closing sufficiently when cold.	Adjust the index of the choke thermostatic (bimetal) coil for richer mixture.
	Choke valve or linkage binding or sticking.	Realign the choke valve or linkage as necessary. If caused by dirt and gum, clean with automatic choke cleaner. DO NOT OIL CHOKE LINKAGE. If parts are replaced, readjust to specifications.
	No fuel in carburetor.	Remove fuel line at carburetor. Connect hose to fuel line and run into metal container. Remove the high tension coil wire from center tower on distributor cap and ground. Crank over engine—if there is no fuel discharge from the fuel line, check for kinked or bent lines. Disconnect fuel line at tank and blow out with air hose, reconnect line and check again for fuel discharge. If none, replace fuel pump. Check pump for adequate flow, as outlined in service manual. If fuel supply is all right, check the following: Inspect fuel filter(s). If plugged replace. If filters are all right, remove air horn or fuel bowl and check for a bind in the float mechanism or a sticking float needle. If all right, adjust float to specifications.
	Engine flooded. To check for flooding remove the air cleaner. With the engine off look into the carburetor bores. Fuel will be dripping off nozzles and/or the carburetor will be very wet.	Check to determine if customer is using proper carburetor unloading procedure. Depress the accelerator to the floor and check the carburetor to determine if the choke valve is opening. If not, adjust the throttle linkage and unloader to specifications.
	Carburetor flooding	Note: Before removing the carburetor air horn, use the following procedure which may eliminate the flooding. Remove the fuel line at the carburetor and plug. Crank and run the engine until the fuel bowl runs dry. Turn off the engine and connect fuel line. Then re-start and run engine. This will often flush dirt past the carburetor float needle and seat. If dirt is in fuel system, clean the system and replace fuel filter(s) as necessary. If excessive dirt is found, remove the carburetor unit. Diassemble and clean. Check float needle and seat for proper seal. If the needle is defective, replace with a matched set. Check float for being loaded with fuel, bent float hanger or binding of the float arm. Adjust float to specifications.
Engine starts hard when hot.	Choke valve not opening completely.	Check for binding choke valve and/or linkage. Clean and free up or replace parts as necessary. DO NOT OIL CHOKE LINKAGE. Check and adjust choke thermostatic coil. Check for choke thermostatic coil binding in well or housing. Check for vacuum leak with integral choke system.
	Engine flooded; carburetor flooding.	See procedure under "Engine cranks, will not start."

TABLE 12-1: CARBURETOR TROUBLESHOOTING GUIDELINES (CONTINUED)

Condition	Possible Cause	Remedy
	No fuel in carburetor.	Check fuel pump. Run pressure and volume test. Check float needle for sticking in seat, or binding float.
	Leaking float bowl.	Fill bowl with fuel and check for leaks.
	Fuel percolation.	Open throttle wide and operate starter to relieve over rich condition.
Engine idles rough and stalls.	Idle mixture adjustment.	Adjust idle mixture screws to lean best idle. Repeat the operation on 2- and 4-barrel carburetors. Turn mixture screws in until idle speed drops 25 rpm on tachometer.
	Idle speed setting.	Reset idle speed per instructions on decal in engine compartment. Check solenoid operation.
	Manifold vacuum hoses disconnected or improperly installed.	Check all vacuum hoses leading to the manifold or carburetor base for leaks, being disconnected or connected improperly. Install or replace as necessary.
	Carburetor loose on intake manifold.	Torque carburetor to manifold bolts (100 in-lb.).
	Intake manifold is loose or gaskets are defective.	Spray carburetor cleaner around manifold legs and carburetor base. If engine rpm changes, tighten or replace the manifold gaskets or carburetor base gaskets as necessary.
	Hot idle compensator not operating (where used).	Normally the hot idle compensator should be closed when engine is running cold and open when engine is hot (approx. 140° F); replace if defective.
	Carburetor flooding.	Correct by using procedure outlined under "Engine cranks, will not start."
Engine starts and stalls.	Engine does not have enough fast idle speed when cold.	Check and re-set the fast idle setting and fast idle cam.
	Choke vacuum diaphragm unit is not adjusted to specification or unit is defective.	Adjust vacuum break to specification. If adjusted correctly, check the vacuum opening for proper operation as follows. On externally mounted vacuum diaphragm unit, connect a piece of hose to fitting on the vacuum diaphragm unit and apply suction preferably by hand vacuum pump or another vehicle. Plunger should move inward and hold vacuum. If not, replace the unit. On the integral vacuum piston unit, remove cover and visually check piston and vacuum channel. If piston is corroded or sticking replace assembly. Note: Always check the fast idle cam adjustment before adjusting vacuum unit.
	Choke coil rod out of adjustment.	Adjust choke coil rod.

TABLE 12-1: CARBURETOR TROUBLESHOOTING GUIDELINES (CONTINUED)

Condition	Possible Cause	Remedy
	Choke valve and/or linkage sticking or binding.	Clean and align choke valve and linkage. Replace if necessary. Readjust if part replacement is necessary.
	Idle speed setting	Adjust idle speed to specifications on decal in engine compartment.
	Not enough fuel in carburetor.	Check fuel pump pressure and volume. Check for partially plugged fuel inlet filter. Replace if dirty. Remove air horn or fuel bowl and check float adjustments.
	Carburetor flooding	Correct by using procedure outlined under "Engine cranks, will not start."
Engine runs uneven or surges.	Fuel restriction	Check all hoses and fuel lines for bends, kinks, or leaks. Check fuel filter. If plugged or dirty, replace.
	Dirt or water in fuel system.	Clean fuel tank and lines. Remove and clean carburetor.
	Fuel level	Adjust float. Check for free float and float needle valve operation.
	Main metering jet defective, loose, or incorrect part.	Replace as necessary. See service manual.
	Power system in carburetor not functioning properly.	Power valve or piston sticking in DOWN position. Free up or replace as necessary. Power valve loose, incorrect gasket or leaking around threads: Tighten or replace as necessary. Leaking diaphragm: Test with hand vacuum pump; replace as necessary.
	Vacuum leaks	It is absolutely necessary that all vacuum hoses and gaskets are properly installed, with no air leaks. The carburetor and manifold should be evenly tightened to specified torque.
Engine hesitates on acceleration.	Defective accelerator pump system. Note: A quick check of the pump system can be made as follows: With the engine off, remove air cleaner and look into the carburetor bores and observe pump stream, while briskly opening throttle valve. A full stream of fuel should emit from pump jet and strike near the center of the venturi area.	Piston type: Remove air horn and check pump cup. If cracked, scored, or distorted, replace the pump plunger. Piston and diaphragm types: Check the pump discharge ball for properly seating and location. The pump discharge ball is located in a cavity next to the pump well. To check for proper seating, remove air horn and gasket and fill cavity with fuel. No leak down should occur. Restake and replace check ball if leaking. Make sure discharge ball, spring, and retainer are properly installed.

TABLE 12-1: CARBURETOR TROUBLESHOOTING GUIDELINES (CONTINUED)

Condition	Possible Cause	Remedy
		Diaphragm type: Check pump discharge as before. Inspect diaphragm; replace if defective. Check pump inlet ball valve clearance. Adjust pump operating lever clearance.
	Dirt in pump passages or pump jet.	Clean and blow out with compressed air.
	Fuel level.	Check for sticking float needle or binding float. Free up or replace parts as necessary. Check and reset float level to specification.
	Leaking air horn to float bowl gasket.	Torque air horn to float bowl using proper tightening procedure.
	Carburetor loose on manifold.	Torque carburetor to manifold bolts. (100 in.-lb.)
No power on heavy acceleration or at high speed.	Carburetor throttle valve(s) not going wide open. (Check by pushing accelerator pedal to floor.)	Adjust throttle linkage to obtain wide open throttle in carburetor.
	Dirty or plugged fuel filter(s).	Replace with a new filter element.
	Power system not operating.	Piston type: Check power piston for free up and down movement. If power piston is sticking check power piston and cavity for dirt, or scores. Check power piston spring for distortion. Clean or replace as necessary. Piston and diaphragm types: Check power valve channel restrictions. Clean if necessary.
	Float level too low.	Check and reset float level to specification.
	Float not dropping far enough into float bowl.	Check for binding float hanger and for proper float alignment in float bowl.
	Main metering jet(s) dirty, plugged or incorrect part.	If main metering jets are plugged or dirty and excessive dirt is in fuel bowl, carburetor should be completely disassembled and cleaned.
Engine backfires.	Choke valve, fully or partially open, binding or sticking.	Free up with choke solvent. Realign or replace if bent.
	Accelerator pump not operating properly.	Remove air cleaner and observe pump discharge. Replace pump cup or diaphragm. Readjust pump to specifications. Restake or replace pump intake or discharge valve.
	Old or dirty (fouled) spark plugs.	Clean or replace spark plugs.
	Old or cracked spark plug wires.	Test with a scope if possible or observe wires on dark night with engine running. Replace wires.
	Partially clogged fuel filter.	Replace filter on regular maintenance schedule.
	Backfire on deceleration. Defective air pump diverter valve.	Check hoses and fittings for tightness and leakage. Disconnect valve signal line. With engine running a vacuum must be felt. With engine idling, hold hand at exhaust port. No air should be felt. If valve or hoses defective, it must be replaced.

TABLE 12-1: CARBURETOR TROUBLESHOOTING GUIDELINES (CONTINUED)

Condition	Possible Cause	Remedy
Secondary system inoperative	Sticking throttle valves.	Readjust secondary throttle valve stop screw. Throttle valves nicked or throttle shaft binding. Repair or replace throttle valve. Check throttle body for warpage. Torque throttle body screws evenly.
	Ruptured or leaking secondary diaphragm.	Inspect diaphragm. Replace or install properly.
	Venturi vacuum ports plugged.	Try cleaning ports with choke solvent or lacquer thinner. It may be necessary to remove the diaphragm assembly and back blow into the venturi.

12–46). In fact, many states require an exhaust analysis with each annual tune-up. It is important that the technician know how to troubleshoot and verify carburetor performance using an analyzer.

Most manufacturers use HC and CO content as a measure of carburetor performance. However, because of the efficiency of feedback carburetors and catalytic converters, four-gas exhaust analyzers are increasingly being used to also monitor oxygen (O) and carbon dioxide (CO_2).

Carbon Monoxide (CO). Carbon monoxide (CO), measured in percent by volume of the exhaust, is a byproduct of incomplete combustion of the fuel mixture. A rich air/fuel mixture, lacking oxygen, generally causes high CO content (Figure 12–47). CO is at its lowest when the air/fuel mixture is 14.7:1 or leaner. CO is probably the best indicator of the air/fuel ratio due to its sensitivity to mixture changes.

Carbon Dioxide (CO_2). Carbon dioxide, measured in percent by volume of the exhaust, is most abundant when the engine operates at maximum efficiency. Low CO_2 readings are caused by mixtures richer or leaner than the optimum 14.7:1 ratio (Figure 12–48). CO_2 is an excellent indicator of the operating efficiency of the engine.

Oxygen (O_2). Oxygen, also measured in percent by volume of the exhaust, is a necessary ingredient for all combustion. The oxygen content of the exhaust gas indicates whether the fuel in the combustion chamber is using all the available oxygen for igniting the mixture. The O_2 reading also indicates whether the exhaust gas sample is being diluted by an air injection system or a leak in the exhaust system. Diluted exhaust samples can cause incorrect CO and HC readings. A lean air/fuel ratio means the O_2 reading will be high. A rich mixture causes a low O_2 reading (Figure 12–49).

Catalytic converters can mask HC and CO readings but do not affect oxygen; therefore, O_2 is valuable in diagnosing mixture problems as well as elec-

FIGURE 12–46 Troubleshooting carburetor performance with an exhaust analyzer

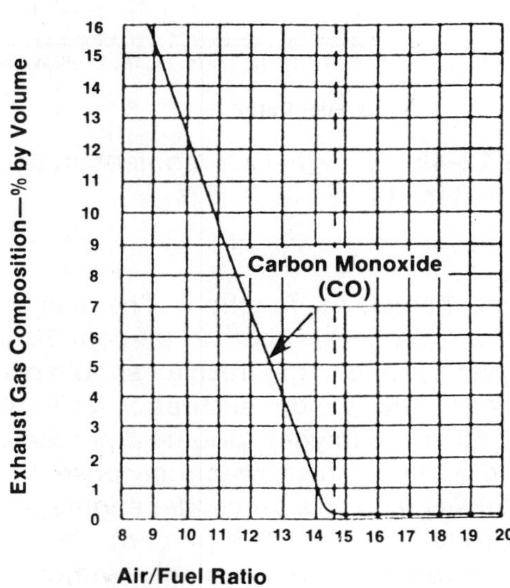

FIGURE 12–47 CO content drops as air/fuel ratio approaches 14.7:1.

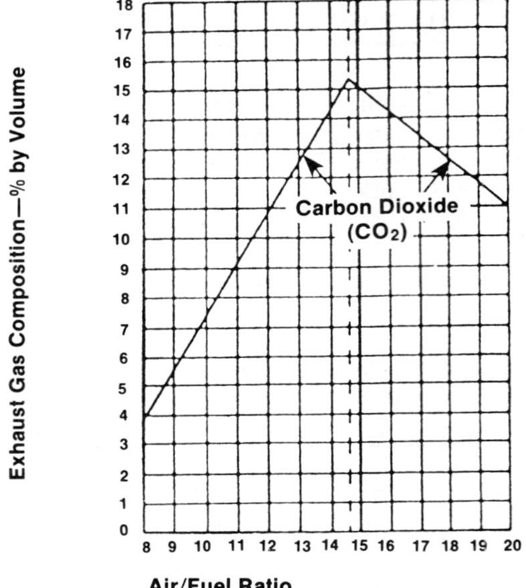

FIGURE 12-48 CO_2 content is highest when air/fuel ratio is 14.7:1.

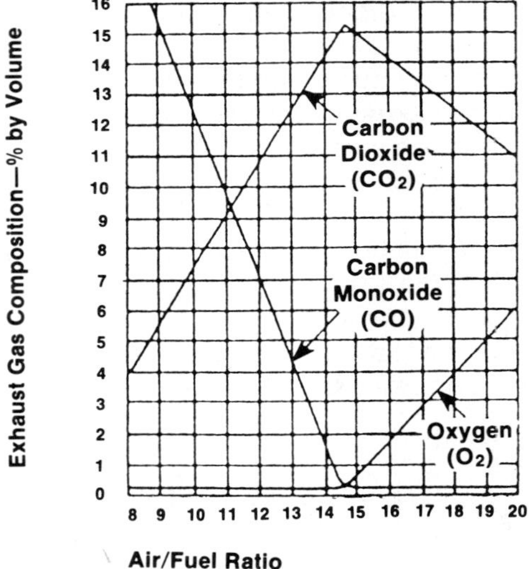

FIGURE 12-49 A high CO_2 content indicates a lean air/fuel mixture.

trical or mechanical malfunctions. The oxygen content of the atmosphere is about 21 percent. Exposing the analyzer's probe to the fresh air should give this reading. If a single cylinder misfires on an 8-cylinder engine, 1/8th of the oxygen (about 2.5 percent) goes directly into the exhaust system because no combustion takes place. In a 6-cylinder engine, a single misfire produces about 3.3 percent O_2. In a 4-cylinder engine, a misfire in a single cylinder would increase the oxygen reading by about 5 percent (1/4th of 20 percent).

Hydrocarbons (HC). Hydrocarbon, measured in parts per million (ppm), is unburned fuel leaving the combustion chamber. HC results when some factor does not allow the air/fuel mixture to burn completely in the combustion chamber (spark plug misfire, over-advanced spark timing, insufficient spark duration, low compression, or extremely lean mixture.

Interpreting Analyzer Readings

The readings of a four-gas analyzer, when taken together, are used to determine whether an air/fuel mixture is lean or rich. For example, CO is fuel that is only *partially* burned due to a lack of sufficient oxygen in the combustion process. With lean mixtures there is usually sufficient oxygen and the result is low CO readings; however, when too much oxygen (and, therefore, too little fuel) is present, proper combustion cannot occur. As a result, unburned fuel is exhausted in the form of hydrocarbons (HC). The O_2 reading will be high because of the excessive amount of air in the cylinder. Lean mixtures appear as very low CO readings, high HC, low CO_2, and high O_2 readings.

SHOP TALK

Catalytic converters can oxidize a great deal of CO and HC before increasing the emissions at the tail pipe. This can delay or mask changes in exhaust gas content readings. Many of the analyzer tests are easier to interpret if the converter is disabled or exhaust gasses are sampled upstream from the converter.

Some manufacturers equip vehicles with a sampling plug upstream from the converter. Locate and remove the plug and insert the analyzer's exhaust probe. Be sure to apply an antiseize compound on the threads before reinstalling the plug.

If the vehicle is not equipped with a sample plug, disable the converter.

1. *Turn the engine off.*
2. *Allow the catalytic converter to cool for at least 10 minutes.*
3. *Restart the engine and perform the necessary tests within 1 minute before the converter "relights."*

TABLE 12-2: INFRARED EXHAUST GAS ANALYSIS PARAMETERS FOR HC, CO, CO_2, AND O_2

Gasses	Idle		1500 rpm		2500 rpm		Condition/ Possible Cause
	Converter*	Nonconverter	Converter*	Nonconverter	Converter*	Nonconverter	
HC (rpm)	0–150	75–250	0–135	50–200	0–75	25–150	Normal gas reading—stable values
CO (%)	1–1.5	.5–3.0	0–1.1	.5–2.0	0–.8	.1–1.5 4 cyl.–3%	
CO_2 (%)	10–12	10–12			11–13	11–13	
O_2 (%)	.1–2.0	.1–2.0	.1–2.0	.1–2.0	.1–1.25	.1–2.0	
HC (ppm)	0–150	75–250	0–135	50–200	0–75	0–100	Rich mixture: Idle mixture too rich; choke set too rich or not opening fully; power valve leaking; float level too high; restricted air cleaner; PCV restricted; contaminated crankcase.
CO (%)	Above 3.0	Above 4.0	Above 3.0	Above 3.5	Above 3.0	Above 3.0	
CO_2 (%)	8–10	8–10			9–11	9–11	
O_2 (%)	0–.5	0–.5	0–.5	0–.5	0–.5	0–.5	
HC (ppm)	0–150	75–250	0–135	50–200	0–75	0–100	Lean mixture: Low float level; idle mixture lean; cruise mixture lean; small air leaks; cracked or disconnected vacuum lines.
CO (%)	0–1.0	0–1.0	0–.8	0–.9	0–.25	0–.75	
CO_2 (%)	8–10	8–10			9–11	9–11	
O_2 (%)	1.5–3.0	1.5–3.0	1.0–2.5	1.0–2.5	1.0–2.0	1.0–2.0	
HC (ppm)	50–850	400–1200	50–850	400–1200	50–750	400–1200	Lean misfire: Severe air leak; bad spark plug or wire; stuck PCV, misadjusted/ defective carb.
CO (%)	0–.3	0–.75	0–.3	0–.75	0–.3	0–.75	
CO_2 (%)	5–9	5–9			6–10	6–10	
O_2 (%)	4–9	4–9	4–9	2–7	2–7	2–7	
HC (ppm)	50–850	Over 1000	50–850	Over 1000	50–750	Over 1000	Misfire: Overadvanced timing; fouled plug wire; EGR stuck open.
CO (%)	.1–1.5	.5–3.0	0–1.1	.5–2.0	0–.8	.1–1.5	
CO_2 (%)	6–8	6–8			8–10	8–10	
O_2	4–12	5–12	4–12	5–12	4–12	5–12	

*Converter readings are taken with the air injection system disabled.

Excessively rich mixtures can cause poor mileage, fouled plugs, and might increase engine wear. A rich fuel mixture is usually indicated by a higher CO reading since a rich mixture is starving for oxygen. HC, CO_2, and O_2 will all be low.

Table 12-2 shows the HC, CO, CO_2, and O_2 readings for normal air/fuel mixtures, rich and lean mixtures, and misfire conditions. Readings are given for systems with and without catalytic converters and at different engine speeds (idle, 1500 rpm, and 2500 rpm). Possible causes for these conditions are also listed. This chart can be used as a troubleshooting guide.

Specific tests can also be performed to diagnose specific carburetor component operation. Typical procedures are given in the following paragraphs.

Idle Mixture Analysis

Today's emission-controlled engines require precise idle mixture adjustments. In states that require vehicles to pass emission test standards, the engine can no longer be adjusted only for "best idle." The car still might not pass stringent idle emission requirements.

FIGURE 12–50 An exhaust analyzer

A majority of vehicle manufacturers provide specified levels of CO at idle as well as recommended methods of adjustment to meet these requirements.

SHOP TALK

Erratic or unsteady readings can indicate internal carburetor or closed-loop problems. If readings are erratic after idle adjustment, mechanical repairs are probably needed.

The following procedure can be used to adjust the idle mixture. Be sure to disconnect or clamp off the main outlet hose from the air pump or to block the "gulp valve" on systems using pulse air. Install and activate the analyzer according to the manufacturer's instructions.

1. Adjust the engine idle rpm to the manufacturer's specifications.
2. Adjust each idle mixture screw 1/8 turn at a time until the maximum rpm is noted.
3. Lean the venturies until there is a 2 percent drop in idle speed. (One percent per venturi.)
4. Return the engine speed to the manufacturer's specified idle with the idle speed adjustment screw or solenoid only. Do not use the idle mixture screws to do this.
5. Observe the CO, CO_2, and HC displays.

The CO_2 reading will be near its highest value when the idle is correctly adjusted. The CO and HC readings noted might be slightly higher than the recommended manufacturer's value, but this does not indicate a problem. Problems are indicated only when the CO and HC readings are excessively high, low or unsteady. It is difficult to say what is too high or too low for a particular vehicle; however, as a general rule, a model vehicle equipped with a catalytic converter will show between 1 and 2 percent CO while the noncatalyst models might show 2 to 3 percent CO.

Before deciding whether a rich or lean mixture exists, gradually bring the engine speed up while observing the CO and HC readings. As engine speed gradually increases, the CO (and HC) readings should decrease slightly. When the engine speed is gradually reduced to idle, the CO (and HC) readings should return to their original settings. Excessive changes in the CO and HC displays or initially high or low readings might indicate carburetor problems.

Propane Enrichment Analysis

An exhaust analyzer can also be used in conjunction with propane enrichment to test carburetor performance (Figure 12–50).

Consult the manufacturer's service manual or emissions decal in the engine compartment for specific instructions for propane enrichment. Then, follow this basic procedure:

1. Warm the engine to its normal operating temperature and adjust the curb idle to the specified rpm.
2. Use propane to enrich the mixture for the highest rpm.
3. While maintaining enrichment, observe rpm; it should be within the manufacturer's specifications for enriched rpm.
4. Stop enrichment and return to idle. Idle rpm should now be within specified curb idle.
5. If not, adjust the idle mixture screws. Curb idle rpm lower than specifications indicates a lean mixture. Curb idle rpm higher than specifications indicates a rich mixture. Adjust the idle mixture screws accordingly until correct curb idle speed is achieved.

Idle Adjustment—Peak CO_2 Method

When a four-gas analyzer is available, using CO_2 as an indicator of the best idle mixture is the best way to set the idle mixture. This is because CO_2 production is highest when the air/fuel mixture is at the stoichiometric ratio of 14.7 to 1.

To adjust idle mixture with the peak CO_2 method, follow this procedure:

1. Disable the air injection system.
2. Lightly bottom the idle mixture screws and back them out two full turns.
3. Start engine and watch the CO, HC, CO_2, and O_2 displays. All should be stable.
4. Adjust the curb idle speed to the manufacturer's specification.
5. Turn the idle mixture screws the same amount and direction 1/8 turn at a time, until the CO_2 reaches its highest value (12 to 15 percent).
6. Adjust the mixture screws *equally* to the leanest setting possible while maintaining the peak CO_2 reading.
7. Reset idle speed as required.
8. Repeat steps 5, 6, and 7 as needed.

Accelerator Pump Test

The accelerator pump can also be tested using an exhaust analyzer.

1. Set the engine speed at fast-idle (approximately 1200 rpm).
2. Allow the CO reading to stabilize.
3. Momentarily open and close the throttle without raising the engine rpm and observe the CO and HC readings.
4. The CO reading should increase up to 2 percent above original CO.
5. The HC reading should increase up to 600 ppm above the original ppm.

If the engine stumbles, and CO reads momentarily lower or remains the same, an accelerator pump problem is indicated. Stop the engine and open the throttle to see if a squirt of fuel is directed into the barrels, or inspect for bent linkage, linkage adjustment, stuck plunger, or defective check ball.

Power Valve Test

The power valve operation can be quickly verified by creating a momentary "near zero" vacuum. The increased airflow and spring tension should lift the power valve from its seat and supplement the fuel flow.

Not only is it important to verify the power valve opening, but equally important to verify whether the power valve reseats properly.

1. Disconnect the accelerator pump linkage and stabilize the engine speed at approximately 1500 rpm using the fast-idle cam.
2. Note the CO (and HC) readings.
3. While holding the fast-idle cam in one hand, quickly open and return the throttle to the same fast idle cam position.

It is important to return to the same fast-idle cam setting in order to check for proper reseating of the power valve. You might wish to repeat this test several times to check for consistency. A momentary increase (1 to 2 percent) in CO and a return to the original reading should also be expected.

IDLE SPEED ADJUSTMENT

As mentioned earlier, most vehicles with computer-controlled fuel systems have no provision for adjusting the idle speed. This is a function of the electronic control module. However, the idle speed on these vehicles should always be checked as part of a tune-up. An idle speed that is not to specifications is a sign that a malfunction, such as an air leak, is occurring somewhere in one of the engine systems. Idle speed is checked using a tachometer.

If idle speed adjustment is possible on the carburetor, be sure to follow the manufacturer's instructions given on the emissions decal. Typical instructions might include the following:

1. Warm the engine to normal operating temperatures.
2. Make sure that the choke plate is open and that the throttle linkage is off fast-idle.
3. Set the parking brake and block the wheels.
4. Turn off all accessories (such as the air conditioner) and close the doors.
5. Shift the transmission into PARK, NEUTRAL, or DRIVE.
6. Vehicles with feedback carburetion might also require that certain connectors be disconnected to keep the carburetor in open-loop.

Then, while watching the tachometer, turn the idle-speed adjusting screw to adjust the idle rpm (Figure 12–51). The idle screw might bear against a tab on the throttle linkage or the plunger of a throttle stop solenoid.

M/C Solenoid Test

As explained earlier in this chapter, a feedback fuel system relies on an oxygen (O_2) sensor in the exhaust system to monitor the air/fuel ratio. Based on the amount of oxygen in the exhaust gasses, the O_2 sensor generates voltage signals that are sent to the vehicle's on-board computer. The computer uses these signals to adjust the air/fuel mixture to the ideal 14.7:1 (stoichiometric) ratio. It does so by controlling the on/off cycling of the mixture control (M/C) solenoid.

FIGURE 12-51 Adjusting idle speed on a Chevrolet Nova

Depending on the manufacturer, the air/fuel ratio can either become richer or leaner as the solenoid's "on" time increases. The amount of time that the solenoid is energized (turned on) is measured either as the duty cycle (expressed as a percentage of the total possible "on" time) or as dwell.

SHOP TALK

Solenoid dwell can also be measured on the 6-cylinder scale of a common dwellmeter. Read the middle of the swinging needles range and disregard any big jumps. Check the dwell at both idle and fast-idle rpms. Manipulating the air/fuel ratio with propane enrichment and watching how the on-board computer and M/C solenoid respond is an excellent way to monitor and diagnose the closed-loop system.

1. Hook up the analyzer's engine harness according to the manufacturer's instructions. Connect the appropriate test lead to the mixture control solenoid test connector (Figure 12-52), wire, or terminal (see the manufacturer's service manual for test connector or terminal identification and location).
2. Start the engine and observe the dwell reading. It should be to specifications given in the manufacturer's service manual.

SHOP TALK

On some carburetors, the dwell can be adjusted by adjusting the idle mixture (Figure 12-53).

3. Use propane to *slowly* enrich mixture and observe dwell or duty cycle readings and exhaust gasses. Note the readings.
4. Stop enrichment.
5. Allow the engine speed to stabilize.
6. Remove a vacuum hose to create an air leak. Note the readings.

As you vary the mixture, the dwell or duty cycle readings should increase or decrease accordingly. The CO reading should momentarily increase or decrease, then restabilize at approximately the original setting. This indicates that the oxygen sensor and the on-board computer sense the change in the air/fuel ratio and are trying to compensate by increasing or decreasing the amount of fuel in the mixture.

If the system responds sluggishly, it might indicate that the oxygen sensor is coated with soot.

FIGURE 12-52 Mixture solenoid test connector

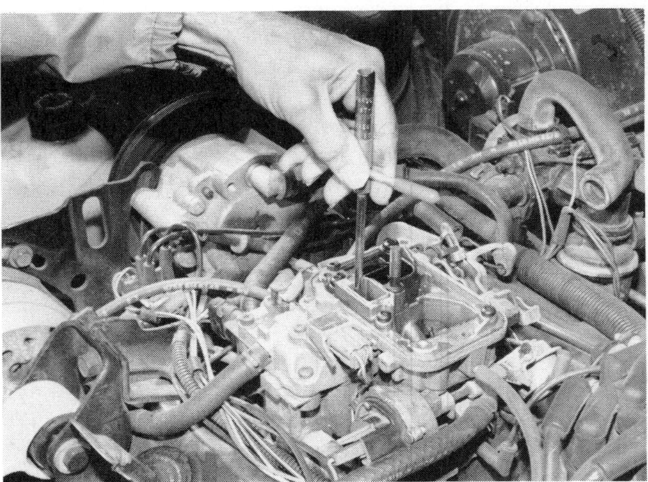

FIGURE 12-53 Adjusting idle mixture on a Rochester Varajet II carburetor

Ground the M/C solenoid lead, run the engine at 2000 rpm for 3 minutes, and retest. If the dwell or duty cycle is not responding, there might be a problem with the circuit between the oxygen sensor and the electronic control module or the mixture control solenoid itself.

CARBURETOR ADJUSTMENTS

Often a carburetor problem cannot be corrected without removing the carburetor from the engine and rebuilding it. Carburetor rebuilding goes beyond the scope of engine tuneup. However, there are other tests and adjustments that can be made without removing the carburetor. Some of those procedures are given here.

Fast-Idle Adjustment

Instructions for setting fast-idle speed is also contained on the emissions decal or in the manufacturer's service manual. Most carburetors have a fast-idle screw (Figure 12-54) that can be adjusted to correct the fast-idle speed. After satisfying any pretest conditions (such as A/C off or engine in gear), place the specified step of the fast-idle cam on the adjusting screw. Turn the screw clockwise to increase rpm and counterclockwise to decrease rpm.

Idle Mixture Adjustment

Prior to 1966, most carburetors allowed the idle mixture to be adjusted. However, in 1966, manufacturers were required by law to provide some means for limiting the adjustability of the idle mixture screws. Modern carburetors have a factory set idle mixture that cannot be adjusted.

FIGURE 12-54 Adjusting a fast-idle screw

On older carburetors that allow for idle mixture adjustment, a simple procedure is as follows:

1. Adjust the idle speed to the specified rpm.
2. Turn the mixture screws to obtain the smoothest idle.
3. Readjust the idle speed to specs.
4. Repeat steps 2 and 3 until the engine idles smoothly at the correct engine rpm.
5. Turn the idle mixture adjustment screw in (in the lean direction) as far as possible without a loss of smoothness. Some procedures require leaning the mixture until there is a definite drop in rpm and loss of smoothness and then backing out the idle mixture screw(s) one quarter.

Idle Adjustment Using Propane Enrichment Method

By the mid-1970s, the combination of catalytic converters and the very lean mixtures made it difficult to properly perform curb idle speed and mixture adjustments. However, these adjustments must be properly made to insure that emission control devices limit CO, HC, and NO_x to specified levels. The solution adapted by Ford, Chrysler, and General Motors is to use propane gas to assist in achieving correct idle settings.

When a controlled amount of propane is injected into the carburetor at idle, there is a direct correlation between the air/fuel ratio and engine rpm gain. If the rpm gain is less than the specifications found on the emissions control decal, the air/fuel ratio is too rich. If the rpm gain is more than specified, the air/fuel ratio is too lean.

Setting idle mixture using the artificial enrichment method will require a commercially available bottle of propane, a metering valve, a length of hose, and an adapter to connect the hose to the carburetor or air cleaner.

SHOP TALK

During the enrichment test, the propane tank must be upright to allow an even flow of gas. The common procedure is to hang the tank from the hood (Figure 12-55).

Although the propane injection method varies slightly from one manufacturer to another (check the service manual), the following procedure can be considered basic:

1. Apply the parking brake and block the wheels. Disconnect the automatic brake release and plug the vacuum connection.
2. Connect a tachometer to the engine.
3. Disconnect the fuel evaporative purge return hose at the engine and plug the connection.
4. Disconnect the fuel evaporative purge hose at the air cleaner and plug the nipple.
5. Disconnect the flexible fresh air tube from the air cleaner duct or adapter. Using a propane enrichment tool, insert the tool hose approximately 3/4 inch of the way into the duct or fresh air tube (Figure 12-56). If necessary, secure the hose with tape and hold the bottle upright to even the propane flow.
6. For vehicles equipped with an air injector system, revise the dump valve vacuum hoses as follows:
   - For dump valves with two vacuum fittings, disconnect and plug the hose(s).
   - For dump valves with one fitting, remove the hose at the dump valve and plug it. Connect a slave hose from the dump valve vacuum fitting to an intake manifold vacuum fitting.
7. Verify that the idle mixture limiter(s) is set to the maximum rich position (counterclockwise against the stop); correct if required.
8. Check the engine curb idle rpm (for A/C-OFF rpm). If necessary, reset to specification.
9. With the transmission in NEUTRAL, run the engine at approximately 2500 rpm for 15 seconds before each mixture check.
10. With the engine idling at normal operating temperature, place the transmission selector in the position specified for the fuel mixture check. Gradually open the propane tool valve and watch for engine speed gain, if any, on the tachometer. When the engine speed reaches a maximum and then begins to drop off, note the amount of speed gain.
11. Compare the measured speed gain to the specified speed gain on the engine decal or specification sheet. If the idle rpm gain is not to specifications, adjust the mixture according to the reset rpm specification.
    - If the measured speed gain is higher than the speed gain specification: Turn the mixture screw(s)/limiter(s) counterclockwise (rich) in equal amounts and simultaneously repeat steps 8 through 10 until the measured speed rise meets the reset rpm specification.
    - If the measured speed gain is lower than the speed gain specification: Turn the mixture screw(s)/limiter(s) clockwise (lean) in equal amounts and simultaneously repeat steps 8 through 10 until the measured speed rise meets the reset rpm specification.

FIGURE 12-55 Propane tank hanging vertically during enrichment test

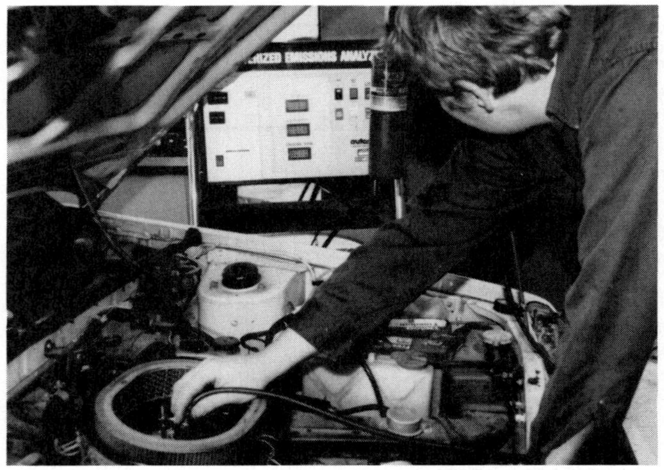

FIGURE 12-56 Check oxygen reading accuracy

CHOKE ADJUSTMENTS

A malfunctioning choke can cause a variety of performance problems, from no-start to missing and power loss at cruising speeds. These problems can be caused by a number of things: choke stuck closed, choke stuck open, inoperative choke pull-off, or malfunctioning bimetal coil.

A visual inspection might identify the trouble spot quickly. Remove the air cleaner, making sure to label each vacuum hose as it is disconnected. If the engine is cold, the choke should be closed. Watch the choke plate as another technician cranks the engine. When the engine starts, the choke pull-off should open the choke slightly (about one quarter of an inch). As the engine warms, make sure that the bimetal coil slowly opens the choke to its fully-open position.

Exact choke servicing and adjustment procedures vary with choke type and carburetor model. In fact, on today's electronically controlled vehicles, the electronic choke is sealed to prevent tampering and misadjusting. Operation of the heating element is checked with a digital volt-ohmmeter. Refer to the manufacturer's service manual for procedures and specifications.

Before making any choke adjustments, move the link by hand to be certain there is no binding or sticking in the linkage, shaft, or choke plate. Use an approved carburetor cleaner to remove any gum deposits that interfere with free operation.

Adjusting the Choke Index

If the choke housing permits adjustment of the choke setting, the choke can be adjusted to richen or lean the air/fuel mixture. With an integral choke, manufacturers recommend specific alignment of the choke index marks. A mark on the plastic choke cover must be lined up with a mark on the metal choke housing or body. This initial choke setting will usually provide choke operation.

Check the thermostat housing "index" setting (Figure 12–57) against the specification on the engine decal or the vehicle's service manual.

For example, if the choke setting is specified as "index," the index mark on the cap should be aligned with the center (index) notch on the housing. A 1R, 2R, or 3R specification means that the cap index

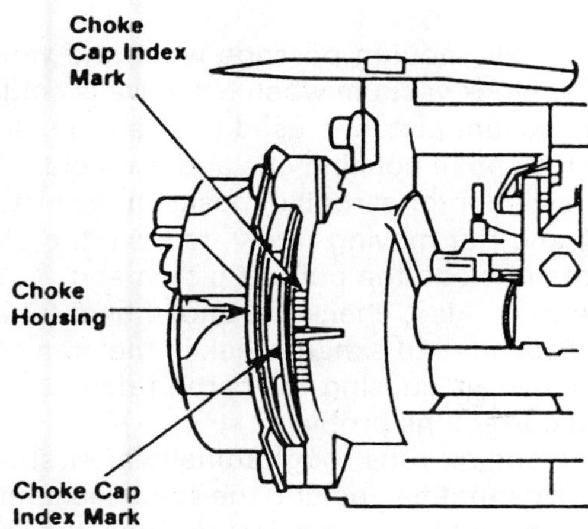

FIGURE 12–57 Choke index adjustment

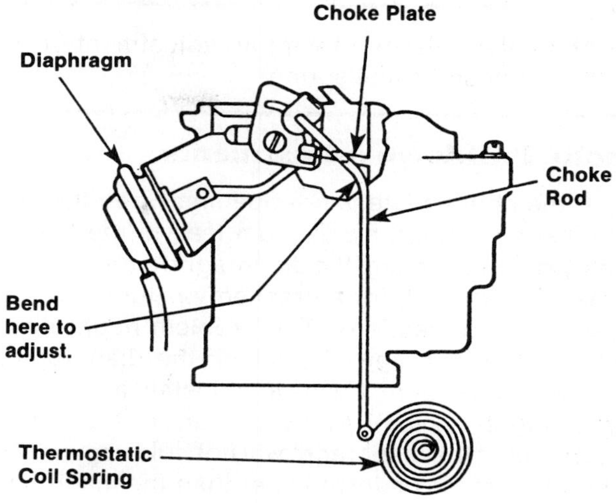

FIGURE 12–58 Ford delayed choke pulldown adjustment

mark should be aligned with the housing marks one, two, or three notches to the right of the center notch. Specs reading 1L, 2L, or 3L require positioning the index mark to the left of the center notch.

To adjust the choke setting, loosen the cap screws and the hot air or coolant tubes (if applicable). Turn the cap to realign the index mark with the specified notch on the housing. Then, retighten the tubes and screws.

Divorced Choke Adjustment

To adjust the remote or divorced choke, proceed in the following manner (Figure 12–58).

1. Disconnect the choke rod from the choke plate and close the plate.
2. Move the rod up or down as directed in the service manual. Check whether or not the rod is above or below the hole in the check plate. The bottom of the rod should align with the top edge of the hole. With choke plates that use a sliding fit, the rod is usually set at the middle of the slot. However, always check the manufacturer's instructions.
3. Bend the choke rod to lengthen or shorten it as required.

CAUTION: Improper bending will cause the choke rod to bind.

---

4. Test for free movement between open and closed choke positions. There must not be any binding or interference.

CAUTION: Never attempt an adjustment on the thermostatic coil spring.

Choke Pulldown Adjustments

If the choke plate is not opening when the engine starts, check for a vacuum leak in the choke qualifying mechanism. If a diaphragm-type vacuum break is used, look for a cracked vacuum hose or loose hose connections. The problem might be a hole in the diaphragm. To check the diaphragm, remove the vacuum hose, and install a vacuum pump (Figure 12-59). Apply vacuum to the pump diaphragm. If the diaphragm will not hold vacuum or if the vacuum leaks down faster than the manufacturer specifies it should, the vacuum break should be replaced.

SHOP TALK

Before many vacuum breaks can be tested with a vacuum pump, a bleed hole must be plugged. To do so place your finger or a piece of tape over the bleed hole. Figure 12-60 shows several vacuum break styles with the recommended procedure for blocking the bleed hole prior to testing.

If the carburetor has a secondary vacuum break, test the thermal vacuum valve or choke vacuum thermal switch. Consult the manufacturer's service manual for temperature specs. Connect a vacuum pump to the lower port and a vacuum gauge to the upper port. Apply the specified vacuum and alternately warm and cool the valve or switch. The valve should permit vacuum passage when the valve is warm and block vacuum when the valve is cold.

If a vacuum piston is used to open the choke, remove the choke housing cap and inspect the piston and cylinder for carbon deposits. If the piston is sticking and not moving freely, remove the choke housing and clean the pull-off piston and vacuum passageways. Also, check the choke heat tube for carbon deposits. An exhaust leak in the area of the choke "stove" is causing the carbon deposits. Locate and correct the problem.

If the engine runs rough initially after startup, the choke cannot be opening the specified amount. To check the qualifying dimension, apply the speci-

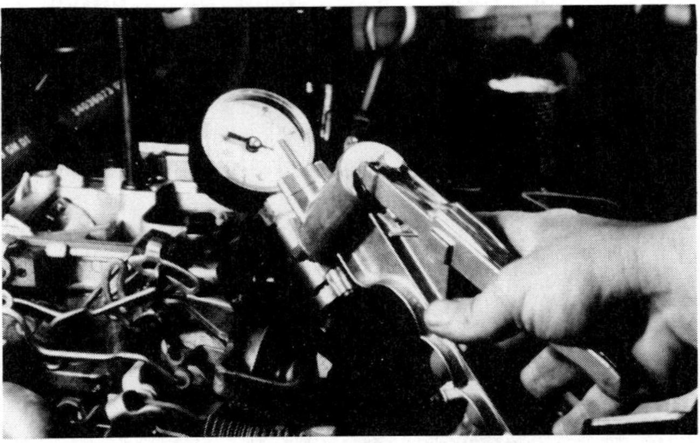

FIGURE 12-59 Checking vacuum break diaphragm with a vacuum pump

PLUGGING AIR BLEED HOLES

Pump Cup or Valve Stem Seal

Tape hole in tube.

Tape end of cover.

BUCKING SPRINGS

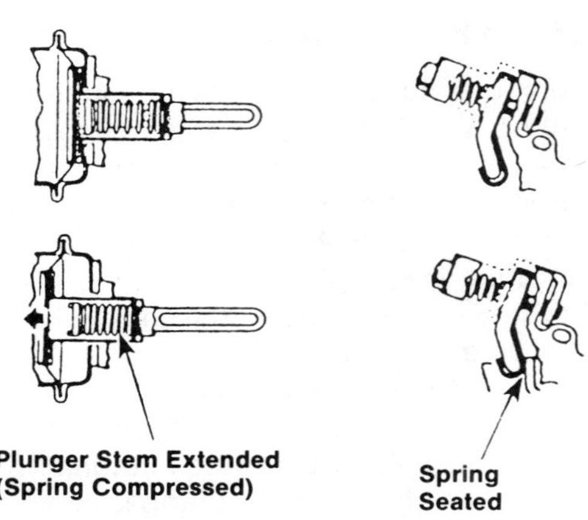

FIGURE 12-60 Vacuum break adjustments

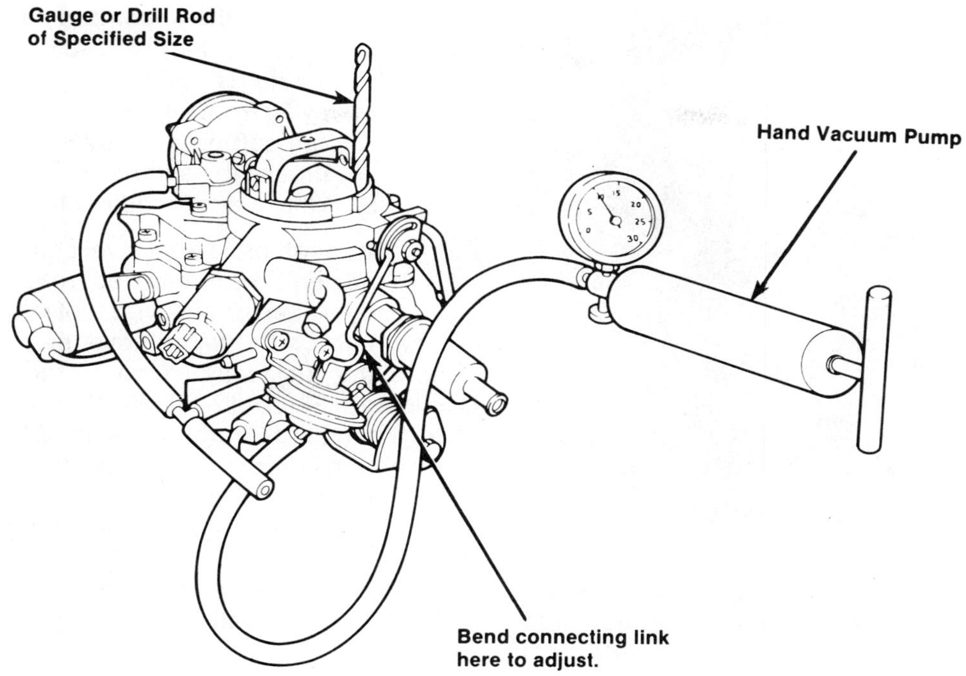

FIGURE 12-61 Choke pulldown adjustment

fied vacuum to the diaphragm, using a vacuum pump. Measure the clearance between the upper edge of the choke plate and the air horn wall, using a drill bit with a diameter equal to the specified dimension (Figure 12-61). To adjust the dimension, bend the choke linkage with a pair of pliers.

Choke Unloader Adjustment

The unloader can be adjusted following this procedure:

1. With the throttle valves wide open, close the choke valve as far as possible without forcing.
2. Use a drill bit with a diameter equal to the specified unload dimension as a gauge and place it between the upper edge of the choke valve and the air horn.
3. Bend the unloader tang on the throttle lever to obtain the dimension as listed in the specifications.

CHECKING THE ACCELERATOR PUMP

If the accelerator pump is not operating properly, the engine will hesitate when the accelerator pedal is depressed.

To check for accelerator pump operation, remove the air cleaner top. Make sure that the actuating linkage is not binding. Start the engine and using a flashlight, look down into the air horn while quickly moving the throttle linkage forward (Figure 12-62). Each time you do so, a stream or squirt of fuel from the discharge tube should be visible. If not, remove the air horn and service the accelerator pump following the specific procedures outlined in the manufacturer's service manual.

FLOAT ADJUSTMENTS

An improperly operating float system can result in a variety of performance problems: flooding, rich

FIGURE 12-62 Checking accelerator pump operation

FIGURE 12-63 Check float condition

fuel mixture, lean fuel mixture, fuel starvation (no fuel), stalling, low-speed engine miss, and high-speed engine miss. Thus a faulty float system can affect engine operations at all speeds.

On some carburetors, the fuel level in the fuel bowl can be checked without removing the air horn assembly from the carburetor. A gauge (Figure 12-63) is placed into the fuel bowl through a vent hole. The level of the fuel in the bowl is an indication of the float condition. Most carburetors, however, require that the air horn assembly be removed from the carburetor to inspect and adjust the float.

An error in float adjustment as small as 1/32 inch can change the air/fuel ratio sufficiently to make other carburetor adjustments difficult, if not impossible. Adjustments include the float level, float drop, float toe, and float adjustment. Since most do not require all four adjustments, check the vehicle's service manual or the rebuilding kit instruction sheet for the adjustment that must be performed.

Float Level Adjustment

A typical float level adjustment is as follows:

1. Remove the upper body assembly and the upper body gasket; install a new gasket before making the adjustment.
2. From cardboard, fabricate a gauge to the specified dimension following the specifications given in the service manual or rebuilding kit. Figure 12-64 shows a typical float gauge. Check the service manual for the exact dimensions.
3. With the upper body inverted, place the fuel level gauge on the cast surface of the upper body (not on the gasket) and measure the vertical distance from the cast surface or the upper body and the bottom of the float (Figure 12-65).
4. To adjust, bend the float operating lever away from the fuel inlet needle to decrease the setting and toward the needle to increase the setting.
5. Check and/or adjust the float drop.

Although most float level checks are performed with the carburetor removed from the vehicle, a wet test can be performed with fuel in the unit. With the bowl filled with fuel, remove the air horn and measure the fuel depth. One method of doing so is shown in Figure 12-66. The float should be adjusted to the manufacturer's specifications in the same manner as for the dry test (without fuel).

Some carburetors have a sight glass in the fuel bowl that makes it possible to check the float level while the engine is operating. Float levels on these carburetors can be adjusted by an external screw or threaded inlet valve. Checking the fuel lever on a

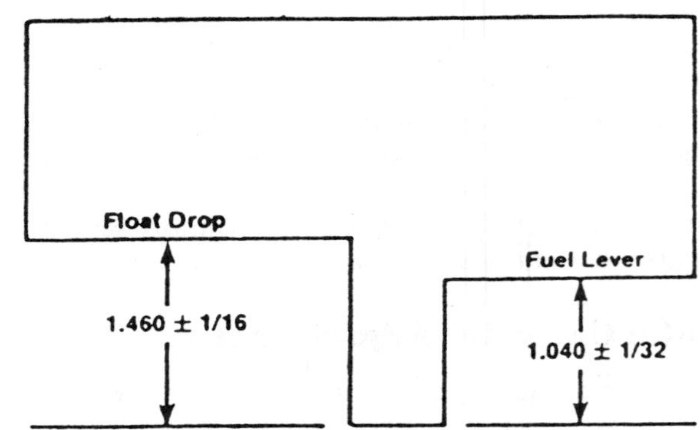

FIGURE 12-64 Float gauge

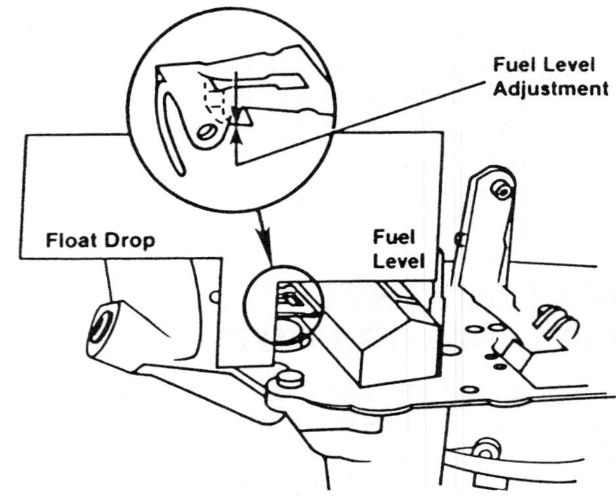

FIGURE 12-65 Testing float level

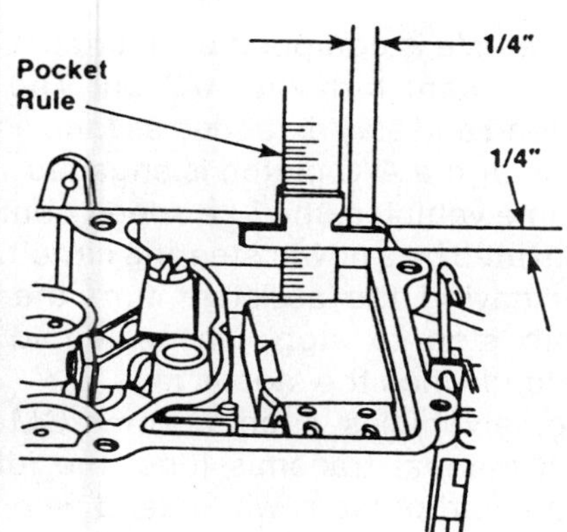

FIGURE 12-66 Testing float level, wet test

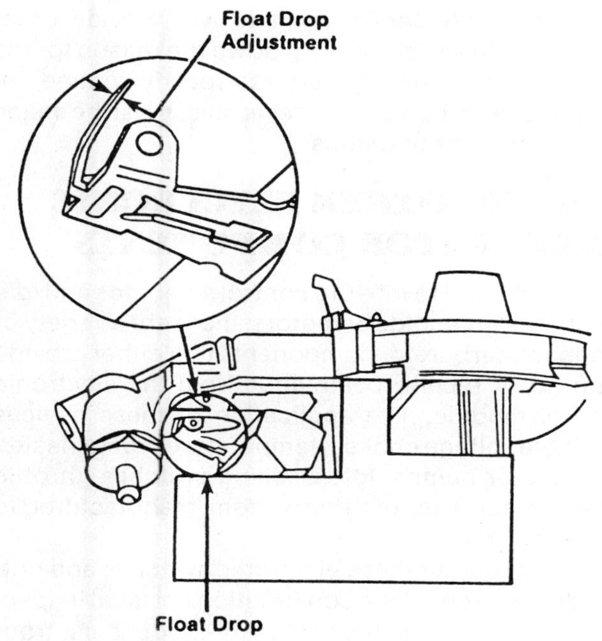

FIGURE 12-67 Checking float drop

carburetor with a sight glass not only aids in adjusting the float level but also in locating float problems.

Float Drop Adjustment

On floats hung on the air horn, a float drop adjustment such as this is required:

1. Fabricate a gauge to the specified dimension as in the float level test (Figure 12-67).
2. With the upper body assembly held upright, place the gauge against the cast surface of the upper body (not on the gasket) and measure the vertical distance between the cast surface of the upper body and the bottom of the float.
3. To adjust, bend the stop tab on the float lever away from the hinge pin to increase the setting and toward the hinge pin to decrease the setting.
4. Reinstall the upper body assembly.

Float Toe Adjustment

In addition to the float level and drop adjustments, some carburetors also have a float toe adjustment. To make this adjustment, the air horn is turned over (no gasket installed) and checked to make sure the float toes are flush with the air horn casting (Figure 12-68A). If the float has dimples (Figure 12-68B), measure from the dimples to the top of the gasket. In either case the float arm is bent to where it meets the float.

After making a toe adjustment, be sure to recheck the float level and drop to see if they have changed.

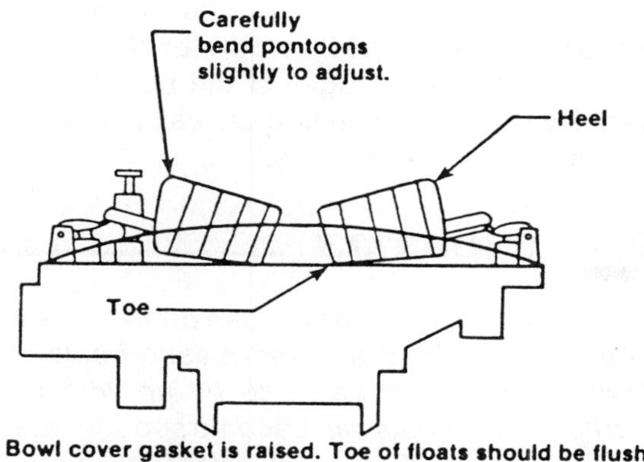

A FLOAT TOE ADJUSTMENT WITHOUT DIMPLE

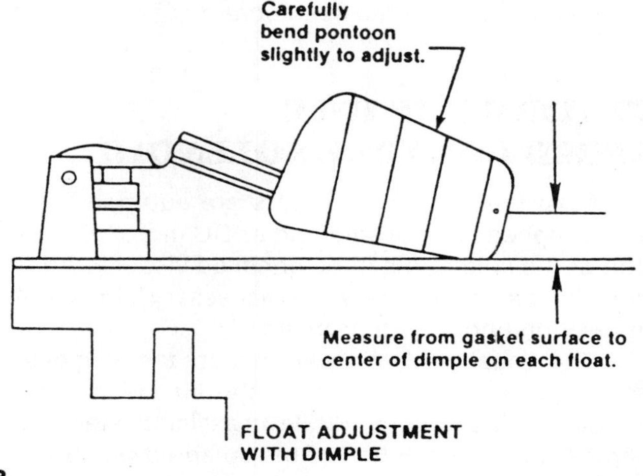

B FLOAT ADJUSTMENT WITH DIMPLE

FIGURE 12-68 Checking float toe

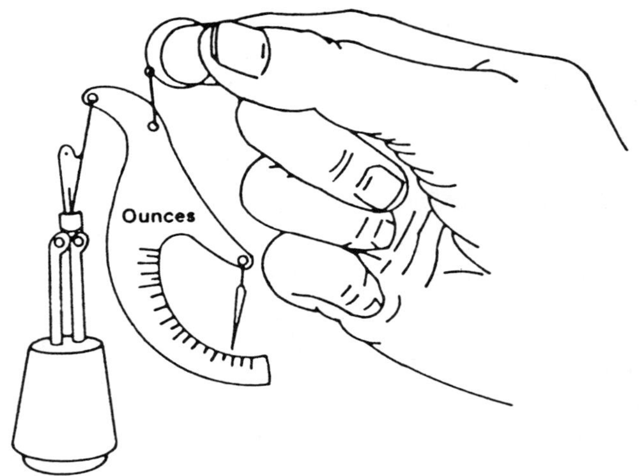

FIGURE 12-69 Weighing a float

Float Alignment Adjustment

The float lung, or pontoon, must be parallel in all cases to the edges of the float bowl or air horn. Badly aligned floats can rub against the float bowl walls and stick with the inlet needle valves open. Straighten by bending the float arms.

SHOP TALK

Problems with the float can often be traced to a heavy float. Brass and plastic floats can develop leaks and begin to fill up with fuel. Nitrile rubber floats can become spongy, soaking up fuel. In all cases, the float loses buoyancy and the fuel level in the bowl exceeds specifications. Always weigh the float (Figure 12-69) when making adjustments, and replace it if it is heavier than specifications.

TESTING THE IDLE SPEED CONTROL SOLENOID

Most feedback carburetors are equipped with an idle speed control solenoid or DC motor. Operation of the solenoid can be checked visually. Additional electrical checks will be necessary if the visual inspection uncovers a problem.

The visual inspection will require the help of an assistant. Begin by removing the air cleaner and installing a tachometer. While an assistant starts the engine, watch the action of the solenoid plunger. It should extend immediately after the engine starts. On some vehicles, it will retract after a certain length of time (10 seconds, for example).

If the vehicle is equipped with air conditioning, have the assistant turn the A/C on. The plunger should extend and stay extended as long as the A/C switch is on or the A/C clutch is engaged.

On some vehicles, the idle speed control solenoid is activated by a power steering circuit. Test the circuit by having the assistant turn the steering wheel from stop to stop. The solenoid plunger should extend when the stop is met.

On some vehicles, such as some GM 1.6L engines with manual transmissions, the idle speed control solenoid plunger will extend when the engine speed exceeds 2500 rpm. This can also be checked visually.

If the idle speed control solenoid fails any of these visual tests (where they are applicable), the manufacturer's service manual will provide other test procedures for tracking down the reason for the malfunction. These procedures specify voltage, resistance, and/or vacuum tests that must be made under specific conditions.

TESTING OTHER ELECTRONIC CARBURETOR COMPONENTS

In addition to mixture control solenoids and idle air control solenoids or motors, there are a variety of feedback carburetor components that either provide input to or receive commands from the electronic control module. The application of these devices (variable voltage chokes, temperature compensated accelerator pumps, idle tracking switches, throttle position sensors, etc.) vary from manufacturer to manufacturer.

Diagnosis of these electronic switches and output devices requires a combination of visual inspections and electrical tests. On late-model cars, troubleshooting is simplified by on-board diagnostics and scan tools that display applicable trouble codes. See Chapter 9 for a description of computer control systems, and self-diagnostic features, and procedures for accessing trouble codes from the diagnostic circuits and computer memory banks. Be sure to follow the step-by-step procedures given in the manufacturer's service manual. These specific troubleshooting procedures are designed to systematically inspect the suspected component, its wiring and connectors, and the electronic control module for faults.

REVIEW QUESTIONS

1. The narrow passageway in a carburetor that speeds up the flow of air intake is called a(n) ______________.

a. air horn
b. barrel
c. venturi
d. all of the above

2. On cold startup, which of the following is true?
   a. The throttle valve and the choke valve are closed.
   b. The throttle valve and the choke valve are open.
   c. The throttle valve is open and the choke is closed.
   d. The throttle valve is closed and the choke is wide open.

3. Which of the following is not part of the metering system?
   a. float
   b. main air bleed
   c. idle feed tube
   d. throttle valve

4. Which of the following draws air from the high-pressure area in the carburetor barrel above the venturi?
   a. idle air bleed
   b. accelerator pump system
   c. main air bleed
   d. choke

5. When does fuel tend to condense on the intake manifold walls?
   a. during cold startup
   b. during heavy acceleration
   c. both a and b
   d. neither a nor b

6. Which of the following could be a cause of rough idling?
   a. defective EGR valve
   b. overheating
   c. clogged fuel filter
   d. all of the above

7. Which of the following adjustments must be corrected before a correct idle speed can be achieved?
   a. vacuum break adjustment
   b. float adjustment
   c. idle mixture adjustment
   d. fast-idle adjustment

8. On which of the following types of choke mechanisms would you expect to find an "indexed" choke setting?
   a. remote choke
   b. integral choke
   c. manual choke
   d. none of the above

9. Which of the following statements about a vacuum break is true?
   a. The vacuum break is operated by ported vacuum.
   b. The vacuum break opens the choke after the engine has started.
   c. All integral chokes use a diaphragm vacuum break.
   d. All of the above.

10. Which of the following is used to speed up the opening of the choke?
    a. choke air modulator
    b. choke delay valve
    c. electric assist heating element
    d. all of the above

11. Which of the following components controls slow-idle speed?
    a. throttle stop solenoid
    b. idle mixture screw
    c. dashpot
    d. none of the above

12. The throttle position solenoid increases idle speed to ____________.
    a. assist in acceleration
    b. assist in cold engine startup
    c. compensate for air conditioner load
    d. all of the above

13. Against what does the fast-idle adjusting screw bear?
    a. main air bleed
    b. throttle plate linkage
    c. fast-idle cam
    d. deceleration valve

14. On vehicles with catalytic converters, propane gas is injected into the carburetor to fine-tune which of the following adjustments?
    a. choke setting
    b. idle mixture
    c. fast-idle speed
    d. all of the above

15. Technician A says that an acceleration pump is necessary because the fuel pump cannot supply fuel rapidly enough when the throttle is suddenly opened. Technician B says that an acceleration pump is needed because the air/fuel mixture gets leaner when the throttle is suddenly opened. Who is correct?
    a. Technician A
    b. Technician B
    c. Both A and B
    d. Neither A nor B

16. During a discussion about carburetors with automatic chokes, Technician A says that some carburetors are designed to have hot water (radiator coolant) flowing through the choke housing. Technician B says that some carburetors are heated electrically. Who is correct?
    a. Technician A
    b. Technician B
    c. Both A and B
    d. Neither A nor B

17. Technician A says that fast idling of a cold engine is generally controlled by a fast-idle circuit in the carburetor. Technician B says that the fast-idle is controlled by a fast-idle cam. Who is correct?
    a. Technician A
    b. Technician B
    c. Both A and B
    d. Neither A nor B

18. Technician A says that an engine generally requires the leanest air/fuel mixture during starting, idling, and full-throttle operation. Technician B says that an engine requires a rich air/fuel mixture during hot weather, high altitude, and high speed operation. Who is correct?
    a. Technician A
    b. Technician B
    c. Both A and B
    d. Neither A nor B

19. Technician A says that the choke thermostatic coil spring tends to close the choke plate when the engine is cold. Technician B says that the choke vacuum diaphragm tends to open the choke after the engine is started. Who is correct?
    a. Technician A
    b. Technician B
    c. Both A and B
    d. Neither A nor B

20. Technician A says that setting the carburetor float level lower than factory specification could cause too lean a mixture. Technician B says that a low float level will cause surging at cruising speeds. Who is correct?
    a. Technician A
    b. Technician B
    c. Both A and B
    d. Neither A nor B

CHAPTER THIRTEEN

TROUBLESHOOTING FUEL INJECTION SYSTEMS

Objectives

Upon completion of this chapter, you should be able to

- Describe the similarities and the differences between fuel injection systems and carburetion.
- Explain the principles of operation of a fuel injection system.
- List the advantages of fuel injection.
- Describe the two distinct methods of fuel injection in use today: electronic fuel injection (EFI) and mechanical or continuous injection system (CIS).
- Explain the differences in point of injection in throttle-body or single-point injection and port or multi-point injection.
- Explain the design and function of EFI components, including fuel delivery system components, system sensors, electronic control unit (ECU), and fuel injectors.
- Explain the design and function of CIS components, including fuel delivery and metering components and airflow components.
- Service fuel injection systems, including conducting preliminary checks, fuel injection and control system checks, fuel system checks, injector checks, oscilloscope checks, injector cleaning and replacement, idle adjustment, and CIS checks and tests.

A few years ago, fuel injection was something shop technicians occasionally read about, rarely saw, and hardly ever serviced. To shop technicians as well as home mechanics, fuel injection was as foreign as the cars it came on. This status prevailed because the American automotive industry built its reputation on big cars with big engines that came factory equipped with carburetors.

Today, the idea of squirting or injecting fuel directly into the intake manifold is quickly becoming the method of choice. America now manufactures cars that are downsized, fuel efficient, have front-wheel drive, and come factory equipped with fuel injection systems.

Learning the operation of fuel injection systems and how to properly diagnose them takes effort, time, and some specialized tools, but the rewards will be apparent. Fuel injection is here to stay and new service opportunities will be for the taking by those having the knowledge and expertise of this fuel delivery concept.

This chapter presents the basics of fuel injection systems, looks at the components that make up these systems and gives guidelines for troubleshooting such systems.

There is one thing in the technician's favor when it comes to understanding fuel injection—similarity of design. The majority of injection systems in use today derived their technology from one company—the Robert Bosch Corporation. It was Bosch who actually perfected and mass-produced diesel fuel injection systems back in 1927. Since that time Bosch has remained the leader in both gasoline and diesel injection systems. Once one or two systems are learned all other systems will be very similar in operation, layout, arrangement, and servicing.

FUEL INJECTION VERSUS CARBURETION

When one considers the general principles of operation, both fuel injection and carburetor systems have similar characteristics. Fuel and air are mixed by an injection system to form a combustible mixture. A carburetor does the same thing. Fuel is sucked into the airflow by a carburetor; it is squirted into the airflow by an injection system. The distinction between an injection system and a carburetor requires closer scrutiny to recognize the differences.

The following list shows the comparable functions of carburetor parts with parts of the injection system.

Fuel Injection System	Carburetor
Throttle switch	Accelerator pump
Fuel pressure regulator	Float
Inlet manifold pressure sensor or airflow sensor	Metering rods
Thermo-timer (switch)	Fast idle cam
Injector valves and electronic control unit	Metering jets and idle fuel system

It is also helpful to contrast the characteristics of the injection system against the carburetor to highlight their unique differences.

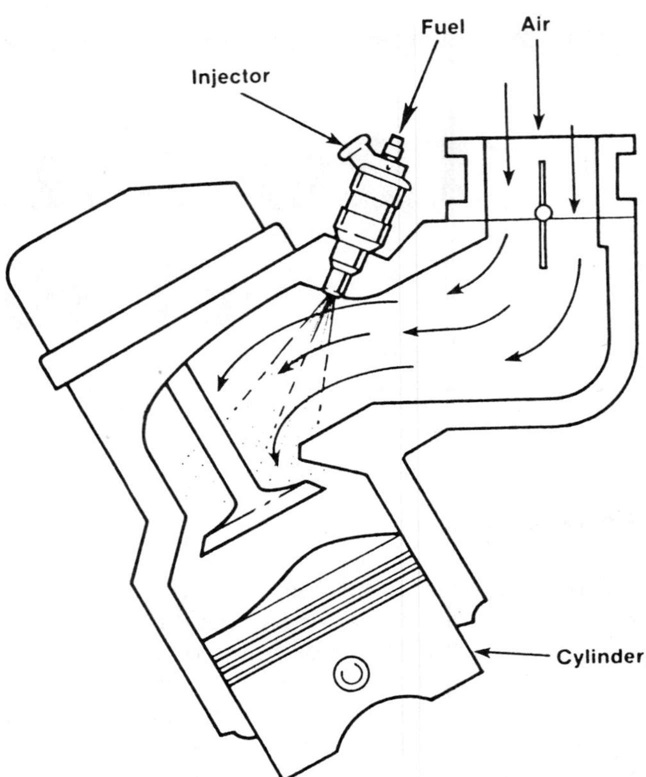

FIGURE 13–1 The mixing of fuel and air in an injection system

PRINCIPLES OF OPERATION

The amount of fuel metered by the injectors depends on several operating conditions. The metering of fuel takes into account variable parameters such as airflow, engine temperature, engine speed, throttle position, and other engine/control factors. For any given set of engine conditions, uniform quantities of fuel are always delivered individually to each cylinder.

As explained in Chapter Twelve, fuel and air are not measured in a carburetor system. A form of fuel metering is maintained at a nearly constant proportion by the manifold vacuum. The fuel mixture (air and fuel) is prepared upstream of the intake manifold and is prepared together for all cylinders. In other words, all cylinders get the same fuel mixture.

AIR/FUEL MIXTURE

In a fuel injection system, fuel droplets are squirted into the inrushing airstream under pressure. The injectors are mounted close to the intake valve preventing the manifold walls from getting wet. All fuel goes into the cylinder. Precise control of the air/fuel ratio is constantly maintained (Figure 13–1).

In a carburetor system, fuel is sucked into the airstream by the vacuum created in the manifold. Because fuel and air are mixed upstream of the manifold, wetting of the manifold walls takes place, thereby upsetting the air/fuel ratio balance (Figure 13–2).

IDLE

In an injection system, fuel delivery is cut off completely when no power is needed. When engine speed comes or is close to an idle speed, fuel delivery is turned on to prevent stalling. This is a big fuel energy saving.

On a carburetor system, fuel is continually mixed into the airstream in predetermined proportions when no power is needed.

ALTITUDE COMPENSATION

Compensation for differing altitudes can be built into the basic controls of an injection system. At higher altitudes, with standard carburetor settings, the engine wastes fuel because of a too-rich mixture; that is, less oxygen intake, but the same quantity of fuel to be mixed.

FUEL CONSUMPTION

An injection system has lower fuel consumption and exhibits a clear advantage at all reasonable speeds. This claim cannot be applied to a carburetor system.

COMPLEXITY

Carburetors are very complex, much more complex than a fuel injection system. Some carburetors can be a maze of mechanical, pneumatic, and hydraulic (fuel) and electrical systems.

Complexity of an injection system is contained within an electronics control unit of which the technician needs to have little knowledge of its internal circuitry. An injection system might seem complex because of the number of sensors it employs. But once operation and diagnostic procedures are understood, servicing becomes simpler and often less expensive than carburetor systems.

ADVANTAGES OF FUEL INJECTION

The move to fuel injection throughout the auto industry has been triggered by the need to produce clean running, fuel efficient vehicles without sacrificing vehicle power or performance. In this regard, fuel injection offers these advantages over carburetor-controlled engines.

1. Improved fuel distribution to each cylinder, thereby providing a more even load distribution between cylinders with less tendency toward detonation because of the leaner air/fuel ratios.
2. Engine power increases on an average by 10 percent. This is due to better volumetric efficiency and is achieved by using larger air inlet passages. The efficiency is further assisted by the fact that cooler fuel is delivered to the injectors and also accounts for better fuel vaporization.
3. A wider range of fuel can be used because of the mechanical atomization of the fuel.
4. Faster acceleration is possible because the atomized fuel is delivered directly to the cylinder.
5. Leaner air/fuel ratios are more easily attained.
6. Provisions are included for fuel shutoff when a vehicle decelerates.
7. Icing is minimized because the fuel is atomized in the cylinder.
8. Higher engine torque, quicker starts, and engine warm-up
9. Fewer exhaust emissions
10. Backfiring in the inlet manifold or throttle body is reduced or practically eliminated.
11. Many pollution control devices normally required by a carburetion system are eliminated.
12. Better fuel economy
13. No choke requirements and no mixture screw adjustments as predominantly used on carburetion systems
14. Accurate control of air/fuel mixture ratios
15. Maintain stoichiometric conditions over a wide range of operating conditions

Although fuel injection (FI) technology has been around since the 1920s, several factors have kept it from gaining widespread popularity until the 1980s. Cheap, plentiful fuel supplies and a lack of concern over engine emissions made the increased efficiencies of fuel injection more of a novelty rather than a necessity. Simple carburetor systems were less expensive, easier to repair, and provided adequate performance.

However, as carburetors evolved into more complex designs to meet increasing demand for fuel economy and emission standard laws differences in

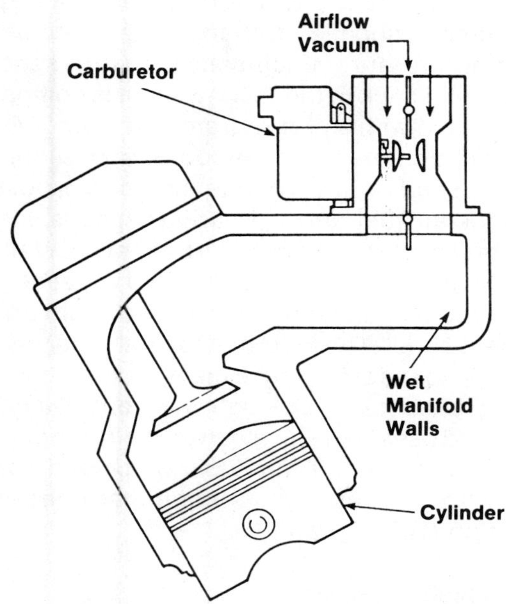

FIGURE 13-2 The mixing of fuel and air in a carburetor system

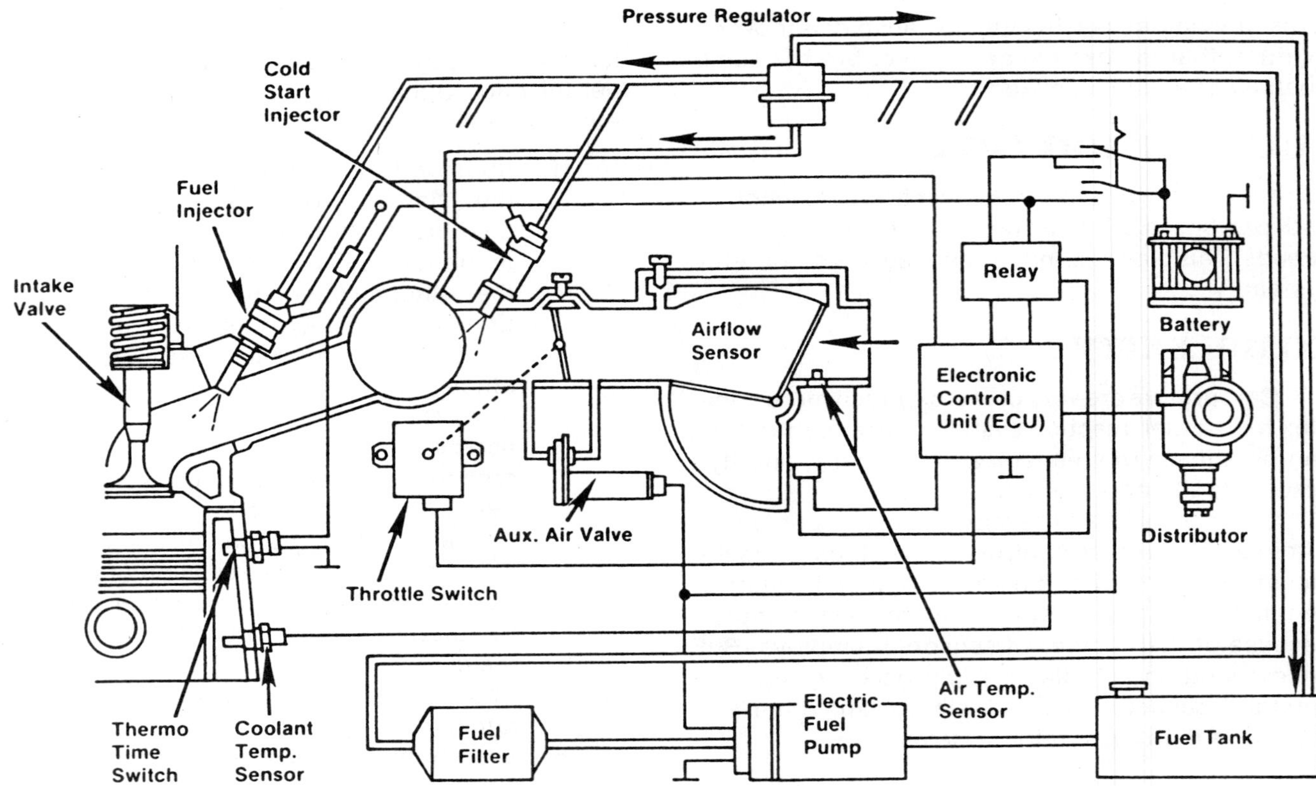

FIGURE 13–3 Schematic view of a fuel injection system that is electronically controlled

price and complexity between FI and carburetor systems have narrowed substantially.

The only true drawback to fuel injection systems is their need for a clean operating environment. Because FI systems must meter extremely precise quantities of fuel, they must have clean fuel as well as clean air to operate properly and efficiently. Dirt is the number one enemy of fuel injection systems.

FUEL INJECTION SYSTEM DESIGN

There are two distinct methods of fuel injection in use today—electronic fuel injection (EFI) and the mechanical or continuous injection system (CIS). Electronic fuel injection is by far the most widely used.

EFI SYSTEM DESIGN

Electronic systems are often called intermittent or pulsed systems, because the fuel flows in short spurts as the injectors turn on and off. An electronic control unit (ECU) determines how long an injector is held open. As the name implies, this system relies on electronic circuitry for its operation. The ECU, with memory capability, also outputs electrical signals to control certain vehicle systems and inputs electrical signals to monitor or sense vehicle operating conditions. The ECU operates in conjunction with several engine and vehicle sensors that are fed a voltage reference signal. Each sensor operates on the resistance principle—a changing temperature or a changing pressure. Each sensor in turn sends a voltage signal back to the ECU where it is compared and computed with a preprogrammed memory system. Based on a reference mode of operation, the ECU then sends out a voltage signal to adjust or change the air/fuel ratio for various speeds, loads, and general operating parameters. If a sensor fails or an abnormal condition exists, the ECU will sense the problem and illuminate or flash a check light on the instrument panel. The ECU has several other names usually tied to the automotive manufacturer. General Motors refers to it as an electronic control module (ECM), Ford uses the term electronic engine control microprocessor control unit (EEC/MCU), and Chrysler refers to their system as computer-controlled combustion (CCC) or an ECU.

EFI System Operation

Most electronic fuel injection systems only inject fuel during part of an engine combustion cycle.

Fuel requirements are measured either by airflow across a sensor or by intake manifold pressure (vacuum). These sensors convert airflow or vacuum parameters into electrical signals that are fed to a microprocessor in the ECU (Figure 13-3). The ECU in turn processes these signals to determine the engine fuel requirements. Once these requirements are determined, the ECU generates an electrical signal to operate the fuel injectors. Basically, the ECU sends a signal that controls the length of time each injector stays open. This open time interval is known as the injector pulse width. This pulsing or intermittent opening and closing of the injectors is a typical characteristic of an electronic fuel injection system. Other sensors mounted within the engine compartment monitor other parameters such as air temperature, coolant temperature, engine speed, and throttle position. These sensors send electrical signals to the ECU that are also used to determine the overall fuel requirements for the engine. The fuel delivery system, consisting of fuel tank, fuel pump, and filter, are typical of any gasoline engine. A fuel pressure regulator is used, however, to control the amount of fuel pressure that is delivered to each injector.

Two modes of operation are characteristic of electronic injection systems. The open loop operational mode uses memory information within the ECU to determine air/fuel ratio, injection timing, etc., as opposed to using "real time" sensor inputs. This mode usually occurs during cold engine operation or whenever a sensor or sensors malfunction. The cold start mode utilizes the thermo-time switch, the auxiliary air, and the cold start injector. The auxiliary air valve provides additional air into the intake manifold. The thermo-time switch operates on coolant temperature and the time it takes for a bimetallic element in the switch to heat up and open the electrical circuit to close the cold start injector.

The other operational mode is the closed loop system. This type of operation utilizes an oxygen sensor to determine air/fuel ratio. This type of operation is also referred to as the feedback system. Basically, this mode of operation limits the amounts of harmful substances in the vehicle's exhaust gasses by controlling the metering of the air/fuel mixture very closely. By constantly measuring the oxygen content of the exhaust gasses, the fuel delivered by the injection system can be adjusted as needed to obtain the proper oxygen output.

CIS SYSTEM DESIGN

The continuous injection system utilizes limited electronics and is basically mechanical in nature. This type of system is often termed CIS because it supplies fuel to the injectors in a continuous, constant flow (Figure 13-4). Fuel flow and/or injector timing is not computer or electronically controlled in the basic CIS system. Fuel flow is controlled by a distributor that feeds fuel to the engine proportional to airflow. In the basic CIS designs, the fuel distributor can be thought of as an ECU that is mechanical in design.

The only electrical component within the CIS system is the fuel pump or, on some systems, the in-tank fuel pump. Fuel is supplied to the injectors in a constant, continuous flow. The volume of fuel is controlled by flow rate and not by injector pulse width. An airflow sensor moves valves that alter the fuel flow. The main attraction of the CIS is the fuel distributor. The distributor is supplied with fuel from the fuel tank and leaves the distributor via "one" fuel line for each injector at a constant predetermined pressure.

The basic CIS has also been equipped with several types of electronic control systems. One is an oxygen (lambda) sensor feedback circuit. The true CIS-E systems use an electromagnetic-controlled fuel regulator to control the fuel mixture in closed-loop operation.

DIFFERENCES IN POINT OF INJECTION

Gasoline electronic fuel injection systems can be further classified by their point of injection. There are two general design types: throttle body or single-point injection, or port or multipoint injection.

THROTTLE BODY FUEL INJECTION

Throttle body injection (TBI) enjoyed great popularity in the mid-1980s as an easy replacement for the carburetor. In this system, fuel is delivered to a carburetor-like throttle body assembly mounted to a conventional intake manifold (Figure 13-5). When fuel is delivered to a central point, the systems are called single-point, or throttle body, even though more than one injector can be used.

The throttle body mounts onto one injector or two injectors in the case of V-type engines. Fuel is supplied to the injector(s) by an electric pump. A regulator built into the throttle body maintains a constant fuel pressure (Figure 13-6).

The throttle body type of injection system is often considered a compromise between a feedback

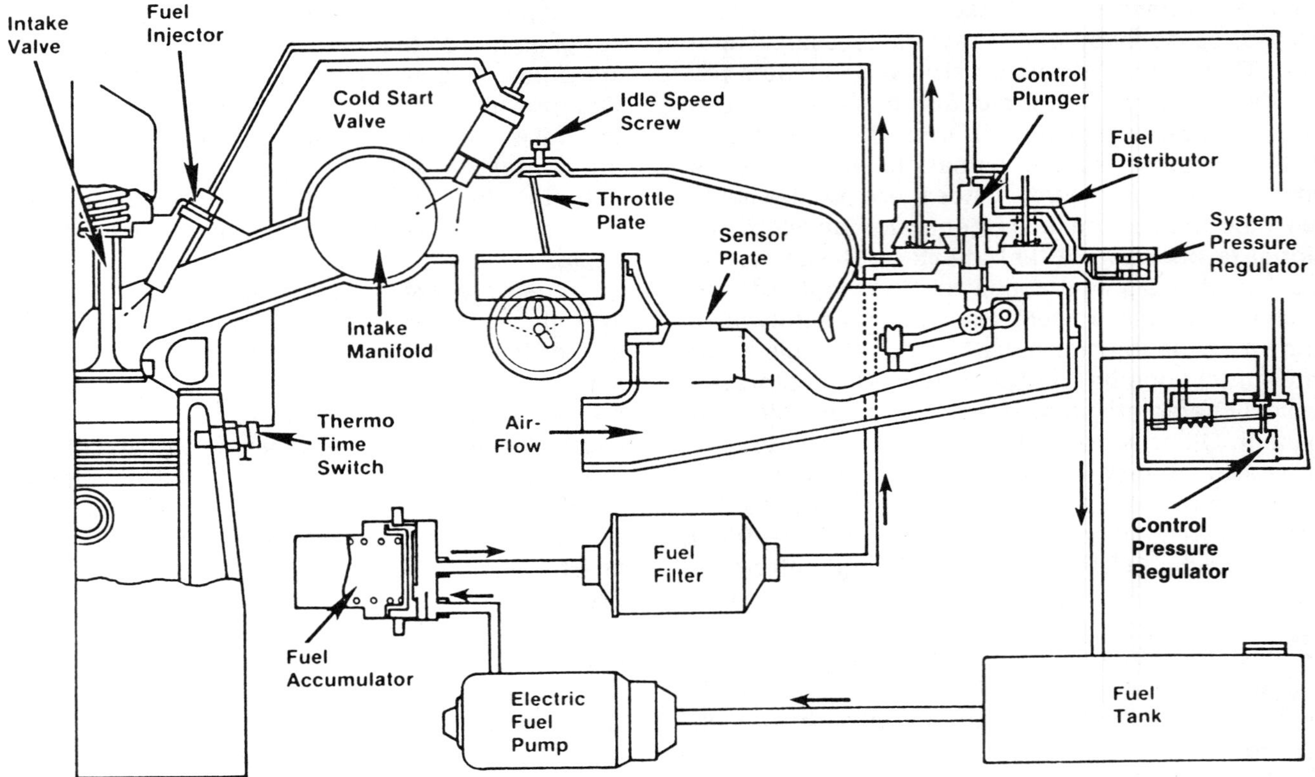

FIGURE 13-4 Schematic view of a continuous injection system that is mechanically controlled

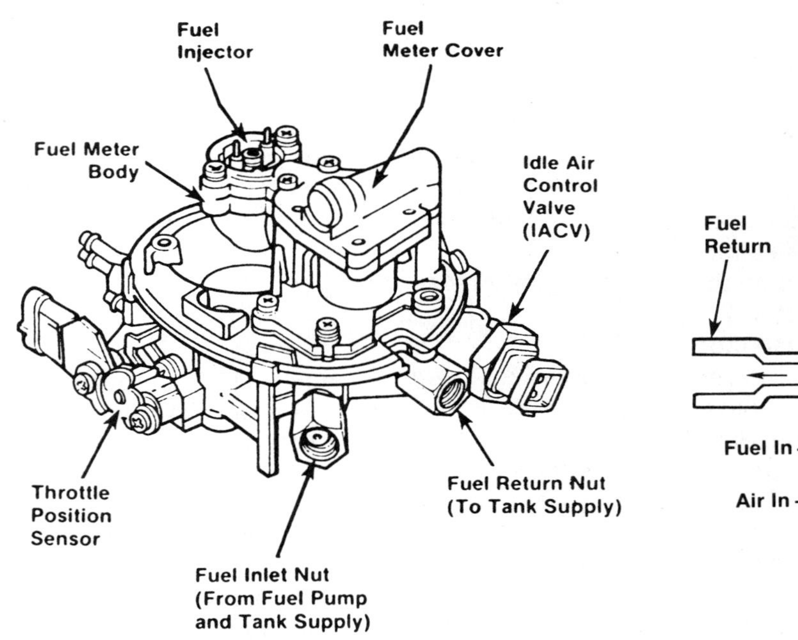

FIGURE 13-5 Throttle body injection (TBI) systems provide more precise fuel control than carburetor systems. Fuel is injected into a carburetor-like assembly mounted on the manifold.

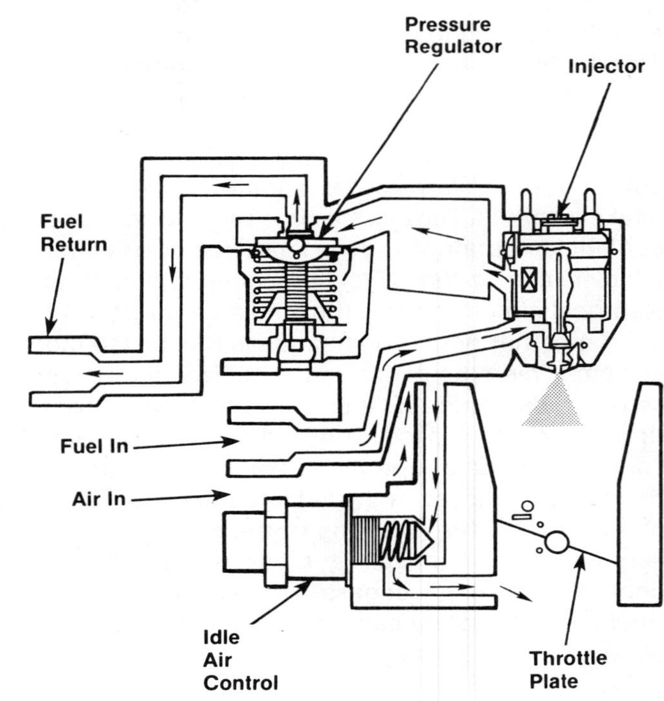

FIGURE 13-6 A regulator built into the throttle body maintains constant fuel pressure.

carburetor and a port injection system, which is described in the following section.

Throttle body systems provide improved fuel metering when compared to feedback carburetors. They are also less expensive and simpler to service. And because they use a maximum of only two injectors, throttle body systems are considerably less expensive than port injection systems, which use an injector for each cylinder.

However, throttle body units are not as efficient as port systems. The disadvantages are primarily manifold related. Like a carburetor system, fuel is still not distributed equally to all cylinders, and a cold manifold may cause fuel to condense and puddle in the chamber. And like a carburetor, throttle body injection systems must be mounted above the combustion chamber level, which eliminates the possibility of "tuning" the manifold design for more efficient operation. Yet when controlled by a computer (or ECU), these systems can produce results approaching port injection systems.

PORT FUEL INJECTION

Port injection systems use one injector at each cylinder. They are mounted in the intake manifold near the cylinder head where they can inject a fine, atomized fuel mist as close as possible to the intake valve (Figure 13-7). Fuel lines run to each cylinder

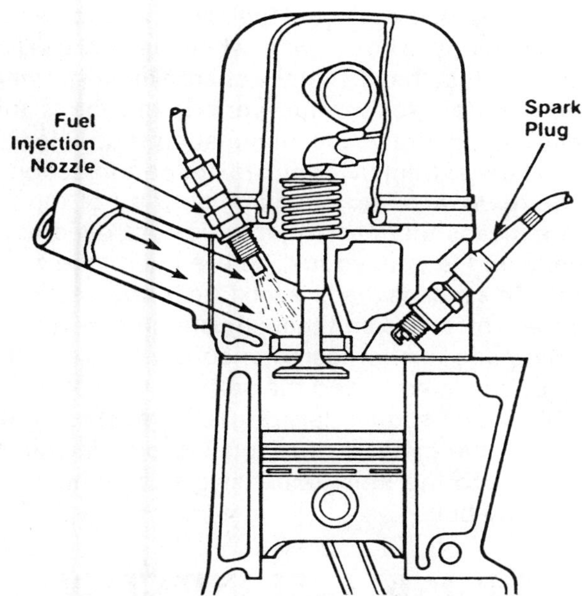

FIGURE 13-7 Port injection system injects fuel into an intake port instead of into a common throttle body assembly. With port injection there is an injector for each cylinder.

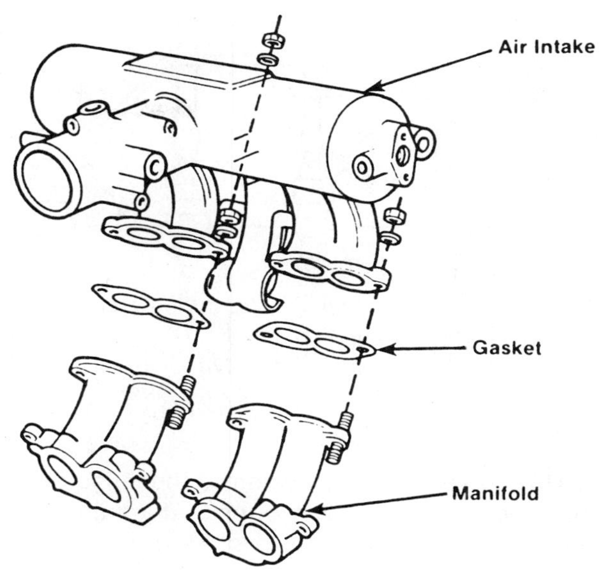

FIGURE 13-8 Port injection systems usually have "tuned" intake manifold runners that optimize low-speed power. TBI does not offer this advantage.

from a fuel manifold, usually referred to as a fuel rail. Since each cylinder has its own injector, fuel distribution is exactly equal. With little or no fuel to wet the manifold walls, there is no need for manifold heat or an early fuel evaporation system. Fuel will not collect in puddles at the base of the manifold. This means that the intake manifold passages can be tuned or designed for better low-speed power availability (Figure 13-8). The port type systems provide a more accurate and efficient delivery of fuel.

Port systems require an additional control system that throttle body injection units do not require. While throttle body injectors are mounted above the throttle plates and are not affected by fluctuations in manifold vacuum, port system injectors have their tips located in the manifold where constant changes in vacuum would affect the amount of fuel injected (at a given pulse width). To compensate for these fluctuations, port injection systems are equipped with fuel pressure regulators that sense manifold vacuum and continually adjust the fuel pressure to maintain a constant pressure drop across the injector tips at all times.

Port Firing Control

While all port injection systems operate using an injector at each cylinder, they do not fire the injectors in the same manner. One of four firing systems are used on domestic vehicles.

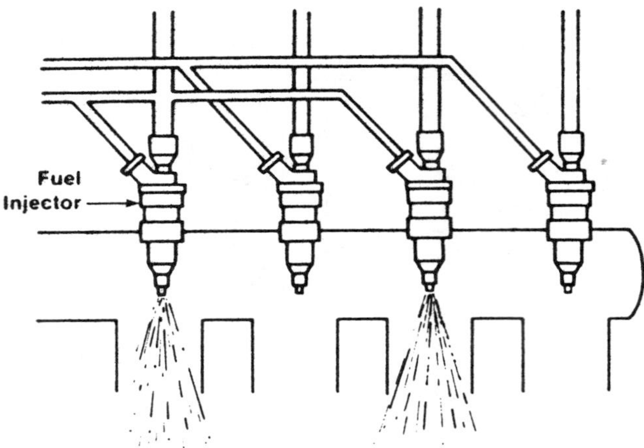

FIGURE 13-9 In port systems, the injectors are often divided into equal groups that are alternately fired with each crankshaft revolution.

1. Grouped single fire
2. Grouped double fire
3. Simultaneous double fire
4. Sequential fire

The terms grouped and simultaneous refer to how the injectors are connected within the system. Some systems fire the injectors in groups; others fire them all together. The terms single and double fire refer to how many firings of each injector are used to make up the fuel charge for one combustion stroke.

Grouped Single Fire. In this type of system, injectors are split into two equal groups. The groups are fired alternately, with one group firing each engine revolution (Figure 13-9). Only one injector pulse is used for each combustion stroke.

Since only two injectors can be fired relatively close to the point where the intake valve is about to open, the fuel charge for the remaining cylinders must stand in the intake manifold for varying periods of time. Also, since a new charge is released only once every two crankshaft revolutions, it is necessary to wait this long before any adjustment can be made in the air/fuel mixture.

Grouped Double Fire. This system also splits the injectors into two equal groups that alternately fire. But in this case, each group fires once each engine revolution. Two injector pulses make up each intake charge. This results in the air/fuel mixture remaining in the manifold for shorter periods of time. Adjustments to the mix can also be made sooner than with single fire systems.

Simultaneous Double Fire. This system fires all of the injectors at the same time once each engine revolution. It offers ease in programming and relatively fast adjustments to the air/fuel mix. The injectors are connected in parallel so the ECU develops just one signal for all injectors. They all open and close at the same time. It simplifies the electronics without compromising injection efficiency. The amount of fuel required for each four-stroke cycle is divided in half and delivered in two injections, one for every 360 degrees of crankshaft rotation.

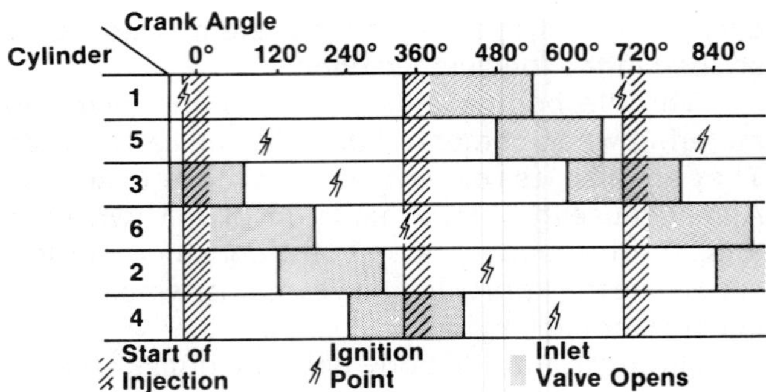

FIGURE 13-10 Charting a simultaneous injection system. Fuel is injected once per crankshaft revolution.

Figure 13-10 shows the intake stroke for each cylinder of a typical 6-cylinder sequential fuel-injected engine. In the firing order of 1-5-3-6-2-4, the first injection (to all cylinders) begins during the intake stroke of cylinder 3. The second pulse begins during the intake stroke of cylinder 4. In each case, the intake valve is open when the injection begins. In the other cases, the injected fuel forms a vapor in the manifold, ready to be drawn in when the intake valve opens. The fact that the intake charge must still wait in the manifold for varying periods of time is the system's major drawback. Simultaneous double fire systems are extremely popular with port injection.

Sequential Fire. Sequential firing of the injectors means that each injector is controlled individually and is opened just before the intake valve opens. This means that the mixture is never static in the intake manifold and that adjustments to the mix can be made almost instantaneously between the firing of one injector and the next.

Obviously, sequential firing is the most accurate and desirable method of regulating port injection. But it is also the most expensive and complex to design and build.

ELECTRONIC FUEL INJECTION SYSTEM COMPONENTS

The fuel injection system must provide the correct air/fuel ratio for all engine loads, speeds, and temperature conditions. Unlike a carburetor, a fuel

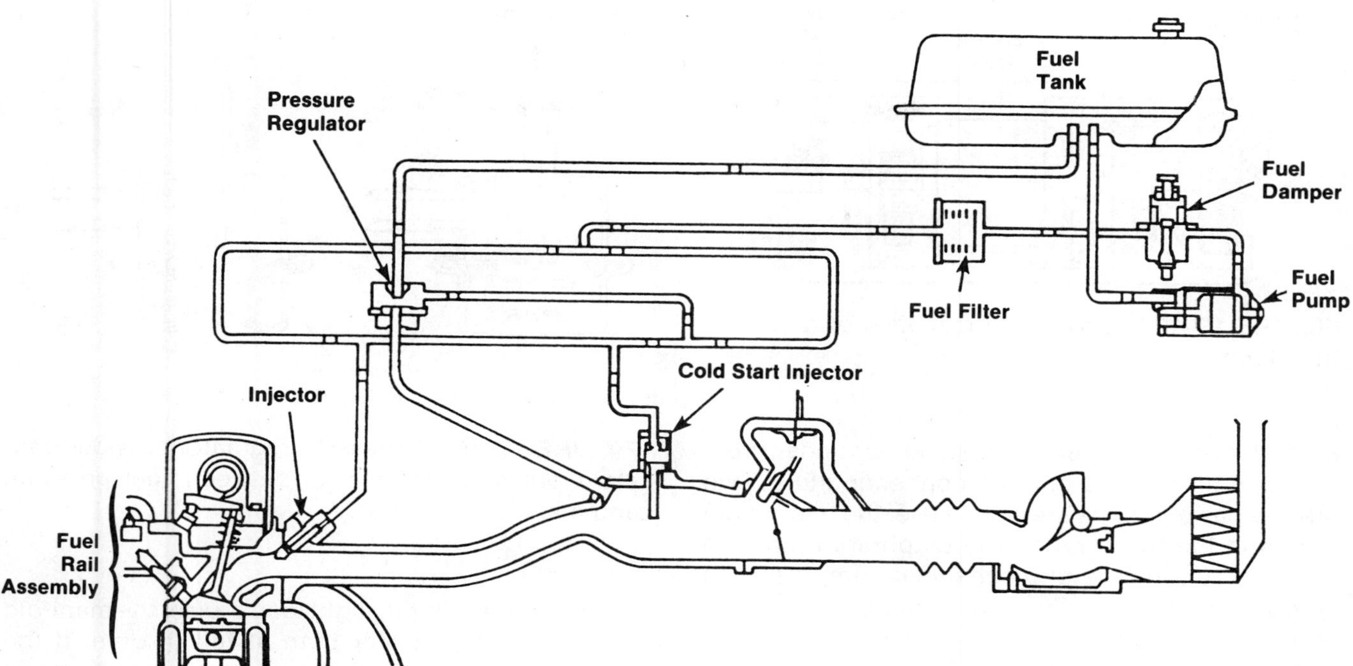

FIGURE 13-11 Fuel circuit used in a typical EFI system

injection system uses the same basic single system to provide these difficult air/fuel ratios. This basic system includes:

- Fuel delivery system
- System sensors
- Electronic control unit (ECU)
- Fuel injectors

FUEL DELIVERY SYSTEM

The fuel delivery system is similar to that used for a carburetion system. This system typically includes such components as an electric fuel pump, fuel filter, pressure regulator, fuel tank, fuel rail assembly, and individual connecting lines to each injector (Figure 13-11). It also includes an injector for cold-start operation.

Fuel Tank and Connecting Lines

The fuel tank and connecting lines are like those of a carburetion system; the main difference is that a return line is used to return excess fuel to the tank. This is accomplished by the pressure regulator.

Fuel Pumps

The fuel pump is a roller-cell pump (Figure 13-12) that is mounted directly on the shaft of the electric motor. The motor runs surrounded by gasoline to cool and lubricate it. An explosive atmosphere is avoided because the mixture in the pump housing is never an ignitable one (restricted air ratio). The pump pressurizes the fuel and thereby prevents air bubbles in the lines and vapor lock. The rotary roller pump consists of rollers on a centrally mounted eccentric that rotates within a housing. The rollers are pushed outboard by springs against the pump housing allowing fuel to flow in behind the rollers. The eccentric continues to rotate thereby compressing the rollers and creating pressure on the fuel behind them. As the eccentric rotates past

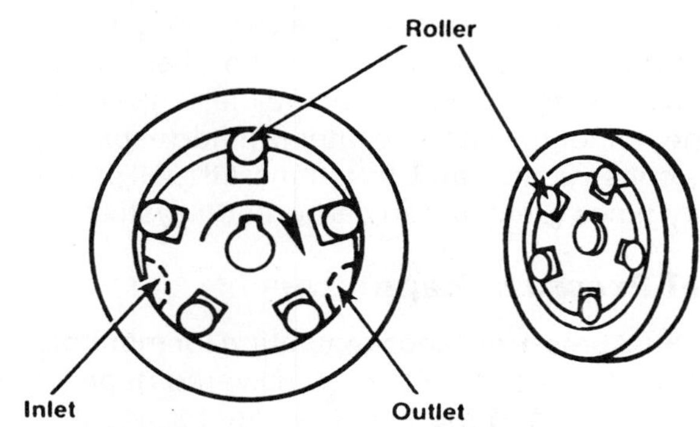

FIGURE 13-12 Operation principles of FI fuel pump. As the rotor spins, centrifugal force presses the rollers against the pump walls, forcing fuel through the pump.

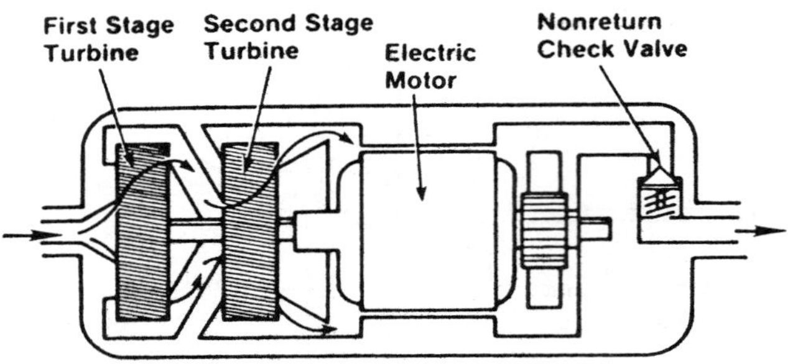

FIGURE 13–13 Cross-sectional view of a turbine fuel pump

an outlet port, the fuel is squeezed out under pressure and discharged into the connecting line. Two other types of pumps are also in use: the diaphragm type and the turbo type. The diaphragm pump is identical to the type of mechanical pump used on carburetion systems. This type of pump is used only as a transfer pump to deliver fuel to a high-pressure pump. The turbo type (Figure 13–13) is similar to a water pump or turbocharger compressor wheel. Fuel is drawn inward at the center of the wheel and forced outward through centrifugal force. The centrifugal force creates the pressure.

Power to the fuel pump is controlled by a relay in the fuel injection electrical system. One circuit delivers fuel pressure during cranking and another circuit delivers fuel when the engine is running and cuts off fuel pressure if the engine stops (even if the ignition is in the run mode).

Fuel Filter

Small contaminant particles can lodge in an injection nozzle and block it partially open or closed. Water can also seep into the system and corrode closely machined parts of the injectors. To eliminate dirt and water contaminants, one or more fuel filters are used in the system and are very similar to those used on carburetion systems. The filter can be located under the hood or near the fuel tank and pump at the rear of the car. The filter is made especially of fine construction and functions at fuel pressures many times greater than carburetion systems.

Fuel Pressure Regulator

Fuel pressure needs regulating for manifold absolute pressure (MAP). Fuel delivered depends on the pressure and the time the injector is open, the pulse width. The pressure difference between the fuel in the injector and the air pressure in the manifold must be constant. To maintain this pressure differential, the fuel pressure must change as the

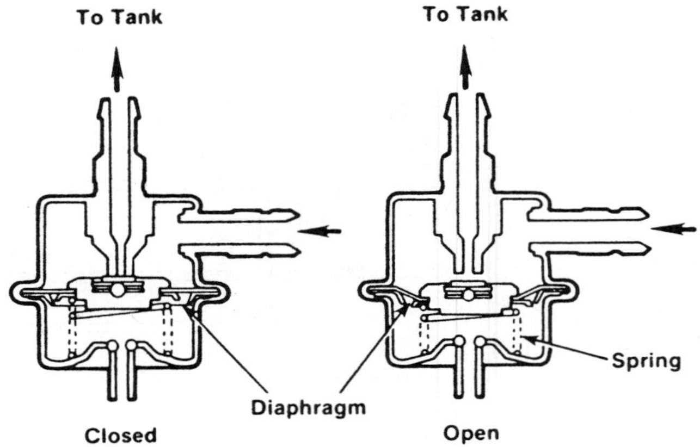

FIGURE 13–14 A pressure regulator provides constant pressure differential between fuel pressure and manifold pressure.

MAP changes. During light load cruise, the manifold pressure will be lower than at full throttle. If the engine is turbocharged, the injector will sometimes have to inject against higher MAP. Figure 13–11 shows the air hose connection between the fuel pressure regulator and the intake manifold. In Figure 13–14, the bottom connection applies the MAP to one side of a diaphragm to regulate fuel pressure. Fuel from the pump, entering the connection, is regulated according to the MAP. Excess fuel is returned to the tank through the return line. This regulator is adjusted at the factory and is not field adjustable. In a typical electronic fuel injection system, the fuel pressure is regulated at a value higher than MAP, expressed as 2.55 bar, 255 kiloPascals (kPa), or 37 pounds per square inch (psi). In normal operation, the regulator acts as a controlled leak, maintaining a constant fuel pressure.

Fuel Rail Assembly

The fuel rail assembly (Figure 13–15) is a mechanical assembly that contains or houses the fuel injectors (usually on a port fuel injection system). On some automotive models the fuel pressure regulator can also be mounted on this assembly.

AIRFLOW SENSORS

The airflow sensor shown in Figure 13–16 measures *airflow,* or air volume. The sensor consists of a spring-loaded flap, potentiometer, damping chamber, backfire protection valve, and idle bypass channel. As air is drawn into the engine, the flap is deflected against the spring. A potentiometer, attached to the flap shaft, monitors the flap movement and

FIGURE 13-15 A fuel rail assembly used on port injected 2.2L Ford engine

produces a corresponding voltage signal (Figure 13-17). The strength of the signal increases as the flap opens. The signal voltage is relayed to the electronic control module.

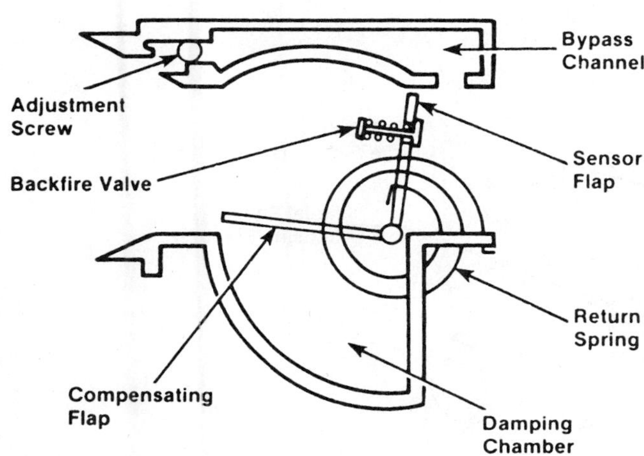

FIGURE 13-16 The airflow sensor flap is equipped with a backfire valve to prevent damage to the sensor.

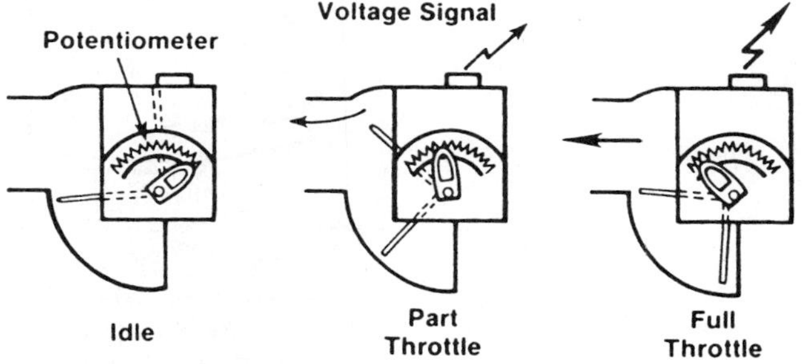

FIGURE 13-17 The voltage signal from the potentiometer varies in response to the amount the airflow sensor flap is deflected.

The electronic control module interprets the signal voltage as a measurement of airflow. However, the potentiometer signal must be modified to compensate for air temperature. Cold air is denser (weighs more) than the same volume of warm air. The denser air needs more fuel to be burned efficiently. The control unit monitors ambient air temperature via an air temperature sensor.

Air Leaks. If air leaks let extra air (sometimes called false air) enter the engine without passing the airflow sensor, the engine cannot run properly. Because the extra air gets in without being measured, it can fool the system. The electronic control unit does not provide fuel for the extra air so the resulting air/fuel mixture is too lean. Checking for leaks is especially important in fuel injection systems that depend on airflow measurement.

Damping Chamber. The curved shape of the airflow sensor shown in Figure 13–17 is the damping chamber. The damping flap in this chamber is on the same shaft as the airflow sensing flap and is also about the same area. As a result, the damping flap smooths out any possible pulsations caused by opening and closing of the intake valves. Airflow measurement can be a steady signal, closely related to airflow as controlled by the throttle.

Backfire Protection. The airflow sensor flap provides for backfire protection with a spring-loaded valve (Figure 13–18). If the intake manifold pressure suddenly rises because of a backfire, this valve releases the pressure and prevents damage to the system.

Idle Bypass. The airflow sensor assemby includes an extra air passage for idle, bypassing the airflow sensor plate. This is seen near the top of Figure 13–18. When the throttle is closed at idle, the opening and closing of intake valves can cause pulsations in the intake manifold. Without the idle bypass, such individual pulsations could cause the flap to shudder, resulting in an uneven air/fuel mixture. The idle bypass smooths the flow of the idle intake air, insuring regular signals to the electronic control unit for regular injections to the engine.

Adjustments. Except in sealed systems, both idle speed and mixture can be adjusted (see pages 386 and 390).

OTHER SENSORS

The signals from the airflow sensor are sometimes called the base pulse, or the minimum fuel to be injected. The base pulse must be modified by the electronic control unit to adjust the injector pulse width (open time) according to engine operating conditions.

Coolant Temperature. The coolant temperature sensor signals the electronic control unit when the engine needs cold enrichment, as it does during warmup. This adds to the base pulse, but decreases to zero as the engine warms up.

Air Temperature. The air temperature sensor signals the electronic control unit when colder outside air is coming in. The colder, heavier air needs more fuel, therefore a longer pulse width signal is sent to the fuel injectors.

Mass Airflow Sensor. The airflow sensor shown in Figure 13–18 measures *air mass.* Mass

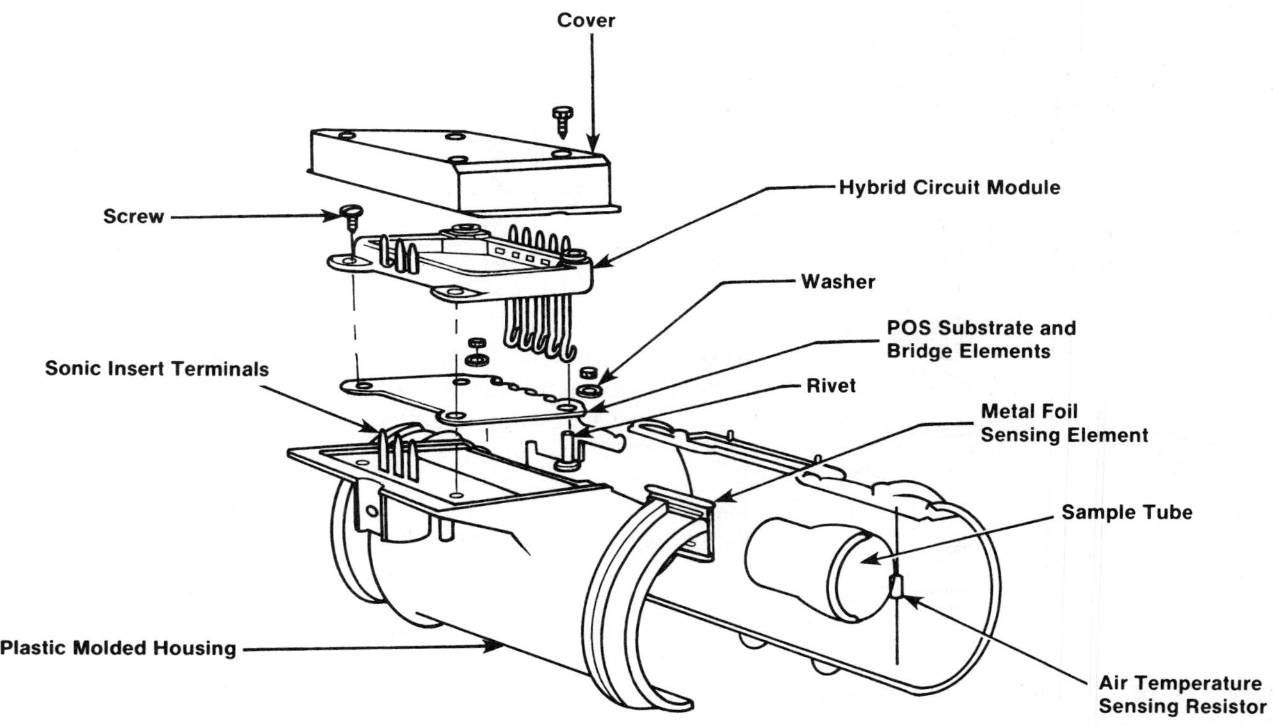

FIGURE 13–18 Mass airflow sensor

equals volume times density, where density is an expression of a material's weight per unit of volume. Variations in temperature, humidity, and altitude affect the oxygen density of a given air mass (more or less molecules per square unit). Monitoring the oxygen in a given volume of air is important, since oxygen is a prime catalyst in the combustion process. From a measurement of mass, the electronic control unit adjusts the fuel delivery for the oxygen content in a given volume of air. The accuracy of air/fuel ratios is greatly enhanced when matching fuel to air mass instead of fuel to air volume.

The primary purpose of the mass airflow sensor is to convert air flowing past a heated sensing element into an electronic signal. The strength of this signal is determined by the energy needed to keep the element at a constant temperature above the incoming ambient air temperature. As the volume and density (mass) of airflow across the heated element changes, the temperature of the element is affected and the current flow to the element is adjusted to maintain the set temperature difference. The varying current flow parallels the particular characteristics of the incoming air (hot, dry, cold, humid, high/low pressure). The electronic control module monitors the changes in voltage to determine air mass and to calculate precise fuel requirements.

There are two basic types of mass airflow sensors: hot wire and hot film. In the first type a very thin wire (about 0.2 mm thick) is used as the heated element. The element temperature is set at 100 degrees Celsius above incoming air temperature. Periodically, the wire is heated to approximately 1000 degrees Celsius for 1 second to burn off any accumulated dust and contaminants.

The second type uses a nickel foil sensor, which is kept 75 degrees Celsius above ambient air temperatures. It does not require a burn off period; thus, it is potentially longer lasting than the hot wire type.

Throttle Position. The switches on the throttle shaft signal the electronic control unit for idle enrichment when the throttle is closed (Figure 13-19). These same throttle switches signal the electronic control unit when the throttle is near the wide-open position to provide full load enrichment.

Engine Speed. The ignition system sends a signal corresponding to engine speed, either from the ignition coil or distributor to the electronic control unit. This signal advises the electronic control unit to adjust the pulse width of the injectors for engine speed. This also times the start of the injection according to the intake stroke cycle.

Cranking Enrichment. The starter circuit sends a signal for fuel enrichment during cranking operations even when the engine is warm. This is independent of any cold-start fuel enrichment demands.

Additional Input Information Sensors. Additional sensors are also used to provide the following information on engine conditions:

- Crankcase position
- Camshaft position
- Timing of ignition spark
- Air conditioner operation
- Gearshift lever position
- Battery voltage
- Amount of oxygen in exhaust gasses
- Emission control device operation

Figure 13-20 shows the inputs and outputs of an onboard computer (ECU) on a typical vehicle.

ELECTRONIC CONTROL UNIT

The heart of the fuel injection system is the electronic control unit (ECU). The ECU is a small computer that is usually mounted within the passenger compartment to keep it away from the heat and vibration of the engine. The ECU includes solid

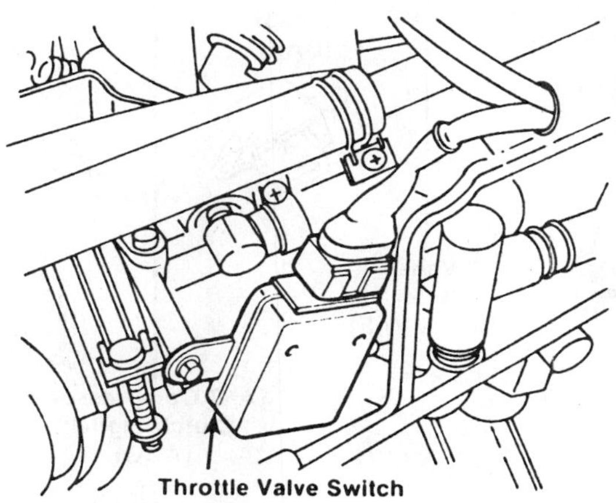

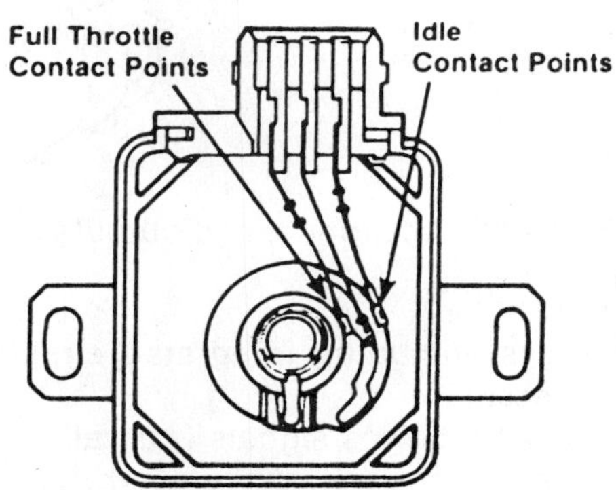

FIGURE 13-19 The throttle position switch signals the ECU when the throttle is wide open or at idle.

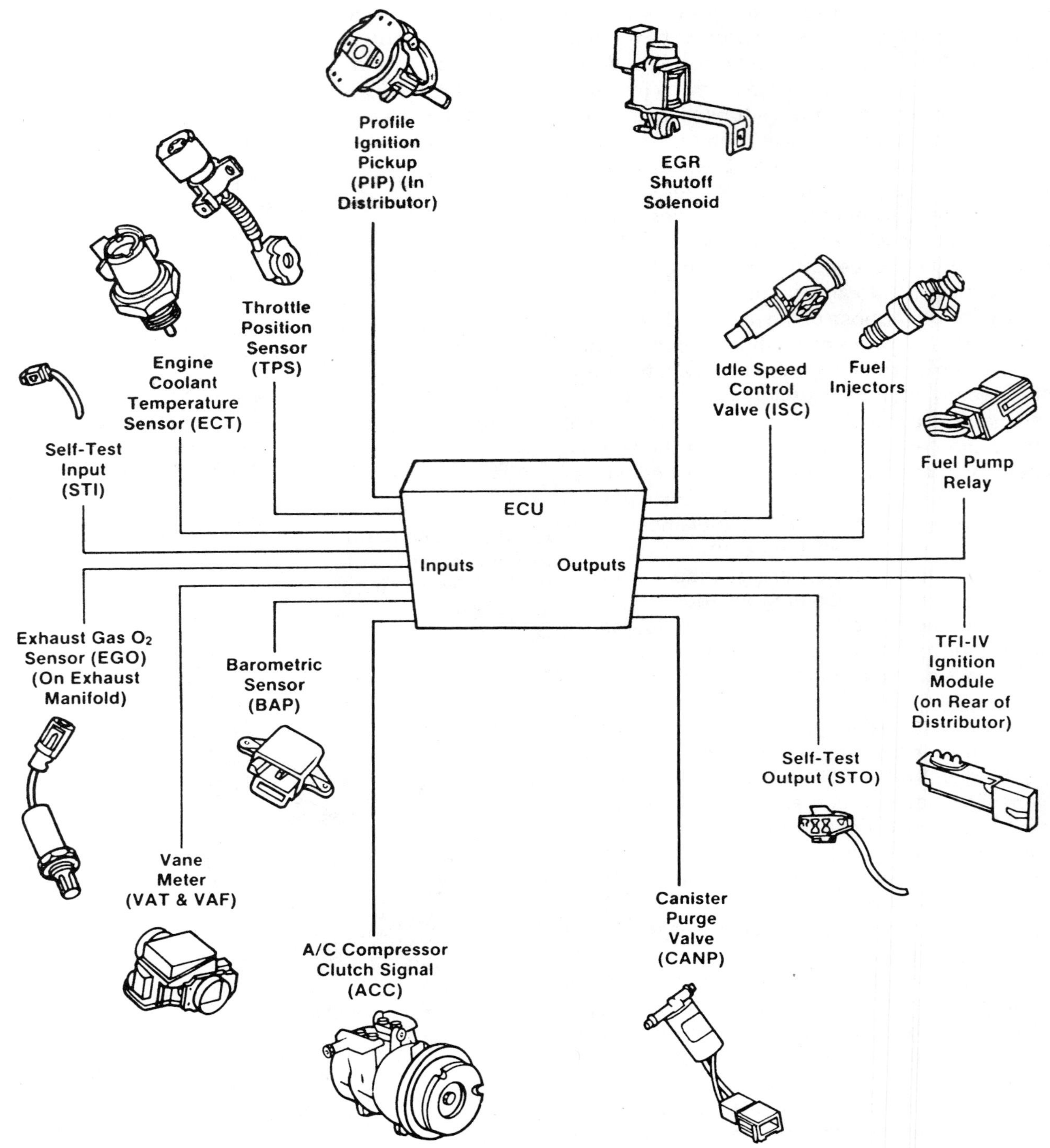

FIGURE 13–20 The inputs and outputs of a typical ECU-controlled vehicle

state devices, including integrated circuits and a microprocessor.

The ECU receives signals from all the system sensors, processes them, and transmits programmed electrical pulses to the fuel injectors. Both incoming and outgoing signals are sent through a wiring harness and a multiple-pin connector.

Electronic feedback in the ECU means the unit is self-regulating, controlling the injectors by the ECU on the basis of operating performance or parameters rather than on preprogrammed instructions. An ECU with a feedback loop, for example, reads signals from the oxygen sensor, varies the pulse width of the injectors, and again reads the signals from the oxygen sensor. This is repeated until the injectors are properly pushed to give the required oxygen output in the exhaust gasses from the engine.

INJECTOR DESIGN AND FUNCTION

Two types of injectors are currently in use (Figure 13-21). In a top-feed injector, fuel is delivered under high pressure (up to 79 psi) to help prevent vaporization. The higher pressure requires a more expensive fuel pump. One disadvantage of this type of injector is that fuel vapors tend to rise into the incoming fuel stream and can prevent or block fuel delivery. Bottom-fed injectors are able to use fuel pressures as low as 10 psi. Fuel vapors flow upward through the injector to a return line that feeds the fuel tank. Using bottom-feed injectors requires a less expensive pump and fuel vapors do not tend to block fuel delivery.

The injector (Figure 13-22) is about the size and shape of a typical spark plug. When the signal from the ECU arrives at the electrical connection, the

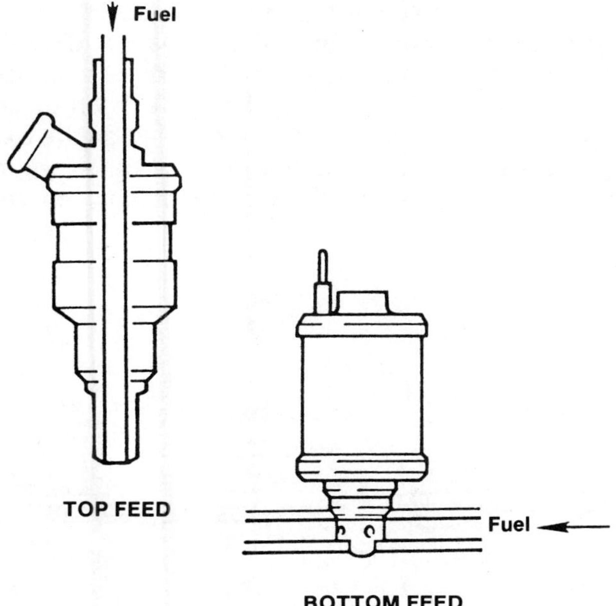

FIGURE 13-21 Examples of both standard, top feed and low-pressure bottom feed injectors

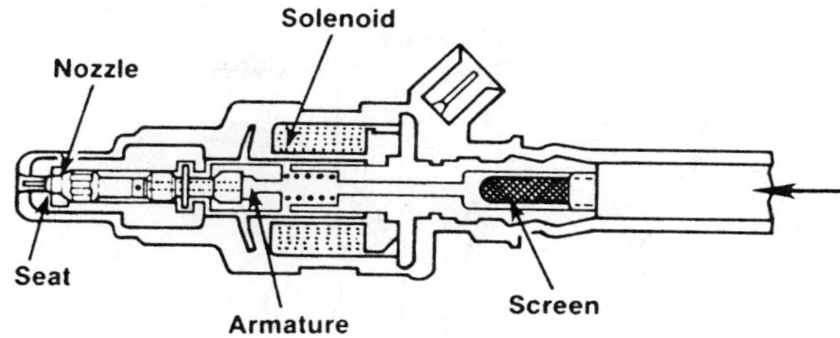

FIGURE 13-22 Cross-sectional view of a typical solenoid-operated injector

solenoid winding pulls the armature, opening the nozzle valve. When the ECU shuts off the signal, the nozzle snaps shut by spring action. At idle, the open time might be as short as 1 millisecond, and at full load as long as 10 millisecond.

In some engines, the injectors operate at less than battery voltage. For each injector, battery voltage is reduced by a ballast resistor, sometimes called a dropping resistor. The resistor protects the injector from voltage surges coming from the alternator and from other surges, or spikes in the electrical system.

Cold Start Operation

In addition to the units already described for normal warm engine operation, additional units are provided for cold starting and warm-up. These include the cold start injector (valve) and its thermo-time switch plus the auxiliary air device (see Figure 13-11).

The cold start injector delivers extra fuel while the cold engine is cranking. When the fuel pump is running, this extra injector is supplied with fuel under pressure. The colder the engine, the more extra fuel will be injected during cranking. When the engine is warm, extra fuel is not injected.

The cold start injector is supplied with power from the starting circuit and is grounded through the thermo-time switch. The switch gets its name because it operates on both temperature and time. The temperature here is coolant temperature and the time involved is modified by an electric heating element in a bimetal switch, also powered by the starting circuit.

After several seconds of cranking, this bimetal switch in the thermo-time switch opens the circuit and stops the cold start fuel injection. The results are as follows: When the engine is cold and the ignition key turned to START, cold start fuel injection is

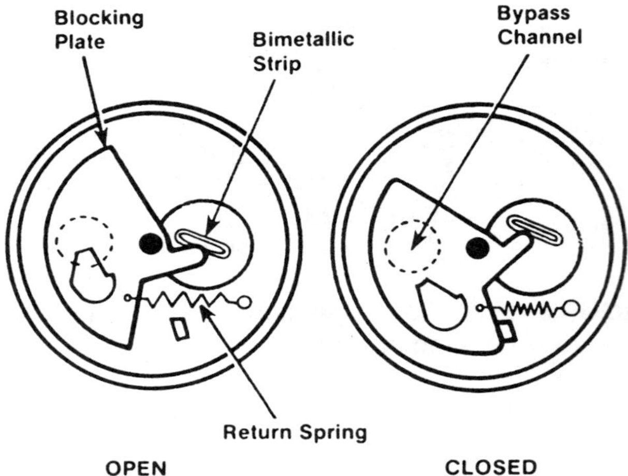

FIGURE 13-23 The auxiliary air device controls an air bypass passage to increase idle speed during cold operation.

delivered. With the engine cold and the ignition key released to RUN, cold start fuel injection switches off. If the engine is warm (about room temperature) and the ignition key is turned to START, there is no cold start fuel injection. When the engine is cold, continued starting or cranking of the engine can cause flooding so the cold start fuel injection switches off. The switch might provide cold start fuel injection for 20 seconds at 0 degrees Fahrenheit. When the engine is warmer, about the temperature of freezing, the switch might allow cold start fuel delivery for about 5 seconds.

When the cold start injector provides extra fuel during cranking, the auxiliary air device provides extra air during start and warm-up. The auxiliary air device is an idle bypass of the throttle, providing extra measured air and, therefore, injected fuel. As a result, the idling engine has additional power to overcome the increased friction of the cold engine. In the bypass hose, the auxiliary air device opens when the engine is cold. Figure 13-23 shows how its operation can be checked. When the plate opens the channel, extra air bypasses the throttle. Opening is controlled by a bimetal strip. As the bimetal heats up, it bends to rotate the blocking plate, gradually blocking the opening. When the device is closed, there is no auxiliary airflow.

The auxiliary air device is usually on the cylinder head or block so it is warm when the engine is warm. In some engines, such as air-cooled engines, lubricating oil warms the auxiliary air device.

Usually from a cold start the auxiliary air must be shut off sooner than it would be by rising engine temperature alone. The bimetal strip is warmed by an electric heating element powered from the run circuit of the ignition switch. This bimetal element is not a switch, but a strip that moves the blocking plate directly. The auxiliary air device is independent of the cold start injector. It is not controlled by the ECU but is continuously powered when the ignition key is set to the RUN position.

When the engine is cold, the passage opens for extra air when the engine starts. When the engine is running and still cold, the passage is open but the heater begins operating to close it gradually, requiring about 8 minutes maximum if start was at 0 degrees Fahrenheit. If the engine is warm at start up, the passage is closed and normal air is delivered for idle.

CLOSED LOOP FEEDBACK SYSTEM

Regulations limiting the amounts of harmful substances in exhaust gasses require manufacturers to control the metering of the air/fuel mixture more closely. The method used by most manufacturers, whether the vehicle is equipped with a carburetor or fuel injection system, is the feedback or closed loop fuel control system.

The principal components are an oxygen sensor, sometimes called a lambda sensor or exhaust gas oxygen sensor (EGO) in the exhaust manifold near the engine (Figure 13-24), additions to the ECU, and the connecting wires.

The oxygen (lambda) sensor measures the oxygen concentration in the exhaust gasses and passes this information to the ECU in the form of a variable voltage signal. The ECU makes a decision on the regulation of the air/fuel mixture based on the oxygen sensor signal (plus the input of other engine sensors). Because the exhaust conditions generate

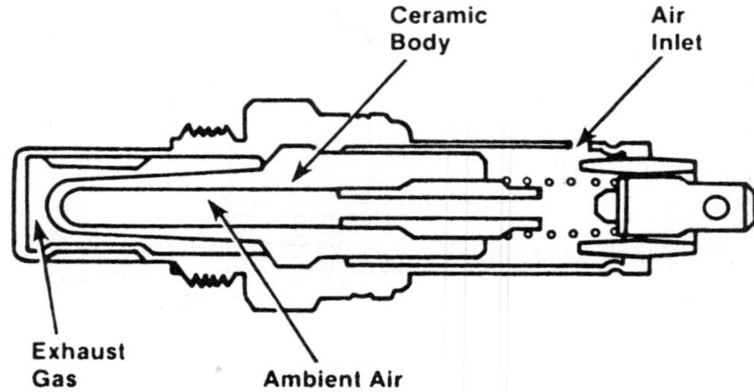

FIGURE 13-24 Cross sectional view of a typical lambda or oxygen sensor that signals the ECU if the fuel mixture is lean or rich.

the information required to modify the exhaust conditions, the system is called closed loop or feedback.

For a feedback system to operate, the air/fuel ratio must be tightly controlled, not too rich or too lean, but just right for the reduction of NO_X and oxidation of HC and CO in the three-way converter. This ideal air/fuel ratio can be achieved with a closed loop or feedback control. By continually measuring the oxygen content of the engine exhaust, the fuel delivered by the injection system can be adjusted as needed.

The oxygen (lambda) sensor in the exhaust manifold (Figure 13-24) tells the ECU if the effective air/fuel mixture being supplied to the engine is too rich or too lean. It does this by measuring the amount of oxygen in the exhaust gasses. The oxygen sensor continually sends this information to the ECU in the form of a switching voltage.

For example, when the incoming mixture is too lean, the oxygen content of the exhaust will be high, and the sensor will put out a low voltage signal. A too-rich mixture will result in a low oxygen content and a high signal voltage. In operation, the change in the oxygen sensor voltage output, even from slightly rich to slightly lean, is so rapid and pronounced that it provides the ECU with a responsive and precise method for monitoring air/fuel ratio.

Closed Loop Operation

The signals from the oxygen sensor are fed to the ECU for processing and modification of the fuel injection pulse width. This changes the engine exhaust, consequently changing the signal from the oxygen sensor. Control continues in this closed loop. Closed loop control continually adjusts the fuel injection so the air/fuel ratio is right for changing conditions.

Closed loop operation depends on several conditions. First, the engine must be warm or the ECU will take away the extra fuel added for the warmup. The oxygen sensor must be hot, about 570 degrees Fahrenheit, or else it will not send reliable signals. Finally, the engine must be between idle and full throttle, or the ECU will take away the extra fuel. When any one of these conditions is not right, the loop is not closed. It is then in open loop, operating on preprogrammed instructions in the ECU. The signals sent to the ECU determine when to switch from closed loop to open loop and back.

Open Loop Operation

In open loop, the ECU disregards signals from the oxygen sensor. Instead, it sends programmed signals for warmup, idle, and full load, similar to the way it does on engines without oxygen sensor control.

In service, several possibilities for open loop operation should be considered. The oxygen sensor is small, so it heats up quickly. At low speeds with no load, it might not heat up enough to operate. Also, the engine is massive, taking up to 10 minutes to warm up and a long time to cool down.

As a result, during engine warmup, the engine operates under an open loop condition for up to 10 minutes. For a hot start after a brief stop, open loop condition might last only 10 seconds. Slow creeping in traffic can also cause an open loop condition as could a loose connection to the coolant temperature sensor.

Advantages and Disadvantages

A properly functioning feedback control loop provides precision control of fuel injection to meet varying operating conditions. It makes possible the use of the three-way catalyst for better emission control, improves driveability, power, and economy. On the other hand, it can cover up engine malfunctions such as a failed EGR or misfiring spark plug. Tests have shown some malfunctions that are not detected by normal maintenance testing. Both operators and service technicians should be alert for signals such as poor driveability, reduced fuel economy, or black exhaust smoke. However, closed loop control is not responsible for hard starting, poor idle, or lack of full power.

COASTING SHUTOFF

Coasting shutoff can be found on a number of engines. It can improve fuel economy as well as reduce emissions of hydrocarbons and carbon monoxide. Fuel shutoff is controlled in different ways depending on the type of transmission (manual or automatic). The ECU makes a coasting shutoff decision based on two input signals:

1. A closed throttle as indicated by the idle switch
2. Engine speed indicated by the signal from the ignition coil

Fuel injection and turbochargers are well suited for each other because the manifold system conveys only air as opposed to the typical air/fuel mixture. This offers flexibility in the installation of the turbocharger and the airflow control sensor unit.

As the car operates at higher altitudes, the thinner air needs less fuel. Altitude compensation in a fuel injection system is accomplished by installing a sensor to monitor barometric pressure. Although the oxygen sensor can compensate for altitude to a slight degree, it can reach the limit of its correction scale, usually about 20 percent on either side of its setting. Signals from the barometric pressure sensor are sent to the ECU to reduce the injector pulse width (or reduce the amount of fuel injected).

CONTINUOUS FUEL INJECTION SYSTEM COMPONENTS

While it is most likely that electronic fuel injection eventually will be the system of choice on future engines, the mechanically controlled continuous or constant injection system (CIS) is still very popular.

Since it first appeared in the early 1970s, CIS has gained an excellent reputation for efficiency and reliability. In the US, CIS operating efficiency even allowed several automakers to pass exhaust emissions tests for several years without the addition of catalytic converters.

More recent variations of the system employ an oxygen (lambda) sensor and an ECU to keep the air/fuel mixture within an even stricter range. Another variation uses an electromagnetic differential fuel pressure regulator controlled by the ECU. But while some electronic control elements are being added to the basic system, most CIS systems in use today are purely mechanically controlled.

The CIS uses fuel under pressure to modulate or change the fuel injection rate. Fuel is delivered continually to the intake manifold and a constant relative fuel pressure is maintained. The fuel delivery rate is not varied by injector pulse width as in electronic systems. Instead the amount of fuel being passed to the injectors is varied. The injectors spray a mist continuously into the intake ports. The amount of fuel contained in the mist is regulated.

The operation and components of a CIS unit is shown in Figure 13-25. Air enters the system through the air filter and is measured by an airflow sensor. Airflow is controlled by the driver, using a regular throttle valve. The air flows through the intake tubes into the combustion chamber.

Fuel is delivered by an electric fuel pump through a fuel accumulator and fuel filter. In the mixture control unit, the movement of the airflow sensor controls how much fuel is delivered through the injectors. In the intake manifold, the injected fuel is delivered to the combustion chamber by intake air when the valve is open. While the intake valve is closed, the fuel vapor is stored in the intake manifold.

In Figure 13-25, the air enters beneath the airflow sensor plate, lifting the plate as it moves toward

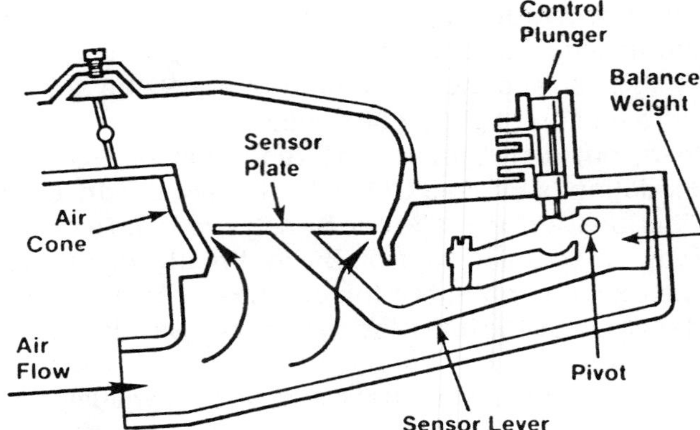

FIGURE 13-25 The CIS airflow sensor is coupled directly to the control plunger in the mixture regulator.

the throttle and intake manifold. This is known as an updraft system. In some systems, however, movement of intake air presses down on the airflow sensor plate. This is called a downdraft system.

FUEL SYSTEM

The fuel system for CIS is similar to that for an electronic system. It includes the fuel tank, fuel pump, accumulator, and filter. The principal difference between the two fuel systems is the control pressure regulator.

The regulator includes a push valve, as shown in Figure 13-26. The regulator with its push valve controls system pressure with the engine running. However, it also retains fuel pressure in the connecting lines when the engine is turned off. This will ensure quick restarting. It also means that the fuel pressure must be relieved before opening any of the fuel lines. Failure to relieve this pressure can cause fuel to be sprayed around the engine compartment, possibly causing a fire or personal injury.

The fuel system also includes a return line to the fuel tank and associated fuel relays to control the fuel pump during starting and running. The relays also shut off the pump, for safety reasons, if the engine should stop with the ignition on.

AIRFLOW MEASUREMENT

Because of the cone shape of the air funnel in the mixture regulator, the airflow sensor rises farther when more air flows into the engine. This movement of the air sensor plate measures the amount of incoming air.

Any air entering the intake system without passing the sensor plate will interfere with the proper air/fuel mixture, causing the engine to run lean. This

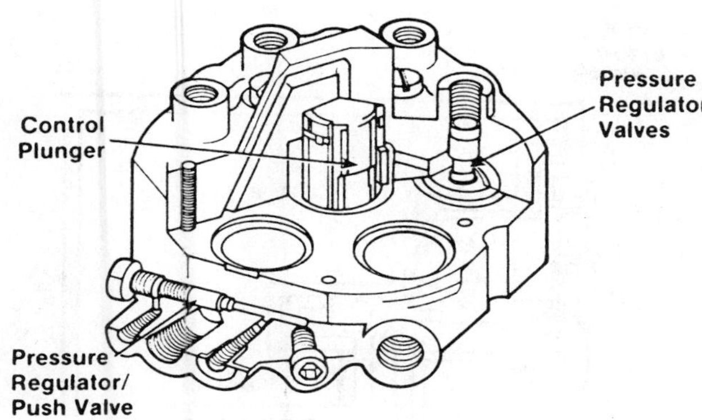

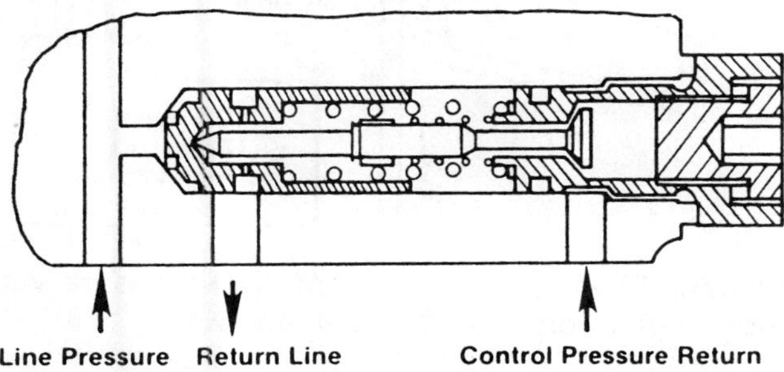

FIGURE 13-26 The push valve maintains line pressure with the engine off to ensure a fast start.

is the same as the electronic system. Proper operation of the CIS depends on having no vacuum leaks.

The airflow sensor (Figure 13-25) is part of the mixture control unit. As the air sensor plate moves because of air entering the engine, it controls the amount of fuel injected by raising the control plunger in the fuel distributor. Although this unit resembles an ignition distributor, it does not rotate and distribute fuel to the cylinder in firing order. Instead, it continuously sends the same amount of fuel to all cylinders at the same time.

The sensor plate moves against the hydraulic counter pressure applied to the top of the control plunger by pressurized fuel. In more recent model years, a small return spring helps the plunger follow rapid movements of the air sensor.

FUEL METERING

The control plunger on the mixture regulator rises and falls in a barrel with metering slits (Figure 13-27). The actual slit is about 0.2 millimeters wide, about twice the thickness of a human hair. The barrel contains one slit for each cylinder in the engine.

Metering by plunger movement is also shown in Figure 13-27. In position 1, the slight airflow at idle lifts the plate only enough to raise the control plunger slightly. The metering slit is opened to allow enough fuel for idle. As airflow increases, more of the metering slit is exposed. At full throttle, position 3, the fuel flow is at maximum.

The fuel control plunger must fit tightly enough in the barrel to minimize leaks, yet it must move freely to respond to delicate movements of the sensor plate. If a small particle of dirt jams the control plunger, it locks the sensor plate in place and the engine will not respond to varying air flows.

Within the mixture control unit, each injector has its own differential pressure valve (Figure 13-28). These valves maintain equivalent fuel flows at each injector by maintaining a constant fuel pressure difference across the metering port.

A steel diaphragm separates the two chambers of each pressure regulator valve. Inside the differential pressure valve, the spring presses down on the diaphragm while, at the same time, the fuel system pressure is pressing up on the lower side of the

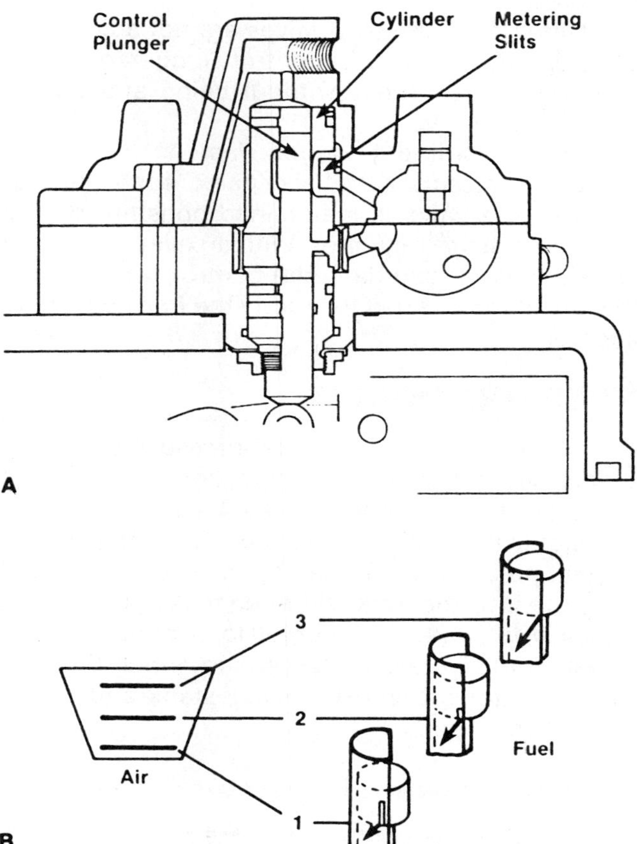

FIGURE 13-27 CIS fuel metering: (A) a control plunger is raised and lowered in a barrel that has one metering slit for each cylinder of the engine; (B) as the plunger is raised in response to increased airflow, it uncovers more of the metering slit.

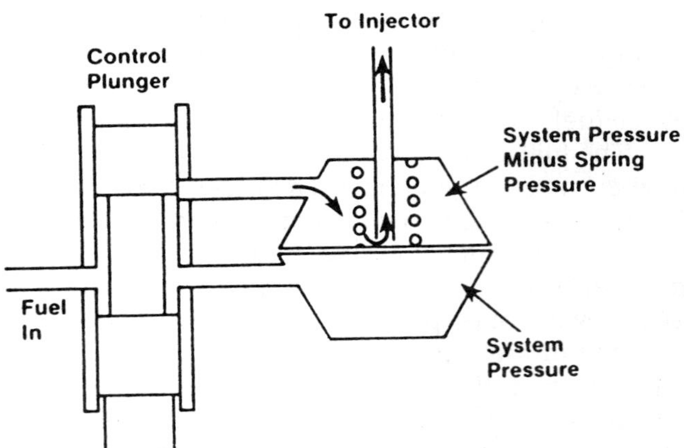

FIGURE 13–28 A pressure regulator valve for each cylinder maintains a constant pressure difference on either side of the slit.

diaphragm. For fuel to flow to the injector, the diaphragm must be deflected slightly against fuel system pressure.

Differential pressure valves are not adjustable. They operate to allow the control plunger to control fuel volume accurately without being affected by pressure changes.

In a CIS without oxygen (lambda) control or feedback, pressure in the lower part of each differential pressure valve is maintained constant by the system pressure regulator. With auxiliary electronics, such as oxygen (lambda) control, for example, this pressure is varied to modify the fuel flow to the injector.

FUEL PRESSURES

The CIS operates on fuel pressure at several different levels. System pressure, determined by the fuel pressure regulator, is 65 to 75 psi.

Control pressure is reduced 50 to 55 psi from system pressure when the engine is warm. Control pressure is further reduced when the engine is cold, typically 19 to 24 psi. Under full load, or under turbo boost, control pressure lessens to enrich the mixture. Injection pressure is not adjustable and ranges from 51 to 60 psi.

Control Pressure Regulator

In a CIS, the control pressure regulator determines the fuel pressure on top of the control plunger to help regulate the air/fuel mixture. Fuel pressure is applied to the top of the control plunger to balance the force from the airflow sensor.

In a typical installation, system pressure will be between 65 and 75 psi. With a warm engine in part-

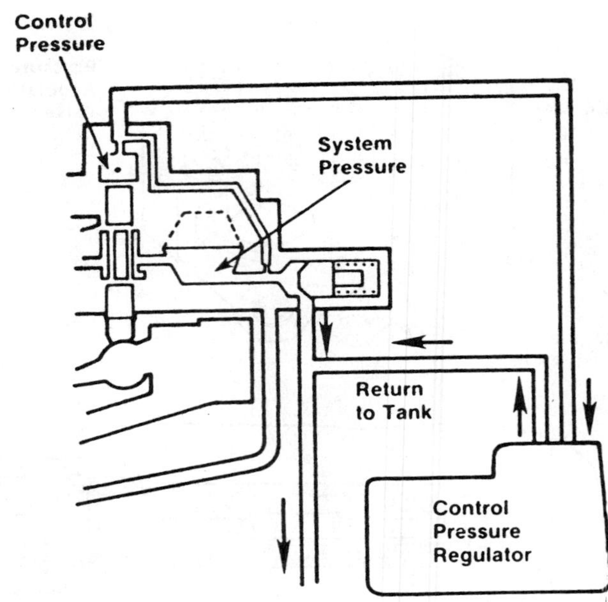

FIGURE 13–29 Altering the control pressure will affect regulation of the air/fuel mixture.

throttle cruise, the control pressure applied to the top of the plunger will be 50 to 55 psi. This control pressure keeps the control plunger down, limiting the amount of fuel passing through each metering slit to its injector.

The key to changing the amount of fuel delivered to the injectors is to reduce the control pressure on top of the control plunger (Figure 13–29). When the control pressure is lower, the control plunger can rise farther under pressure from the airflow sensor, permitting more fuel to pass to the injectors, thereby enriching the mixture.

Cold Start Provisions

Mixture adaption in the CIS, as for cold start operations, is similar to the electronic fuel injection system. Refer to the electronic fuel injection system description for cold start operation (includes the cold start valve, thermo-time switch, coolant temperature switch, and auxiliary air device).

Warm-up Enrichment

The mechanism for changing control pressure is the control pressure regulator (Figure 13–30). It is usually located on the engine block so it is sensitive to engine temperatures. In some engines, its only purpose is to enrich the mixture during warm-up. It is sometimes called a warm-up regulator. The control pressure regulator reduces control pressure by increasing the return flow to the tank. The control

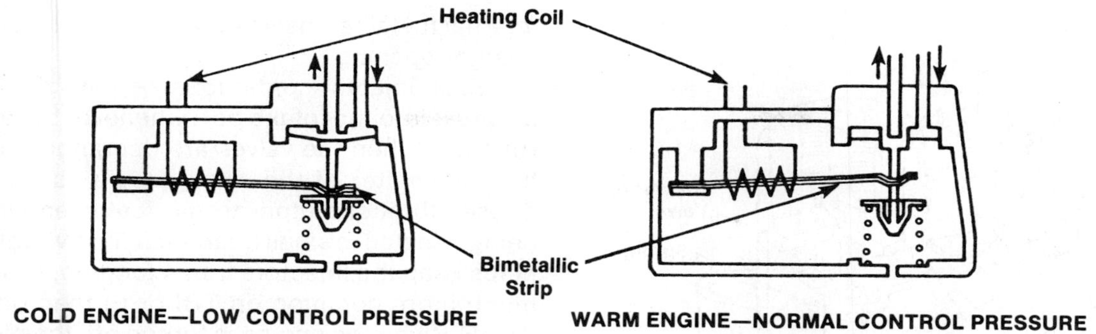

FIGURE 13-30 The cold pressure regulator reduces control pressure during cold engine operation to create a richer mixture.

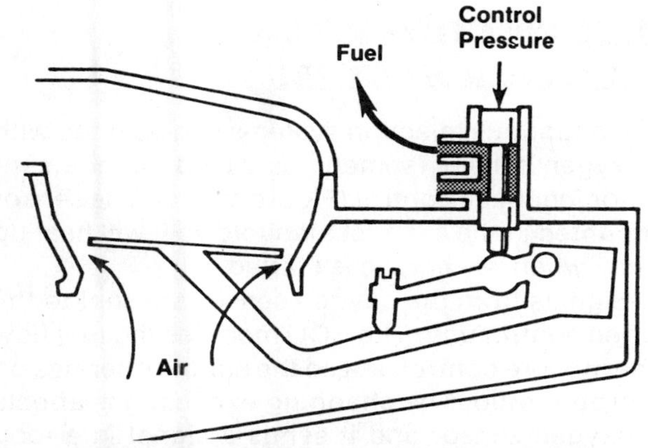

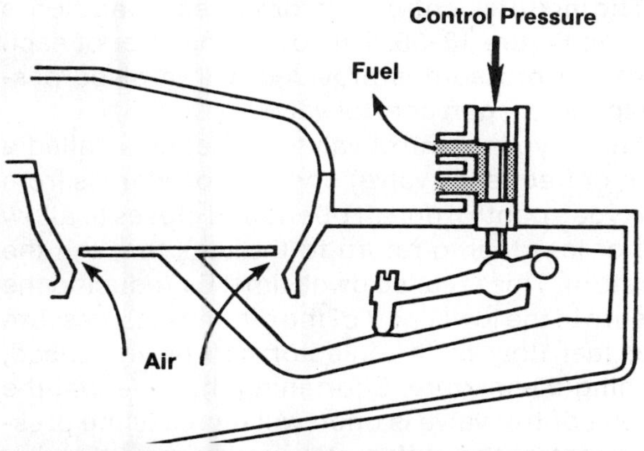

FIGURE 13-31 Lower control pressure allows the control plunger to rise farther, exposing more of the metering slit and allowing more fuel flow.

pressure regulator on the engine block is connected to the mixture control unit by hoses. The opening of the valve diaphragm is resisted by the valve spring and bimetal spring. The bimetal spring is electrically heated by a resistance coil, enabling the warm-up regulation to be precisely tailored to engine characteristics.

When the engine is cold, the control pressure regulator is also cold (Figure 13-31A). The bimetal spring bends down against the valve spring, allowing the valve diaphragm to open farther and return more fuel to the tank. This lessens the control pressure, allowing the plunger to rise and admit more fuel, enriching the mixture.

The control pressure regulator is not adjustable. High readings could mean a restricted return line. A low reading when warm could mean a break in the ground circuit or current supply. Otherwise, incorrect readings mean the control pressure regulator should be replaced. The colder the engine, the lower the control pressure, allowing the plunger to rise to enrich the mixture (Figure 13-31B).

When the engine is started, the electric heating element begins to heat the bimetallic strip. As the bimetallic strip rises, the valve spring raises the valve diaphragm, reducing the return flow (Figure 13-30). This gradually increases the control pressure on top of the control plunger; therefore, the airflow sensor plate cannot lift the control plunger as far. This cuts down the amount of fuel delivered to the injectors and leans the fuel mixture as the engine warms up.

Control Under Other Conditions

Some control pressure regulators can do more than enrich the mixture for the engine during warm-up. Full load enrichment can also be provided by a control pressure regulator (Figure 13-32).

During normal part-load cruise, the vacuum from the intake manifold applied through the port draws air from the control pressure regulator. The atmospheric air presses the valve diaphragm shut, increasing the control pressure and keeping the mixture lean. When the throttle is open to full load, the manifold absolute pressure (MAP) rises, pressing down on the diaphragm. This lowers the valve

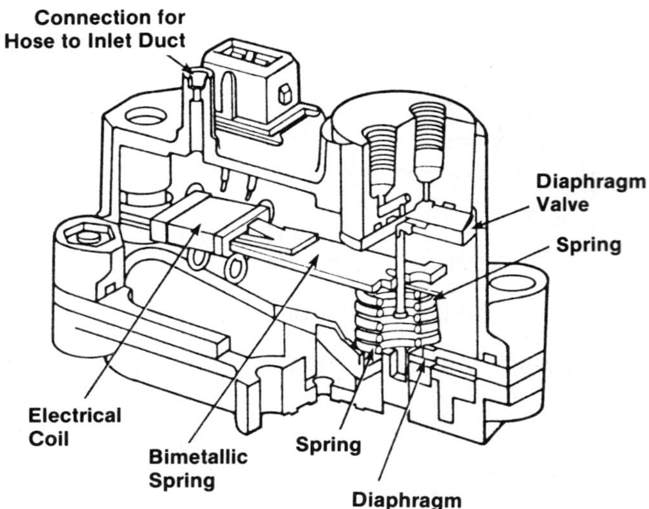

FIGURE 13-32 Some control pressure regulators are connected to the intake manifold vacuum to allow mixture enrichment during full throttle operation.

diaphragm, bleeding off the control pressure and enriching the fuel mixture for full-load operation.

The control pressure regulator can also compensate for thinner air at high altitude. It can also enrich the mixture when manifold absolute pressure increases because of boost from a turbocharger or supercharger.

Fuel Injectors

Each fuel injector (Figure 13-33) is pressed into its manifold opening and sealed with an O-ring. The internal fuel filter insures that clean fuel reaches the injector opening.

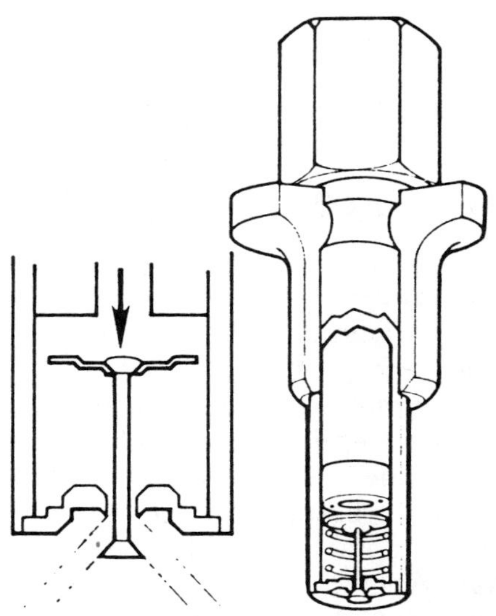

FIGURE 13-33 Design of a typical CIS fuel injector

Each injector continuously sprays finely atomized fuel into the intake port whenever the engine is running. When the valves are operating, the vibration or chatter of the valve needle can be heard. These vibrations atomize the fuel even when it is being injected in small quantities. This vibration also helps keep the injectors from clogging. Clogging is much more common on TBI or PI than on CIS or CIS-E. When the engine is turned off, the pin closes tightly under spring pressure. This traps fuel under pressure in the line, preventing vapor lock and insuring a quick start the next time.

OXYGEN CONTROL FEEDBACK SYSTEM

Continuous injection systems can be fitted with an oxygen sensor (sometimes called lambda sensor) for feedback control (Figure 13-24). The sensor is mounted in the exhaust manifold so it will heat up rapidly when the engine is started.

Signals from the oxygen sensor are sent to the oxygen control unit. The ECU modifies the fuel flow in the mixture control unit so the engine operates on the proper ratio. The changing exhaust gas affects the oxygen sensor and it sends a signal in a loop through the mixture control unit to the engine. This closed loop operation is shown in Figure 13-34.

The mixture control unit for system operation is shown in Figure 13-35. The lower chamber of each differential pressure is provided with a bleed passage to the oxygen control valve.

The oxygen control valve (sometimes called a timing or frequency valve) operates on signals from the oxygen control unit. It opens and closes to allow more or less fuel to return to the tank through the fuel return. This is called dwell time. By reducing the pressure in the lower part of the differential pressure valve, fuel flow to the injector can be increased, enriching the mixture. Shortening the time that the oxygen control valve is open will increase the pressure beneath the differential pressure valve diaphragm. This will lessen the amount of fuel injected, leaning the mixture.

Normally, the oxygen control valve is open about 50 percent of the time. Based on a series of signals from the sensor from oxygen-rich to oxygen-lean, the control valve continually cycles from being open about 40 percent of the time to being open about 60 percent of the time averaging about 50 percent.

The oxygen control unit switches to open loop during conditions when the oxygen sensor is cold or

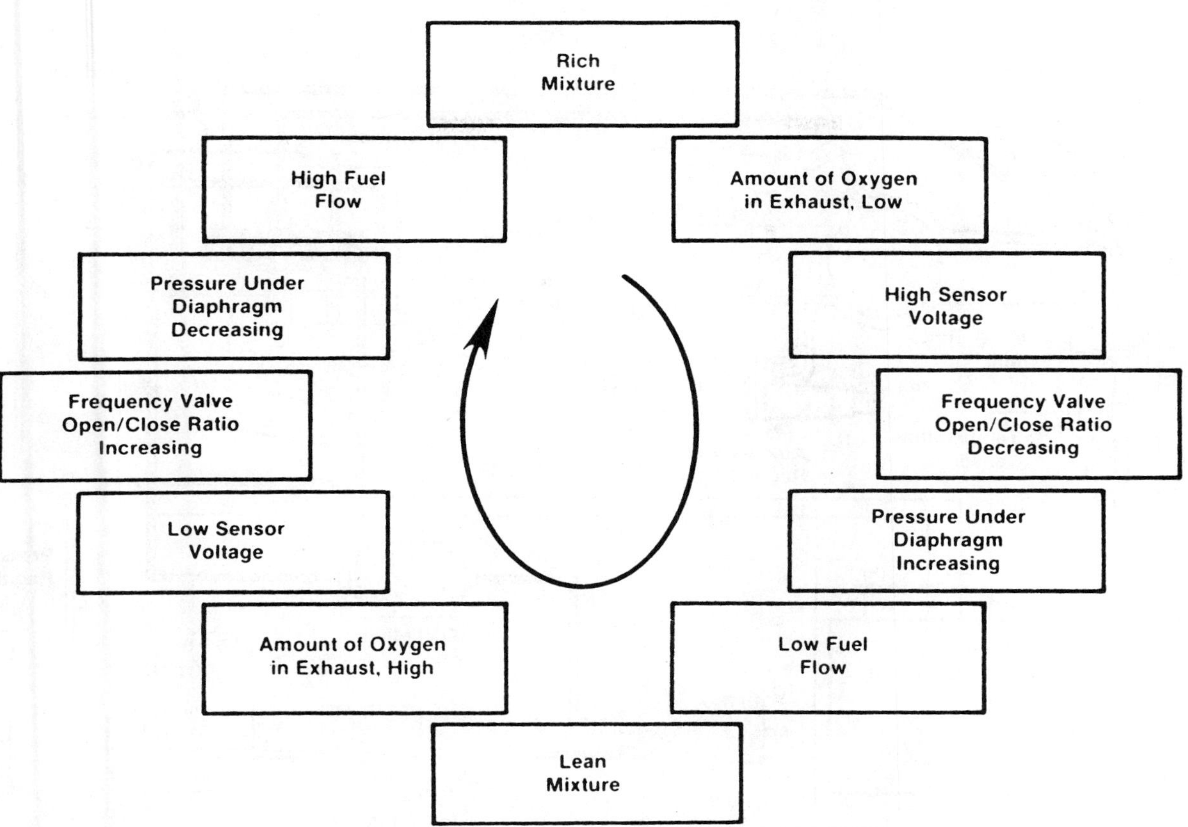

FIGURE 13-34 Closed loop operation refers to the way various components respond to each other to maintain optimum air/fuel proportioning.

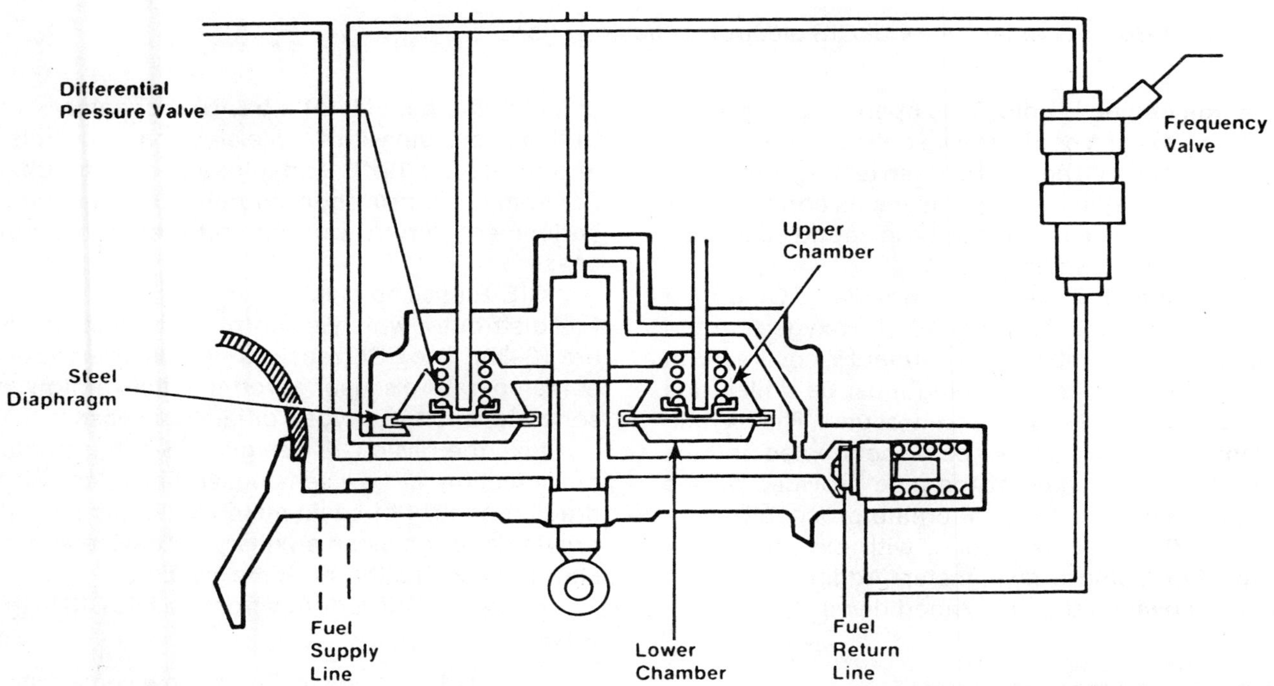

FIGURE 13-35 The frequency valve lowers the pressure in the lower portion of the pressure regulator valves to alter the fuel quantity to the injectors in response to the oxygen sensor.

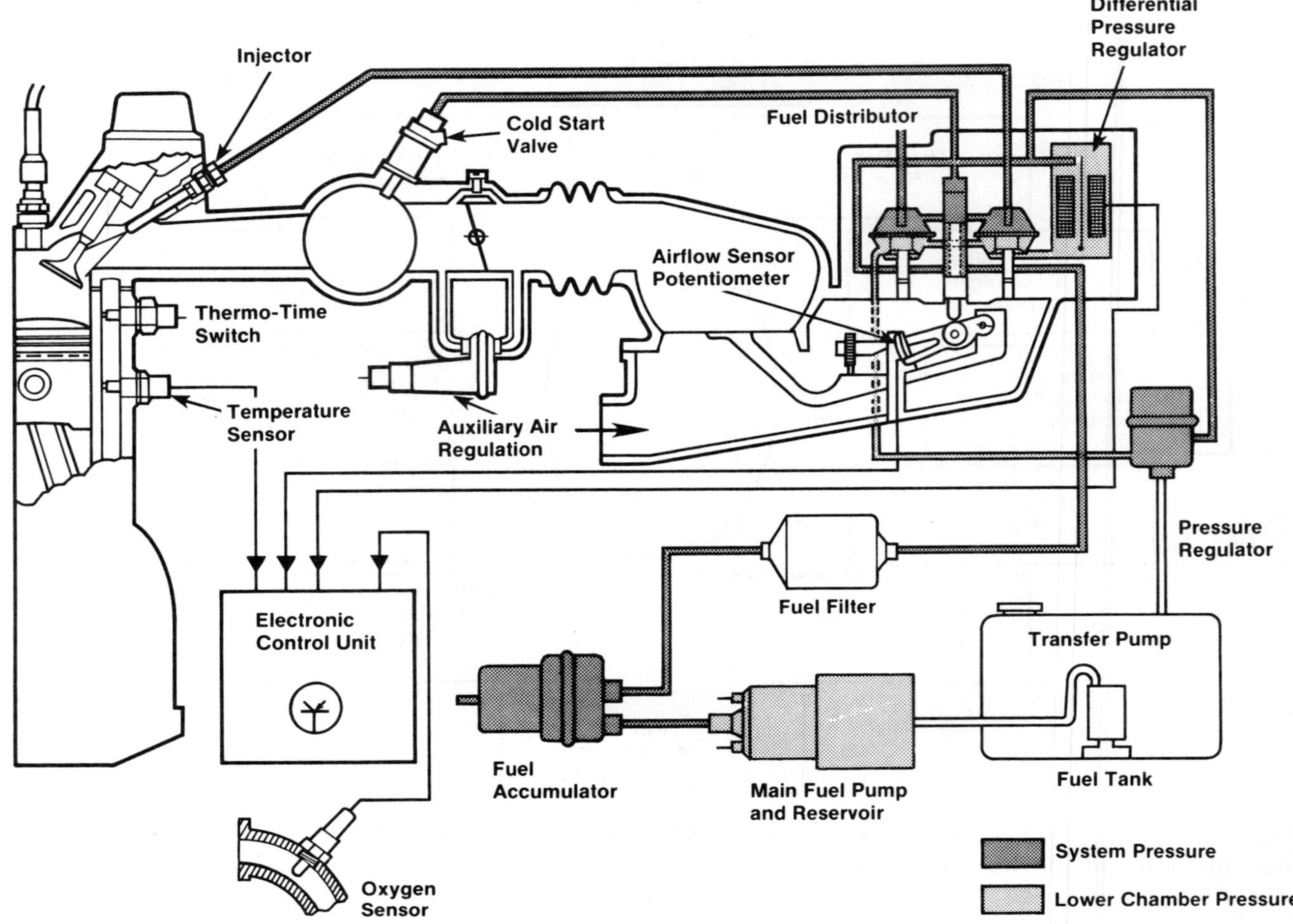

FIGURE 13-36 Schematic view of CIS electronic fuel injection

when the engine is cold. This open-loop operation holds the oxygen control valve open a fixed amount of time (around 60 percent). When testing the operation of the oxygen control unit and its control valve, it is possible to hear the change in sound caused by the change in open time.

Adjusting the carbon monoxide (CO) output level for those models equipped with oxygen control is different. If the mixture adjustment is covered with a tamper-proof plug, this plug must be drilled and removed from the mixture control unit. The oxygen (lambda) sensor wire can be disconnected and the exhaust sample taken at the pipe provided on the exhaust manifold. As an alternate procedure, mixture is adjusted closed loop, with oxygen sensor connected, using a dwellmeter. Additional details will be shown on the underhood decal.

CIS-E COMPONENTS

As mentioned earlier, CIS systems can also be equipped with certain electronic controls. They combine the benefits of a basic mechanically controlled FI system with simple electronic controls for enrichment, cutoff, and closed loop feedback. Economy is improved through the use of minimum enrichment during warmup and fuel cutoff during coast.

CIS-E uses an electrohydraulic actuator in the fuel distributor which is controlled by an ECU (Figure 13-36). The ECU receives signals from the coolant temperature sensor, throttle switch, airflow sensor plate, and the oxygen or lambda sensor.

Like the basic CIS design, CIS-E has continuous injection, airflow sensing, and mechanical (hydraulic) control of basic metering. It has cold start provisions such as an auxiliary air device and cold start valve with a thermo-time switch.

However, CIS-E differs from a basic CIS in three ways.

1. **Airflow Sensor.** The airflow sensor mechanism includes a potentiometer, which signals the position and movement of the plate for acceleration enrichment.

2. **System Pressure Regulator.** The system pressure regulator maintains a constant system pressure. It also relieves electric fuel pump pressure and maintains pressure in the system for easy restarts by controlling return fuel flow from the fuel distributor.
3. **Electrohydraulic Actuator.** The electrohydraulic actuator is an electromagnetic differential pressure regulator on the fuel distributor. It operates a plate valve.

System Operation

The system pressure regulator maintains a constant system of primary pressure. Constant system pressure is applied to the control plunger to counter the force of the airflow sensor mechanism. There is no control pressure and no control pressure regulator.

The electrohydraulic actuator provides enrichment reducing the pressure below the diaphragm of each differential pressure valve. This has the effect of increasing the pressure at each metering slit. In turn, this increases the amount of fuel delivered. For fuel cutoff during coasting or for rpm limitation, the electrohydraulic action is reversed. A decrease in the pressure at the metering slits cuts off delivery of fuel.

SERVICING FUEL INJECTION SYSTEMS

Troubleshooting fuel injection systems requires a systematic step-by-step test procedure. With so many interrelated components and sensors controlling fuel injection performance, a hit-or-miss approach to diagnosing problems can quickly become frustrating, time-consuming, and costly.

Most fuel injection systems are integrated into engine control systems. The self-test modes of these systems are designed to help in engine diagnosis. Unfortunately, when a problem upsets the smooth operation of the engine, many mechanics automatically assume that the computer (ECU) is at fault. It seems that the presence of a computer frequently short circuits the technician's good judgment and years of experience, which usually results in an incorrect diagnosis and a damaged reputation.

In the vast majority of cases, complaints about driveability, performance, fuel mileage, roughness, or hard starting or no-starting are due to something other than the computer itself. (Although many problems are caused by sensor malfunctions that can be traced using the self-test mode.)

But even before condemning sensors as bad, remember that weak or poorly operating engine components can often affect sensor readings and result in poor performance. For example, a sloppy timing chain or bad rings or valves will reduce vacuum and cylinder pressure, resulting in a lower exhaust temperature. This can affect the operation of a perfectly good oxygen or lambda sensor, which must heat up to approximately 600 degrees Fahrenheit before functioning in its closed loop mode. A new timing chain can raise tailpipe temperature 125 degrees Fahrenheit and ensure optimum closed loop performance.

Intake manifold leaks are another example of a problem that causes a sensor, the MAP sensor in this case, to adjust engine operation to less than ideal conditions.

One of the basic rules of electronic fuel injection servicing is as follows: You cannot adjust EFI to match the engine, you have to make the engine match EFI. In other words, make sure the rest of the engine is sound before condemning the fuel injection and engine control components.

PRELIMINARY CHECKS

The best way to approach a problem on a vehicle with electronic fuel injection is to treat it as though it had no electronic controls at all. As the above examples illustrate, any engine is susceptible to problems that are unrelated to the control system itself. And unless all engine support systems are hooked up and operating correctly, the control system will not operate as designed.

Before proceeding with specific fuel injection checks and electronic control testing, be certain of the following:

- The battery is in good condition, fully charged, with clean terminals and connections.
- The charging and starting systems are operating properly.
- All fuses and fusible links are intact.
- All wiring harnesses are properly routed with connections free of corrosion and tightly attached.
- All vacuum lines are in sound condition, properly routed, and tightly attached. For example, a dead miss in one cylinder can be caused by a leaking vacuum hose. The extra air leans out the air/fuel mixture to the point where it cannot burn. Also check for plugged lines.
- The PCV system is working properly and maintaining a sealed crankcase.

- All emission control systems are in place, hooked up and operating properly. For example, a frequent cause of hesitation is a faulty heated air intake system, and the most common reason for detonation or pinging is an inoperative EGR setup.
- The level and condition of the coolant/antifreeze is good and the thermostat is opening at the proper temperature.
- The secondary spark delivery components are in good shape with no signs of crossfiring, carbon tracking, corrosion, or wear.
- The base timing and idle speed are set to specs.
- The engine is in good mechanical condition.
- The gasoline in the tank is of good quality and has not been substantially cut with alcohol or contaminated with water. Testing the specific gravity of the fuel is a good method of ensuring that poor fuel has not fouled the injection system. Specific gravity of gasoline should be at least 0.735.

FUEL INJECTION AND CONTROL SYSTEM COMPONENT CHECKS

In any FI system three things must occur for the system to operate.

1. An adequate air supply must be supplied for the air/fuel mix.
2. A pressurized fuel supply must be delivered to properly operating injectors.
3. The injectors must receive a trigger signal from the control computer.

If all of the preliminary checks listed above do not reveal a problem, proceed to test the electronic control system and fuel injection components.

Some older control systems require involved test procedures and special test equipment, but most newer designs have a self-test program designed to help diagnose the problem. These self-tests perform a number of checks on components within the system. Input sensors, output devices, wiring harnesses, and even the electronic control computer itself may be among the items tested.

The results of the testing are converted into two-digit trouble codes (Figure 13–37) that the technician may read using special test equipment, a test light, or an analog voltmeter. The meaning of trouble codes vary from manufacturer to manufacturer, year to year, and model to model, so it is important to have the appropriate shop manuals.

Always remember that trouble codes only indicate the particular circuit in which a problem has been detected; they do not pinpoint individual components. So if a code indicates a defective lambda or oxygen sensor, the problem could be the sensor itself, the wiring to it, or its connector. Trouble codes are not a signal to replace components. They signal that a more thorough diagnosis is needed in that area.

More information on troubleshooting using electronic control systems can be found in Chapter 9. The following sections outline general troubleshooting procedures for the most popular EFI and CIS designs in use today. Port and throttle body trouble-

Code	Circuit Affected or Possible Cause
12	No distributor reference pulses to the ECM. This code is not stored in memory and will flash only while the fault is present. Normal code with ignition on, engine not running.
13	Oxygen Sensor Circuit. The engine must run up to 4 minutes at part-throttle under road load before this code will set.
14	Shorted coolant sensor circuit. The engine must run 2 minutes before this code will set.
15	Open coolant sensor circuit. The engine must run 5 minutes before this code will set.
21	Throttle Position Sensor (TPS) circuit voltage high (open circuit or misadjusted TPS). The engine must run 10 seconds at specified curb idle speed before this code will set.
22	Throttle Position Sensor (TPS) circuit voltage low (grounded circuit or misadjusted TPS). Engine must run 20 seconds at specified curb idle speed to set code.
23	M/C solenoid circuit open or grounded.
24	Vehicle speed sensor (VSS) circuit. The vehicle must operate up to 2 minutes at road speed before this code will set.
32	Barometric pressure sensor (BARO) circuit low.

FIGURE 13–37 Example of ECU trouble codes and their meaning

TABLE 13-1: PORT TYPE SYSTEM TROUBLESHOOTING

Problem	Cause	Remedy
Preliminary checks		Check fuel system for fuel leaks
		Check battery state of charge
		Check all wiring and connections
		Check cooling system level
		Check ignition system
		Check air cleaner and preheat system
		Check fuel system pressure
		Check fuel lines for restrictions
		Check vacuum hoses for leaks and restrictions
Hard start—cold or rough idle—cold	CTS	Check coolant level or replace sensor
	Fuel pressure bleed down	Check for fuel leak or defective fuel pump
	Cold start injector	Check cold start injector, service or replace as required
	Leaking manifold gasket or base gasket	Replace defective gasket
	ACT/MAT Sensor	Replace defective ACT/MAT sensor
	Wrong PCV Valve	Replace PCV valve
	Warm-up regulator	Replace warm-up regulator
	Injector	Check injectors for variation in spray pattern, clean or replace injectors as required
	Mass air flow sensor	Check air flow meter, fuel pump contacts
	Pressure regulator	Check pressure regulator for setting and bleed down
Hesitation or surging—hot or cold	CTS	Check coolant level or replace sensor
	Low fuel system pressure	Check fuel filter and fuel pump, service or replace as required.
	Restricted air intake system	Check air cleaner and preheat system, service or replace as required
	TPS defective or not adjusted correctly	Check TPS, adjust or replace as required
	Mass air flow sensor	Check air flow meter, fuel pump contacts
	ACT/MAT sensor	Replace defective ACT/MAT sensor

TABLE 13-1: PORT TYPE SYSTEM TROUBLESHOOTING (CONTINUED)

Problem	Cause	Remedy
	Air leak in air intake system	Check gaskets, hoses and ducting, service or replace as required
	Defective oxygen sensor	Replace oxygen sensor
	Defective computer	Replace computer
Hard start—hot	Bleeding injector	Inspect injector for dripping, service or replace as required.
	Leaking intake manifold gasket or base gasket	Replace defective gasket
	MAP sensor	Check MAP sensor and vacuum hose, service or replace as required.
	Pressure regulator	Check pressure regulator for setting and bleed down, service or replace as required.
Rough idle—hot	MAP sensor	Check MAP sensor and vacuum hose, service or replace as required
	CTS	Check coolant level or replace sensor
	TPS	Check TPS, adjust or replace as required
	Injector	Check injector for variation in spray pattern, clean or replace injector as required
	Oxygen sensor	Replace oxygen sensor
	Defective computer	Replace computer
	ISC/IAC	Check idle speed control device, service or replace as required
Stalling	ISC/IAC	Check idle speed control device, service or replace as required
	TPS	Check TPS, adjust or replace as required
	MAP Sensor	Check MAP sensor and vacuum hose service or replace as required
Poor power	Dirty injector	Check injector spray pattern, clean or replace injector as required
	Fuel pump	Check fuel pump pressure, replace fuel pump
	Fuel pump pick up strainer	Check strainer, replace as required
	Fuel filter	Check fuel filter, replace as required
	Pressure regulator	Check pressure regulator for setting and bleed down, service or replace as required.

CTS—Coolant Temperature Sensor
MAT—Manifold Air Temperature Sensor
TPS—Throttle Position Sensor
MAP—Manifold Absolute Pressure Sensor
ACT—Air Charge Temperature Sensor
ISC—Idle Speed Control
IAC—Idle Air Control

TABLE 13-2: THROTTLE BODY FUEL INJECTION SYSTEM TROUBLESHOOTING

Problem	Cause	Remedy
Preliminary checks		Check fuel system for fuel leaks
		Check battery state of charge
		Check all wiring and connections
		Check cooling system level
		Check ignition system
		Check air cleaner and preheat system
		Check fuel system pressure
		Check fuel lines for restrictions
		Check vacuum hoses for leaks and restrictions
Hard start—cold or rough idle—cold	CTS	Check coolant level or replace sensor
	Fuel pressure bleed down	Check for fuel leak or defective fuel pump
	Leaking manifold gasket or base gasket	Replace defective gasket
	MAT Sensor	Replace defective MAT sensor
	Wrong PCV valve	Replace PCV valve
Stalling, hesitation, surging—hot or cold	CTS	Check coolant level or replace sensor
	Low fuel system pressure	Check fuel filter and fuel pump, service or replace as required
	Restricted air intake system	Check air cleaner and preheat system, service or replace as required.
	TPS defective or not adjusted correctly	Check TPS, adjust or replace as required
Hard start—hot	Bleeding injector	Inspect injector for dripping, service or replace as required
	Leaking intake manifold gasket or base gasket	Replace defective gasket
	MAP sensor	Check MAP sensor and vacuum hose, service or replace as required
Rough idle—hot	MAP sensor	Check MAP sensor and vacuum hose, service or replace as required.
	CTS	Check coolant level or replace sensor
	TPS	Check TPS, adjust or replace as required
Stalling	ISC/IAC	Check idle speed control device, service or replace as required
	TPS	Check TPS, adjust or replace as required

TABLE 13-2: THROTTLE BODY FUEL INJECTION SYSTEM TROUBLESHOOTING (CONTINUED)

Problem	Cause	Remedy
Poor power	Dirty injector	Check injector spray pattern, clean or replace injector as required
	Fuel pump	Check fuel pump pressure, replace fuel pump
	Fuel pump pick up strainer	Check strainer, replace as required.
	Fuel filter	Check fuel filter, replace as required.

CTS—Coolant Temperature Sensor
MAT—Manifold Air Temperature Sensor
TPS—Throttle Position Sensor
ISC—Idle Speed Control
IAC—Idle Air Control
MAP—Manifold Absolute Pressure Sensor
ACT—Air Charge Temperature Sensor

shooting is summarized in Tables 13-1 and 13-2. A listing of common CIS problems and their solutions begins on page 387.

AIR SYSTEM CHECKS

In an injection system (particularly designs that rely on airflow meters or mass airflow sensors), all the air entering the engine must be accounted for by the air measuring device. If it is not, the air/fuel ratio will become overly lean resulting in a performance complaint that resembles a vacuum leak on carbureted vehicles. For this reason, cracks or tears in the plumbing between the airflow sensor and throttle body are potential air leak sources that can affect the air/fuel ratio.

During a visual inspection of the air control system, pay close attention to these areas, looking for cracked or deteriorated ductwork. Also make sure all induction hose clamps are tight and properly sealed. Look for possible air leaks into the crankcase, for example, dipstick tube and oil filter cap. Any extra air entering the intake manifold through the PCV system will not be measured either and can upset the delicately balanced air/fuel mixture at idle (Figure 13-38). This is especially true on turbo-equipped vehicles and many mass airflow systems where the amount of air entering an unsealed oil dipstick tube can affect speed by as much as 100 rpm or more.

When looking for the cause of a performance complaint that relates to poor fuel economy, erratic performance/hesitation or hard starting, make the following checks to determine if the airflow sensor (Figure 13-39) (all types except CIS) is at fault. Start by removing the air intake duct from the airflow sensor to gain access to the sensor flap. Check for binding, sticking, or scraping by rotating the sensor flap (evenly and carefully) through its operating range. It should move freely, make no noise and feel smooth. Next, turn the ignition on (do not start the engine) and move the flap toward the open position. The electric fuel pump should come on as the flap is opened. If it does not, turn off the ignition, remove the sensor harness, and check for specific resistance values with an ohmmeter at each of the sensor's terminals.

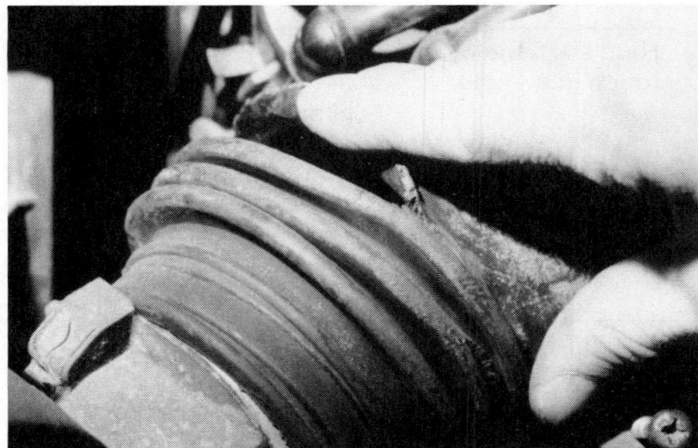

FIGURE 13-38 Any air leaks allowing air to enter the cylinder without being accounted for by the airflow sensor will result in a lean mix and poor performance.

On some models it is possible to check the resistance values of the potentiometer by moving the air flap, but in either case a shop manual is needed to identify the various terminals and to look up resistance specifications.

On systems that use a mass airflow meter or manifold pressure sensor to measure airflow, other than checking for good electrical connections, there are no actual physical checks. However, a hand-held scan tool can be plugged into the diagnostic connector on some models to check for proper voltage values.

FUEL SYSTEM CHECKS

If the air control system is in working order, move on to the fuel delivery system. It is important to always remember that fuel injection systems operate at higher fuel pressure levels than a carbureted system. This fuel pressure must be relieved before any fuel line connections can be broken. Spraying gasoline (under a pressure of 35 psi or more) on a hot engine creates a real hazard when dealing with a liquid that has a flash point of minus 45 degrees Fahrenheit.

In place of the "old-shop-towel-around-the-fuel line" method of absorbing pressurized fuel when opening connections in the fuel system, Chrysler, Ford, and GM recommend the following pressure relief procedures:

- *Chrysler.* After removing the fuel cap to relieve pressure in the tank (which, by the way, applies to all systems, makes, and models), remove the electrical connection from an injector and, using two jumper wires, ground one injector wire and connect the other to battery voltage. This will energize the injector allowing pressure to bleed off. However, to avoid permanently damaging the injector coil (by overheating), it is important not to energize any one injector for more than 10 seconds.
- *GM.* The recommended procedure requires removing the fuel pump fuse located in the underdash fuse panel, starting the vehicle, and allowing it to run until it dies. After it dies, crank it over for a few more seconds and, if it does not start, the lines are empty and it is safe to proceed. Afterward, do not forget to replace the fuse.
- *Ford.* Connect a fuel pressure gauge to the Schrader valve service fitting located in the fuel rail on Ford's EFI systems and on the throttle body on their CFI systems. Use the bleed-off hose that is designed as a part of most fuel injection pressure gauges to relieve fuel pressure into an appropriate container. This procedure applies to all fuel-injected systems that provide a service fitting for gauge hookup. If planning to connect the gauge for additional testing anyway, this procedure could save time.

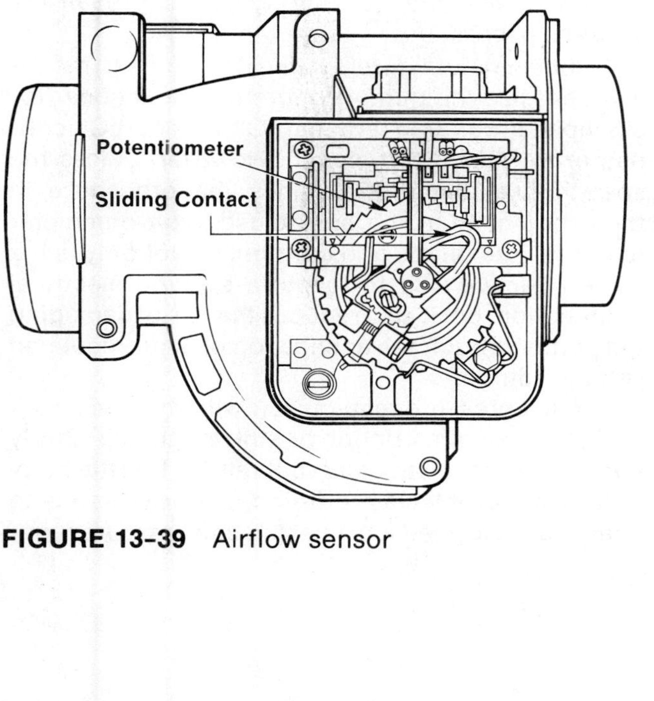

FIGURE 13-39 Airflow sensor

If none of the above procedures apply to the particular vehicle, a "generic" alternative is to apply 20 to 25 inches of vacuum to the externally mounted fuel pressure regulator (with a hand vacuum pump connected to the manifold control line of the regulator), which will bleed fuel pressure back into the tank (Figure 13-40). If the vehicle does not have an exter-

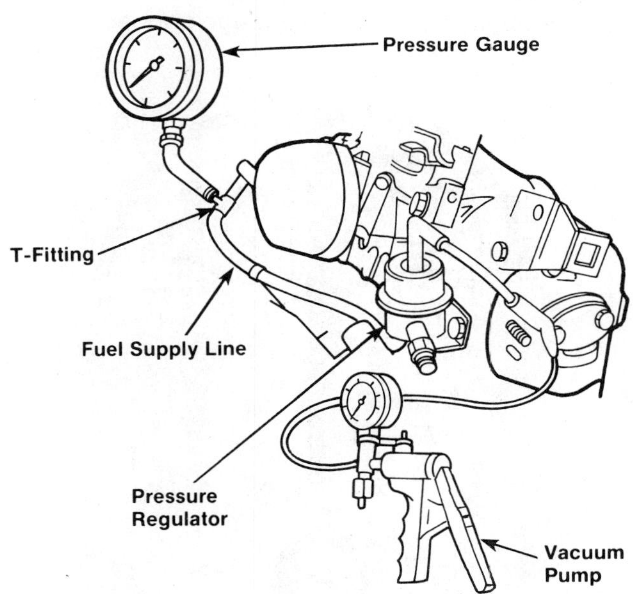

FIGURE 13-40 With an externally mounted fuel pressure regulator it is possible to bleed off system pressure into the fuel tank using a hand vacuum pump.

nally mounted regulator, or a service valve or injector to energize, use the shop towel to catch excess fuel, while slowly breaking a fuel line connection. This method should only be used when all other possibilities have been tried.

Fuel Delivery

When dealing with an alleged fuel complaint that is preventing the vehicle from starting, the first step (after spark, compression, etc., have been verified) is to determine if fuel is reaching the cylinders (assuming of course there is fuel in the tank and that it is gasoline). Checking for fuel delivery is simple on throttle body systems. Remove the air cleaner, crank the engine, and watch the injector for signs of a spray pattern. If a better view of the injector's operation is required, an ordinary strobe light does a great job of highlighting the spray pattern.

Checking for fuel on a port-injected system, however, is more of a challenge since the injector's location precludes a visual examination. Fortunately, there are alternatives to getting the needed information without actually seeing what is going on inside.

Pressure Gauges. The fuel system is checked using a fuel pressure gauge and various system fittings (Figure 13-41).

Attach the gauge by connecting it to the service fitting (usually a Schrader type) that is located in the fuel rail or, in lieu of a service fitting, use one of the adapters or T-fittings in the gauge kit. If the system must be opened to make the gauge connection, remember to release any residual pressure first. Also, make sure the connection is made upstream from the pressure regulator or pressure readings will be inaccurate.

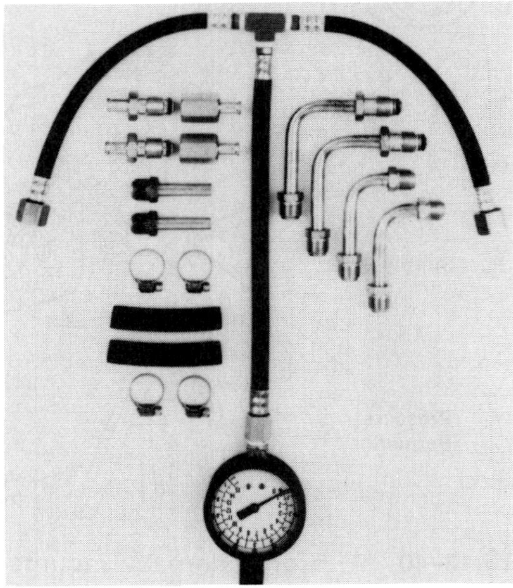

FIGURE 13-41 Typical fuel pressure gauge kit, complete with adapters

Once the pressure gauge is hooked up, bleed the air out of the gauge line, switch the ignition on, and listen for the operation of the fuel pump. Most injection systems have a 1 or 2 second priming period to pressurize the system for starting so there should be a gauge reading. If there is no reading but the service manual indicates there should be, crank the engine. If pressure builds while the engine is cranking, a problem exists in the portion of the pump circuit that controls the priming function.

If a gauge reading is not present with the key in any position, verify the operation of the fuel pump. Start by checking the fuel pump fuse. If the fuse is not blown, turn the ignition switch on and check for operating voltage at the pump connector with a voltmeter or test light (keep in mind that voltage will only appear for 1 or 2 seconds with the ignition in the "on" position, so it might be necessary to crank the engine during this test). If there is no indication of voltage at the connector, use a wiring diagram to help search for damaged wires, and loose or dirty connections (pay close attention to all grounds), and make certain control devices such as relays or safety cutoff switches are working (Figure 13-42).

If there is a normal voltage reading at the pump connector, connect an ohmmeter to the pump wiring harness and see if there is continuity there. If the answer is no, make the one more check for continuity at the pump terminals. Lack of continuity at the pump terminals means the pump is defunct and will have to be replaced. With continuity at the pump and voltage at the connector, the problem has to lie in the wiring harness between the pump and pump connector. All that remains is to locate and repair the wires as necessary.

Some manufacturers make it possible to manually energize the pump from remote locations (for example, grounding terminal "G" on the ALCL connector for GM C-3 systems). However, unless the specific manufacturer outlines the procedure in their shop manual, it is best to use a volt-ohmmeter to check for voltage and continuity. Not only is it a more effective way to locate a specific electrical malfunction, but it also reduces the risk of damaging the pump through the application of an unregulated voltage source.

If fuel pressure remains low with the pump running, look for a restriction or kink in the fuel supply line between the tank and fuel rail (or throttle body unit). If no problems are found, the next step is to check for a clogged fuel filter(s), bad pressure regu-

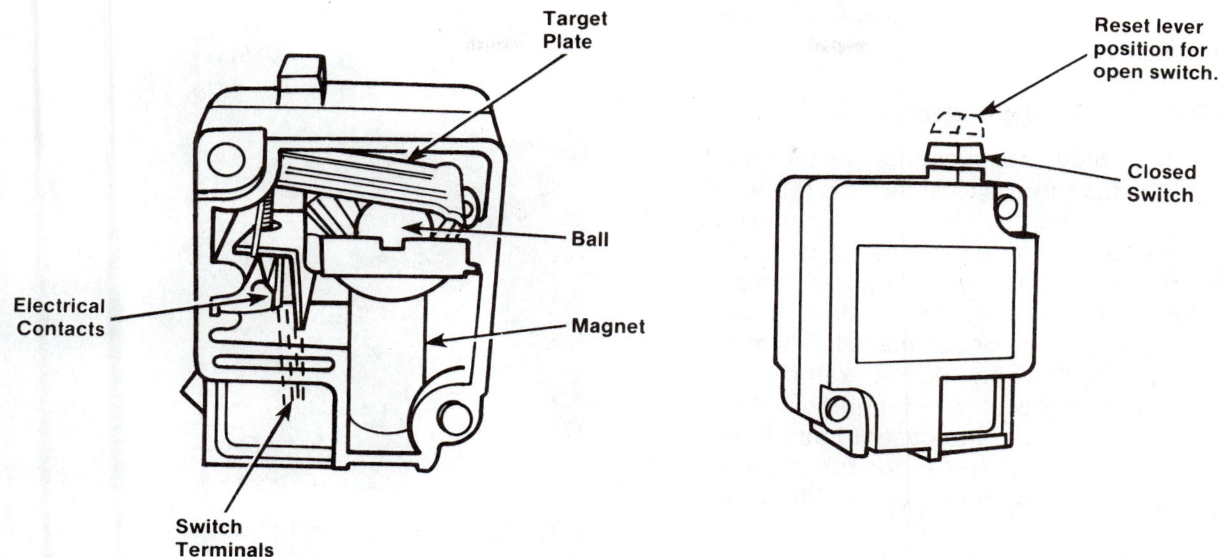

FIGURE 13-42 Fuel pumps are designed with safety cut-off switches. An overly sensitive switch can trip unexpectedly and become the source of no fuel delivery. On this Ford design, be sure the reset button is pushed down.

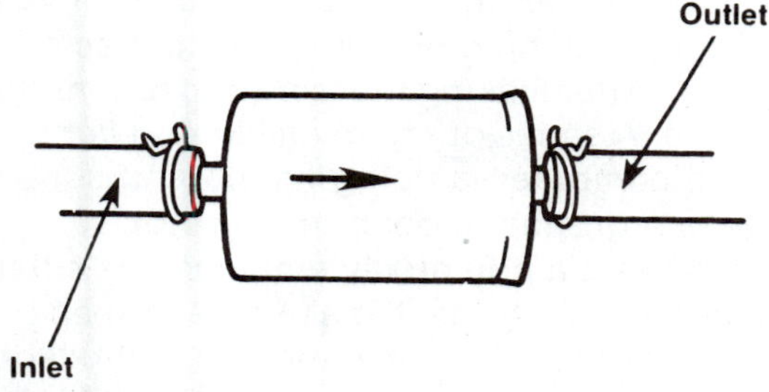

FIGURE 13-43 Always install in-line fuel filters so the fuel flows through them in the direction indicated by the arrow on the filter.

lator, and/or an improperly operating fuel pump (supplying an inadequate fuel volume). There is one final item to consider—an injector that is stuck open. This problem is covered in the next section, so for now assume the injectors are sealing properly.

To determine whether or not the in-line fuel filter is at fault, install the pressure gauge between the fuel filter hose and fuel line. Energize the pump and note the pressure. A correct pressure reading indicates the need for a new filter (Figure 13-43). However, if the pressure is still low, try slowly pinching off the fuel return hose while the pump is running. If pressure increases, suspect a faulty pressure regulator. No pressure change means the problem is a clogged in-tank fuel filter or a faulty fuel pump unit.

Before pulling the tank to replace the pump and/or in-tank filter, confirm a bad regulator by applying 15 to 20 inches of vacuum to the regulator diaphragm (on manifold vacuum-controlled units only). A vacuum reading that does not hold indicates a leaking regulator.

INJECTOR CHECKS

Up to this point, if the vehicle seems to have all the necessary ingredients but still will not run, about all that is left on the fuel side to cause problems are the injectors themselves.

Although fuel injection systems are often associated with terms like performance, power and economy, the fuel injector itself is nothing more than a solenoid-actuated fuel valve. Its operation is quite basic in that as long as it is held open and the fuel pressure remains steady, it will keep on delivering fuel until it is told to stop.

Because all fuel injectors operate in a similar manner, fuel injector problems will tend to exhibit the same failure characteristics. The main difference is that in a TBI design, generally all cylinders will suffer if the injector(s) malfunctions, whereas in port systems the loss of one injector is not as crucial.

An injector that does not open will cause hard starts on port-type systems and an obvious no-start on single-point TBI designs. An injector that is stuck partially open will cause loss of fuel pressure (most noticeably after the engine is stopped and restarted within a short time period) and flooding due to raw fuel dribbling into the engine. In addition to a rich running engine, a leaking injector will also cause the engine to "diesel" or "run on" when the ignition is turned off.

Checking Voltage Signals

When an injector is suspected as the cause of a problem, the first step is to determine if the injector is receiving a signal (from the control unit) to fire. Fortunately, determining if the injector is receiving a voltage signal is easy and requires simple test equipment. Unfortunately, the location of the injector's electrical connector can make this simple voltage check somewhat difficult. For example, on some recent Chevrolet 2.8 liter V-6 engines, the cover must be removed from the cast aluminum plenum chamber that is mounted over the top of the engine before the injector can be accessed (Figure 13–44).

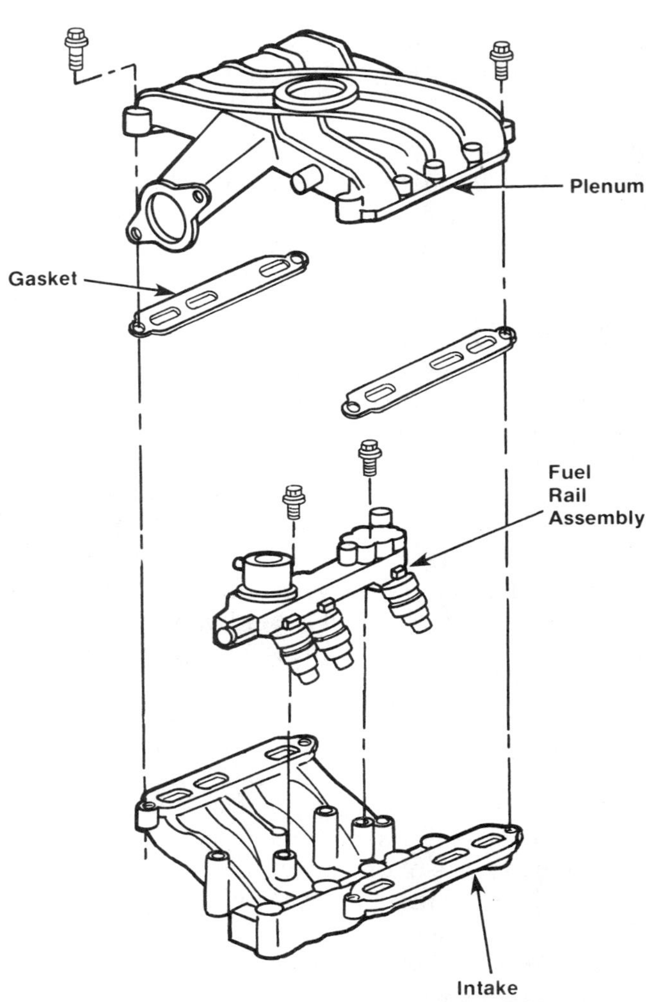

FIGURE 13–44 To access the injectors on the Chevrolet 2.8L V-6, the air plenum must be removed. Remember to torque screws to prevent air leaks.

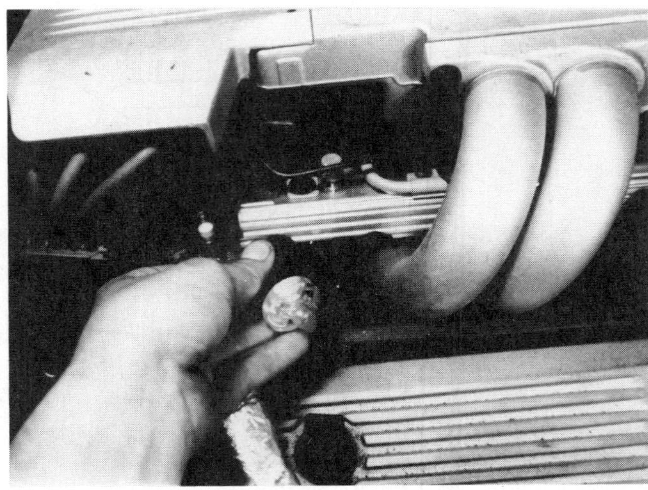

FIGURE 13–45 A "noid" light is a handy testing device. When plugging into an injector connector receiving an electrical signal, it will pulse rapidly.

Once the injector's electrical connector has been removed, check for voltage at the injector using an ordinary test light or one of those convenient "noid" lights that plug right into the connector (Figure 13–45). After making the test connections, crank the engine. A series of rapidly flickering lights indicates the computer is doing its job and supplying voltage or a ground to open the injector.

To prevent a potentially embarassing situation when performing this test, make sure to keep off the accelerator pedal. On some models, fully depressing the accelerator pedal activates the "clear flood" mode. In "clear flood," the voltage signal to the injectors is automatically cut off. Technicians unaware of this will waste time tracing a phantom problem.

If sufficient voltage is present after checking each injector, check the electrical integrity of the injectors themselves. Use an ohmmeter to check each injector winding for shorts, opens, or excessive resistance. Compare resistance readings to the specifications found in the service manual.

Injector Balance Test

If the injectors are electrically sound, perform an injector pressure balance test. This diagnostic test will help isolate a clogged or dirty injector that may be opening and closing properly, but not allowing adequate fuel to pass through.

Performing an injector balance test is quite simple. All that is needed is to energize each injector while using the fuel pressure gauge to monitor the fuel pressure drop. However, to ensure accurate results and prevent damage to the injector, an elec-

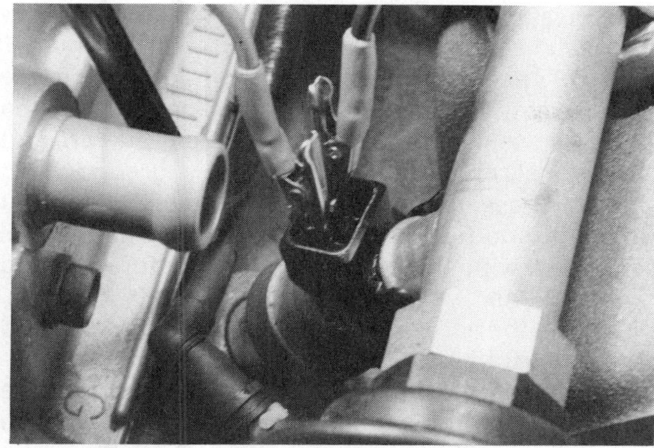

FIGURE 13-46 To perform an injector balance test, connect an electronic injector triggering device to the fuel injector terminals.

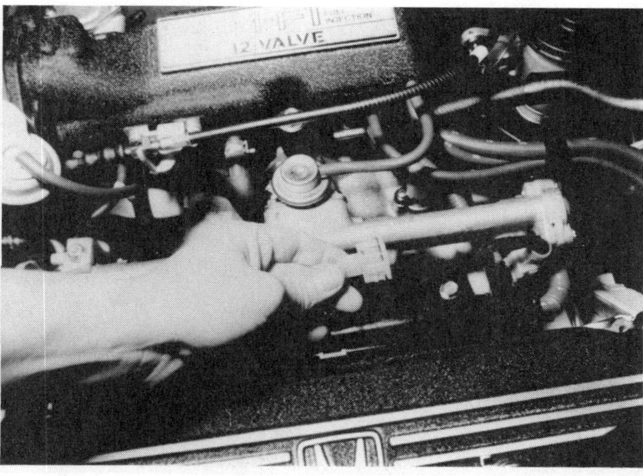

FIGURE 13-47 To perform a power balance test, disconnect the injector leads one at a time and watch for rpm drop.

tronic injector tester is needed to energize the injectors. The tester is designed to safely pulse each injector for a controlled length of time—something that CANNOT BE DONE by applying 12 volts to the injector terminal.

With the pressure gauge connected, pull the connectors off each injector (to prevent oil dilution and flooding), and attach the tester's lead to one of the injectors (Figure 13-46). Next, attach the tester's power leads to the vehicle's battery. Turn on the ignition until pressure builds to the maximum, then switch off the key. See that the reading holds. If the tester has various pulse settings (one pulse at 500 milliseconds, 50 pulses at 10 milliseconds, or 100 pulses at 5 milliseconds), select one, watch the gauge, hit the tester's activation button, and record the fuel pressure reading after the needle stops pulsing.

The difference between the maximum and minimum reading is obviously the pressure drop. Next, do the other injectors, running the pressure up again for each, and compare drop readings. Ideally, each injector should drop the same amount when opened. A variation of 1.5 to 2 psi (10 kPa) or more is cause for concern. If there is no pressure drop or a low pressure drop, suspect a plugged injector. A higher than average pressure drop is indicative of a rich condition. If there are inconsistent readings, the nonconforming injector(s) will either have to be cleaned or replaced (see page 384).

Injector Power Balance Test

The injector pressure balance test described above is a good method of checking injectors on vehicles with no-start problems. When the vehicle will run, several other tests are possible. One such test, the injector power balance test, is an easy method of determining if an injector is causing "missing."

To perform this test, first hook up both a tachometer and fuel pressure gauge to the engine. Once all gauges are in place, start the engine and allow it to reach operating temperature. As soon as the idle stabilizes, unplug each injector (Figure 13-47), one at a time, and note the rpm drop and pressure gauge reading. To ensure accurate test results, on many electronic systems it may be necessary to disconnect some type of idle air control device to prevent the computer from trying to compensate for the unplugged injector.

After recording the rpm and fuel pressure drop, reconnect the injector, wait until the idle stabilizes, and move on to the next one, noting the rpm and pressure drop each time. If an injector does not have much effect on the way the engine runs, chances are it is either clogged or electrically defective. To avoid any guesswork, though, it is a good idea to back up the power balance test with an injector resistance check, pressure balance test, and/or "noid" light check.

ALTERNATE CHECKS

If the injector's electrical leads are difficult to access, an injector power balance test will be hard to perform. As an alternative, start the engine and use a technician's stethoscope to listen for correct injector operation (Figure 13-48). A good injector will make a rhythmic clicking sound as the solenoid is energized and de-energized several times each sec-

FIGURE 13–48 Using a technician's stethoscope to check injector operation

ond. If a clunk-clunk instead of a steady click-click is heard, chances are the problem injector has been found. Cleaning or replacement is in order. If a stethoscope is not handy, use a thin steel rod, wooden dowel, or fingers to feel for a steady "on/off" pulsing of the injector solenoid.

Another way to isolate an offending cylinder when injector access is limited is to perform the more traditional cylinder balance test and disable the spark plugs instead of the injectors (remember to bypass the idle air control). Following the same warm-up and idle stabilizing procedures discussed above, watch the rpm drop and note any change in idle quality as each plug is shorted. If a lazy or dead cylinder is located, concentrate efforts on the portion of the fuel or ignition system pertaining to that cylinder (assuming, of course, no mechanical problems are present).

CAUTION: Any time cylinders are shorted during a power balance test, make the readings as quickly as possible. Prolonged operation of a shorted cylinder will cause excessive amounts of unburned fuel to accumulate inside the catalytic converter and increase the risk of premature converter failure.

OSCILLOSCOPE CHECKS

An oscilloscope can be used to monitor the injector's pulse width and duty cycle when an injector-related problem is suspected. As covered earlier in the chapter, the pulse width is the time in milliseconds that the injector is energized. The duty cycle is the percentage of on-time to total cycle time.

To check the injector's firing voltage on the scope, a typical hookup involves connecting the scope's positive lead to the injector supply wire and the scope's negative lead to an engine ground. Even though these connections are considered typical, it is still a good idea to read the instruction manual provided with the test equipment before making connections.

With the scope set on the low voltage scale and the pattern adjusted to fill the screen, a "square"-shaped voltage signal should be present with the engine running or cranking. If the voltage pattern reads higher than normal, excessive resistance in the injector circuit is indicated. Conversely, a low-voltage trace indicates low-circuit resistance. If the pattern forms a continuous straight line, it means the injector is not functioning due to an open circuit somewhere in the injector's electrical circuit.

INJECTOR CLEANING

Since a single injector can cost up to several hundred dollars, arbitrarily replacing injectors when they are not functioning properly, especially on multiport systems, can be an expensive proposition. If injectors are electrically defective, replacement is the only alternative. However, if the vehicle is exhibiting rough idle, stalling or slow or uneven acceleration, the injectors may just be dirty and require a good cleaning.

In recent years, a combination of declining fuel quality and poor injector designs has led to a substantial increase in the frequency of clogged injectors (mainly on port-type). Fortunately, a number of aftermarket injector cleaning systems are available. In addition to restoring lost power and fuel economy, injector cleaning can also create a profitable new service opportunity.

Before covering the typical cleaning systems available and discussing how they are used, several cleaning precautions are in order. First, never soak an injector in cleaning solvent. Not only is this an ineffective way to clean injectors, but it will most likely destroy the injector in the process. Also, never use a wire brush, pipe cleaner, toothpick, or other cleaning utensil to unblock a plugged injector. The metering holes in injectors are eyed to precise tolerances. Scraping or reaming the opening will result in a clean injector that will no longer be an accurate fuel-metering device.

Always use an approved on-the-car cleaning system or injector cleaning bench to effectively clean clogged injectors. On-the-car cleaners are by far the most popular and practical method.

The basic premise of all injection cleaning systems is similar in that some type of cleaning chemical is run through the injector in an attempt to dissolve deposits that have formed on the injector's tip. The methods of applying the cleaner can range from single shot, premixed, pressurized containers to self-mix, self-pressurized chemical tanks resembling "bug sprayers." The premixed, pressurized container systems are fairly simple and straightforward to use since the technician does not need to mix, measure, or otherwise handle the cleaning agent. The easiest way to describe this cleaning method is to compare it to the charging of an A/C system with individual cans of R-12.

Other systems require the technician to assume the role of chemist and mix up a desired batch of cleaning solution for each application. The chemical solution then is placed in a holding container and pressurized by hand pump or shop air to a specified operating pressure.

The more advanced units feature electrically operated pumps neatly packaged in roll-around cabinets that are quite similar in design to an A/C charging station (Figure 13-49).

Regardless of the cleaning system used, the standard procedure for cleaning the injectors is as follows:

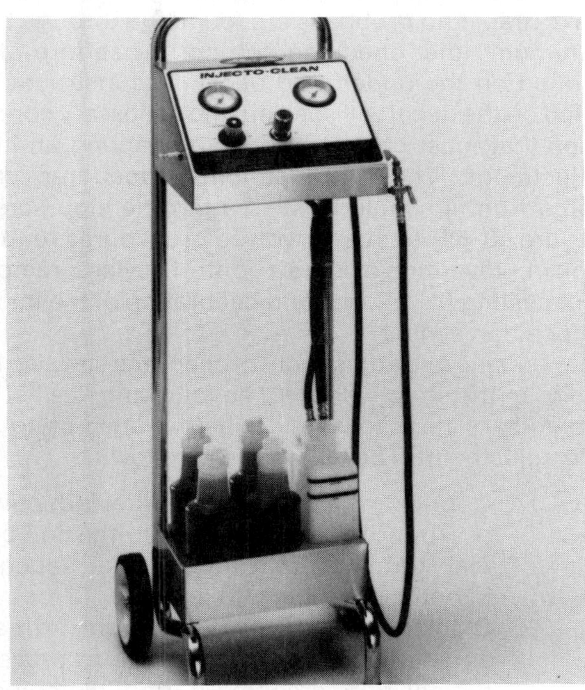

FIGURE 13-49 Typical fuel injection cleaning system for on-car cleaning

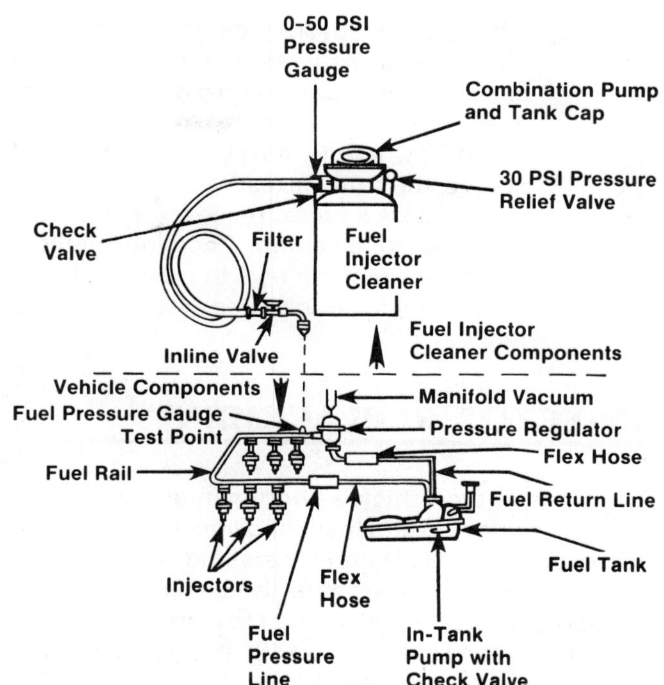

FIGURE 13-50 Typical setup for on-car injector cleaning

1. Attach the cleaner's service hose to the fuel rail following the same procedure used when hooking up the fuel pressure gauge.
2. Disable the fuel pump as per the car manufacturer's instructions (for example, pull fuel pump fuse, disconnect lead at pump, etc.), and clamp off the fuel pump return line at the flex connection to prevent the cleaner from seeping into the fuel tank (Figure 13-50). For an extra margin of safety, clamp off the inlet line also.
3. Before starting the engine, open the cleaner's control valve one-half turn or so to prime the injectors and then start the engine.
4. If available, set and adjust the cleaner's pressure gauge to the operating pressure of the injection system (generally between 25 to 45 psi) and let the engine idle at 2000 rpm for 10 to 15 minutes or until the cleaning mix has run out.
5. If the engine is still running, shut it off, remove the cleaning setup, and enable the fuel pump.
6. After removing the clamping devices from around the fuel lines, start the car and let it idle for 5 minutes or so to remove any leftover cleaner from the fuel lines.

7. On the more severely clogged cases, the idle improvement should be noticeable almost immediately. With more subtle performance improvements, an injector balance test will verify the cleaning results. Of course, the balance test will not mean anything unless a pressure measurement was taken before cleaning. Once the injectors are clean, recommend the use of an in-tank cleaning additive or a detergent-laced fuel.

INJECTOR REPLACEMENT

Consult the vehicle's shop manual for instructions on removing and installing injectors that need to be replaced. But before installing the new one, always check to make sure the sealing ring is in place (Figure 13–51). Also, prior to installation, lightly lubricate the sealing ring with engine oil or automatic transmission fluid (avoid using silicone grease, which tends to clog the injectors) to prevent seal distortion or damage.

If a locking, or retaining ring is used, make sure it is open and then, using a slight twisting/rocking motion, push downward to fully seat the injector and lock it in. Failure to follow this procedure could result in a massive vacuum leak or dangerous fuel spill. On models where the injectors are closely connected to, or part of the fuel rail, pay special attention to injector installation. When trying to seat multiple injectors at once, putting one in cocked or not fully seating the sealing ring is very easy to do.

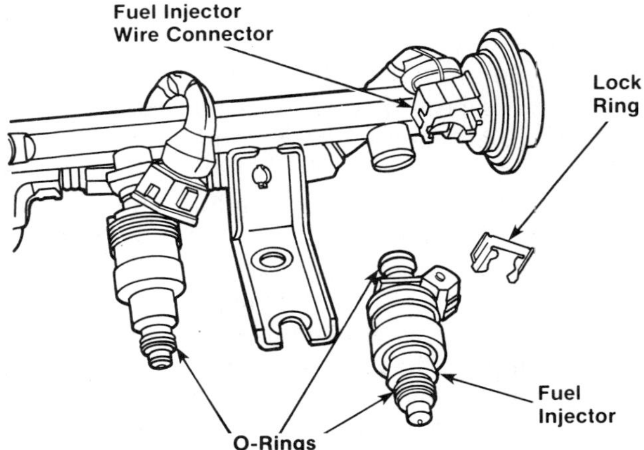

FIGURE 13–51 Before replacing the injector, inspect the sealing O-rings for any signs of damage and replace as needed. Be sure any retaining rings are open.

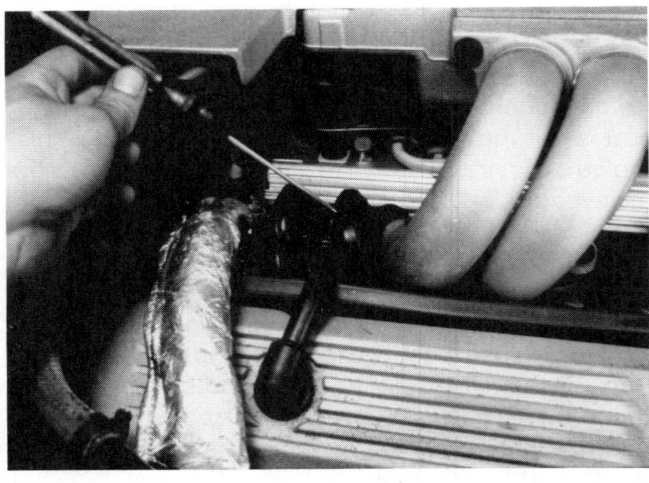

FIGURE 13–52 On many models, a tamper-proof idle stop screw plug must be removed before adjustments can be made.

IDLE ADJUSTMENT

In a fuel injection system, idle speed is regulated by controlling the amount of air that is allowed to bypass the airflow sensor or throttle plate(s). When presented with a car that tends to stall, especially when coming to a stop, or idles too fast, look for obvious problems like binding linkage and vacuum leaks first. If no problems are found, go through the minimum idle checking/setting procedure described on the underhood decal. The instructions listed on the decal will spell out the necessary conditions that must be met prior to attempting an idle adjustment. These adjustment procedures can range from a simple twist of a throttle stop screw (Figure 13–52) to more involved procedures requiring circumvention of idle air control devices, removal of casting plugs, and/or recalibration of the throttle position sensor.

Specific idle adjustment procedures can also be found in the shop manual. The following is a list of preliminary steps to consider before attempting an idle adjustment. They include the following:

1. Connecting a tachometer either inductively around an ignition wire, to the coil primary terminals or at the tach filter, depending on availability and access.
2. Blocking the drive wheels (front or rear) and setting the parking brake to prevent the car from creeping during idle adjustment, especially if the adjustment must be made with the transmission in gear.

3. Checking and adjusting base ignition timing before adjusting engine idle.
4. Making all adjustments with the engine at normal operating temperature to prevent cold-start enrichment and fast idle speed devices from influencing idle.
5. Following the car maker's specific procedures with respect to position of the air cleaner (on or off); procedure for disconnecting and plugging the vapor purge canister; and operation of the cooling fan, A/C or headlights. With respect to this last point, failure to compensate for the added load of high current-drawing items may result in a car that stalls when the accessories are in use.

The idle speed should always be adjusted before adjusting the air/fuel mixture. The mixture is adjusted by turning the mixture adjustment screw clockwise to decrease the bypass air, which enriches the mixture. This also increases the CO in the exhaust. Backing out the mixture screw increases the bypass air, leaning the mixture and reducing CO. Because some electronic fuel injection engines run so lean, the CO meter might not react to the mixture adjustment screw unless special enrichment procedures are followed. Instructions are usually given in the engine manual for the particular application.

CIS CHECKS AND TESTS

Servicing mechanically controlled continuous injection systems requires the same logical approach and preliminary checks used in troubleshooting throttle body and port EFI systems. CIS systems are normally very reliable, and should not be blamed for a problem until the preliminary ignition and engine system checks given earlier have been checked out.

The following list summarizes the major problems associated with CIS mechanically controlled fuel injection.

- Difficult cold starting may be due to a malfunction cold-start valve or thermo-time switch.
- Rough running or stalling during warm up is cause to suspect the auxiliary air regulator, system pressure, and the idle speed and mixture adjustments.
- Surging or missing when the engine is warm points to a problem in system pressure, warm control pressure, the fuel distributor and injectors, and the idle speed and mixture settings.
- Hard hot starting may be caused by a faulty hot start relay (if present) improper air sensor plate height, incorrect idle speed and mixture setting, or faulty leakdown/cut-off pressure.

The following service tips apply to most CIS vehicles in use today.

DIFFICULT COLD START

To test the thermo-time switch, it is necessary to first make sure that the coolant is below 35 degrees Celsius (95 degrees Fahrenheit).

1. Remove the coil high-tension wire from the distributor, then connect it to the ground.
2. Unplug the connector from the cold start valve, attach a test light to the connector's terminals, then crank the engine. The light should go on for several seconds, then go out. In cases where this does not happen, the thermo-time switch or its circuit is faulty.

   Continuing to operate the starter on a car with a hot-start relay will cause the light to blink on and off at regular intervals. Before it is possible to check the cold-start valve itself, it is necessary to make the fuel pump run constantly. On a typical model without air sensor wiring, remove the fuel pump relay and bridge the appropriate pair of relay board terminals with a fused jumper wire to energize the pump. (Refer to the vehicle shop manual for proper terminals.)
3. Unplug the cold-start valve's electrical connector, remove the valve from the intake manifold, place its nozzle end into a container, connect a jumper between one of the valve's terminals and ignition coil terminal, and a second jumper between the valve's other terminal and ground. Turn the ignition on and look for a steady, cone-shaped fuel spray pattern. Shut the ignition off, wipe the nozzle, and watch it for a minute. No drops should appear.

ROUGH RUNNING/STALLING

The auxiliary air regulator keeps rpm up while the engine is cold by providing a path for air to bypass the throttle plate. A heating coil connected to

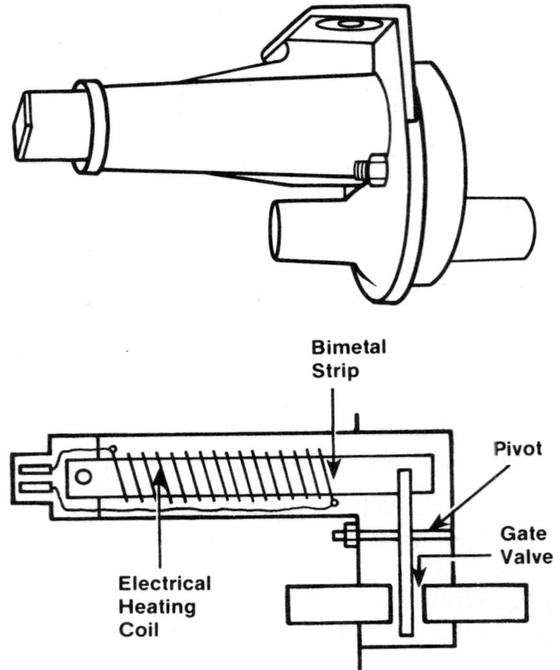

FIGURE 13-53 The auxiliary air regulator in a CIS design provides extra air during warmup.

the fuel pump and control pressure regulator circuit warms a bimetal strip, and the deflection of the strip gradually closes the regulator's gate valve, cutting off the flow of extra air (Figure 13-53).

To test it with the engine cold, unplug its electrical connector, start the engine and let it idle, then pinch either hose that is connected to the unit. Idle speed should decrease slightly. Plug the connector back in, and let the engine run until warm, then pinch either hose again. This time, there should be no change in idle rpm.

If the idle speed does slow down, remove the electrical connector and use a test light across the connector's terminals. If the light is on with the engine idling, the auxiliary air regulator is faulty.

FUEL PRESSURE PROBLEMS

As in EFI systems, CIS requires high fuel pressure to operate. Improper pressures or pressure leaks can cause a number of problems such as rough idling, stalling, surging or missing, or hard start problems.

As covered earlier in this chapter, a fuel pressure gauge of sufficient capacity is needed to make pressure checks. Use the gauge to check the three key pressures:

1. The system pressure provided by the fuel pump
2. The control pressure that opposes the airflow plate and controls the air/fuel ratio
3. The rest pressure, which can reveal the presence of a leak that allows the system to bleed down while the engine is shut off. Hot-starting problems are the result.

Begin by checking system and control pressures. If the system pressure provided by the fuel pump matches specifications, make any changes needed in the control pressure by changing shims in the pressure relief valve.

To conduct a rest pressure test, begin by taking a pressure reading with the test valve open. After the system has reached control pressure, shut off the ignition and fuel pump and time any pressure drop. Although the specs vary for different cars, a severe pressure drop will be obvious and happen quickly—in under 5 minutes. To be sure though, let the pressure drop for 20 minutes and compare the drop to specs.

If the pressure drop is greater than allowed, repeat the test with the test valve closed. If the system now holds fuel pressure, the warm-up regulator is at fault. Replace it. If the pressure drop is the same as before, check the rest of the system.

The check valve in the fuel pump outlet is a likely culprit. Build up pressure in the system, shut off the pump and immediately pinch closed the fuel line from the pump outlet to the accumulator. If the system holds pressure, replace the check valve, which is available as a detail part.

Another possible source of pressure leakage is the accumulator. Run the pump to build up control pressure in the system and then remove the screw from the back end of the accumulator. Insert a piece of wire in the hole to measure the depth inside the accumulator, with the pump running. If it is out of spec, replace the accumulator.

If the check valve and accumulator are good, examine the O-ring seal in the pressure regulator relief valve. If it is nicked or deteriorated, fuel pressure will not hold.

FUEL DISTRIBUTION PROBLEMS

Surging or missing can also be caused by unequal fuel distribution to the injectors. Ideally, CIS supplies all the cylinders with the same amount of fuel. If not, poor idle quality and part throttle performance will be the symptoms. To find out if all the injectors are providing equal amounts of gasoline, the car makers say to use two sets of graduated measuring tubes and a special fixture that depresses the air sensor plate. To use a typical setup, proceed as follows:

1. Unplug the connector from the auxiliary air regulator to keep it from heating up.
2. Remove the injectors from the cylinder head and place them in the containers.
3. Install the special air sensor fixture.
4. Activate the fuel pump as previously described.
5. Push the fixture's sleeve down until it stops, then turn the adjusting screw until the magnetic end of the sleeve touches the bolt in the center of the air sensor plate.
6. Rotate the adjusting screw to move the sensor until at least one nozzle just starts to deliver fuel, then turn off the pump and empty the measuring tubes.
7. Lift the fixture's sleeve to its first stop, which simulates the idle position.
8. Run the pump until one of the tubes is filled to the 20-milliliter mark, and stop the pump. There should be a cone-shaped spray pattern visible, and there should not be more than a 3 milliliters difference between the amount of fuel in the highest and lowest tube. Empty the containers again.
9. Lift the fixture's sleeve to its last stop, which simulates full throttle.
10. Run the pump until one of the tubes is filled to the 80-milliliter mark, and stop the pump. Now, the difference between the highest and lowest reading should not be more than 8 milliliter, and the cone-shaped spray pattern should be visible again.

If there is too much variation in fuel volume, swap the injectors of the two lines with the largest difference and repeat the test. If the same injector again flows too little fuel, it is faulty. On the other hand, if the same fuel line that supplied insufficient fuel in the first test does so again even though connected to a different injector, the problem can be found in the fuel distributor.

It is possible for the plunger in the fuel distributor to be worn or scored, allowing fuel to push past it rather than flowing to the injectors. If all else fails, remove the distributor. Then carefully remove the plunger and inspect it. If there is any hint of damage, replace the fuel distributor and plunger assembly. It is an expensive component but is handled as a core.

SHOP TALK

The special measuring tubes and fixture are not absolutely necessary. A similar check can be made using any type of graduated container for each injector, and moving the air sensor by hand.

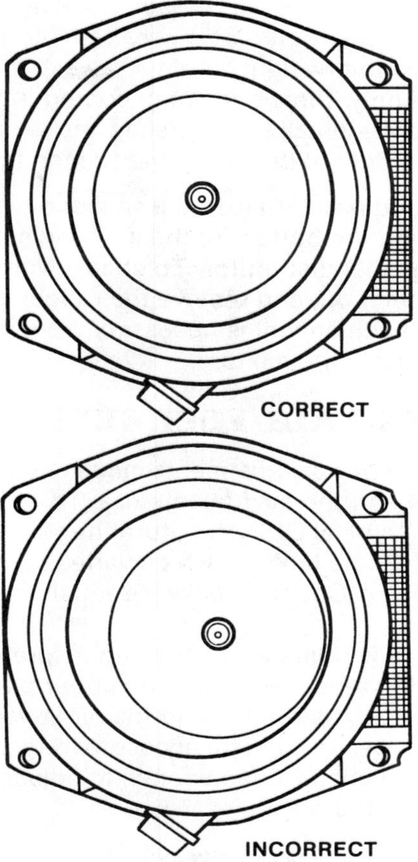

FIGURE 13–54 If an airflow sensor is off-center in its bore, air can leak by, causing a hot-start problem.

AIRFLOW SENSOR ADJUSTMENT

An airflow sensor plate that is off-center in its bore (Figure 13–54) can allow air to leak by, causing severe hot-start problems.

Check the air sensor plate for drag by using a magnet on the plate bolt to move the air sensor plate up and down with the pump on. If there is drag, or if the plate does not appear to be concentric with the air cone, it should be centered. To do so, perform the following procedures:

1. Remove the bolt that holds the air sensor plate to the level, coat the bolt threads with an anaerobic locking compound, and reinstall the bolt finger tight.
2. Move the plate as necessary to allow a 0.10 millimeter (0.004 inch) feeler gauge to be inserted at any point around its circumference, tighten the bolt, and recheck it with the feeler gauge.
3. If the plate cannot be centered, and the lever appears to be too far to one side,

remove the air sensor housing and the damper bolt that is screwed into the lever counterweight.

4. Apply thread locking compound to the bolt, install it, then slide the lever to the middle of the pivot, and tighten the bolt.

A related point: At rest, the sensor plate should be even with the bottom of the air cone at the side nearest the fuel distributor. To check, warm up the engine, shut it off, and remove the rubber connecting boot. If adjustment is necessary, bend the lever stop clip. Reset the mixture.

IDLE AND AIR/FUEL SETTINGS

The two basic adjustments made to the CIS are idle speed and the air/fuel mixture. To set the idle speed on a typical CIS setup, turn the screw next to the throttle plate linkage, which varies the size of a bypass drilling. Counterclockwise equals faster (Figure 13–55).

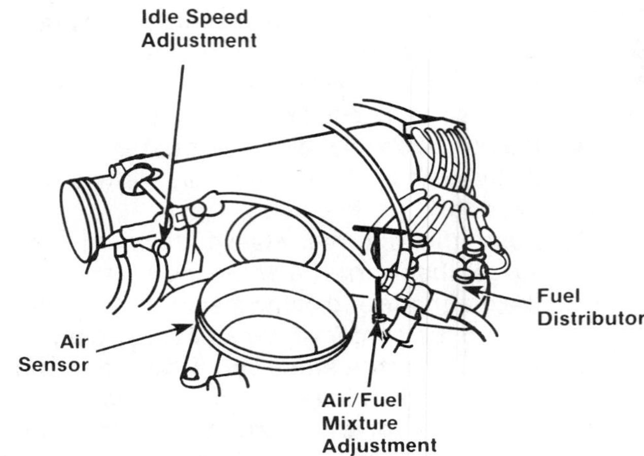

FIGURE 13–55 Location of idle speed and air/fuel mix screw on a typical CIS system

Mixture is adjusted by turning a screw, which bears on the air sensor lever. A rubber plug is between the fuel distributor and the air sensor cone, which must be removed for access to the screw. Use a long, 3-millimeter Allen wrench to adjust—clockwise richens the blend.

CIS-E IDLE CONTROL

In a CIS-E engine the control computer uses signals from the engine temperature sensor, air-plate potentiometer, throttle switch, and distributor to decide mixtures for special purposes. It controls idle speed with the idle-speed stabilizer, holding a higher speed for a colder engine and gradually slowing it down as the engine warms. Idle specs are given in percent of dwell to rpm, and turning the idle screw, as on most domestic cars, will not change the idle speed because the idle stabilizer will change to correct it.

Idle speed is programmed into the computer and cannot be changed. Some systems include an increased idle setting when the air-conditioning is on to provide better compressor function and prevent stalls.

While the idle speed cannot be adjusted with an idle stabilizer, the dwell should be checked to see that it is within range. Changing the idle speed screw will correct incorrect dwell. Typical warm idle dwell is 30 percent. Note that that figure is in percent, not in degrees. Most of the dwellmeters used for contact-point ignition are calibrated in degrees of distributor shaft rotation. For example, GM eight-cylinder distributors are set to 30 degrees. That was 30 degrees each for eight cylinders in 360 degrees of distributor shaft rotation. The points were closed for 240 degrees of that time, or 66 percent of the time. That is a 66 percent dwell.

When using an older dwellmeter calibrated in degrees, the best way to get the meter in the center of its range to check for 30 percent dwell is to set it on the four-cylinder scale and then adjust for 27 degrees (30 percent of the maximum possible 90 degrees).

If the idle speed is wrong and it is not possible to get the adjustment correct, check the idle-speed stabilizer. If it works properly, the problem is in either the connecting wires or the computer itself. Remember, the air-conditioning can turn up the idle, but headlights, heater, and other alternator loads should not reduce it.

ADDITIONAL CIS-E CHECKS

On CIS-E engines a feedback circuit test is used to check the operation of the electromagnetic differential pressure regulator on the fuel distributor. Electromagnets are almost 100 percent efficient, so the current used is very small, measured in milliamps (thousandths of an amp). To check circuit current, a high-quality digital ammeter wired in series and a special adapter for the regulator is needed. The key to whether the feedback system is working properly is whether the regulator current hunts back and forth. It may be necessary to hold the throttle open to 2000 rpm for a few seconds to warm the oxygen sensor.

Idle mixture adjustment on CIS-E is done through a hole in the fuel distributor/air flap housing using a long 3-millimeter Allen wrench. Set up the fuel distributor regulator and an ammeter on the exhaust analyzer on the exhaust tap (which is generally covered with a blue cap). It is very critical to have a secure seal around the exhaust probe or the readings can be false. An exhaust leak can also fool the oxygen sensor into giving false lean readings and driving the system rich.

The key is to set both regulator current and exhaust carbon monoxide (CO) in spec. Some vehicle service procedures call for adjusting mixture in closed loop by setting the parameters of the fluctuating current and checking against CO. Others call for unplugging the oxygen sensor (hence open loop) and adjusting for CO and checking against current fluctuation range. Obviously, these are two ways of doing the same thing, but the shop manuals will list specifications for one method or the other.

REVIEW QUESTIONS

1. The throttle switch in a fuel injection system corresponds in function to what component in a carburetor?
   a. float
   b. accelerator pump
   c. metering rods
   d. fast idle cam

2. Which of the following statements concerning the operation of a fuel injection system is incorrect?
   a. In an injection system, air and fuel are measured together.
   b. In an injection system, fuel delivery is cut off completely when no power is needed.
   c. An injection system has lower fuel consumption.
   d. Compensation for differing altitudes can be built into the basic controls of an injection system.

3. Which of the following is *not* an advantage that fuel injection offers over carburetion?
   a. leaner air/fuel ratios
   b. better fuel economy
   c. no choke requirements
   d. lower engine torque

4. What factor is the number one enemy of fuel injection systems?
   a. extreme cold
   b. moisture
   c. dirt
   d. lack of lubrication

5. In most electronic fuel injection systems, fuel requirements are measured by ____________.
   a. airflow across a sensor
   b. intake manifold pressure
   c. vacuum
   d. all of the above

6. The length of time that an injector stays open is called ____________.
   a. intermittent system
   b. pulsed system
   c. injector pulse width
   d. open loop mode

7. In a continuous injection system, fuel flow is controlled by ____________.
   a. a distributor
   b. a metering valve
   c. a fuel pump
   d. an airflow sensor

8. When diagnosing a hard cold start condition on a car with a CIS system, Technician A tests the cold start valve operation. Technician B tests the thermo-time switch. Who is correct?
   a. Technician A
   b. Technician B
   c. Both A and B
   d. Neither A nor B

9. When servicing the fuel delivery system of an EFI, Technician A also adjusts the fuel pressure regulator. Technician B adjusts the electronic control unit. Who is right?
   a. Technician A
   b. Technician B
   c. Both A and B
   d. Neither A nor B

10. What EFI component houses the fuel injectors?
    a. fuel pressure regulator
    b. ECU
    c. fuel rail assembly
    d. damping chamber

11. The signals from the airflow sensors in an EFI are sometimes called ______________.
    a. base pulse
    b. injector pulse width
    c. intermittent system
    d. loop mode

12. What units are provided for cold starting and warm-up?
    a. cold start injector valve
    b. thermo-time switch
    c. auxiliary air device
    d. all of the above

13. To discover the cause of an engine "miss," Technician A performs an injector power balance test. Technician B tests the ignition system on an oscilloscope. Who was correct?
    a. Technician A
    b. Technician B
    c. Both A and B
    d. Neither A nor B

14. Which of the following problems with components outside the fuel injection system can cause sensors to adjust engine operation to less than ideal conditions?
    a. intake manifold leaks
    b. sloppy timing chain
    c. bad valve or ring
    d. all of the above

15. Which of the following CIS-E component is not in a basic CIS system?
    a. potentiometer
    b. system pressure regulator
    c. electrohydraulic actuator
    d. all of the above

16. Technician A relieves fuel pressure on GM cars by connecting a pressure gauge to a Schrader valve. Technician B does so by disabling the fuel pump and running the car until it dies. Who is correct?
    a. Technician A
    b. Technician B
    c. Both A and B
    d. Neither A nor B

17. Which of the following adjustments can be made to a CIS?
    a. idle speed
    b. air/fuel mixture
    c. both a and b
    d. neither a nor b

CHAPTER FOURTEEN

UNDERSTANDING EMISSION CONTROL SYSTEMS

Objectives

Upon completion of this chapter, you should be able to:

- Describe the inspection and replacement of PCV system parts.
- Name the components of an evaporative emission control system.
- Explain the air temperature emission control system inspection and testing.
- Name the components of the manifold heat control system and what they do.
- Describe the application of the EFE control.
- List the EGR valves and controls.
- Explain air injection system.

Federal standards of HC, CO, AND NO_X have been established for these pollutants. The exception to these standards are a few high-altitude western states and California. Because there is less oxygen at high altitudes to promote combustion, emission standards at high altitudes are slightly less strict. California's standards allow less pollution than federal standards.

Two basic types of emission control systems are used in modern vehicles:

1. **Precombustion Control Systems.** Most of the pollution control systems used today prevent emissions from being created in the engine, either during or before the combustion cycle. The common precombustion control systems are as follows:
   - *Positive Crankcase Ventilation (PCV).* First used in the early 1960s, the PCV system removes gasses that blow by the pistons into the crankcase. Originally, these gasses (HC, CO, and NOx) were vented to the air by a road draft tube. Now, they are recirculated to the induction system.
   - *Engine Modification Systems.* These systems appeared in the 1960s as a group of engine modifications designed to improve combustion and reduce HC-CO in the exhaust. It included a heated primary air system, carburetor design changes, engine "breathing" refinements, and some spark timing controls.
   - *Evaporative Control Systems.* In the 1960s, evaporative control systems were used to trap raw gas vapors from the fuel tank (and later the carburetor bowl) and route them to the air cleaner when the engine ran. In the 1970s the system was refined to a sealed housing system to better control emissions and purge them to the intake manifold in specific engine models.
   - *Exhaust Gas Recirculating (EGR) Systems.* EGR is strictly a control for NOx in the exhaust gasses. It reduces NO_X by diluting the air/fuel mixture with some exhaust gas, which does not burn.
2. **Post-Combustion Control Systems.** Post-combustion control systems clean up the exhaust gasses after the fuel has been burned. Secondary air or air injector systems appeared in the mid 1960s. Their function is to pull fresh air into the exhaust to reduce HC and CO to harmless water

vapor and carbon dioxide by chemical (thermal) reaction with oxygen in the air. In the 1970s, catalytic converters were added to help this process. Some catalysts now reduce NOx as well as HC-CO.

Most emission control devices have some kind of engine vacuum control. These controls are used to block out the operation in certain modes:

- Systems that put air or fuel vapors into the air/fuel induction system do not operate at idle, when the engine is cold, and sometimes when the intake air is cold. The exception is PCV, which operates whenever the engine is running.
- EGR systems are usually blocked out at idle, since mixture dilution then causes rough idle. They have various controls to vary the gas flow in other models.
- Thermactor (secondary air) systems are usually blocked out from the exhaust when it is rich (idle, deceleration, cold engine, WOT, etc.) to prevent backfire and damage to the catalysts.

The control devices are principally vacuum diaphragms (Figure 14–1). Engine vacuum from various sources is switched on or off to the diaphragms. The switching devices can be in the air cleaner (controlled by inlet air temperature), in the water jacket (controlled by coolant temperature), or incorporated into solenoids that are controlled by the electronic control assembly (ECA). The principal sources for the vacuum muscle are manifold vacuum, spark port or "S" vacuum, and "E" port (EGR) vacuum. "S" and "E" vacuum sources are used to lock out certain systems at curb idle when they are essentially zero. Venturi vacuum is used only occasionally.

PCV SYSTEMS

The crankcase of an engine must "breathe" to purge itself of fuel and water vapors. If these vapors are not purged, they can thin the oil or form sludge. Water forms in the crankcase when a cold engine condenses it from the air. In fact, it has been estimated that every gallon of gasoline burned forms more than a gallon of water. During the last part of the engine's combustion stroke, some unburned fuel and products of combustion—water vapor, for instance—leak past the engine's piston rings into the crankcase. This leakage results in the following:

- *High Pressures in the Engine Combustion Chamber.* This condition is created by the normal compression stroke in the engine under operation.
- *Necessary Working Clearance of Piston Rings in Their Grooves.* Without normal ring clearance, the engine's piston rings do not

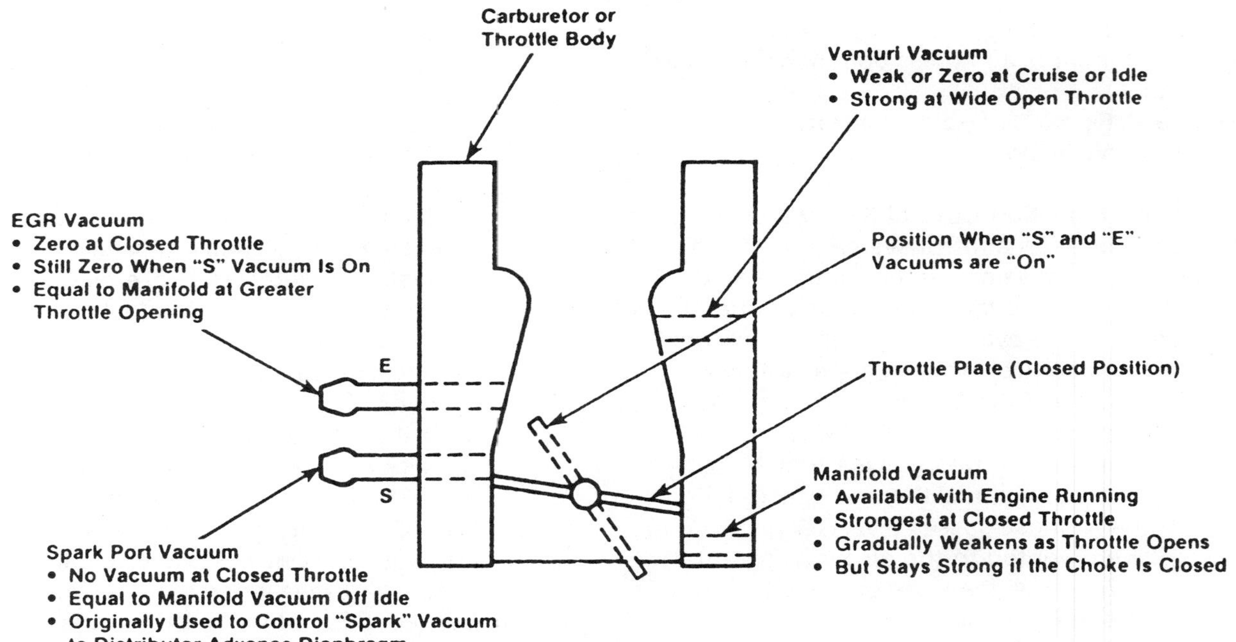

FIGURE 14–1 Emission control devices that are used principally are vacuum diaphragms.

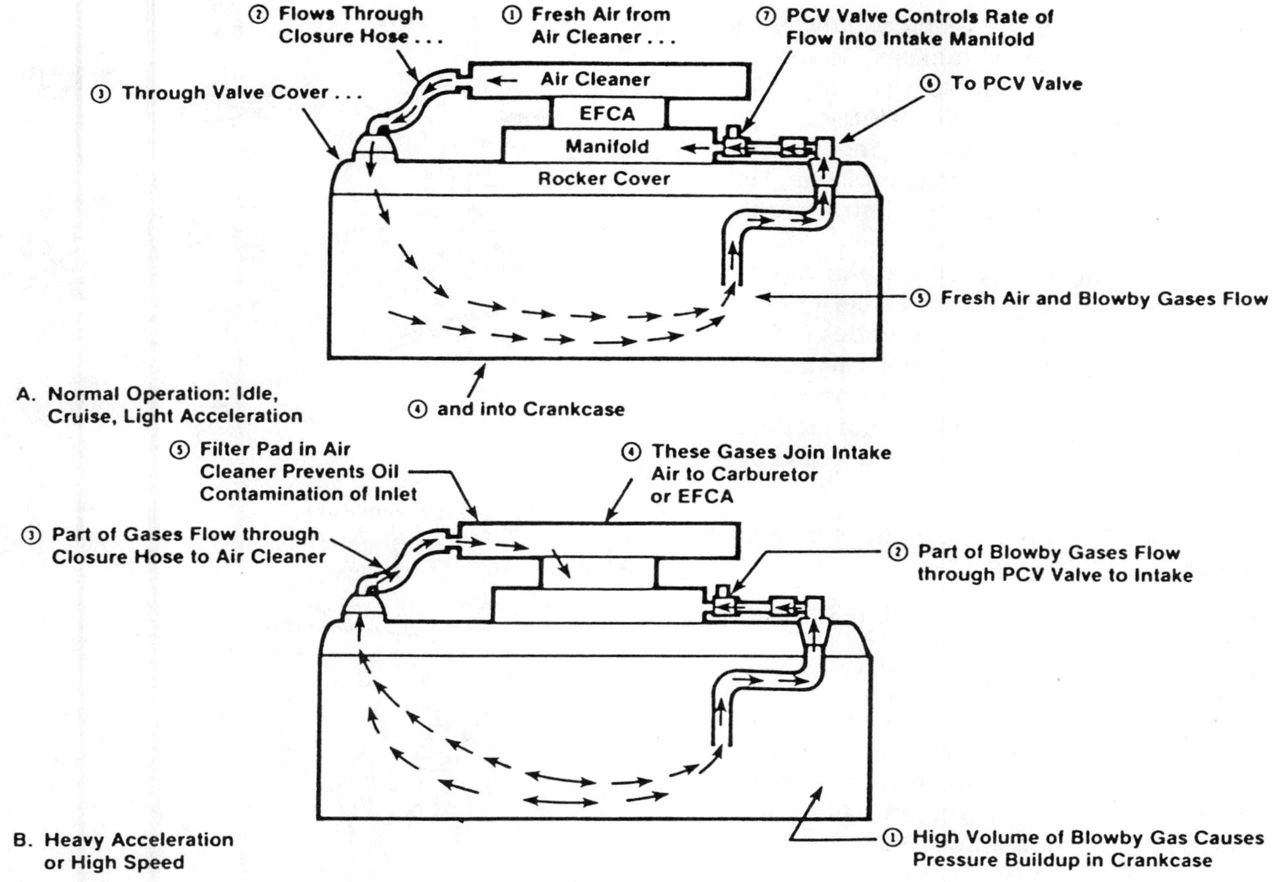

FIGURE 14-2 (A) Fresh air flow during normal operation: idle, cruise, and light acceleration; (B) during heavy acceleration or high speed.

have room to expand from heat, which is created by normal engine operation, and seal properly against the cylinder walls.

- *Normal Shifting of Piston Rings in Their Grooves.* Sometimes this shift lines up the clearance gaps of two or more rings, which is a normal condition. As the piston rings continue to turn in their grooves, the situation corrects itself.
- *Reduction in Piston Ring Sealing Contact Area.* This occurs as the piston moves up and down in the cylinder.

Leakage into the engine crankcase is called *blowby*. Blowby must be removed from the engine before it condenses in the crankcase and reacts with the oil, which forms sludge. Sludge, if allowed to circulate with engine oil, will corrode and accelerate wear of pistons, piston rings, valves, bearings, and other internal working parts of the engine.

Because the air/fuel mixture in an engine never completely burns, blowby carries some unburned fuel into the crankcase. If unremoved, the unburned fuel dilutes the crankcase oil. When oil is diluted with gasoline, it does not lubricate the engine properly, which causes excessive wear.

Combustion gasses that enter the crankcase are removed by a positive crankcase ventilation (PCV) system, which uses engine vacuum to draw fresh air through the crankcase. This fresh air, which dissipates the harmful gasses, enters through the air filter on top of the carburetor or through a separate PCV breather filter located on the inside of the air filter housing (Figure 14-2).

Because the vacuum supply for the PCV system is from the engine's intake manifold, the airflow through this system must be controlled in such a way that it varies in proportion to the regular air/fuel ratio being drawn into the intake manifold through the carburetor. Otherwise, the additional air that is drawn into the system would cause the air/fuel mixture to become too lean for efficient engine operation.

The PCV is a highly engineered device that has two major functions:

1. It prevents the emission of blowby gasses from the engine crankcase to the atmosphere. These gasses were once vented through a road draft tube. Now they are recirculated to the engine intake and burned during combustion. Thus, it is able to sense vacuum and control crankcase airflow.
2. It scavenges the crankcase of vapors that could dilute the oil and cause it to deteriorate or that could build undesirable pressure in the crankcase. Fresh air from the air cleaner mixes with these vapors and makes them flow to the air/fuel intake. It has the ability to regulate the air/fuel ratio over all operating ranges.

Thus, the PCV system benefits the vehicle's driveability by

- Eliminating harmful crankcase gasses
- Reducing air pollution
- Promoting fuel economy

Recirculated gasses in the system are a combustible mixture that becomes fuel for the engine when added to the air/fuel mixture entering the intake manifold from the carburetor fuel injectors. Consequently, an inoperative PCV system could shorten the life of the engine by allowing harmful blowby gasses to remain in the engine, causing corrosion and accelerating wear.

PCV SYSTEM DIAGNOSIS AND SERVICE PROCEDURES

No adjustments can be made to the PCV system. Service of the system involves a careful inspection, operation, and replacement of faulty parts. When replacing a PCV valve, match the part number on the valve with the vehicle maker's specifications for the proper valve. If the vehicle cannot be identified, refer to the part number listed in the manufacturer's service manual.

PCV System Inspection

The first step in PCV servicing is a visual inspection. As shown in Figure 14-3, the PCV valve can be located in several places. The most common location is in a rubber grommet in the valve or rocker arm cover (Figure 14-3A). Or it can be installed in the middle of the hose connections (Figure 14-3B), as well as installed directly in the intake manifold (Figure 14-3C).

Once the PCV valve is located, make the system inspection by using the following procedure:

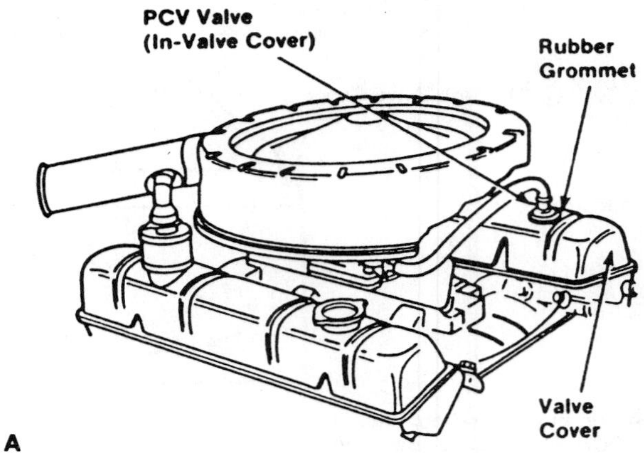

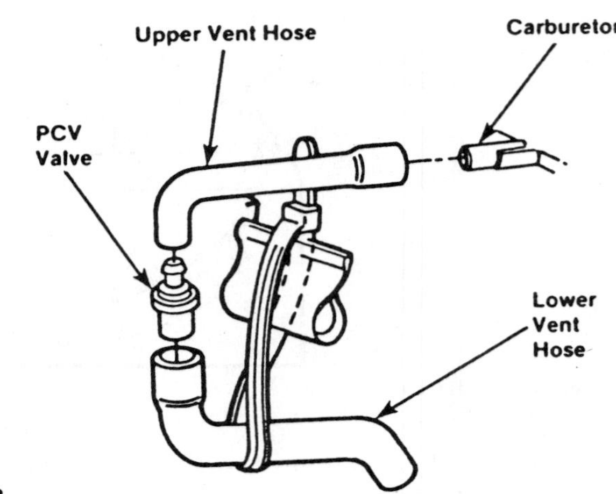

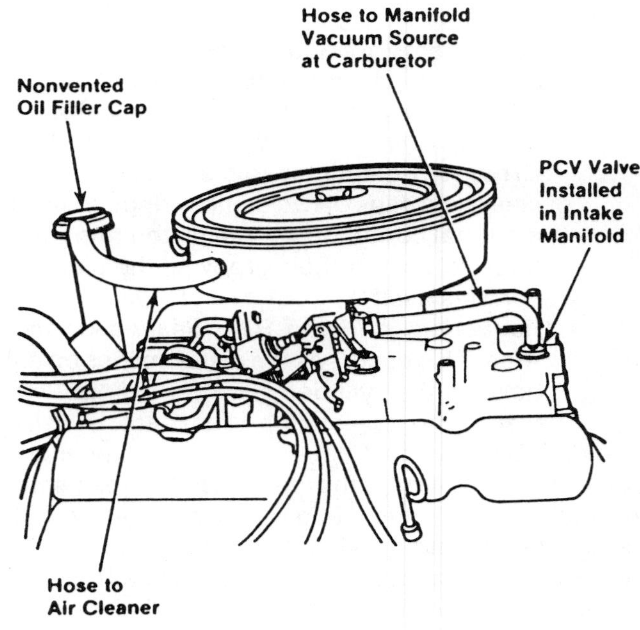

FIGURE 14-3 Various locations of the PCV valve

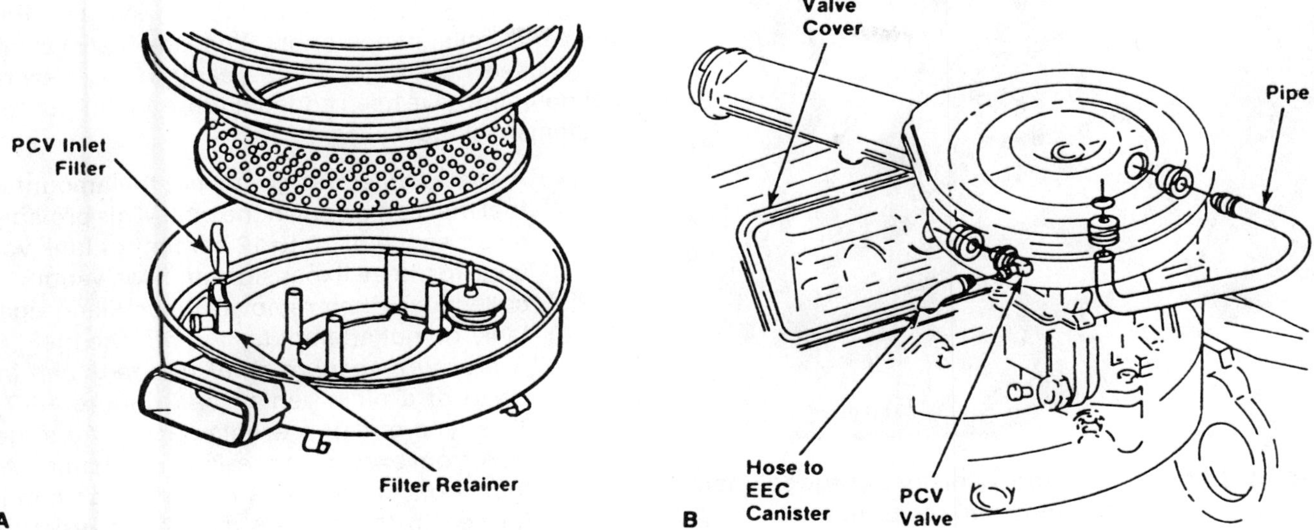

FIGURE 14–4 When inspecting the PCV system, check the air filter in the (A) air cleaner or in the (B) air hose.

1. Make sure all the PCV system hoses are properly connected and that they have no breaks or cracks.
2. Remove the air cleaner and inspect the carburetor or fuel injector air filter. Crankcase blowby can clog these with oil. Clean or replace such filters.
3. Check the crankcase inlet air filter located in the air cleaner (Figure 14–4A), in the air hose (Figure 14–4B), or in the crankcase inlet air cleaner in the valve corner. As the filter does its job, it gets dirty. If it is oil soaked, it is a good indication that the PCV valve is not working the way that it should. When the filter becomes so dirty that it restricts the flow of clean air to the crankcase, it can cause the same problems as a clogged PCV valve. So be sure to check this particular filter when checking the PCV system and replace it as necessary.
4. Inspect for dirt deposits that could clog the passages in the manifold or carburetor base. These deposits can prevent the system from functioning properly, even though the PCV valve, valve filter, and hoses might not be clogged.

Functional Checks of PCV Valve

If an engine is idling roughly, check for a clogged PCV valve or plugged hose. Check the PCV part number to be sure the correct PCV is installed. Perform a functional check of the PCV valve by using the following procedure:

1. Disconnect the PCV valve from the valve cover, intake manifold, or hose.
2. Start the engine and let it run at idle. If the PCV valve is not clogged, a hissing will be heard as air passes through the valve.
3. Place a finger over the end of the valve to check for vacuum. If there is little or no vacuum at the valve, check for a plugged or restricted hose. Replace any plugged or deteriorated hoses.
4. Turn off the engine and remove the PCV valve. Shake the valve and listen for the rattle of the check needle inside the valve. If the valve does not rattle, replace it.

SHOP TALK

When diagnosing poor idle condition, check out the PCV system very carefully because it can affect the vehicle's idling. Service technicians must remember that an engine operated without any crankcase ventilation can be damaged. Therefore, it is important to replace the PCV valve and air cleaner breather at proper intervals. Periodically inspect the hoses and clamps and replace any that show signs of deterioration. An engine that runs too richly during warmup but runs well if the PCV valve is removed should be checked for crankcase lube oil contamination and the cause of the contamination should be investigated (for example, short trips, fouled plugs, faulty thermostat, or leaking injectors).

FIGURE 14-5 This vapor canister's purge solenoid is operated by the engine control computer.

Proper operation of the PCV system depends on a sealed engine. Remember that the crankcase is sealed by the dipstick, valve cover, gaskets, and sealed filler cap. If oil sludging or dilution is noted and the PCV system is functioning properly, check the engine for a possible cause and correct it to ensure that the system will function as intended.

EVAPORATIVE EMISSION CONTROL SYSTEM

The fuel evaporative emission system (Figure 14-5) reduces the amount of raw fuel vapors (HC) that are emitted into the air from the fuel tank and carburetor (if so equipped). While different manufacturers have slight variations and names for their evaporative emission control (EEC) systems they are very similar in operation. Since the first systems were used nationwide in the early 1970s, several refinements have been added (Figure 14-6). Current systems include:

- *A special filler design* to limit the amount of fuel that can be put in the tank. This provides an air space of 10 to 12 percent of tank volume for heat expansion and for vapors to collect until their removal. When filling a gas tank, do not attempt to "top off" the fuel.
- *A pressure/vacuum relief fuel tank cap* instead of a plain vented cap (Figure 14-7). This type of cap prevents damage to a fuel tank from excessive pressure or vacuum. As fuel cools, it contracts. A partial vacuum is formed in the tank. Atmospheric pressure acting on the walls of a tank could collapse it. A vacuum relief valve opens to allow outside air into the tank. This action balances pressure against the outside surfaces of a tank. When a fuel tank is heated by outside temperatures, fuel and vapors expand, producing pressure that could rupture fuel tank seams. Normally, this pressure buildup is slight and pressure is dissipated throughout a vapor-recovery system. However, if hoses or tubes become clogged or accidentally pinched, pressure could damage a tank. A pressure-relief valve opens to allow excessive pressure to escape.
- *A vapor separator* in the top of the fuel tank (Figure 14-8). This device permits fuel vapors to pass into connecting hoses. The vapor/fuel separator collects droplets of liquid

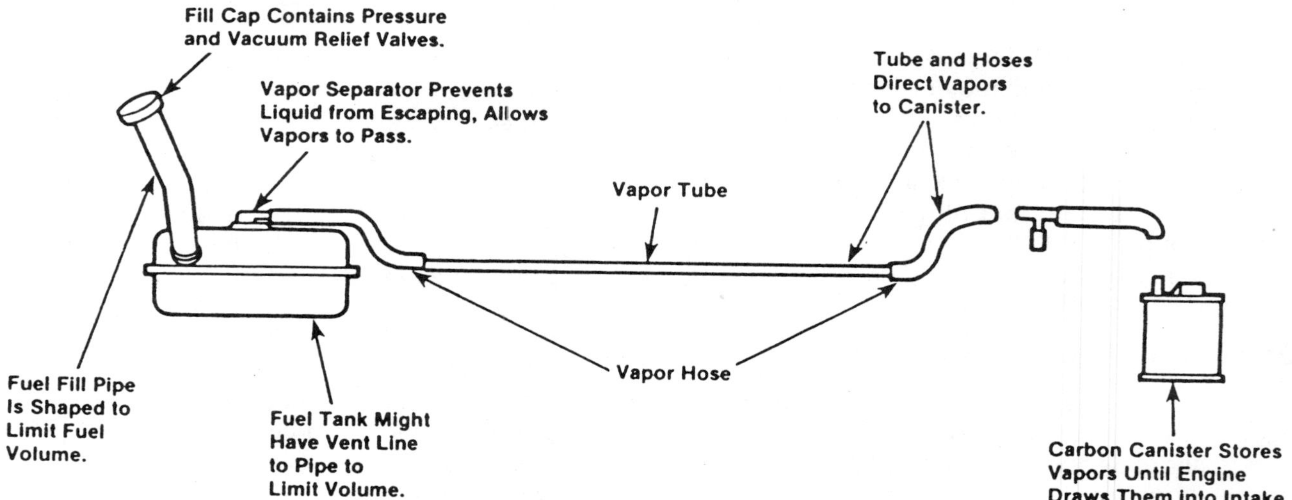

FIGURE 14-6 Refinements of the EEC system

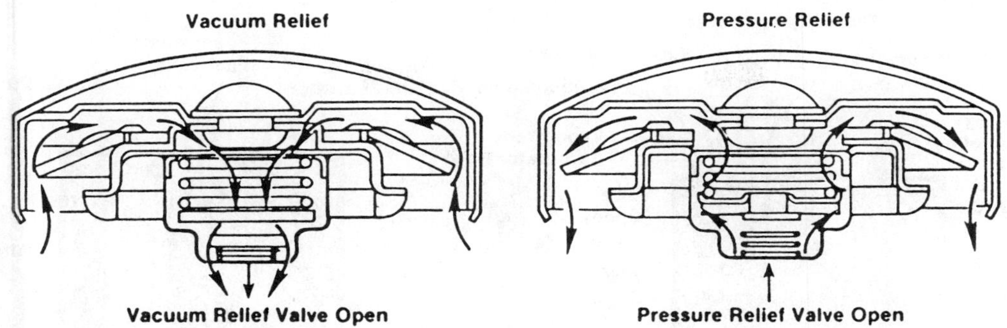

FIGURE 14-7 Sealed fuel tank cap

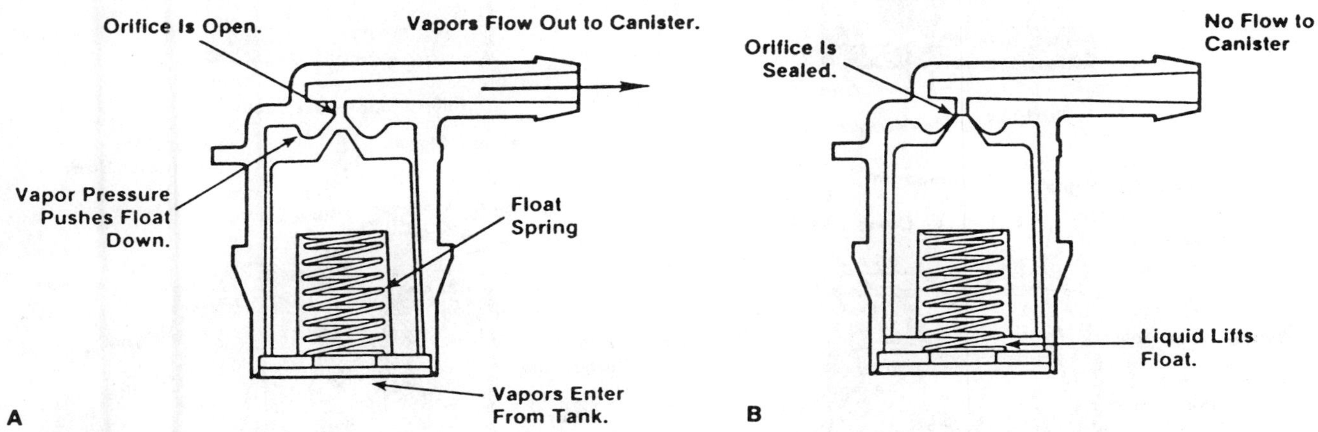

FIGURE 14-8 (A) Normal operation of vapor separator; (B) with liquid in separator

fuel and directs them back into the tank. In some cases the liquid vapor separator is located outside the fuel tank.

- *A domed fuel tank* in which its upper portion is raised. Fuel vapors rise to this upper portion and collect.
- *Check, or one-way, valves* keep vapors confined. When an engine runs, vacuum and/or electrically operated valves open. Fresh air and vapors are drawn into the intake manifold.
- *Hoses and tubes* connect parts of a vapor-recovery system. Special fuel/vapor rubber tubing must be used. Ordinary rubber hoses would rot and crack quickly from the effects of fuel. Metal tubing is used under a vehicle.

The most obvious part of the evaporative emission control system is the charcoal (carbon) canister in the engine compartment. This canister (Figure 14-9) that stores the vapors is located in or near the engine compartment. It is connected to the air cleaner, carburetor, or throttle body so as to purge the vapors when the engine is running. The following are parts of the charcoal canister:

- Two canisters are sometimes used with both single and dual tanks, dual fuel bowls or large capacity tanks.
- Storage medium is activated carbon (charcoal).
- Dust cap and vapor purge connections are interchangeable.
- Canister purge control is usually accomplished with a vacuum-operated or electrically operated valve. The valve opens during specific operating conditions; when the engine can accept the extra fuel vapor.
- Fuel bowl vent control is part of the system on carbureted engines (Figure 14-10). The carburetor bowl is vented to the bowl vent line. Bowl vapors thus flow into the canister when the purge is not operating. Several types of valves are used on various engines to control the bowl venting.

In 1978, new government emission standards led to a group of changes called sealed housing evaporative determination (SHED) or evaporative emission SHED system. The principal features of SHED, carried through to current models, are:

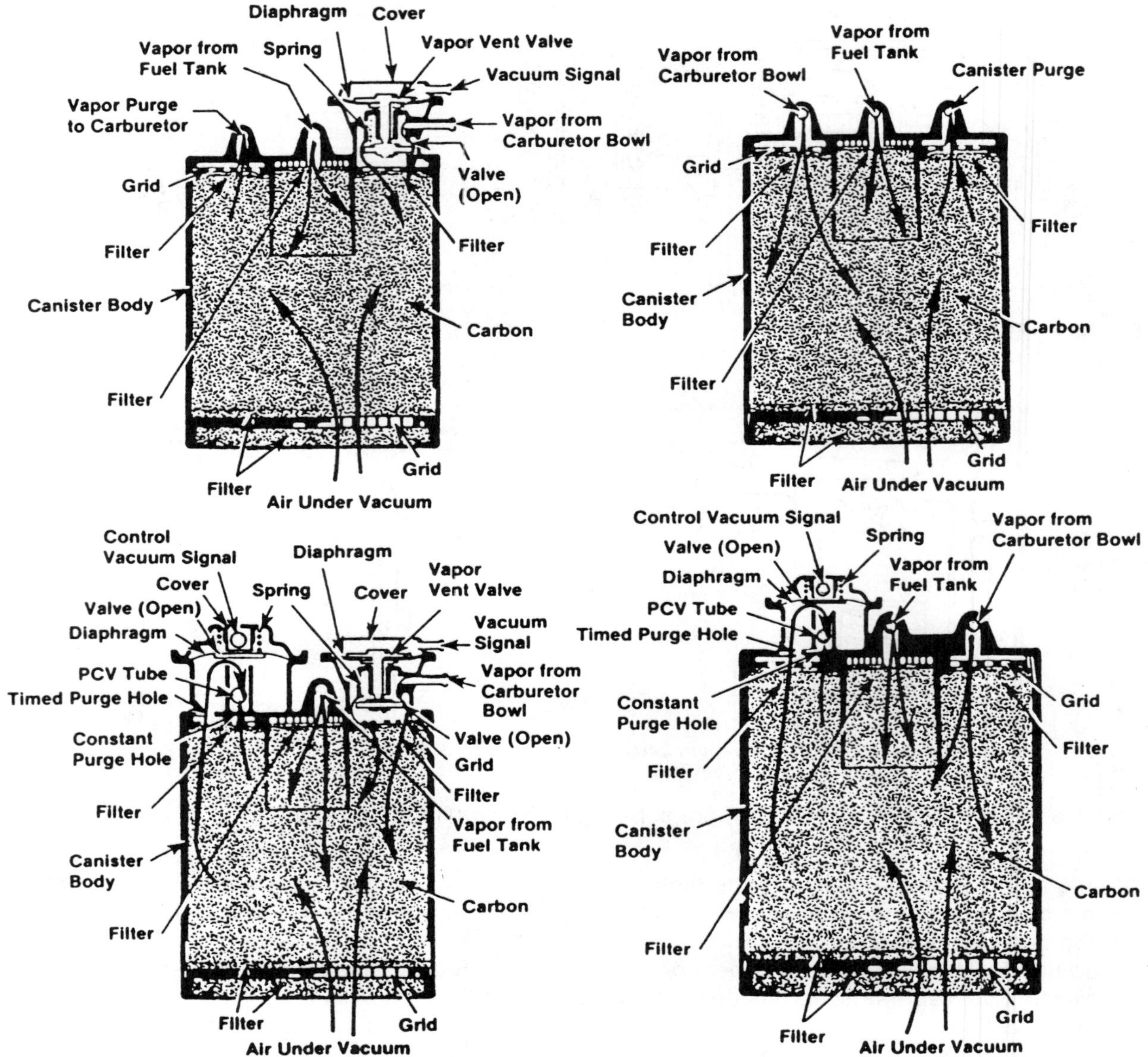

FIGURE 14-9 Various canister designs

- Vapors from the gas tank and carburetor fuel bowl are routed to and stored in the carbon canister.
- The carbon canister is purged while the engine is running.

Fuel vapors are absorbed onto the surfaces of the charcoal granules. When the vehicle is restarted, vapors are sucked into the carburetor or intake manifold to be burned in the engine. Canister purging varies widely with make and model. In some instances a fixed restriction allows constant purging whenever there is manifold vacuum. In others, a staged valve provides purging only at speeds above idle. Generally, the canister purge valve is normally closed (Figure 14–11). It opens the inlet to the purge outlet when vacuum is applied. Some units incorporate a thermal-delay valve so the canister is not purged until the engine reaches operating temperature. Purging at idle or with a cold engine creates other problems, such as rough running and increased emissions because of the additional vapor added to the intake manifold. Typical purge valve mountings are shown in Figure 14–12.

CANISTER TESTING

Check the canister to make sure that it is not cracked or otherwise damaged (Figure 14–13). Also make certain that the canister filter is not completely

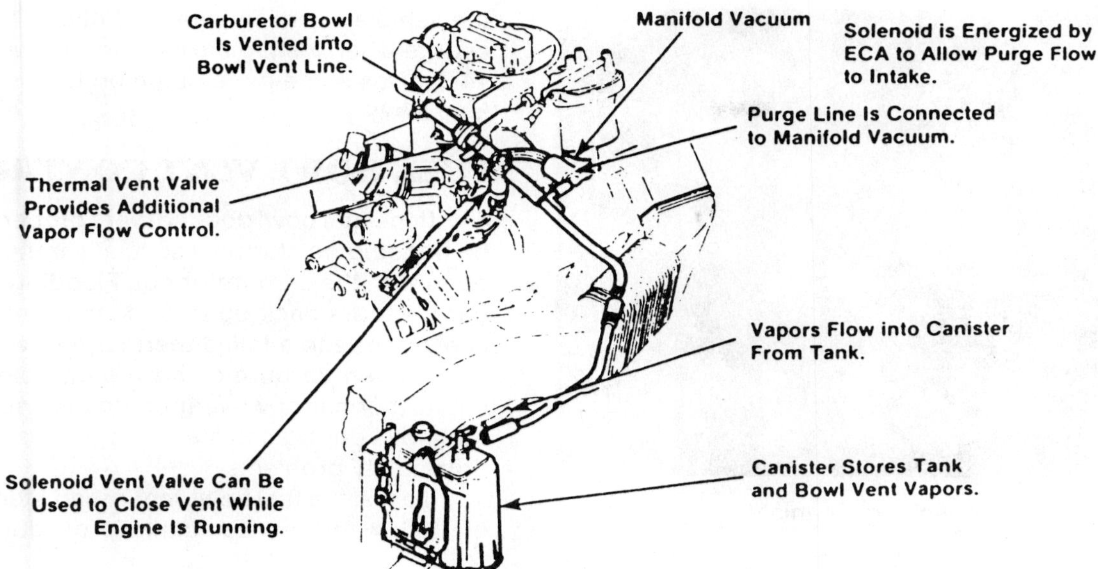

FIGURE 14-10 Typical carbureted engine vapor control

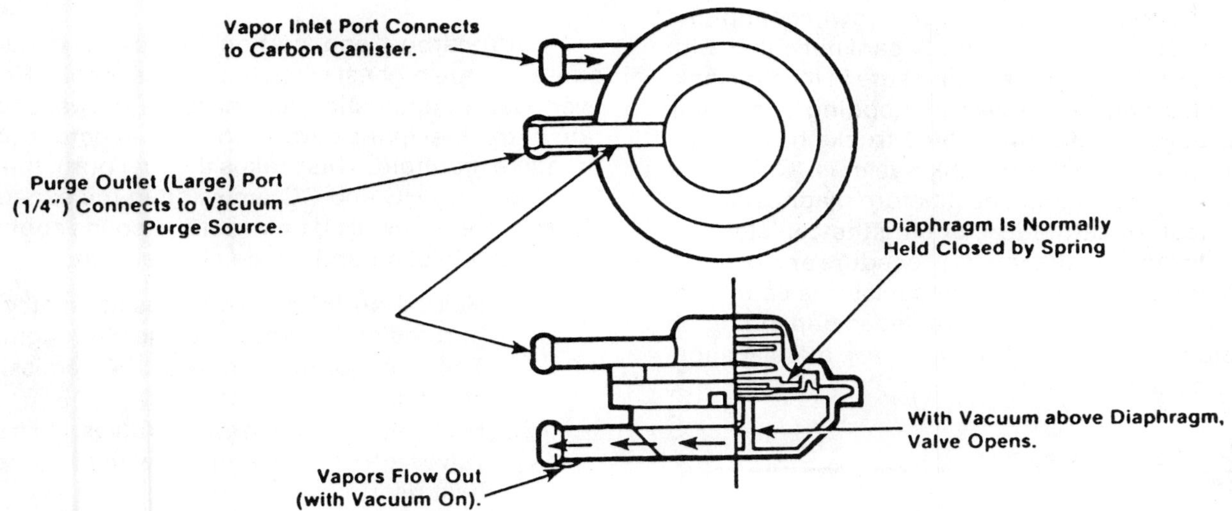

FIGURE 14-11 Canister purge valve operation

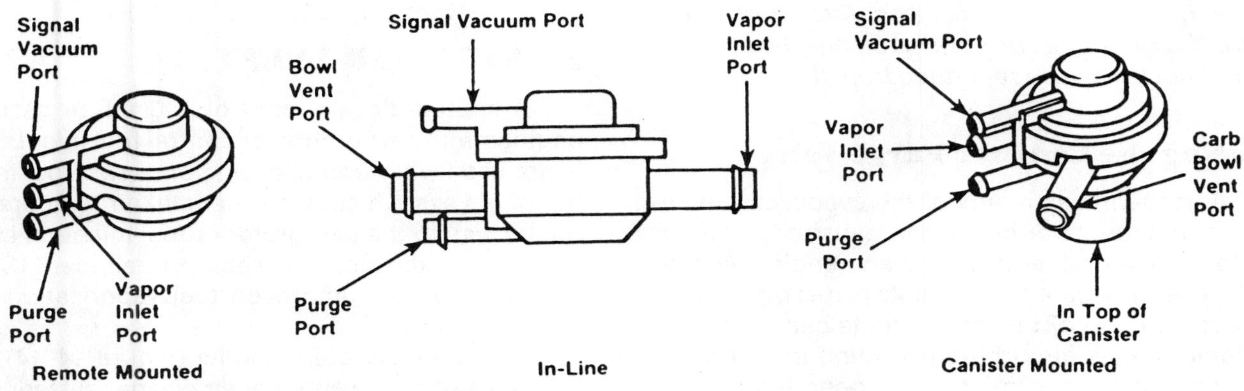

FIGURE 14-12 Typical purge valve mountings

FIGURE 14-13 Checking a canister

saturated. Remember that a saturated charcoal filter can cause symptoms that can be mistaken for carburetor problems. Rough idle, flooding, and other conditions associated with poor bowl venting can indicate a canister problem. A canister filled with liquid or water causes back pressure in the fuel tank. It can also cause richness or flooding symptoms during purge or startup. (Some trucks have intentionally pressurized fuel tank systems. Check the calibration and engine decal before diagnosing.)

To test for saturation, unplug the canister momentarily during a diagnosis procedure and observe the engine's operation. If the canister is saturated, either it or the filter must be replaced depending on its design. That is, some models have a replaceable filter, others do not.

SHOP TALK

Sometimes a partially saturated canister can be evacuated by removing it from its mounting area (leaving the hoses connected), inverting it, and running the engine at high idle for 3 to 5 minutes. Be sure to replace the canister filter if it can be separated from the canister.

Testing the Canister Purge Valve

A vacuum leak in any of the evaporative emission components or hoses can cause starting and performance problems as can any engine vacuum leak. It can also cause complaints of fuel odor. Incorrect connection of the components can cause rich stumble or lack of purging (resulting in fuel odor). To conduct a vacuum-on, valve-open test consult the vehicle's service manual. Follow the same procedure when testing the canister purge valve. Table 14-1 gives some more common EEC system problems.

FUEL BOWL VENT CONTROL

If the fuel bowl does not vent properly, there can be flooding on startup, especially if the vent fails to open with the carburetor hot. Flooding can also occur if vapors back up into the carburetor from the canister due to a failed thermal vent valve (TVV). A failed open carburetor bowl vent solenoid or vacuum/thermal bowl vent can cause suction in the fuel bowl during canister purging, resulting in lean driveability problems.

To test the fuel bowl vent or canister purge shutoff, follow the service manual procedure.

AIR TEMPERATURE EMISSION CONTROL

Hydrocarbon and carbon monoxide exhaust emissions are highest when the engine is cold. However, warm combustion air improves the vaporization of the fuel in the carburetor, fuel injector body, or intake manifold. This makes the fuel burn, therefore reducing HC and CO emission in the exhaust. Three systems are used on various gasoline engines to heat the inlet air and/or the air/fuel mixture.

1. **Heated Air Inlet.** This is used on all carbureted and central fuel injected engines. The idea originated with EC emission control.
2. **Manifold Heat Control Valves.** These valves are used on engines with a manifold heat riser.
3. **Early Fuel Evaporation (EFE) Heater.** Its function is basically the same as the manifold heat control valve, except that a heater warms the air.

HEATED AIR INLET

A heated air inlet control is used on gasoline engines with carburetion or central fuel injection. It is not used with turbocharging or ported fuel injection. This system controls the inlet air temperature on the way to the carburetor or fuel injection body. By warming the air as necessary, it reduces HC and CO emissions by improved fuel vaporization and faster warmup.

The principal components (Figure 14-14) and functions of a conventional air cleaner system are:

TABLE 14-1: EXHAUST GAS RECIRCULATION SYSTEM PROBLEMS

Condition	Possible Cause	Remedy
Engine idles abnormally roughly and/or stalls	EGR valve vacuum hoses misrouted.	Check EGR valve vacuum hose routing. Correct as required.
	Leaking EGR valve	Check EGR valve for correct operation.
	Failed EGR valve gasket or loose EGR attaching bolts	Check EGR attaching bolts for tightness. Tighten as required. If not loose, remove EGR valve and inspect gasket. Replace as required.
	Improper vacuum to EGR valve at idle	Check vacuum from throttle body EGR port with engine at stabilized operating temperature and at curb idle speed. If vacuum is more than 1.0 in. Hg refer to throttle body idle diagnosis and EGR check chart for computer command control system.
Engine runs roughly on light throttle acceleration and has poor part load performance.	EGR valve vacuum hose misrouted.	Check EGR valve vacuum hose routing. Correct as required.
	Check for loose valve.	Torque valve.
	Sticky or binding EGR valve	Clean EGR passage of all deposits. Remove EGR valve and inspect. Replace as required.
	Wrong or no EGR gasket(s) and/or spacer	Check and correct as required. Install new gasket(s), install spacer (if used), torque attaching parts.
Engine stalls on decelerations.	Control valve blocked or airflow restricted.	Check internal control valve function per service procedure.
	Restriction in EGR vacuum line	Check EGR vacuum lines for kinks, bends, etc. Remove or replace hoses as required. Check EGR valve for excessive deposits causing sticky or binding operation. Replace valve.
	Sticking or binding EGR valve	Remove EGR valve and replace.
Part-throttle engine detonation[1]	Insufficient exhaust gas recirculation flow during part-throttle accelerations	Check EGR valve hose routing. Check EGR valve operation. Repair or replace as required. Check EGR passages and valve for excessive deposit. Clean as required. Check EGR per service procedure.
Engine starts but immediately stalls when cold.[2]	EGR valve hoses misrouted.	Check EGR valve hose routings.

[1]Nonfunctioning EGR valve could contribute to part-throttle detonation. Detonation can be caused by several other engine variables. Perform ignition and throttle body related diagnosis.
[2]Stalls after start can also be caused by throttle body problems.

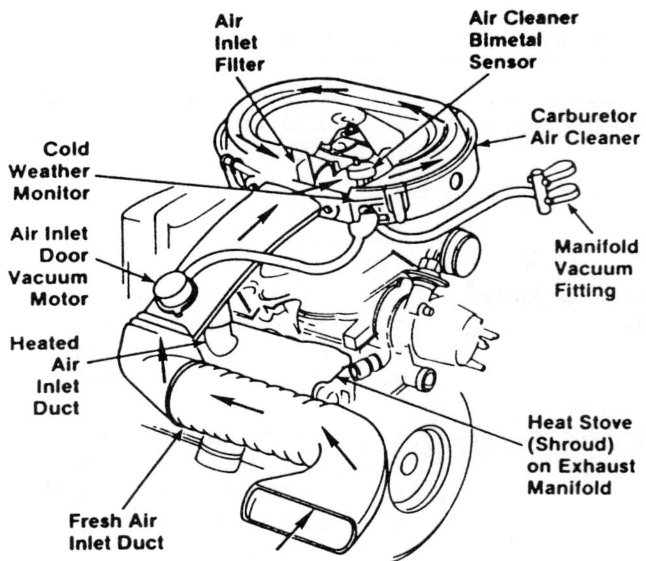

FIGURE 14-14 Typical system with conventional air cleaner

- *Heated Air Inlet Duct.* Directs air that has been warmed by the heat stove (shroud) on the intake manifold to the snorkel of the air cleaner.
- *Air Inlet Door Vacuum Motor.* Controls a flapper door inside the snorkel to admit manifold heated air or fresh air (or a mixture of both) into the air cleaner.
- *Air Cleaner Bimetal Sensor.* Regulates vacuum to the vacuum motor to determine the position of the air door. It is sensitive to air cleaner temperature.
- *Cold Weather Modulator (CWM).* Traps vacuum to the motor if manifold vacuum drops off due to the throttle opening while the air cleaner is cold.

The air cleaner bimetal sensor, which is installed in the air cleaner body (or air horn), senses the air cleaner temperature. Depending on the calibration, the sensor can be set to operate at 75 degrees Fahrenheit, 90 degrees Fahrenheit, or 105 degrees Fahrenheit. It is what controls the operating modes shown in Figure 14–15.

Some vacuum control systems use a retard delay valve (Figure 14–16) instead of a cold weather modulator. The difference is that the retard delay valve traps the vacuum for a few seconds when the throttle opens. Its function is the same: to prevent a change in the air door position if vacuum drops off because the throttle opens.

With a remote air cleaner, the functions are the same. As illustrated in Figure 14-17, however, the components are located differently:

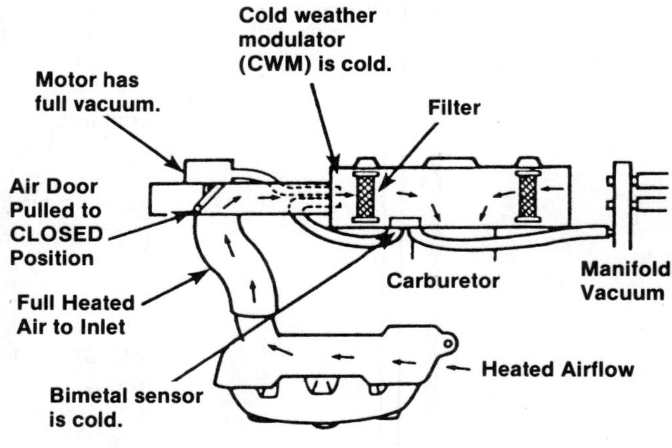

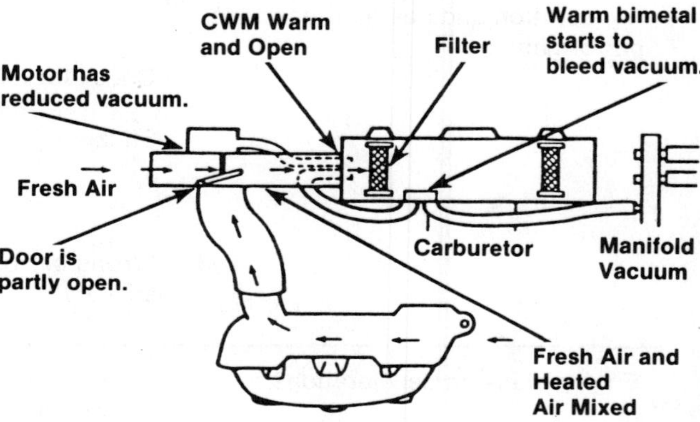

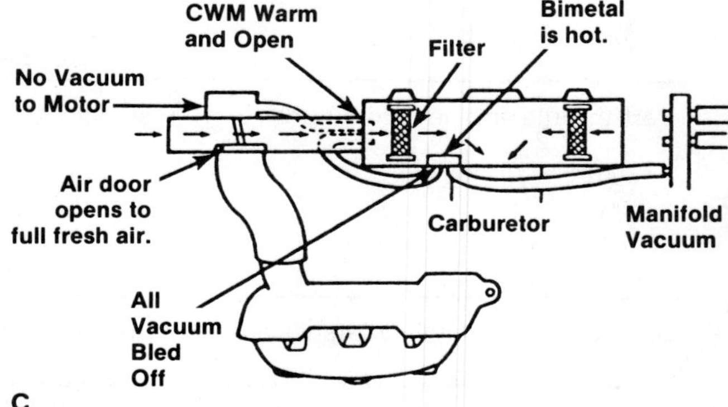

FIGURE 14-15 Air cleaner bimetal sensor at (A) cold start-up; (B) modulating partial warm-up; (C) hot engine/hot ambient air

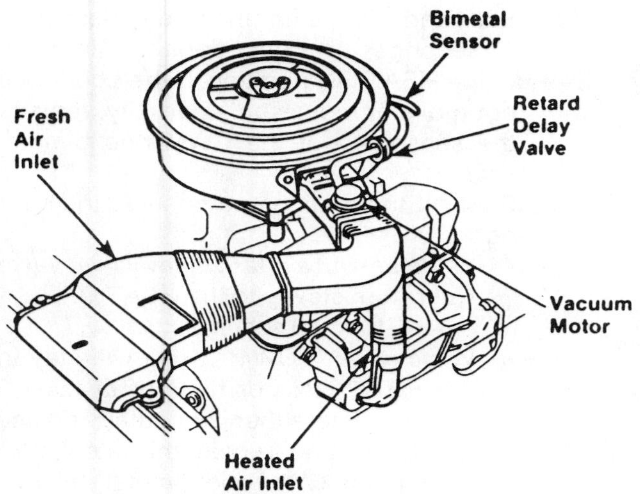

FIGURE 14-16 Conventional air cleaner with retard delay valve

FIGURE 14-18 Air cleaner bimetal sensor operation

- The bimetal sensor is in the air horn assembly.
- The air inlet door vacuum motor and door are in the air cleaner assembly instead of the inlet snorkel.

The bimetal sensor (Figure 14-18) is installed in the air cleaner body. It senses air cleaner temperature and starts to bleed off vacuum near its setting. The sensing element is a bimetal spring, which is linked to a sensing valve. Several temperature settings are available. The sensor is calibrated to provide a specific output vacuum to the air door motor as it warms to its temperature setting. The calibration is based on 16 inches of source vacuum.

Heated Air Inlet System Inspection and Performance Test

To perform an inspection and performance test on a typical heated air inlet system (Figure 14–19), proceed as follows:

1. Apply the parking brake and block the wheels.
2. Remove the air cleaner cover and element. Inspect the heated air duct for proper in-

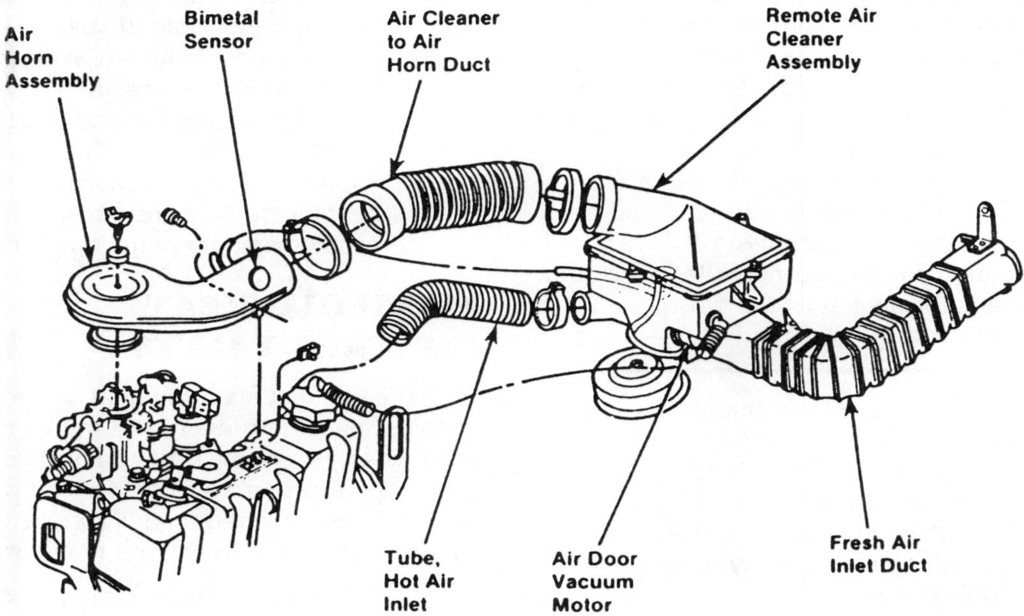

FIGURE 14-17 Remote air cleaner

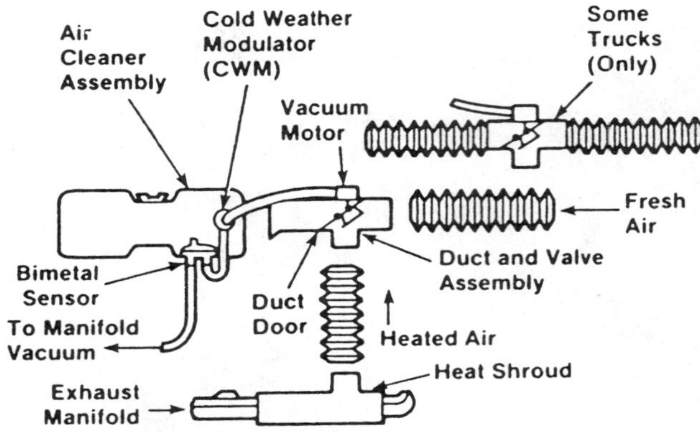

FIGURE 14-19 Typical heated air inlet system

stallation and/or damage. Service as required.

3. Remove components as necessary to ensure that the duct door is in the OPEN TO FRESH AIR position. If the door is in the CLOSED TO FRESH AIR position, check for binding and sticking. Service or replace as required.
4. Check the vacuum source and integrity of the vacuum hoses to the bimetal sensor, cold weather modulator (CWM), and vacuum motor.
5. Start the engine. If the duct door has moved to the HEAT ON position (closed to fresh air), proceed to step 7. If the door stays in the HEAT OFF position (closed to warm air), place a finger over the bleed of the bimetal sensor. The air duct door must move rapidly to the HEAT ON position. If the door does not fully move to this position, stop the engine and test the vacuum motor. Repeat this step with a new vacuum motor. The ambient air must be at least 60 degrees Fahrenheit during this test.
6. With the engine off, cool the bimetal sensor and CWM by spraying with liquid from a small can of refrigerant R-12 with a proper adapter for 20 seconds after the liquid contacts the sensor and CWM.

WARNING: Do not cool the bimetal sensor while the engine is running. If refrigerant R-12 is drawn into the intake system while the engine is running, poisonous phosgene gas will be expelled into the test area. Perform this test only in a well-ventilated area. Wear goggles when using refrigerant.

---

7. Start and run the engine briefly (less than 15 seconds). The duct door should move to the HEAT ON position. If the door does not move or moves only partially, replace the sensor. Cool the CWM and bimetal sensor.
8. Shut off the engine and observe the duct door:
   - *Vehicles with a retard delay valve:* Valve will return slowly to the HEAT OFF position (10 to 30 seconds).
   - *Vehicles with CWM:* Valve will stay in the HEAT ON position for at least 2 minutes. If less than 2 minutes, replace the CWM and repeat this step after cooling the CWM and bimetal sensor.

Solutions to common heated air inlet problems are given in Table 14-2.

Troubleshooting the Heated Inlet Control System

In a modern gasoline engine, emission controls that have any effect on the mixture necessarily have an effect on driveability, since all the systems are calibrated to operate as one integral system. Problems with the heated air inlet fall into the following groups:

- Obstructions to airflow, causing hard starting, performance problems, and poor fuel economy
- Vacuum loss, causing cold driveability problems since the system goes to full fresh air
- Air leaks, which "bypass" the control and usually cause cold driveability problems
- Vacuum trap failure (CWM or retard delay valve), which causes stumble on cold acceleration because the engine suddenly gulps in cold air
- Failure to switch to fresh air, which can overheat the mixture and cause warm driveability problems and detonation

MANIFOLD HEAT CONTROL VALVES

The exhaust manifold heat control valve routes exhaust gasses to warm the intake manifold heat riser when the engine is cold. This heats the air/fuel mixture in the intake manifold and improves ventilation. The result is reduced HC and CO emission in the exhaust. The two general types of valves are:

1. **Vacuum-Operated (Figure 14-20).** Some V-8 engines use a vacuum-operated valve,

TABLE 14-2: EEC SYSTEM PROBLEMS AND THEIR REMEDIES

Condition	Possible Cause	Remedy
Gasoline odor	Carburetor bowl vent solenoid or vacuum valve (VBVV or V/TBVV) failed in closed or restricted condition. Disconnected/damaged hoses to carburetor fuel bowl vent or canister. Fill cap gasket damaged or relief valve leaks. Vapor separator or hose leaks. Canister damaged or filled with liquid (gas or water). Kinked hose from vapor separator to canister. Carburetor bowl vent line blocked or restricted.	Inspect system very carefully. Correct faulty situation or replace faulty components.
Gasoline odor Possible flooding in some cases.	Purge line blocked. Purge valve failed to close. Disconnected or leaking vacuum signal port on purge valve. Purge solenoid failed to close or is not energized. Purge valve damaged by belts or pulleys. Canister loaded with liquid fuel.	Inspect system very carefully. Correct faulty situation or replace faulty components.
Hard starting Poor idle Poor performance Purging whenever engine runs/cranks	Purge valve or solenoid failed to open. Leaking signal port on purge shutoff valve.	Check purge valve or solenoid and replace if necessary.
Engine vacuum leak, no purging; canister loaded with liquid fuel	Purge line leak downstream of valve.	Repair leak.
Lean hesitation Vacuum in bowl during purge Poor cruise performance	Bowl vent valving failed to open.	Replace valve.

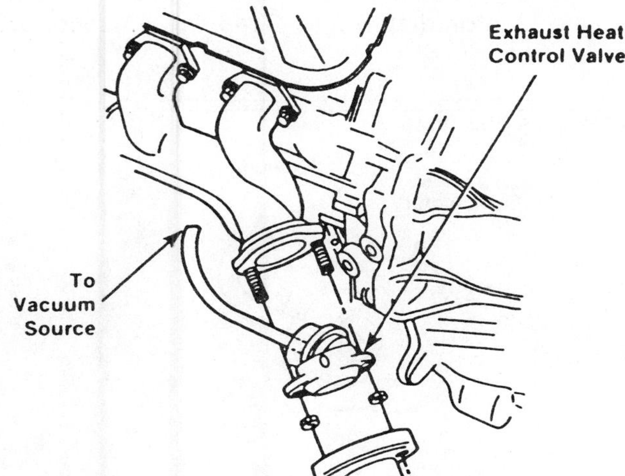

FIGURE 14-20 Vacuum-operated valve

which is bolted between the left-side exhaust manifold and exhaust pipe. The vacuum diaphragm connects to the manifold vacuum through a ported vacuum switch (PVS). On EEC-controlled engines, the vacuum system also includes an electric solenoid-operated vacuum valve.

2. **Thermostat-Operated (Figure 14-21).** A thermostat-operated valve is used mostly on 6-cylinder engines. It has the same function as the vacuum-operated valve. It operates as follows:
   - *Cold Engine.* The thermostat closes the valve to block exhaust gas flow from the manifold, which forces the gas to flow up through the heat riser and then to the exhaust pipe.

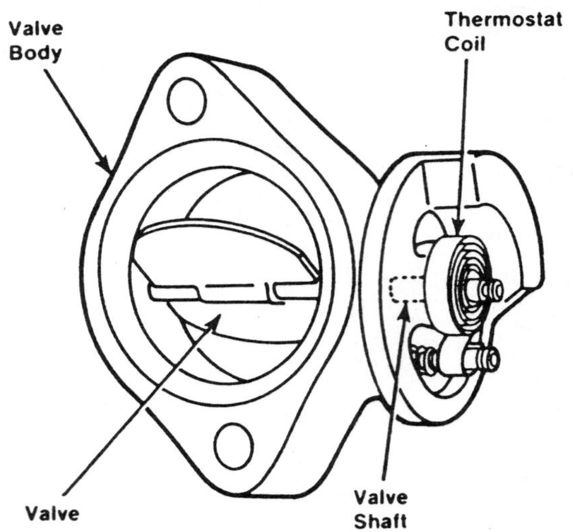

FIGURE 14-21 Thermostat-operated valve

- *Warm Engine.* The thermostat opens the valve to a position that seals off the heat riser passage. Exhaust gasses flow directly to the exhaust pipe.

Some V-8 engines use a more complicated manifold heat control valve, which is called a power heat control valve (Figure 14-22). It works similarly to the vacuum-controlled EFE, but is designed specifically to work with a minicatalyst as well as to preheat the air/fuel mixture for improved cold engine driveability. A vacuum actuator keeps the power heat control valve closed during warmup. All right-side exhaust gas travels up through the intake manifold crossover to the left side of the engine. Then, all exhaust gas from the engine passes through a miniconverter just down from the left manifold. This converter warms up rapidly because it is small and close to the engine. Its rapid warmup reduces exhaust emissions. As the engine and main converter warm up, a coolant-controlled engine vacuum switch (CCEVS) closes, which cuts vacuum to the actuator and allows the valve to open. Exhaust gas flows through both manifolds into the exhaust system and main converter.

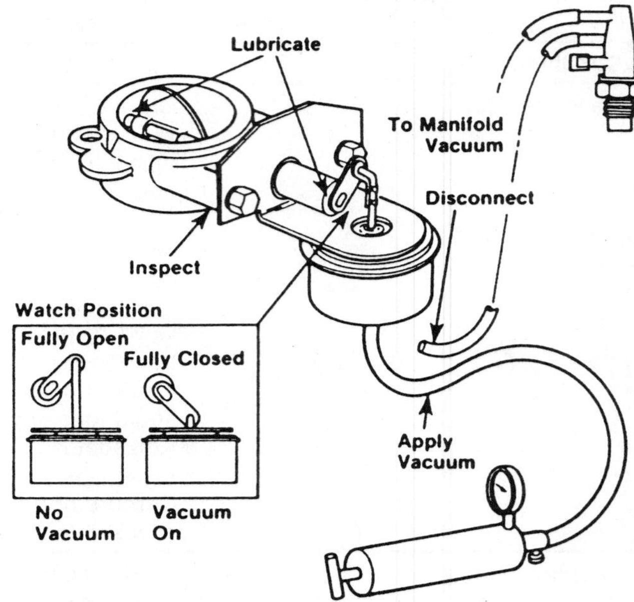

FIGURE 14-23 Testing a vacuum-operated valve

Testing and Maintenance of a Vacuum-Operated Valve

To check a vacuum-operated valve, proceed as follows (Figure 14-23):

1. Inspect the valve for any abnormal condition. Repair or replace it as necessary.
2. Disconnect the hose from the PVS.
3. Apply 15 inches of vacuum to the vacuum motor diaphragm. Trap for 60 seconds.

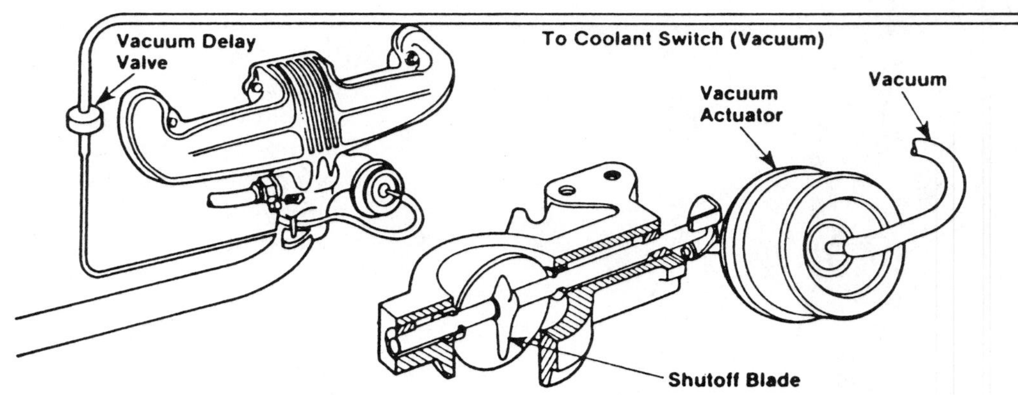

FIGURE 14-22 Power heat control valve

The valve must leak no more than 2 inches of vacuum in 60 seconds.

4. Watch the position of the vacuum motor stem. It must go to the fully closed position with the vacuum on and to the fully open position when the vacuum is released.
5. If necessary, lubricate the shaft with a graphite lubricant.

Functional Testing a Vacuum-Operated Valve System

To check the function of a vacuum-operated valve, proceed as follows:

1. With the engine off, note the position of the valve motor stem. It should be fully extended (Figure 14-24).
2. Start the engine cold. The stem should pull into the vacuum motor housing.
3. If the stem does not move, stop the engine. Attach a vacuum pump to the vacuum motor diaphragm and apply 15 inches of vacuum. The valve stem must pull in and stay in with the vacuum trapped. It may not leak more than 2 inches in 60 seconds.
4. Lubricate the valve shaft if the stem sticks when pulling in or releasing.
5. If the valve and motor are all right, check the vacuum system. Look for a leaking or restricted hose; repair as necessary. Check the hoses for proper routing.
6. If the valve closes with the engine started, let it warm. When the coolant warms, it should close.

If the valve operates with a separate vacuum source, but not from the PVS, test the PVS as described later in this chapter.

Because of its location, the exhaust heat control valve can stick if the shaft is not serviced regularly with graphite lubricant or heat control solvent. Sticking closed can cause overheating, detonation, and hot performance problems. Sticking cold can cause poor idle or poor performance cold. A vacuum loss usually causes the valve to fail closed, resulting in cold performance conditions.

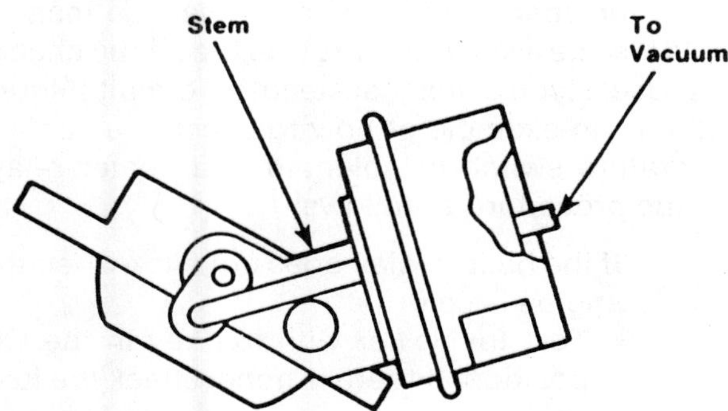

FIGURE 14-24 Valve motor stem should be fully extended.

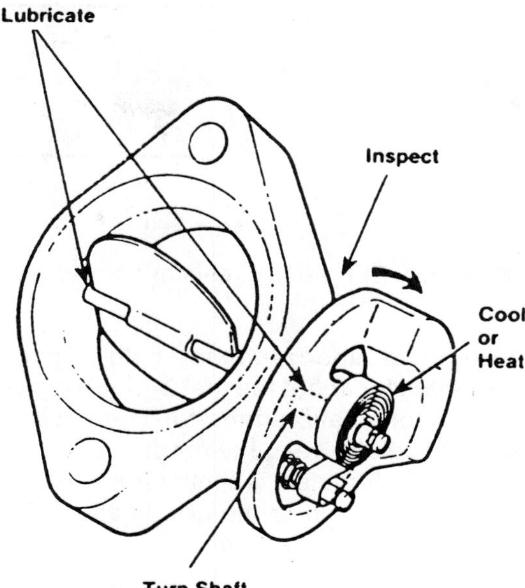

FIGURE 14-25 Inspect the valve assembly for damage.

Thermostat-Operated Heat Control Valve

Although a thermostat-operated heat control valve is not as common as the vacuum type, some vehicles have them. To inspect and service this type of valve, proceed as follows:

1. Inspect the valve assembly for damage (Figure 14-25). Replace or repair it as necessary.
2. Turn the valve shaft by hand to see if it is free. It must turn freely and return to the CLOSED position when cold.
3. Cool or heat the thermostat as required to check for proper opening or closing.
4. Lubricate the valve with a graphite lube.

EARLY FUEL EVAPORATION (EFE) CONTROL

The early fuel evaporation (EFE) heater contains a resistance grid that heats the mixture from the primary venturi of the carburetor (Figure 14-26). Its purpose is the same as a manifold heat control

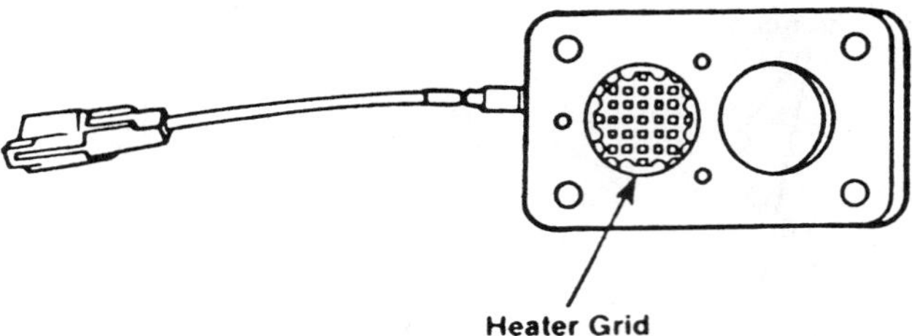

FIGURE 14-26 EFE heater resistance grid

valve: to improve vaporization in a cold engine. The heater operates for about the first 2 minutes, permitting leaner choke calibrations for improved emissions without cold driveability problems.

The basic EFE system is similar from one engine to the next. In addition to the grid heater, EFE has two other important components:

1. **Coolant Temperature Switch.** The EFE temperature switch or solenoid mounts to the engine, usually on the bottom of the intake manifold (Figure 14-27). The switch is closed when its temperature is below a specified temperature (generally between 130 to 150 degrees Fahrenheit). The switch opens as the engine coolant temperature goes above the specified temperature (see the service manual for the exact temperature).

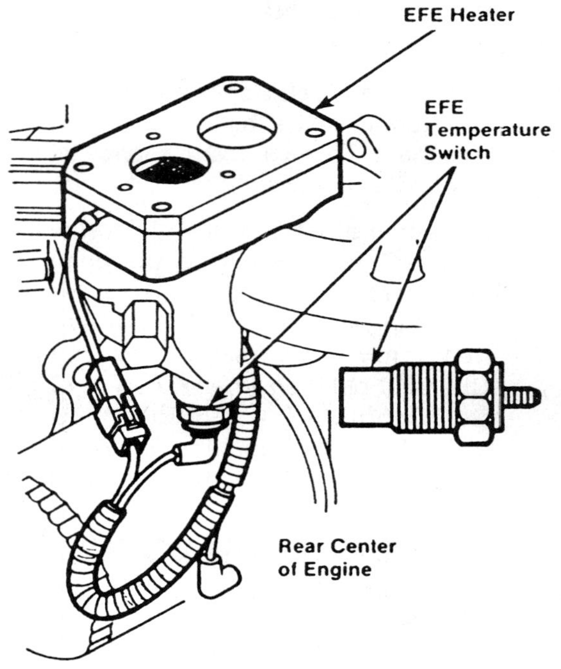

FIGURE 14-27 The EFE temperature switch is mounted on the engine.

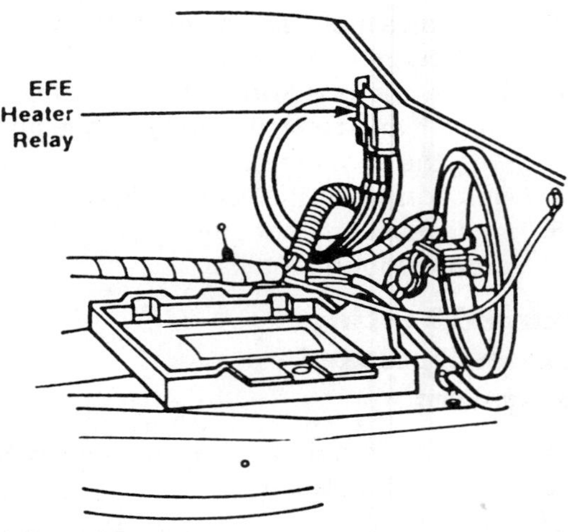

FIGURE 14-28 The EFE heater really is usually mounted on the body of the vehicle.

2. **EFE Heater Relay.** The temperature switch controls the EFE relay or valve. It powers the EFE heater when the temperature switch is cold and closed. After the engine has warmed up and the EFE is no longer needed, the relay deenergizes and the grid heater turns off. The EFE heater relay usually mounts on the body of the vehicle (Figure 14-28).

EFE Heater Check

To understand the EFE heater system operation, look at the electrical diagram shown in Figure 14-29.

An open circuit in EFE will cause no heating of the mixture on a cold engine resulting in performance problems the first minute or two of cold operation. If the heater is powered continually for some reason, it can cause warm engine driveability problems and possibly overheating and detonation.

To perform a typical EFE electrical system check, use a 12-volt test light and jumper wires as needed for open-circuit testing. A service manual should also be available to provide the circuit checkpoint. Use a typical manual electrical circuit (Figure 14-29) as an example of how to check the coolant temperature switch or solenoid and heater relay. Start the procedure as follows:

1. If the heater relay does not click when the engine is cold:
   - Test for power with the key in the ON position. If there is none, check the fuse link.
   - Remove the wire at point 45 in Figure 14-29 and jump to ground. Replace the

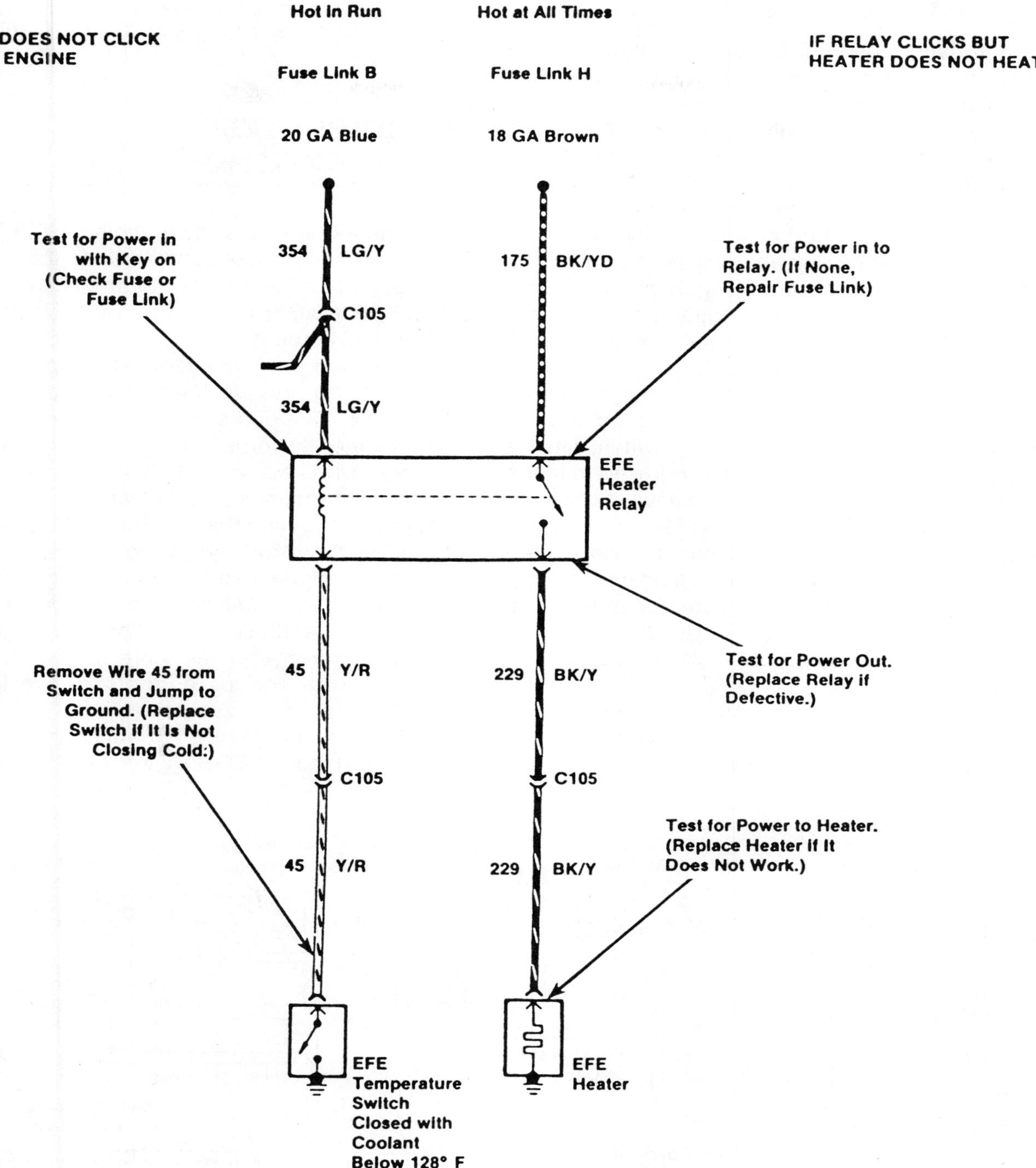

FIGURE 14-29 Testing an electric circuit

switch if it is closing when the engine is cold.

2. If the relay clicks but the heater does not, check the following:
   - Test for power at the heater relay. If there is none, repair the fuse link.
   - Test for power out of the relay. Replace the relay if it is defective.
   - Test for power to the heater. Replace the heater if it does not work.
3. If the heater is on constantly, check the following:
   - Test to be sure that the coolant switch fails to close.
   - Test to be sure that there is no short to ground in the electrical circuit.

Vacuum EFE Controls

Although most EFE systems are similar to the one just described, some use vacuum in place of an electrical arrangement. The method of testing a vacuum system is basically the same as the heated air inlet controls described earlier in the chapter.

Before testing the coolant temperature vacuum switches, carefully inspect the vacuum lines. A cracked, pinched, or burned hose can cause an EFE malfunction. Then, with the engine running cold (coolant at room temperature), apply vacuum from a pump to the inlet port of the thermal vacuum switch. Check for vacuum at the outlet port. If there is none, replace the thermal vacuum switch.

With the engine warmed up, the thermal vacuum switch must be closed. Remember that on engines that use an oil temperature sensitive switch, it might take up to 10 miles of cold weather driving to reach the temperature at which the switch will close.

To check an EFE relay or valve for proper operation, inspect the valve and linkage for obvious damage. Apply at least 10 inches of vacuum from a vacuum source of the EFE valve actuator motor. The valve should move freely to the CLOSED position, and the vacuum diaphragm should hold vacuum for at least 1 minute.

If the valve binds, lubricate it with special heat valve lubricant following directions on the can. If the lubricant will not free the valve, it must be replaced. If the diaphragm lets the vacuum leak down in less than a minute, replace the vacuum actuator.

EXHAUST GAS RECIRCULATING (EGR) SYSTEMS

The exhaust gas recirculating (EGR) systems (Figure 14–30) in use today reduce the amount of oxides of nitrogen emitted by the exhaust system. The EGR system dilutes the air/fuel mixture with controlled amounts of exhaust gas. Since exhaust gas does not burn, this reduces the peak combustion temperatures. At lower combustion temperatures, very little of the nitrogen in the air combines with oxygen to form NO_x. Most of the nitrogen is simply carried out with the exhaust gases. For driveability/performance, it is desirable to have the EGR valve opening (and the amount of gas flow) proportional to the throttle opening. Driveability is also improved on most applications by shutting off the EGR when the engine is started up cold, at idle, and at full throttle. Since the NO_x control requirements vary on different engines, there are several different systems with various controls to provide these functions.

When first introduced to emission control, there were two types of EGR systems that regulated the

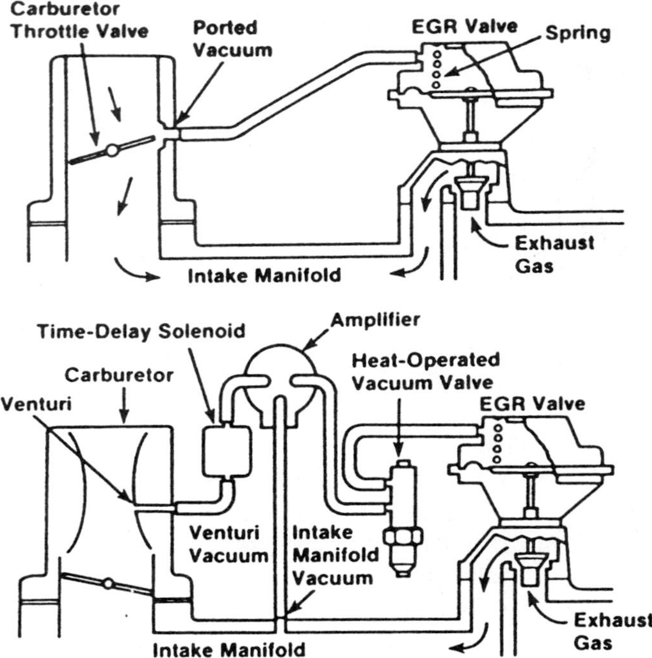

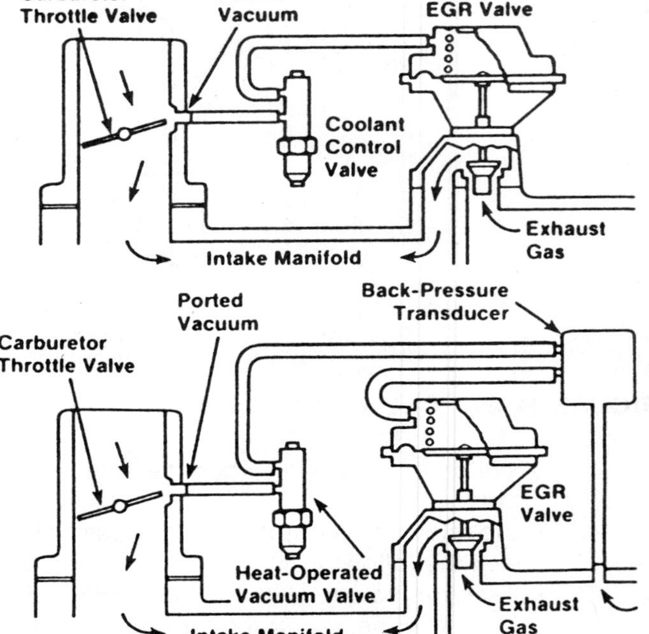

FIGURE 14-30 Summary of EGR variations

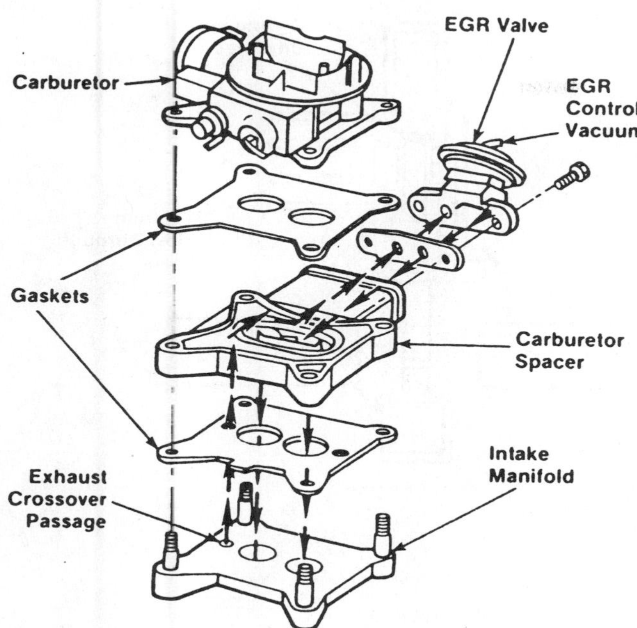

FIGURE 14-31 How the EGR system works

amount of exhaust gasses entering the intake system. One was the floor jet system, which utilized two orificed jets built into the intake manifold. The other was the valve-controlled system, which utilized a vacuum-operated valve. In the mid-1970s, the floor jet system was abandoned by the auto industry and the valve-controlled system became the universally accepted system.

This system uses a vacuum-operated EGR valve to regulate the exhaust gas flow into the intake manifold. Exhaust crossover passages under the intake manifold channel the exhaust gas to the valve. (Some in-line engines route the exhaust gas to the valve through an external tube.) Typical mounting of the EGR valve is either on a plate under the carburetor or directly on the manifold. Figure 14-31 illustrates how the basic valve system operates:

1. The EGR valve is a vacuum-operated, flow control valve. On most systems, it is attached to a carburetor spacer.
2. The carburetor spacer is sandwiched between the carburetor and intake manifold.
3. Gaskets are used above and below the spacer to seal the EGR system and the carburetor-to-manifold air/fuel flow.
4. A small exhaust crossover passage in the intake manifold admits exhaust gasses to the spacer. These gasses flow through the spacer to the inlet port of the EGR valve.
5. Opening the EGR valve by control vacuum at the diaphragm allows exhaust gasses to flow through the valve and back to another port of the spacer.
6. Here, the exhaust gas mixes with the air/fuel mixture, leaving the carburetor and then entering the intake manifold. The effect is to dilute or lean-out the mixture so that it still burns completely but with a reduction in combustion chamber temperatures.

TYPES OF EGR VALVES

Several types of control valves are used in EGR systems. The more common types are:

- Ported EGR valve
- Remote back pressure transducer EGR valve
- Positive integral back pressure transducer EGR valve
- Negative integral back pressure transducer EGR valve

The EGR system works when the engine reaches operating temperature or when the engine is operating under conditions other than idle or wide-open throttle. EGR systems include various functions that control the operation of the EGR valve. Some applications use cold engine EGR lockout and wide-open throttle EGR lockout. Basically, cold EGR lockout is necessary to keep the EGR valve closed during cold engine operation. Wide-open throttle EGR lockout might be required to keep the EGR valve closed when the engine is under maximum load. The following are various controls that relate directly to the EGR system:

- *Thermal Vacuum Switch (TVS).* Senses the air temperature in the carburetor air cleaner to control vacuum to the EGR valve (Figure 14-32). When the engine reaches operating temperature, the TVS opens to supply vacuum to the EGR valve. This opens the EGR valve for exhaust gas recirculation.

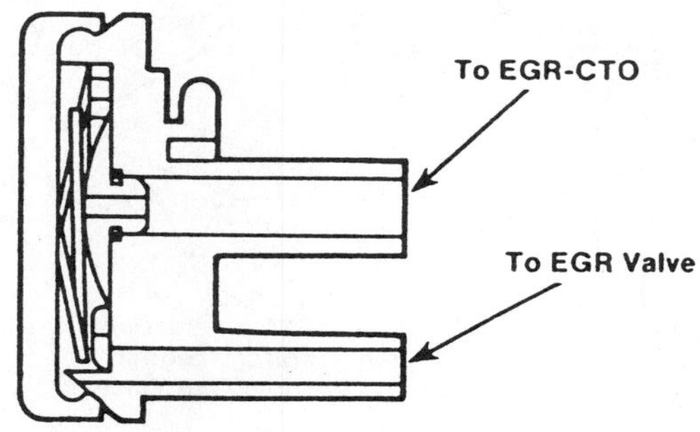

FIGURE 14-32 Thermal vacuum switch

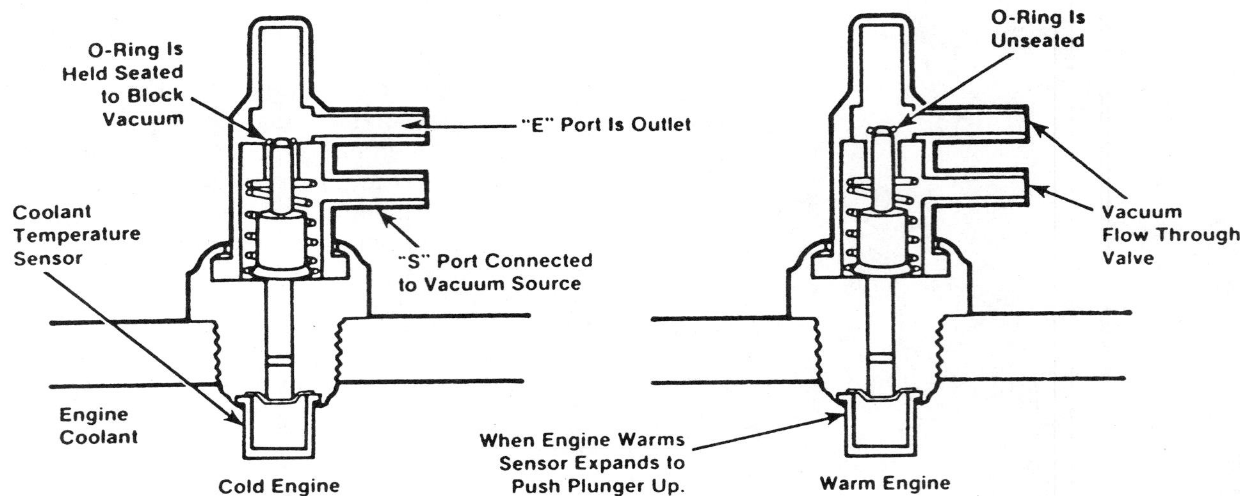

FIGURE 14-33 How the two-port PVS switch works

- *Ported Vacuum Switch (PVS).* Senses the coolant temperature to control vacuum to the EGR valve (Figure 14-33). The PVS operates in the same manner as the TVS, except it senses the coolant temperature instead of the air temperature. That is, the PVS function is to cut off vacuum to the EGR valve when the engine is cold and connects the vacuum to the EGR valve when the engine is warm.
- *Coolant Controlled Exhaust Gas Recirculation (CCEGR).* Operation of the CCEGR is the same as the PVS, but another manufacturer uses and labels it differently (Figure 14-34).
- *Venturi Vacuum Amplifier (VVA).* Some EGR systems use the VVA so that the carburetor

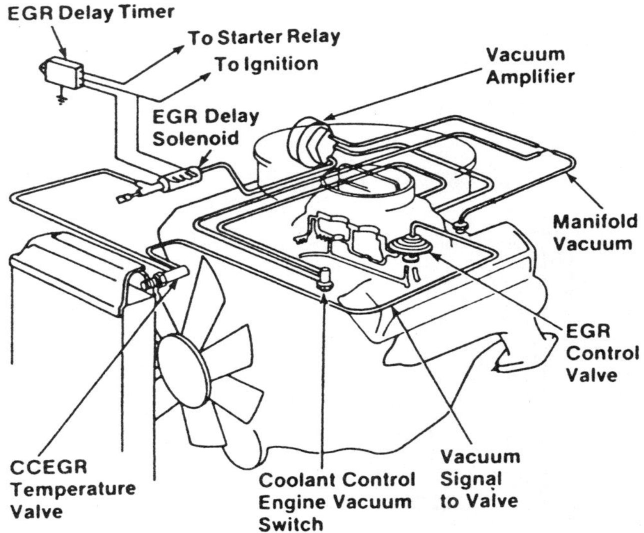

FIGURE 14-34 EGR system

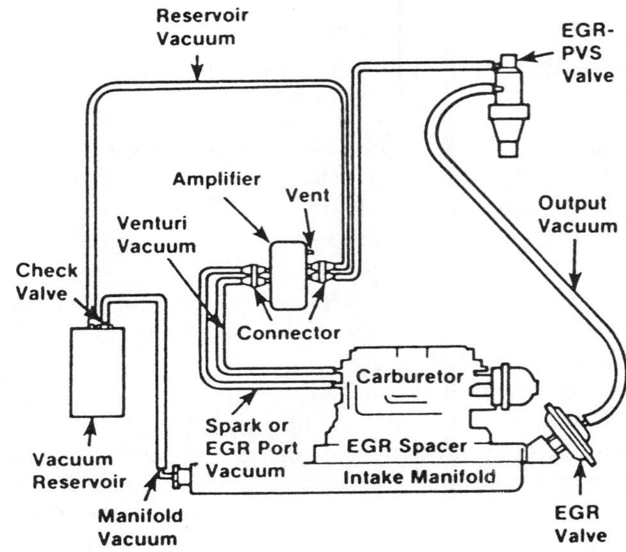

FIGURE 14-35 Venturi vacuum port

venturi vacuum can control the EGR valve operation (Figure 14-35). Venturi vacuum is more desirable because it is in proportion to the airflow through the carburetor. Since the venturi vacuum is a relatively weak vacuum signal, the VVA converts it to a strong enough signal to operate the EGR valve. The VVA system uses the manifold vacuum for the strength and venturi vacuum for the control signal.

- *EGR Delay Timer Control.* Some vehicles have an EGR delay system, which consists of an electrical timer that connects to an engine-mounted solenoid. Together, the purpose of the delay timer and solenoid is to prevent

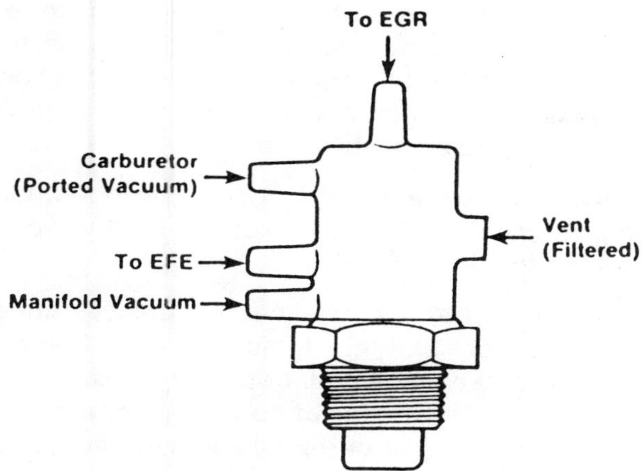

FIGURE 14-36 Early fuel evaporation/thermal vacuum switch

EGR operation for a predetermined amount of time after warm engine startup. On cold engine startups, the TVS and/or PVS valve override the delay timer.

- *Early Fuel Evaporation/Thermal Vacuum Switch (EGR—EFE/TVS).* In most common applications the EFE uses a valve that increases the exhaust gas flow under the intake manifold during cold engine operation through a crossover passage to heat up the incoming air/fuel charge (Figure 14-36). The EFE is vacuum-operated and controlled by a TVS that applies vacuum to the EFE valve when the coolant temperature is low. Once the engine reaches operating temperature, the TVS will block off vacuum to the EFE and direct it to the EGR valve for EGR operation.
- *Wide-Open Throttle Valve (WOT).* Some applications use the WOT valve where it is desirable to cut off EGR flow at wide-open throttle (Figure 14-37).

These various EGR system controls represent some of the common controls currently used by automobile manufacturers. Control devices used in the various systems might have different labels but actually complete the same function within the EGR system. For further information on EGR system controls, it is advisable to consult the service manual that pertains to each vehicle being tested or serviced.

EGR SYSTEM DIAGNOSIS AND SERVICE

Manufacturers calibrate the amount of EGR gas flow for every engine. If there is too much or too little, it can cause performance problems by changing the engine breathing characteristics. Also, with the little EGR flow, the engine can overheat and detonate. Typical problems that show up in ported EGR systems are:

- *Rough Idle.* Possible causes are an EGR valve stuck open, PVS fails to open, dirt on the valve seat, or loose mounting bolts. Loose mounting will cause a vacuum leak and a hissing noise. Diagnose a stuck valve by a functional test and visual inspection.
- *Surge, Stall, or Will Not Start.* Probable cause is the valve stuck open.
- *Detonation (Spark Knock).* Any condition that prevents proper EGR gas flow can cause detonation. This includes a valve stuck closed, leaking valve diaphragm, restrictions in flow passages, EGR disconnected, or a problem in the vacuum source. Detonation on engines with high spark advance is serious enough to destroy them.
- *Lead Poisoning.* If using leaded gas improperly, it can cause deposits on the seat and valve. This will restrict flow, causing detonation and possibly overheating.
- *Poor Fuel Economy.* This is an EGR condition only if it relates to detonation or other symptoms of restricted or zero EGR flow.

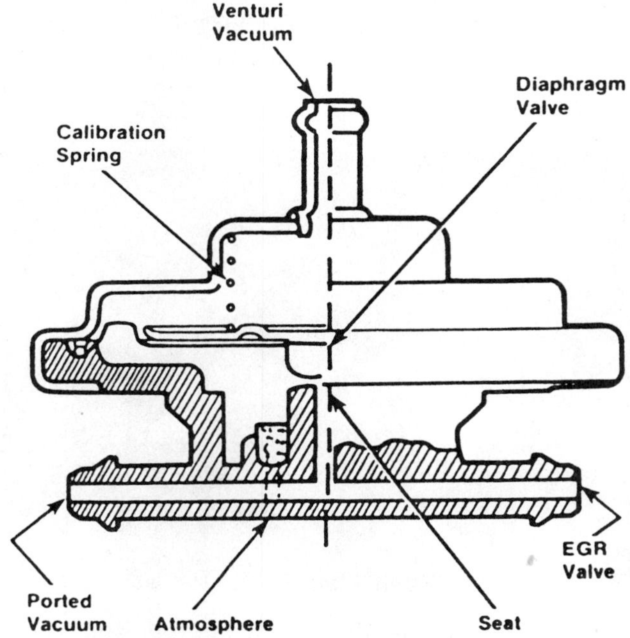

FIGURE 14-37 Wide-open throttle valve

EGR VALVES AND SYSTEMS TESTING

Test the EGR valve by using a vacuum gauge or a hand vacuum pump. Follow these procedures for using either piece of test equipment.

Vacuum Gauge. Use a vacuum gauge to check for excessive exhaust back pressure by using the following procedure:

1. Disconnect a vacuum line connected to an intake manifold port.
2. Put a vacuum gauge between the disconnected vacuum line and the intake manifold port.
3. Connect a tachometer.
4. Start the engine and gradually increase speed to 2000 rpm with the transmission in NEUTRAL.
5. The reading from the manifold vacuum gauge (Figure 14–38) should be above 16 inches of vacuum. If not, there could be an excessive back pressure in the exhaust system. To verify an excessive back pressure problem or a vacuum leak, perform the following:
   - Turn off the engine.
   - Disconnect the exhaust system at the exhaust manifold.
   - Repeat steps 4 and 5.
   - If 16 inches of vacuum is still not on the gauge, there is a blockage in the exhaust manifold. Visually inspect the exhaust manifold for the blockage.
   - If the reading is 16 inches of vacuum, the blockage is either in the muffler, exhaust pipe(s), or catalytic converter. If the problem is a damaged converter, inspect the muffler to be sure that converter debris is not in the muffler, causing an obstruction.

FIGURE 14–38 Reading from the manifold vacuum gauge should be above 16 inches mercury

Hand Vacuum Pump. Use a hand vacuum pump to check the operation of the EGR valve by following these procedures:

1. Check all vacuum lines for correct routing. Ensure that they attach securely. Replace cracked, crimped, or broken lines.
2. With the engine at normal operating temperature, be certain there is no vacuum to the EGR valve at idle.
3. Install a tachometer.
4. On EFI engines (multi-point injection), disconnect the throttle air bypass valve solenoid.
5. Remove the vacuum supply hose from the EGR valve port and plug the hose.
6. Start the engine and idle with the transmission in NEUTRAL, then observe the engine idle speed. If necessary, adjust idle speed to the emission decal specification.
7. Slowly apply 5 to 10 inches of vacuum to the EGR valve vacuum port, using a hand vacuum pump.
8. If any of the following occurs when applying vacuum to the EGR valve, replace the valve.
   - Engine does not stall.
   - Idle speed does not drop more than 100 rpm.
   - Idle speed does not return to normal (±25 rpm) after removing vacuum.
9. If the EGR valve is operating properly, unplug and reconnect the EGR valve vacuum supply hose.
10. Reconnect the throttle air bypass valve solenoid, if removed.

Typical EGR problems and their remedies are given in Table 14–3.

AIR INJECTION SYSTEMS

One of the earliest components used to control HC and CO was the air pump. Formal names of these systems included air injection reaction (AIR) by GM, thermactor emission (TE) by Ford, air guard by American Motors, and air injection system by Chrysler. The name used by auto mechanics in the field was smog pump.

TABLE 14-3: HEATER AIR INLET SYSTEM PROBLEMS AND THEIR REMEDIES

Condition	Possible Cause	Remedy
Cranks normally but starts hard Hesitates or stalls on acceleration Lack of power Poor fuel economy	Debris in snorkel or air door Collapsed air duct Dirty air filter	Check for restricted airflow and correct the problem.
Hesitates on cold acceleration Lack of power Poor fuel economy	Vacuum leak Bimetal stuck in bleed-off Vacuum motor diaphragm leak Air door stuck open.	Correct to be sure system has heated air.
Warm driveability problems in cold ambient air conditions.	Heated air duct leaking or not installed Air cleaner not sealed tight Broken or leaking duct or air cleaner	Correct all fresh air leaks in the system.
Lack of power in warm engine Might overheat Might detonate	Air door stuck open Sensor fails to bleed off when it warms	Check air door motor and sensor. Repair or replace as necessary.
Poor acceleration cold Hesitates or stumbles on acceleration	No vacuum trap to air door motor (at full throttle) Check valve leaks in cold weather modulator Retard-delay valve leaks.	Check air door motor, CWM, and retard-delay valve; replace if faulty.

The basic air injection system, regardless of name, forces air from an air pump, through air manifolds, into the cylinder head exhaust ports or exhaust manifold. The fresh air or oxygen promotes after-burning of any combustibles remaining in the exhaust gasses flowing from the combustion chamber. Carburetor calibrations are generally slightly richer, which also permits improved driveability.

Earlier this chapter states that air injection is a post-combustion emission system. Actually the typical air injector system has elements of both post- and precombustion systems. That is, the air injection system introduces extra air (oxygen) into the exhaust stream of the engine. In most cases, combustion in the cylinders is limited by the amount of oxygen available to sustain burning; so when introducing extra air into the hot exhaust system, all the remaining fuel oxidates, or burns. Since air injection helps the process of combustion, it qualifies as a precombustion device. But the fact that the combustion process continues in the exhaust system (and requires additional components) qualifies air injection as a post combustion system. However it is classified, air injection is an effective system that economically reduces HC and CO emissions.

The typical air injector system used in most vehicles consists of a belt-driven vane pump (Figure 14-39); a vacuum-operated diverter valve to vent pump air to the atmosphere during deceleration so the combustion of a rich fuel mixture and oxygen does not cause backfiring; a pressure relief valve that allows excess pump output to escape; a one-way check valve that allows air into the exhaust but prevents exhaust from entering the pump in the event the belt breaks; and the hoses and nozzles necessary to distribute and inject the air (Figure 14-40).

On most air pump-equipped models, the system has the ability to switch airflow from the exhaust manifold to the catalytic converter. During engine warmup, air injects directly into the exhaust manifold. But once the engine is warm, the extra air in the manifold would affect EGR operation, so air injection switches downstream to the converter, where it aids the converter in oxidizing emissions.

A vacuum-operated valve does the actual switching. A thermal vacuum switch controls vacuum to the valve. When the coolant is cold, it signals the switching valve to direct air to the exhaust manifold. Then when the engine warms to normal operating temperature, the thermal vacuum switch signals the switching valve to reroute the air to the converter. A relief valve in the switching valve vents extra air to the atmosphere. When air pump output rises with engine speed, the air pressure overcomes the relief valve spring pressure.

An antibackfire valve prevents backfire in the exhaust. There are two types of antibackfire valves, both of which the intake manifold vacuum controls. All late-model engines use an air bypass system

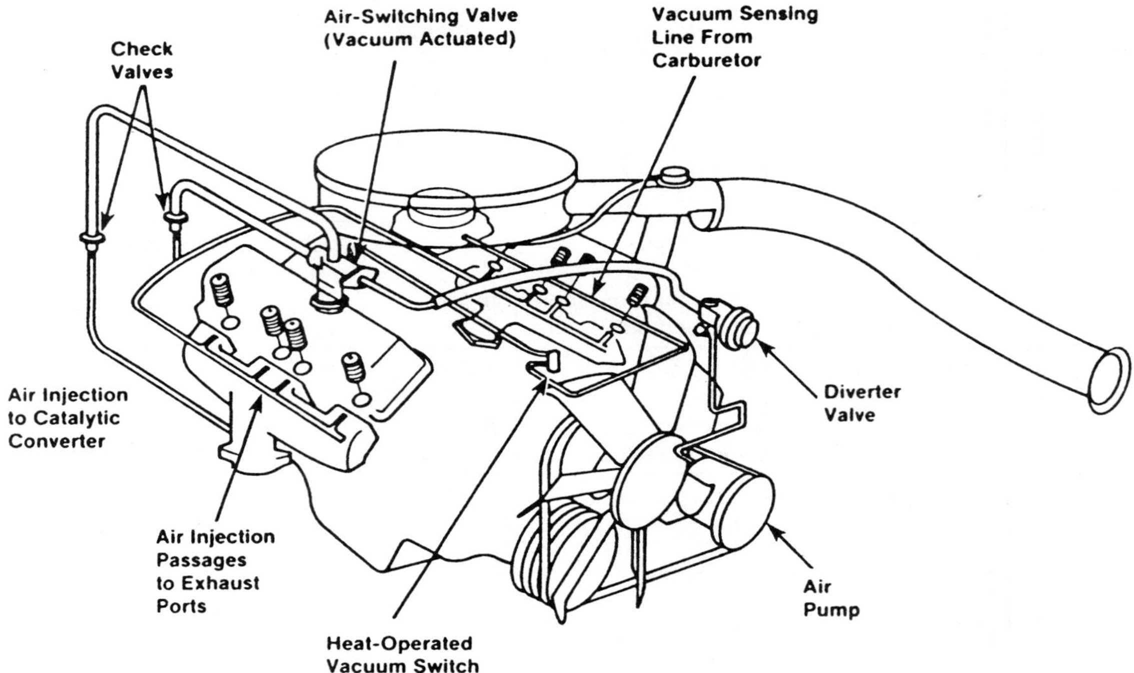

FIGURE 14-39 Engine-driven air pump

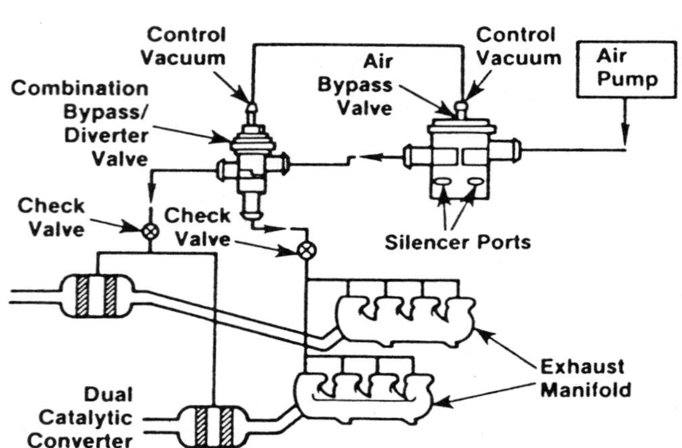

FIGURE 14-40 Typical managed air system

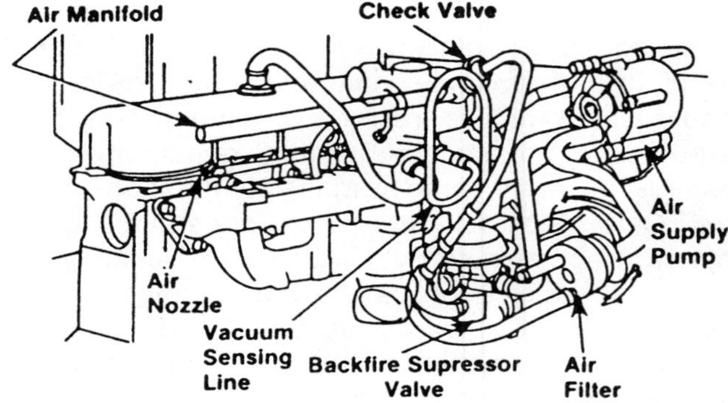

FIGURE 14-41 Air injection system with a bypass type of backfire-suppressor valve

(Figure 14–41), which momentarily bypasses or dumps the air pump output at the start of deceleration when manifold (engine) is highest. The diverter or deceleration bypass valve (Figure 14–42) directs airflow through a muffling device, which usually mounts on the air pump. After 1 or 2 seconds, the valve closes and again supplies air to the exhaust manifold.

In some newer systems, there is a vacuum differential valve (VDV), which is used in conjunction with an air bypass valve. The VDV shuts off the signal to the bypass valve momentarily whenever intake manifold vacuum rises or drops sharply. When the vacuum signal is gone, a spring inside the bypass opens a vent port and dumps the air pressure. Normal vacuum overrides the vent spring and air flows into the exhaust ports.

Older model engines and many imports use an air gulp type antibackfire valve, which accomplishes the same task in a slightly different manner. The gulp valve also monitors engine vacuum and, when it gets a strong signal, instead of dumping the air into the atmosphere, it routes pump air into the intake manifold leaning the mixture enough to prevent unwanted explosions in the exhaust system. Some gulpers do not use pump air. They simply open a passage to the air cleaner on deceleration and let the

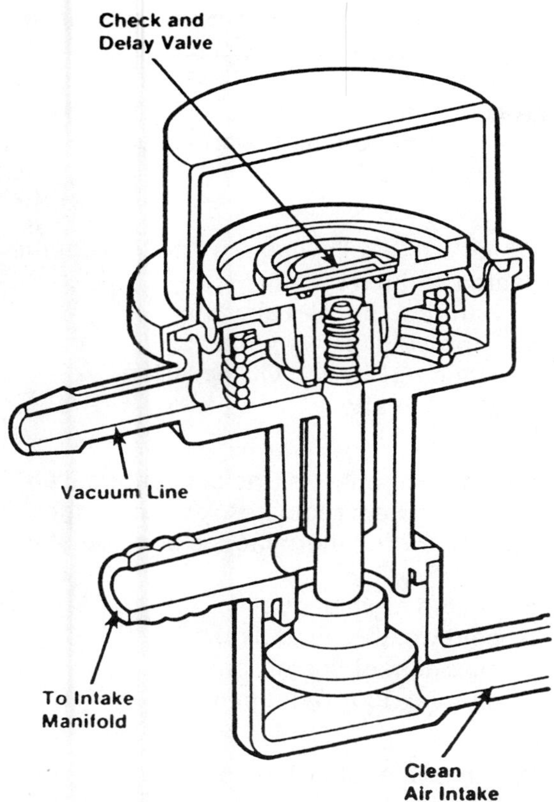

FIGURE 14-42 The deceleration valve should prevent exhaust backfire during deceleration.

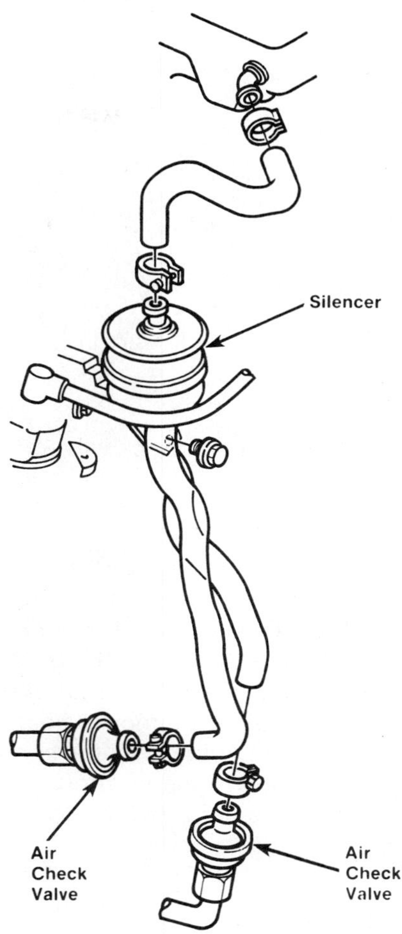

FIGURE 14-43 Typical pulse air system

engine, with its high vacuum, take as deep a breath as it can to lean the mixture.

In the late 1970s, the aspirator air system replaced the mechanical air pump. This system draws air into the system from the air cleaner. The aspirator valve, which contains a spring-loaded one-way valve, seats in a hose that runs between the air cleaner and the exhaust manifold. Negative pressure pulses in the exhaust system cause the aspirator valve to open and allow fresh air to mix with the exhaust gasses. When positive pulses occur in the exhaust, the one-way aspirator valve closes to prevent the hot exhaust gasses from backing up in the hoses.

Several other pulse air systems are in use. In one operation, the system uses the natural pulses already mentioned in this chapter to "pull" air from the air cleaner into the catalytic converters. The typical pulsed air system operates in the following manner:

- An air check valve and silencer (Figure 14-43) connects in-line between the air cleaner and the catalytic converter.
- When pressure in the exhaust system is greater than the pressure in the air cleaner, the reed valve (Figure 14-44) inside the air check valve closes.
- When the pressure in the exhaust system is less than the pressure in the air cleaner, the reed valve opens and air draws into the catalytic converter.
- The incoming oxygen reduces the hydrocarbons and carbon monoxide content of the exhaust gasses by continuing the combustion of unburned gasses in the same manner as the conventional system (with an air pump).

Air injector systems rarely cause problems. If one should cause a difficulty, check the vehicle's service manual.

SPARK ADVANCE SYSTEMS

Spark advance systems have been in use since the earliest gasoline engines. It was discovered that

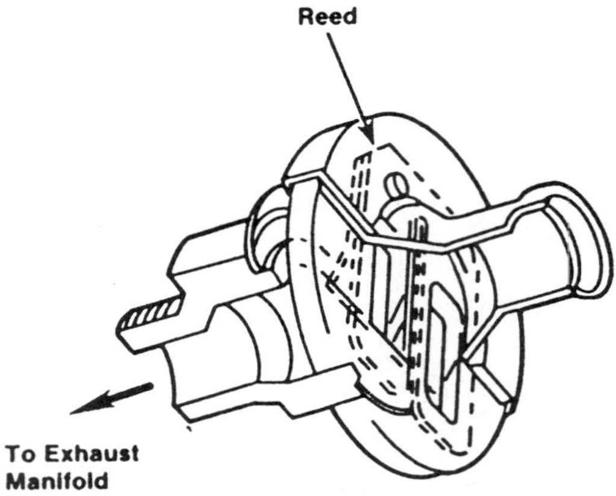

FIGURE 14-44 Typical air check valve

the proper timing of the ignition spark helped the car engine to reduce exhaust emission and develop more power output. Throughout the years each car manufacturer developed slightly different spark timing controls according to engine requirements and emission standards for each model year, but the systems and devices all operate on the same principles. Complete details on the operations and servicing spark advance systems can be found in Chapters 7 and 8.

REVIEW QUESTIONS

1. A faulty PCV valve that has failed to close will result in which of the following conditions?
   a. a dilution of the air/fuel mixture, resulting in rough idle
   b. trapped gasses in the crankcase combine with oil to create sludge
   c. an overly lean air/fuel mixture, resulting in hard, cold startup.
   d. All of the above

2. A vehicle exhibits flooding symptoms during startup. Technician A says the problem might be that the canister is filled with liquid or water. Technician B says the problem might be a failed open carburetor bowl vent solenoid. Who is right?
   a. Technician A
   b. Technician B
   c. Both A and B
   d. Neither A nor B

3. A carburetor bowl vent valve that has failed to close will result in ____________ .
   a. flooding
   b. lean conditions
   c. sludge buildup in the crankcase
   d. all of the above

4. Which of the following are functions of the heated air inlet system?
   a. warming the inlet air
   b. improving fuel vaporization
   c. reducing HC and CO emissions
   d. all of the above

5. Which component of the conventional air cleaner system regulates vacuum to the vacuum motor to determine the position of the air door?
   a. heated air inlet duct
   b. cold weather modulator
   c. air cleaner bimetal sensor
   d. air inlet door vacuum motor

6. Which of the following components is not a part of the early fuel evpaoration (EFE) control system?
   a. power heat control valve
   b. resistance grid
   c. coolant temperature switch
   d. EFE relay

7. Which of the following devices delays the introduction of exhaust gasses into the intake manifold until after the engine warms up?
   a. thermal vacuum switch
   b. ported vacuum switch
   c. delay timer control
   d. all of the above

8. During a check of the intake manifold vacuum, the vacuum reading at 2000 rpm was lower than the reading at idle. Technician A says the problem is an EGR valve failed in the closed position. Technician B says the problem is excessive exhaust back pressure. Who is probably correct?
   a. Technician A
   b. Technician B
   c. Both A and B
   d. Neither A nor B

9. Which of the following conditions can cause detonation?
   a. disconnected EGR
   b. valve stuck closed
   c. leaking valve diaphragm
   d. all of the above

10. When verifying an excessive exhaust back pressure problem, the reading is 16 inches of vacuum. Technician A says the blockage might be in the muffler. Technician B says the blockage might be in the catalytic converter. Who is right?
    a. Technician A
    b. Technician B
    c. Both A and B
    d. Neither A nor B

11. Which EGR control senses the air temperature in the carburetor air cleaner to control vacuum to the EGR valve?
    a. ported vacuum switch
    b. venturi vacuum amplifier
    c. delay timer control
    d. thermal vacuum switch

12. Which of the following is a part of a pulse air injection system?
    a. air pump
    b. antibackfire valve
    c. aspirator valve
    d. all of the above

13. Where is an air gulp type antibackfire valve likely to be found?
    a. import vehicles
    b. older engines
    c. both a and b
    d. neither a nor b

14. In which of the following systems could a vacuum leak affect drivability?
    a. EGR system
    b. evaporative emission control system
    c. PCV system
    d. all of the above

15. Leakage into the engine crankcase is called ____________ .
    a. blowby
    b. sludge
    c. combustion
    d. raw vapor

APPENDIX

GLOSSARY OF TERMS

A

Actuator A device that delivers motion in response to an electrical signal.

Additive As used with reference to automotive oils, a material added to the oil to give it certain properties. For example, a material added to engine oil to lessen its tendency to congeal or thicken at low temperature.

Advance Spark or ignition advance. Causing the ignition spark to occur earlier to compensate for faster engine operation or slower fuel combustion.

Air A gas containing approximately 4/5 nitrogen, 1/5 oxygen, and some carbonic gas.

Air Bypass Valve A valve in the air pump system. At high engine vacuum, it vents pressurized air from the air pump to the atmosphere in order to prevent backfiring. At other times it sends air to the exhaust manifold; on cars with a three-way catalyst, it sends air only to the oxidation catalyst when the engine warms up. Also known as the **anti-backfire valve** or **diverter valve.**

Air Charge Temperature Sensor A thermistor sensor responsible for inputting to the processor the temperature of an air stream in the air filter or intake manifold. If air is cold, signals choke to let off slowly. Alters engine speed after choke is off, and below a certain temperature dumps thermactor air for catalyst protection.

Air Cleaner A device for filtering, cleaning, and removing dust from the air admitted to a unit, such as an engine or air compressor.

Air Cleaner Bimetal Sensor Senses the temperature of incoming fresh air. Bleeds off vacuum when air cleaner is warm.

Air Cleaner Duct and Valve Vacuum Motor Opens and closes the air cleaner duct to provide heated or unheated air to the engine dependent upon the temperature of the incoming air.

Air Control Valve A vacuum-controlled valve in the thermactor system that diverts air pump air to either the upstream (exhaust manifold) or downstream (underbody catalyst) air injection points as required; a diverter valve or a combination bypass/diverter valve.

Air/Fuel Mixture A ratio of the amount of air that is mixed with fuel before it is burned in the combustion chamber.

Air Gap The space between spark plug electrodes, motor and generator armatures, field shoes, etc.

Air Lock A bubble of air trapped in a fluid circuit that interferes with normal circulation of the fluid.

Air Pump A device to produce a flow of air at higher than atmospheric pressure. Normally referred to as a thermactor air supply pump.

Alloy A mixture of different metals such as solder, which is an alloy consisting of lead and tin.

Altitude Compensation System An altitude barometric switch and solenoid used to provide better driveability at more than 4,000 feet above sea level.

Aluminum A metal noted for its lightness and often alloyed with small quantities of other metals for automotive use.

Ambient Temperature Temperature of air surrounding an object.

Ammeter An electrical meter used to measure current flow in amperes.

Ampere The unit of measure of current flowing in an electrical circuit.

Amplifier A circuit or device used to increase the voltage or current of a signal.

Annealing A process of softening metal. For example, the heating and slow cooling of a piece of iron.

Antibackfire Valve Often called a gulp valve, the antibackfire valve is located downstream from the air bypass valve. Its purpose is to divert a portion of the thermactor air to the intake manifold when it is triggered by intake manifold vacuum on deceleration. The valve operates only during periods of sudden decrease in intake manifold pressure.

Antifreeze A material such as alcohol, glycerin, etc., added to water to lower its freezing point.

Arcing Electrical energy jumping across a gap. Arcing across ignition points causes pitting and erosion.

Atmospheric Pressure The weight of the air at sea level; about 14.7 pounds per square inch; less at higher altitudes.
Automotive Emissions Gaseous and particulate compounds that are emitted from a car's crankcase, exhaust, carburetor, and fuel tank (hydrocarbons, nitrogen oxide, and carbon monoxide).

B

Backfire Suppressor Valve A device used in conjunction with the early design thermactor exhaust emission system. Its primary function is to lean out the excessively rich fuel mixture, which follows closing of the throttle after acceleration. Allows additional air into the induction system whenever intake manifold vacuum increases.
Barometric Pressure A sensor or its signal circuit that sends a varying frequency signal to the processor relating actual barometric pressure.
Battery A cell or group of cells connected together to deliver electrical energy (in the automobile battery, by chemical action).
Boiling Point The temperature at atmospheric pressure at which bubbles or vapors rise to the surface and escape.
Bore The diameter of a hole, such as a cylinder; also to enlarge a hole as distinguished from making a hole with a drill.
Boring Bar A stiff bar equipped with multiple cutting bits, which is used to bore a series of bearings or journals in proper alignment with each other.
Boss An extension or strengthened section, such as the projections within a piston, which support the in piston por piston pin bushings.
Bowl Vent Port Port in a carburetor that vents fumes and excess pressure from the bowl to maintain atmospheric pressure.
Breaker Arm The movable member of a pair of breaker points.
Breaker Cam The multilobed cam rotating in the ignition distributor, which interrupts the primary circuit to induce a high-tension spark for ignition.
Breaker Plate The plate in the distributor that supports the breaker points.
Breaker Points Contact points in the distributor that close and open the ignition primary circuit. (Also called ignition points, distributor points, or points.)
British Thermal Unit A measurement of the amount of heat required to raise the temperature of 1 pound of water 1 degree Fahrenheit.
Burnish To smooth or polish by the use of a sliding tool under pressure.
Bushing A removable liner for a bearing.
Bypass An alternate path for a flowing substance.
Bypass Valve Valve that opens under certain conditions to permit a flow of liquid or gas by some alternate to its normal route. See **Air Bypass Valve.**

C

Calibrate To determine or adjust the graduation or scale of any instrument giving quantitative measurements.
Caliper (Inside and Outside) An adjustable tool for determining the inside or outside diameter by contact and retaining the dimension for measurement or comparison.
Cam The rotating member of the distributor with lobes that open and close the breaker points as many times per distributor-shaft revolution as there are cylinders in the engine.
Cam Angle Also known as dwell period. Referring to an ignition distributor, the number of degrees of rotation of the ignition distributor shaft during which the contact points are closed.
Cam Ground Piston A piston ground to a slightly oval shape which, under the heat of operation, becomes round.
Camshaft The shaft containing lobes or cams to operate the engine valves.
Canister A container in an evaporative emission control system that contains charcoal to trap vapors from the fuel system.
Canister Purge Shutoff Valve A vacuum-operated valve that shuts off canister purge when the thermactor air diverter valve dumps air downstream.
Canister Purge Solenoid Electrical solenoid or its control line. Solenoid opens valve from fuel vapor canister line to intake manifold when energized. Controls flow of vapors between carburetor bowl vent and carbon canister.
Canister Purge Valve Valve used to regulate the flow of vapor from the charcoal canister to the engine.
Carbon A common nonmetallic element that is an excellent conductor of electricity. It also forms in the combustion chamber of an engine during the burning of fuel and lubricating oil.
Carbon Dioxide Compressed into solid form, this material is known as dry ice and remains at a temperature of 109 degrees. It goes directly from a solid to a vapor state.
Carbonize The process of carbon formation within an engine, such as on the spark plugs and within the combustion chamber.
Carbon Monoxide Gas formed by incomplete combustion; colorless, odorless, very poisonous.

Carburetor A device for automatically mixing fuel in the proper proportion with air to produce a combustible gas.
Case Harden To harden the surface of steel.
Catalyst A compound or substance that can speed up or slow down the reaction of other substances without being consumed itself. In an automatic catalytic converter, special metals (i.e., platinum, palladium) are used to promote more complete combustion of unburned hydrocarbons and a reduction of carbon monoxide.
Catalytic Converter A muffler-like component in the exhaust system that promotes a chemical reaction that converts certain air pollutants in the exhaust gases into harmless substances.
Cathode Ray Tube An electronic tube with an electron gun at the rear that projects a stream of electrons to a spot on the screen end of the tube. The height of the spot on the screen is controlled by the height of the voltage applied to the tube.
Centigrade A measurement of temperature used principally in foreign countries, zero on the centrigrade scale being 32 degrees on the Fahrenheit scale.
Centrifugal Force A force that tends to move a body away from its center of rotation. An example is supplied by whirling a weight attached to a string.
Chamfer A bevel or taper at the edge of a hole.
Charging Circuit The alternator (or generator) and associated circuit used to keep the battery charged and to furnish power to the automobile's electrical systems when the engine is running.
Chase To straighten up or repair damaged threads.
Chassis Dynamometer A machine for measuring the power delivered to the drive wheels of a vehicle.
Check Valve A gate or valve that allows passage of gas or fluid in one direction only.
Chilled Iron Cast iron on which the surface has been hardened.
Chromium Steel An alloy of steel with a small amount of chromium to produce a metal which is highly resistant to oxidation and corrosion.
Circuit The complete path of an electrical current, including the generating device. When the path is unbroken, the circuit is closed and current flows. When the circuit continuity is broken, the circuit is open and current flow stops.
Clearance The space allowed between two parts, such as between a journal and bearing.
Clockwise Rotation Rotation in the same direction as the hands of a clock.
Closed Circuitry A circuit that is uninterrupted from the current source and back to the current source.
Closed Loop A system that feeds back its output to the input side of the processor, which monitors the output and makes corrections as necessary.
Closed Loop Mode Mode in which the ECA operates with EGO sensor feedback.
Closed System Crankcase ventilation system that vents crankcase pressure and vapors back into the engine where they are burned during combustion rather than venting to the atmosphere.
Coefficient of Friction A measurement of the amount of friction developed between two surfaces pressed together and moved one on the other.
Coil A spiral winding of electrical wire. In the automobile ignition system, a step up transformer (pair of coils) that changes lower voltage and high current in the primary coil winding to high voltage and low current in the secondary coil windings.
Cold Manifold An intake manifold to which the exhaust gas is not applied for heating purposes.
Cold Weather Modulator A vacuum modulator located in the carburetor air cleaner on some models. Prevents the air cleaner duct door from opening to nonheated intake air when fresh air is below 55 degrees Fahrenheit. Looks like a **Temperature Vacuum Switch,** differentiated by color codes.
Combination Thermactor Air Bypass and Air Diverter Valve Combines the function of a normally closed air bypass valve and an air control valve in one integral valve.
Combustion The process of burning.
Combustion Space or Chamber In automobile engines, the volume of the cylinder above the piston with the piston on top center.
Compound A mixture of two or more ingredients.
Compression The reduction in volume or the squeezing of a gas. As applied to metal, such as a coil spring, compression is the opposite of tension.
Compression Ratio The volume of the combustion chamber at the end of the compression stroke as compared to the volume of the cylinder and chamber with the piston on bottom center.
Compression Rings The upper rings on a piston designed to hold the compression in the cylinder.
Computed Timing The relationship of spark plug firing to crankshaft position expressed in crankshaft degrees. On some vehicles, the crankshaft position is determined by the crankshaft position (CP) sensor. On the others, this function is controlled by the profile ignition pickup (PIP) sensor.
Computer A device that takes information, processes it, makes decisions, and outputs those decisions. Computers are generally large and not portable (see **Microcomputer**).
Condensation The process of a vapor becoming a liquid; the reverse of evaporation.

Condenser An electrical device used to store the excess coil energy above spark-plug firing needs and discharge this excess energy back through the primary ignition circuit. By providing a place for primary current to flow while the points are opening, the condenser prevents arcing at the points.
Concentric Two or more circles have a common center.
Conductor Any material that permits electrical current to flow easily. A nonconductive material is called an insulator.
Connecting Rod Rod that connects the piston to the crankshaft.
Contraction A reduction in mass or dimension; the opposite of expansion.
Convection A transfer of heat by circulating heated air.
Conventional Oxidation Catalyst Catalyst that acts on two major pollutants: hydrocarbons and carbon monoxide.
Coolant The liquid that circulates in an engine cooling system.
Corrode To eat away gradually as if by gnawing, especially by chemical action such as rust.
Counterbore To enlarge a hole to a given depth.
Counterclockwise Rotation Rotating the opposite direction of the hands on a clock. The same as anticlockwise rotation.
Countersink To cut or form a depression to allow the head of a screw to go below the surface.
Crankcase The housing within which the crankshaft and many other parts of the engine operate.
Crankcase Breather A port or tube that vents fumes from the crankcase. An inlet breather allows fresh air into the crankcase.
Crankcase Dilution Under certain conditions of operation, unburned portions of the fuel find their way past the piston rings into the crankcase and oil reservoir where they dilute or thin the engine lubricating oil.
Cranking Circuit The starter and its associated circuit, including battery, relay (solenoid), ignition switch, neutral start switch (on vehicles with automatic transmission), and cables and wires.
Crankshaft The main shaft of an engine which in conjunction with the connecting rods changes the reciprocating motion of the pistons into rotary motion.
Crankshaft Counterbalance A series of weights attached to or forged integrally with the crankshaft so placed as to offset the reciprocating weight of each piston and rod assembly.
Crude Oil Liquid oil as it comes from the ground.
Current The movement of electrical energy. Generally considered to be the movement of electrons, or negatively charged particles, within the conductor. Measured in amperes.
Cylinder Head A detachable portion of an engine fastened securely to the cylinder block, which contains all or a portion of the combustion chamber.
Cylinder Sleeve A liner or tube interposed between the piston and the cylinder wall or cylinder block to provide a readily renewable wearing surface for the cylinder.

D

Dead Center The extreme upper or lower position of the crankshaft throw at which the piston is not moving in either direction.
Deceleration A decrease in speed.
De-energized Having the electric current or energy source turned off.
Degree Abbreviated "deg." or indicated by a small "°" placed alongside a figure; can be used to designate temperature readings or to designate angularity, one degree being 1/360 part of a circle.
Delay Valve Vacuum restriction used to retard or delay the application of a vacuum signal.
Delay Valve Two-Way A delay valve that functions in both directions
Density Compactness; relative mass of matter in a given volume.
Detergent A compound of a soap-like nature used in engine oil to remove engine deposits and hold them in suspension in the oil.
Detonation As used in an automobile, indicates a too rapid burning or explosion of the mixture in the engine cylinders. It becomes audible through a vibration of the combustion chamber walls and is sometimes confused with a "ping" or spark knock.
Dial Gauge A type of micrometer wherein the readings are indicated on a dial rather than on a thimble.
Diaphragm A flexible partition or wall separating two cavities.
Die One of a pair of hardened metal blocks for forming metal into a desired shape or a thread die for cutting external threads.
Die Casting An accurate and smooth casting made by pouring molten metal or composition into a metal mold or die under pressure.
Diode An electronic device that permits flow of electricity in only one direction. In automotive application, used at the alternator to change alternating current into direct current.
Direct Current Current flowing in one direction only.

Disable A type of microcomputer decision that results in an automotive system being deactivated and not permitted to operate.
Distortion A warpage or change in form from the original shape.
Distributor The mechanism in an ignition system that opens and closes the primary circuit through the distributor breaker points and directs the secondary high voltage to the spark plugs at the correct time for firing.
Distributor Terminal That terminal of the ignition coil to which primary current flows from the breaker point circuit of the distributor, permitting the coil's magnetic field to saturate and collapse as the points close and open.
Distributor Vacuum Vent Valve Required by some engines to prevent fuel migration to distributor advance diaphragm and to act as a spark advance delay valve.
Diverter Valve Valve that directs secondary air downstream to mid-bed port of three-way catalyst. See **Air Control Valve.**
Dowel Pin A pin inserted in matching holes in two parts to maintain those parts in fixed relation one to the other.
Draw Filing A method of filing wherein the file is drawn across the work while held at a right angle to the length of the file.
Drill A tool for making a hole or to sink a hole with a pointed cutting tool rotated under pressure.
Drive-Fit Also known as a force-fit or press-fit. This term is used when the shaft is slightly larger than the hole and must be forced in place.
Drop Forging A piece of steel shaped between dies while hot.
Dwell The period during which the breaker points are closed. Measured in degrees of distributor shaft rotation.
Dwell Period Also known as cam angle.
Dwell Section That segment of the ignition pattern on a scope from the point at which breaker points close to the point at which they open.
Dynamometer A machine for measuring the actual power produced by an internal combustion engine.

E

Early Fuel Evaporation A device to heat the air fuel mixture entering the intake manifold when the engine is cold.
Eccentric One circle within another circle wherein both circles do not have the same center. An example of this is a cam on a camshaft.
EGR Exhaust Gas Recirculation.
EGR Act EGR solenoid pressure valve assembly.
EGR Control Solenoid Electrical solenoid or its control line. Solenoid switches engine manifold vacuum to operate EGR valve when solenoid is energized.
EGR Cooler Assembly Heat exchanger using engine coolant to reduce exhaust gas temperature.
EGR Valve Position Sensor Potentiometric sensor or its signal line used in electronically controlled EGR systems. Sensor wiper position is proportional to EGR valve pintle position. This allows the electronic control assembly (ECA) to determine actual EGR flow at any point in time.
EGR Vent Solenoid Electrical solenoid or its control line. Solenoid normally vents EGRC vacuum line. When EGRV is energized, EGRC can open the EGR valve.
EGR Venturi Vacuum Amplifier Device that uses relatively weak venturi vacuum to control a manifold vacuum signal to operate the EGR valve. Contains a check valve and relief valve that open whatever venturi vacuum signal is equal to or greater than manifold vacuum.
Electrode The firing terminals that are found in a spark plug.
Electromechanical Refers to a device that incorporates both electronic and mechanical principles together in its operation.
Electronic Pertaining to the control of systems or devices by the use of small electrical signals and various semiconductor devices and circuits.
Electronic Control Assembly A vehicle computer consisting of a calibration assembly containing the computer memory and thus its control program and processor assembly, which is the computer hardware.
Electronic Engine Control A computer-directed system of engine control; it can be controlled by one or more sensors. As a rule it controls engine timing and/or fuel system (with the FBC system). The typical electronic engine control system operates on the circuit theory. Once the ignition switch is turned on, EEC is in operation.
Enable A type of microcomputer decision that results in an automotive system being activated and permitted to operate.
Energized Having the electrical current or electrical source turned on.
Engine As used in automobiles, the term applies to the prime source of power generation used to propel the vehicle.
Engine Coolant Temperature Sensor Refers to thermistor sensor or its signal line. Sensor is immersed in engine coolant. Provides engine coolant temperature information to ECU, which is used to

alter spark advance and EGR flow during warm-up or overheat condition. (Replaces cooling and EGR PVS.)

Engine Displacement The sum of the displacement of all the engine cylinders.

"E" Port A carburetor source for EGR vacuum.

Evaporative Emission Controls Emission control system responsible for preventing fuel vapors from entering the atmosphere, primarily from the fuel tank and carburetor.

Evaporative Emission Shed System A system for containment of fuel vapors (evaporative emissions) introduced in 1978. Annual improvements have modified this system.

Exhaust Gas Check Valve Allows thermactor air to enter exhaust manifold but prevents reverse flow in the event of improper operation of other components.

Exhaust Gas Oxygen Sensor Exhaust Gas Oxygen Sensor or its signal line. Sensor changes its output voltage as exhaust gas oxygen content changes when compared to the oxygen content of the atmosphere. Constantly changing voltage signal is sent to the processor for analysis and adjustment to the air/fuel ratio.

Exhaust Gas Recirculation A procedure in which a small amount of exhaust gas is readmitted to the combustion chamber to reduce peak combustion temperatures and thus reduce nitrogen oxide emissions.

Exhaust Heat Control Valve A valve that routes hot exhaust gases to the intake manifold heat riser during cold engine operation. The valve can be thermostatically controlled or vacuum operated. In many vehicles, this valve is controlled by the signal from the ECA.

Exhaust Manifold That part of an engine through which exhausted or spent gases flow to the exhaust system.

Exhaust Pipe The pipe connecting the engine to the muffler to conduct the exhausted or spent gases away from the engine.

Exhaust Valve A valve that permits the exhausted or spent gases to escape from a chamber.

Expansion An increase in size. For example, when a metal rod is heated, it increases in length and perhaps also in diameter; expansion is the opposite of contraction.

F

Fahrenheit A scale of temperature measurement ordinarily used in English-speaking countries. The boiling point of water is 212 degrees Fahrenheit as compared to 100 degrees Celsius.

Feedback Carburetor Actuator A computer-controlled stepper motor that varies the carburetor air/fuel mixture.

Feeler Gauge A metal strip or blade finished accurately with regard to thickness. It is used for measuring the clearance between two parts. Such gauges ordinarily come in a set of different blades that are graduated in thickness by increments of 0.001 inch.

Ferrous Metal Metal that contains iron or steel and is therefore subject to rust.

File To finish or trim with a hardened rasp or file.

Fillet A rounded filling between two parts joined at an angle.

Filter (Oil, Water, Gasoline, Etc.) A unit containing an element, such as a screen of varying degrees of fineness. The screen or filtering element is made of various materials depending on the size of the foreign particles to be eliminated from the fluid being filtered.

Firing Section That segment of the ignition pattern on a scope from the point at which the breaker points open, initiating the firing of the spark plug, to the point at which the spark extinguishes.

Firing Spike The vertical line appearing on the scope as the plug begins to fire, representing the voltage required to jump the spark plug gap.

Fit A kind of contact between two machined surfaces.

Flange A projecting rim or collar on an object for keeping it in place.

Floating Piston Pin A piston pin that is not locked in the connecting rod or the piston but is free to turn or oscillate in both the connecting rod and piston.

Flutter or Bounce As applied to engine valves, refers to a condition wherein the valve is not held tightly on its seat during the time the cam is not lifting it.

Flywheel A heavy wheel in which energy is absorbed and stored by means of momentum.

Foot-Pound This is a measure of the amount of energy or work required to lift 1 pound to 1 foot.

Force-Fit Also known as a press-fit or drive-fit. This term is used when the shaft is slightly larger than the hole and must be forced in place.

Forge To shape metal while hot and plastic by hammering.

Four-Cycle Engine Also known as Otto cycle, wherein an explosion occurs every other revolution of the crankshaft; a cycle being considered as half a revolution of the crankshaft. These strokes are intake, compression, power, and exhaust.

Freewheeling A mechanical device that engages the driving member to impart motion to a driven

member in one direction but not the other. Also known as an overrunning clutch.
Fuel Knock Same as **Detonation.**
Fuel (Rich, Lean) A qualitative evaluation of air/fuel ratio based on the voltage from the exhaust gas oxygen sensor.
Fuel Tank Vapor Valve A valve mounted in the top of the fuel tank responsible for venting excess vapor and pressure from the fuel tank into the evaporative emission control system.
Fuel-Vacuum Separator Used to filter waxy hydrocarbons from carburetor ported vacuum to protect the vacuum delay and distributor vacuum controls.
Fulcrum The support on which a lever turns in moving a body.

G

Gap A break in the continuity of a circuit. The distance between the electrodes of a spark plug. The distance between breaker points at the high point on the distributor cam.
Gasket Anything used as a packing, such as a nonmetallic substance, placed between two metal surfaces to act as a seal.
Glaze As used to describe the surface of the cylinder, an extremely smooth or glossy surface such as a cylinder wall highly polished over a long period of time by the friction of the piston rings.
Glaze Breaker A tool for removing the glossy surface finish in an engine cylinder.
Grind To finish or polish a surface by means of an abrasive wheel.
Ground The negatively charged side of a circuit. A ground can be a wire, the negative side of the battery or even the vehicle chassis.
Ground Circuit The return side of an electric circuit.
Gum In automotive fuels, this refers to oxidized petroleum products that accumulate in the fuel system, carburetor, or engine parts.

H

Hall Effect A process in which current passes through a small slice of semiconductor material at the same time as a magnetic field to produce a small voltage in the semiconductor.
Harmonic Balancer A device to reduce the torsional or twisting vibration that occurs along the length of the crankshaft used in multiple cylinder engines. Also known as a vibration damper.
H.C. High compression.
Heated Air Inlet System System that operates during cold weather and cold start. Its purpose is to bring warm, filtered air into the engine to control the volume of air entering the engine, vaporize the fuel better, and reduce hydrocarbon and carbon monoxide emissions.
Heated Exhaust Gas Oxygen Sensor An EGO sensor with a heating element.
Heat Riser The passage in the manifold between the exhaust and intake manifold.
Heat Treatment A combination of heating and cooling operations timed and applied to a metal in a solid state in a way that will produce desired properties.
Helical Shaped like a coil of wire or screw thread.
Hg Chemical symbol for mercury. Also, a reference to amount of vacuum, i.e., "inches of mercury."
High Tension High voltage. In automotive ignition systems, voltages (up to 40 kilovolt) in the secondary circuit of the system as contrasted to the low, primary circuit voltages (nominally 6 or 12 volts).
Hone An abrasive tool for correcting small irregularities or differences in diameter in a cylinder, such as an engine cylinder or brake cylinder.
Hot Spot Refers to a comparatively thin section or area of the wall between the inlet and exhaust manifold of an engine, the purpose being to allow the hot exhaust gases to heat the comparatively cool incoming mixture. Also used to designate local areas of the cooling system which have attained above average temperatures.
HP Horsepower, the energy required to lift 550 pounds 1 foot in 1 second.
Hydrocarbon A chemical composition, made up of hydrogen and carbon, that is a component of exhaust emissions.
Hydrocarbon Engine An engine using petroleum products, such as gas, liquefied gas, gasoline, kerosene, or fuel oil, as a fuel.
Hydrogen Highly flammable elemental gas. Chemical symbol is H.

I

I.D. Inside Diameter.
Idle Refers to the engine operating at its slowest speed with a vehicle not in motion.
Idle Vacuum Valve This device can be used in conjunction with other vacuum controls to dump thermactor air during extended periods of idle. Provides protection to catalyst.
Ignition In internal-combustion gasoline engines, the process of igniting the air/fuel mixture in

the combustion chamber by means of an electrical spark from a spark plug.
Ignition Distributor An electrical unit that provides a means for conveying the secondary or high tension current to the spark plug wires as required.
Ignition Module Signal The signal produced by the ECA that controls the ignition module on and off time and, therefore, the spark plug timing and coil dwell.
Ignition System The means for igniting the fuel in the cylinders.
IHP Indicated horsepower developed by an engine and a measurement of the pressure of the explosion within the cylinder expressed in pounds per square inch.
Induction The transfer of electricity by magnetism rather than direct flow through a conductor. When current flows through a wire or coil, it builds up a magnetic field. If another conductor moves through this field (or the field moves through another conductor), a voltage is induced in the other conductor. This is the principle of the ignition coil and also of the engine analyzer's induction pickups.
Inertia A physical law that tends to keep a motionless body at rest or also tends to keep a moving body in motion; effort is thus required to start a mass moving or to retard or stop it once it is in motion.
Infrared Analyzer Instrument used to measure exhaust emissions.
Inhibitor A material to restrain or hinder some unwanted action, such as a rust inhibitor, which is a chemical added to cooling systems to retard the formation of rust.
Injector Electronic fuel injection solenoid (one of two) or its control line. Solenoid, when energized, allows fuel to flow into throttle body and then into one plane of the dual-plane intake manifold.
Inlet Valve or Intake Valve A valve that permits a fluid or gas to enter a chamber and seals against an exit.
Input Information provided to a microcomputer to allow accurate control of a system.
Input Conditioner A device or circuit that conditions or prepares an input signal for use by a microcomputer.
Intake Manifold or Inlet Pipe The tube used to conduct the gasoline and air mixture from the carburetor to the engine cylinders.
Integrated Circuit A small semiconductor device with circuitry that can perform numerous functions.
Intermediate Section That segment of the ignition pattern on a scope from the point at which the spark extinguishes to the point at which the breaker points close.

J

Journal A bearing within that a shaft operates.

K

Keep Alive Memory A series of vehicle battery-powered memory locations in the microcomputer that allow the microcomputer to store information on input failure, identified in normal operations for use in diagnostic routines. Keep Alive Memory adopts some calibration parameters to compensate for changes in the vehicle system.
Key A small block inserted between the shaft and hub to prevent circumferential movement.
Keyway or Key Seat A groove or slot cut to permit the insertion of a key.
Knock A general term used to describe various noises occurring in an engine; can be used to describe noises made by loose or worn mechanical parts, preignition, detonation, etc.
Knock Sensor A device designed to vibrate at approximately the same frequency as the engine knock frequency. The knock sensor provides information on engine knock to an electronic engine control microcomputer.

L

Lapping The process of fitting one surface to another by rubbing them together with an abrasive material between the two surfaces.
Liner Usually a thin section placed between two parts, such as a replaceable cylinder liner in an engine.
Liter A metric measure equal to 2.11 pints.

M

Magnet Any body with the property of attracting iron or steel.
Magnetic Field The area surrounding the poles of a magnet that is affected by its attraction or repulsion forces.
Malleable Casting A casting that has been toughened by annealing.
Manifold A pipe with multiple openings used to connect various cylinders to one inlet or outlet.
Manifold Absolute Pressure Sensor A pressure sensitive disk capacitor sensor used to measure air pressure inside the intake manifold.
Manifold Control Valve A thermostatically operated valve in the exhaust manifold for varying heat to

the intake manifold with respect to the engine temperature.

Manifold Vacuum Vacuum generated below the throttle plates of a carburetor. Vacuum present in the intake manifold. Manifold vacuum is high at idle and lowers as the throttle plates open.

Mechanical Efficiency (Engine) The ratio between the indicated horsepower and the brake horsepower of an engine.

Microcomputer A device that takes information, processes it, makes decisions, and outputs those decisions. Microcomputers are generally small and portable and are located inside a processor.

Micrometer A measuring instrument for either external or internal measurement in thousandths and sometimes tenths of thousandths of inches.

Microprocessor An integrated circuit within a microcomputer that controls information flow within the microcomputer. Also called the Central Processing Unit (CPU).

Microprocessor Control Unit Integral part of electronically controlled feedback carburetor using TWC catalyst. Various sensors monitor mode conditions. MCU is widely used on Ford-built vehicles for the control of air/fuel ratios.

Mill To cut or machine with rotating tooth cutters.

Millimeter One millimeter is the metric equivalent of 0.039370 of an inch or one inch being the equivalent of 25.4 mm.

Misfiring Failure of an explosion to occur in one or more cylinders while the engine is running; can be continuous or intermittent failure.

Monolithic Substrate The ceramic honeycomb structure used as a base to be coated with a metallic catalyst material for use in the catalytic converter.

Motor This term should be used in connection with an electric motor and should not generally be used when referring to the engine of an automobile.

Muffler A chamber attached to the end of the exhaust pipe that allows the exhaust gases to expand and cool. It is usually fitted with baffles or porous plates and serves to subdue much of the noise created by the exhaust.

N

Needle Bearing An antifriction bearing using a great number of small-diameter rollers of greater length.

Negative A pole from which electricity flows. Having an excess of electrons (negatively-charged particles). Opposite of **Positive.**

Neutral Drive Switch A sensor that provides information on transmission status to the ECA.

Nitrogen An elemental gas that is inert. Seventy-eight percent of the air is nitrogen.

Nitrogen Oxides A compound formed during the engine's combustion process when oxygen in the air combines with nitrogen in the air to form the nitrogen oxides, which are agents in photochemical smog.

Nonferrous Metals This designation includes practically all metals that contain no iron or very little iron and are therefore not subject to rusting.

Normally Closed Refers to a switch or solenoid that is closed when no control or force is acting on it.

Normally Open Refers to a switch or solenoid that is open when no control force is acting on it.

Normal Pattern A scope pattern that represents ignition system operation of an engine completely and accurately tuned to manufacturer's specifications.

O

Octane Number A unit of measurement on a scale intended to indicate the tendency of a fuel to detonate or knock.

Octane Selector A calibrated device for adjusting the timing of the ignition distributor in accordance with the characteristics of the fuel in use.

Ohm A unit of electrical resistance opposing current flow. Resistance varies in different materials and with temperature.

Ohmmeter An instrument that measures resistance of a conductor in units called ohms.

Open Circuit An electrical circuit whose path has been interrupted or broken either accidentally (e.g., a broken wire) or intentionally (e.g., a switch turned off).

Open Loop An electronic control system in which sensors provide information, the microcomputer gives orders, and the output actuators obey the orders without feedback to the microcomputer.

Open System Descriptive term for crankcase emissions control system, which vents to the atmosphere.

Oscillation A flow of electricity rapidly changing direction.

Oscilloscope A cathode ray tube with controls and associated circuitry to present on its screen wave forms or patterns for observation and measurement of voltage with respect to time.

Output Driver A transistor in the output control area of the processor that is used to turn various actuators on and off.

P

Particulate A tiny bit of solid matter found in exhaust gases.

Paraded A scope display mode that presents cylinder patterns horizontally one after the other in firing order, beginning with No. 1 plug at the left.

Pattern In automotive ignition system service, designates the wave form on the screen of a scope.

Peen To stretch or clinch over by pounding with the rounded end of a hammer.

Petcock A small valve placed at various points in a fluid circuit usually for draining purposes.

Petroleum A group of liquid and gaseous compounds composed of carbon and hydrogen that are removed from the earth.

Phosphor-Bronze An alloy consisting of copper, tin, and lead sometimes used in heavy-duty bearings.

Photochemical The action of light (photo) on air pollutants (chemicals), which creates smog.

Piston A cylindrical part closed at one end, which is connected to the crankshaft by the connecting rod. The force of the explosion in the cylinder is exerted against the closed end of the piston causing the connecting rod to move the crankshaft.

Piston Collapse A condition describing a collapse or a reduction in diameter of the piston skirt due to heat or stress.

Piston Displacement The volume of air moved or displaced by moving the piston from one end of its stroke to the other.

Piston Head That part of the piston above the rings.

Piston Lands Those parts of a piston between the piston rings.

Piston Pin The journal for the bearing in the small end of an engine connecting rod, which also passes through piston walls; also known as a **Pin Boss.**

Piston Ring An expanding ring placed in the grooves of the piston to provide a seal to prevent the passage of fluid or gas past the piston.

Piston Ring Expander A spring placed behind the piston ring in the groove to increase the pressure of the ring against the cylinder wall.

Piston Ring Gap The clearance between the ends of the piston ring.

Piston Ring Groove The channel or slots in the piston in which the piston rings are placed.

Piston Skirt That part of the piston below the rings.

Piston Skirt Expander A spring or other device inserted in the piston skirt to compensate for collapse or decrease in diameter.

Polarity The particular state, either positive or negative, with reference to the two poles or to electrification.

Poppet Valve A valve structure consisting of a circular head with an elongated stem attached in the center which is designed to open and close a circular hole or port.

Port In engines, the openings in the cylinder block for valves, exhaust and inlet pipes, or water connections. In two-cycle engines, the openings for inlet and exhaust purposes.

Ported Vacuum Vacuum tapped from a passage just above the throttle plate in a carburetor; S-port vacuum.

Ported Vacuum Switch A temperature-actuated switch that changes vacuum connections when the coolant temperature changes. (Originally used to switch spark port vacuum; now used for any vacuum switching function that requires coolant temperature sensing.)

Porting As generally applied to racing engines, the enlarging, matching, streamlining, and polishing of the inside of the manifolds and valve ports to reduce the friction of the flow of gases.

Positive The positive plate of an electrolytic cell. The terminal of a battery toward which electrons are presumed to flow. Opposite of **Negative.**

Positive Crankcase Ventilation System An emission control system that routes engine crankcase fumes into the intake manifold or air cleaner, where they are drawn into the cylinders and burned along with the air/fuel mixture.

Positive Crankcase Ventilation Valve A valve that controls the flow of vapors from the crankcase into the engine.

Preheating The application of heat as a preliminary step to some further thermal or mechanical treatment.

Preignition Ignition occurring earlier than intended. For example, the explosive mixture being fired in a cylinder as by a flake or incandescent carbon before the electric spark occurs.

Press-Fit Also known as a force-fit or drive-fit. This term is used when the shaft is slightly larger than the hole and must be forced in place.

Primary Pertaining to that circuit of a transformer that induces current. The low-voltage circuit in an ignition system.

Primary Wires The wiring circuit used for conducting the low tension or primary current to the points where it is to be used.

Processor The on-board computer that receives data from a number of sensors and other electronic components. Based on data programmed into the processor's memory, the processor generates output signals to control various engine functions.

Program A set of detailed instructions a microcomputer follows when controlling a system.

PSI A measurement of pressure in pounds per square inch.

Pulse Air System A type of exhaust emission control system that uses the natural exhaust pulses in a turned pipe. A reed-type check valve responds to the

negative pressure pulses and permits air to be drawn into the exhaust system.
Purge Control Valve Valve used to control the release of fuel vapors from the charcoal canister into the engine.
Purge Solenoid Device used to control the operation of the purge valve in evaporative control emission systems.
Pushrod A connecting link in an operating mechanism, such as the rod interposed between the valve lifter and rocker arm on an overhead valve engine.

Q

Quenching A process of rapid cooling of hot metal by contact with liquid, gases, or solids.

R

Race As used with reference to bearings, a finished inner and outer surface in which or on which balls or rollers operate.
Race Cam A type of camshaft for race cars that increases the lift of the valve, increases the speed of valve opening and closing, increases the length of time the valve is held open, etc. Also known as full, three-quarter, or semi-race cams, depending upon design.
Radiation The transfer of heat by rays, such as heat from the sun.
Random Access Memory A type of memory which is used to store information temporarily.
Raster Scope display mode; identical to stacked. (See **Stacked.**)
Ratio The relation or proportion that one number bears to another.
Read A microcomputer operation wherein information is retrieved from memory.
Read Only Memory A type of memory used to store information permanently. Information cannot be written to ROM; as the name implies, information can only be read from ROM.
Ream To finish a hole accurately with a rotating fluted tool.
Reciprocating A back-and-forth movement, such as the action of a piston in a cylinder.
Reed Valve Check valve used on some secondary air applications. Prevents reverse flow of air from exhaust manifold to intake air cleaner.
Reference Voltage A voltage provided by a voltage regulator to operate potentiometers and other sensors at a constant level.
Relay A switching device operated by a low-current circuit that controls the opening and closing of another circuit of high current capacity.
Relief The amount one surface is set below or above another surface.
Relieving As applied to racing engines, the removal of some metal from around the valves and between the cylinder and valves to facilitate flow of the gases.
Resistance The opposition offered by a substance or body to the passage of electric current through it.
Resistor A current-consuming piece of metal wire or carbon inserted into an electrical circuit to decrease the flow.
Retard When used with reference to an ignition distributor, it means to cause the spark to occur at a later time in the cycle of engine operation; opposite of **Spark Advance.**
Rocker Arm In an automobile engine, a lever located on a fulcrum or shaft, one end of the lever being on the valve stem, the other being on the pushrod.
Rockwell Hardness A scale for designating the hardness of a substance.
Roller Bearing An inner and outer race upon which hardened steel rollers operate.
Rotary Engine An engine construction wherein the crankshaft remains stationary and the cylinder spins around it as in a pinwheel.
Rotary Valve A valve construction in which ported holes come into and out of register with each other to allow entrance and exit of fluids or gases.
RPM Revolutions per minute.
Rubber An elastic vibration-absorbing material of either natural or synthetic origin.
Running-Fit Where sufficient clearance has been allowed between the shaft and journal to allow free running without overheating.

S

SAE Society of Automotive Engineers.
SAE Steels A numerical index system used to identify composition of SAE steel. Basically the first digit indicating the type to which the steel belongs; thus 1 indicates a carbon steel, 2 a nickel steel, etc. The second digit indicates the approximate percentage of the predominant alloying element. Usually the last two or three digits indicate the approximate average carbon content in points or hundredths of 1 percent. Thus "SAE 2340 steel " indicates a nickel steel of approximately 3 percent nickel and 0.40 percent carbon.
SAE Thread A table of threads set up by the society of Automotive Engineers and determines the number of threads per inch. For example, a 1/4-inch-diameter rod with an SAE thread would have 28 threads per inch.

Safety Relief Valve A spring-loaded valve designed to open and relieve excessive pressure in a vessel when it exceeds a predetermined safe point.

Sampling The act of periodically collecting information, as from a sensor. A microcomputer samples input from various sensors in the process of controlling a system.

Saturation The point reached when current flowing through a coil or wire has built up the maximum magnetic field.

Saybolt Test A method of measuring the viscosity of oil with the use of a viscosimeter.

Scale A flaky deposit occurring on steel or iron. Ordinarily used to describe the accumulation of minerals and metals in an automobile cooling system.

Score A scratch, ridge, or groove marring a finished surface.

Seat A surface, usually machined, upon which another part rests or seats; for example, the surface upon which a valve face rests.

Secondary Pertaining to that side of a transformer in which current is induced (by the primary). The high-voltage (high-tension) part of the ignition system.

Seize When one surface moving upon another scratches, it is said to seize. An example is a piston score or abrasion in a cylinder due to lack of lubrication or overexpansion.

Semiconductor General term for transistors, integrated circuits, and other electronic devices made of material, such as silicon, that are used to moderate the flow of electricity.

Sensor A device that measures an operating condition and provides an input signal to a microcomputer.

Short Circuit A usually unintentional grounding of a circuit such that current is diverted from part of the normal circuit by following a shorter (or easier) path to the current source. Commonly called a short.

Shrink Fit Where the shaft or part is slightly larger than the hole in which it is to be inserted. The outer part is heated above its normal operating temperature or the inner part chilled below its normal operating temperature and assembled in this condition; then upon cooling an exceptionally tight fit is obtained.

Signal A voltage condition that transmits specific information in an electronic system.

Sleeve Valve A reciprocating sleeve or sleeves with ported openings placed between the piston and cylinders of an engine to serve as valves.

Sliding Fit Where sufficient clearance has been allowed between the shaft and journal to allow free running without overheating.

Slip-In Bearing A liner made to extremely accurate measurements that can be used for replacement purposes without additional fitting.

Sludge As used in connection with automobile engines, it indicates a composition of oxidized petroleum products along with an emulsion formed by the mixture of oil and water. This forms a pasty substance and clogs oil lines and passages and interferes with engine lubrication.

Smog Air pollution created by the reaction of nitrogen oxides to sunlight.

Solder An alloy of lead and tin used to unite two metal parts.

Soldering To unite two pieces of metal with a material, such as solder having a comparatively low melting point.

Solenoid A wire coil with a movable core that changes position by means of electromagnetism when current flows through the coil.

Solenoid Vent Valve Energized by the ignition switch to control fuel vapor flow to the canister. When the ignition is off, the valve is open.

Solid State Any electronic circuit that does not use tubes can be considered solid state. The term refers to circuits that use transistors, integrated circuits, and/or other semiconductors.

Solvent Simply, a solution that dissolves some other material. For example, water is a solvent for sugar.

Spacer Entry EGR System Exhaust gases are routed directly from the exhaust manifold through a stainless steel tube to the carburetor base.

Spark An electrical current possessing sufficient pressure to jump through the air from one conductor to another.

Spark Advance When used with reference to an ignition distributor, means to cause the spark to occur at an earlier time in the cycle of engine operation; opposite of **Retard.**

Spark Delay Valve A valve that operates spark vacuum advance during rapid acceleration from idle or from speeds below 15 miles per hour and to cut off spark advance immediately on deceleration. Has an internal sintered orifice to slow air in one direction, a check valve for free airflow in opposite direction, and a filter.

Spark Gap The space between the electrodes of a spark plug through which the spark jumps. Also a safety device to provide an alternate path for the current when it exceeds a safe value.

Spark Knock Same as **Preignition.**

Spark Line The normally horizontal line of a scope pattern that represents the firing of the plug once the plug gap has been bridged by the spark. Also called the firing line.

Spark Output (Spout) The output signal from the processor that triggers the TFI-IV Module to fire the ignition coil.
Spark Plug A device inserted into the combustion chamber of an engine containing an insulated central electrode for conducting the high-tension current from the ignition distributor. This insulated electrode is spaced at a predetermined distance from the shell or side electrode in order to control the dimensions of the gap that the spark will be jumping across.
S-Port A special carburetor port to take off ported vacuum.
Stacked A scope display mode that presents cylinder patterns vertically one above another in firing order with No. 1 plug at the bottom. Also called raster display.
Stamping A piece of sheet metal cut and formed into the desired shape with the use of dies.
Standard Thread Refers to the U.S.S. table of the number of threads per inch. For example, a 1/4-inch-diameter standard thread would have 20 threads per inch.
Steel Casting Cast iron to which varying amounts of scrap steel have been added.
Stellite An alloy of cobalt, chrome and tungsten which is often used for exhaust valve seat inserts. It has a high melting point, good corrosion resistance and unusual hardness when hot.
Stoichiometric Chemically correct. An air/fuel mixture is considered stoichiometric when it is neither too rich nor too lean; stoichiometric ratio is 14.7 parts of air for every part of fuel.
Strategy A plan the microcomputer follows to achieve a desired goal.
Stress The force or strain to which a material is subjected.
Stroke In an automobile engine, the distance moved by the piston.
Stroking As applied to racing engines, remachining the crankshaft throws off center to alter the stroke.
Studs A rod with threads cut on both ends, such as a cylinder stud that screws into the cylinder block on one end and has a nut placed on the other end to hold the cylinder head in place.
Suction Suction exists in a vessel when the pressure is lower than the atmospheric pressure; also see **Vacuum.**
Superimposed A scope display mode that presents the patterns of all cylinders superimposed on one another.
Switch A device used to open, close, or redirect the current in an electrical circuit.

T

Tachometer A device for measuring and indicating the rotative speed of an engine.
Tap To cut threads in a hole with a tapered, fluted, threaded tool. In England, term for drain cock.
Tappet The adjusting screw for varying the clearance between the valve stem and the cam. Can be built into the valve lifter in an engine or installed in the rocker arm on an overhead valve engine.
Temperature Vacuum Switch Located in the carburetor air cleaner to control vacuum to the EGR valve or other diaphragm. A cold air lockout. Responds to the temperature of the inlet air heated by the exhaust manifold. Looks like a cold weather modulator, differentiated by color.
Test Lead A wire or cable from an engine analyzer connected directly or by induction pickup to specific points in an ignition system to sense and transmit electrical signals to the analyzer.
Thermactor An air injection type of exhaust emission control system.
Thermactor Air Bypass Solenoid Electrical solenoid or its control line. Solenoid switches engine manifold vacuum. Vacuum bypasses thermactor air to atmosphere.
Thermactor Air Control Solenoid Vacuum Valve Assembly Used on thermactor air control systems, the valve assembly consists of two normally open solenoid valves with vents.
Thermactor Air Control Valve Combines a bypass (dump) valve with a diverter (upstream/ downstream) valve. Controls the flow of air in response to vacuum signals to its diaphragms.
Thermactor Air Diverter Solenoid Electrical solenoid or its control line. Solenoid switches engine manifold vacuum. Vacuum switches thermactor air from downstream (past EGO sensor in exhaust system) to upstream (before EGO sensor) when solenoid is energized.
Thermal Reactor Exhaust manifold design using heat and air to burn unburned hydrocarbons in the exhaust gases to reduce air pollution. Requires a secondary air source to the exhaust system.
Thermal Vacuum Switch See **Temperature Vacuum Switch.**
Thermal Vent Valve Temperature-sensitive valve assembly located in the canister vent line. Closes when engine is cold and opens when hot. Prevents fuel tank vapors from being vented through the carburetor fuel bowl when fuel tank heats up before engine compartment.
Thermistor A resistor that changes its resistance with temperature.

Thermo Efficiency A gallon of fuel contains a certain amount of potential energy in the form of heat when burned in the combustion chamber. Some of this heat is lost and some is converted into power. The thermal efficiency is the ratio of work accomplished compared to the total quantity of heat contained in the fuel.

Thermostat A heat controlled valve used in the cooling system of an automobile engine to regulate the flow of water between the cylinder block and the radiator or used in the electrical circuit of the car heating system to control the amount of heat supplied to the passengers.

Thick Film Integrated Module An electronic module that is mounted on the distributor and fires the ignition coil.

Three-Way Catalyst A catalyst designed to simultaneously convert hydrocarbons, carbon monoxide, and oxides of nitrogen into relatively inert substances. Only effective when the exhaust gases are chemically correct; that is, when the air/fuel ratio of the mixture burned in the cylinders is 14:6:1. Also see **Stoichiometric.**

Throttle Position Sensor Potentiometric sensor or its signal line. Sensor wiper position is proportional to throttle position. Defines engine operation mode: closed, partly open, or wide open throttle; also throttle angle rate for accelerator pump function (replaced carburetor spark/ERG port.) There are two types of throttle positioners: rotary and linear.

Throw With reference to an automobile engine, usually the distance from the center of the crankshaft main bearing to the center of the connecting rod journal.

Timer A feature of microcomputers that allows them to measure time intervals, such as "time since engine start" or "time in neutral idle."

Timing Chain Chain used to drive camshaft and accessory shafts of an engine.

Timing Gears Any group of gears that are driven from the engine crankshaft to cause the valves, ignition, and other engine-driven apparatus to operate at the desired time during the engine cycle.

Tolerance A permissible variation between the two extremes of a specification of dimensions.

Torque An effort devoted toward twisting or turning.

Torque Wrench A special wrench with a built-in indicator to measure the applied force.

Transformer An electrical device that transforms electrical voltage from high to low or from low to high (as in an ignition coil). Note that the total energy is not changed because current increases proportionately to voltage decrease and vice-versa (see **Watt**).

Transistor A solid semiconducting combination of chemical elements used in place of vacuum (radio) tubes. As used in automotive electronic ignition systems, the transistor is a means of furnishing greater current to the ignition coil, resulting in greater secondary voltage to fire the spark plugs while at the same time requiring less current through the breaker points.

Turbulence A disturbed or disordered irregular motion of fluids or gases.

U

Upper Cylinder Lubrication A method of introducing a lubricant into the fuel or intake manifold to permit lubrication of the upper cylinder, valve guides, etc.

V

Vacuum The lack of air or air pressure. There are three type of vacuums important to engine and component function: manifold, ported, and venturi. The strength of these vacuums depends on throttle opening, engine speed, and load.

Vacuum Advance Advances ignition timing with relation to engine load conditions. This is achieved by using engine intake manifold vacuum to operate the distributor diaphragm.

Vacuum Check Valve A one-way valve used to retain vacuum signal in a line after vacuum source is gone; prevents back flow.

Vacuum Control Valve A ported vacuum switch. Controls vacuum to other emission devices during engine warm-up. The two-port types simply open when engine coolant reaches their pre-determined calibration temperatures. The four-port types open likewise, since they are nothing more than two-port types in one housing. The three-port types switch the vacuum source to the center port from the top or the bottom ports. Most VCV's respond to a sensing bulb immersed in engine coolant by utilizing a wax pellet principle.

Vacuum Differential Valve A valve used in the Thermactor system with a catalyst to cut off vacuum to the air bypass valve during deceleration.

Vacuum Gauge An instrument designed to measure the degree of vacuum existing in a chamber.

Vacuum Regulator Provides constant vacuum output from manifold when vehicle is at idle. Switches to engine vacuum off idle.

Vacuum Regulator Valve Two-Port Vacuum regulator that provides a constant output signal when the input level is greater than a preset level. At a lower input vacuum, the output equals the input.

Vacuum Regulator Valve Three- and Four-Port Used to control the vacuum advance to the distributor.

Vacuum Reservoir Stores vacuum and provides "muscle vacuum." Prevents rapid fluctuations and sudden drops in a vacuum signal, such as during acceleration.

Vacuum Restrictor Used to control the flow rate and/or timing in actions to different emission control components.

Vacuum Retard Delay Valve A valve that delays a decrease in vacuum at the distributor vacuum advance unit when the source vacuum decreases. (Can be used to delay release of vacuum from any diaphragm; a "momentary" trap for vacuum.)

Vacuum Vent Valve Controls the induction of fresh air into a vacuum system to prevent chemical decay of the vacuum diaphragm that can occur on contact with fuel.

Valve A valve is a device for opening and sealing an aperture.

Valve Clearance The air gap allowed between the end of the valve stem and the valve lifter or rocker arm to compensate for expansion due to heat.

Valve Face The part of a valve that mates with and rests on a seating surface.

Valve Grinding A process of lapping or mating the valve seat and valve face usually performed with the aid of an abrasive; also called **Valve Lapping.**

Valve Head The portion of the valve upon which the valve face is machined.

Valve-in-Head Engine Same as overhead valve engine.

Valve Key or Valve Lock The key, washer, or other device that holds the valve spring cup or washer in place on the valve stem.

Valve Lifter A pushrod or plunger placed between the cam and the valve on an engine; often adjustable to vary the length of the unit.

Valve Margin On a poppet valve the space or rim between the surface of the head and the surface of the valve face.

Valve Overlap An interval expressed in degrees where both valves of an automobile engine cylinder are open at the same time.

Valve Seat The matched surface upon which the valve face rests.

Valve Spring A spring attached to a valve to return it to the seat after it has been released from the lifting or opening means.

Valve Stem That portion of a valve that rests within a guide.

Valve Stem Guide A bushing or hole in which the valve stem is placed that allows only lateral motion.

Vane Airflow Meter A sensor with a movable vane connected to a potentiometer calibrated to measure the amount of air flowing through the sensor. The output signal is relayed to the ECA which then translates the information into the amount of air flowing into the engine.

Vane Air Temperature Sensor Located in the vane airflow meter, it senses the temperature of the air flowing into the engine. Changes of temperature result in changes in a resistive element in the sensor.

Vapor Lock A condition wherein the fuel boils in the fuel system forming bubbles that retard or stop the flow of fuel to the carburetor.

Vehicle Emission Control Information Decal Critical specifications for servicing emission systems. Decal located in engine compartment.

Venturi Vacuum Amplifier Used with some EGR systems so that carburetor venturi vacuum can control EGR valve operation. Venturi vacuum is desirable because it is proportional to the airflow through the carburetor.

Vibration Damper A device to reduce the torsional or twisting vibration that occurs along the length of the crankshaft used in multiple cylinder engines; also known as a harmonic balancer.

Viscosimeter An instrument for determining the viscosity of an oil by passing a certain quantity at a definite temperature through a standard size orifice or port. The time required for the oil to pass through expressed in seconds gives the viscosity.

Viscosity The resistance to flow or adhesiveness characteristics of an oil.

Volatility The tendency for a fluid to evaporate rapidly or pass off in the form of vapor. For example, gasoline is more volatile than kerosene because it evaporates at a lower temperature.

Volt A unit of measurement of electromotive force. One volt of electromotive force applied steadily to a conductor of one ohm resistance will produce a current of one ampere.

Voltage The total electromotive force applied to a given circuit, expressed in volts.

Volume The measure of space expressed as cubic inches, cubic feet, etc.

Volumetric Efficiency A combination between the ideal and actual efficiency of an internal combustion engine. If the engine completely filled each cylinder on each induction stroke the volumetric efficiency of the engine would be 100 percent. In actual operation, however, volumetric efficiency is lowered by the inertia of the gases, the friction between the gases and the manifolds, the temperature of the gases, and the pressure of the air entering the carburetor. Volumetric efficiency is ordinarily increased by the use of large valves, ports, and manifolds and can be further increased with the aid of a supercharger.

Vortex A whirling movement or mass of liquid or air.

W

Watt A unit of power equal to the rate of work represented by a current of 1 ampere under a pressure of 1 volt. (Formula: Watts = amperes × volts.)

White Metal An alloy of tin, lead, and antimony having a low melting point and a low coefficient of friction.

Winding A single complete turn of the wire wound into a coil or transformer.

Wringing-Fit Wherein the clearance is less than for a running or sliding fit and the shaft will enter the hole by means of twisting and pushing by hand.

Wrist Pin The journal for the bearing in the small end of an engine connecting rod that also passes through piston walls; also known as a piston pin.

Write A microcomputer operation wherein information is sent to and stored in memory.

INDEX

Complete Engine Performance and Diagnostics

Student Technician's Shop Manual

Delmar Publishers Inc.®

NOTICE TO THE READER

Publisher does not warrant or guarantee any of the products described herein or perform any independent analysis in connection with any of the product information contained herein. Publisher does not assume, and expressly disclaims, any obligation to obtain and include information other than that provided to it by the manufacturer.

The reader is expressly warned to consider and adopt all safety precautions that might be indicated by the activities described herein and to avoid all potential hazards. By following the instructions contained herein, the reader willingly assumes all risks in connection with such instructions.

The publisher makes no representations or warranties of any kind, including but not limited to, the warranties of fitness for particular purpose or merchantability, nor are any such representations implied with respect to the material set forth herein, and the publisher takes no responsibility with respect to such material. The publisher should not be liable for any special, consequential or exemplary damages resulting, in whole or in part, from the readers' use of, or reliance upon, this material.

Text, Design, and Production by
Scharff Associates, Ltd.
RD 1 Box 276
New Ringgold, PA 17960

Scharff Staff
Production Manager: Marilyn Strouse-Hauptly
Staff Writers: Dave Caruso, John Corinchock, and Keith Mullen
Text Design: Barbara J. Gould
Cover Design: Eric Schreader
Logo Design: Ed Foulk

Delmar Staff
Editor-in-Chief: Mark W. Huth
Associate Editor: Joan Gill

For more information, address Delmar Publishers Inc.
3 Columbia Circle, Box 15-015
Albany, New York 12212-5015

Printed in the United States of America
Published simultaneously in Canada
by Nelson Canada,
a division of International Thomson Limited

10 9 8 7 6 5

Library of Congress Cataloging-in-Publication Data

Scharff, Robert.
Complete engine performance and diagnostics.

Includes index.
1. Automobiles—Motors—Maintenance and repair.
I. Motor service (Chicago, Ill. : 1951) II. Title.
TL210.S35 1989 629.25'04 89-7832
ISBN 0-8273-3579-2
ISBN 0-8273-3581-4 (shop manual)
ISBN 0-8273-3580-6 (instructor's guide)

CONTENTS

CHAPTER ONE

INTRODUCTION TO AUTOMOTIVE TUNE-UPS

Objectives

Upon completion of this chapter, you should be able to:

- State the objectives of a tune-up.
- Explain the changes that have affected tune-up procedures.
- Outline the basic steps in a quality tune-up.
- List the sources of information needed when performing a tune-up.
- Explain safe tune-up practices.

PRACTICE QUESTIONS

1. Which of the following components are replaced when tuning up a late model vehicle?
   a. spark plugs
   b. breaker points
   c. condenser
   d. all of the above

2. A vehicle suffers from poor startup and hesitation. Technician A says such performance problems can possibly be traced to a defective valve in the emission system. Technician B says they can possibly be traced to a vacuum leak in the emission system. Who is correct?
   a. Technician A
   b. Technician B
   c. Both A and B
   d. Neither A nor B

3. Which of the following emission components was not introduced to reduce the level of HC emissions?
   a. air injector reactor
   b. vapor canister
   c. EGR valve
   d. catalytic converter

4. Which of the following combustion byproducts are not subject to emission levels established by the Clean Air Act?
   a. oxygen
   b. carbon monoxide
   c. oxides of nitrogen
   d. hydrocarbons

5. What is denoted by the electrical diagram shown in Figure 1–1?
   a. motor
   b. switch
   c. ground
   d. coil

6. What is denoted by the electrical diagram shown in Figure 1–2?
   a. wires crossed
   b. contact points

Figure 1–1

Figure 1–2

c. spark gap
d. circuit breaker

7. Which of the following troubles will prevent you from tuning a vehicle properly?
   a. low battery charge
   b. low cylinder compression
   c. loose vacuum hoses
   d. all of the above

8. Where can you find information on the routing of vacuum hoses in an evaporative emission control system?
   a. VECI decals
   b. electrical diagrams
   c. troubleshooting trees
   d. all of the above

9. Which of the following is not ordinarily found on the VECI decal?
   a. initial timing setting
   b. fast-idle speed
   c. mileage information
   d. carburetor adjustment instructions

10. Technician A says that being an ASE certified mechanic will help her find a job. Technician B says being ASE certified will help him make more money. Who is correct?
    a. Technician A
    b. Technician B
    c. Both A and B
    d. Neither A nor B

SHOP ASSIGNMENT 1

Identify the components of a typical exhaust gas recirculation (EGR) system. Refer to Figure 1-3 as needed.

SHOP ASSIGNMENT 2

Locate the VECI decal on a late-model vehicle. Study the vacuum diagram and trace the vacuum hoses on the vehicle. As you do so, check all hose connections for proper fit and look for cracks and other defects in the hoses.

SHOP ASSIGNMENT 3

For a late-model car, record the following information, using the VIN, VECI decal, and service manual as sources of information:

Make ______________________

Model ______________________

Model year ______________________

Displacement ______________________

Type of ignition system ______________________

Type of fuel system ______________________

Base idle speed ______________________

Fast idle speed ______________________

Base timing ______________________

Advance timing ______________________

SHOP ASSIGNMENT 4

Using an owner's manual as a source of information, record the maintenance intervals for the following components:

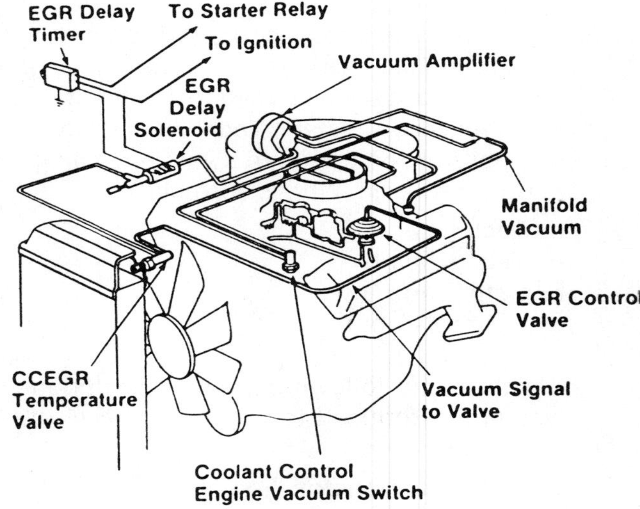

Figure 1-3

Spark plugs ______________________

Air filter ______________________

Fuel filter ______________________

Oil filter ______________________

PCV valve ______________________

Ignition points ______________________

Coolant ______________________

Crankcase oil ______________________

Ignition timing ______________________

Idle speed ______________________

Spark plug cables ______________________

CHAPTER TWO

ENGINE TUNE-UP TOOLS AND EQUIPMENT

Objectives

Upon completion of this chapter, you should be able to:

- Name the diagnostic tools and equipment used in engine tune-ups.
- Describe the basic application and use of diagnostic tools.
- Explain the importance of trained technicians skilled in the use of state-of-the-art electronic and computerized test equipment.

PRACTICE QUESTIONS

1. The tool shown in Figure 2–1 is used to check ______________.
   a. fuel pressure
   b. manifold vacuum
   c. cylinder compression
   d. all of the above

2. Which of the following tools would be used to test the operation of a choke pull-off diaphragm?
   a. fuel injector cleaner
   b. vacuum pump
   c. tach/dwellmeter
   d. oscilloscope

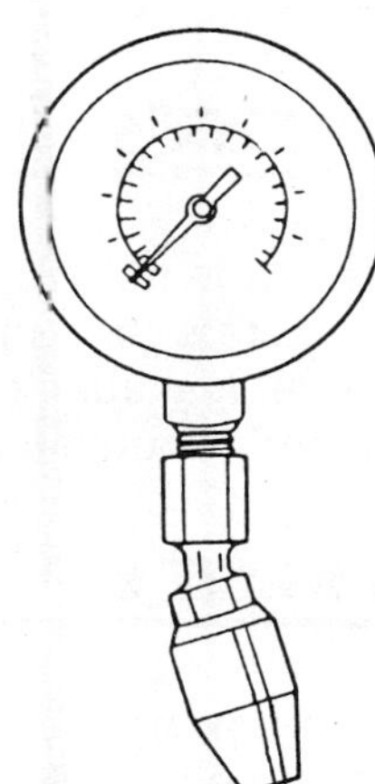

Figure 2–1

3. A circuit tester is not used for ______________.
   a. measuring amperage in a circuit
   b. testing a circuit for voltage
   c. searching for open circuits
   d. all of the above

4. Technician A uses a digital multimeter to test the voltage output of an oxygen sensor. Technician B favors an analog-type meter. Who is correct?
   a. Technician A
   b. Technician B
   c. Both A and B
   d. Neither A nor B

5. The tool shown in Figure 2–2 is being used to test ______________.
   a. injector pressure drop
   b. fuel pump discharge volume
   c. for ultrasonic vacuum leaks
   d. none of the above

6. The arrow in Figure 2–3 is pointing to a/an ______________.
   a. air bleed screw
   b. magnetic timing probe receptacle

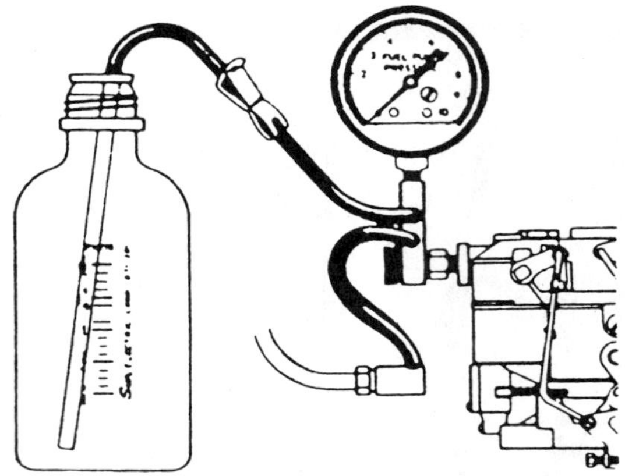

Figure 2-2

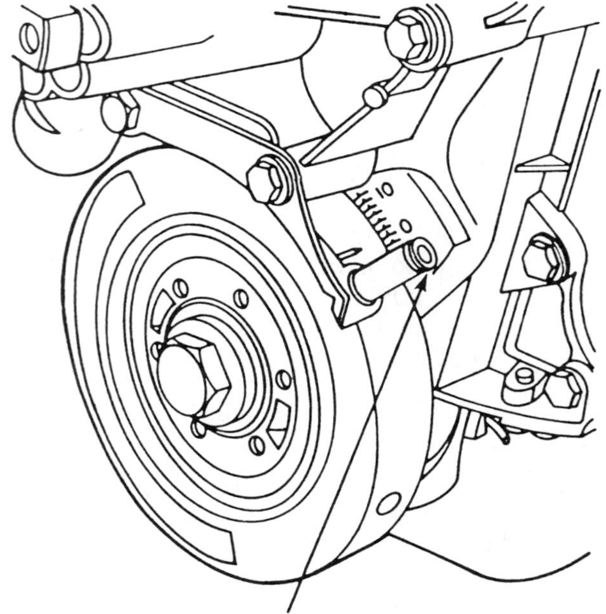

Figure 2-3

c. dip stick
d. vacuum pump

7. Technician A wants to see a digital volt/ohmmeter to test voltage in a starter circuit. Technician B wants to use the same volt/ohmmeter to test current flow to the alternator. Who made the correct choice of tools?
   a. Technician A
   b. Technician B
   c. Both A and B
   d. Neither A nor B

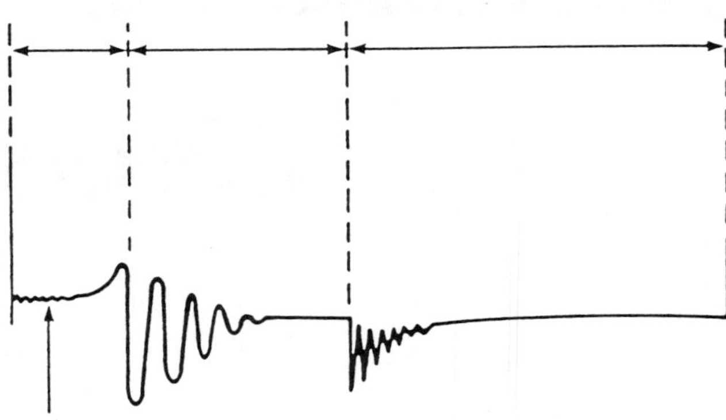

Figure 2-4

8. The arrow in Figure 2-4 is pointing to the ______________.
   a. firing line
   b. spark line
   c. dwell section
   d. intermediate section

9. A microcomputer scan tool can be used to test for ______________.
   a. cylinder compression
   b. exhaust emissions
   c. fuel flow rates
   d. none of the above

10. Technician A want to use an engine analyzer to test HC levels in the exhaust. Technician B wants to use the engine analyzer to check the ignition timing advance. Who is correct?
    a. Technician A
    b. Technician B
    c. Both A and B
    d. Neither A nor B

SHOP ASSIGNMENT 5

Identify and explain how to use various compression and vacuum testers, including the compression gauge, cylinder leakage tester, vacuum gauge, vacuum pump, and vacuum leak detector.

SHOP ASSIGNMENT 6

Identify and explain how to use various pieces of fuel system test equipment, including the pressure gauge, fuel injector cleaner, and fuel injector pulse tester.

SHOP ASSIGNMENT 7

Identify and explain how to use various pieces of electrical test equipment, including the circuit tester, voltmeter, ohmmeter, ammeter, volt/amp tester, and multimeter.

SHOP ASSIGNMENT 8

Identify and explain how to use various pieces of ignition system test equipment, including the tach-dwellmeter, digital tachometer, timing light, magnetic timing probe, electronic ignition module tester, and oscilloscope.

CHAPTER THREE

ENGINE PERFORMANCE TESTS

Objectives

Upon completion of this chapter, you should be able to:

- Explain the principle of compression ratio and how it relates to engine torque and horsepower.
- Prepare an engine for a dry compression test, perform the test, and interpret the results.
- Explain when a wet compression test is needed, perform the test, and interpret the results.
- Prepare an engine for a cylinder leakage test, perform the test, and interpret the results.
- Perform an engine vacuum test and interpret the results.
- Use a vacuum gauge to test a PCV system and an exhaust system.
- Explain the significance of engine noise as a diagnostic aid.
- Perform a cylinder power balance test.
- Describe the three ways that the valve clearance is adjusted.
- Perform a valve adjustment on a running engine and a stationary engine.

PRACTICE QUESTIONS

1. The timing belt should be replaced every ______________.
   a. 30,000 to 40,000 miles
   b. 50,000 to 60,000 miles
   c. 25,000 miles
   d. three years, regardless of mileage

2. A vehicle suffers from loss of compression and poor driveability. Technician A says that leaking valves is a possible cause of the problem. Technician B says that a leaking cylinder head gasket is a possible cause of the problem. Who is correct?
   a. Technician A
   b. Technician B
   c. Both A and B
   d. Neither A nor B

3. In addition to a compression gauge, what other special tool is required to do a compression test?
   a. stethoscope
   b. remote starter switch
   c. leakage tester
   d. vacuum gauge

4. The wet compression test should not be performed ______________.
   a. on horizontally opposed cylinder engines
   b. before the engine is disassembled for inspection
   c. using a compression gauge with a threaded adapter
   d. all of the above

5. When connecting an indicator light prior to performing a cylinder leakage test as shown in Figure 3–1, one of the leads is grounded and the other is attached to the ______________.
   a. primary battery terminal
   b. distributor's secondary terminal on the coil
   c. distributor's primary terminal on the coil
   d. number 1 cylinder spark plug hole

6. A vacuum test produces a low but steady reading, between 12 and 15 inches of mercury. Technician A says that this can be caused by the

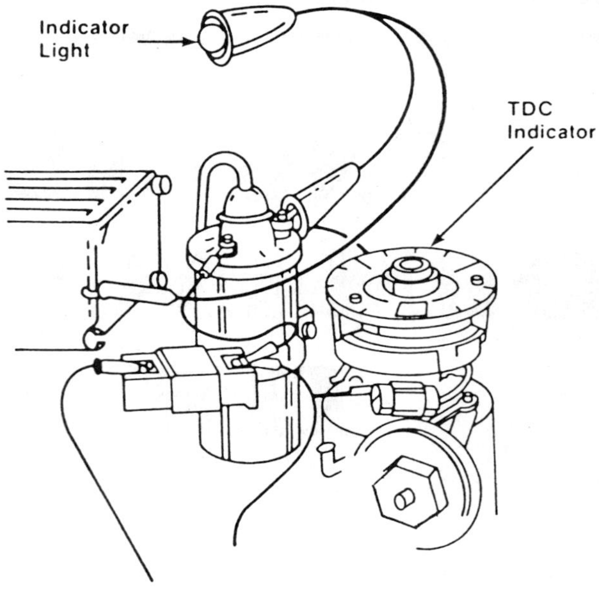

Figure 3-1

ignition timing being too far advanced. Technician B says that the spark plug in one of the cylinders is not firing. Who is correct?

a. Technician A
b. Technician B
c. Both A and B
d. Neither A nor B

7. Ring noise can be heard ______________.

a. when the engine is cold
b. when the engine is idling
c. only at speeds over 50 mph
d. during acceleration

8. Figure 3-2 shows examples of worn ______________.

a. connecting-rod bearings
b. crankshaft journals

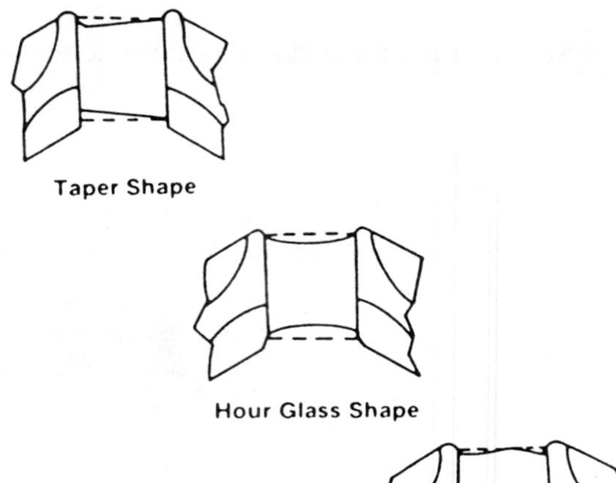

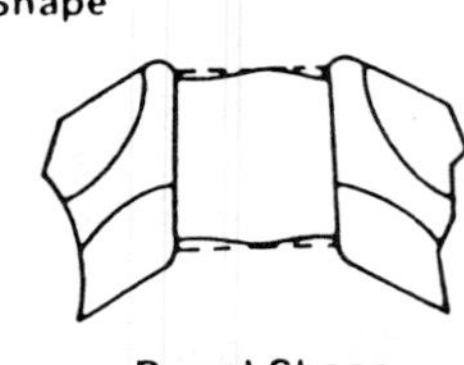

Figure 3-2

c. vibration dampers
d. rocker arms

9. Before performing a power balance test on a vehicle with a catalytic converter, Technician A shorts each spark plug at 30 second intervals. Technician B does not perform power balance tests on vehicles with catalytic converters; he does a compression test instead. Who is correct?

a. Technician A
b. Technician B
c. Both A and B
d. Neither A nor B

10. Adjustable rocker arms are found on ______________.

a. overhead valve engines with the camshaft in the engine block
b. overhead cam engines
c. all General Motors engines
d. all Ford engines

JOB SHEET

SHOP ASSIGNMENT 9
PERFORM A DRY COMPRESSION TEST

NAME ______________________ **STATION** ______________________ **DATE** ______________

Tools and Materials

Spark plug socket and ratchet
Compressed air
Screwdrivers
Remote starter switch
Compression gauge
Pencil and paper
Adjustable wrench

Protective Clothing

Safety goggles or glasses with side shields

Procedure

1. Turn on the engine; when it reaches its normal operating temperature, turn it off.

   Task completed ____

2. Disconnect the spark plug cables from the plugs.

   Task completed ____

3. Use compressed air to clean all foreign matter out of the plug wells.

   Task completed ____

4. Remove all the spark plugs.

   Task completed ____

   a. Set the plugs on a workbench or other clean surface in the order in which they were removed.

   Task completed ____

5. Remove all plug gaskets or tubes from the cylinder head.

   Task completed ____

6. Remove the air cleaner from the carburetor.

   Task completed ____

   a. Block the choke and throttle plate in the wide-open position using a screwdriver or similar tool.

   Task completed ____

7. Remove the ignition coil's high-tension wire from the distributor cap and ground it.

   Task completed ____

8. Connect a remote starter switch to the starter relay or solenoid by attaching one of its leads to the positive battery terminal and the other to the relay or solenoid S-terminal.

Task completed ____

9. Does the compression gauge being used have a threaded adapter?

Yes ____ No ____

   a. If yes, thread the adapter into the plug hole, fingertighten, and connect the gauge to the adapter.

   Task completed ____

   b. If no, insert the rubber tip into the plug hole and hold it firmly in place.

   Task completed ____

10. Turn the engine over four full compression strokes, watching the gauge needle at all times. Record the first and fourth gauge readings.

Task completed ____

11. Did the needle rise steadily with each stroke, with the final reading falling well within specifications?

Yes ____ No ____

   a. If yes, the cylinder compression is fine. Go to step 12.

   Task completed ____

   b. If no, a problem exists. Locate the source, then go to step 12.

   Task completed ____

12. Open the vent valve to release the compression pressure.

Task completed ____

13. Disconnect and remove the gauge.

Task completed ____

14. Repeat the procedure on the other cylinders.

Task completed ____

PROBLEMS ENCOUNTERED: ____________________

____

____

INSTRUCTOR'S COMMENTS: ____________________

____

____

JOB SHEET

SHOP ASSIGNMENT 10
PERFORM A CYLINDER LEAKAGE TEST

NAME ______________ **STATION** ______________ **DATE** ______________

Tools and Materials

Spark plug socket and ratchet
Compressed air
Screwdrivers
Radiator coolant (if applicable)
Leakage tester and whistle
Test adapter hose
TDC indicator
Indicator light
Adjustable wrench
Jumper lead
Chalk

Protective Clothing

Safety goggles or glasses with side shields

Procedure

1. Check the coolant level and fill if needed.

   Task completed ____

2. Turn on the engine; when it reaches its normal operating temperature, turn it off.

   Task completed ____

3. Disconnect the spark plug cables from the plugs.

   Task completed ____

4. Use compressed air to clean all foreign matter out of the plug wells.

   Task completed ____

5. Remove all the spark plugs.

   Task completed ____

   a. Set the plugs on a workbench or other clean surface in the order in which they were removed.

   Task completed ____

6. Remove all plug gaskets or tubes from the cylinder head.

   Task completed ____

7. Remove the air cleaner from the carburetor.

   Task completed ____

a. Block the choke and throttle plate in the wide-open position using a screwdriver or similar tool.

Task completed ____

8. Disconnect the PCV hose from the crankcase.

Task completed ____

9. Install a test adapter hose in the number 1 cylinder spark plug hole.

Task completed ____

a. Connect a tester whistle to the adapter hose.

Task completed ____

10. Using a wrench on the crankshaft pulley nut or bolt, slowly rotate the engine in the normal direction; this indicates the beginning of a compression stroke.

Task completed ____

11. Continue the rotation until the timing mark on the crankshaft pulley lines up with the engine-timing pointer on the timing chain cover.

Task completed ____

a. Remove the tester whistle from the adapter hose.

Task completed ____

12. Use a jumper lead to connect the coil-to-distributor secondary cable to a good ground.

Task completed ____

13. Remove the distributor cap and rotor, then mount a TDC indicator on the distributor shaft.

Task completed ____

a. Mark a chalk reference point on the engine that lines up with the appropriate cylinder marking on the TDC indicator.

Task completed ____

14. Connect an indicator light by attaching one of its leads to the distributor's primary terminal on the coil, and the other lead to a ground. Turn on the ignition switch.

Task completed ____

15. Connect the tester to the adapter hose.

Task completed ____

a. Does the gauge show more than 20 percent leakage? If so, look for air leaking from the carburetor, tail pipe, or crankcase, and for air bubbles in the radiator.

Yes ____ No ____

16. Disconnect the tester from the adapter hose.

Task completed ____

a. Resume rotating the engine until the next appropriate cylinder mark on the TDC indicator lines up with the chalk mark on the engine. A glowing indicator light means that the piston is in firing position.

Task completed ____

17. Remove the adapter from the previously tested cylinder and install it in the plug hole of the next cylinder in the firing order.

Task completed ____

18. Repeat steps 15, 16, and 17 on each of the remaining cylinders.

Task completed ____

PROBLEMS ENCOUNTERED: ____

____

____

INSTRUCTOR'S COMMENTS: ____

____

____

JOB SHEET

SHOP ASSIGNMENT 11
TEST AN EXHAUST SYSTEM FOR RESTRICTION USING A VACUUM GAUGE

NAME ______________________ **STATION** ______________________ **DATE** ______________

Tools and Materials

Vacuum gauge
Tachometer
Length of hose
Clamp (if applicable)

Protective Clothing

Safety goggles or glasses with side shields

Procedure

1. Connect a vacuum gauge to a nonrestricted port on the intake manifold using a length of hose.

   Task completed ____

2. If necessary, clamp the hose to further dampen the needle.

   Not applicable ____ Task completed ____

3. Connect a tachometer to the engine.

   Task completed ____

4. Run the engine until it reaches its normal operating temperature.

   Task completed ____

5. Slowly accelerate the engine until it reaches 2000 rpm. The gauge needle should drop a bit, then rise sharply.

   Task completed ____

6. Close the throttle quickly. The needle should return to the normal idle reading as quickly as it rose.

   Task completed ____

7. If the needle did not respond as it should, check for muffler and tail pipe damage, and make sure the manifold heat control valve is not stuck or frozen.

   Not applicable ____ Task completed ____

PROBLEMS ENCOUNTERED: ____

____

____

INSTRUCTOR'S COMMENTS: ______________________________

____

____

JOB SHEET

SHOP ASSIGNMENT 12
PERFORM A CYLINDER POWER BALANCE TEST

NAME ______________________ **STATION** ______________________ **DATE** ______________

Tools and Materials

Screwdrivers
Plug
Jumper lead
Engine analyzer

Protective Clothing

Safety goggles or glasses with side shields

Procedure

1. Does the engine have an air/fuel mixture feedback control, as shown in Figure 3-3?

   Yes ____ No ____

   a. If yes, disconnect and plug either the air pump hose going to the catalytic converter or the downstream hose between the air switching valve and the check valve.

   Task completed ____

   b. If no, disconnect and block the air pump on the valve side.

   Task completed ____

2. Can the electric cooling fan be bypassed?

   Yes ____ No ____

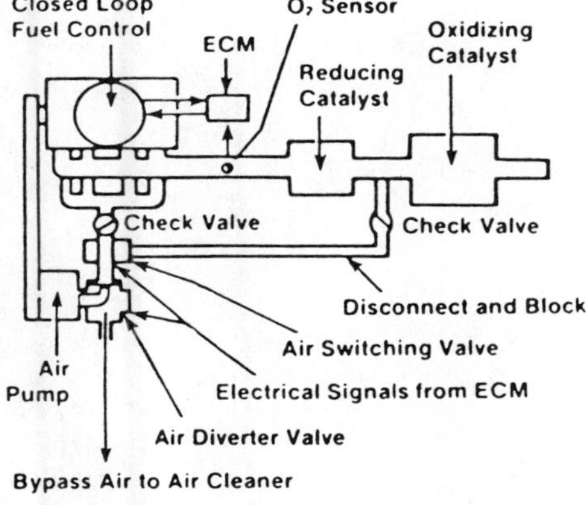

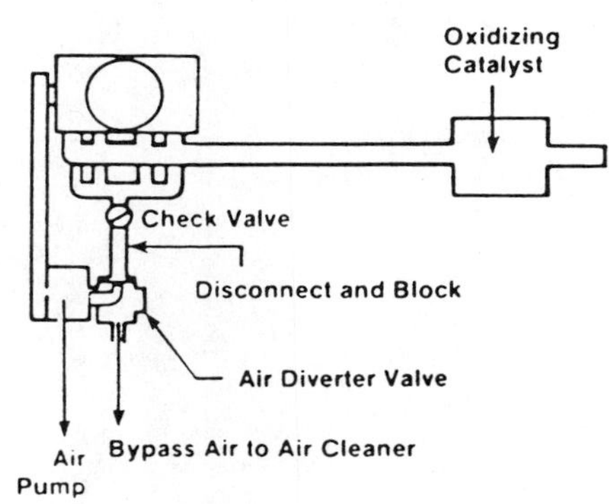

Figure 3-3

a. If yes, override the controls of the fan by jumper wiring them so that the fan runs constantly.

Task completed ____

b. If no, disconnect the fan.

Task completed ____

3. Connect the engine analyzer's leads, referring to the service manual for specific instructions.

Task completed ____

4. Turn on the engine and let it reach its normal operating temperature. Achieve a speed of approximately 1000 rpm.

Task completed ____

5. As each cylinder is shorted out, is there a noticeable drop in engine speed?

Yes ____ No ____

a. If yes, the engine is in sound mechanical condition.

Task completed ____

b. If no, further testing must be done to determine the exact source of the problem.

Task completed ____

PROBLEMS ENCOUNTERED: ____

____

____

INSTRUCTOR'S COMMENTS: ____

____

____

JOB SHEET

SHOP ASSIGNMENT 13
MAKE A VALVE CLEARANCE CHECK AND ADJUSTMENT ON A RUNNING ENGINE

NAME ______________ **STATION** ______________ **DATE** ______________

Tools and Materials

Fender covers
Screwdrivers
Feeler gauge
Engine flush
Adjustable wrench (if applicable)
Valve cover gaskets

Protective Clothing

Safety goggles or glasses with side shields

Procedure

1. Protect both fenders with covers.

   Task completed ____

2. Turn on the engine; when it reaches its normal operating temperature, turn it off.

   Task completed ____

   a. Remove the valve covers.

   Task completed ____

3. Check the valve springs and valve stems for damage.

   Task completed ____

4. Check that the oil-drain passages in the cylinder heads are open.

   Task completed ____

   a. Remove any sludge built up on the valve-train components.

   Task completed ____

5. Turn on the engine and run it at its slowest idle. Using a feeler gauge and a front-to-rear or rear-to-front pattern, check the valves for clearance. The feeler gauge strip should pass through the gap with a slight drag.

   Task completed ____

   a. If the strip must be forced, or if the engine starts to miss, the clearance is too small. Turn the adjustment screw in until a slight drag is felt.

   Not applicable ____ Task completed ____

b. If the strip passes through easily, or if a jerking sensation is felt, the clearance is too large. Turn the adjustment screw out until a slight drag is felt.

Not applicable ____ Task completed ____

c. If the adjustment screw has a separate locknut, recheck the clearance after tightening the nut.

Not applicable ____ Task completed ____

6. Replace the valve covers and any other accessories that were removed.

Task completed ____

7. Turn on the engine and check the valve covers for oil leaks.

Task completed ____

a. Check that the oil is up to the recommended level.

Task completed ____

PROBLEMS ENCOUNTERED: ____

____

____

INSTRUCTOR'S COMMENTS: ____

____

____

CHAPTER FOUR

BATTERY TESTING AND SERVICING

Objectives

Upon completion of this chapter, you should be able to:

- Perform safety and working precautions that must be taken when inspecting, testing, and charging storage batteries.
- Perform the proper steps for routine battery inspections, cleaning, testing, and replacement.
- Describe the function and operation of battery fast and slow chargers and how they differ.
- Properly jump-start vehicle with a discharged battery using a booster battery.

PRACTICE QUESTIONS

1. The hydrometer shown in Figure 4–1 indicates a ______________.
   a. sulfated battery
   b. low specific gravity
   c. high specific gravity
   d. both a and b

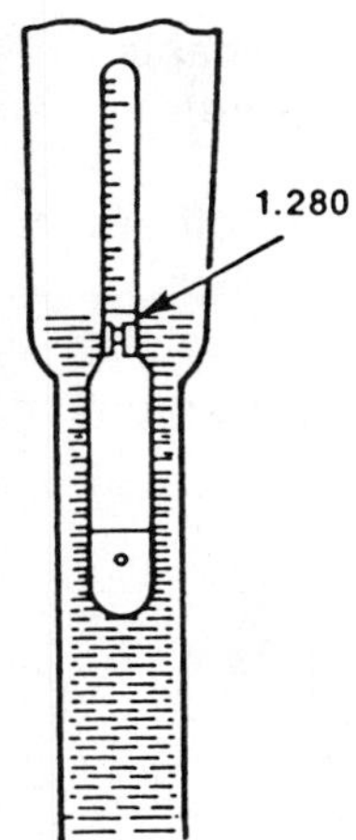

Figure 4–1

2. A cold cranking amps rating specifies the minimum amps available at ______________.
   a. 32° F
   b. 0° F and -20° F
   c. 0° F and 20° F
   d. 0° F and 32° F

3. Why can overfilling a battery be harmful?
   a. It weakens the concentration of sulfuric acid.
   b. The plates' grids deteriorate more rapidly.
   c. It causes the active materials in the battery to harden.
   d. All of the above.

4. When testing with a built-in hydrometer, if the green dot is not visible and has a dark appearance as shown in Figure 4–2, what does it mean?
   a. The battery is sufficiently charged.
   b. The battery must be charged.
   c. The battery must be replaced.
   d. The battery needs water.

5. A cadmium-probe test reveals that all the cells vary less than five divisions on the top scale, but some test in the red section on the lower scale. Technician A says this means the battery is in good condition and sufficiently charged. Technician B says it means the battery is in good condition but has a low charge. Who is correct?
   a. Technician A
   b. Technician B
   c. Both A and B
   d. Neither A nor B

6. Which of the following statements is incorrect?
   a. Spring-type battery cables are removed using a box wrench or cable-clamp pliers.
   b. A stiff bristle brush is ideal for removing heavy corrosion buildup on a battery.

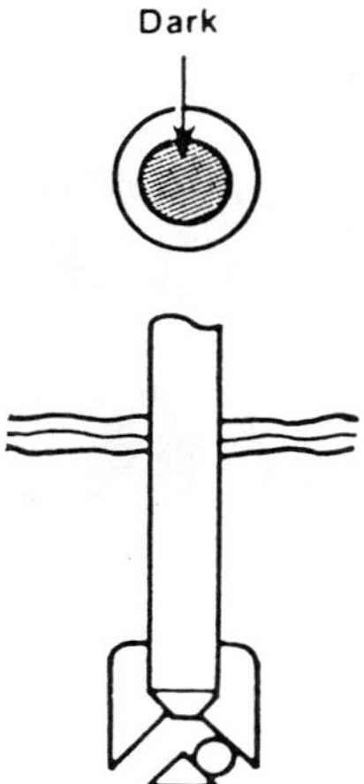

Figure 4–2

c. Felt washers treated with corrosion-resistant compound can be installed over battery terminals.
d. Always grip the cable when loosening a battery connector nut.

7. A specific gravity difference of more than 50 points between cells is ____________.
   a. a good sign
   b. common with maintenance-free batteries
   c. a sign of a defective battery
   d. both a and b

8. An open circuit voltage test reveals a charge below 75 percent of full charge. Technician A responds by recharging the battery. Technician B recharges the battery as well, then performs a capacity test. Who is correct?
   a. Technician A
   b. Technician B
   c. Both A and B
   d. Neither A nor B

9. What test is used to determine if an unsealed battery is too sulfated to accept a charge?
   a. three-minute charge test
   b. cadmium-probe test
   c. capacity test
   d. alternative capacity test

10. A battery should not be charged if its built-in hydrometer registers ____________.
   a. clear
   b. light yellow
   c. both a and b
   d. neither a nor b

JOB SHEET

SHOP ASSIGNMENT 14
PERFORM AN ALTERNATIVE CAPACITY TEST

NAME ______________ **STATION** ______________ **DATE** ______________

Tools and Materials

Voltmeter
Screwdriver

Protective Clothing

Safety goggles or glasses with side shields

Procedure

1. Make sure the battery is at or near full charge.

   Task completed ____

2. Make sure the starting circuit and starter are in good condition.

   Task completed ____

3. Connect the voltmeter positive lead to the positive battery terminal, and the voltmeter negative lead to the negative battery terminal.

   Task completed ____

4. Crank the engine over continuously for 15 seconds.

   Task completed ____

5. After cranking the engine, observe the voltmeter reading. Is it above the minimum specified load test voltage?

   Yes ____ No ____

   a. If yes, the battery and cranking circuit are in good condition.

      Task completed ____

   b. If no, either the battery is in poor condition or the starting circuit is drawing too much current. Both a three-minute charge test and a starting system load test must be performed.

      Task completed ____

PROBLEMS ENCOUNTERED: ______________________________

____

____

INSTRUCTOR'S COMMENTS: ______________________________

____

JOB SHEET

SHOP ASSIGNMENT 15
PERFORM A CADMIUM-PROBE TEST

NAME ______________________ **STATION** ______________________ **DATE** ______________

Tools and Materials

Specific tester
Battery charger (if applicable)
Pencil and paper

Protective Clothing

Safety goggles or glasses with side shields

Procedure

1. Remove the vent caps and check the electrolyte level in each cell. Must water be added to the battery?

   Yes ____ No ____

   a. If yes, charge the battery at a slow rate for 5 to 10 minutes to mix the electrolyte. Proceed to step 2.

   Task completed ____

   b. If no, proceed to step 3.

   Task completed ____

2. Remove the surface charge from the battery by turning on the headlights for 1 minute. If the vehicle was not operated within the last eight hours, this step is not necessary.

   Not applicable ____ Task completed ____

3. Be sure the headlights and all accessories are turned off.

   Task completed ____

4. Place the tester's red probe in the positive cell next to the positive terminal, and place the blade probe in the second cell. Record the reading on the meter.

   Task completed ____

5. Move the red probe to the second cell and the black probe to the third cell. Record the reading.

   Task completed ____

6. Move the red probe to the third cell and the black probe to the fourth cell. Record the reading.

   Task completed ____

7. Continue in the same manner until all the cells have been tested.

Task completed ____

8. Does the reading of any two cells vary five divisions or more on the top scale, regardless of the section into which they fall on the lower scale?

Yes ____ No ____

   a. If yes, the battery must be replaced.

Task completed ____

   b. If no, proceed to the next step.

Task completed ____

9. Do all the cells vary less than five divisions on the top scale, with all of them in the green section on the lower scale?

Yes ____ No ____

   a. If yes, the battery is in good condition and sufficiently charged.

Task completed ____

   b. If no, proceed to the next step.

Task completed ____

10. Do all the cells vary less than five divisions on the top scale, with at least some of them in the red section on the lower scale?

Yes ____ No ____

   a. If yes, the battery is in good condition but needs to be charged.

Task completed ____

   b. If no, proceed to the next step.

Task completed ____

11. Are any of the cells in the recharge-retest section of the top scale, with the rest in the first four divisions?

Yes ____ No ____

   a. If yes, the battery must be recharged and the surface charge removed before it can be properly tested.

Task completed ____

   b. If no, the test is complete.

Task completed ____

PROBLEMS ENCOUNTERED: ______________________________

____

____

INSTRUCTOR'S COMMENTS: ______________________________

____

____

JOB SHEET

SHOP ASSIGNMENT 16
RECHARGE A BATTERY

NAME ______________________ **STATION** ______________________ **DATE** ____________

Tools and Materials

Battery charger
Voltmeter (if necessary)
Pyrometer
Baking soda or ammonia
Rags

Protective Clothing

Safety goggles or glasses with side shields

Procedure

1. Check the electrolyte level and add water if needed.

   Task completed ____

2. Connect the charger leads to the battery as shown in Figure 4-3.

   Task completed ____

3. Check the specific gravity, the electrolyte temperature, and the voltage across the terminals periodically during charging.

   Task completed ____

4. The battery is fully charged when all the cells are gassing freely and specific gravity has not increased for three hours. The fully charged specific gravity should be 1.260 to 1.280. At that point, disconnect the charger.

   Task completed ____

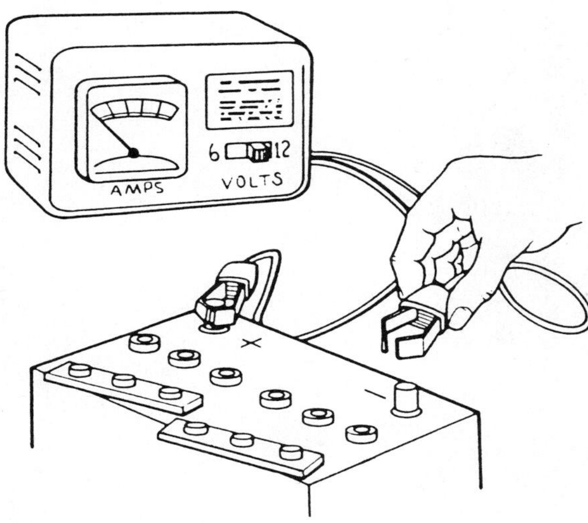

Figure 4-3

5. Flush the battery top with baking soda or ammonia solution.

Task completed ____

6. Dry the battery, then check the electrolyte level and add water if needed.

Task completed ____

PROBLEMS ENCOUNTERED: ____________________

____

____

INSTRUCTOR'S COMMENTS: ____________________

____

____

CHAPTER FIVE

STARTING SYSTEM TESTING AND DIAGNOSIS

Objectives

Upon completion of this chapter, you should be able to:

- List components of the starter system, starter circuit, and control circuit.
- Explain the different types of magnetic switches and starter drive mechanisms.
- Explain how a starter motor operates.
- Describe the operation and function of an overriding clutch.
- Explain how to perform and interpret the following tests: battery load test, cranking voltage test, cranking current test, insulated circuit resistance test, starter relay bypass test, ground circuit resistance test, and control circuit voltage and resistance tests.
- Explain how to replace a starter.

PRACTICE QUESTIONS

1. Figure 5-1 illustrates the two distinct circuits that make up a typical starting system. Which of the following statements is true?
   a. The dashed lines show the control circuit, while the solid lines show the starter circuit.
   b. The dashed lines show the starter circuit, while the solid lines show the control circuit.
   c. The dashed lines can indicate either the starter or control circuit, depending on whether the system is charging or discharging.
   d. None of the above.

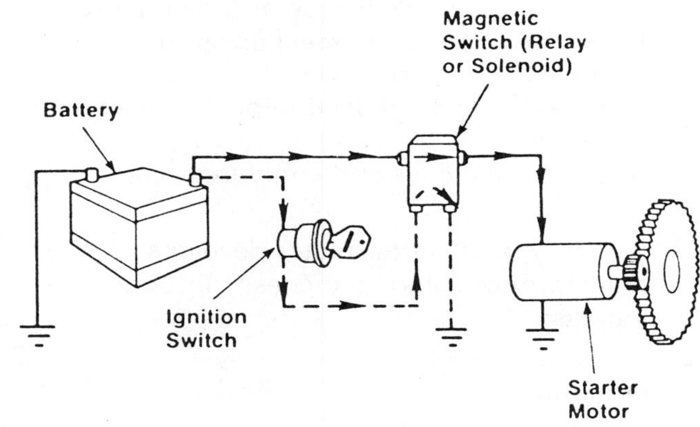

Figure 5-1

2. The purpose of the magnetic switch (either a relay or solenoid) is to ____________.
   a. regulate the amount of current flowing through the starter circuit
   b. open the control circuit and close the starter circuit
   c. allow the starting motor to engage the flywheel
   d. both a and b

3. Figure 5-2 shows a ____________.
   a. relay with a hinged armature
   b. fusible link before a short circuit
   c. fusible link after a short circuit
   d. relay after a short circuit

4. In a solenoid-actuated direct drive starting system, ____________.

Figure 5-2

a. the solenoid is mounted directly on top of the starter
b. the solenoid is mounted as close to the battery as possible to keep cables as short as possible
c. a spring is used to help align the pinion teeth with the teeth on the flywheel
d. both a and c

5. Which part of the starter motor is the arrow pointing to in Figure 5-3?
   a. commutator
   b. solenoid
   c. armature windings
   d. shaft

6. The one area in which starter motors vary greatly is in ____________.
   a. design of the electric motor using magnetic fields to produce torque or twisting force
   b. design of the drive mechanism used to engage the flywheel
   c. types of housing used to protect the starter motor
   d. use of field coils and armatures

7. Which type of starter motor develops its maximum torque at startup and less torque as speed increases?
   a. series motors
   b. shunt motors
   c. compound motors
   d. all of the above

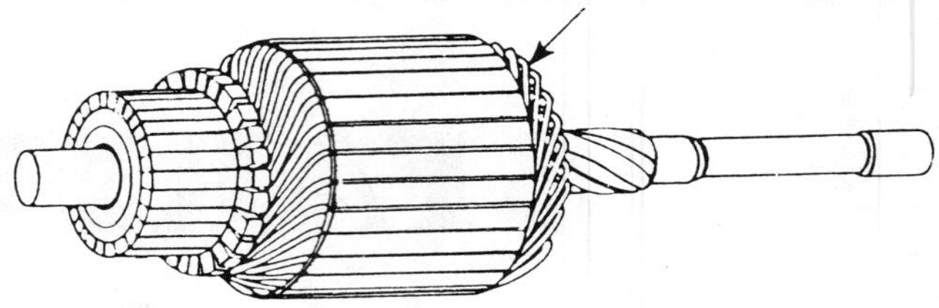

Figure 5-3

8. Which of the following drive mechanism designs is found exclusively on Ford models, has the drive mechanism as an integral part of the motor, and has the drive pinion engage and flywheel before the motor is energized?
   a. permanent magnet planetary drive
   b. movable pole shoe
   c. solenoid-actuated drive
   d. solenoid-actuated reduced drive

9. To disable a standard ignition system, Technician A removes the primary lead from the distributor center tower and grounds the lead using a jumper wire. Technician B removes the secondary lead from the distributor center tower and grounds it using a jumper wire. Who is correct?
   a. Technician A
   b. Technician B
   c. Both A and B
   d. Neither A nor B

10. The function of a starter safety switch is to ____________.
    a. prevent vehicles from being started with the transmission in gear (automatics) or the clutches engaged (standards)
    b. prevent the vehicle from being started with the steering wheel in the locked position
    c. prevent a voltage overload at the starter relay
    d. all of the above

JOB SHEET

SHOP ASSIGNMENT 17
PERFORM A CRANKING VOLTAGE TEST ON A LATE-MODEL CHRYSLER VEHICLE

NAME ______________________ **STATION** ______________________ **DATE** ______________

Tools and Materials

Remote starter switch
Voltmeter

Protective Clothing

Safety goggles or glasses with side shields

Procedure

1. Bypass the ignition switch by connecting the remote starter switch leads between the battery terminal at the starter relay and the terminal.

   Task completed ____

2. Connect the voltmeter negative lead to a ground.

   Task completed ____

3. Connect the voltmeter positive lead to the proper test point as shown in Figure 5-4.

   Task completed ____

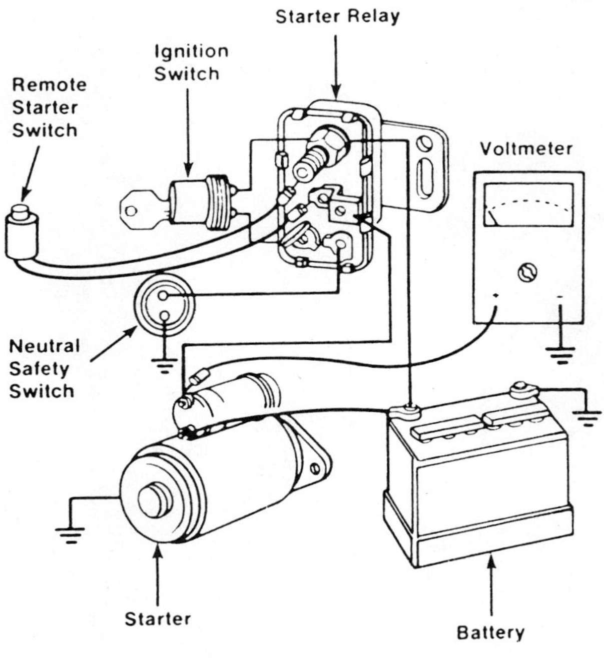

Figure 5-4

4. Crank the engine and note the voltmeter reading. Is it below specifications, even if the battery is fully charged?

Yes ____ No ____

a. If yes, a cranking current test should be performed to determine where the problem lies.

Task completed ____

b. If no, but the motor still cranks poorly, the starter motor is the problem.

Task completed ____

PROBLEMS ENCOUNTERED: ____

____

____

INSTRUCTOR'S COMMENTS: ____

____

____

JOB SHEET

SHOP ASSIGNMENT 18
PERFORM A CRANKING CURRENT TEST USING A CONVENTIONAL AMMETER

NAME ______________ **STATION** ______________ **DATE** ______________

Tools and Materials

Voltmeter
Ammeter
Carbon pile
Remote starter switch

Protective Clothing

Safety goggles or glasses with side shields

Procedure

1. Disable or bypass the ignition using a remote starter switch.

   Task completed ____

2. Connect the voltmeter positive lead to the positive battery terminal, and the voltmeter negative lead to the negative battery terminal.

   Task completed ____

3. Set the carbon pile to its maximum resistance (open).

   Task completed ____

4. Connect the ammeter positive lead to the positive battery terminal and the ammeter negative lead to one lead of the carbon pile.

   Task completed ____

5. Connect the other lead of the carbon pile to the negative battery terminal.

   Task completed ____

6. Crank the engine and note the voltmeter reading.

   Task completed ____

7. Stop cranking the engine. Adjust the carbon pile until the voltmeter reading matches the reading taken in step 6.

   Task completed ____

8. Note the ammeter reading and set the carbon pile back to open.

   Task completed ____

9. As a rule of thumb, readings for 8-cylinder engines should be approximately 280 amps; 6-cylinder engines approximately 240 amps; and 2- to 5-cylinder engines about 210 amps.

   Task completed ____

PROBLEMS ENCOUNTERED: ______________________________

____

____

INSTRUCTOR'S COMMENTS: ______________________________

____

____

JOB SHEET

SHOP ASSIGNMENT 19
PERFORM AN INSULATED CIRCUIT RESISTANCE TEST ON A GM VEHICLE

NAME ______________________ **STATION** ______________________ **DATE** ______________

Tools and Materials

Remote starter switch
Voltmeter

Protective Clothing

Safety goggles or glasses with side shields

Procedure

1. Disable or bypass the ignition using a remote starter switch.

   Task completed ____

2. Connect the voltmeter positive lead to the positive battery terminal post or nut.

   Task completed ____

3. Connect the voltmeter negative lead to the terminal at the starter.

   Task completed ____

4. Crank the engine and note the voltmeter reading. Is it within specifications?

   Yes ____ No ____

   a. If yes, the insulated circuit does not have excessive resistance. Perform a ground circuit test.

      Task completed ____

   b. If no, and the voltage loss is above specifications, move the volt lead on the starter progressively toward the battery. Crank the engine at each of the test points indicated in Figure 5-5.

      Task completed ____

   c. If no, and there is a decrease in voltage, inspect for damaged wiring or faulty connections.

      Task completed ____

PROBLEMS ENCOUNTERED: ____

____

____

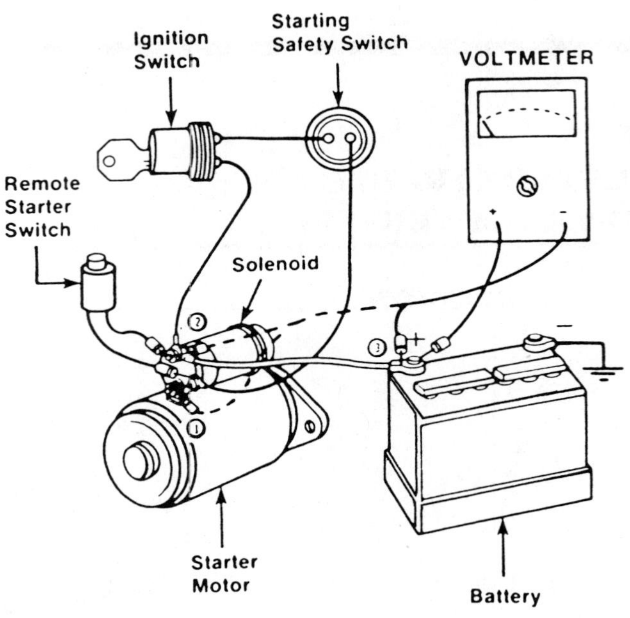

Figure 5-5

INSTRUCTOR'S COMMENTS: ____

____

____

JOB SHEET

SHOP ASSIGNMENT 20
REMOVE AND REPLACE A STARTER MOTOR

NAME ______________________ **STATION** ______________________ **DATE** ______________

Tools and Materials

Appropriate wrenches
Support brackets (if applicable)
Appropriate screwdrivers
Hoist or safety stands
Tags

Protective Clothing

Safety goggles or glasses with side shields

Procedure

1. Disconnect the battery ground cable.

   Task completed ____

2. Raise and support the vehicle using the hoist or safety stands.

   Task completed ____

3. If required to access the starter, turn the front wheels, disconnect the tie-rods, and remove the wheel.

   Task completed ____

4. Loosen or remove the exhaust pipes and other components that block access to the starter.

   Task completed ____

5. Disconnect and tag all wires from the starter motor or solenoid.

   Task completed ____

6. Remove the heat shields and support brackets.

   Task completed ____

7. Remove all mounting bolts and shims that secure the starter to the engine, and lift out the starter.

   Task completed ____

8. If they are being used, attach support brackets to the new or serviced starter motor.

   Task completed ____

   a. If possible, attach the heat shield before mounting.

   Task completed ____

9. Install the motor on the engine using all the mounting bolts.

Task completed ____

10. Check for proper flywheel engagement.

Task completed ____

11. Reconnect all wires to the solenoid or motor terminals.

Task completed ____

12. Remount or tighten any removed or loosened parts disassembled previously.

Task completed ____

13. Lower the vehicle and reconnect the battery ground cable.

Task completed ____

14. Test the starter for proper operation.

Task completed ____

PROBLEMS ENCOUNTERED: ____

____

____

INSTRUCTOR'S COMMENTS: ____

____

____

CHAPTER SIX

OPERATION AND TESTING OF CHARGING SYSTEMS

Objectives

Upon completion of this chapter, you should be able to:

- Explain how AC systems work, and describe their advantages over DC systems.
- Explain half- and full-wave rectification, and how they relate to alternator operation.
- Identify the different types of AC voltage regulators.
- Perform a charging system inspection.
- Perform a general charging system test using a charging-starting-battery analyzer.
- Describe the five basic charging system tests.

PRACTICE QUESTIONS

1. In a DC charging system, energy is produced by _______________.
   a. a series of rotating conductors
   b. a series of rotating regulators
   c. a single regulator
   d. the cutout relay

2. What conducts current to the rotor?
   a. slip rings and brushes
   b. stator
   c. cutout relay
   d. coil

3. When a diode is placed in a simple AC circuit, _______________.
   a. one-half of the voltage is blocked
   b. all of the voltage is blocked
   c. full-wave rectification occurs
   d. none of the voltage is blocked

4. To keep alternator output at a constant maximum, even when the rotor speed increases, Technician A raises the field current. Technician B lowers the field current. Who is correct?
   a. Technician A
   b. Technician B
   c. Both A and B
   d. Neither A nor B

5. Isolated field circuits are found mainly in _______________.
   a. vehicles with electromagnetic voltage regulators
   b. vehicles with solid-state voltage regulators
   c. foreign vehicles
   d. Chrysler vehicles

6. Which of the following statements is incorrect?
   a. Integrated circuit regulators are nonserviceable.
   b. Most AC systems include a field relay along with the regulator.
   c. Today's transistorized regulators use an electromagnetic field relay.
   d. The majority of electromagnetic regulators are double-contact units.

7. The _______________ is mainly a battery condition monitor.
   a. voltmeter
   b. indicator light

c. ammeter
d. engine analyzer

8. Technician A diagnoses a whining alternator as being caused by a shorted diode. Technician B says that the likely cause is a defective stator. Who is correct?
   a. Technician A
   b. Technician B
   c. Both A and B
   d. Neither A nor B

9. When should the carbon pile be turned on?
   a. never
   b. at all times
   c. at all times, except when testing is being done
   d. only when testing is being done

10. Figure 6-1 illustrates a typical 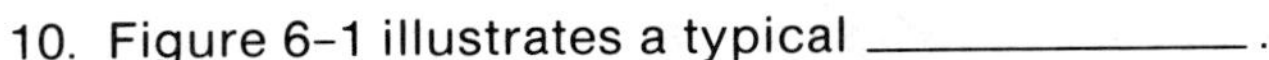.
    a. ammeter
    b. regulator
    c. stator
    d. heat sink

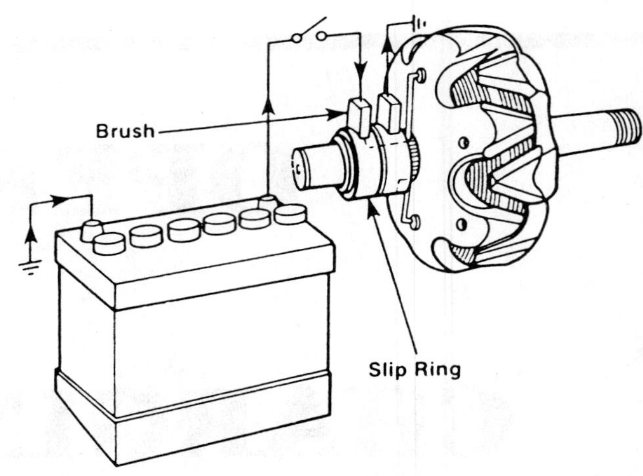

Figure 6-1

JOB SHEET

SHOP ASSIGNMENT 21
CONDUCT A CHARGING SYSTEM INSPECTION

NAME ______________ **STATION** ______________ **DATE** ______________

Tools and Materials

Belt tension gauge
Battery charger (if necessary)
Hydrometer
Inductive probe
Appropriate wrenches

Protective Clothing

Safety goggles or glasses with side shields

Procedure

1. Check for proper pulley alignment.

   Task completed ____

2. Inspect the alternator drive belt with a belt tension gauge.

   Task completed ____

   a. Also check the belt for fraying, cracking, or glazing, and replace if needed.

   Task completed ____

3. Check the battery for proper electrolyte level and specific gravity.

   Task completed ____

   a. If necessary, charge the battery.

   Task completed ____

4. Clean and tighten all battery posts and cable clamps.

   Task completed ____

5. Use an inductive probe to inspect the wiring and connections. Especially look for short circuits, open grounds, or high resistance.

   Task completed ____

6. Check the alternator and regulator mountings for loose and/or missing bolts.

   Task completed ____

7. Listen for noisy belts, bad bearings, or bad diodes (whirring sound).

   Task completed ____

PROBLEMS ENCOUNTERED: ____

____

____

INSTRUCTOR'S COMMENTS: ____

____

____

JOB SHEET

SHOP ASSIGNMENT 22
USE A CSB ANALYZER TO TEST A CHARGING SYSTEM

NAME ______________________ **STATION** ______________________ **DATE** ______________

Tools and Materials

CSB analyzer

Protective Clothing

Safety goggles or glasses with side shields

Procedure

1. Connect the analyzer to the vehicle following the directions in the operator's manual and the example shown in Figure 6-2.

   Task completed ____

2. Set the test selector switch to the charging test position.

   Task completed ____

3. Turn on the ignition switch and record the ammeter reading.

   Task completed ____

4. Start the engine and adjust it to approximately 2000 rpm (or whatever speed is called for by the manufacturer).

   Task completed ____

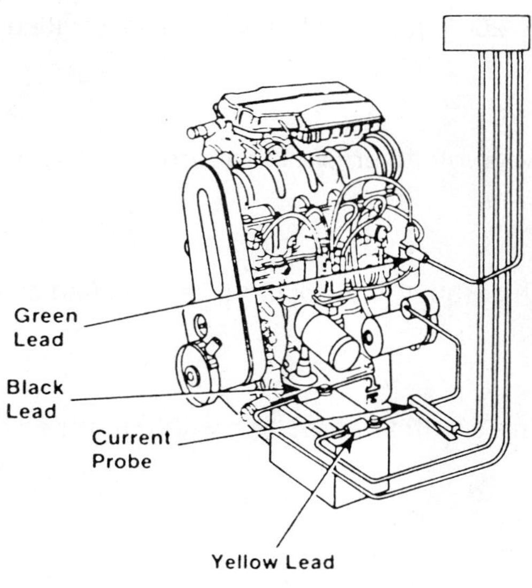

Figure 6-2

5. Adjust the carbon pile to achieve the maximum ammeter reading, while keeping the system voltage above 12 volts.

Task completed ____

6. Is the ammeter reading within 10 percent of specifications?

Yes ____ No ____

   a. If yes, go on to the next step.

   Task completed ____

   b. If no, go on to step 11.

   Task completed ____

7. Set the test selector switch to the regulator test position.

Task completed ____

8. Adjust the engine speed to 2000 rpm (or whatever speed is called for by the manufacturer).

Task completed ____

9. When the voltmeter stops rising, note the reading. Is it within specifications?

Yes ____ No ____

   a. If yes, the charging system is in good condition.

   Task completed ____

   b. If no, replace the regulator and repeat steps 7 to 9.

   Task completed ____

10. Turn off the engine and disconnect the field wire from the alternator.

Task completed ____

11. Connect the field lead to the alternator field terminal.

Task completed ____

12. Start the engine and adjust it to 2000 rpm (or whatever speed is called for by the manufacturer).

Task completed ____

13. Adjust the carbon pile to achieve a voltmeter reading approximately 2 volts less than the system voltage.

Task completed ____

14. Switch the field selector switch to the A or B position, as specified by the manufacturer.

Task completed ____

15. Adjust the carbon pile to achieve a voltmeter reading within the specified range.

Task completed ____

16. Record the ammeter reading.

Task completed ____

a. Return the carbon pile to the open position, then turn off the engine.

Task completed ____

17. Add the ammeter reading from step 3 to the reading made in the previous step. Is the total within 10 percent of specifications?

Yes ____ No ____

a. If yes, go on to the next step.

Task completed ____

b. If no, remove the alternator for further testing.

Task completed ____

18. Examine the wiring between the alternator and regulator. Are repairs necessary?

Yes ____ No ____

a. If yes, make them. Then repeat steps 12 to 17.

Task completed ____

a. If no, replace the regulator.

Task completed ____

PROBLEMS ENCOUNTERED: ____________________

____

____

INSTRUCTOR'S COMMENTS: ____________________

____

____

CHAPTER SEVEN

TROUBLESHOOTING BREAKER POINT IGNITION SYSTEMS

Objectives

Upon completion of this chapter, you should be able to:

- Name the components of the primary and secondary ignition system circuits, and describe the function of each.
- Explain dwell and its significance to breaker point ignitions.
- Explain the importance of reach, heat range, and air gap to spark plugs.
- Describe the difference between centrifugal and vacuum spark advance.
- Perform the full range of primary and secondary circuit tests using an oscilloscope, voltmeter, and ohmmeter.
- Perform breaker point maintenance, including aligning and setting points.
- Remove, inspect, regap, and install spark plugs.
- Explain the procedural differences involved in setting static and dynamic timing.

PRACTICE QUESTIONS

1. The two main types of ignition systems are ______________.
   a. breaker point and conventional
   b. electronic and solid-state
   c. primary and secondary
   d. breaker point and electronic

2. The ______________ controls the flow of low-voltage current from the battery to the coil primary winding.
   a. ignition switch
   b. primary resistor
   c. ballast resistor
   d. none of the above

3. A pitted set of breaker points has a built-up metal mound on the positive point. Technician A says that this means the condenser capacity is too high. Technician B says that it means the condenser capacity is fine. Who is correct?
   a. Technician A
   b. Technician B
   c. Both A and B
   d. Neither A nor B

4. Figure 7-1 illustrates a typical ______________.
   a. ballast primary resistor
   b. resistance wire primary resistor
   c. primary resistor that is part of the coil
   d. moveable primary resistor

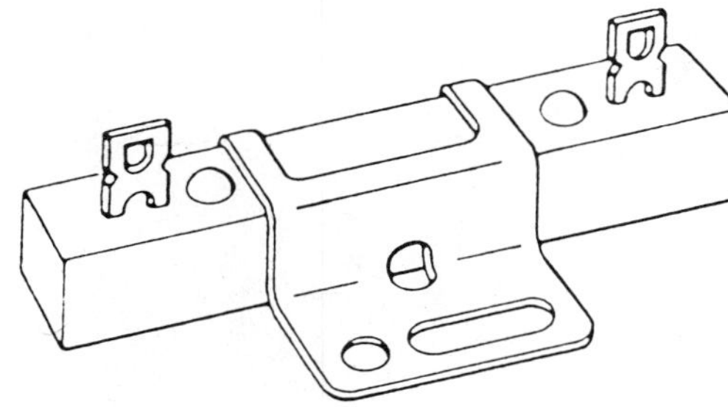

Figure 7-1

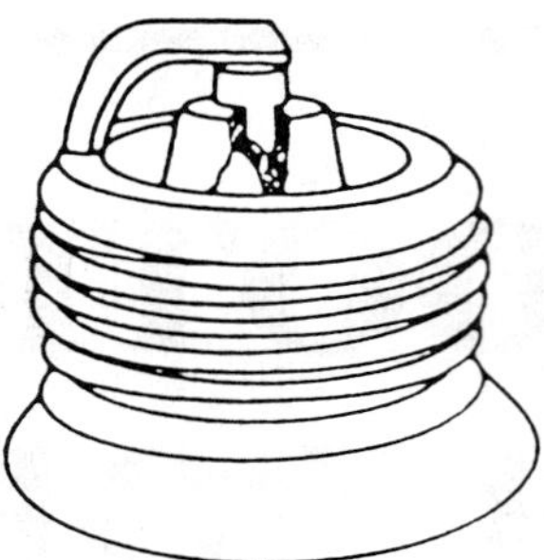

Figure 7-2

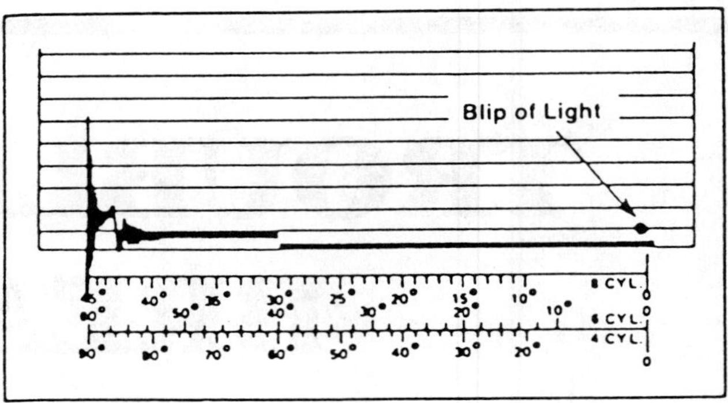

Figure 7-3

5. The spark plug shown in Figure 7-2 has probably experienced ____________.
   a. preignition damage
   b. overheating
   c. turbulence burning
   d. splash fouling

6. Which part of a spark plug acts as the heat conductor?
   a. steel shell
   b. ceramic core
   c. plastic shell
   d. electrodes

7. The discharge across the spark plug air gap can be _____ _______.
   a. capacitive or ionized
   b. hot or cold
   c. centrifugal or inductive
   d. capacitive or inductive

8. Most oscilloscopes display ____________ different patterns for both the primary and secondary circuits.
   a. nine
   b. six
   c. eighteen
   d. three

9. Technician A begins an available coil voltage test by connecting the voltmeter positive lead to the positive coil terminal and the voltmeter negative lead to a ground. Technician B connects the voltmeter positive lead to the positive coil terminal and the voltmeter negative lead to the negative coil terminal. Who is correct?
   a. Technician A
   b. Technician B
   c. Both A and B
   d. Neither A nor B

10. Figure 7-3 illustrates ____________.
    a. point bounce
    b. point arcing
    c. an intermittently open condenser
    d. condenser series resistance

JOB SHEET

SHOP ASSIGNMENT 23
PERFORM A BREAKER POINT VOLTAGE DROP TEST

NAME ______________________ **STATION** ______________________ **DATE** ______________

Tools and Materials

Voltmeter
Remote starter switch
Feeler gauge

Protective Clothing

None required

Procedure

1. Hook up the voltmeter by connecting the positive lead to the negative coil terminal and the negative lead to a ground.

   Task completed ____

2. Connect a remote starter switch.

   Task completed ____

   a. Bump the engine until the breaker points close.

   Task completed ____

3. Turn the ignition switch on and note the voltmeter reading. Is it less than 0.2 volt?

   Yes ____ No ____

   a. If yes, the points are in good condition.

   Task completed ____

   b. If no, go on to the next step.

   Task completed ____

4. Inspect the points for loose wiring connections; faulty ground connections; poor alignment; burned, pitted, or oxidized contacts; and excessive wear.

   Task completed ____

5. Adjust or replace the points as needed.

   Task completed ____

   a. Repeat the test procedure.

   Task completed ____

PROBLEMS ENCOUNTERED: ____________________

____

____

INSTRUCTOR'S COMMENTS: ____________________

____

____

JOB SHEET

SHOP ASSIGNMENT 24
PERFORM A CURRENT DRAW TEST

NAME ______________ **STATION** ______________ **DATE** ______________

Tools and Materials

Ammeter
Screwdriver

Protective Clothing

None required

Procedure

1. Turn the ignition switch off.

   Task completed ____

2. Disconnect the positive coil primary wire from the coil.

   Task completed ____

3. Hook up the ammeter by connecting the positive lead to the positive coil primary wire and the negative lead to the positive coil primary terminal.

   Task completed ____

4. Turn the ignition switch on and either start the engine or close the breaker points (depending on the manufacturer's recommendations).

   Task completed ____

5. Is the ammeter reading less than the manufacturer's specifications?

   Yes ____ No ____

   a. If yes, go on to the next step.

   Task completed ____

   b. If no, go on to step 7.

   Task completed ____

   c. If there is no reading at all, look for an open somewhere in the primary circuit.

   Not applicable ____ Task completed ____

6. Check for a discharged battery; high resistance in the primary wiring or coil primary winding; or loose or corroded battery connections.

   Task completed ____

7. Check for a shorted coil or resistor, or the wrong type of coil or resistor.

   Task completed ____

PROBLEMS ENCOUNTERED: ______________________________

____

____

INSTRUCTOR'S COMMENTS: ______________________________

____

____

JOB SHEET

SHOP ASSIGNMENT 25
SET NEWLY ALIGNED BREAKER POINTS

NAME ________________ **STATION** ________________ **DATE** ____________

Tools and Materials

Feeler gauge
Remote starter switch
Screwdriver

Protective Clothing

None required

Procedure

1. Remove the distributor cap and rotor.

   Task completed ____

2. Connect a remote starter switch.

   Task completed ____

   a. Bump the engine until the rubbing block is exactly on a high point of the cam.

   Task completed ____

3. Slide a clean feeler gauge of the proper thickness between the points. It should slide between the contacts with a slight drag.

   Task completed ____

4. Make any necessary adjustment to achieve the proper gap dimension.

   Task completed ____

5. Tighten the holddown screw and recheck the gap.

   Task completed ____

   a. Make any further adjustments that may be needed.

   Task completed ____

PROBLEMS ENCOUNTERED: ____

____

____

INSTRUCTOR'S COMMENTS: ____

____

____

JOB SHEET

SHOP ASSIGNMENT 26
SET IGNITION TIMING DYNAMICALLY

NAME ______________ **STATION** ______________ **DATE** ______________

Tools and Materials

Tachometer
Timing light
Screwdrivers
Service manual

Protective Clothing

None required

Procedure

1. Make sure that the dwell setting is correct.

   Task completed ____

2. Disconnect the distributor vacuum lines, if called for in the service manual.

   Not applicable ____ Task completed ____

3. Connect a tachometer and timing light.

   Task completed ____

4. Start the engine. Run it at normal operating temperature, or the speed called for in the service manual.

   Task completed ____

5. Observe the position of the timing marks.

   Task completed ____

   a. If adjustment is necessary, loosen the distributor holddown clamp and bolt. Advance the timing by rotating the distributor in the opposite direction of the rotor rotation; set it back by rotating the distributor in the direction of the rotor rotation.

   Not applicable ____ Task completed ____

6. Tighten the holddown clamp and bolt, then recheck the timing.

   Task completed ____

7. Reconnect the vacuum lines and readjust the idle, if necessary.

   Not applicable ____ Task completed ____

   a. If the idle speed was adjusted, recheck the timing.

   Not applicable ____ Task completed ____

PROBLEMS ENCOUNTERED: ______________________________

____

____

INSTRUCTOR'S COMMENTS: ______________________________

____

____

CHAPTER EIGHT

TROUBLESHOOTING ELECTRONIC IGNITION SYSTEMS

Objectives

Upon completion of this chapter, you should be able to:

- Identify the limitations of a conventional ignition system.
- Name the basic parts of an electronic ignition system and explain how they work.
- Explain the similarities and differences between different electronic switching devices.
- Describe the early electronic systems used on domestic cars.
- Explain the use of an oscilloscope to diagnose electronic ignition system performance.
- Implement electronic troubleshooting techniques to test individual electronic components.

PRACTICE QUESTIONS

1. Figure 8-1 shows a ____________.
   a. metal detection sensor
   b. magnetic pulse generator
   c. Hall-effect sensor
   d. photoelectric sensor

2. Which of the following is an advantage of electronic ignition systems over conventional systems?
   a. more reliable performance at all speeds
   b. reduced exhaust emissions
   c. high secondary voltage
   d. all of the above

3. Electronic ignition systems are ____________ to interrupt the coil's primary circuit.
   a. an NPN-type transistor
   b. breaker points
   c. a toggle switch
   d. all of the above

4. Figure 8-2 shows a ____________.
   a. metal detection sensor
   b. magnetic pulse generator
   c. Hall-effect sensor
   d. photoelectric sensor

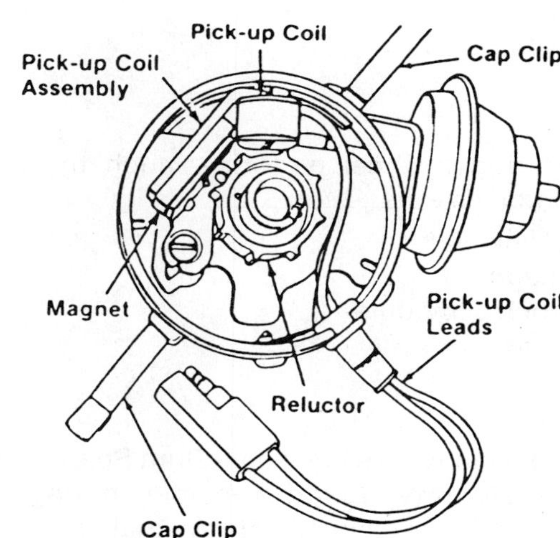

Figure 8-1

5. Technician A says that electronic ignition systems have always had computer-controlled spark advance. Technician B says that early electronic ignition systems had centrifugal and vacuum advance mechanisms. Who is correct?
   a. Technician A
   b. Technician B
   c. Both A and B
   d. Neither A nor B

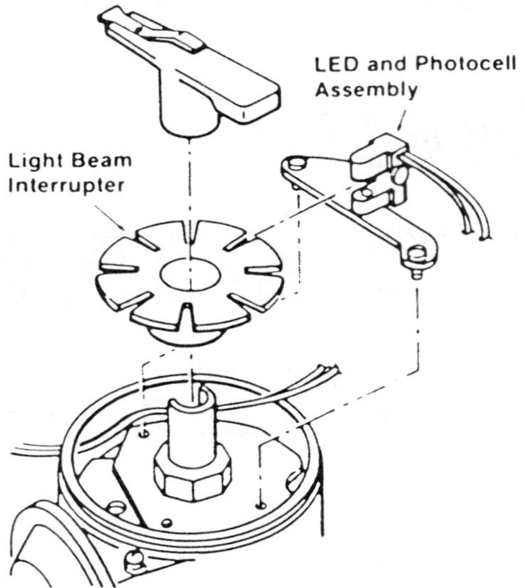

Figure 8-2

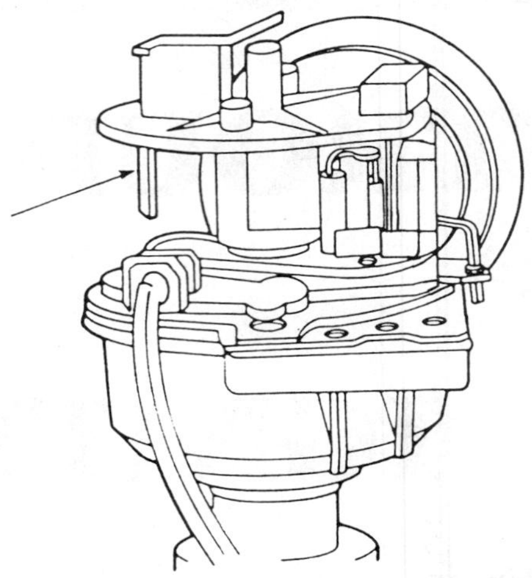

Figure 8-3

6. On the Hall-effect distributor shown in Figure 8-3, what is the arrow pointing to?
   a. rotor
   b. permanent magnet
   c. reluctor
   d. pick-up coil

7. The trigger wheel performs much the same function as a ____________.
   a. crankshaft
   b. magnet
   c. control module
   d. cam

8. Technician A says that some Dura Spark II ignition systems retard ignition timing during hard acceleration. Technician B says that some Dura Spark II systems retard ignition timing during startup. Who is correct?
   a. Technician A
   b. Technician B
   c. Both A and B
   d. Neither A nor B

9. An E-core coil ____________.
   a. is filled with oil
   b. has high primary resistance
   c. generates high secondary voltages
   d. all of the above

10. Technician A says that the electronic control module is housed inside the distributor on GM's HEI system. Technician B says that the HEI distributor also contains the pick-up coil assembly. Who is correct?
    a. Technician A
    b. Technician B
    c. Both A and B
    d. Neither A nor B

JOB SHEET

SHOP ASSIGNMENT 27
TEST COIL POLARITY

NAME ______________ **STATION** ______________ **DATE** ______________

Tools and Materials

Oscilloscope

Protective Clothing

None required

Procedure

1. Set the pattern selector of the oscilloscope to the display position.

   Task completed ____

2. Set the function selector of the oscilloscope to secondary.

   Task completed ____

3. Set the pattern height control of the oscilloscope to the O–25 kV scale.

   Task completed ____

4. Start the engine.

   Task completed ____

5. Do the firing lines of the display pattern extend downward?

   Yes ____ No ____

   a. If yes, the polarity is reversed. Check the coil primary connections to see if they are reversed.

   Task completed ____

   b. If no, and the firing lines extend upward instead, the coil polarity is correct.

   Task completed ____

PROBLEMS ENCOUNTERED: ____

____

____

INSTRUCTOR'S COMMENTS: ____

____

____

JOB SHEET

SHOP ASSIGNMENT 28
TEST SPARK PLUG FIRING VOLTAGE

NAME ______________________ **STATION** ______________________ **DATE** ______________

Tools and Materials

Oscilloscope
Pencil and paper

Protective Clothing

None required

Procedure

1. Set the pattern selector of the oscilloscope to the display position.

   Task completed ____

2. Set the function selector of the oscilloscope to secondary.

   Task completed ____

3. Set the pattern height control of the oscilloscope to the 0–25 kV scale.

   Task completed ____

4. Connect the oscilloscope ground lead and the primary, secondary, and trigger pickups to the vehicle.

   Task completed ____

5. Start the engine and adjust the speed to 1000 rpm.

   Task completed ____

6. Note the firing voltage of all the cylinders for height and uniformity.

   Task completed ____

   a. Record the highest and lowest firing voltage.

   Task completed ____

7. Was one or more of the firing voltages uneven, low, or high?

   Yes ____ No ____

   a. If yes, consult an appropriate source for possible causes.

   Task completed ____

   b. If no, the spark plug firing voltage is fine.

   Task completed ____

PROBLEMS ENCOUNTERED: ______________________________

____

____

INSTRUCTOR'S COMMENTS: ______________________________

____

____

JOB SHEET

SHOP ASSIGNMENT 29
PERFORM A ROTOR AIR GAP VOLTAGE DROP TEST

NAME ______________ **STATION** ______________ **DATE** ______________

Tools and Materials

Oscilloscope
Pencil and paper
Grounding probe

Protective Clothing

None required

Procedure

1. Set the pattern selector of the oscilloscope to the display position.

   Task completed ____

2. Set the function selector of the oscilloscope to secondary.

   Task completed ____

3. Set the pattern height control of the oscilloscope to the 0–25 kV scale.

   Task completed ____

4. Start the engine and adjust the speed to 1000 rpm.

   Task completed ____

5. Observe the height of the firing voltages, and record the height and firing order of any abnormal cylinder.

   Task completed ____

6. Turn off the engine and remove the spark plug wire of the abnormal cylinder from the distributor cap.

   Task completed ____

7. Connect one end of the jumper lead to a ground and the other end to the large portion of a grounding probe.

   Task completed ____

   a. Place the other end of the grounding probe in the distributor cap tower terminal.

   Task completed ____

8. Start the engine and adjust the speed to 1000 rpm.

   Task completed ____

9. Observe the firing voltage of the abnormal cylinder(s) recorded in step 5. Is there a slight drop in the firing voltage?

   Yes ____ No ____

a. If yes, the rotor and distributor cap are fine.

Task completed ____

b. If no, and the voltage either remains high or drops below specifications, the rotor or distributor cap may be defective. Inspect both and replace as needed.

Task completed ____

PROBLEMS ENCOUNTERED: ______________________________

____

____

INSTRUCTOR'S COMMENTS: ______________________________

____

____

JOB SHEET

SHOP ASSIGNMENT 30
TEST FOR A BREAK IN SECONDARY CIRCUIT INSULATION

NAME ____________________ **STATION** ____________________ **DATE** ______________

Tools and Materials

Oscilloscope
Grounding probe

Protective Clothing

None required

Procedure

1. Set the pattern selector of the oscilloscope to the display position.

   Task completed ____

2. Set the function selector of the oscilloscope to secondary.

   Task completed ____

3. Set the pattern height control of the oscilloscope to the 0–25 kV scale.

   Task completed ____

4. Connect one end of the jumper lead to a ground and the other end to the large portion of a ground probe.

   Task completed ____

5. Start the engine and adjust the speed to 1000 rpm.

   Task completed ____

6. Run the grounding probe around all the secondary circuit components, one by one.

   Task completed ____

7. While scanning with the grounding probe, observe the firing lines on the scope. Do they decrease in height when the secondary voltage arcs across to the probe?

   Yes ____ No ____

   a. If yes, there is an insulation break somewhere in the circuit.

   Task completed ____

   b. If no, the secondary circuit insulation does not have any breaks.

   Task completed ____

PROBLEMS ENCOUNTERED: ______________________________

____

____

INSTRUCTOR'S COMMENTS: ______________________________

____

____

CHAPTER NINE

COMPUTER-CONTROLLED IGNITION SYSTEMS

Objectives

Upon completion of this chapter, you should be able to:

- Explain how a computer-controlled ignition system computes spark advance based on engine operation conditions.
- Describe the components and operation of most domestic computer-controlled ignition systems.
- Perform test procedure on most domestic computer-controlled ignition systems.
- Explain the operation of a distributorless ignition system and perform tests on GM's DIS systems.

PRACTICE QUESTIONS

1. Figure 9-1 is an example of a ____________.
   a. spark control computer
   b. throttle position transducer
   c. vacuum transducer
   d. dual pick-up distributor

2. When presented with a complaint by a vehicle owner, Technician A begins his diagnostic routine by inspecting the computer system. Technician B starts by looking for loose, corroded, or damaged electrical connections. Who is correct?
   a. Technician A
   b. Technician B
   c. Both A and B
   d. Neither A nor B

3. The Hall-effect pick-up assembly ____________.
   a. was used on four-cylinder engines before 1978
   b. was used on 1976 to 1979 six- and eight-cylinder engines
   c. has been used on six- and eight-cylinder engines since 1976
   d. has been used on four-cylinder engines since 1978

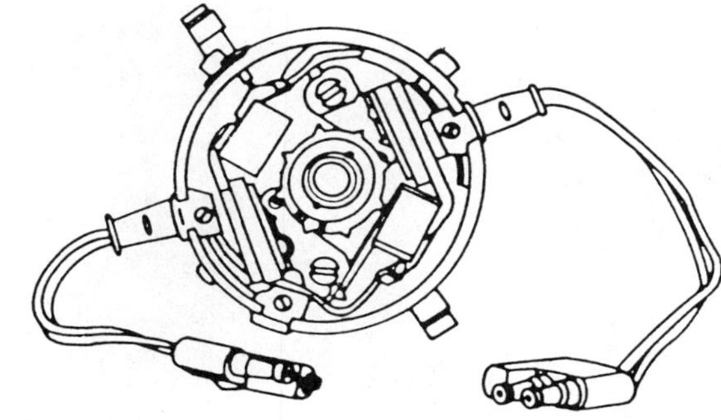

Figure 9-1

4. On a Chrysler modular control system, a fault code stored in the computer's memory will remain there until ____________.
   a. the computer no longer sees a problem and erases it automatically
   b. the battery is physically disconnected
   c. the computer reverts to a limp-in mode
   d. both a and b

5. A vehicle with an ESC system is experiencing advanced idle, and the idle does not vary when engine rpm is increased. Technician A says this could mean that the SCC has failed. Technician B says it means that the system could be in the limp-in mode. Who is correct?

a. Technician A
b. Technician B
c. Both A and B
d. Neither A nor B

6. Which of the following statements is incorrect?
   a. Ford's EEC-II system uses 7-, 9-, and 10-pin connectors to attach the ECA to the system.
   b. Ford's EEC-IV system has no crankshaft position sensor.
   c. The EEC-IV system has self-diagnostic capabilities through a self-test connector.
   d. All of the above.

7. A faulty TFI module can cause ______________.
   a. rough idling
   b. braking problems
   c. clogged fuel injectors
   d. both b and c

8. On the EEC-IV self-diagnostic system, which code is the same as a hard fault?
   a. on-demand
   b. dynamic response
   c. memory
   d. all of the above

9. Technician A performs the wiggle test with the vehicle engine running. Technician B performs the wiggle test with the vehicle engine off. Who is correct?
   a. Technician A
   b. Technician B
   c. Both A and B
   d. Neither A nor B

10. On a 1978 MISAR-ESC model vehicle, the CHECK IGNITION light comes on during normal operation, when the reference timing connector is not grounded. Technician A says this means there is a problem in the charging system voltage. Technician B says it means the control assembly is faulty. Who is correct?
    a. Technician A
    b. Technician B
    c. Both A and B
    d. Neither A nor B

JOB SHEET

SHOP ASSIGNMENT 31
CHECK THE TIMING ADVANCE ON A CHRYSLER SPARK CONTROL SYSTEM

NAME ______________________ **STATION** ______________________ **DATE** ______________

Tools and Materials

Advance timing light
Appropriate screwdrivers

Protective Clothing

None required

Procedure

1. Connect an advance timing light to the engine.

   Task completed ____

   a. Turn on the engine and allow it to warm up thoroughly.

   Task completed ____

2. Run the engine at idle speed for 1 minute, then remove the vacuum hose from the SCC.

   Task completed ____

3. Check the position of the timing mark. It should be at the base timing position specified on the VECI decal.

   Task completed ____

4. Accelerate the engine to 2000 rpm and observe the timing mark again. Did it advance significantly?

   Yes ____ No ____

   a. If yes, the timing advance is fine.

   Task completed ____

   b. If no, go on to the next step.

   Task completed ____

5. Disconnect the coolant switch and recheck the timing.

   Task completed ____

6. If there is still no advance, inspect the carburetor switch wiring for a short circuit to ground. In the event the switch is grounded, the SCC will not adjust the timing.

   Task completed ____

7. If the carburetor and coolant switches checked out fine, inspect the throttle and vacuum transducers.

Task completed ____

8. If steps 5, 6, and 7 revealed no problems, inspect the SCC; it may be faulty.

Task completed ____

PROBLEMS ENCOUNTERED: ______________________________

____

____

INSTRUCTOR'S COMMENTS: ______________________________

____

____

JOB SHEET

SHOP ASSIGNMENT 32
PERFORM A KEY-ON/ENGINE-OFF TEST ON A FORD EEC-IV SYSTEM

NAME ______________________ **STATION** ______________________ **DATE** ______________

Tools and Materials

Jumper wires
Voltmeter
Pencil and paper

Protective Clothing

None required

Procedure

1. Locate the self-test output connector under the hood.

   Task completed ____

2. Place one end of a jumper lead in the pin 4 slot of the output connector, and the other end to the negative voltmeter lead.

   Task completed ____

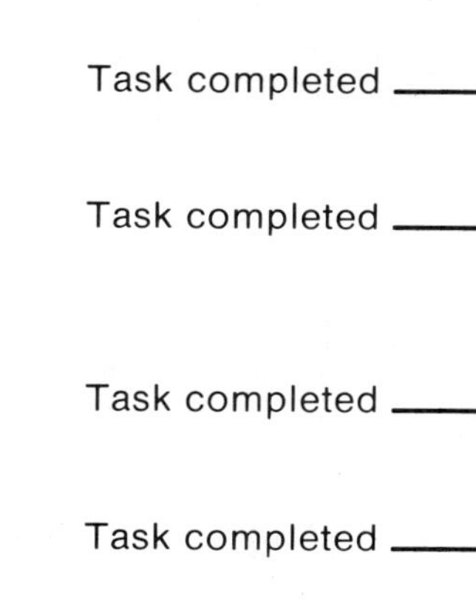

3. Connect the positive voltmeter lead to the positive battery terminal.

   Task completed ____

4. Use a second jumper wire to bridge the pin 2 slot of the output connector and the self-test input connector, as shown in Figure 9–2.

   Task completed ____

5. Set the voltmeter on the 0-15 scale.

   Task completed ____

6. Place the ignition switch in the RUN position, but do not start the engine yet.

   Task completed ____

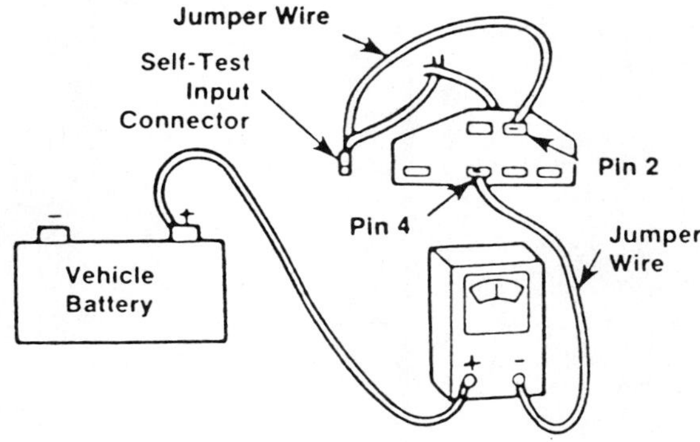

Figure 9–2

7. When the fast codes are finished, the first usable code will be displayed. Does the voltmeter needle sweep once, pause two seconds, then sweep again?

Yes ____ No ____

a. If yes, this indicates a code 11, meaning the system is fine.

Task completed ____

b. If no, the code for the particular problem will be counted out on the voltmeter. Use it to locate the fault.

Task completed ____

PROBLEMS ENCOUNTERED: ______________________________

____

____

INSTRUCTOR'S COMMENTS: ______________________________

____

____

JOB SHEET

SHOP ASSIGNMENT 33
MAKE A DETONATION CHECK ON A FORD MCU SYSTEM

NAME ____________________ **STATION** ____________________ **DATE** ____________

Tools and Materials

Appropriate screwdrivers
Jumper wire
3/4" wrench (if applicable)

Protective Clothing

None required

Procedure

1. Check and set the initial timing.

Task completed ____

   a. Disconnect the knock sensor.

Task completed ____

2. Disconnect the two-wire ignition module, and use a jumper wire to connect the two wires as shown in Figure 9-3.

Task completed ____

3. When the wires were connected, did the timing retard 15 to 20 degrees?

Yes ____ No ____

   a. If yes, reconnect the two-wire ignition module and set the engine to 2000 rpm.

Task completed ____

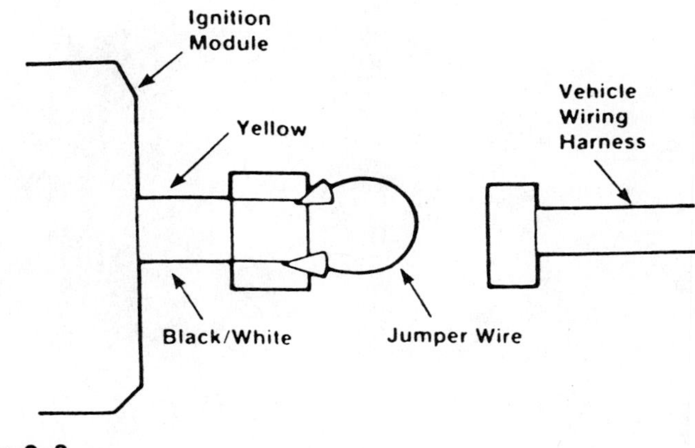

Figure 9-3

b. If no, replace the ignition module.

Task completed ____

4. Disconnect the zone vacuum switch hose.

Task completed ____

5. As the hose was disconnected, did the timing retard less than 5 degrees?

Yes ____ No ____

a. If yes, check the yellow wire between the MCU and the ignition module for shorts or opens. If the wire is fine, replace the MCU and recheck the timing.

Task completed ____

b. If no, and the timing retards more than 5 degrees, go on to the next step.

Task completed ____

6. Reconnect the knock sensor and run the engine up to 2400 rpm.

Task completed ____

7. Tap on the intake manifold with a 3/4-inch wrench. If the timing does not retard, replace the knock sensor and recheck the timing.

Task completed ____

8. If the timing still does not retard, replace the MCU.

Task completed ____

PROBLEMS ENCOUNTERED: ____________________

____

____

INSTRUCTOR'S COMMENTS: ____________________

____

____

JOB SHEET

SHOP ASSIGNMENT 34
CHECK THE TIMING ADVANCE ON A GM MISAR-ESC SYSTEM

NAME ____________ **STATION** ____________ **DATE** ____________

Tools and Materials

Ohmmeter
Timing light
Appropriate screwdrivers

Protective Clothing

None required

Procedure

1. Start the engine and check the ignition timing.

   Task completed ____

2. Is the timing advanced to about 35 degrees BTDC with the manifold vacuum at 18 inches or more?

   Yes ____ No ____

   a. If yes, go on to the next step.

   Task completed ____

   b. If no, recheck the base timing and the coolant sensor operation. If they check out fine, or if the timing remains at about 15 degrees, replace the control assembly and then recheck the timing.

   Task completed ____

3. Remove the control assembly vacuum hose.

   Task completed ____

4. When the hose was removed, did the ignition timing drop to 15 degrees?

   Yes ____ No ____

   a. If yes, the controller is okay.

   Task completed ____

   b. If no, go on to the next step.

   Task completed ____

5. Use an ohmmeter to check the coolant temperature sensor resistance. Is the resistance within specifications?

   Yes ____ No ____

a. If yes, replace the controller and recheck the timing.

Task completed ____

b. If no, replace the sensor.

Task completed ____

PROBLEMS ENCOUNTERED: ______________________________

____

____

INSTRUCTOR'S COMMENTS: ______________________________

____

____

CHAPTER TEN

SERVICING THE FUEL SUPPLY SYSTEM

Objectives

Upon completion of this chapter, you should be able to:

- Define the fuel delivery system components and their functions, including the fuel tank, fuel lines, filters, and pump.
- Conduct inspecting and servicing procedures for a fuel system.
- Define the components and their functions in an electronic engine control (EEC) fuel delivery system.

PRACTICE QUESTIONS

1. Which of the following statements is incorrect?
   a. Even when a gas tank is empty, it can still contain enough gas fumes to cause an explosion.
   b. Class B fire extinguishers are specially intended for use on gasoline fires.
   c. Damaged fuel tanks should always be welded.
   d. A good motor fuel should provide good mileage under various driving conditions.

2. A typical compression ratio for a post-1975 vehicle would be ____________.
   a. 4.5 to 1
   b. 12.5 to 1
   c. 8.2 to 1
   d. 3.2 to 1

3. Which of the following factors is likely to promote engine knock?
   a. advancing the ignition timing
   b. rich fuel mixture
   c. low air temperature when it enters the cylinder
   d. all of the above

4. A vehicle is experiencing hard starting, hesitation, and stumbling during warmup. Technician A says this can be caused by the fuel not vaporizing readily. Technician B says the problem stems from the fuel vaporizing too quickly. Who is correct?
   a. Technician A
   b. Technician B
   c. Both A and B
   d. Neither A nor B

5. Some electric fuel pumps have a(n) __________ that shuts off the fuel pump if the vehicle is involved in a collision or rolls over.
   a. one-way check valve
   b. inertia switch
   c. surge plate
   d. additional cam lock

6. Steel fuel line tubing can be substituted with ____________.
   a. copper tubing
   b. aluminum tubing
   c. both a and b
   d. neither a nor b

7. A fuel line is leaking, but only in one area. Technician A says the entire fuel line must be replaced. Technician B says repairing the leaking section is sufficient. Who is correct?
   a. Technician A
   b. Technician B
   c. Both A and B
   d. Neither A nor B

8. Which type of in-carburetor gasoline filter has a special element to remove metallic contamination before it reaches the carburetor?
   a. screw-in
   b. sintered bronze

c. ceramic
d. pleated paper

9. ______________ can lead to a vapor lock condition.

a. Low pressure
b. High temperatures
c. Both a and b
d. Neither a nor b

10. On the gasoline filter shown in Figure 10-1, the arrow is pointing to the ______________.

a. vacuum relief valve
b. fuel pick-up tube
c. charcoal canister
d. check valve

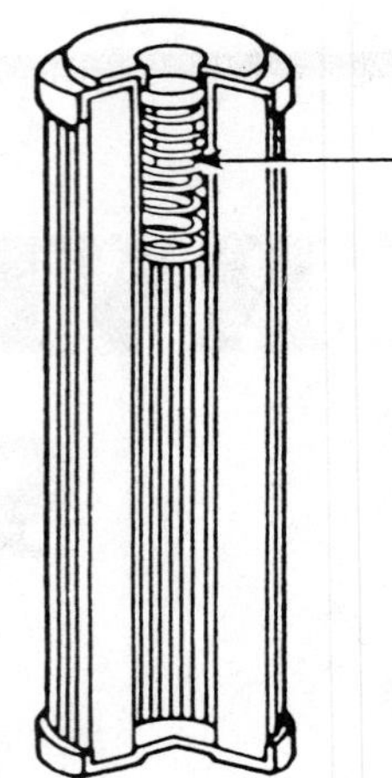

Figure 10-1

JOB SHEET

SHOP ASSIGNMENT 35
REPLACE A FUEL TANK

NAME ______________________ **STATION** ______________________ **DATE** ______________

Tools and Materials

Appropriate wrenches
Siphon hose
Masking tape
Appropriate screwdrivers
Safety can

Protective Clothing

Safety goggles or glasses with side shields

Procedure

1. Disconnect the negative battery terminal.

   Task completed ____

2. Remove the fuel tank filler cap and completely drain the tank.

   Task completed ____

3. Disconnect the fuel line at the tank that runs to the fuel pump.

   Task completed ____

4. Connect a siphon hose to the tank and draw the fuel into a safety can as shown in Figure 10-2.

   Task completed ____

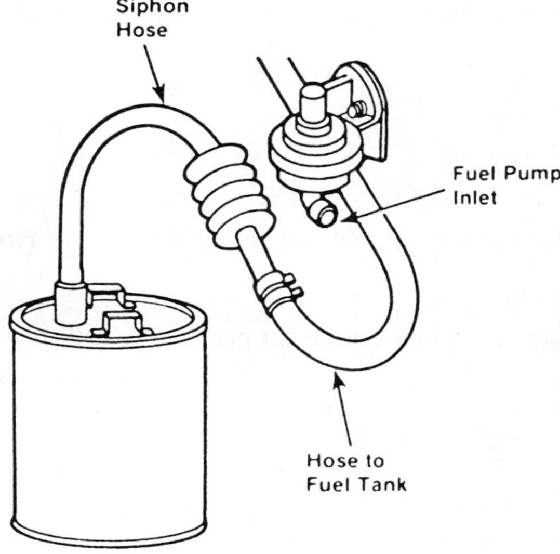

Figure 10-2

5. Use masking tape on each line to ensure correct reinstallation.

Task completed ____

a. Disconnect the vent lines and the plugged wire connected to the sending unit.

Task completed ____

6. Unfasten the filler from the tank. Remove the screws around the outside of the filler neck near the cap if it is a one-piece tube; for a three-piece unit, remove the neoprene hoses after loosening the clamp.

Task completed ____

7. Loosen the bolts holding the tank straps to the vehicle until they are about two threads from the end.

Task completed ____

8. Hold the tank against the underchassis with one hand while removing the bolts. Lower the tank to the ground, making sure all wires and tubes are unhooked.

Task completed ____

9. Before installing the new or repaired tank, make sure all the rubber or felt tank insulators are in place.

Task completed ____

10. Put the tank straps in place and position the tank. Loosely fit the straps around the tank, but do not tighten them yet.

Task completed ____

11. Make sure all hoses, wires, and vent tubes are connected properly, and check the filler neck for alignment and insertion into the tank.

Task completed ____

12. Tighten the strap bolts to secure the tank.

Task completed ____

13. Install the vent line, sending unit wires, ground wire, filler tube, and any other tank accessories.

Task completed ____

14. Fill the tank with fuel.

Task completed ____

a. Examine for leaks, especially around the filler neck and pick-up assembly.

Task completed ____

15. Reconnect the battery and check the fuel gauge for proper operation.

Task completed ____

PROBLEMS ENCOUNTERED: ____________________

____

____

INSTRUCTOR'S COMMENTS: ____________________

____

____

JOB SHEET

SHOP ASSIGNMENT 36
REPLACE AN IN-CARBURETOR GASOLINE FILTER

NAME ________________ **STATION** ________________ **DATE** ____________

Tools and Materials

Appropriate wrenches
Filter
Service manual

Protective Clothing

None required

Procedure

1. Disconnect the fuel line from the carburetor fuel inlet nut. Use two wrenches, one to hold the nut and the other to turn the fuel line fitting.

   Task completed ____

2. Remove the inlet nut from the carburetor.

   Task completed ____

   a. Retain the inlet nut gasket, but discard the used filter.

   Task completed ____

3. Consult the service manual for the correct replacement filter.

   Task completed ____

4. Install the new filter, inlet nut gasket, and inlet nut.

   Task completed ____

5. Reconnect the fuel line, making sure to tighten securely.

   Task completed ____

6. Turn the engine on and check the fuel line connections for leaks.

   Task completed ____

PROBLEMS ENCOUNTERED: ____

____

____

INSTRUCTOR'S COMMENTS: ____

____

____

JOB SHEET

SHOP ASSIGNMENT 37
TEST A FUEL PUMP FOR PRESSURE AND CAPACITY

NAME ______________________ **STATION** ______________________ **DATE** ______________

Tools and Materials

Fuel filter (if applicable)
Appropriate wrenches
Appropriate screwdrivers
Pressure gauge
Flexible hose
Restrictor
Graduated fuel container

Protective Clothing

Safety goggles or glasses with side shields

Procedure

1. Inspect the fuel filter and replace if necessary.

   Task completed ____

2. Remove the air cleaner assembly.

   Task completed ____

3. Disconnect the fuel inlet line or the fuel filter at the carburetor.

   Task completed ____

4. Connect a pressure gauge, restrictor, and flexible hose between the filter and carburetor as shown in Figure 10-3.

   Task completed ____

5. Position the hose and restrictor so the fuel can be discharged into a graduated container.

   Task completed ____

6. Turn on the engine and run it at the specified idle rpm.

   Task completed ____

   a. Open the restrictor momentarily to vent the system into the container, then close the restrictor.

   Task completed ____

7. After allowing the pressure to stabilize, take the reading. Is it within specifications?

   Yes ____ No ____

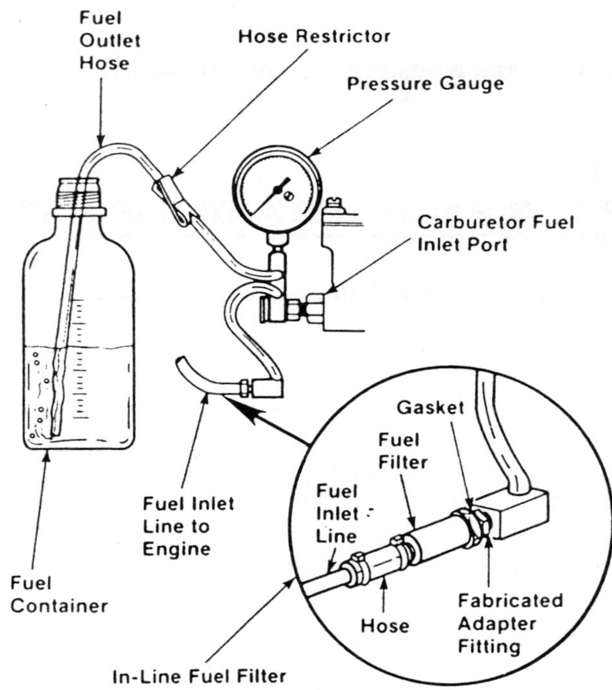

Figure 10-3

a. If yes, the pressure is fine. Go on to the next step.

Task completed ____

b. If no, and the fuel lines and filter are in good condition, replace the pump. Then go on to the next step.

Task completed ____

8. Before testing for capacity, run the engine at the specified idle rpm.

Task completed ____

9. Open the restrictor for the specified time to let the fuel discharge into the graduated container, then close the restrictor.

Task completed ____

10. Did at least one pint of fuel discharge?

Yes ____ No ____

a. If yes, the capacity is fine.

Task completed ____

b. If no, go on to the next step.

Task completed ____

11. Repeat steps 8, 9, and 10 using an auxiliary fuel supply and a new fuel filter.

Task completed ____

12. If the capacity now meets specifications, check for a restriction in the fuel supply from the tank and for improper tank ventilation.

Task completed ____

PROBLEMS ENCOUNTERED: ____

____

____

INSTRUCTOR'S COMMENTS: ____

____

____

CHAPTER ELEVEN

SERVICING THE VACUUM/AIR SUPPLY SYSTEMS

Objectives

Upon completion of this chapter, you should be able to:

- Define vacuum and explain how it affects and assists vehicle operation and driveability.
- Define the air intake system components and their functions, including air intake ductwork, air cleaner/filter assembly, and intake air temperature controls.
- Conduct inspection and servicing procedures for a vacuum system, air intake system, and exhaust system.
- Explain the operation of a turbocharger.

PRACTICE QUESTIONS

1. The complete absence of pressure is known as ______________.
   a. atmospheric pressure
   b. ambient air
   c. induction
   d. perfect vacuum

2. In which of the following systems is vacuum used?
   a. fuel induction
   b. emission control
   c. air intake and fuel
   d. all of the above

3. A customer complains that his or her engine is running roughly. Technician A says the source of the problem could be a leaking vacuum hose. Technician B says that a defective vacuum hose would not affect engine operation. Who is correct?
   a. Technician A
   b. Technician B
   c. Both A and B
   d. Neither A nor B

4. The vacuum gauge reading in Figure 11-1 probably indicates a ______________.
   a. choked catalytic converter or muffler
   b. manifold leak
   c. carburetor or injector adjustment
   d. leaking head gasket

5. In Figure 11-2, what part of the air cleaner assembly is the arrow pointing to?
   a. snorkel
   b. air filter element
   c. vacuum motor
   d. air duct

6. In most air intake systems, air to the engine is controlled by a ______________ valve or valves.
   a. butterfly
   b. heat riser
   c. vacuum retard-delay
   d. none of the above

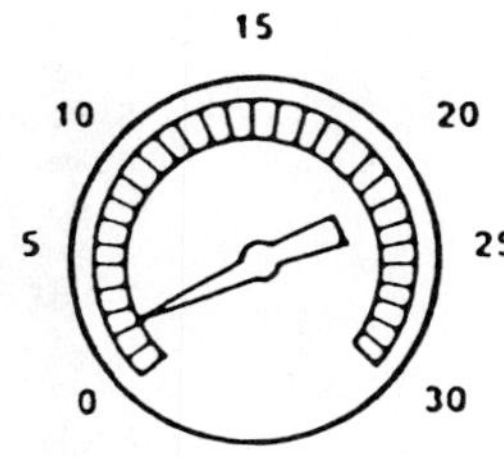

Figure 11-1

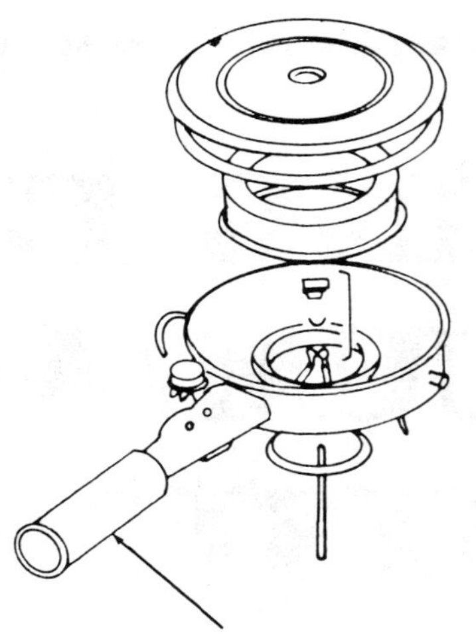

Figure 11-2

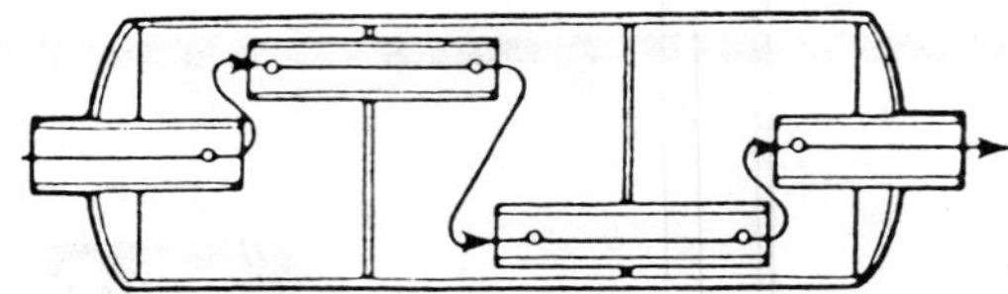

Figure 11-3

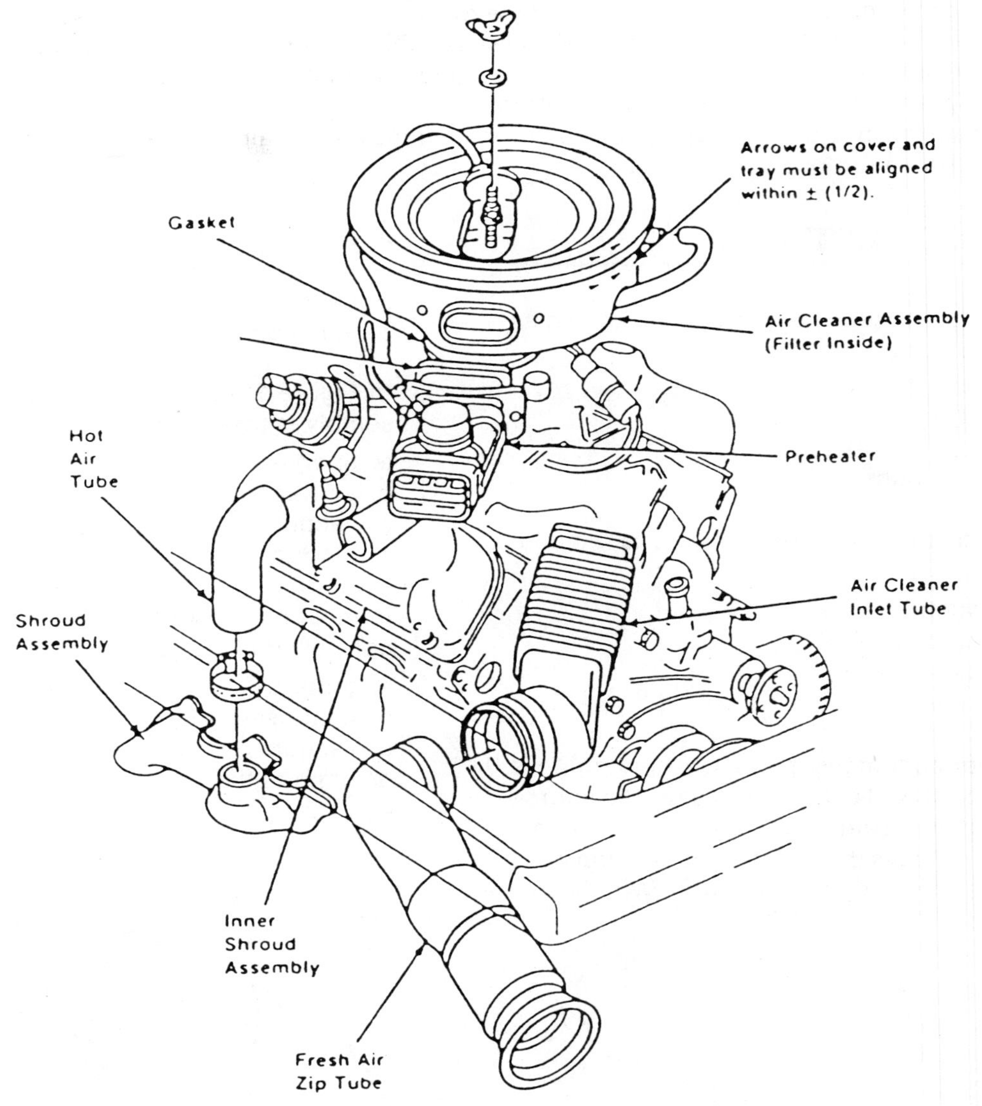

Figure 11-4

7. When selecting a replacement air filter, Technician A picks one with a wire outer screen. Technician B chooses one with an expanded metal outer screen. Who is correct?
   a. Technician A
   b. Technician B
   c. Both A and B
   d. Neither A nor B

8. Figure 11-3 illustrates a typical ____________ muffler.
   a. rear-mounted
   b. reverse-flow
   c. straight-through
   d. perforated

9. Technician A uses compressed air to clean heavy-duty air filter elements, while Technician B uses water. Who is correct?
   a. Technician A
   b. Technician B
   c. Both A and B
   d. Neither A nor B

10. The air filter unit used on most cars and light-duty trucks is located ____________.
    a. outside the air cleaner housing
    b. inside the air cleaner housing
    c. in a remote location on the engine
    d. none of the above

SHOP ASSIGNMENT 38

Identify the components of a typical air intake system. Refer to Figure 11-4 as needed.

SHOP ASSIGNMENT 39

Identify various exhaust system tools, including the air chisel, chain pipe cutter, pipe cutter, muffler cutter, shaper, expander, and hanger removal tool.

JOB SHEET

SHOP ASSIGNMENT 40
CLEAN A HEAVY-DUTY AIR FILTER ELEMENT

NAME ______________ **STATION** ______________ **DATE** ______________

Tools and Materials

Compressed air (if applicable)
Filter cleaner
Hose and water supply
Light

Protective Clothing

None required

Procedure

1. Use a flow of water or compressed air to remove any loose dust.

   Task completed ____

2. Soak the element in a filter cleaner solution for at least 15 minutes, occasionally agitating it to loosen the dirt particles.

   Task completed ____

3. Remove the element from the solution and rinse it by hosing water through it from the clean side to the dirty side until the water comes out clean.

   Task completed ____

4. Let the element air dry completely.

   Task completed ____

5. Use a light to inspect the element for pinholes, ruptures, or damaged or missing gaskets. If any of these conditions exist, replace the element.

   Task completed ____

PROBLEMS ENCOUNTERED: ______________________________

____

____

INSTRUCTOR'S COMMENTS: ______________________________

____

____

JOB SHEET

SHOP ASSIGNMENT 41
REPLACE AND START UP A TURBOCHARGER

NAME ______________ **STATION** ______________ **DATE** ______________

Tools and Materials

Appropriate wrenches
Service manual
Gaskets and seals
Torque wrench
Engine oil
Appropriate screwdrivers

Protective Clothing

Safety goggles or glasses with side shields

Procedure

1. Loosen all bolts, connections, and gaskets to remove the old turbocharger.

   Task completed ____

2. Consult the service manual to make sure the new turbocharger is the correct one.

   Task completed ____

3. Install the new gasket and seals, making sure they are properly positioned.

   Task completed ____

4. Install the new turbocharger and spin the impeller-turbine to check for binding.

   Task completed ____

   a. Torque the fasteners to specifications.

   Task completed ____

5. Change the engine oil and flush the oil lines.

   Task completed ____

6. If oil problems caused the old unit to fail, check the oil supply pressure in the feed line to the turbocharger.

   Not applicable ____ Task completed ____

7. Fill the oil filter with clean oil to minimize cranking time.

   Task completed ____

8. Leave the oil drain line disconnected at the turbocharger and crank the engine without starting it until oil flows out of the turbocharger drain port.

   Task completed ____

9. Connect the oil drain line.

Task completed ____

a. Start the engine and operate it at low idle for a few minutes before running it at higher speeds.

Task completed ____

PROBLEMS ENCOUNTERED: ______________________________

____

____

INSTRUCTOR'S COMMENTS: ______________________________

____

____

CHAPTER TWELVE

CARBURETOR TESTING AND SERVICING

Objectives

Upon completion of this chapter, you should be able to:

- Explain the purpose and functions of a carburetor.
- Identify the main systems and components of a carburetor.
- Recognize carburetor-related performance problems.
- Troubleshoot and adjust a carburetor for optimum performance and efficiency.

PRACTICE QUESTIONS

1. Which of the following is not part of the carburetor float circuit?
   a. idle limiter cap
   b. fuel inlet fitting
   c. fuel bowl
   d. float

2. On the external vent system illustration shown in Figure 12-1, what component is the arrow pointing to?
   a. charcoal canister
   b. check valve
   c. fuel inlet
   d. accelerator pump

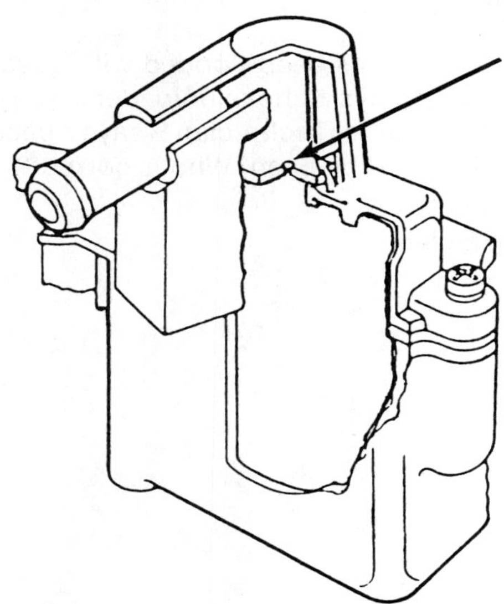

Figure 12-1

3. The metering systems of a carburetor are designed to function properly ______________.
   a. at all times
   b. only at full-throttle operation
   c. only when the fuel level in the bowl is at a specific level
   d. only when the secondary throttle plates are open

4. Figure 12-2 illustrates a ______________ carburetor.
   a. variable venturi
   b. feedback
   c. single-stage
   d. two-stage

5. As carburetors developed, what replaced mechanical metering rods?
   a. venturi
   b. power valve

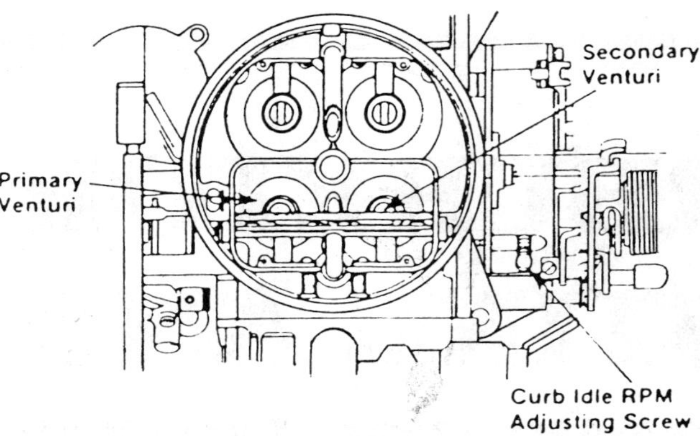

Figure 12-2

c. free-floating lever
d. discharge jet

6. ____________ have a qualifying diaphragm instead of a vacuum piston.
   a. Some integral chokes
   b. All divorced chokes
   c. Both a and b
   d. Neither a nor b

7. A cold engine has been flooded with gasoline. Technician A says a choke unloader is required to start the engine. Technician B says a deceleration valve must be used. Who is correct?
   a. Technician A
   b. Technician B
   c. Both A and B
   d. Neither A nor B

8. Which of the following statements is incorrect?
   a. The sole function of the throttle position solenoid is to control idle speed when the transmission is engaged.
   b. A variable venturi increases in size as engine demands increase.
   c. For the three-way catalyst on a feedback carburetor to work efficiently, the air/fuel ratio must be maintained very close to 14.7 to 1.
   d. Both a and b

9. An engine starts hard when it is hot. Technician A says a possible cause of this problem is a leaking float bowl. Technician B says a possible cause is that the choke coil rod is out of adjustment. Who is correct?
   a. Technician A
   b. Technician B
   c. Both A and B
   d. Neither A nor B

10. An improperly operating float system can result in ____________.
    a. fuel starvation
    b. rich fuel mixture
    c. lean fuel mixture
    d. all of the above

JOB SHEET

SHOP ASSIGNMENT 42
MAKE A PROPANE ENRICHMENT ANALYSIS

NAME ______________ **STATION** ______________ **DATE** ______________

Tools and Materials

Tachometer
Propane

Protective Clothing

None required

Procedure

1. Warm the engine to its normal operating temperature.

   Task completed ____

   a. Adjust the curb idle to the specified rpm.

   Task completed ____

2. Use propane to enrich the air/fuel mixture for the highest rpm. The rpm should be within the manufacturer's specifications for enriched rpm.

   Task completed ____

3. Stop the enrichment and return the engine to idle. Is the idle rpm within the specified curb idle?

   Yes ____ No ____

   a. If yes, the idle mixture is fine.

   Task completed ____

   b. If no, adjust the idle mixture screws as needed to achieve the correct curb idle speed.

   Task completed ____

PROBLEMS ENCOUNTERED: ______________________________

____

____

INSTRUCTOR'S COMMENTS: ______________________________

____

____

JOB SHEET

SHOP ASSIGNMENT 43
MAKE A CARBURETOR IDLE ADJUSTMENT USING THE PEAK CO_2 METHOD

NAME ______________________ **STATION** ______________________ **DATE** ______________

Tools and Materials

Four-gas exhaust analyzer

Protective Clothing

None required

Procedure

1. Hook up the four-gas exhaust analyzer as per the manufacturer's instructions.

   Task completed ____

2. Disable the air injection system.

   Task completed ____

3. Lightly bottom the idle mixture screws and back them out two full turns.

   Task completed ____

4. Turn on the engine and observe the CO, HC, CO_2, and O_2 displays. They should be stable.

   Task completed ____

5. Adjust the curb idle speed to the manufacturer's specification.

   Task completed ____

6. Turn the idle mixture screws the same amount and direction, 1/8 turn at a time, until the CO_2 reaches its highest value.

   Task completed ____

7. Turn the screws equally to the leanest setting possible, while still maintaining the peak CO_2 reading.

   Task completed ____

8. Reset the idle speed as required.

   Task completed ____

9. Repeat steps 6, 7, and 8 as needed until the ideal idle mixture is achieved.

   Task completed ____

PROBLEMS ENCOUNTERED: ______________________________

____

____

INSTRUCTOR'S COMMENTS: ______________________________

____

____

JOB SHEET

SHOP ASSIGNMENT 44
ADJUST A DIVORCED CHOKE

NAME ______________________ **STATION** ______________________ **DATE** ______________

Tools and Materials

Appropriate screwdrivers
Service manual

Protective Clothing

None required

Procedure

1. Disconnect the choke rod from the choke plate.

   Task completed ____

   a. Close the choke plate.

   Task completed ____

2. Move the choke rod up or down as specified in the service manual. On choke plates with a sliding fit, it is usually set at the middle of the slot.

   Task completed ____

3. Bend the choke rod to lengthen or shorten it as required, as shown in Figure 12-3.

   Task completed ____

4. Check for free movement between the open and closed choke positions.

   Task completed ____

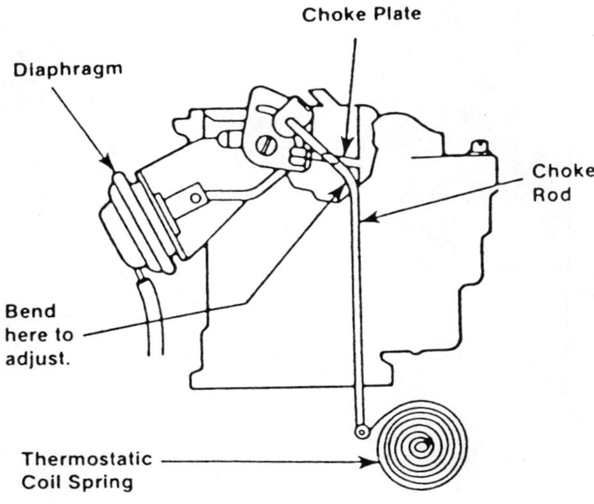

Figure 12-3

PROBLEMS ENCOUNTERED: ______________________________

____

____

INSTRUCTOR'S COMMENTS: ______________________________

____

____

JOB SHEET

SHOP ASSIGNMENT 45
MAKE A CARBURETOR FLOAT LEVEL ADJUSTMENT

NAME ______________________ **STATION** ______________________ **DATE** ______________

Tools and Materials

Appropriate screwdrivers
Gasket
Cardboard
Service manual

Protective Clothing

None required

Procedure

1. Remove the upper body assembly and gasket.

   Task completed ____

   a. Install the new gasket.

   Task completed ____

2. Use cardboard to make a fuel level gauge like the one shown in Figure 12-4. Make it to the specified dimensions found in the service manual.

   Task completed ____

3. With the upper body inverted, place the gauge on the cast surface of the body (not on the gasket).

   Task completed ____

   a. Measure the vertical distance from the upper body to the bottom of the float.

   Task completed ____

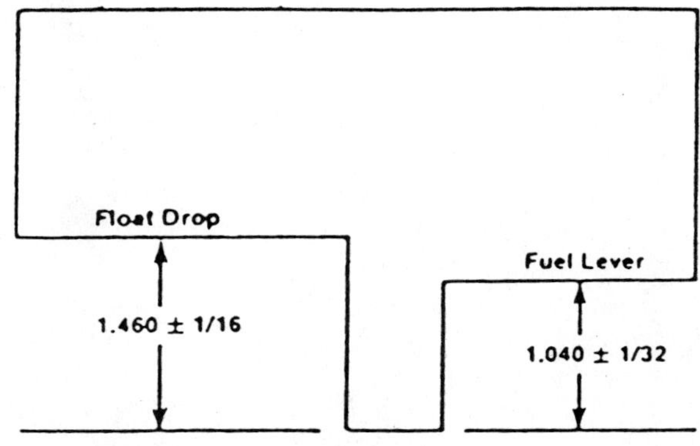

Figure 12-4

4. If necessary, adjust the float drop by bending the float operating lever away from the fuel inlet needle to decrease the setting and toward the needle to increase the setting.

Not applicable ____ Task completed ____

PROBLEMS ENCOUNTERED: ______________________________

____

____

INSTRUCTOR'S COMMENTS: ______________________________

____

____

CHAPTER THIRTEEN

TROUBLESHOOTING FUEL INJECTION SYSTEMS

Objectives

Upon completion of this chapter, you should be able to:

- Describe the similarities and the differences between fuel injection systems and carburetion.
- Explain the principles of operation of a fuel injection system.
- List the advantages of fuel injection.
- Describe the two distinct methods of fuel injection in use today: electronic fuel injection (EFI) and mechanical or continuous injection system (CIS).
- Explain the differences in point of injection in throttle-body or single-point injection and port or multi-point injection.
- Explain the design and function of EFI components, including fuel delivery system components, system sensors, electronic control unit (ECU), and fuel injectors.
- Explain the design and function of CIS components, including fuel delivery and metering components and airflow components.
- Service fuel injection systems, including conducting preliminary checks, fuel injection and control system checks, fuel system checks, injector checks, oscilloscope checks, injector cleaning and replacement, idle adjustment, and CIS checks and tests.

PRACTICE QUESTIONS

1. The carburetor float corresponds to the ____________ of the fuel injection system.
   a. throttle switch
   b. thermo-timer switch
   c. airflow sensor
   d. none of the above

2. Which of the following statements are incorrect?
   a. Fuel and air are not measured in a carburetor system.
   b. Precise control of the air/fuel ratio is constantly maintained in a fuel injection system.
   c. Fuel injection produces less exhaust emissions than carburetion.
   d. A fuel injection system requires more pollution control devices than a carburetion system.

3. The only electrical component within the CIS system is the ____________.
   a. throttle plate
   b. fuel pump
   c. airflow sensor
   d. distributor

4. In Figure 13-1, what part of the throttle body is the arrow pointing to?
   a. throttle plate
   b. throttle position sensor
   c. injector
   d. pressure regulator

5. The airflow sensor potentiometer shown in Figure 13-2 indicates ____________.
   a. idle
   b. part throttle
   c. full throttle
   d. none of the above

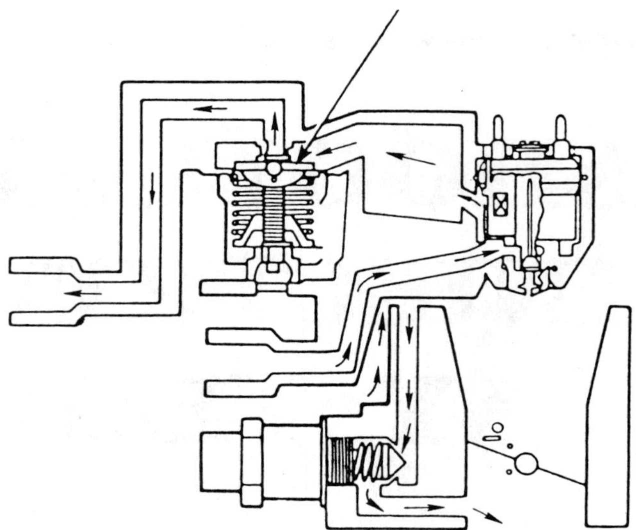

Figure 13-1

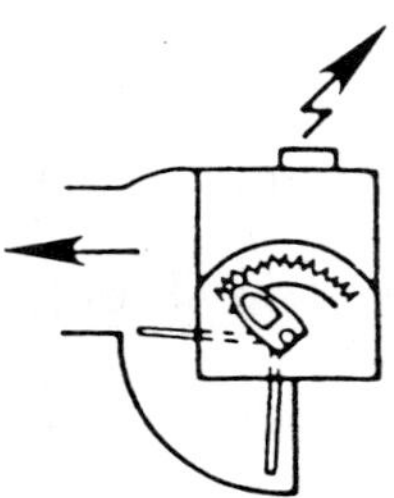

Figure 13-2

6. During warmup, the engine operates under an open loop condition for _____________.
   a. up to 10 seconds
   b. up to 10 minutes
   c. up to 10 miles
   d. none of the above

7. Clogged injectors are more common on _____________.
   a. CIS
   b. CIS-E
   c. PI
   d. both a and b

8. Before beginning specific fuel injection checks and electronic control testing, Technician A makes sure the base timing and idle speed are set to specifications. Technician B makes sure all fuses and fusible links are intact. Who is right?
   a. Technician A
   b. Technician B
   c. Both A and B
   d. Neither A nor B

9. A vehicle with throttle body fuel injection is experiencing stalling, surging, and hesitation; the problem occurs when the engine is hot or cold. Technician A says a possible cause is a restricted air intake system. Technician B says a possible cause is a defective MAP sensor. Who is right?
   a. Technician A
   b. Technician B
   c. Both A and B
   d. Neither A nor B

10. Prior to installing a new fuel injector, the sealing ring should be lubricated with _____________.
    a. engine oil
    b. automatic transmission fluid
    c. silicone grease
    d. both a and b

JOB SHEET

SHOP ASSIGNMENT 46
PERFORM A FUEL INJECTOR POWER BALANCE TEST

NAME ____________ **STATION** ____________ **DATE** ____________

Tools and Materials

Tachometer
Fuel pressure gauge
Pencil and paper

Protective Clothing

None required

Procedure

1. Hook up a tachometer and fuel pressure gauge to the engine.

   Task completed ____

2. Turn on the engine and let it reach its normal operating temperature.

   Task completed ____

3. When the idle stabilizes, unplug one of the injectors.

   Task completed ____

   a. Record the rpm drop and the pressure gauge reading.

   Task completed ____

4. Reconnect the injector, wait until the idle stabilizes, and move on to the next injector. Repeat on all of the injectors.

   Task completed ____

5. If any injector fails to cause a noticeable drop in rpm and fuel pressure when it is unplugged, examine it; chances are it is either clogged or defective.

   Task completed ____

PROBLEMS ENCOUNTERED: ____________

____

____

INSTRUCTOR'S COMMENTS: ____________

____

____

JOB SHEET

SHOP ASSIGNMENT 47
CLEAN FUEL INJECTORS

NAME ______________________ **STATION** ______________________ **DATE** ______________

Tools and Materials

Injector cleaning system
Appropriate wrenches
Appropriate screwdrivers
Clamps

Protective Clothing

Safety goggles or glasses with side shields

Procedure

1. Attach the cleaning system service hose to the fuel rail.

Task completed ____

2. Disable the fuel pump following the vehicle manufacturer's instructions.

Task completed ____

3. To prevent cleaning solution from getting into the fuel tank, clamp off the fuel pump return line at the flex connection as shown in Figure 13-3.

Task completed ____

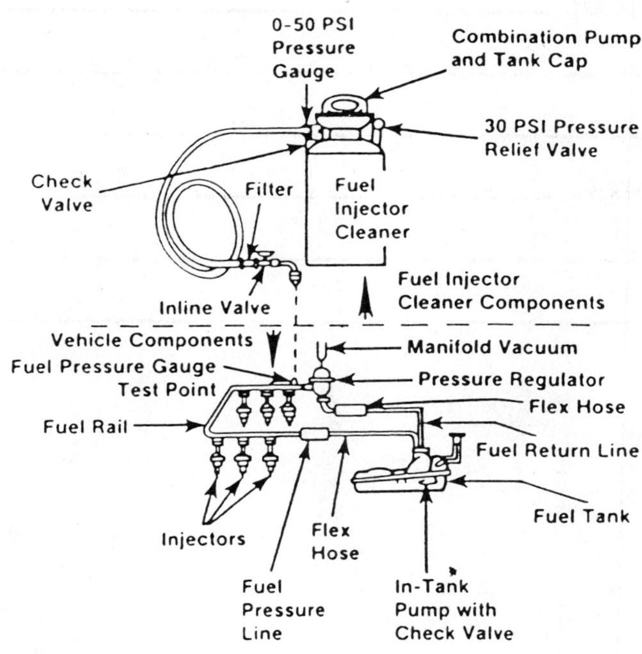

Figure 13-3

a. Clamp off the inlet line for added protection.

Task completed ____

4. Open the cleaning system's control valve about one-half turn to prime the injectors.

Task completed ____

5. Start the engine.

Task completed ____

6. If available, set and adjust the cleaning system's pressure gauge to the operating pressure of the injection system (approximately 25 to 45 psi).

Not applicable ____ Task completed ____

7. Let the engine idle at 2000 rpm for 10 to 15 minutes, or until the cleaning solution has run out.

Task completed ____

8. Turn off the engine and disconnect the cleaning system.

Task completed ____

a. Reconnect the fuel pump.

Task completed ____

9. Remove the clamps from the fuel lines.

Task completed ____

10. Start the car and let it idle for about 5 minutes to remove any leftover cleaning solution from the fuel lines.

Task completed ____

PROBLEMS ENCOUNTERED: ____

____

____

INSTRUCTOR'S COMMENTS: ____

____

____

JOB SHEET

SHOP ASSIGNMENT 48
MAKE AN AIRFLOW SENSOR ADJUSTMENT

NAME ______________________ **STATION** ______________________ **DATE** ______________

Tools and Materials

Appropriate wrench
Anaerobic locking compound
Feeler gauge
Appropriate screwdriver

Protective Clothing

None required

Procedure

1. Remove the bolt securing the air sensor plate to the level.

   Task completed ____

2. Coat the bolt threads with anaerobic locking compounds.

   Task completed ____

3. Reinstall the bolt fingertight.

   Task completed ____

4. Move the plate as needed to insert a 0.004 inch feeler gauge at any point around its circumference.

   Task completed ____

5. Tighten the bolt and recheck it with the feeler gauge.

   Task completed ____

6. If the plate cannot be centered, and the level appears to be too far to one side, remove the air sensor housing and the damper bolt that is screwed into the lever counterweight.

   Not applicable ____ Task completed ____

7. Apply anaerobic locking compound to the bolt, reinstall it, then slide the lever to the middle of the pivot and tighten the bolt.

   Not applicable ____ Task completed ____

PROBLEMS ENCOUNTERED: ____

____

____

INSTRUCTOR'S COMMENTS: ___

CHAPTER FOURTEEN

UNDERSTANDING EMISSION CONTROL SYSTEMS

Objectives

Upon completion of this chapter, you should be able to:

- Describe the inspection and replacement of PCV system parts.
- Name the components of an evaporative emission control system.
- Explain the air temperature emission control system inspection and testing.
- Name the components of the manifold heat control system and what they do.
- Describe the application of the EFE control.
- List the EGR valves and controls.
- Explain air injection system.

PRACTICE QUESTIONS

1. The ______________ system is strictly a control for NO_X in the exhaust gasses.
   a. exhaust gas recirculating
   b. evaporative control
   c. engine modification
   d. positive crankcase ventilation

2. Thermactor systems are usually blocked out ______________.
   a. during deceleration
   b. at idle
   c. when the engine is cold
   d. all of the above

3. In Figure 14-1, the arrow is pointing out one of the various locations of the ______________.
   a. PCV valve
   b. vapor canister
   c. vapor purge connection
   d. remote air cleaner

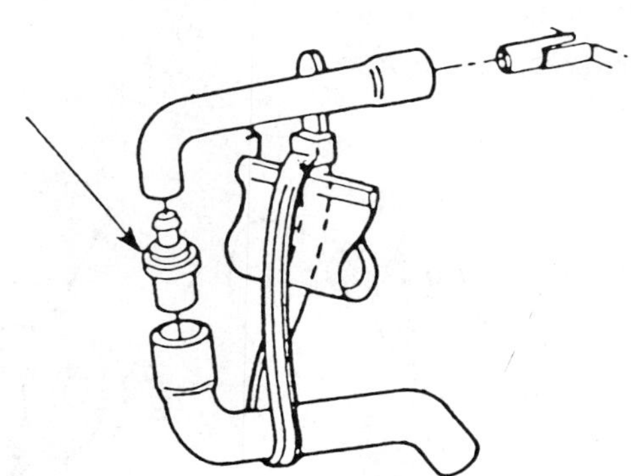

Figure 14-1

4. An engine is idling roughly. Technician A checks for a plugged hose. Technician B checks for a clogged PCV valve. Who is correct?
   a. Technician A
   b. Technician B
   c. Both A and B
   d. Neither A nor B

5. Figure 14-2 shows a(n) ______________ purge valve.

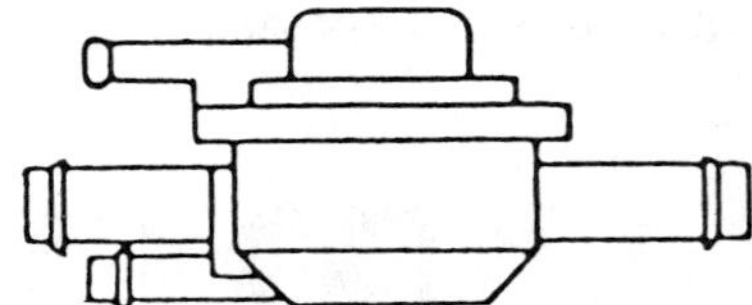

Figure 14-2

a. in-line
b. remote mounted
c. canister mounted
d. thermostat-controlled

6. Which of the following statements is incorrect?
   a. The bimetal sensor is installed in the air cleaner body.
   b. Thermostat-operated heat control valves are found mainly on V-8 engines.
   c. A possible cause of hard starting and poor performance is a leaking signal port on the purge shutoff valve.
   d. A power heat control valve is designed specifically to work with a minicatalyst.

7. ______________ EGR lockout might be required to keep the EGR valve closed when the engine is under maximum load.
   a. Wide-open throttle
   b. Cold
   c. Both a and b
   d. Neither a nor b

8. A vehicle cranks normally but starts hard, and hesitates or stalls on acceleration. Technician A says a possible cause of this problem is a vacuum leak in the heated air inlet system. Technician B says a possible cause is a collapsed air duct. Who is correct?
   a. Technician A
   b. Technician B
   c. Both A and B
   d. Neither A nor B

9. In the typical pulse air system shown in Figure 14-3, what component is the arrow pointing to?
   a. silencer
   b. air check valve
   c. check and delay valve
   d. air manifold

10. Which component of the typical air injector system is designed to prevent backfiring?
    a. vane pump
    b. one-way check valve
    c. pressure relief valve
    d. diverter valve

Figure 14-3

JOB SHEET

SHOP ASSIGNMENT 49
PERFORM A FUNCTIONAL CHECK OF A PCV VALVE

NAME ______________ **STATION** ______________ **DATE** ______________

Tools and Materials

Appropriate wrench

Protective Clothing

None required

Procedure

1. Disconnect the PCV valve from the valve cover, intake manifold, or hose.

   Task completed ____

2. Start the engine and run it at idle speed.

   Task completed ____

3. Is a hissing sound heard as air passes through the PCV valve?

   Yes ____ No ____

   a. If yes, the PCV valve is not clogged.

   Task completed ____

   b. If no, the PCV valve is clogged.

   Task completed ____

4. Check for vacuum by placing a finger over the end of the PCV valve.

   Task completed ____

5. Is there little or no vacuum present at the valve?

   Yes ____ No ____

   a. If yes, check the hoses for restrictions and/or deterioration, and replace if needed.

   Task completed ____

   b. If no, the hoses are in good condition.

   Task completed ____

6. Turn off the engine and remove the PCV valve.

   Task completed ____

7. Shake the valve. Can the rattle of the check needle inside the valve be heard?

   Yes ____ No ____

a. If yes, the valve is in good condition.

Task completed ____

b. If no, the valve must be replaced.

Task completed ____

PROBLEMS ENCOUNTERED: ____

____

____

INSTRUCTOR'S COMMENTS: ____

____

____

JOB SHEET

SHOP ASSIGNMENT 50
CHECK A VACUUM-OPERATED HEAT CONTROL VALVE

NAME ____________________ **STATION** ____________________ **DATE** ____________

Tools and Materials

Vacuum pump
Graphite lubricant (if applicable)

Protective Clothing

None required

Procedure

1. Inspect the valve for abnormal conditions, and repair or replace as needed.

   Task completed ____

2. Disconnect the hose from the PVS.

   Task completed ____

3. Use a vacuum pump to apply 15 inches of vacuum to the vacuum motor diaphragm.

   Task completed ____

   a. Trap the vacuum for 60 seconds. The valve should leak no more than 2 inches of vacuum in this time.

   Task completed ____

4. Observe the position of the vacuum motor stem. It should go to the fully closed position with the vacuum on and to the fully open position when the vacuum is closed, as shown in Figure 14-4.

   Task completed ____

5. If necessary, use a graphite lubricant to lubricate the valve shaft.

   Not applicable ____ Task completed ____

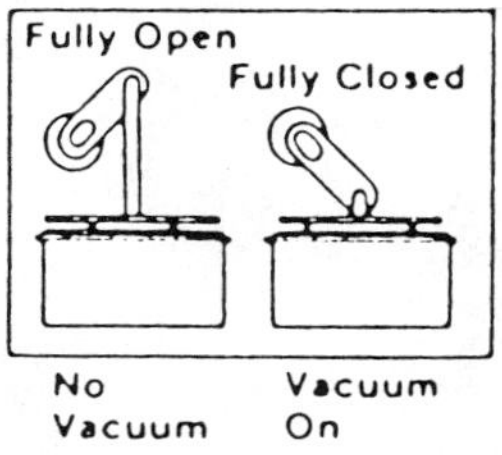

Fully Open
Fully Closed
No Vacuum
Vacuum On

Figure 14-4

PROBLEMS ENCOUNTERED: ______________________________

____

____

INSTRUCTOR'S COMMENTS: ______________________________

____

____

JOB SHEET

SHOP ASSIGNMENT 51
INSPECT AND SERVICE A THERMOSTAT-OPERATED HEAT CONTROL VALVE

NAME ____________________ **STATION** ____________________ **DATE** ____________

Tools and Materials

Graphite lubricant
Cold and/or hot water

Protective Clothing

None required

Procedure

1. Inspect the valve assembly for damage, and repair or replace as needed.

   Task completed ____

2. Turn the valve shaft by hand to test if it is free, as shown in Figure 14-5. It should turn freely and return to the closed position when cold.

   Task completed ____

3. Place the thermostat in cold or hot water as required to check for proper opening or closing.

   Task completed ____

4. Lubricate the valve with graphite lubricant.

   Task completed ____

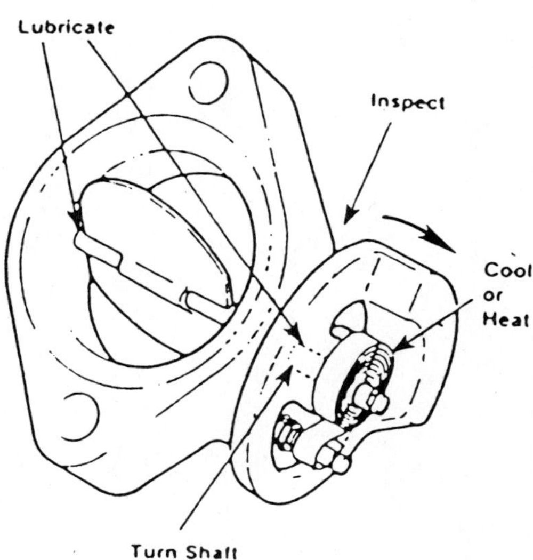

Figure 14-5

PROBLEMS ENCOUNTERED: ____

____

____

INSTRUCTOR'S COMMENTS: ____

____

____

JOB SHEET

SHOP ASSIGNMENT 52
CHECK THE OPERATION OF AN EGR VALVE

NAME ______________ **STATION** ______________ **DATE** ______________

Tools and Materials

Tachometer
Vacuum pump
Appropriate screwdrivers
Appropriate wrenches

Protective Clothing

None required

Procedure

1. Inspect all vacuum lines to make sure they are correctly routed and attached securely.

   Task completed ____

   a. Replace any cracked, crimped, or broken lines.

   Not applicable ____ Task completed ____

2. Turn on the engine and let it reach its normal operating temperature, then make sure there is no vacuum to the EGR valve at idle.

   Task completed ____

3. Install a tachometer.

   Task completed ____

4. If working with an EFI engine, disconnect the throttle air bypass valve solenoid.

   Not applicable ____ Task completed ____

5. Remove the vacuum supply hose from the EGR valve port, then plug the hose.

   Task completed ____

6. Run the engine at idle with the transmission in NEUTRAL.

   Task completed ____

   a. If necessary, adjust the idle speed to the emission decal specification.

   Not applicable ____ Task completed ____

7. Using a vacuum pump, slowly apply 5 to 10 inches of vacuum to the EGR valve vacuum port.

   Task completed ____

8. Replace the EGR valve if any of the following occurs when vacuum is applied: the engine does not stall, the idle speed does not drop more than 100 rpm, and the idle speed does not return to normal after removing the vacuum.

Not applicable ____ Task completed ____

9. If the EGR valve is operating properly, unplug and reconnect the vacuum supply hose.

Not applicable ____ Task completed ____

10. If it was removed, reconnect the throttle air bypass valve solenoid.

Not applicable ____ Task completed ____

PROBLEMS ENCOUNTERED: ____

____

____

INSTRUCTOR'S COMMENTS: ____

____

____

ANSWER KEYS

ANSWERS TO PRACTICE QUESTIONS

Chapter 1

1. a	3. c	5. d	7. d	9. c
2. c	4. a	6. c	8. a	10. c

Chapter 2

1. c	3. a	5. b	7. a	9. d
2. b	4. a	6. b	8. b	10. c

Chapter 3

1. b	3. b	5. c	7. d	9. a
2. c	4. a	6. d	8. b	10. a

Chapter 4

1. b	3. a	5. b	7. c	9. a
2. b	4. b	6. a	8. b	10. c

Chapter 5

1. a	3. c	5. c	7. a	9. b
2. b	4. d	6. b	8. b	10. a

Chapter 6

1. a	3. a	5. d	7. c	9. c
2. a	4. b	6. c	8. a	10. c

Chapter 7

1. d	3. d	5. c	7. d	9. a
2. a	4. a	6. b	8. d	10. b

Chapter 8

1. b	3. a	5. b	7. d	9. c
2. d	4. d	6. c	8. b	10. c

Chapter 9

1. d	3. d	5. c	7. a	9. c
2. b	4. d	6. a	8. a	10. b

Chapter 10

1. c	3. a	5. b	7. a	9. c
2. c	4. a	6. d	8. a	10. d

Chapter 11

1. d	3. a	5. d	7. c	9. c
2. d	4. b	6. a	8. b	10. b

Chapter 12

1. a	3. c	5. b	7. a	9. a
2. b	4. d	6. c	8. a	10. d

Chapter 13

1. d	3. b	5. c	7. c	9. a
2. d	4. d	6. b	8. c	10. d

Chapter 14

1. a	3. a	5. a	7. a	9. a
2. d	4. c	6. b	8. b	10. d